Becoming an Aircraft Stress Engineer

A guide that bridges university with the Aerospace industry.

by

***Stephane Couture**, B. Eng.*

First edition

Version 1.6

Edited and Published by

AircraftStress Publishing

2023

DISCLAIMER

This book "BECOMING AN AIRCRAFT STRESS ENGINEER" is to be strictly used for academic and entertainment purposes and is not intended to be used to design actual aircraft or be referred to in actual professional analyses or reports. It is simply to show some of the analysis approaches and steps that the author used during his career by using fictitious data and fictitious aircraft structures examples.

The publisher and the author do not guarantee the contents provided in this book. Professional structural engineers must use up-to-date and accepted references for methodologies and allowables.

The author and the publisher are not responsible for any injury or damage or financial losses or loss of lives for using the data or the information in the book.

Legal Deposit to the Library and Archives Canada

Couture, Stéphane
Becoming an Aircraft Stress Engineer: A guide that bridges university with the aerospace industry - 1st ed.

ISBN: 978-1-7782450-0-8 Hard cover

1. Aircraft structures 2. Structural analysis 3. Aerospace engineering

A Secured PDF version of the book is also available here:

Aircraftstress Publishing
https://aircraftstress.com
info@aircraftstress.com

Cover design by Elisabeth Couture.

First edition 2023.

This book is dedicated to my mother *(Sammy)* thanks to whom I was able to follow my dreams.

I also dedicate this book to all the children of the world, hoping that they will also be able to realize their dreams.

CONTENTS

FOREWORD

One of my grand-fathers was a mechanic, owned a garage as a business and was quite inventive in creating and building new machines to meet certain needs. My other grand-father was a welder on ships and loved those huge boats. My father was a machinist and would build anything and everything to the thousandth of an inch! One of my uncles was building and driving race-cars. He was also a heck of a good mechanic, maybe better than he was a driver... I sure spent a lot of hours in garages when I was a child watching them weld, machine parts, and clean pistons.

All these people were talking about mechanical systems, mechanical structures, engines, etc. They would build them but who was designing those vehicles I wondered? Who decided how thick the cylinder head of an engine needed to be so that it doesn't break under the pressure from the "explosions" in the cylinders? They could manage to make things work by trial and error or by looking at other machines but they could not draw them or even less calculate or design them.

That's what I wanted to do. I wanted to design vehicles and, although nobody in my family or in my surrounding owned an airplane or even boarded one, I always wanted to know how an airplane flew and how it was built.

After high-school, I went to a technical college in aeronautics to study aircraft and electronics. At the time, I thought I knew *enough* about mechanical systems already but didn't know anything about electrical and electronic circuits. So I studied Avionics. I liked it although I realized toward the end of that program that I was only going to repair radios and electrical wirings not design them. I didn't know what an engineer was, or even less, what it was doing and started to ask around to my teachers until one of them told me that I had to go to University to become an engineer in order to design those vehicles. That person was quick at telling me that the engineering program was very difficult and that I wouldn't probably succeed because of the much higher requirements than for becoming a technician... Other "friends" told me that I was dreaming in color as I was from a rural community and that nobody in my family was a professional! As if that was a prerequisite for admission...

Luckily, my mom encouraged me and, since I had good marks (and a lot of drive), I was admitted and started the four year engineering program at l'École Polytechnique of the Université de Montréal in the electrical engineering specialty. After just a few classes, where the professor would describe what a resistor was and how a transistor worked, I felt I was losing my time since I had already seen all that in the tech program. So I "switched" over to my original passion: mechanical engineering.

I liked the classes of physics and thermodynamics but I sure preferred the stress analysis classes. Being able to predict when a part was going to fail was something of a mystery to me, similar in a way as predicting the future! I was learning about metals and heat treatments. Then after some time, I realized I had acquired the knowledge that my late father and the other members of my family, all gone by then, would certainly have loved to possess. That was awesome!

After four years in the mechanical engineering program, as all of my buddies at the tech school were starting to buy houses and raise kids, I got my bachelor's degree; with high marks I must add. As my mom was struggling to make ends meet, I didn't go for a master's degree but went on and got a job as a junior structural engineer in a civil engineering firm to help the family. In 1987, engineering jobs (especially in aerospace) were somewhat scarce and my bank account was empty. I also needed to pay back the $22,000 student loan. So, there I was, analyzing steel and wooden building structures. That was pretty cool but it wasn't flying. I was also pretty much left on my own as there were no really experienced engineers to guide me and my boss was rarely in the office or was too busy.

As "funny" as life can be sometimes, after a few months there, I met a young intern whose father was working for a prestigious aerospace company. Next thing I know, I am offered a job as a Liaison Engineer for that company.

I didn't know what a "Liaison" engineer was but whatever it was, I was going to be in Aerospace engineering and I would be designing aircraft, I told myself. The whole aerospace system was new to me. I felt that I had learned very little in school or, that what I had learned was obsolete or not applicable to aerospace. Note that it all came back to me later.

In those days, engineers wore white shirts and ties, and could smoke in the office! One could easily tell who the engineer on the production line was. Then, again, I realized that, although I had learned a lot about drawings, aluminum alloys, rivets and manufacturing procedures, I wasn't calculating anything. We were restoring the original structural integrity with relatively small repairs so, no analyses were needed. When more extensive repairs were required, then we would pass it on to a "**stressman**" that would perform the necessary calculations and let us know if it worked or not.

After a couple of months, I started looking around the engineering department to try to find a group where I could do just that. I looked into the weights group, stress group, design group and the propulsion group to realize that the stress group was supporting all of the other groups when analyses were necessary. In the stress group, I could analyze a part of structure, a part from the electrical system, a part from the engine support, etc. So I asked for a transfer. The transfer was denied as they didn't need anybody in that group at the moment.

There were *old chaps* in the Liaison group that had been doing Liaison engineering during their entire career. One of them explained to me that: "Once you're tagged as a liaison engineer, you remain as such all of your life". This old man would have cut off both of my legs from under me and it wouldn't have been more devastating. I then started to look into the stress reports for whenever there was a repair to do on the aircraft, I would learn what load goes into that part and how it was analyzed. At the same time, I was looking for another job as a stressman in other companies as it seemed that my career in this one was sealed for the next thirty-five years.

"Cemeteries are filled with irreplaceable people..."

Then, after a year, it happened. Again, life plays funny tricks on you, I suppose... One of my friend got a job as a *draftsman* in a very small European based aerospace company that had just started in Montreal and was doing a design package for a much larger company. He contacted me and told me they were looking for a stress analyst. With his help, I got my first job as a *stressman*. During the interview, the person, that was later going to be my lead, told me that the job implied very simple beam calculations. I was put on the wing package of a small commercial jet!.. That small company was certainly looking more for somebody cheap than for somebody experienced and that lead had obviously never stressed a wing, or not much else for that matter. Boy, what a difference that was. At last I could analyze parts and interact with the designers. Have a word to say in the way the structure was going to look like. I would create stress notes and sign drawings. The real deal!

However, that large company didn't take too long to realize that their subcontractor's stress engineer (me) wasn't too experienced. Therefore, upon their request, I was then re-located on site of that large company to be supervised and guided by more experienced fellows. After a year, they offered me to work for them permanently. I guess they must have liked what I had been doing...

After that job, and for the rest of my career, all of the other ones were variations on the same theme. I got to work on the cockpit, fuselage, doors, empennage, engine pylon, flight controls, dynamic components of helicopters, water and waste systems, cargo systems, and a bunch of other secondary structures.

The analyses I was performing on those aircraft structures were static, fatigue "durability", and damage tolerance. From all three, I always preferred the static part of the analysis as it is when the parts get designed "sized" and you can control (sort of) all of the thicknesses and the type of material.

Although, in some companies, the fatigue analysis is done at the same time as the static analysis and can thus drive design changes as well. As for Damage Tolerance Analysis (*DTA*), there is no control on the design. The objective is usually to calculate Inspection Intervals and very rarely does it lead to a design change.

During those 30 years, I had "touched" pretty much all parts of an aircraft, airplanes and helicopters. I had been a direct employee for the first quarter of that time and a contractor for the remaining. I changed jobs more than a dozen times and went to work for most of the small and large commercial aerospace companies (some start with the letter "A" or "B") all across North America. And since I had acquired a truck load of expertise from all of those companies, quite often, I would be bringing some knowledge to the less experienced engineers and I would become a source of information for them in the field.

"You make the job as interesting as you want it to be"

Through my journey, I met a lot of young engineers that would come to me for advice since their boss was simply not available or wasn't easy to talk to. I just loved showing those brilliant young motivated minds how to perform analyses of aircraft structures the best I could. As in martial arts, some apprentices have become better than the master, and that's the way it should be. But practically, the number of new aircraft analysts that I could "train" was limited. With the help of this book, my hope is to reach as many new **aircraft stress engineers** as possible in order to pass on the knowledge to them.

As I get into some form of retirement, I didn't want all that information to be lost like I've seen too many times in this industry, and in others for that matter. So much time and money is wasted each time the wheel gets re-invented. So let's keep going forward and eventually we'll get somewhere no engineer has gone before...

This book has three parts. ***Part I*** is about my own experience and what I think is required to be proficient as an aircraft stress engineer. ***Part II*** is about a "Quiz" that I had been through during a job interview. The first chapter shows the complete stress analysis of the "bad" design and, the second, the complete stress analysis of a better design. Finally, ***Part III*** presents two complete detailed analysis examples of fictitious but realistic structures. The first chapter covers a cabinet installation in an airplane, while the second chapter covers details of a wing rear spar.

Through-out the book, I have incorporated some quotes that I've heard and read over time. Those quotes have an important meaning to me as an engineer and relate some words of wisdom based on my experience. Some are somewhat funny and some are more philosophical thoughts. I sure hope you'll enjoy reading them.

As my *Liaison engineering* boss told me when I started: "You make the job as interesting as you want it to be". So I encourage the reader to go even deeper into this field, to keep investigating, to read more reference books comparing analysis methodologies (hand calculations and FEA) and validate them through test results and especially, to keep your interest and curiosity always active regarding those wonderful flying machines.

Stef Couture

November 2020

ABBREVIATIONS, ACRONYMS AND NOMENCLATURE

a	largest dimension of a panel
A	Area
b	smallest dimension of a panel
BL	Buttock Line, along the lateral *(Y)* axis
brg or br	bearing
c.g.	center of gravity
C.L..	Center Line
d	distance
D	Diameter
DAR	Designated Airworthiness Representative (FAA)
DER	Designated Engineering Representative (FAA)
DFEM	Detailed Finite Element Model
DT	Damage Tolerance or Diagonal Tension
DTA	Damage Tolerance Analysis
E	Young's Modulus
f	applied stress
F	Force or allowable stress
FBD	Free Body Diagram
FEA	Finite Element Analysis
FEM	Finite Element Model
FS	Fuselage Station, along the longitudinal *(X)* axis or, Front Spar
G	Shear modulus
GAG	Ground-Air-Ground
GFEM	Global Finite Element Model
GPFB	Grid Point Force Balance
in.	inch
I_i	Second moment of inertia around the *(i)* axis.
ksi	1,000 pound square inch
l	length
lbs	pounds
LE	Leading edge
LHS	Left Hand Side *(Port)*
msi	1,000,000 pound square inch
M	Bending Moment
MRB	Material Review Board
MS	Margin of Safety
N_i	Acceleration factor along *(i)* axis
OML	Outer Mold Line
psi	pound square inch
P	Applied force
q	Shear flow

R	Reaction force or radius
RHS	Right Hand Side (Starboard)
R/S	Rear Spar
s	Fastener spacing (pitch), also noted as *p*
S.C.	Shear Center
t	thickness
T	Torque
TE	Trailing Edge
w	distributed load
WL	Water Line, along the vertical *(Z)* axis
WS	Wing Station, along the elastic axis of a wing (usually following the front spar)

Indices

a	axial
b	bending
c	compression or, column
cr	critical *(compression allowable)*
l	lower or, longitudinal
o	original or, at point *(o)*
pb	plastic bending
r	radial
s	shear
t	*t*ension or, transverse
u	ultimate, or upper
x	Along the *(X)* axis
y	yield, or along the *(Y)* axis
z	Along the *(Z)* axis

#	pounds
'	temporary value
"	inches

REFERENCES

You may be working for a large company that already has all the necessary tools for a stress engineer to perform analyses. However, since it is not the case for all companies, stress engineers must sometimes carry their own tools on the job. The following non exhaustive list of books is considered by myself to be the essentials that an aerospace engineer will need. During my career, I've used some more often than others but yet, all have something that others don't and thus they somehow complete each other.

Other than for the references containing the material allowables, the edition of each one is not really important as they pretty much contain the same information that we need in all editions. If you're lucky enough to get your hands on old yellowed copies of these books, then all the better. Books that have belonged to former stress engineers usually have highlighted equations, corrections, and have hand written comments in them. They are worth gold!

Recommended References for the aircraft stress engineer:

1. Abbott I.H. and von Doenhoff A.E. *Theory of wing sections.* (1959).
2. Bruhn, E.F. *Analysis and Design of Flight Vehicle Structures*. (1973).
3. Composite Material Handbook-17; *CMH-17*. https://www.cmh17.org/
4. Engineering Sciences Data Unit; IHS *ESDU*.
5. Flabel J.C. *Practical Stress Analysis for Aircraft Design Engineers* (1987)
6. Madier, D. *Practical Finite Element Analysis for Mechanical Engineers* (2020)
7. McCombs, W.F. *McCombs Bruhn Supplement*. (2004).
8. Metallic Materials Properties Development and Standardization; *MMPDS*.
9. Niu, Michael C.Y. *Airframe Stuctural Design*. (1988).
10. Niu, Michael C.Y. *Airframe Stress analysis and Sizing*. (1997).
11. Niu, Michael C.Y. *Composite Airframe structures – Practical design information and data*. 2nd ed. (1993).
12. Oberg, E. *Machinery's Handbook* (1914)
13. Pilkey Walter.D. *Formulas for Stress, Strain, and Structural Matrices.* (1994).
14. Pilkey Walter.D. *Peterson's Stress Concentration Factors.* 2nd ed. (1997).
15. Shigley J.E. and Mischke C.R. *Mechanical Engineering Design*. 5th ed. (1989).
16. Timoshenko S. *Theory of Plates and Shells.* 2nd ed. (1987)
17. Young, Warren C. *Roark's Formulas for Stress & Strain.* 6th ed. (1989).

Public Domain References:

The following references can be accessed at this website: https://ntrs.nasa.gov/search

18. Ramberg,W., Osgood,W.R. *NACA-TN-902 Description of stress-strain curve by 3 parameters* (1943).
19. Kuhn,P., Peterson,J.P., Levin,L.R. *NACA-TN-2661 A summary of diagonal tension Part I* (1952).
20. Gerard,G., Becker,H. *NACA-TN-3781 Handbook of structural stability, buckling of flat plates Part I* (1957).
21. Sullins, R.T., Smith, G.W., Spier, E.E., *NASA CR-1457 Manual for structural stability analysis of sandwich plates and shells* (1969).

References available from other sources:

22. Air Force Flight Dynamic Laboratory *AFFDL-TR-69-42 Stress analysis manual* (1967)
23. Air Force – Navy – Civil Aircraft design. *ANC-5; Strength of metal aircraft elements* (1951).
24. Hi-Lites and Hi-Loks Fasteners http://www.lisi-aerospace.com.
25. Code of Federal Regulations https://www.ecfr.gov/current/title-14/chapter-I/subchapter-C/part-25.
26. Cherry Aerospace fasteners https://www.fsirivet.com.
27. Blogs https://aircraftstress.com/blog/
28. Bolts MS calculations https://ntrs.nasa.gov/archive/nasa/casi.ntrs.nasa.gov/20120003667.pdf

Note: As web sites change through time, the links in this book to internet web sites cannot be guaranteed.

ACKNOWLEDGMENTS

I wish to thank the following persons, some of them for their help on this book and, all of them for their support and friendship during my career as an aerospace engineer.

Not only are they all great engineers but, more importantly, they are my friends. We probably shared more hours next to each other in an engineering office than the time we spent with our respective families...

In alphabetical order: Lisa Chamber, Kathy DeLong, Walter Faessler, Tam Ha, Patrick McConnell, Mike Ozerov, Lowell Plough, Wayne Pollard, Serge Radacovici, Julian Sharpe, Nate Warfield, Mike Wolff and Bosko Zdanski.

Special thanks to Maryse Couture, Alain Gendron, Serge Lascelle, Bilal Merachli, Gavin Rowlandson, Martin Savard, and Conrad Young, for proof reading the manuscript and for their precious comments.

These people are the ones I look up to in the aircraft stress analysis field and in life. Not only have the greatest airplane and helicopter companies benefited from their expertise and professionalism but they also gladly share their knowledge with other fellow engineers.

And finally, many thanks to my wife Manon for her patience and encouragement during all those months (years) that I've spent sitting in front of a computer screen writing this book. Thanks also to my daughters Elisabeth and Marianne for their encouragements. All of you are the women of my life and I can never be grateful enough for your support, your love, and your trust in me and my project(s).

"No one is king in his own country."

PART I

The journey to a proficient career

1. INTRODUCTION

What is the most important thing in the aerospace business?

- Safety
- Weight
- High pay/profit
- Glamour

Well, most managers will tell you: **It's the schedule!**

Because without a followed schedule, there are no aircraft, then no sales, hence we don't have to worry about any of the points above!

Most of the time, the simplest way to meet the schedule (and the budget), there is no magic trick to this, is by having people that know what they're doing. Pretty obvious will you say! But surprisingly, as I progressed through my career, it appeared to me that, as the years went by, less and less people were competent in their field (unfortunately, that seems to be applicable to many fields…).

Throughout the years, experienced people slowly but steadily disappeared. They retired, got early retirement (got terminated) or simply died. Then, those people were replaced by young less expensive engineers. To my opinion, for some accountants, all engineers are equal and therefore, they ask themselves the question: –*Why pay more for the experienced and expensive analysts?* So companies may see the opportunity to increase their profit margin by reducing the staff and/or by replacing the old with the young (some are almost ready to work for free to follow their passion) fresh out of engineering schools.

The experienced analysts will tell you that, in stress analysis, it takes five years to **know** what you're doing and then another five more years to **understand** what you're doing. What they don't say though is that it is true only if the young analysts are mentored or have a boss that could take the time to guide them.

Unfortunately, without generalizing, this is getting more and more rare as companies try to function with the minimum manpower and/or don't want to invest in a "green" engineer because they suspect that, after they would get good and autonomous, they might be leaving for another company for a higher pay or, will leave to become contractors for an even higher pay. So the young engineers that don't want to change job, because they like where they are, could find themselves somewhat penalized in their progression towards becoming skillful stress analysts.

Quite often, as the experienced people disappear, the remaining young ones get promoted into lead positions or even as supervisors after only a few years and with very little hands-on exposure to real design work. They might get more responsibility (without having the real control), do more hours and won't necessarily get a higher pay. They then don't have anybody to go to for advice and nobody to challenge them, so they call the shots upon themselves based on their best knowledge. Most incompetent people don't know they are incompetent! Therefore, it is up to the stress analysts themselves to perfect their knowledge and expertise by getting it from other sources.

In this book, I have tried to relate some of my experiences by providing complete analysis examples that show the reader all the necessary steps that I would take along with my thoughts. This can serve as a guide or inspiration to whoever is new to the field. No, it doesn't necessarily make me a guru or that I know all there is to know about stress analysis and that I never make mistakes. To the contrary, mistakes are essential to the learning process.

This work is meant to be used as a map to guide the aircraft stress engineer through the process of writing a complete stress report while making sure nothing is missed.

Sometimes, a lot of times actually, I feel that we "stressmen" are the police. The guardians of the truth! Newton's laws representatives! During the release of a drawing, where people from ten different groups will co-sign the release sheet, the stress engineer's signature pretty much guarantees the adequacy (substantiation) of the part or the assembly drawn on that drawing. To me this is the most important signature but, don't tell the other groups...

We (stress analysts) have to make sure that the design makes sense (works as intended), is well balanced (sum of the forces equals zero), is as light as possible and more important, can be **analyzed**. In this book, I will try to make the reader realize or determine by simple calculations or by logic alone if a load path makes sense or not quickly. However, don't get me wrong, when the designer creating the part is experienced, the stress analyst merely confirms that the design is adequate and may only be optimizing it to reduce its weight.

Most young graduates nowadays solely rely, it seems, on FEA "Finite Element Analysis". As you'll see, this is often a much longer process than doing simple hand calculations. Providing the FEM is well constructed with good boundary conditions, elements, properties, etc. and providing that a thermal, linear, non-linear buckling is performed, the young analyst could most likely be covered. Sometimes designers will create a concept in 3D and then the analyst will go off for days or weeks before getting back to the designers, with changes of course... That's when you don't make friends! Designers cannot wait weeks for the stressman to get back to them. Sometimes, it's a small change like a fastener type change but sometimes, the whole concept has to be scrapped and redone with better load paths.

In which case, it is most important that the designer(s) and the analyst know that they are in the same team and that they have the exact same goal, which is making a good aircraft.

That's where your social skills have to be at their best. Avoid sending text messages or lengthy emails. Go talk to the person and try to develop a very good relationship. I have as many designer friends as I have analyst friends. By being quick at spotting bad designs (flaws), you will earn respect and a good reputation from the designers, the chief designer, your peers, your boss, and also, from the manager and V.P.

Then of course, an email with a sketch of the suggested change is required and serves two purposes: 1- An image is worth a thousand words and, 2- You might be asking for many changes or suggestions, therefore you want to keep a record of the changes you ask for. I personally use *PowerPoint*™ with sketches, or CAD screen shots showing the WAS and the NOW. Number your changes or suggestions. Title the file with a change number, the date and the drawing or installation number.

These records are also for your protection as it is a proof that you asked for those changes since it is possible that you or the designer (or both) may very well forget about it over a long weekend. And send that file by email to the designer and copy (cc) your lead and the designer's lead. Also create a folder on the shared network with all your changes (*PowerPoint* files) at a location where anybody in engineering working on your program can access in the case the email gets "lost" or the recipient is off on vacation and someone else has to pick-up the ball.

Also, make a list in a spreadsheet of all of your change requests and a check box next to each request to check or highlight green when the change is implemented. You get the idea. I would recommend you use this system or develop your own but the important thing here is to keep a good track of your change requests.

Some people may probably argue that such a system is a waste of time and that we don't have the time. They might tell you that the development phase is in continuous evolution and that we just have to cope with it one problem at a time. To my opinion, we can never be too prepared.

Granted that we can never avoid encountering obstacles but, by organizing ourselves, those obstacles will be easier to overcome. A working environment is always less stressful when we organize and get prepared than when we constantly react.

"Getting organized is NEVER EVER a waste of time"

I can't count the number of times that just a day or two of organizing and planning on how things were going to be done would have saved the company months of delayed schedules and possibly millions of dollars.

I have seen designers that were hard to keep up with as they kept changing things without letting me know and therefore, I had been working on the wrong configuration of structure for days and sometimes weeks.

Now if you believe that getting organized and taking the time to create a good communication with all parties involved is a waste of time, now try to imagine the following. How much time and money it would take to redesign, say, an avionic rack (for example) so it is stiff enough not to get into resonance with the windmilling frequency now that the electrical group has run their wire bundles through it that will have to move and that the primary structure person has to add an Intercostal between two Frames requiring the hydraulic lines to be re-located? *Ouch!* Now that's what makes the schedule slip and the budget bust, not the few days of preparation and organization!

When movies are created, a script is made then a storyboard is drawn which contains every single shot that is going to be taken during filming. It is a dry run that allows debugging and getting ready for the day when the expensive big stars show up on set. Steven Spielberg says he'll work on a storyboard for two years before starting filming. I don't think he considers that preparation phase to be a waste of time, to the contrary.

On the other hand, some will say: Chaos create cash! Good for contractors as they'll get overtime to fix problems, but not so good for the schedule and the budget.

One of the chief designers on a certain program would sometimes tell me, after I would tell him that his design was inadequate (read: crap) just by looking at it, "Prove to me that it is not acceptable". That person didn't trust my expertise obviously. I guess when you're incompetent you tend to think that everybody is also incompetent. So I would tell him: "Why not putting my time and energy into substantiating a design that I am confident will work instead".

We can never be 100% sure that a design or a part will be structurally adequate. To be absolutely sure, we would have to have 100% confidence in the equations and our assumptions that we use, in the material, in the FEM, in the loads from the loads group, etc. But, even if we are only 90% sure about our analysis, we have to show total confidence towards everybody else. We have to have the confidence of a car salesperson! It is always a good idea to ask one of our peers that we trust for a second opinion before dropping the bomb on the designer that he or she will have to change that fuselage Frame from an extrusion to a machined frame!

I sure hope that by the end of the book, you will have gained the ability to determine load paths, do quick and good FBD's (Free Body Diagrams), you will have polished your analysis skills and, more importantly, you will have learned how to develop your social skills. I also hope that you will have acquired the confidence to make good sales pitches on your way to becoming an excellent **aircraft stress engineer**.

Please note that the masculine form may be used in some places to simply lighten the text and is not in any form discriminatory as I have an equal respect for both genders.

Now let's get to it shall we?

2. GOAL

Becoming an Aircraft Stress Engineer is not an aircraft design manual. Although extensive fictitious but realistic analysis examples are shown, the goal is to guide the reader to evolve from someone having a pretty good knowledge in stress analysis towards a fully autonomous aircraft stress engineer capable of analyzing any structure and communicating with his peers efficiently.

Therefore, the intent of this book is not to replace a bachelor's degree in engineering or to duplicate the already existing really good reference books like: *Bruhn, Peery, Roark, Niu, Flabel, ESDU,* etc. It focuses on the engineer (the individual) and is to be your analysis companion. The reference that will actually show you the steps on becoming a proficient structural engineer in the field of aircraft (or other vehicles). Through many components stress analysis examples, its purpose is to share the tips and "secrets" of the trades and the words of wisdom that can only be learned through many years of exposure among experienced analysts in the field of aircraft design.

As you go through the book, you may find that you do not agree with some of the assumptions that were made or, that you would have used some other methodology. This is perfectly legitimate (and encouraged) since nobody possesses the absolute truth. One could say that there are almost as many ways to analyze a part as there are stress engineers.

I wrote this book while imagining that I am sitting right next to a new hire (you the reader) and that, as we're looking at a component of aircraft structure on a screen (or a drawing on paper) together, I am suggesting methods and commenting on the different analysis approaches using the different tools (references) we have available.

No calculation tools are provided. Since the book is purely informative and not aimed at being an actual design tool, no spreadsheet or codes or the like is provided. Furthermore, since most of you are most likely more up-to-date and better programmers or coders than I am, it is the reader's responsibility to make his own calculation tools using his favorite software, or use the ones provided by the company that he's working for.

"Only those who do nothing never make mistakes."

Although the analysis examples are mainly for aluminum parts, the exact same steps or procedures are to be used with quasi-isotropic composite structures. Once the properties are known, the analysis is pretty much the same thing as for metallic parts. Therefore, whether the aircraft is made out of aluminum alloy, graphite-epoxy, wood or *unobtainium* (This is the term used by some procurement people for a very rare and expensive material that is very difficult to obtain; I don't think it even exists on this planet!), the approach is the same.

It is hoped that, with this book, the new (and not so new) stress analyst will be confident knowing what steps to take when analyzing aircraft structures and writing stress reports. Also, more importantly perhaps, he will have gained a general knowledge on how to thrive in the engineering stress department of an aircraft manufacturer. The results are: better stress analysts, better aircraft designs and therefore, better performances leading to better aircraft sales! Everybody wins.

As a second thought, the real reason for me to write this book is to form better stress analysts so that, when my family and I will be traveling when I'm retired, we will then be boarding the aircraft that YOU my friend will have analyzed, and I want to feel safe! Just kidding!

3. PREREQUISITES

Like any mechanic, dentist or electrician, the aerospace stress engineers need tools in order to do their job. I consider this book to be like a recipe book, but in order for a chef to cook, pots, pans and a stove must be available. Those are the reference text books. See the chapter on *References* for a list of essential and strongly recommended references (tools). This guide was written expecting the reader to already have access to most of those books.

The reader is assumed to be an undergraduate student or a newly graduated engineer in aeronautical, aerospace, or in mechanical engineering and has already acquired the main knowledge of the basic stress analysis methods. Regardless on the marks you had in the structures classes, the important thing is to like it and to want to learn more about it in order to improve yourself.

Therefore, I won't be showing how to calculate a second moment of Inertia or perform any mathematical demonstration as I expect the reader to have gone through that in school. It is now time to put that knowledge to profit in creating something concrete...but light!

In the case that structures stress analysis is not your favorite class, that's fine. Just know that there are many things that an engineer can do besides stress analysis. There are many opportunities to develop one's skills in other engineering groups like: design, weights, flight tests, materials, propulsion, systems, etc. This book however only focuses on making a career in the stress department of an aircraft manufacturer or a repair and maintenance facility.

It is the utmost importance to do something we like as a career (and in life) otherwise, it will be very long until retirement.

Some aerospace engineers I know are also pilots and fly Cessna's, some are mechanics and rebuild old cars or old airplanes, and others are hobbyists who design and build their own radio controlled airplanes. All those people are not necessarily better engineers but they have, to my humble opinion, a more practical view about structures and mechanical concepts than others.

Don't get me wrong. It is certainly not a prerequisite to get your hands dirty with motor grease to be a good stress analyst. I'm just saying that by experimenting with real structures, this is to me the best way to really get a feel and a good understanding on the behavior of materials. Tests are the best!

"Perseverance is more effective than good marks" (Stef)

To my opinion, one of the greatest qualities of a person, regardless of his profession or cultural background, are open-mindedness and respect! That goes of course for stress engineers as any other professions. When reviewing someone else's work, for instance, one can find that it's not good or that there are errors. I recommend keeping humble and have an open mind towards the person that did the work. Because, for all I know, YOU could be wrong or you could have interpreted the words incorrectly. Remember, we all make mistakes and I can only recommend you to treat people the same way that you would want to be treated.

I will finally say that one of the best quality for this job is the thirst to learn. I believe that learning new methodologies, new designs, new tools and software, new solutions to problems, is what makes the job interesting and, especially, makes better engineers.

4. WHAT IS A STRESS ANALYST?

A stress analyst is the same as a stress engineer, a structural engineer, a stress-man or, as called in France: *un calculateur* (a calculator)! The maths required to do this job is high school level, maybe college, but, with the use of the pre-programmed computer tools, the stress analyst nowadays merely puts numbers into some cells in a spreadsheet. The trick though is to know what equations to use and what numbers to type in!

When I was studying to get my *Private Pilot License* for single engine airplanes, the instructor would define the skills distribution of a pilot as follows:

- Flying the plane: 25%
- Communication: 25%
- Navigation: 25%
- Weather judging: 25%

Therefore, a mistake in any of the four skills above could possibly ruin your day. An analogy can be done for the stress analyst by listing the hypothetical skills distribution as follows:

- Performing calculations: 25%
- Good communication: 25%
- Good attitude and health: 25%
- Good (clean) looking (nice attire): 25%

Like for the pilot, a mishap in any of the four skills above could lead to a very bad day or a bad career. The following are details on how each of those four bullets can be addressed.

4.1 CALCULATIONS

To perform analyses, the stress analyst need tools. You need references that can be some of the books listed in the *References* chapter, or the Design Criteria Manual for the aircraft your are working on, or any other documentation provided by the aircraft manufacturer company. Of course, you'll need a computer along with an electronic spreadsheet program and, CAD and FEA software.

Some companies will want you to do the final stress report as you're sizing the parts while others will let you do *stress notes* to speed up the drawing release process while the final stress report will be done at a later time, possibly by somebody else. I prefer doing my stress notes in a spreadsheet first before writing the final stress report. It allows me to: 1- Use already made calculations templates, 2- Perform iterations easily for non-linear material behavior or for optimization, 3- easily share my files with others.

> *"P goes on top, and, don't push on a rope"*

Large companies tend to have their own calculation templates that are already formatted to be part of the final report. Although, I do encourage the stress engineer to always double check any tool or template by comparing the results with hand calculations or his own tool. It's always good to know what a particular tool does, what reference it uses, and what assumptions are made before blindly using any "black box" and put our name on an analysis that uses it.

The analysis methodologies tend to be standardized within a company so that all stress reports look alike as if made by the same person. It also helps the DER feel reassured when reviewing stress reports made by different analysts having the same approach and analysis philosophy.

4.2 COMMUNICATIONS

When working in a team, a good communication between individuals and all the different parties is essential.

For the communication skills, I recommend taking classes or reading books about *conversational hypnosis*. Some people call that science: *Neuro-Linguistic Programming* (NLP). The best sales persons, politicians, managers, and leaders use that method.

You're probably doing it naturally without knowing it. Don't worry, this isn't anything that will make people do stupid things, although it could, but that wouldn't be too professional would it?

A very simplistic description of the method is to have the person with which you want to communicate (the designer you want to ask to change his whole design, for example) to have him to talk about his passions or what he loves doing. Then you simply listen to him attentively while looking at him straight in the eyes with a smile. You have to seem genuinely interested on what he's talking about for it to work. Who knows, you might learn a thing or two about many subjects as a secondary benefit.

Then, while the person is still floating on a cloud from talking about his favorite subject, you gently and slowly bring in the request. Chances are, he'll be more than happy to do the change for you as you somehow just became his BFF (Best Friend Forever)...

It'll take some practice to master the technique but after the first year of interacting with many people in this way, you will have become a master conversationalist and have more "friends" than you could hope for!

> *"Good communication is not important, it is essential!"*

I remember arguing for almost an hour about some analysis methodology with someone who's mother tongue wasn't English (which is quite common in Aerospace) only to realize after a while that we were talking about the very same thing! So, after we realized how silly this was, we laughed and went back to our desks remaining good friends. I was glad that neither of us had lost his temper and that each of us gave each other the chance to explain his point of view respectfully.

Sometimes, for people with different backgrounds, the choice of words or the way they see things can be different. It doesn't mean that their understanding of a particular subject is incorrect. Therefore, some patience is definitely needed to avoid jumping to conclusions and say or do something we could regret.

4.3 ATTITUDE AND HEALTH

Always be polite and respectful in all situations but be firm. Try to stay cool, especially when confronted to people with hot heads. You're better than that.

I always admired people that were responsible and adult enough to admit they were wrong. It takes a great deal of courage to say that we made a mistake or that our point of view about something was incorrect. Those people earned my respect and I tend to trust them more than others for that.

My respect and loyalty goes also to those leaders that will never put down someone from their team in front of others when that person had made a mistake. They would talk to that individual in private to make sure that the error doesn't happen again but would rather take the blame upon themselves when having to explain what happened. On the opposite, they will gladly publicly recognize the merit of one team member's achievement without taking the credit themselves.

> *"A team leader is only as good as the weakest member of his team."*

Stay in physical shape! Just keep in mind that, for airplanes to perform well, the stress analyst tries to use the material having the highest strength to weight ratio possible so that the parts are as light as can be and yet be sufficiently strong. That is also applicable to ourselves. If we want to be able to intellectually and physically perform for many years, we have to take care of our body and our mind. Doctors recommend a good diet, a daily 30 to 60 minute walk and a positive attitude.

The structure analysis job requires us to sit on a chair for 8, and sometimes up to 12 hours a day. So, I find it absolutely essential to do some form of physical activities before or after work and during the weekends. Otherwise, you're sure to get back pain eventually and weak legs! Believe me, I know...

4.4 THE LOOK

Most of you have probably raised an eyebrow by reading the fourth bullet from the skills distribution. Well, let me tell you that, by experience, the better dressed the person, the more that person is to be taken seriously and listened to. You can never be too well dressed, although don't push it! The people in the office may be surprised on your first day at the job if you wear, for example, a tie and a jacket but then, they'll get used to it and you will have created your own reputation of being a professional. Wear a nice perfume and please, don't drink coffee and smoke unless you don't intend to talk to anybody.

You could be the best stressman but if you have a bad personal hygiene, wear worn out clothes or cheap shoes, people unconsciously will consider you as a second or third class engineer. I know this is silly and that we should never judge a book by its cover but, whether we want it or not, humans tend to judge right away by how we look. Having a scientific mind, I know very well that how we dress or what car we drive has absolutely nothing to do with how good people are in their own field and yet, I can't help it. I wouldn't want to hire a lawyer that drives a 10 year old car and wears cheap cotton pants because I would unconsciously assume that he is poor because he surely doesn't get too many cases and thus ... is not a good lawyer.

Engineers wearing nice clothes tend to go up in a company quicker. That is a fact! If you dress like a manager, people will see you as a manager and it won't be any surprise to them when you become their boss. Dress for the job you want.

> *"Dress for the job you ultimately seek."*

4.5 REAL LIFE STORIES

Friday afternoon at 4 o'clock, the manager comes to your desk and tells you that there is a big dent on the side of the aircraft, and then he says: "Don't spend too much time on this, (but) just let me know if it's ok to fly with this discrepant condition this weekend!" Well the quick answer to that is: NO! But the answer that he may want is a quick YES! Otherwise, there is no 5 minute jobs!

First, you need to go see the damage for yourself. It is usually worse than what the manager implied. Then, when discussing with the inspector that reported the problem, I usually ask him: "What are your intentions? Are you planning on removing and replacing the part or you want to repair it?" When I tell them what the repair would look like, they usually opt to replace the part since a brand new aircraft with a patch doubler in plain sight having two rows of fasteners all around won't most likely be accepted by the customer who paid millions of dollars for it..."

Note that, to some production managers, the schedule is more important than anything else. So the total responsibility bears on you to ensure that safety is the prime directive. Remember that YOUR NAME will appear against that decision not the manager's or anybody else that will most likely try to put pressure on you to sign: ACCEPTABLE AS IS!.

Sometimes, even when the analysis showed a positive margin of safety for the non-conforming part (not to the drawing), I often got away with the following disposition:

"*Part is structurally acceptable to engineering as is. Condition with proposed repair must be approved by Quality Assurance manager for aesthetic acceptability*". Quite often, the part was removed and replaced. Why didn't they go to this guy first? Maybe they somehow knew that it would look like hell, or they didn't care, who knows. When I got bombarded by many people pushing me to accept the condition as is, I would simply reply: *"Would you buy a multi-million aircraft with such a repair? I know I wouldn't accept a $25,000 car with a visible patch but again, it might be just me..."*

When somebody, usually not your boss (as he knows better), asks you how long it will take for you to do your analysis, I always tell them a worst case time-frame. Therefore, if I do it quicker, I look good, if I do it in time, I still look good.

> *"Arguing with an engineer is like wrestling with a pig in the mud! After three hours you realize that the darn thing loves it!"*

Someone once told me: "When doing a job time estimate, consider including a good buffer since your manager is most likely going to cut it in two anyway. We all have to justify our jobs I guess". Well, from my experience, the vast majority of bosses and managers are very good (fortunately) and will listen to you and trust your judgment. So it's always preferable to be as realistic and honest as possible.

Do NOT assume anything! I recall a large program where many sub-contractors were involved. In that case, (not the actual event) company *A* was designing the forward fuselage and company *B* was designing the aft fuselage. Both appeared to have "assumed" that the other was designing the joint connecting the two assemblies together and guess what happened! The joint analysis fell in the crack, as we say, as nobody did it.

The contract may have not been clear on this however but, a simple and quick phone call would have removed any doubt and saved the prime manufacturer a lot of time and money! When in doubt, ask! You'll be surprised on how many people assume things on the basis that they think they are obvious to everybody or that somebody ***else*** must have taken action. Again, this is another assumption...

5. DISCOVERING THE PASSION

It all started when I was a kid and was watching "*The Thunderbirds*™" television show that was playing in the 60's. I was fascinated with all the different aircraft and other vehicles and was hoping that one day I would be able to develop such fine flying machines. I clearly remember a particular episode when they had to transport a huge rocket across a suspended bridge. The structural engineer that had designed the bridge warned that the bridge could support the weight of the rocket across the full span of the bridge but wasn't designed to support that weight at a concentrated location on the span.

I was still very far from understanding shear and moment diagrams but I could understand the principle of a distributed load versus a concentrated load intuitively grasping that it made sense somehow that if the same weight is spread over a greater surface, then it wouldn't be as critical as if it were all in one spot.

In the story, as expected, as the truck carrying the huge rocket was slowly moving across the suspended bridge span, the cables supporting the span started failing and, as expected, the *Thunderbirds*™ were called to the rescue. Absolutely beautiful!

I believe that this TV show sparked my passion for aircraft and structural analysis. My father was a machinist and could do pretty much anything out of metal, but he didn't do anything that looked like an aircraft structure. I didn't know anybody who owned a plane or was a pilot and I had never boarded an airplane before the age of 20, so I was curious to see and learn how they did it. To me, what differentiated an aircraft structure from another structure was the shiny aluminum color, the rivets and the lightening holes! Those round holes seemed to be the trademark of any structure that flew.

Nowadays, even the architects will borrow those features and incorporate them into their visible building structures to give it a futuristic look and not necessarily for the intended reason which was to reduce the weight.

I was so fascinated by those features that I would "invent" and draw spaceships with beams having lightening holes in them. That looked so cool!

Then I went to university and it seemed that I was learning more about maths than about structures. It is only when I started working for an aircraft company that I really got to see the real airframe structures and how simplistic the structure analysis methods were. Somehow, that disappointed me a bit as I was expecting some very complex calculations. I guess it's like any magic trick, once you learn how it's done, the "magical" aspect of it disappears and then, what used to be a feeling of admiration and curiosity, becomes a feeling of slight disappointment, along with a satisfaction and a peace of mind however.

For engineers, there is always a feeling of satisfaction, of relief or peace whenever we understand the root cause of a problem. After that, it is much easier to find a solution to that problem. A passion for aviation is not a prerequisite to become an aircraft analyst. Actually a lot of the people I worked with, if not most of them, were surprisingly not really passionate about those flying machines. However, most seemed to be sharing one thing: curiosity! You could tell from the spark in their eyes whenever there was an intriguing problem or a new mechanical concept to be understood.

On a couple of occasions during my career, I have questioned myself and explored the idea of changing career and become a teacher, a science fiction novel writer and even a film maker. Within the science fiction field, I could then invent all sorts of spaceships (having lightening holes) and somewhat see them "fly". As the years went by however, as I was working on one aircraft program after another and was having a steady income, it was becoming more and more difficult for me to find the courage to make such a drastic change.

During my years in university, I recall that all of us in the mechanical engineering program were in our low twenties except one student that was close to forty. This guy had worked as a car mechanic for many years and, after reflecting on his life and his passion, had come to realize that he wanted to understand how vehicles were designed and be part of it. I don't remember the details but I remember that he had to make big sacrifices to get where he was.

I truly admire those people that dare changing career, sometimes more than once during their life, in order to pursue their dreams and ambitions. It takes courage and a lot of will, often despite the comments of discouragement we might get from our family and friends.

It is only at the end of my career as a stress engineer that I decided to write and "teach" at the same time by creating this book.

Don't get me wrong, I loved doing my job! However, I must admit that when I started working in aerospace, I thought, naively perhaps, that eventually I'd be working on weird vertical take-off supersonic transport aircraft and spaceships made out of *unobtainium*. That is probably why I was attracted to science fiction since I could have somehow "materialized" my expectations.

Regularly questioning ourselves about whether we, for instance, are in the right job, live in the right location and are with the right person, is, to me, essential and a sign of an healthy mind. It is only when we sincerely answer YES to all those questions that we can really be happy.

As I'm writing this book, some aircraft companies are currently developing super sonic new aircraft and designing space rockets. So I guess the chance of working on those vehicles will be for you my friend. I believe that great advancements in aerospace lie ahead for this generation of aircraft engineers.

Nevertheless, that flame, that passion is what kept me going for all of those years. The need to learn new things all the time. Things like learning how to create some codes in a spreadsheet to calculate Margins of Safety for each of the thousands of load cases, learning how a hydraulic actuator works, learning how a certain part failed on one of the aircraft in the fleet, etc. My mom likes to say: "A day without learning anything new is a day wasted".

So don't waste your time! Learn as much as you can and share this knowledge with others. That's the most valuable gift you can make.

> *"A day without learning something is a wasted day"*
>
> (Sammy)

6. SO YOU WANNA BE A STRESSMAN?

One day this drafter (designer) comes to me and asks: *How long does it take to learn PATRAN™, I want to become a stressman?* I knew there were two answers to that question and he wouldn't like any of them. I wanted to avoid the long answers so I answered with a question: *Why do want to become a stressman? – Because it pays more!* he said.

That was a legitimate reason. However, the company we were both working for at the time would only consider engineers as stress analysts only if they had a bachelor's degree as a minimum. So I told him that even if he was to spend hundreds of hours trying to learn that tool, he would be doing that in vain as the company wouldn't probably pay him more. So I couldn't hold what I had in my mind any longer and told him: *All stress analysts here did a four year degree in mechanical or aeronautical engineering as a minimum, and a good number of them have a master's degree as well. Most of us learned Patran over many years through classes and on the job but most of our calculations are done by hand and not with that software. To do stress analyses requires the proper education and academic training to know what you're doing. So good luck with that!*

I don't think he quite grasped what I meant by having to know what you're doing. To him, that job was easy. He would make pretty pictures using that FEA software which would simply spit-out the answer and then, he would sign-off drawings. Piece of cake.

Things done by a professional always look easy to a person not familiar with that particular art. Only somebody that has done that exact same thing can recognize the difficulty and all the hours of training required to make it look easy. That is applicable to any field, whether it's in sports, arts or engineering.

> *"Teaching by example is not the best way to teach, it is the* ***only*** *way"*
> (Albert Schweitzer)

Although the calculations are not very complicated, a lot of knowledge is required behind those calculations. It would be comparable to a doctor that prescribed medications. Writing the prescription is relatively easy. I say relatively because it seems very hard for some of them to represent the letters of the alphabet in their readable form... But knowing what the patient's problem is and what treatment to do, **is** the reason why they spend so much time in school and in training.

In structural analysis, the young graduates learned some tools in school but they are somewhat lacking supervised hands-on experience before they hit the job market. Some will do internships in some companies (an aircraft company, hopefully) but are often given a task that won't require too much time from their mentor. The mentors have their own job to do and don't have much time for supervising adequately an undergraduate student or even a recently graduated engineer, or anybody in fact.

This book certainly doesn't replace a mentor. However, for the lack of a better way, I think that it does close a gap between the newly graduate and the fully autonomous stress analyst. The reader is literally taken by the hand over complete analyses of actual parts from start to finish. From the determination of the loads, the Free Body Diagrams, the assumptions, the tools used and the presentation, the reader is guided all along the way as if a mentor were sitting next to him. A video would have been good also but a book is quicker and easier to consult.

7. DO'S AND DON'TS

Here are some design guidelines that, when followed, will make for a better aircraft, easier to maintain and should also get rid of a lot of durability issues. You'll also be a more respected person.

7.1 DO NOT ASSUME ANYTHING!

Throughout the book you'll be reading something like:

- *Assuming a simply-supported beam*, or
- *Assuming that the stiffness is the same as...*, or
- *By conservatively assuming a* ***60-40*** *load distribution on the lugs*, etc.

Nevertheless, I'm telling you this: **DO NOT ASSUME ANYTHING!** Well, although this might seem contradictory, it is not as I'll explain here.

For some reason, the verb: "***to assume***" has been used everywhere in stress analysis reports which should really had been the verb: "***to hypothesize***". This is when we are not exactly sure about a particular condition and therefore we play **safe** (and save time) by making a hypothesis (assumption) that we know for sure is conservative (safe).

For example, in the case of a column that is attached to a structure at its top and bottom, I'm **assuming** the column to be pinned at both ends to be safe. Since I don't know exactly what the end restraints (rotational stiffness) are, unless I were to do some tests or some very detailed and time consuming FEM, I know that by assuming the column to be pinned at both ends is safe since the calculated compression load capability is smaller than if ANY end restraint value were to be used.

Your hypotheses must always be stated clearly and you shall stick to them. Don't change your hypotheses on the go because you got a negative MS. Be consistent! If you do change them either because you think it was too conservative, because you realize a different hypothesis was used on another program, or because the DER didn't like it, you should be redoing the analysis completely from the start showing only that one hypothesis.

Now, what is the thing about **not assuming anything**? Here are some examples of cases where no assumption should have been made:

- I'm analyzing the forward fuselage and an outside company is analyzing the mid fuselage. Since neither my boss nor our design criteria mentioned anything about the interface joint analysis between the Fwd Fuse and the Mid Fuse, that outside company must certainly be analyzing that joint...

- The designer I normally work with is working on an extended version of aircraft *model A* called *model B*. Since that designer didn't mention anything to me about any differences or changes in the thicknesses or materials between the two models for the structure I'm involved with, then everything must be the same...

- I work very hard to do the best job I can. To be the best stress engineer in the company. I read books on stress-analysis at home sacrificing quality time I could spend with my family, I do the extra mile to make sure that nothing falls into the cracks (nothing is missed), then my boss will surely notice that and reward me accordingly when the evaluation time comes...

All the above are **assumptions** that you must **NOT** make! Whether it is to save the company you work for time and money, to ensure the integrity of the structure is adequate or, that you get paid according to your knowledge, skills and efforts, there are times when you need to speak up. Of course it requires more efforts and yes you may possibly cause quite a stir but you won't gain anything by taking the path of inaction.

To summarize, a simple rule to remember is: if the consequences of being wrong in the assumption you make could be costly or unsafe, then don't make it! Take action! Go ask questions, interrogate the digital model, maintain a good communication with your boss and other people.

To my opinion, it's better to be thorough and be a bit late than to be quick and miss important points. Because in the end, if the company wins, you win and vice versa.

7.2 FASTENERS AND JOINTS

- Joints must be balanced. There must be an equal amount and same size of fasteners on both sides of a joint. Newton's 3rd law! Ideally the joint should be mirrored exactly.

- Rivets (solid and blind) should always be working in shear. A small tension load is fine but if the tension load is above 100 *lbs* (or 1/4 of the shear strength), consider using Hi-Lites or changing the design to make them work in shear. Then if the Hi-Lites have a significant tension load above 1000 *lbs* (or 1/2 of the shear strength), use tension application collars. Specifically if the tension load is cyclic like at each *GAG* cycle.

- Use fasteners sizes according to the thickness of the parts being fastened together. A joint should be bearing critical not shear critical. Shear critical joints will cause a "zipper" effect. That is once the first fastener fails in shear for a certain ultimate load, the rest of the other fasteners in the joint will then fail in sequence as the sum of the remaining fasteners cannot sustain that same applied load.

- Leave enough room for oversized fasteners or inter-pitched ones. Some designs scallop the edge of a part by going around the fasteners. That may save a little weight but seriously limits repair possibilities. Consider removing weight somewhere else.

- Do not mix fasteners types and sizes in a same joint. When mixed, the load will be taken by the larger diameter and stronger material fasteners. For example, aluminum rivets mixed with titanium Hi-Lites will barely take any load. Also, the mechanic installing those fasteners has less chance of errors if all the fasteners are the same in a joint.

- Lugs (and joints) are not the place to try to save weight. However, try to have a MS above 0.25 including the fitting factor.

- When bushings are installed by interference fit into a part with a different material, one must check the stress due to the interference at both temperature extremes (hot and cold). For instance, make sure that, for a steel bushing installed in an aluminum lug, the hoop stress in the lug doesn't get too high at cold temperature thus reducing the lug's fatigue life due to the GAG cycles. Also, make sure that there is always an interference at high temperature to prevent loose bushings conditions.

- When two parts with two different materials are fastened together, ensure to check the thermal loads at both temperature extremes. You possibly may want to consider combining this thermal load/stress with the one from the appropriate/most critical flight load conditions.

- Use a "3D" bolt group tool for groups of fasteners of a dozen fasteners or so maximum. The theory behind such a tool (ref. 2.), assumes that the parts involved are rigid. As the number of fasteners increases, the part size is also increased and a certain deformation from the part has to be considered somehow. That is especially true for long and narrow bolt groups. In which case, the tool can still be used but a stiffness reduction factor must be used for the fasteners that are further away from the bolt group center.

- Checking fasteners and bolts is necessary but quite often the parts that those bolts are connecting have more critical modes of failure like tension fitting/clips, bearing, bending, tension and shear. Or may allow undesirable excessive deformations.

- In a clevis (double lugs), assume a 60/40 distribution instead of a 50/50 load distribution to take into account slight geometrical differences that would cause one lug to be a bit more loaded than the other.

- It is a good practice to check a lug for a side load by assuming a value of 5 to 10 percent of the total applied load.

- In a fitting or a complex joint, a fitting factor of 1.15 should be applied in the margin of safety calculation as specified in the (ref. 25. *FAR §25.625)*.

- Try to avoid using Monel solid rivets in the case where the shear load is too high for an aluminum rivet. Monel rivets are hard and therefore are difficult to install with a bucking bar or a squeezer. Often, the parts being joined get deformed or damaged during installation. Use Hi-Lites instead if possible. However, monel solid rivets (no aluminum) must be used with titanium due to material compatibility.

- Hi-Lites and Hi-Loks fasteners should always be installed in an interference fit condition as recommended by the manufacturer.

- Always use a minimum of two fasteners in any joint. Never only one! One fastener is a pivot and can only cause trouble. Although, one screw or bolt can be used for struts.

- Don't use set (socket head cap) screws as structural fasteners. Set screws may look cool but the head often shears off at installation. Use standard *NAS1131* type screws or equivalent.

- All screws should have a minimum shear strength of 95 *ksi* and a tensile strength of 160 *ksi*. This is also applicable to bolts. However, bolts with a 75 *ksi* shear strength can be used in less structural applications. A fastener is considered a bolt when its diameter is at least 1/4 inch.

- Never have threads in bearing.

- When using high strength tension bolts, the installation torque is usually quite high and so is the bolt preload. Be mindful of that when selecting the washers. Do not use too thin aluminum washers as they might deform (get squashed like pancakes) under the very high pressure causing the bolt to lose its preload over time.

- When checking (analyzing) screws or bolts, check the shank for a combined tension (due to the preload), shear and bending. Then check the minimum section at the threads due to the tension only. I do not recommend applying a 1.5 ultimate factor to the bolt preload as I deem it too conservative.

- Shear pins are tube-like bolts that are designed to break at a certain shear load. Calculate the MS using the shear load only and the average shear area without considering any bending and that should be pretty close to the tested value.

- In a clevis, ensure that bolts don't yield or show permanent deformations under limit loads. Otherwise, the mechanic may not be able to remove and replace it.

- Plastic bending may be used in bolt analysis but should be combined with the average shear. If you have to use plastic bending to get a positive MS for a bolt, chances are that the bolt would yield under limit load which is not acceptable. See above.

- In a fail-safe design, more than one bolt is used. For instance there could be two "tubes" that fit one inside the other that are called bolts along with a usually small (1/4 inch) bolt at the center to keep everything together. The 1/4 *in.* bolt doesn't take any load and each of the tube takes its share of the load according to its relative stiffness. Do not consider the two tubes as one. That would lead to a non-conservative allowable. For a fail-safe design, one bolt (usually the bigger one) is assumed failed and the other bolt is checked under the limit load using the ultimate allowable.

7.3 REPAIRS

Fastener pitch shall be between 4D and 6D. Providing there is no inter-rivet buckling, try maximizing the pitch to allow for the next size fastener to be installed in case of a repair. Note that 3D is low as pitch and should not occur in high stress areas. Sometimes, in case of a low pitch, one could plug one of the two holes using a "soft" rivet *(MS20426A)* double flush on both sides. Plug the two parts separately. In this case, an oversize fastener could be installed in the remaining (good) hole to compensate for the missing rivet. If need be, install oversized rivets on either sides of the plugged hole but do not install a Hi-Lite for example as it will attract more load next to the plug and we want to try to keep the load away from that area.

> *"If it ain't broke', don't fix it!"*

7.4 FRAMES AND RIBS

In a frame or in any machined beam, the web should be thinner than the flanges. The web mostly works in shear and is subjected to buckling. Since buckling is not a mode of failure, most of the beam's material should be located in the caps, where crippling would occur.

In a significantly curved frame or beam, make the inner cap straight between the radial stiffeners. The portion of the inner cap (not fastened to a stabilizing skin) behaves like a column. If the column is already curved, its compression allowable will be much lower.

Always have an integral stiffener whenever a load changes direction in order to take the "kick" load.

Don't allow too small fillet radii. There are plenty of other locations to remove weight. Try to have the fillet radii as generous as possible in function of the thickness. Generally, a 1/16" fillet radius is too small and may lead to fatigue issues.

When a frame or a beam is curved and/or has a varying cross-section, calculate the section properties from section cuts that are normal to the average neutral axis of the beam. This will provide the smallest height for the section. However, actual flange thicknesses shall be used and not the thickness obtained from the angled section. As the section in this case may not necessarily be normal to the OML, the tension *(P)* and shear *(V)* must be transformed into the section's coordinate system.

7.5 SOCIAL BEHAVIOR

Don't be afraid of losing your job! I know, we all need money and we all have responsibilities like a family, a mortgage, a car payment, etc. However, we should never be worried about losing our job or by getting laid off. Believe me, no matter how good or "indispensable" you think you might be, or how much you've laughed at your boss's jokes, if a company has to reduce its staff, it will! So why worry? Whether we worry or not about something, remember that this will absolutely have no effect on the outcome. It might just keep you awake at night, prevent you from concentrating at work, make you do mistakes and make you feel miserable at home.

All aircraft companies have ups and downs. It is cyclic. Since it normally takes 4 or 5 years to certify an aircraft, the duration of a cycle is about that same time; for commercial aircraft companies anyway. Through the years, I've never encountered a company that didn't go through the cycle of hiring a lot of people and, a couple of years later, firing a lot of people. So much for a "permanent" job. For that reason (and others), I've always preferred being a contractor and surf the waves of the companies that were on their up phase of the cycle. Swiftly shifting wave when the cycle was getting to its down phase (usually when the overtime is cut as a first warning of the coming beheading).

Don't get me wrong, some people do stay for the same company for 35 years and manage to go through the cycles of ups and downs without work interruption. However, sometimes during the down times, they are asked to do other tasks that may be more clerical than what they would normally like to do, that is: doing stress analysis.

Another reason why you shouldn't be afraid of losing your job is to remain true to who you are, to your values and to what you believe is right. Unfortunately, some people won't speak up when they see something they don't like or that is plain bad (design, analysis method, working conditions, etc.) in fear of delaying the schedule, not getting a raise or a promotion or a bonus or, getting fired. The one that speaks up is either considered as the black sheep in a group or as the hero.

7.5.1 RAISING THE RED FLAG WHEN NECESSARY

Regardless if you are voluble or a quiet person, to be ethical in the profession, an engineer MUST speak up and raise a "red" flag if one encounters something that affects safety or possible problems. So if you see something that makes you uncomfortable, speak to your lead or to another lead that you trust and send him an email for the record. Send also a copy of that email to your manager or even the V.P. if need be. Always cover your "6" by putting those findings in writing. As some are better at blaming others than taking action.

I know of one particular stress engineer who quit because he believed that a design wasn't airworthy and that his boss would go ahead with the flight test no matter what. Some people said he was chicken but I thought he was the bravest person among all of us. There might have been other reasons as well as to why that engineer left, we'll never know. However, you certainly don't have to leave your job if you are made aware of something wrong but you certainly can't ignore it and, as a result, you have to let your superiors know about it.

If you want to be able to look at yourself in the mirror then take action!

Young engineers are not experienced enough to see a possible problem and, old ones are all too often disabused or disinterested to care or to fight anymore. Engineering is no politics! There shouldn't be any dealing to do or threats to endure. We shouldn't have to worry about how we look to our superiors by telling them the truth about unforeseen delays or problems.

Employees (contractors included) should be rewarded for bringing up something they are fairly certain could be of a problem to the safety, the schedule, the budget, the company's reputation, goals and objectives. I have seen some people (section chiefs, managers, project engineers) that were hiding delays, problems, lack of expertise or staff so that their boss would be happy about them. This kind of behavior has led to millions if not billions of dollars in additional costs or lack of revenues.

Because as certain as the sun will rise tomorrow, eventually the truth of the matter always ends-up coming out! And what would have cost $1.00 to fix at the time will cost $10.00 or more later.

Always stay polite and respectful but say what you have to say. Be firm and confident. It's always good to seek the opinion of somebody more experienced to make sure before dropping the bomb. You'll earn respect from others and yourself. If you get tagged as the trouble maker, then you shouldn't probably be working there as that company (or one of your superiors) is not professional enough!

Aerospace engineers have knowledge, expertise and the intellectual power to solve pretty much any problem. Most of us are class "A" engineers and somewhat, sometimes, we end up working for people with less or no engineering skills. This shouldn't be a problem, but they shouldn't come and tell us how to do our job! Elon Musk, the CEO of *Tesla Motors*™, said that he hires brilliant engineers not to tell them what to do but so that THEY can tell him what to do.

7.5.2 THE THREE PHASES OF A SALES PITCH

Sometimes, for some reason, a boss, a salesperson, or a brother-in-law **really** wants something from you. They may want you to do overtime during a long weekend, to buy their gizmo, or to lend them your new car... There are usually three different behavioral phases that might occur during the interaction "discussion" of the sales pitch. Those will only occur if you say **no** of course.

The description below shows the typical pattern that, after you read it, will be easily recognizable and, hopefully, will give you more power over the situation if it ever occurs.

1. The friendly phase: The "salesperson" is very happy (that's a hint right there because it is usually not the case...). He will flatter you, give you some goodies (coffee, pen, attention, etc.), tell you how good and brilliant you are and that, whatever he wants you to do, is going to be a great opportunity for you. Then you say: *–NO thank you!*

2. The drama phase: After your refusal, he gets very sad. Then, he will try to get you to have pity over someone (usually himself) or have some remorse because the world is going to end if you don't say yes; and that will be because of you, of course! *–Mmm not interested!*

3. The insults phase: After you say: *–Thanks but **no** thanks* for a third time, he might get angry and treat you with disdain. Then, all of a sudden, you're not to him that nice and intelligent person anymore. He might say that you are selfish, stupid and even an a$$hole...

If and when such a situation should arise, stand your ground. Say **no** politely but firmly. You don't have to provide any excuses or justifications and most of all, don't feel bad. Just smile and walk away. You'll earn their respect, or at least they'll know you're not the sheep they expected you to be and they'll be asking somebody else from now on for their favors. But most importantly, you'll earn self-respect.

Fortunately, only a very small minority of people would act like that towards people/employees. Pretty much all the bosses that I've had were very respectful and friendly.

Of course, you may want to work during that long weekend and that's fine. As long as you're *OK* with that, then that's what matters. For instance, by accepting those kind of special requests may help you, for example, getting a Friday off once in a while. I think it's fair.

8. ANALYSIS ROAD MAP

While this book is a very good guide to the less experienced stress engineer and can be pretty useful, a good friend of mine suggested that I'd do a guide for the guide. So I thought about it and, he was right, there should be some kind of a condensed tool, like a map, that could be consulted rapidly in order to quickly direct the stress analyst into the right direction for the analysis of the part that needs to be performed. A map and a compass for the stress analysis steps to take... Great idea!

I had seen some people do what we call *Mind Maps* in order to help the brain better remember information. This is something visual, colorful, with sketches and doodles. The one I'm providing, shown on the next page, is more like a radial flow chart that I call a: *Stress Analysis Road Map.*

Here is how the *Stress Analysis Road Map* works:

1. As always, the very first thing to do is to draw the Free Body Diagram by applying the external loads to the part or the assembly that you want to analyze. The FBD MUST be balanced. Note that there shall be no moving parts included in the FBD like linkages, struts, hinges and doors, or the like. Force vectors must be along the longitudinal axis of struts and cables. Contact forces must always be applied normal to the surface.

2. Section cuts are made at the locations where the analyst thinks they are critical and at many locations when in doubt. They are made through holes and cutouts, through joints and all locations where the section properties change, when applicable.

3. Determine the internal loads at each section cuts. The FBD at each section must balance and the loads must be extracted at the neutral axis of the section.

4. Calculate the axial (tension and compression), bending, and shear stresses at the most critical points for each section. The most critical points are generally the furthest from the section's c. of g.

5. At this point, you are ready to use the *Road Map* to help determine the analysis steps necessary to get the different Margins of Safety.

6. Every part or assembly can essentially be idealized as a beam. From that beam, the part can be dissected into, somewhat isolated, caps or straps, columns, webs and flanges. Since the most critical type of internal load is compression, because it leads to instability at stress levels sometimes much lower than the yield strength, you'll notice from the *Road Map* that the tension checks to perform are far less extensive than for compression.

7. By looking at a particular isolated component (column, flange, etc), and by knowing the stresses at the stress points of the component's section, the analyst can follow the *Road Map* to know what should be analyzed.

8. For example, for a wing spar, the upper cap being in compression at the section that we're looking at, the upper cap can be considered as a column simply supported (end restraints) at the ribs. The column may be considered long and a distributed air load (and an eccentricity) can cause some bending in the column. Thus, a beam-column analysis must be done in order to get a MS for the general stability.

9. In addition to that MS, other checks must be done for the local stability like: free flange and crippling.

The examples detailed in *Parts II* and *III* use most of the different steps shown in the *Road Map.*

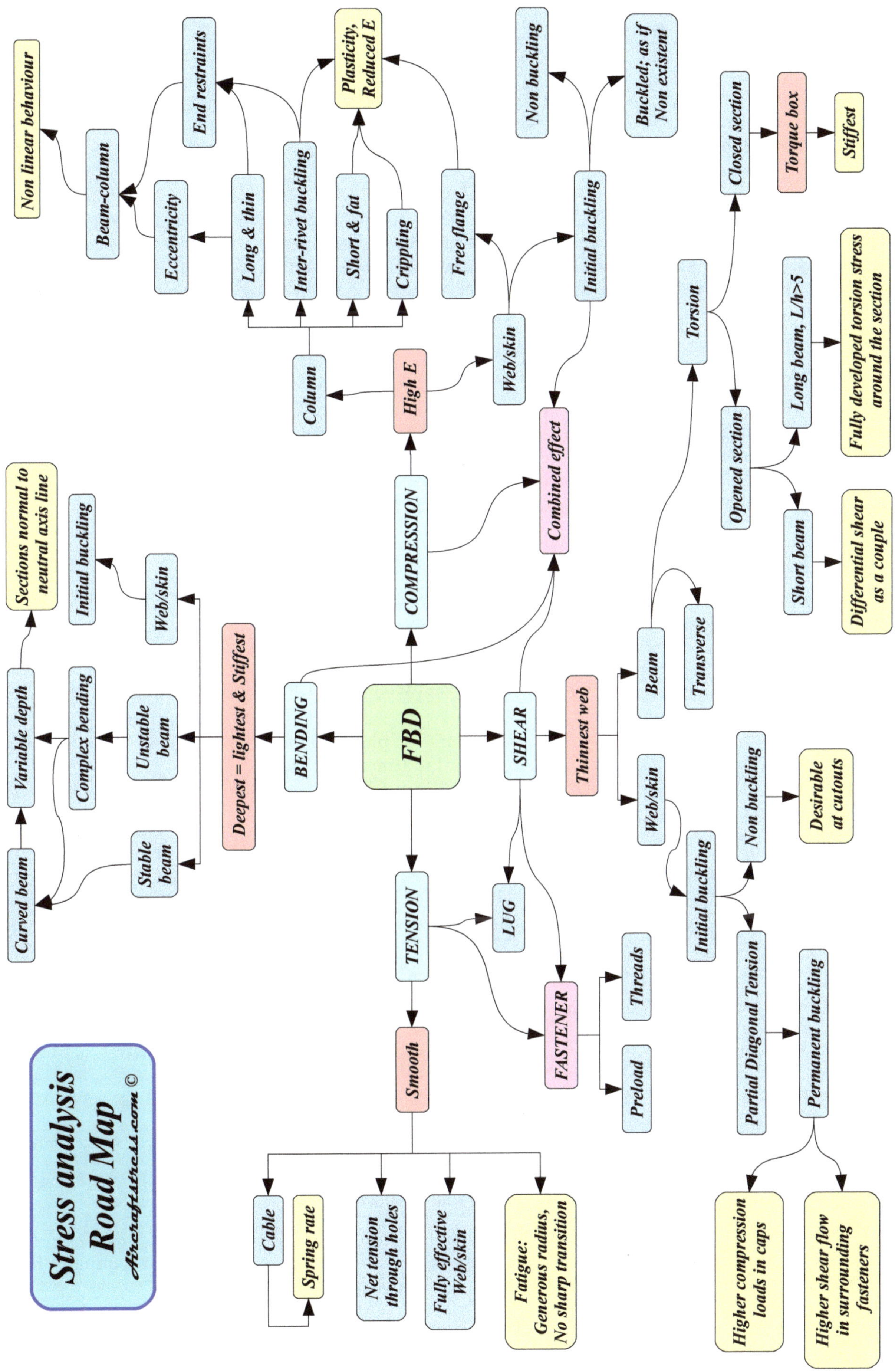
Stress analysis
Road Map
Aircraftstress.com ©
FBD
BENDING
COMPRESSION
TENSION
SHEAR
Deepest = lightest & Stiffest
Stable beam
Unstable beam
Web/skin
Curved beam
Variable depth
Complex bending
Initial buckling
Sections normal to neutral axis line
High E
Column
Web/skin
Long & thin
Inter-rivet buckling
Short & fat
Crippling
Eccentricity
Beam-column
Non linear behaviour
End restraints
Plasticity, Reduced E
Free flange
Initial buckling
Non buckling
Buckled; as if Non existent
Combined effect
Smooth
Cable
Spring rate
Net tension through holes
Fully effective Web/skin
Fatigue: Generous radius, No sharp transition
LUG
FASTENER
Threads
Preload
Thinnest web
Beam
Transverse
Web/skin
Initial buckling
Non buckling
Desirable at cutouts
Partial Diagonal Tension
Permanent buckling
Higher compression loads in caps
Higher shear flow in surrounding fasteners
Torsion
Closed section
Torque box
Stiffest
Opened section
Long beam, L/h>5
Fully developed torsion stress around the section
Short beam
Differential shear as a couple

9. THE STRESS REPORT

You may be working on the detailed design phase of an aircraft program for years making spreadsheets, detailed FEM's, stress notes, signing off drawings but, although essential, those are not the end products of a stress analyst. For stress engineers working on an aircraft to be certified, their final product is the *stress report* or *certification report*.

All the work that was done before (by other people quite often) has to be revisited to create the final stress report. Years ago, that report used to be much smaller than the stress notes where only the most critical MS were shown. But, lately, it appears that companies want a very complete stress report. One reason I think is that the stress report will also be used by MRB people to support the production of that aircraft and will need all the analyses they can get to help speed-up their response time.

Since I only dealt with the FAA and *Transport Canada*, I couldn't know what other certifying agencies in other countries want to see in a stress report. However, I'm pretty sure that the information is pretty much the same even if the requirements may slightly vary. That general information is detailed in the following pages.

9.1 THE MARGIN OF SAFETY PRESENTATION

Figure 9-1 shows a typical MS presentation in a stress report. There are four blocks which contain essential information to the reader which summarizes the analysis performed. Each block is explained as follows:

- *Margin of Safety:* This number is always presented with two decimals. If that number is greater than "1.00", then it is usually presented as: > **1.0**. This number shall **NEVER** be negative!
- *Part, location and load level*: The part name (or part number), the section or the location in aircraft coordinates along with the load level (Ultimate or Yield).
- *Load case and material*: The most critical loading condition from which the loads were extracted. The material of the part analyzed.
- *Mode of failure*: This is the analysis methodology used to derive the MS. It could be: Column buckling, plastic bending, net tension, etc.

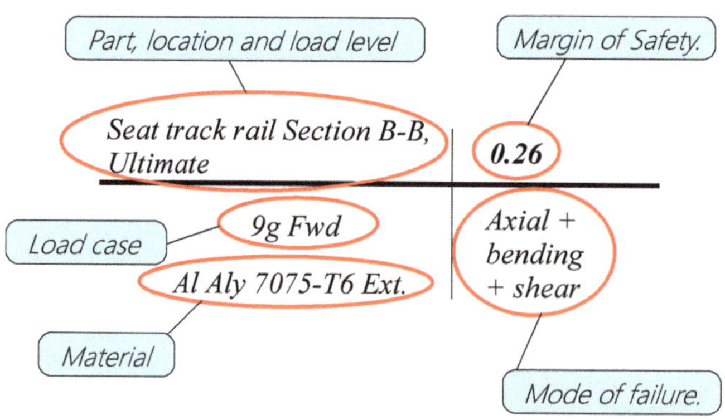

Figure 9-1 *– Margins of Safety presentation.*

Of course, there are variations to the way the MS is presented as every aircraft company has its own format. However, the basic information should be the same.

9.2 STRESS REPORT TEMPLATE

Every analysis must be done by following the same steps, whether it is a simple shear clip or a complex wing box assembly. Again, the order of the steps to present the analysis may vary from one company to another but it is mainly as shown below.

1. Description
2. Summary of minimum margins of safety
3. Loads
4. Material allowables
5. Section properties
6. Assumptions
7. Stress analysis
8. Connections
9. Conclusion/recommendations

Each step is detailed as follows. Parts *II & III* of this book show many examples using those steps.

9.2.1 DESCRIPTION

We start with the larger picture, like the whole airplane, and then we show detailed pictures of the part to be substantiated. So usually, a picture of the airplane showing in which area the part to be analyzed is located. The following information must be shown:

- Picture of the part
- Name and Part Number
- Its use and how it works — tell a little "story"
- Operating conditions/temperature
- Dimensions
- Orientation in the global coordinate system
- Material and temper

9.2.2 SUMMARY OF MINIMUM MARGINS OF SAFETY

This is very important in order for a reader to know, not only what is the minimum MS but, what is the critical mode of failure at a glance. This table is often used in order to evaluate if a weight reduction analysis on that part would be worth it. In general, all the MS are shown even if they are very large (>1.0). We show as much information as possible. See *Table 13-1* as an example.

- Nomenclature
- Part number
- Material and temper
- Mode of failure
- Critical load case
- MS
- Page number

9.2.3 LOADS

In this chapter, one should describe the assumed load paths.

- Explain what is the most critical load case; flight, emergency landing, thermal, etc. Ultimate or Limit loads.
- Free Body Diagram of the most critical load case(s) in all three planes if necessary. The sum of the forces and moments MUST be zero.
- Arrows representing the reaction forces and moments are shown with a slash, or a different color than the applied loads.

9.2.4 MATERIAL ALLOWABLES

- The latest material allowables from your company or from (ref. 8.) shall be used.
- Ensure to use the allowables for the expected worst operating temperature.
- Ensure to use the allowables for the grain direction *(L. LT. ST)* that matches the part at the analysis location.
- Use joints allowables whenever possible for fasteners allowables.

9.2.5 SECTION PROPERTIES

- The sections are made at the most critical locations.
- Sections are made normal to the part's elastic axis while keeping the minimum thickness of the web and flanges.

9.2.6 ASSUMPTIONS

All assumptions regarding the load paths or how the part behaves must clearly be stated at the beginning.

9.2.7 STRESS ANALYSIS

The analysis must be done in a logical way. From the most general to the most detailed. For example, one must perform a column buckling check (general instability) of a strut in compression first before analyzing the rod-ends and the bolts.

- The analysis should flow from top to bottom without having to jump too much.
- Every formula must be clearly written along with its reference.
- Before using a reference, make sure that it follows the politics of your department.
- The formula must then be shown with the numbers so that anybody can verify the accuracy of the answer.
- Loads and stresses are usually shown without any decimals. We can hardly justify showing a stress of 34,567.891 *psi* when the allowable is rounded off to the 1,000 *psi (1 ksi)*!
- For each mode of failure analysis, show one detailed analysis example for the most critical load case and the location, then show a table that summarizes the results for the rest of the hundreds or thousands of load cases.

9.2.8 CONNECTIONS

Every means of attachment shall be covered whether it's rivets, bolts, welds or bonding.

- Use joint allowables when available
- Fasteners shall be checked in shear and tension in combination.
- Bearing stress in all parts involved

9.2.9 CONCLUSION/RECOMMENDATIONS

This is the important statement to make. Does the part meet the requirements or not. If not, make the necessary recommendations to a re-design.

- In stress reports, ALL parts shall meet the requirements
- Every single part in a *Parts List* must be covered somewhat; making sure no part falls into a "crack" (is forgotten!)
- Parts can be substantiated by analysis showing a proper MS
- Parts can be *Passed by Comparison* to another part which has the exact same geometry and has higher material allowables for the same applied load, or same geometry and material but lower loads.
- Parts can be *Passed by engineering judgment* if it is obvious that any analysis would lead to a very high MS.

9.3 PURPOSE

Stress reports are essential if one wants to certify an aircraft. It is required by the certifying authority to provide certification documents (stress reports in our case) that demonstrate that this new aircraft meets all of the airworthiness requirements (FARs in the USA).

Although I recommend "telling a story" to explain to a reader what the analysis is about, this is not a novel! Be concise by trying not to repeat the same thing.

It doesn't matter much how well or not the stress notes were made or who worked on them, what matters is that your name will appear on the cover of the stress report. So it is important to shine and do the best work possible.

> *"Margins of Safety are not negotiable, but loads could be..."*

"There are three important things in a job:

1- The pay,

2- How interesting the job is,

3- A good working environment.

One could possibly do without point 1- or 2-, but without 3-, a job simply sucks!"

PART II

The QUIZ job interview

and

its improved structure

10. QUIZ INTERVIEW

During a job interview, the person I was meeting, which was going to be my boss later on, didn't talk much. He said *Hi* and simply showed me a hand sketched of a small structure and he asked me to simply say what checks I would do to analyze that structure. Then, he told me that if I were to see any possible improvement to the design to let him know afterwards. The applied ultimate load *"P"* is 100 lbs and the ends at points *D* and *E* are assumed clamped. The material is *Al Aly 2024-T42 Clad* 0.050" thick.

From what I remember, it looked something like *Figure 10-1*.

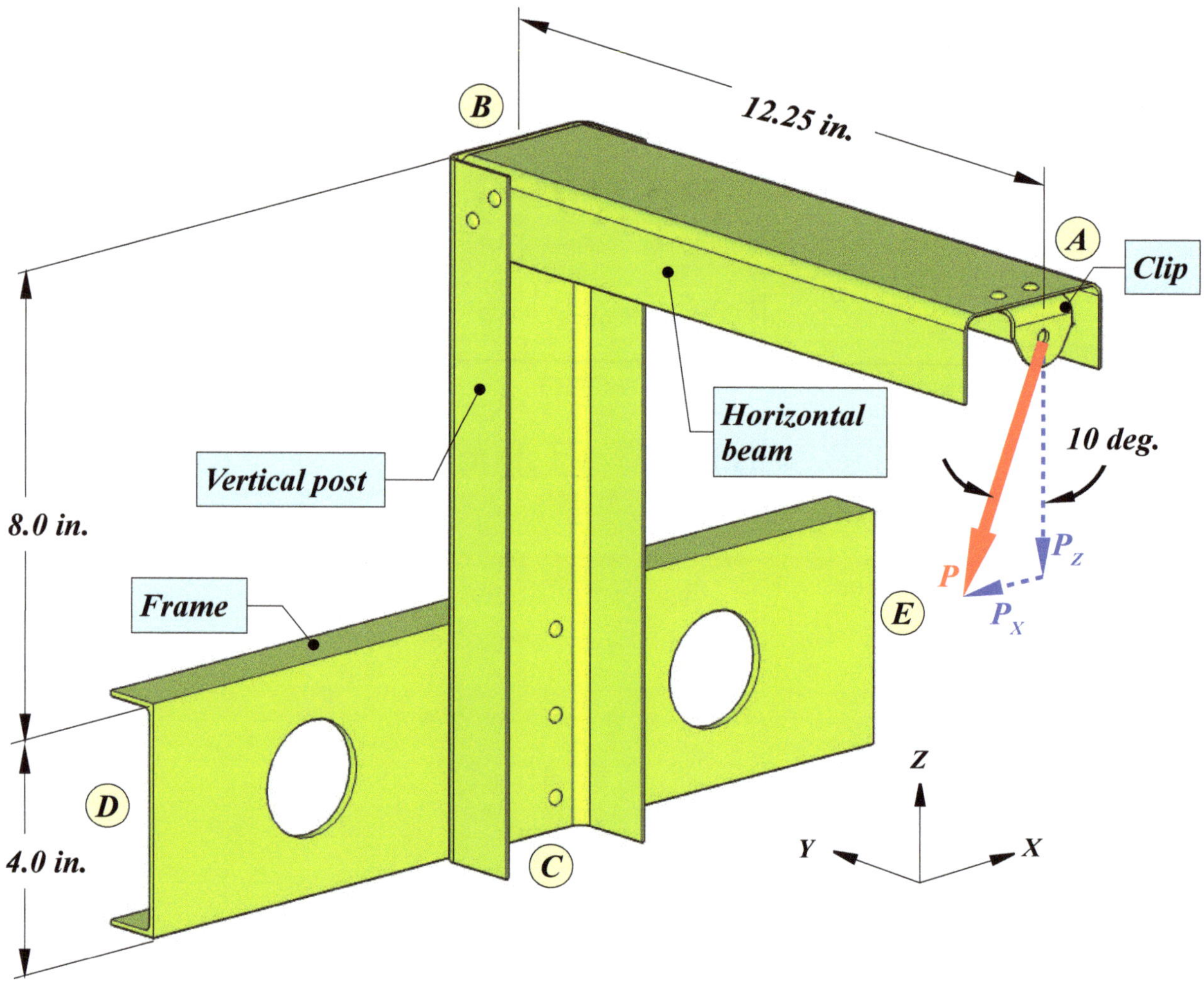

Figure 10-1 *– Isometric view of the skill test structure.*

Before reading any further, do the same exercise and just think about what checks you would need to perform in order to ensure that the complete structure is adequate for the applied loads. So I went ahead and said something like:

Well I'd check the lug and the clip at point A then the bending and the torsion in the horizontal beam and checking the free flanges in compression buckling, the beam-column and the torsion in the vertical post, the rivets at the bottom (point C). I would check the small rivets at the joints in point B but those are probably not going to be adequate in shear or in bearing in the relatively thin part.

I certainly would recommend adding a strut that would connect point A to the base at point C!

At that point, I could see a slight grin on his face. He then appeared to be a bit more friendly and the rest of the interview was just small talks.

I got the job and, later on, I remember having a group meeting where my boss (the same guy as in the interview), showed that same little sketch on the screen and asked everybody how they would analyze this. Since I was one of the newest member of the group, I assumed that not everybody in our group had seen that sketch before. People had pretty much the same answer as I did. He then went on to admit that, by simply using that little sketch when doing job interviews, he could see right away the real stress analysts from the *"wannabes"*. More than half the engineers he would interview couldn't provide the expected answers. It was very effective as I recalled that our group was pretty sharp and that there was not too many *"dead wood"* that could hide under a rock for many months doing...time.

So, did you have pretty much the same list of checks as shown above? If not, don't worry. By the end of this book you'll know what to check and how to analyze any structure.

It takes time (experience) for an analyst to be capable of knowing, or at least to have a pretty good idea, if a structure makes sense or not just by looking at it. You'd still have to perform the analysis obviously but to the least, you won't be spending a lot of time analyzing something that would give you huge negative margins of safety.

10.1 EXTERNAL LOADS

The following analysis shows how this design is adequate (or not) in the case you're not sure or, that your boss insists that you do the calcs (since they probably want to release that drawing today - they might get disappointed...).

So you looked at the design of the example in *Figure 10-1* and you're not sure whether it's acceptable or not just by looking at it. One trick I use in trying to evaluate if a structure is adequate or not is by converting the applied load by an object I know like a car (3,000 *lbs*), a truck (10,000 *lbs*) or, a small person (100 *lbs*). By looking at the structure from *Figure 10-1*, could you safely suspend a 100 *lbs* person at point *A* from a helicopter a hundred feet above ground? Mmm, maybe, maybe not. Personally, I would like to have a certain buffer. A 1.5 ultimate factor would be a minimum.

However, although most of the load is vertical "P_Z" and will simply introduce bending into the beam *A-B* and the post *B-C*, one can see that the horizontal force component "P_X" would cause a twisting moment "M_Z" in the post *B-C*. The post being an open section, it is very weak in torsion.

In stress analysis, we are not only concerned about the stress but also (and more importantly I think) about deflections. In our case, you could show positive margins of safety in torsion shear stress but what about the deflection due to the twist? Is the structure stiff enough to react that load without deforming so much that the structure becomes un-useable or that the load possibly re-distributes somewhere else?

Getting back to our Quiz, the applied forces components are thus:

$$P_X = P\sin(10) = -100\sin(10) = -17.4\ lbs$$
$$P_Z = P\cos(10) = -100\cos(10) = -98.5\ lbs$$

The design shown in *Figure 10-1* actually sucks and I'll show you why! So we'll look at it closely and see how we could improve it. But before everything else, a Free Body Diagram *(FBD)* has to be done.

Figure 10-2 shows the *FBD* in the *X-Z* plane.

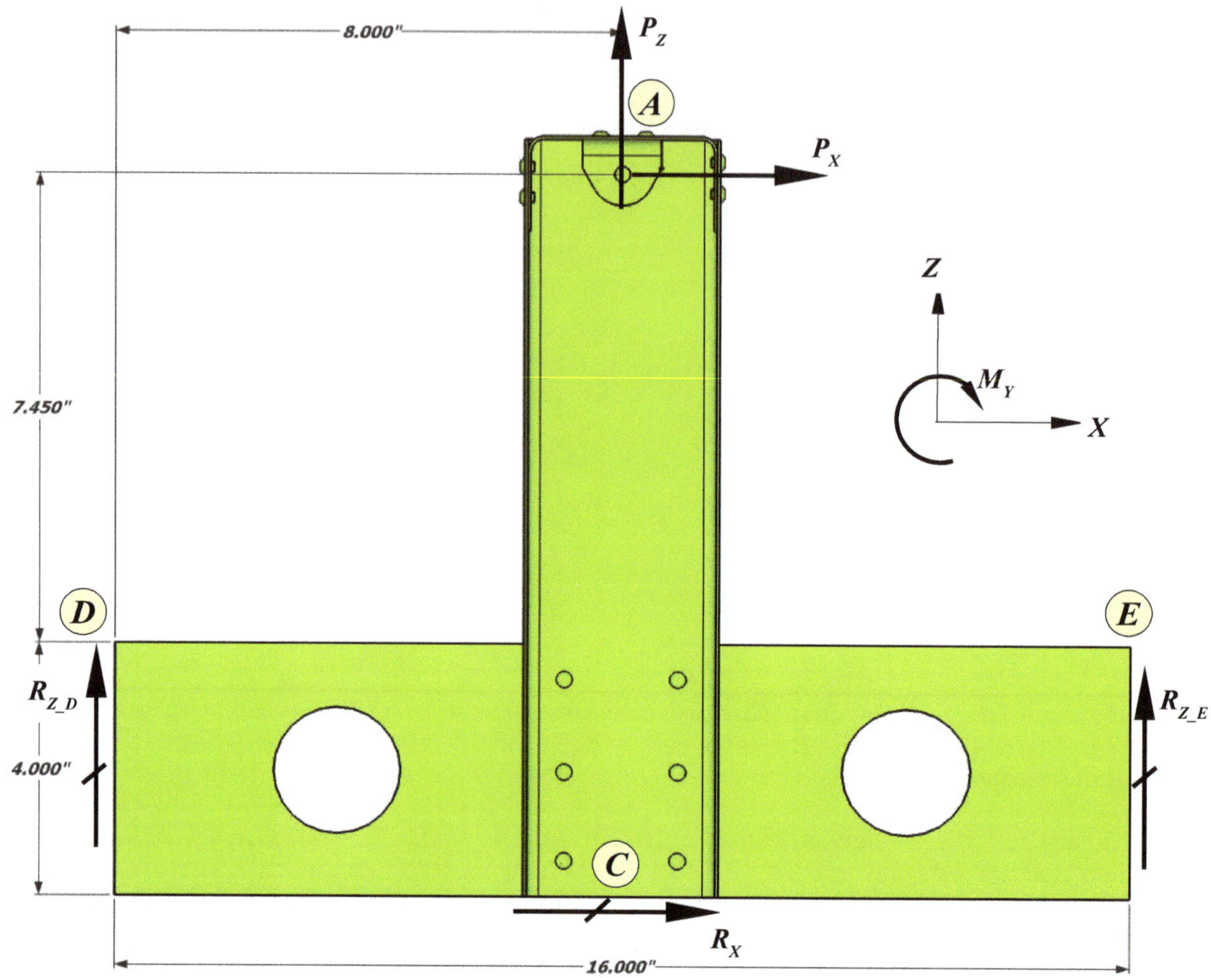

Figure 10-2 – *Free Body Diagram of complete structure View X-Z.*

The load *"P_X"* is assumed reacted at the base of the Frame by *"R_X"* as if the Frame was fastened to a fuselage skin. The points *D* and *E* are assumed to react only shear loads in the *Y* and the *Z* directions as well as the moment around the *X* axis. The reactions at points *D*, *E* and at the Frame base are calculated below by doing the sum of the moments and forces. Reaction vectors have cross lines.

$$R_{Z-D} = -\left[\frac{P_Z}{2} + \frac{M_Y}{16}\right] = -\left[\frac{-98.5}{2} + \frac{-17.4(7.45+4.0)}{16}\right] = 62\ lbs$$
$$R_{Z-E} = -\left[P_Z + R_{Z-D}\right] = -\left[-98.5 + 62\right] = 36.5\ lbs$$
$$R_X = -P_X = -(-17.4) = 17.4 lbs$$

Note that all the force vectors are shown in the positive directions of the Global Coordinate System (G.C.S.) and that the signs of the forces are according to the G.C.S. used and not to tension or compression forces.

Figure 10-3 shows the *FBD* in the *X-Y* plane.

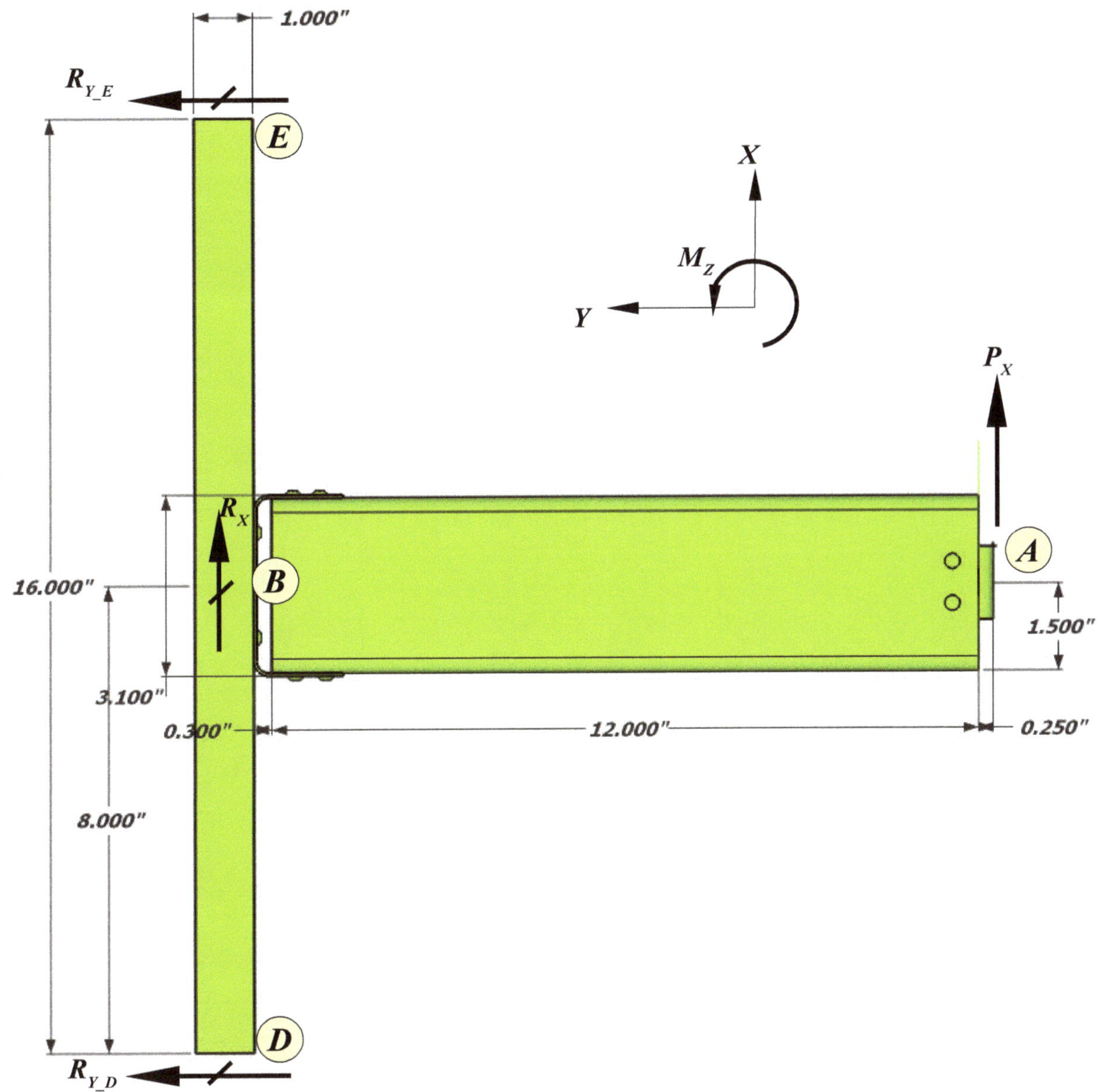

Figure 10-3 *– Free Body Diagram of complete structure View X-Y.*

As shown previously, the load "P_X" is assumed reacted in the middle of the Frame base flange "R_X". Points *D* and *E* are assumed to react the shear loads in the *Y* direction. The reactions are calculated below by doing the sum of the moments and forces. R_{Y_D} and R_{Y_E} are assumed reacted by the top flange of the Frame only.

$$R_{Y-D} = \frac{M_Z}{16} = \frac{P_X(12+0.25+0.30+1.0/2)}{16} = \frac{-17.4(13.05)}{16} = -14.2\ lbs$$

$$R_{Y-E} = -R_{Y-D} = -(14.2) = 14.2\ lbs$$

$$R_X = -P_X = -(-17.4) = 17.4\ lbs$$

Figure 10-4 shows the *FBD* in the *Y-Z* plane.

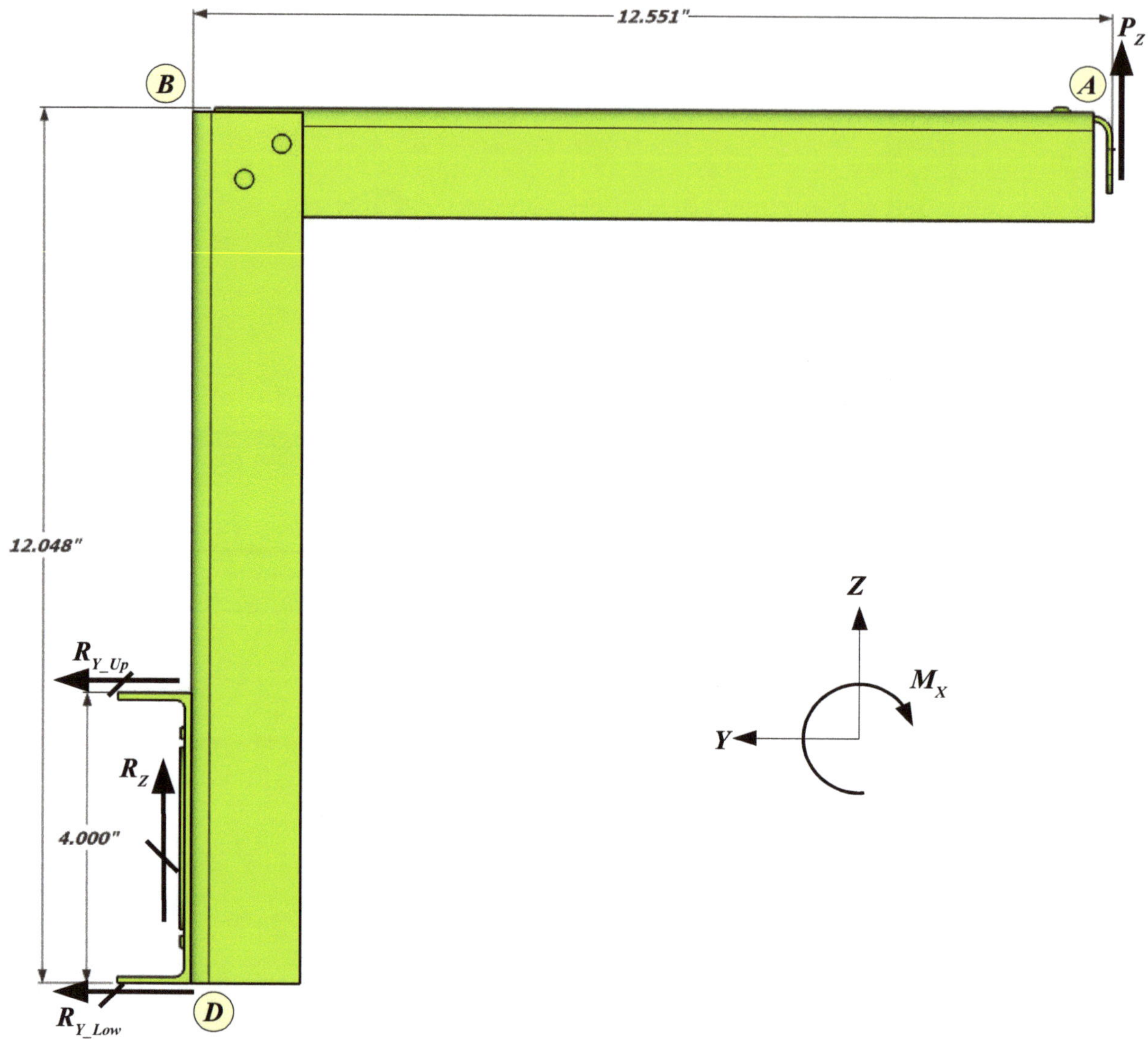

Figure 10-4 – *Free Body Diagram of complete structure View Y-Z.*

The "M_X" moment is assumed reacted as couples (R_{Y_Up} and R_{Y_Low}) as 50% at points *D* and 50% at point *E*, since the post is exactly located in the middle of the Frame, while the vertical shear reactions "R_Z" at those points is reacted at the Frame web.

$$R_{Y-Up} = \frac{\frac{-P_Z}{2}(12.55)}{4} = \frac{\frac{-(-98.5)}{2}(12.55)}{4} = 155\ lbs$$

$$R_{Y-Low} = -R_{Y-Up} = -(155) = -155\ lbs$$

$$R_X = -P_X = -(-17.4) = 17.4\ lbs$$

Figure 10-5 shows the Free Body Diagram of the complete structure along with the actual forces. The *FBD* is essential to comprehend the load path and clarify the assumptions made regarding the boundary conditions.

Now that we've determined the loads at the boundaries (external loads), we will look at each part and create an *FBD* for each of them by extracting the internal loads.

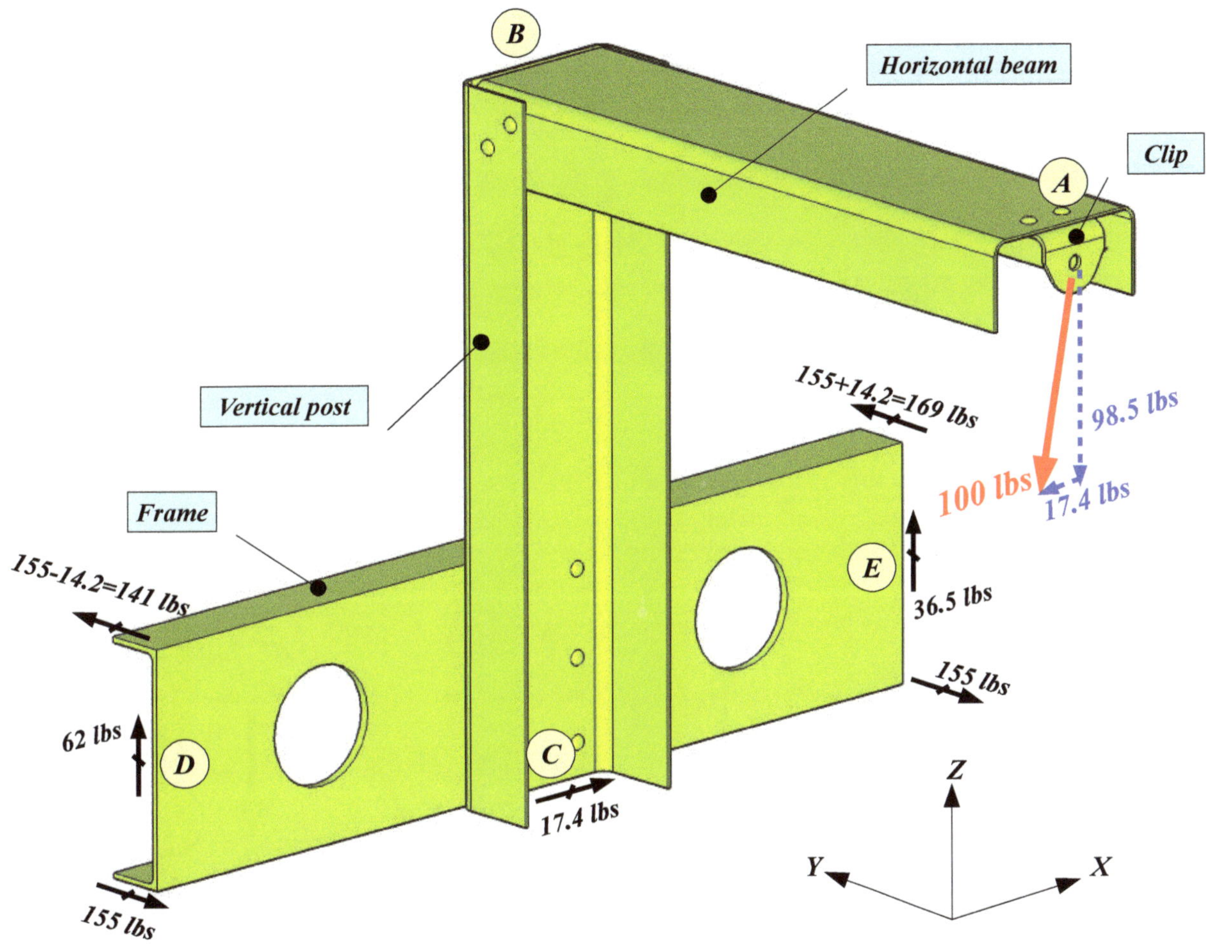

***Figure 10-5** – Free Body Diagram of complete structure, ISO View.*

10.2 CLIP

The clip is used to connect whatever is pulling onto the structure. It could be a cable, a strut, etc. In our case, it is made out of *Al Aly 7075-T62 Clad* 0.071 *in*. thick. It is connected to the horizontal beam with *qty (2) MS20470AD5* (BJ5) rivets. *Figure 10-6* shows the *F.B.D.* of the Clip.

To analyze the clip, we will assume that the rivets are pinned and thus can't react any moment locally. Thus the moment "M_X" due to P_Z times the horizontal distance to the line of rivets (0.695") is reacted by a couple between the rivets and the end of the part ($R_{Z1,2}$ and R_{Z3}). Also, due to the load P_X, a torque moment "M_Y" will be reacted by the couple R_{Z1} and R_{Z2}. The vectors of the reactions forces are shown as expected, but I could be wrong...

10.2.1 LOADS

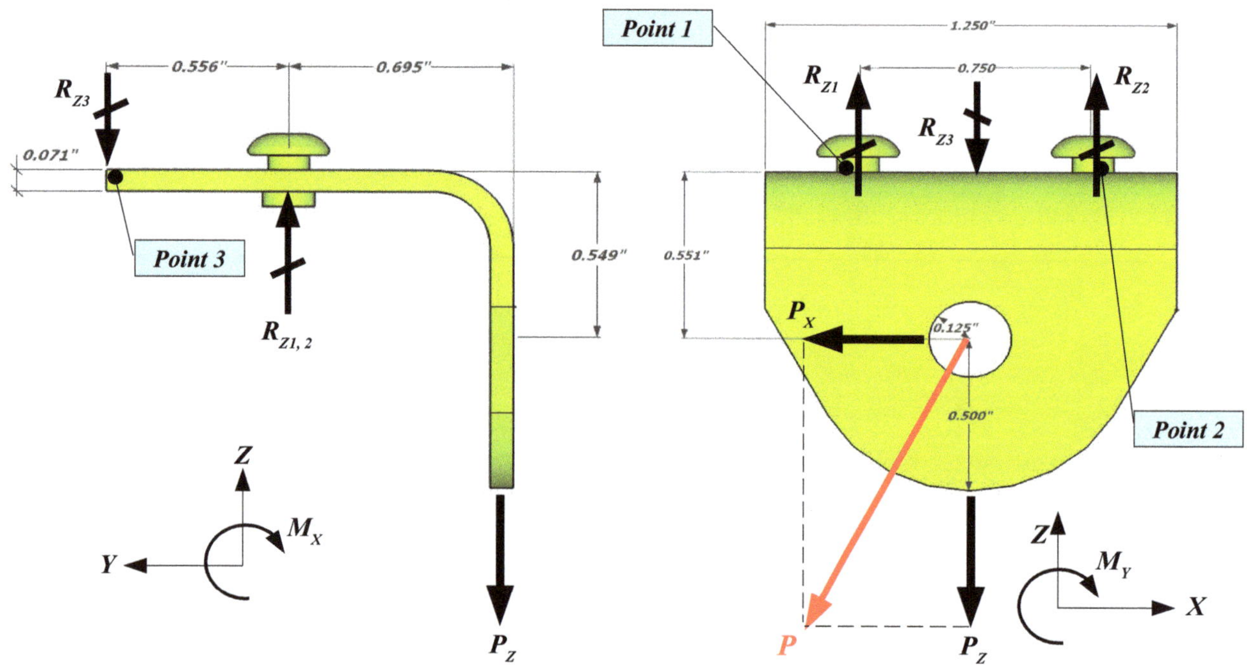

Figure 10-6 *– Clip dimensions and FBD.*

The reactions in the *YZ* plane are calculated by doing the sum of the moments at point *3* and the sum of the forces in the *Z* direction as follows.

$$\sum M_{X3} = P_Z\,(0.556+0.695) - R_{Z1,2}\,(0.556) = 0$$

$$R_{Z1,2} = \frac{98.5\,(0.556+0.695)}{0.556} = 221.6\ lbs$$

$$\sum F_Z = R_{Z1,2} - P_Z - R_{Z3} = 0$$

$$R_{Z3} = 221.6 - 98.5 = 123\ lbs$$

The reactions in the *XZ* plane are calculated by doing the sum of the moments at point *1* and the sum of the forces in the *Z* direction as follows.

$$\sum M_{Y\,1} = (P_Z + R_{Z\,3})\left(\frac{0.75}{2}\right) + P_X\,(0.551) - R_{Z\,2}\,(0.75) = 0$$

$$R_{Z\,2} = \frac{(98.5 + 123)\left(\frac{0.75}{2}\right) + 17.4(0.551)}{0.75} = 123.5\ lbs$$

$$\sum F_Z = R_{Z1} + R_{Z2} - P_Z - R_{Z3} = 0$$

$$R_{Z\,1} = -123.5 + 98.5 + 123 = 98\ lbs$$

As a verification, $R_{Z1} + R_{Z2} = 98 + 123.5 = 221.5\ lbs$ which equals to $R_{Z1,2}$ calculated earlier (221.6 *lbs*) for the *YZ* plane.

We can notice that the results of all the reaction load values calculated are positive, which means that the drawn force orientation vectors are indeed as assumed. Sometimes I prefer doing that than, as in section *10.1*, draw all force vectors in the positive global coordinate system. One simply has to stick to the assumed directions when calculating the sum of the forces and moments.

In this particular case, I preferred drawing the forces in the directions that I thought they were going to be. The main reason for it was that, at point 3, there is no fastener. The reaction force can only exist there when there is a contact. Otherwise, if the applied load P_Z were to be in the opposite direction (Up instead of down), then the Clip would tend to pivot around the line of rivets (points 1 and 2) and no contact would occur at point 3. There would be a contact between the edge of the Beam *A-B* and the Clip instead which would lead to a totally different *FBD*.

The Clip is located underneath the Horizontal Beam thus causing the rivets to be in tension. Even if the tension loads in the rivets aren't very high, having tension loads in rivets is undesirable. Since the applied load is downward, it would be preferable that the Clip be located on top instead.

10.2.2 MATERIAL ALLOWABLES

The actual and latest material allowables for all metallic materials can be obtained from the latest (ref. *8.)*. The **Clip** is made out of *Al Aly 7075-T62 Clad*, 0.071 *in.* thick sheet. *Table **10-1*** shows fictitious values for that material.

Table 10-1 *- 7075-T62 Clad 0.071 in. fictitious material allowables*

Aluminum Alloy 7075-T62 Clad sheet
Fictitious Values, NOT FOR DESIGN

	0.071" [ksi]
F_{tu}	75
F_{ty}	65
F_{cy}	62
F_{su}	42
F_{bru} $(e/D = 2.0)$	135
F_{bry} $(e/D = 2.0)$	100
$E, 10^3$	10
μ	0.33
n	20

10.2.3 FASTENERS ALLOWABLES

Table 10-2 shows fictitious ultimate shear allowables "P_{SU}" and ultimate tension allowables for 1/8", 5/32" and 3/16" *MS20470AD* solid rivets. Note that this data is fictitious, it is thus required to use the latest allowables possible from either (ref. 8.) or from the test results of the company you work for in your analysis. *Table 10-3* shows fictitious values for single shear correction factors *"SF"*.

Table 10-2 *- MS20470AD Solid rivets ultimate shear & tension allowables.*

	MS20470AD Al Aly 2117-T3 $F_{su} = 30$ ksi		
	4	**5**	**6**
Dia. [in.]	0.1285	0.159	0.191
P_{su} [lbs]	**388**	**596**	**862**
P_{tu} [lbs]*	**375**	**430**	**520**

** in 0.050" 2024-T3 & 7075-T6 sheets.*

Table 10-3 *- Shear Correction Factor "SF".*

	Diameter of rivet [in]			
Thickness [in]	3/32"	1/8"	5/32"	3/16"
0.025	0.972	–	–	–
0.032	1.000	0.964	–	–
0.036		0.980	–	–
0.040		0.996	0.964	–
0.045		1.000	0.98	–
0.050			0.996	0.972
0.063			1.000	1.000

10.2.4 SECTION PROPERTIES

The section at the rivet line is shown in *Figure 10-7*.

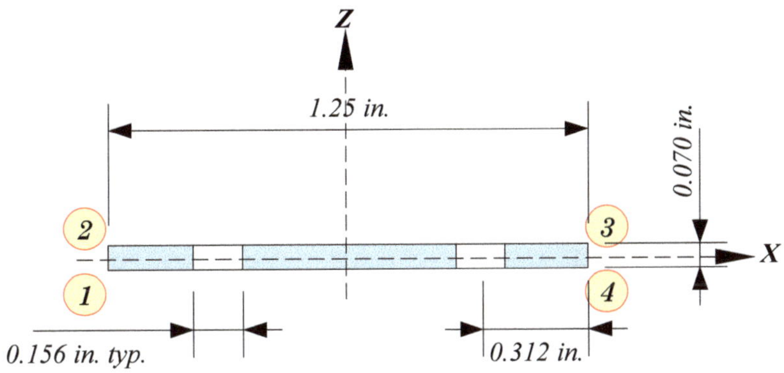

Area = 0.06566 in^2
I_X = 0.000027 in^4
I_Z = 0.009209in^4
I_{XZ} = 0.0000 in^4

Figure 10-7 *– Clip Section at rivets 1 & 2.*

10.2.5 STRESS ANALYSIS

Bending

From *Figure 10-6*, the maximum bending moments occur at the rivet line *(points **1** & **2**)* and are calculated as: $M_X = P_Z(d)$, $M_Y = P_X(h)$ and $M_Z = P_X(d)$. For d = ***0.695*** *in*, h = 0.551−0.071/2 = 0.515 *in.*, P_X = 14.7 *lbs* and P_Z = 98.5 *lbs*, then we obtain:

$$M_X = 98.5(0.695) = 68.5\ in.-lbs$$
$$M_Y = T = 17.4(0.515) = 9\ in.-lbs$$
$$M_Z = 17.4(0.695) = 12\ in.-lbs$$

The bending stresses are calculated at the corners of the section.

$$f_b = \frac{M_X(z)}{I_X} + \frac{-M_Z(x)}{I_Z}$$

For
M_X = 68.5 in-lb
M_Z = -12 in-lb
I_X = 0.000027 in^4
I_Z = 0.009209 in^4

Calculation example at point *1*:

$$f_{b1} = \frac{M_X(z_1)}{I_X} + \frac{-M_Z(x_1)}{I_Z}$$
$$f_{b1} = \frac{(68.5)(-0.035)}{0.000027} + \frac{-(-12)(-0.625)}{0.009209}$$
$$f_{b1} = -88{,}796 - 814 = -89{,}611\ psi$$

Bending stresses at all corners:

Table 10-4 *- Bending stresses at corners*

Points	x [in]	z [in]	f_b [psi]
1	-0.625	-0.035	**-89,611**
2	-0.625	0.035	87,982
3	0.625	0.035	89,611
4	0.625	-0.035	-87,982

As seen above, the most critical location is at point *1* where the compression stress is maximum and exceeds the allowables (75 *ksi*). A plastic bending analysis could be done to demonstrate the Clip good. However, as we will see later, the design should be changed due to other reasons as well. Therefore, this concept is simply shown as being non-adequate for now.

Shear

The transverse shear exists in both the *X* and the *Z* directions and are calculated as follows:

For $V_X = P_X = 17.4\ lbs$, $V_Z = P_Z = 98.5\ lbs$ and, $A = 0.06566\ in^2$

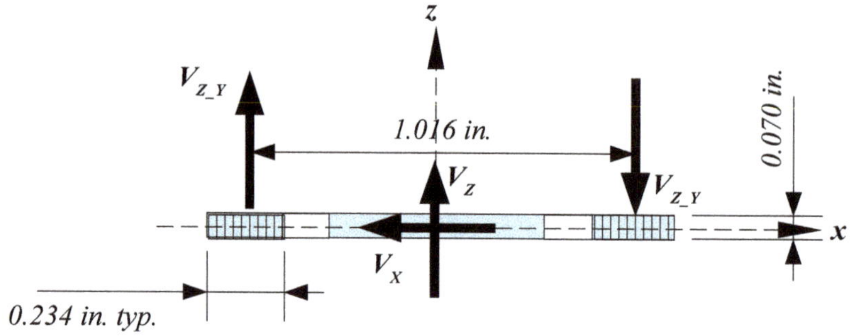

Figure 10-8 *– Shear forces at Clip Section at rivets 1 & 2.*

A torque "M_Y" exists in the Clip due to the vertical offset between the load "P_X" application point and the shear center of the section which, in this case, corresponds to the center of gravity. Because of that torsion, one would be tempted to evaluate the torsional shear stress by using the equation:

$$\tau = \frac{T}{J}t$$

However, that would be too conservative as the torsional shear stress doesn't have time to fully develop due to the short distance between the free end (where the load is applied) and the restrained end (at the rivets *1 & 2*). A distance of about **four** to **six** times the section's width (or largest dimension) is needed to justify using the above equation.

What happens is that, for short distances (below four times the width), the torque is mainly reacted as differential transverse shear. See *Figure 10-8* where the differential shear load "V_{Z_Y}" is shown as assumed applied at the center of the most outboard material section area "A_S" of 0.234 * 0.071 = 0.0166 in^2. That shear load is calculated as:

$$V_{Z-Y} = \frac{T}{arm} = \frac{9}{1.016} = 9\ lbs$$

Therefore, the transverse shear stresses are calculated as follows:

$$f_{sX} = \frac{V_X}{A} = \frac{17.4}{0.06566} = 265\ psi$$

$$f_{sZ} = \frac{V_Z}{A} + \frac{V_{Z-Y}}{A_S} = \frac{98.5}{0.06566} + \frac{9}{0.0166} = 1{,}500 + 542 = 2{,}042\ psi$$

The vertical shear stress is calculated at the location where V_Z and $V_{Z\text{-}Y}$ are in the same direction (points 1 & 2) as this will be the maximum stress.

Since the shear stresses calculated are orthogonal to each other, they are combined as follows:

$$f_{S-TOTAL} = \sqrt{f_{SX}^2 + f_{SZ}^2} = \sqrt{265^2 + 2{,}042^2} = 2{,}059\ psi$$

The resulting shear is relatively low but yet must be combined to the high bending stress.

Margin of safety

The bending stress is combined with the shear stress in order to get the MS. The following interaction equation is normally accepted when combining an axial (or bending) stress with a shear stress at yield or ultimate:

$$R_b^2 + R_S^2 = 1$$

Thus the MS is calculated as:

$$MS = \frac{1}{\sqrt{R_b^2 + R_S^2}} - 1$$

Where R_b and R_S are respectively the bending and shear stress ratios calculated as follows:

$$R_b = \frac{f_b}{F_{tu}} = \frac{|-89{,}611|}{75{,}000} = 1.195$$

$$R_s = \frac{f_s}{F_{su}} = \frac{2{,}059}{42{,}000} = 0.049$$

The bending stress ratio is greater than ONE thus indicating a negative *MS*. *Ftu* is used in this case as the allowable. It is a common practice to use *Ftu* even when there is a compression stress knowing that *Fcu* is normally a bit larger. However, that is only applicable to very stable cross sections (when no risk of buckling or warping exists). In which case, *Fcy* would have to be the allowable, just like for the crippling analysis.

The shear stress ratio is quite small and therefore could have been neglected without any impact on the *MS*.

I could have used the plastic bending approach and possibly get a positive *MS*, but I figure that, if I have to resort to plastic bending to show a sheet metal design good, something is probably wrong. I'm not saying not to use plastic bending here, it could be legitimate to use that approach in some circumstances. I simply try to avoid it when I can.

Hence the MS is:

$$MS = \frac{1}{\sqrt{1.195^2 + 0.049^2}} - 1 = -0.16$$

10.2.6 TENSION CLIP

Recommendation: The Clip should be relocated on top to reduce the bending moment.

Another (much quicker) type of analysis could be done for the Clip by considering it as a tension clip. See below.

From *Figure 10-6*, one can see that the Clip works as a Tension Clip (or Angle Clip). A very fast check can be done to see if that Angle Clip can take the applied tension load of 98.5 *lbs* (Downward in our case). Each major aircraft company has its own set of graphics but they provide more or less the same answer.

For formed *2024* and *7075* Aluminum Alloy Angle Clips, the only information we need is:

- The Clip thickness; $t = 0.071$ *in.*
- The distance from the load to the fastener; $e = 0.695$ *in.* ≈ 0.7 *in.*

Figure 10-9 shows an approximation of a tension clip allowable curve. From it we get: P_{all} = **280 *lbs*** ultimate. This is more than twice the applied tension load.

Of course, at failure, the Clip acts more like a strap. It is completely deformed and yielded and, usually, the mode of failure is the fastener head that pops-off due to the prying effect. This allowable is applicable for a 1.0 *in.* wide Clip and should be multiplied by the number of fasteners that are reacting the applied load, two in our case **or**, by the ratio of the actual width over 1.0 *inch*.

However, although this method would show the Clip acceptable, I wouldn't recommend using it to justify that the design is good for the following reason:

- We don't know the type of fasteners used in the static tests from which the graph is based. I seriously doubt that they used *BJ5* rivets, they most likely used steel bolts!
- The excessive deformation that we should expect from the stretching of the Clip could render the design unusable.
- The torsion in the clip due to the P_X load isn't considered.

Therefore, for those reasons, this design is still considered inadequate.

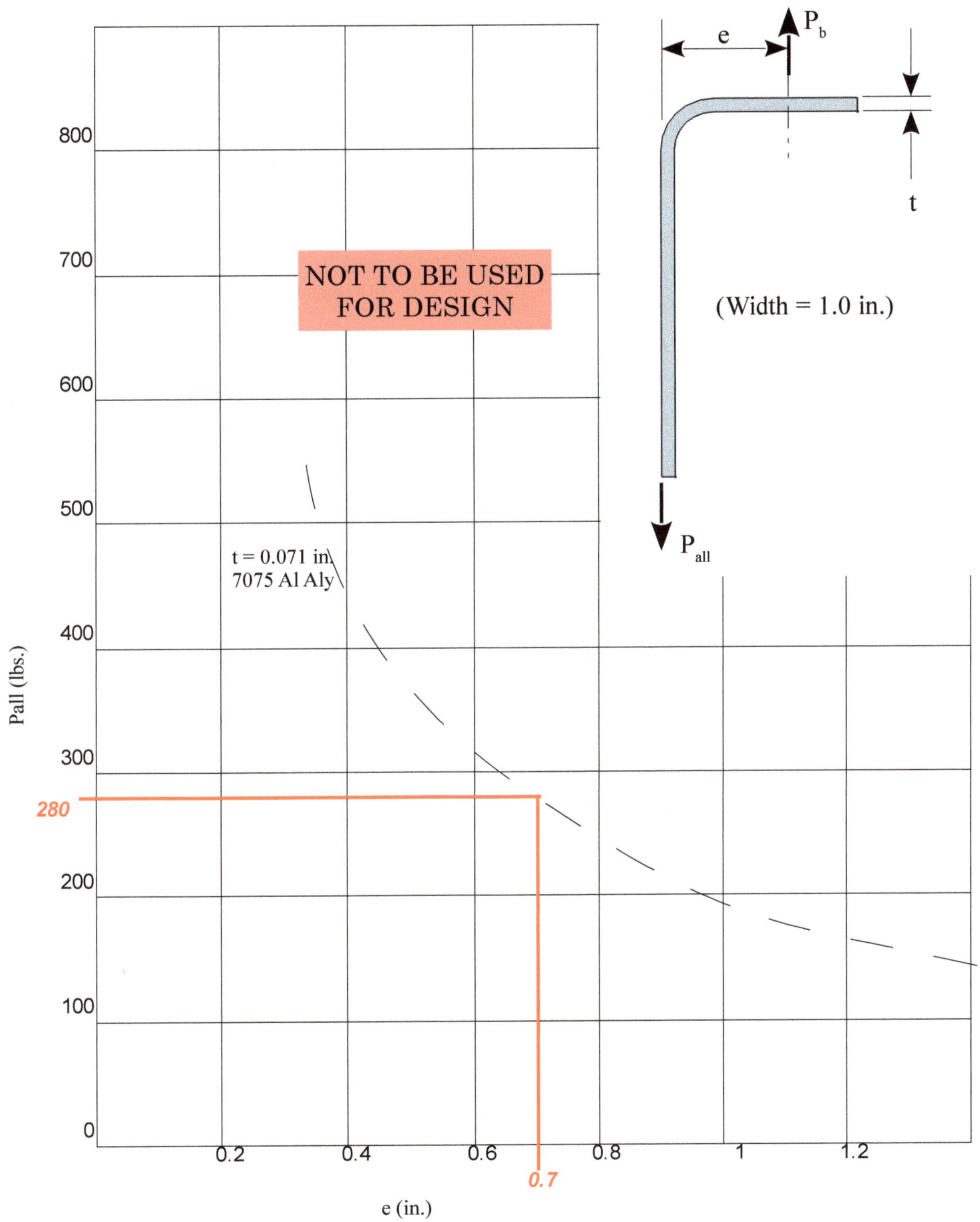

Figure 10-9 *– Angle Clip allowable approximation.*

10.2.7 LUG

At first glance, the designer did a good job for the Lug. The hole is relatively large (0.25 *in.* dia.) and at 0.50 *inch*, it is far enough from the edge at two times the bolt diameter (*2D*). In this case, there is no bushing but if there was, we would have to consider the hole size without the bushing.

Since the loaded hole is at *2D* from the edge, we could simply do a bearing check and call it a day. Let's do it below.

The bearing load allowable "P_{bru}" is calculated as:

$$P_{bru} = F_{bru} \cdot D \cdot t = 135{,}000 \cdot (0.25) \cdot (0.071) = 2{,}396 \ lbs$$

For:
$t = 0.071$ in
$e = 0.50$ in
$D = 0.25$ in
$e/D = 2$
$F_{bru} = 135$ ksi

This is way higher than the 100 *lbs* ultimate applied load. Therefore the lug is *Passed by engineering judgment*. But for comparison, let's do the complete lug analysis.

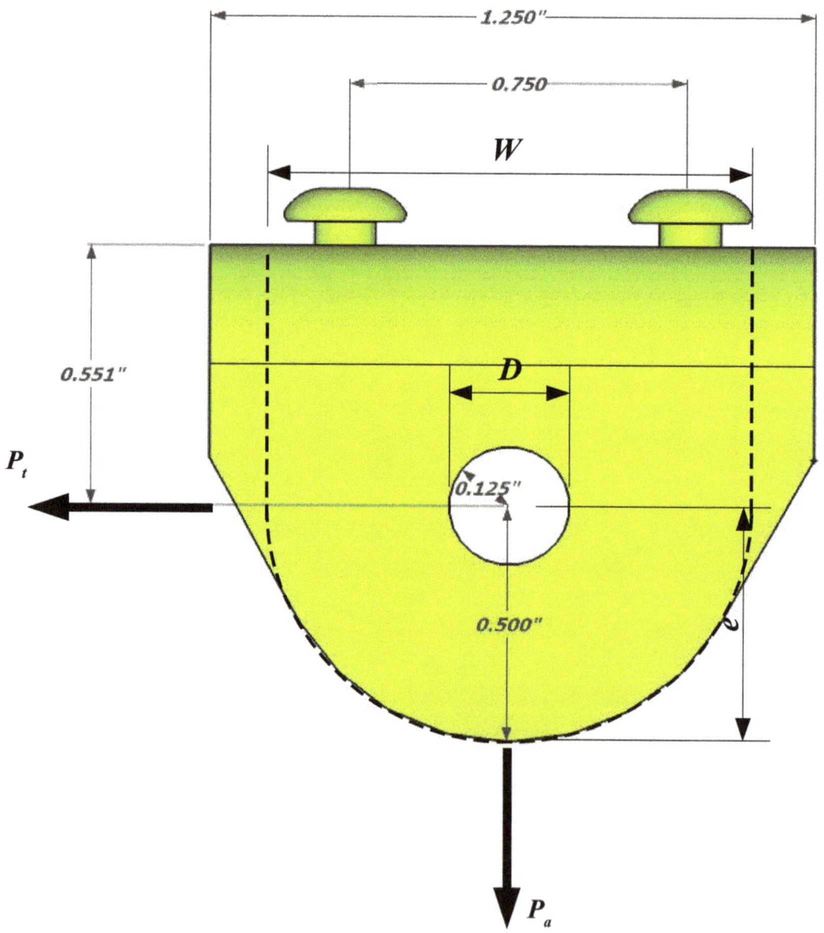

Figure 10-10 *– Angle Clip Lug geometry.*

Lug axial net section strength

For conservatism and to simplify the analysis, the lug geometry is assumed as the dotted line shown on *Figure 10-10*.

I'd rather have some additional conservatism in lugs (as in joints in general) and an MS of 0.25 minimum than trying to save weight in those locations. The joints are typically areas of concern in durability and DTA. There are plenty of other places to optimize the weight than at lugs.

Since I'm considering the lug as a joint, I would also use a *1.15* Fitting Factor. Make sure to verify with your boss, with your company's Design Criteria Manual, and your DER to get their approval before using any factor. One important thing, make sure you use the same approach throughout the stress report. If this leads to negative MS in some locations during the sizing phase, consider re-designing.

The axial tension allowable "P_u" is calculated as:

$$P_u = K_t\, F_{tu}\, A_t$$

The tension efficiency factor "K_t" is obtained from *Figure B-6* using curve *1*. For $W/D = 1.0 / 0.25 = 4$, we find a K_t of 0.9.

Since that the material grain direction of sheet metal parts is almost never specified on the drawings, we always have to assume that it is in the weakest direction (L or LT).

For
$W = 1.0$ in.
$D = 0.25$ in.
$t = 0.071$ in.
$F_{TU} = 75$ ksi

The net tension area "A_t" is calculated as:

$$A_t = (W - D)t = (1.0 - 0.25)\ 0.071 = 0.05325\ in.^2$$

Therefore, the tension allowable is:

$$P_u = (0.9) * 75{,}000 * 0.05325 = 3{,}594\ lbs$$

Shear out bearing Strength

Like for the precedent axial net section check, to be consistent, the lug geometry is assumed as the dotted line shown on *Figure 10-10*.

Ultimate:

The shear out-bearing ultimate allowable "P_{bru}" is calculated as:

$$P_{bru} = K_{bru}\, F_{tu}\, A_{br}$$

The shear-bearing efficiency factor "K_{bru}" for axially loaded lugs is obtained from *Figure B-7*. For $D/t = 0.25 / 0.071 = 3.5$ and $e/D = 2$, we find a value for K_{bru} of 1.9.

The bearing area "A_{br}" is calculated as:

$$A_{br} = D \cdot t = 0.25(0.071) = 0.01775\ in.^2$$

Therefore, the shear out-bearing allowable is:

$$P_{bru} = (1.9) * 75{,}000 * 0.01775 = 2{,}529\ lbs$$

*Note that this value is close to the value calculated on page **62**.*

Yield:

The shear bearing yield allowable "P_{bry}" is calculated as:

$$P_{bry} = K_{bry}\ F_{ty}\ A_{br}$$

The shear-bearing yield efficiency factor "K_{bry}" for axially loaded lugs is obtained from *Figure B-8.* For $t/D = 0.071 / 0.25 = 0.28$ and $e/D = 2$, we find a K_{bry} of 1.48.

For
$A_{br} = 0.01775\ in^2$
$F_{TY} = 65\ ksi$

Therefore, the shear bearing yield allowable is:

$$P_{bry} = (1.48) * 65{,}000 * 0.01775 = 1{,}707\ lbs$$

Since $P_{bru} / P_{bry} = 2{,}529 / 1{,}707 = 1.48$ and thus < 1.5, then no *Yield MS* needs to be calculated.

Lug Strength under transverse loading

Using the same lug geometry as the dotted line shown on *Figure **10-9*** for consistency and conservatism.

Ultimate:

The transverse ultimate allowable "P_{tu}" is calculated as:

$$P_{tu} = K_{tu}\ F_{tu}\ A_{br}$$

The tension efficiency factor "K_{tu}" for transversely loaded lugs is obtained from *Figure B-9* curve *2* for 7075-T6 plate ≤ 0.5 *in*. For $A_{av} / A_{br} = 0.0283 / 0.01775 = 1.6$ (using the maximum value of 1.4 in the graph), we find a value for K_{tu} of 0.68.

The average area "A_{av}" is calculated using *Figure **10-11*** as:

$$A_{av} = \frac{6}{\frac{3}{A_1} + \frac{1}{A_2} + \frac{1}{A_3} + \frac{1}{A_4}}$$

Where, in our case;

$$A_2 = A_3 = \frac{(W-D)t}{2} = \frac{(1.0-0.25)(0.071)}{2} = 0.02662\ in.^2$$

$$A_1 = A_4 = \frac{(W-D\cos(45))t}{2} = \frac{(1.0-0.25\cos(45))(0.071)}{2} = 0.02922\ in.^2$$

Therefore;

$$A_{av} = \frac{6}{\frac{3}{0.02922} + \frac{1}{0.02662} + \frac{1}{0.02662} + \frac{1}{0.02922}} = 0.0283\ in.^2$$

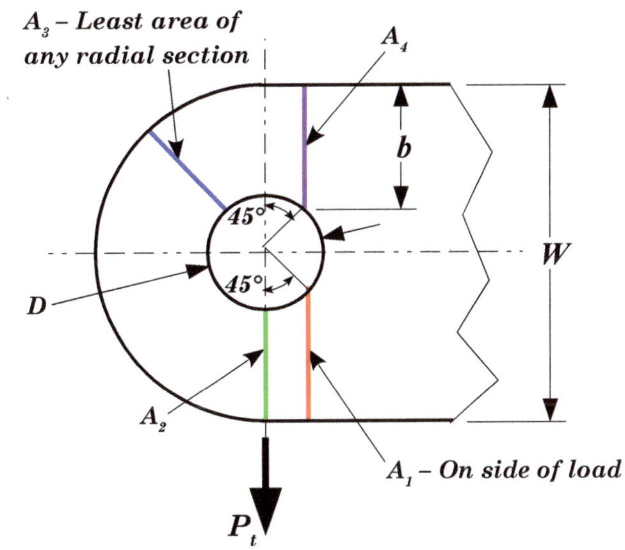

Figure 10-11 *– Lug section areas for transverse loading.*

Therefore, the transverse ultimate allowable is:

$$P_{tu} = (0.68) * 75{,}000 * 0.01775 = 905 \; lbs$$

> *For*
> *$A_{br} = 0.01775 \; in^2$*
> *$F_{TU} = 75 \; ksi$*
> *$K_{TU} = 0.68$*

<u>Yield:</u>

The transverse yield allowable *"P_{ty}"* is calculated as:

$$P_{ty} = K_{ty} \, F_{ty} \, A_{br}$$

The tension efficiency factor *"K_{ty}"* for transversely loaded lugs is obtained from *Figure B-9* curve *1* for all Aluminum at yield. For $A_{av} \,/\, A_{br}$ = 0.0283 / 0.01775 = 1.6 (using the maximum value of 1.4 in the graph), we find a value for K_{ty} of 1.28.

Therefore, the transverse yield allowable is:

$$P_{ty} = (1.28) * 65{,}000 * 0.01775 = 1{,}476 \; lbs$$

> *For*
> *$A_{br} = 0.01775 \; in^2$*
> *$F_{TY} = 65 \; ksi$*
> *$K_{TY} = 1.28$*

Here, the ultimate allowable is lower and thus more critical than the yield allowable.

<u>Margin of safety</u>

Since we have a combination of axial and transverse loading (oblique loading), the following interaction equation is used at yield and ultimate:

$$R_a^{1.6} + R_{tr}^{1.6} = 1$$

Thus the MS is calculated as:

$$MS = \frac{1}{\left(R_a^{1.6} + R_{tr}^{1.6}\right)^{0.625}} - 1$$

Where R_a and R_{tr} are respectively the axial and transverse load ratios calculated as follows:

$$R_a = \frac{P_a}{Min(P_u\ ;\ P_{bru})} = \frac{98.5}{Min(3{,}594\ ;\ 2{,}529)} = 0.039$$

$$R_{tr} = \frac{P_t}{P_{tu}} = \frac{17.4}{905} = 0.019$$

Hence the MS is:

$$MS = \frac{1}{\left(0.039^{1.6} + 0.019^{1.6}\right)^{0.625}} - 1 => 1.00$$

10.2.8 COMMENTS

A modification of the assembly design has to be done in order to show a positive MS for the Clip in bending. As specified in page *59*, by installing the Clip on top of the beam, the bending moment would be reduced and the rivets loads would be reduced also. However, the material and the geometry of the Clip could remain exactly the same.

The MS calculated is larger than 1.0. Therefore, that translates to a design which is more than two times stronger than required. So at that point, there is no need to show the actual *MS*.

A statement such as: *Passed by Inspection* or *Passed by Engineering Judgment* could also have been used instead of calculating the actual large *MS*.

"Two wrongs don't always make a right."

10.3 HORIZONTAL BEAM

The Horizontal beam transfers the large vertical load (Z) and the small fore-aft load (X) to the vertical post. It is made out of *Al Aly 2024-T42 Clad 0.050 in.* thick. It is connected to the vertical post with *qty (4) BJ5* rivets. *Figure 10-12* shows the *F.B.D.* of the Horizontal beam.

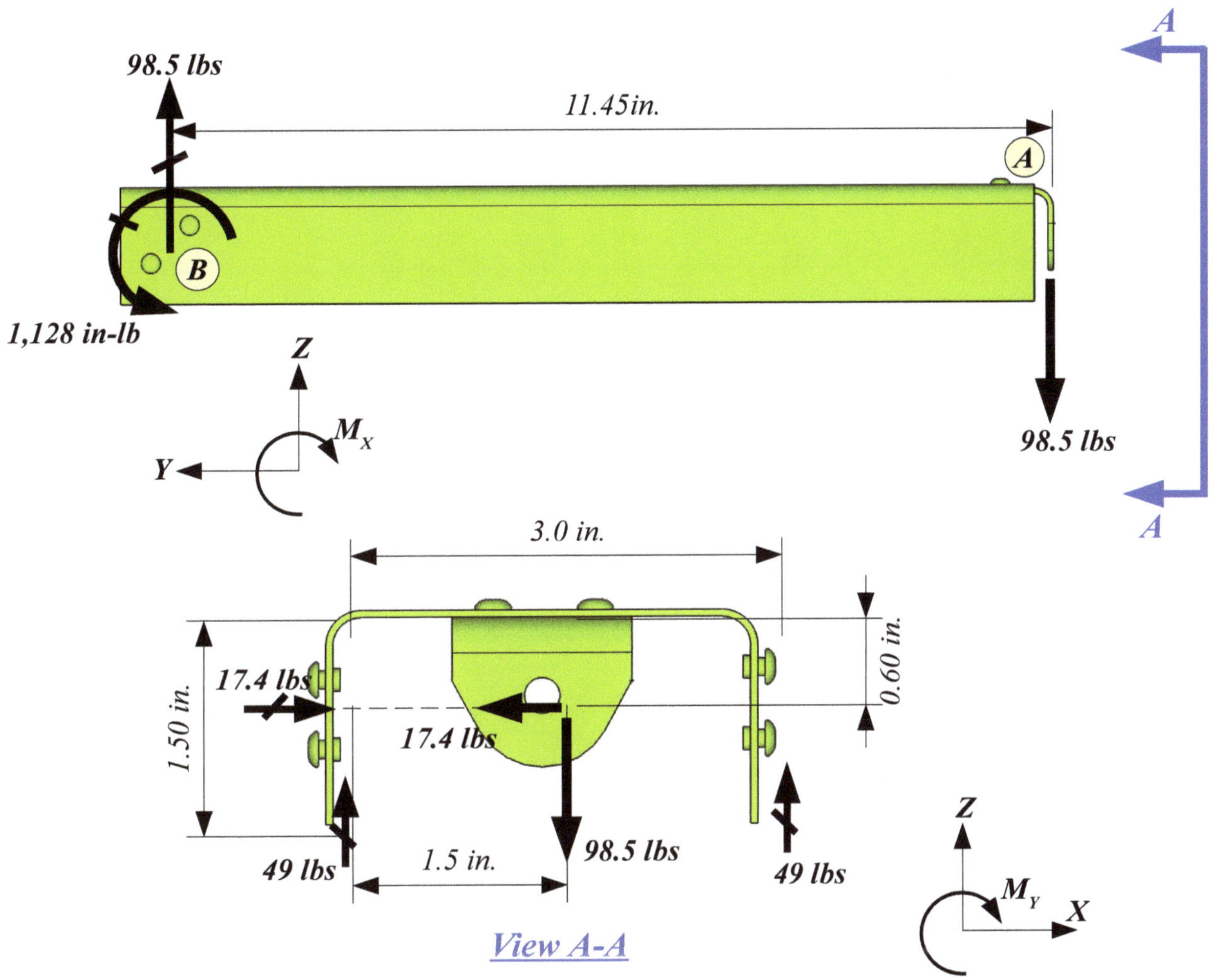

***Figure 10-12** – Horizontal beam dimensions and FBD.*

Here, the location of the reaction *X* load (17.4 *lbs*) is assumed right in line with the applied load. This simplifies the calculations but mainly makes us realize that this load path is not so good since this load is pushing onto the beam flange in its out-of-plane (weak/soft) direction.

There should be a shear clip connecting the web of the Horizontal beam to the web of the Vertical post. In which case, the location of the *X* load reaction would be 0.60 *inch* above the applied load thus changing slightly the vertical reactions at the rivets.

Recommendation: A shear clip should be added at the connection to the vertical post.

10.3.1 LOADS

Shear and moment diagram

The shear and moment diagrams of the Horizontal beam can be seen in *Figure 10-13* and *Figure 10-14*. The beam is cantilevered. The applied load is located at the free end (lug of the Clip) and the "clamped" end is considered located at the center of gravity of the rivet group. The moment at that location is thus calculated as $M_X = P\,(l)$.

> When I'm analyzing an assembly of many parts, I like using a local coordinate system *(ijk)* to analyze the detail parts. By doing that, I know that any axial and bending stresses are along the "*i*" axis, and the shear loads are in the "*jk*" plane. It's a matter of preference.

Plane i-k:

For $P_k = 98.5\ lbs$ and $l = 11.45\ in.$

$$M_j = P_k(l) = -98.5\,(11.45) = -1{,}128\ in.-lb$$

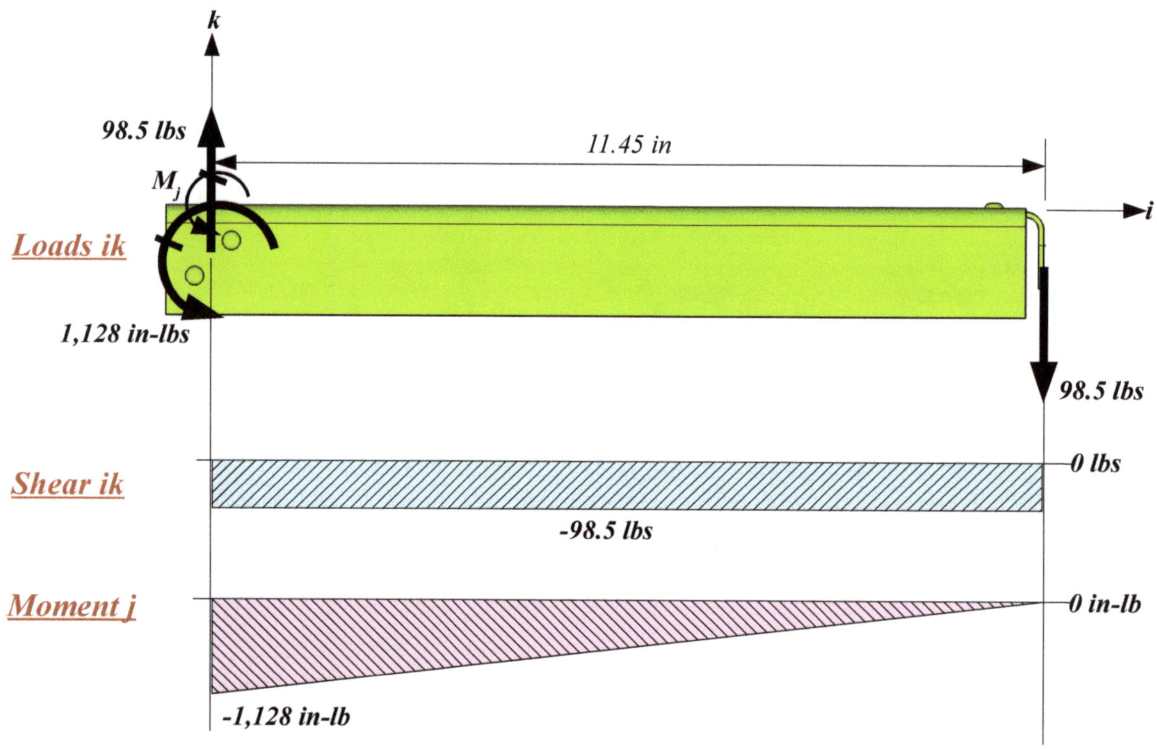

Figure 10-13 *– Horizontal beam Shear and moment diagram.*

> Of course, in reality, the exact shear and moment curves are a bit more complex at the fastener locations. But the simplification shown is sufficient for the analysis performed.

Plane i-j:

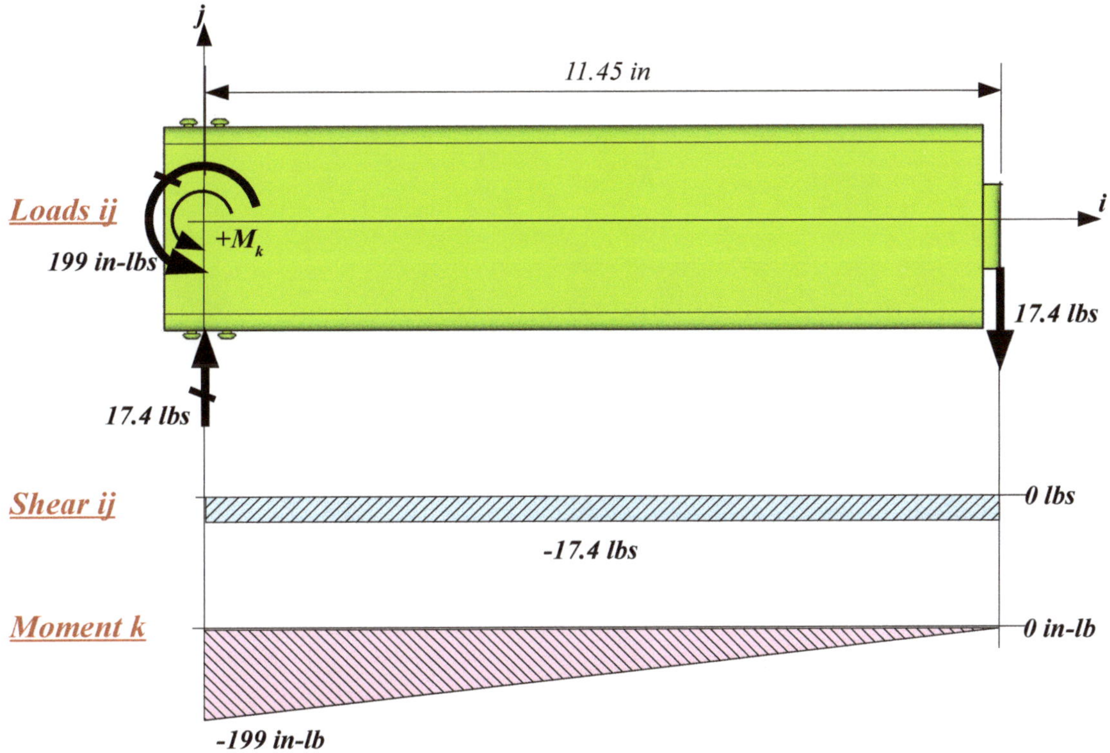

Figure 10-14 *– Horizontal beam Shear and moment diagram*

10.3.2 MATERIAL ALLOWABLES

The **Horizontal beam** is made out of *Al Aly **2024-T42** Clad, **0.050** in.* thick sheet. *Table **10-5*** shows fictitious values for that material.

Table 10-5 *- 2024-T42 Clad 0.050 in. fictitious material allowables.*

*Aluminum Alloy **2024-T42** Clad sheet*
Fictitious Values, NOT FOR DESIGN

	0.050" [ksi]
F_{tu}	55
F_{ty}	35
F_{cy}	37
F_{su}	35
F_{bru} (e/D = 2.0)	115
F_{bry} (e/D = 2.0)	70
E, 10^3	
Primary	10
Secondary	9.5
μ	0.33
n	20

10.3.3 SECTION PROPERTIES

The Horizontal beam has a constant cross-section. Therefore, only one section needs to be calculated. The section properties are obtained from a *3D CAD* drawing software. See *Figure 10-15.*

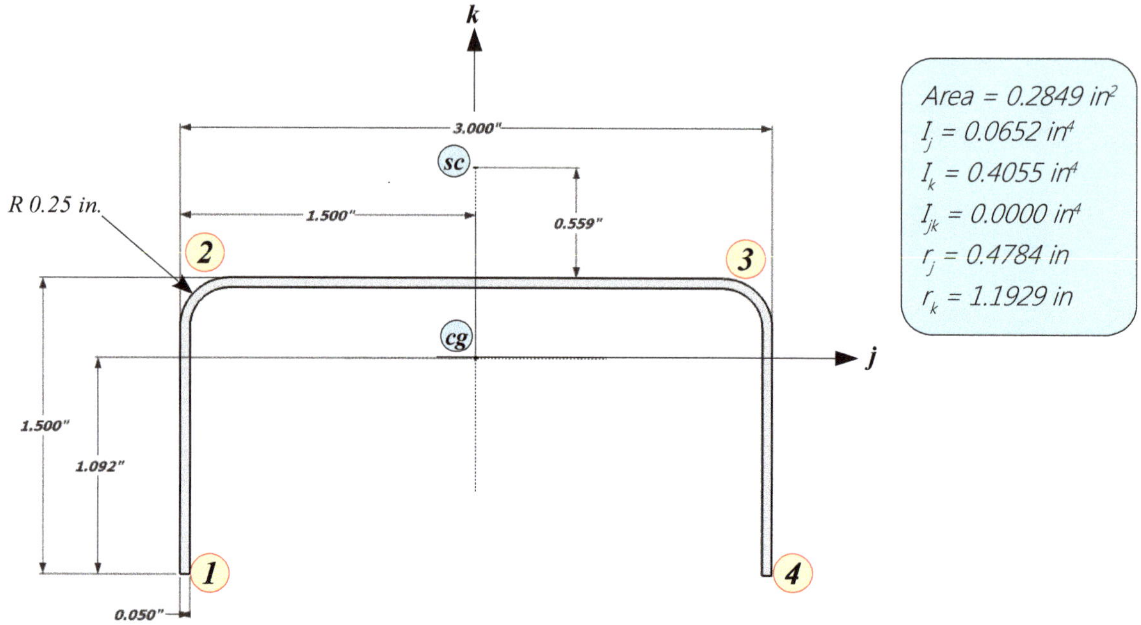

Figure 10-15 *– Horizontal beam section properties.*

However, the distance *"e"* of the shear center *"SC"* is calculated for Channels. The calculations are shown in *Figure 10-16.* Therefore, the distance from the top of the web to the shear center is: $e - t/2$ = 0.584 – 0.050/2 = 0.559 *in*. The torsional stiffness *"K"* and the warping constant "C_w" of the section are calculated as shown below.

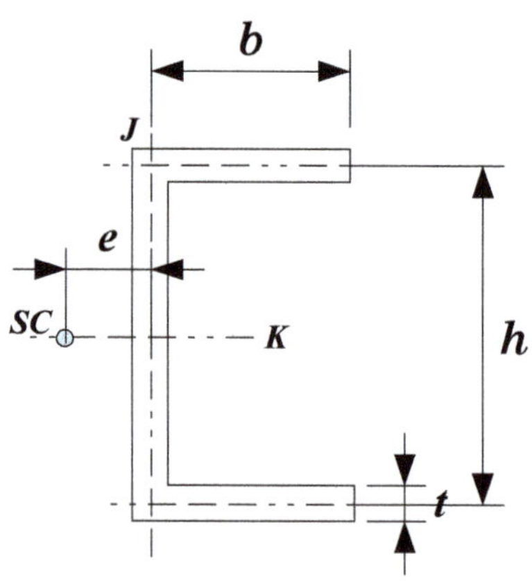

$e = b^2h^2t/4I_k$

For

$b = 1.475$ in

$h = 2.950$ in

$t = 0.050$ in

$I_k = 0.4055$ in⁴

Hence

$e = 0.584$ in

$$K = \frac{t^3}{3}(h + 2b)$$

$$C_w = \frac{h^2\,b^3\,t}{12}\,\frac{2\,h + 3\,b}{h + 6\,b}$$

And

$K = 2.46E^{-4}$ in⁴

$C_w = 0.9914$ in⁶

Figure 10-16 *– Horizontal beam shear center position*

10.3.4 STRESS ANALYSIS

There is no axial load in the beam. Just shear and bending. Therefore, the following checks have to be performed:

- Free-flange buckling
- Shear buckling
- Shear strength
- Deflections (See Section *10.7)*
- Fasteners (See Section *10.6.1)*

Bending stress

The bending stress is calculated at the location where the bending moment is the largest. That is at the center of the rivet group (end *B*). It is calculated at the top of the web and at the edge of the free flanges at points *1* through *4* as shown on *Figure 10-15*. The results of the bending stress calculated at each of the four points are shown in *Table 10-6*.

Since the section is symmetrical ($I_{jk} = 0$), no complex bending needs to be used.

The bending stress is thus calculated as:

$$f_b = -\frac{M_j(k)}{I_j} + \frac{-M_k(j)}{I_k}$$

For
M_j= -1,128 in-lb
M_k= -199 in-lb
I_j = 0.0652 in^4
I_k = 0.4055 in^4

Calculation example at point *1*:

$$f_{b1} = -\frac{M_j(k)}{I_j} + \frac{-M_k(j)}{I_k}$$

$$f_{b1} = -\frac{(-1{,}128)(-1.092)}{0.0652} + \frac{-(-199)(-1.50)}{0.4055}$$

$$f_{b1} = -19{,}628 \ psi$$

The bending stresses at all corners are:

Table 10-6 *- Bending stresses at corners.*

Points	j [in]	k [in]	f_b [psi]
1	-1.5	-1.092	**-19,628**
2	-1.5	0.408	6,323
3	1.5	0.408	7,795
4	1.5	-1.092	-18,156

As seen above, the most critical location is at point *1* where the compression stress is maximum at the free edge. Hence, a free flange buckling analysis will be performed.

10.3.5 FREE FLANGE BUCKLING

Since there is no axial load in the Horizontal beam, then the stress is zero at the section center of gravity. This corresponds also to the neutral axis. Therefore, considering the flange to start from the free edge and stop at the beginning of the bending radius, the variation of the stress goes from compression at the free edge and some tension stress at the bend.

For conservatism, the flange is considered simply supported at all sides except at the free edge. For flat plate buckling in bending, the bending stress allowable is calculated as:

$$F_{b\,cr} = \frac{k_b\,\eta_c\,\pi^2\,E}{12\left(1-\mu^2\right)}\left(\frac{t}{b}\right)^2$$

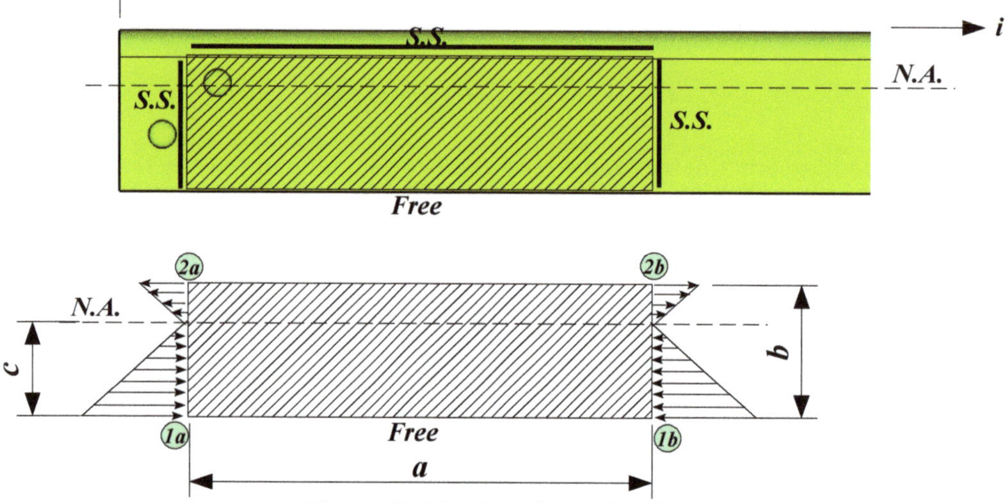

Figure 10-17 *– Free flange buckling in bending*

From *Figure 10-17*, the length *"a"* is arbitrarily shown since it is not required in this analysis. Since the beam is cantilevered, the bending stress at point *1a* is different (higher) than the one at point *1b*. The same for *2a* and *2b*. The maximum (at points *1a* and *2a*) are used for conservatism.

In this case, $b = 1.25$ *in.* and $c = 1.092$ *in.* Thus $b/c = 1.25 / 1.092 = 1.14$. Hence from *Figure B-1* the value for k_b read is: 0.56. The secondary Young's modulus in compression *"E_{c_2}"* is used from *Table 10-5.*

From the above equation:

$$\frac{F_{b\,cr}}{\eta_c} = \frac{k_b\,\pi^2\,E}{12\left(1-\mu^2\right)}\left(\frac{t}{b}\right)^2$$

$$\frac{F_{b\,cr}}{\eta_c} = \frac{(0.56)\,\pi^2\,9.5*10^6}{12\left(1-0.33^2\right)}\left(\frac{0.050}{1.25}\right)^2 = 7{,}856\ psi$$

For
$k_b = 0.56$
$b = 1.25$ in
$t = 0.050$ in
$\mu = 0.33$
$E_{c\_2} = 9.5*10^6$ psi

Since the critical stress is relatively low, the true buckling stress is the same since it is in the elastic portion of the stress-strain curve. The inelastic correction factor *"η_c"* is 1.0 in this case.

Because of the pure Aluminum cladding on the beam material, the secondary elastic modulus is used and, since we are mainly concerned about the side in compression, then the compression secondary Young's modulus *"E_{C_2}"* is used in the buckling allowable equation. See (ref. 8) for more details.

Margin of safety

The bending stress in compression is much higher at point *1* than is the tension stress at point *2*. Hence, knowing that the allowable in tension "F_{TU}" is much higher than the allowable in compression "F_{bcr}", there is no need to check the tension side.

Thus the MS is calculated as:

$$MS = \frac{F_{bcr}}{f_{b1}} - 1 = \frac{7{,}856}{|-19{,}628|} - 1 = -0.60$$

The purpose of this analysis is to merely demonstrate the weakness in this design. Therefore, I didn't consider the impact of the vertical shear as it would have made the *MS* even more negative.

Recommendation: The Horizontal beam should be flipped such that the free flange is in tension and that the compression stress is located at the strong beam corners.

10.3.6 SHEAR BUCKLING

Due to the side load "P_K", a constant shear exists in the web of the Horizontal beam. It is assumed that the beam flanges don't pick-up any of that shear so that only the web carries the shear in the *i-j* plane. On *Figure 10-18*, the shear panel is conservatively assumed simply-supported on all four sides.

For $P_j = V = 17.4\ lbs$, the shear stress "f_s" is calculated as: $f_s = V/A_{web} = 17.4 / (3.0 * 0.05) = 116\ psi$.

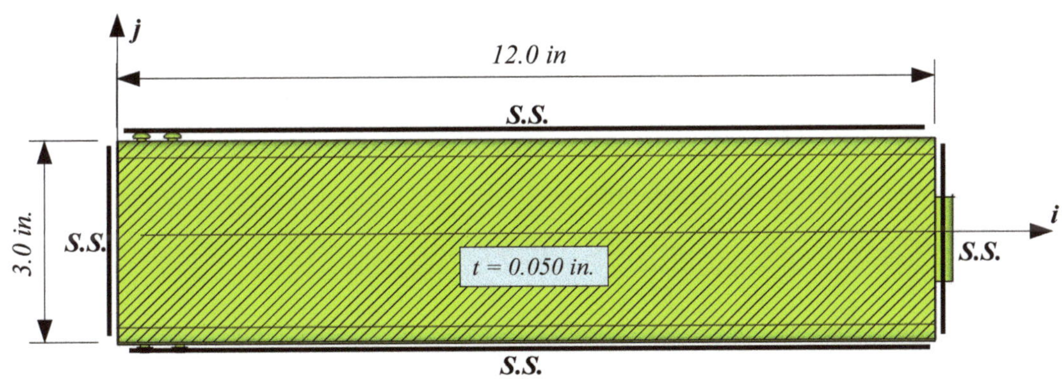

***Figure 10-18** – Web shear buckling.*

For shear panels to work as intended, they MUST be framed on all sides. They are usually thinner than the frames so that their internal shear flow is pretty much constant across the width.

In this design, the Horizontal beam web is framed by the two flanges on the two long sides but there isn't really anything to frame it at the Clip side and at the fasteners' side. Thus this analysis would be somewhat un-conservative, or better yet, the design should be changed to incorporate some framing.

Using (ref. 20.) for flat plate buckling in shear, the shear stress allowable is calculated as:

$$F_{S\,cr} = \frac{k_S\,\eta_S\,\pi^2\,E}{12\left(1-\mu^2\right)}\left(\frac{t}{b}\right)^2$$

From *Figure 10-18*, the dimension *"a"* is 12.0 *in.* and the dimension *"b"* is 3.0 *in.* Note that for shear buckling, the value for *"b"* is always the smallest of the two dimensions.

Thus a/b = 12.0 / 3.0 = 4. Hence from *Figure A-2*, the value for k_s read for hinged edges is: 5.8. The primary Young's modulus in tension *"E"* is used from *Table 10-5*. Because the material is Clad, a reduction factor "λ" is to be applied to the thickness. From (ref. 20.) for 2024 we get: $\lambda = 0.9$ since the total cladding thickness for 0.050" thick sheets is considered to be 10%.

For
k_s = 5.8
a/b = 4
λ = 0.9
t = 0.050 in
μ = 0.33
*E = 10.0*10⁶ psi*

From the above equation:

$$\frac{F_{s\,cr}}{\eta_s} = \frac{k_s\,\pi^2\,E}{12\left(1-\mu^2\right)}\left(\frac{\lambda\,t}{b}\right)^2$$

$$\frac{F_{s\,cr}}{\eta_s} = \frac{(5.8)\,\pi^2\,10.0*10^6}{12\left(1-0.33^2\right)}\left(\frac{0.90*0.050}{3.0}\right)^2 = 12{,}045\ psi$$

Since this stress is low, the true shear buckling stress is the same as it is in the elastic portion of the stress-strain curve. The inelastic correction factor "η_c" is 1.0 in this case.

Using the thickness reduction factor "λ" along with the bare material properties *"E"* is found to be more conservative (lower allowable) than using the secondary elastic modulus "E_2" without the reduction factor. One could use both I guess to be very conservative but that would be "double-dipping" to my opinion. The elastic modulus for tension is used for conservatism since it is slightly lower than the one for compression.

<u>Margin of safety</u>

Thus the MS in shear buckling is calculated as:

$$MS = \frac{F_{s\,cr}}{f_s} - 1 = \frac{12{,}045}{116} - 1 = >1.0$$

10.3.7 SHEAR STRENGTH

The section of the Horizontal beam is opened and has its shear center "*SC*" at a distance "*d*" of 0.600 + 0.559 = 1.159 *in.* from the P_X load (17.4 *lbs*) application point. See *Figure 10-12* and *Figure 10-15*. Therefore, the torque applied at the Clip end is:

$$T = P_X * d = 17.4 * 1.159 = 20.2\ in.-lb$$

The maximum shear stress due to torsion is calculated as follows for a cantilevered Channel beam having one end free to twist and warp and the other end free to warp but not twist. Section properties are from *Figure 10-16*. *Figure 10-19* shows a model of the beam to evaluate the maximum shear stress due to torsion. The shear stress is located everywhere on the surface of the section.

$$G = \frac{E}{2(1+\mu)}$$

$$G = \frac{10.0\,E^6}{2(1+0.33)} = 3.76\,E^6\ psi$$

For
$T = 20.2$ in-lbs
$K = 2.46\ E^{-4}\ in^4$
$G = 3.76\ E^6$ psi
Hence
$\theta' = 0.021839$

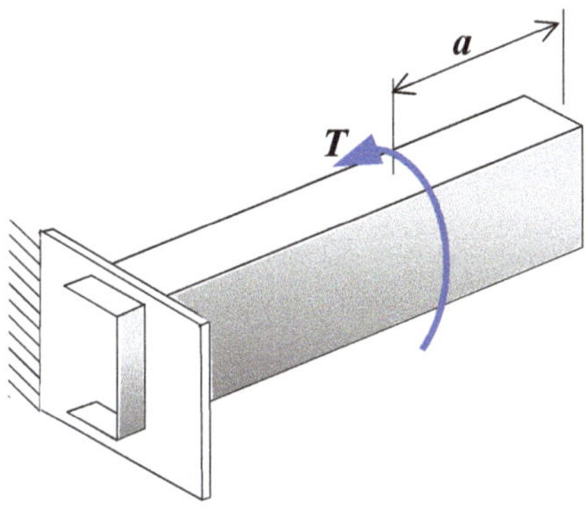

$$Max\ \tau_1 = t\ G\theta'$$
(at the surface everywhere)

where for $a = 0$

$$\theta' = \frac{T}{KG} = \frac{20.2}{2.46\times10^{-4}\ 3.76\times10^{6}} = 0.021839$$

so

$$Max\ \tau_1 = (0.050)\ 3.76\,E^6(0.021839) = 4{,}106\ psi$$

Figure 10-19 – *Cantilevered beam with one end free to twist and warp and one end free to warp but not twist.*

The total shear stress "f_{s_Total}" is then calculated by summing up the torsional stress "$Max\ \tau_1$" and the transverse shear stress "f_s" calculated on page *73*.

Hence

$$f_{s\ Total} = Max\ \tau_1 + f_s = 4{,}106 + 116 = 4{,}222\ psi$$

<u>Margin of safety</u>

Thus for a shear allowable "F_{SU}" of 35 *ksi*, the MS in shear is calculated as:

$$MS = \frac{F_{SU}}{f_{s\ Total}} - 1 = \frac{35{,}000}{4{,}222} - 1 = >1.0$$

The shear stress due to torsion calculated above is for a fully developed torsion stress. That is at a distance sufficiently far from the restrained (clamped) end. So that stress is maximum at the free end and would get somewhat smaller as we get closer to the clamped end since the torque would be reacted by differential shear in the vertical flanges.

10.4 POST

The vertical Post transfers the load from the horizontal beam to the Frame. It is made out of *Al Aly 2024-T42 Clad* sheet 0.050 *in.* thick. It is connected to the horizontal beam with *qty (4) BJ5* rivets and to the Frame with *qty (6) BJ5* rivets. *Figure 10-20(a)* shows the *F.B.D.* of the vertical Post while *Figure 10-20(b)* shows the *F.B.D.* at the center of gravity of the Post's section to determine the internal loads.

10.4.1 LOADS

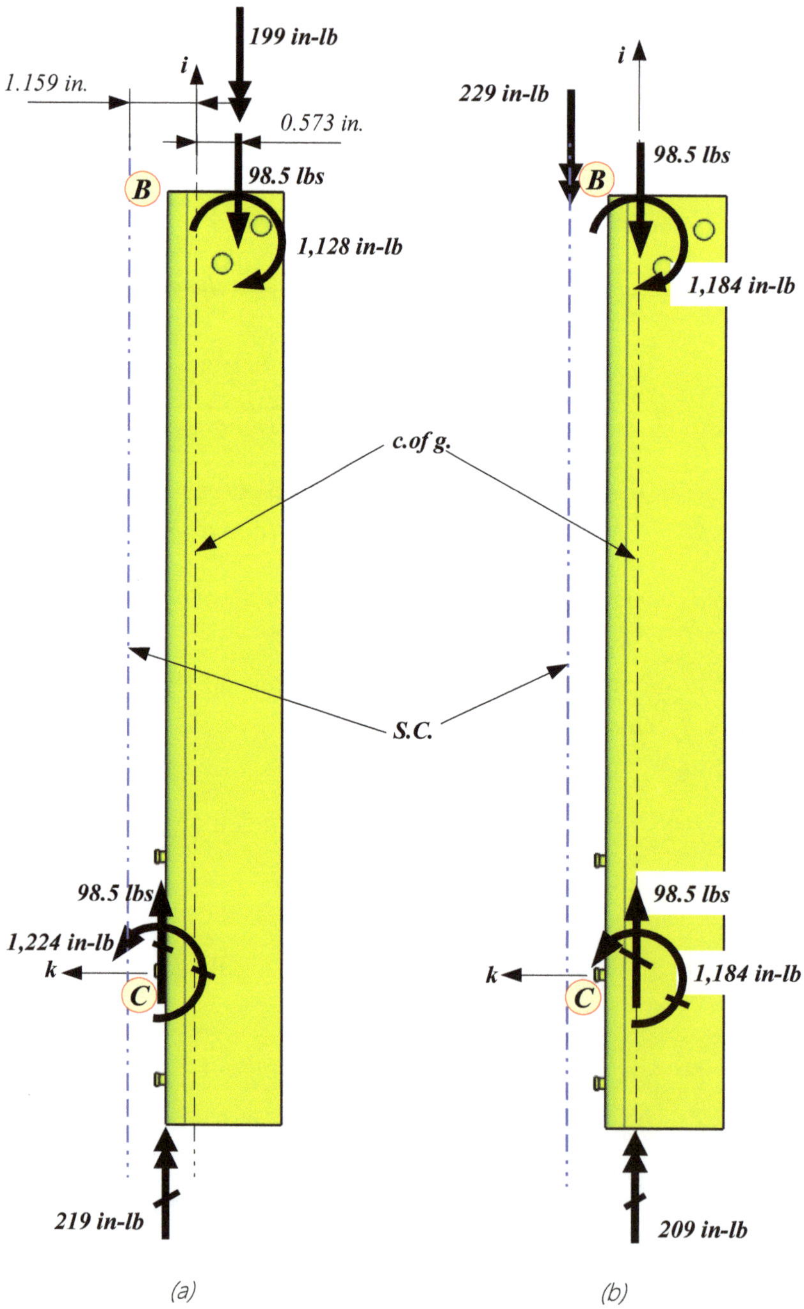

Figure 10-20 – *(a) F.B.D of Post in i-k plane, (b) F.B.D. at section's center of gravity for internal loads.*

As shown on *Figure 10-20(a)*, the loads at end *(B)* are applied at the center of the rivet group. The loads are then reacted at end *(C)* at the center of the rivet group. However, in order to draw the shear and moment diagram of the Post, the *F.B.D.* is re-calculated at the Post's section centroid as shown on *Figure 10-20(b)*. The new F.B.D. is shown in *Figure 10-21*.

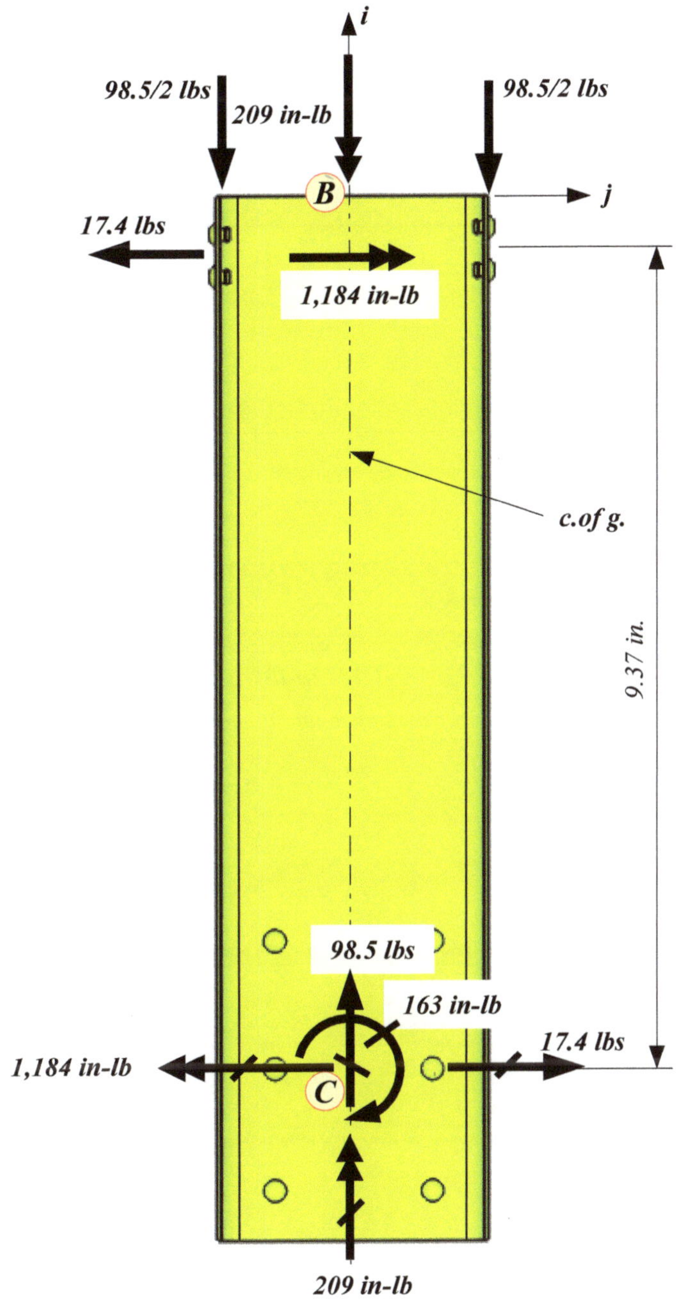

***Figure 10-21** – F.B.D. in i-j plane at section's center of gravity (centroid).*

Shear and moment diagram

The shear and moment diagrams of the Vertical Post can be seen in *Figure 10-22* and *Figure 10-23*. The beam is cantilevered off the Frame and is considered clamped at the rivet group centroid. The applied load is located at the top end.

Plane i-k:

From *Figure 10-13* the moment at point *(B)* is 1,128 *in-lbs*. For P_i = 98.5 *lbs* and an offset *"e"* between the centroid of the rivets and the centroid of the Post's section of 0.573 *in*, then the applied moment is calculated as:

$$M_j = M_{jB} - P_i(e) = -1{,}128 - 98.5\,(0.573) = -1{,}184\ in.-lb$$

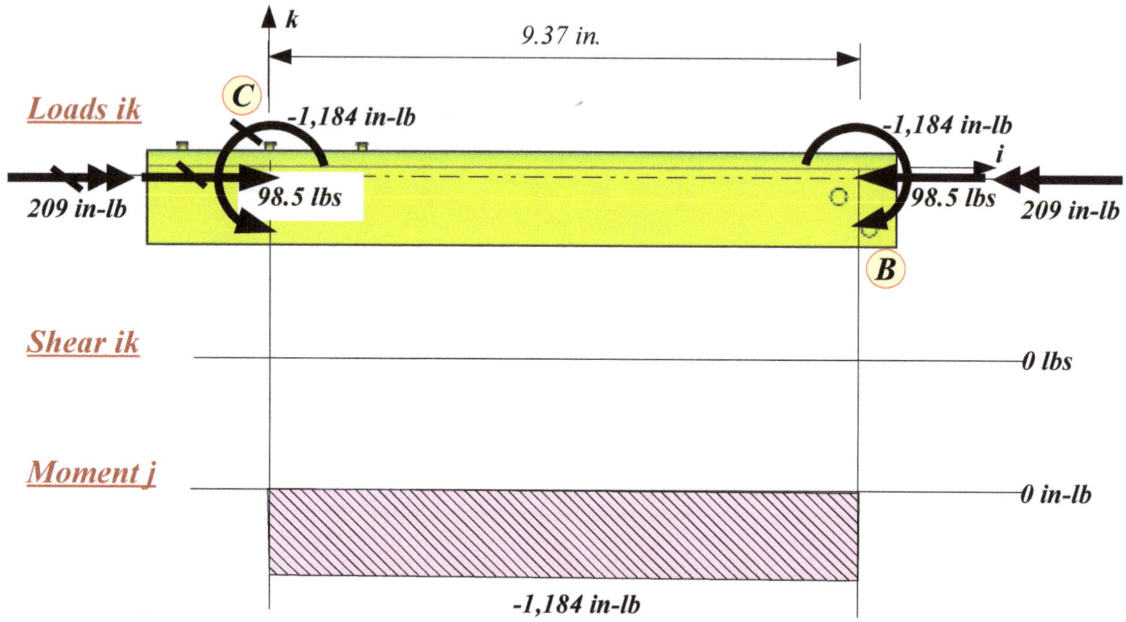

Figure 10-22 – *Vertical Post Shear and moment diagram in plane i-k.*

Plane i-j:

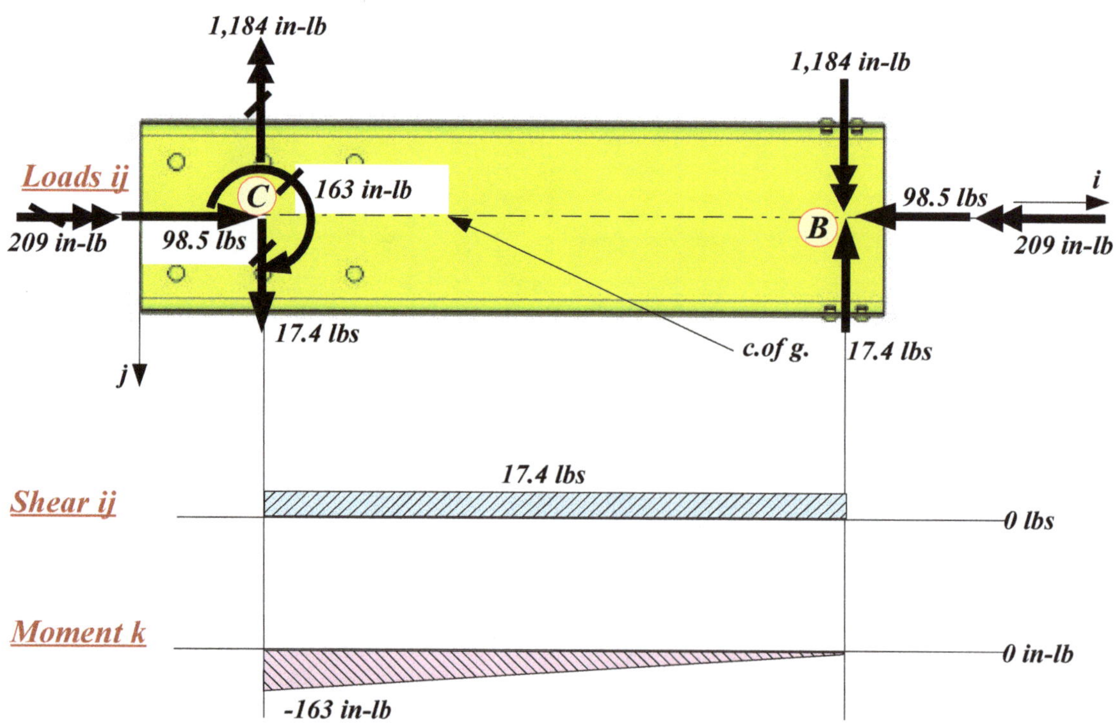

Figure 10-23 – *Vertical Post Shear and moment diagram in plane i-j.*

By assuming the end moments to be located at the centroids of the rivets groups, this approach is found conservative and simpler (quicker) than considering the exact reaction at each fastener at this point. The detailed fastener loads will be calculated later in the Connections chapter.

10.4.2 MATERIAL ALLOWABLES

The **Vertical post**, is made out of *Al Aly 2024-T42 Clad, 0.050 in.* thick sheet. *Table 10-7* shows fictitious values for that material.

Table 10-7 *- 2024-T42 Clad 0.050 in. fictitious material allowables.*

Aluminum Alloy 2024-T42 Clad sheet
Fictitious Values, NOT FOR DESIGN

	0.050"
	[ksi]
F_{tu}	55
F_{ty}	35
F_{cy}	37
F_{su}	35
F_{bru} $(e/D = 2.0)$	115
F_{bry} $(e/D = 2.0)$	70
E, 10^3	
Primary	10
Secondary	9.5
μ	0.33
n	20

10.4.3 SECTION PROPERTIES

The vertical Post has a constant cross-section. Therefore, only one section needs to be calculated. The section properties are obtained from a *3D CAD* drawing software. It is slightly larger than the section of the Horizontal beam. See *Figure 10-24.*

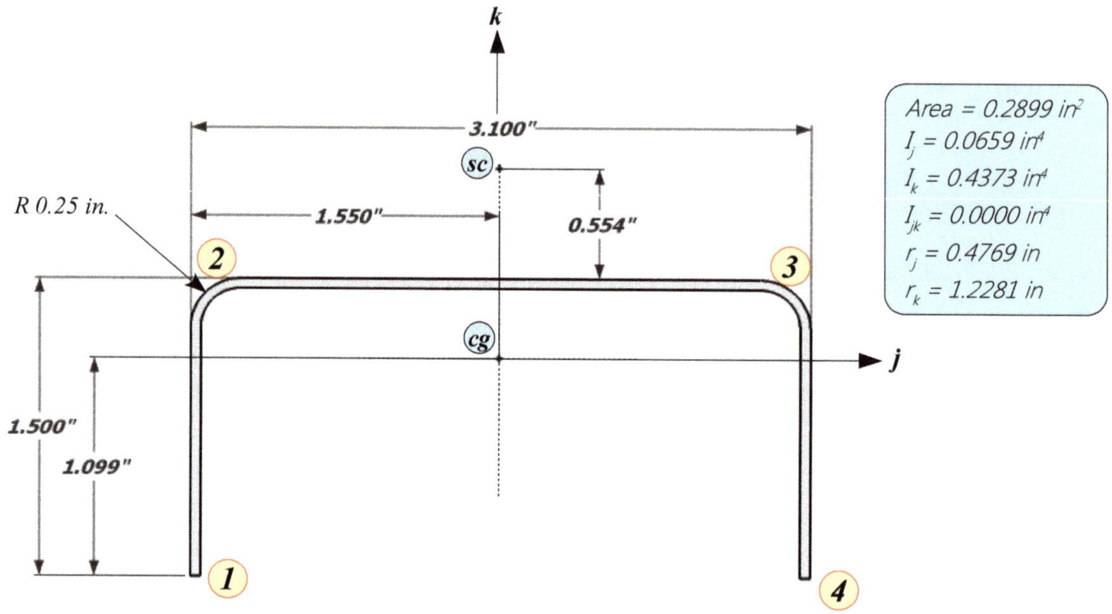

Figure 10-24 *– Vertical Post section properties.*

However, the distance *"e"* of the shear center *"SC"* is calculated for Channels. The calculations are shown in *Figure 10-25.* Therefore, the distance from the top of the web to the shear center is: $e - t/2 = 0.579 - 0.050/2 = 0.554$ *in*. The torsional stiffness *"K"* and the warping constant "C_w" of the section are calculated below.

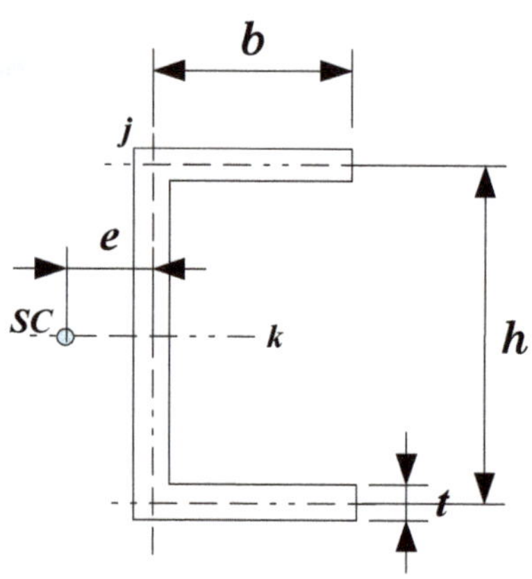

$e = b^2h^2t/4I_k$

For
$b = 1.475$ in
$h = 3.050$ in
$t = 0.050$ in
$I_k = 0.4373$ in^4

Hence
$e = 0.579$ in

$$K = \frac{t^3}{3}(h + 2b)$$

$$C_w = \frac{h^2\,b^3\,t}{12}\,\frac{2\,h + 3\,b}{h + 6\,b}$$

And
$K = 2.50E^{-4}$ in^4
$C_w = 1.0088$ in^6

Figure 10-25 *– Vertical Post shear center position*

10.4.4 ANALYSES

There is an axial compression load and internal moments in the Post. Thus, the following checks have to be performed:

- Column buckling
- Free-flange buckling
- Shear buckling
- Shear strength
- Out-of-Plane bending
- Deflections (See section *10.7)*
- Fasteners (See section *10.6.2)*

10.4.5 COLUMN BUCKLING

The Vertical Post is considered clamped at its base (point *"C"*) and Free where the load is applied (point *"B"*). It is submitted to a 98.5 *lbs* ultimate compression load. The Euler column buckling allowable load is calculated as:

$$P_E = \frac{\pi^2 (EI)}{(L')^2}$$

where L' is the effective length calculated as:

$$L' = \frac{L}{\sqrt{c}}$$

Where "L" is the column length and "c" is the end fixity coefficient. In the case shown in *Figure 10-26,* the end fixity coefficient "c" for a free-fixed column is obtained from *Figure B-2*.

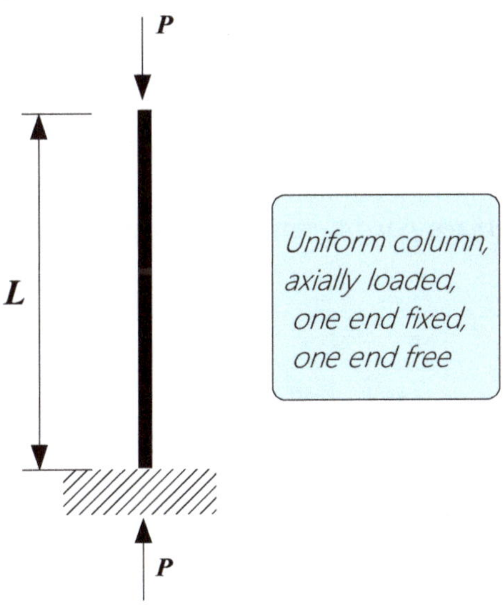

For

$L = 9.37\ in.$

$c = 0.25$

$\frac{1}{\sqrt{c}} = 2$

thus

$L' = 2L = 2*9.37 = 18.74\ in.$

Figure 10-26 *– Vertical Post column model.*

Hence, for $E = E_{C2} = 9.5 * 10^6\ psi$, $I = min(I_j, I_k) = I_j = 0.0659\ in^4$,

then

$$P_E = \frac{\pi^2\ 9.5*10^6 (0.0659)}{(18.74)^2} = 17{,}594\ lbs$$

This corresponds to a stress of: $F_E = P_E / A$ = 17,594 / 0.2899 = 60,690 *psi*. This is way above the yield stress allowable (F_{cy} = *38 ksi*) and is therefore unrealistic.

> The secondary compression elastic modulus was used because the part is in compression and the material is cladded. The Euler buckling is quite high since the column is relatively short with a value for L'/ρ = 18.74 / 0.4769 = 39.

Now, to obtain a more realistic compression load allowable "P_C", one must use the actual elastic modulus at the point of the allowable load. That is the tangent modulus "E_T".

This can be done by using the actual stress-strain curve of the material. In order to do that we will be using the Ramberg-Osgood *(three-parameter)* methodology that uses a parameter (shape factor) value "*n*", the stress value at 70% of *E* "$F_{0.7}$" and the elastic modulus "*E*". See (ref. 18., 20.) for detailed explanations on that methodology.

For *Al Aly 2024-T42 Clad* 0.050" thick sheet, the actual value for "*n*" can be found in the latest edition of (ref. *8.*). In this book, I'm assuming a value of 20 for all aluminum alloys. This is a good approximation for a typical aluminum alloy when the actual data is unavailable.

As for the $F_{0.7}$ value, a simple way to determine it is to use the stress-strain curve approximated here for a cladded aluminum alloy in *Figure 10-27* for the purpose of the demonstration. In this case, from the approximated stress-strain and the tangent modulus curves, the stress value at 70% of the primary *E* "$F_{0.7}$" value is obtained.

For $E = 10.0 * 10^6$ *psi* , then $E_{0.7}$ = 70% * $E = 7.0 * 10^6$ *psi*. We then find a value for $F_{0.7}$ of 37.5 *ksi*.

> Because the cladding material is made out of almost pure aluminum, it is very weak (very low yield strength). This can be seen by looking at the tangent modulus curve where a transition between the primary modulus and the secondary modulus can clearly be noticed.

Now that we have the value for "$F_{0.7}$", the allowable compression load "P_C" can be calculated as follows. Using the *three-parameter* equation for the tangent modulus from (ref. 17), the compression stress allowable "F_C" is obtained by iteration. The *three-parameter* equation is:

$$\frac{E_t}{E} = \frac{1}{1 + \frac{3}{7} n \left(\frac{F_C}{F_{0.7}} \right)^{n-1}}$$

using the Euler equation:

$$F_C = \frac{\pi^2 E_t}{\left(\frac{L'}{\rho} \right)^2}$$

introducing the three-parameter equation into the Euler equation we get:

$$F_C = \frac{\pi^2}{\left(\frac{L'}{\rho} \right)^2} \frac{E}{1 + \frac{3}{7} n \left(\frac{F_C}{F_{0.7}} \right)^{n-1}}$$

Solving by iteration for "F_C" using a *Goal seek* we get:

$$F_C = \frac{\pi^2}{(39)^2}\,\frac{9.5*10^6}{1+\frac{3}{7}(20)\left(\frac{33{,}166}{37{,}500}\right)^{20-1}} = 33{,}166\ psi$$

For
$L' = 18.74$ in
$\rho = 0.4769$ in
$L'/\rho = 39$
$F_{0.7} = 37.5$ ksi
$n = 20$
$E = E_{C2} = 9.5*10^6$ psi

The allowable compression load is thus, $P_C = F_C * A = 33{,}166 * 0.2899 = 9{,}614\ lbs$.

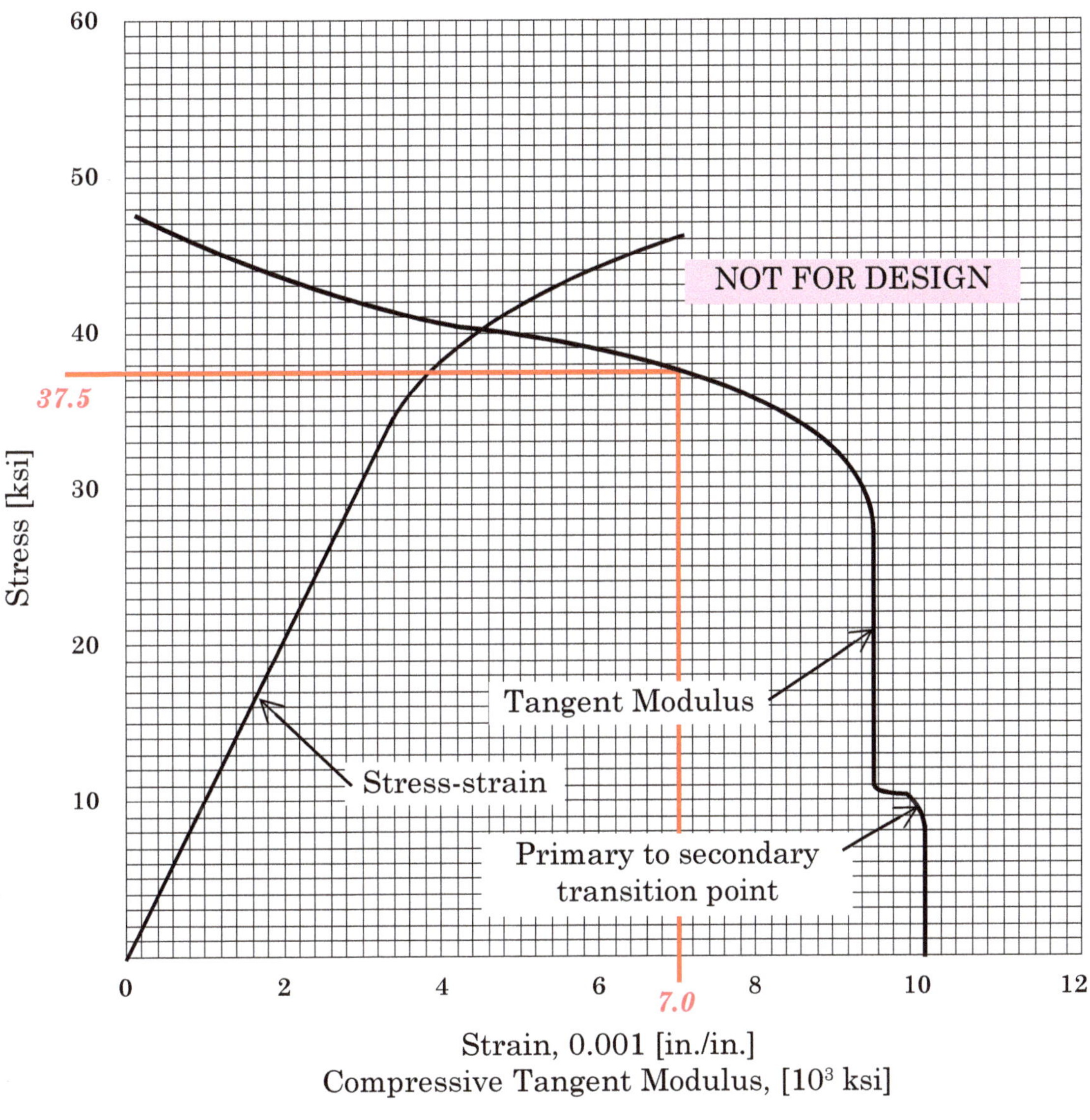

Figure 10-27 *– Approximation of Stress-strain and Tangent Modulus curves for Al Aly 2024-T42 Clad.*

This is still way higher than the applied 100 *lbs* or so. However, the bending will have to be taken into account to get the MS. Especially, because of the secondary bending effect, an "amplified" bending moment will be calculated using the approximate method as shown below. M_o is the primary bending moment while M is the "amplified" bending moment.

$$M = \frac{M_o}{1 - \frac{P}{P_{CR}}}$$

From the above equation, the "amplifying" factor is calculated as:

$$\frac{M}{M_o} = \frac{1}{1 - \frac{P}{P_{CR}}}$$

For $P = 98.5$ *lbs* and $P_{CR} = 9{,}614$ *lbs*, hence:

$$\frac{M}{M_o} = \frac{1}{1 - \frac{98.5}{9{,}614}} = 1.01$$

Therefore, because the applied axial load is small, the maximum bending moment is amplified by only about 1%. This factor will be applied to both M_j and M_k.

Thus the amplified values becomes:

- $M_j = 1.01 * -1{,}184 = -1{,}196$ *in-lb*,
- $M_k = 1.01 * -163 = -165$ *in-lb*,

10.4.6 STRESS ANALYSIS

Bending stress

The bending stress is calculated at the location where the bending moment is the largest. That is at the center of the rivet group (*end (C)*). It is calculated at the top of the web and at the edge of the free flanges at points *1* through *4* as shown in *Figure 10-24*. The results of the bending stress calculated at each of the four points are shown in *Table 10-8*.

Since the section is symmetrical ($I_{jk} = 0$), no complex bending needs to be used.

The bending stress is thus calculated as:

$$f_b = -\frac{M_j(k)}{I_j} + \frac{-M_k(j)}{I_k}$$

Calculation example at point *1*:

For
$M_j = -1{,}196$ in-lb
$M_k = -165$ in-lb
$I_j = 0.0659$ in^4
$I_k = 0.4373$ in^4

$$f_{b1} = -\frac{M_j(k)}{I_j} + \frac{-M_k(j)}{I_k}$$

$$f_{b1} = -\frac{(-1{,}196)(-1.099)}{0.0659} + \frac{-(-165)(-1.55)}{0.4373}$$

$$f_{b1} = -20{,}530 \; psi$$

Axial stress

The axial stress is calculated as $f_c = P_i / A = -98.5 / 0.2899 = -340\ psi$

Total stresses at all corners (bending + axial):

Table 10-8 *- Total stresses at corners.*

Points	*j* [in]	*k* [in]	f_b [psi]	f_a [psi]	f_{Total} [psi]
1	-1.55	-1.099	**-20,530**	-340	**-20,870**
2	-1.55	0.401	6,693	-340	6,353
3	1.55	0.401	7,862	-340	7,523
4	1.55	-1.099	-19,361	-340	-19,700

As seen above, the most critical location is at point *1* where the compression stress is maximum at the free edge. Hence, a free flange buckling analysis will be performed.

10.4.7 FREE FLANGE BUCKLING

Since there is an axial load in the Vertical Post, then the location where the stress is zero (neutral axis) does **not** correspond to the section center of gravity. The axial stress in the Vertical Post is added to the bending stress. Therefore, considering the flange to start from the free edge and stop at the beginning of the bending radius, as shown on *Figure 10-28*, the variation of the stress goes from compression at the free edge and some tension stress at the bend.

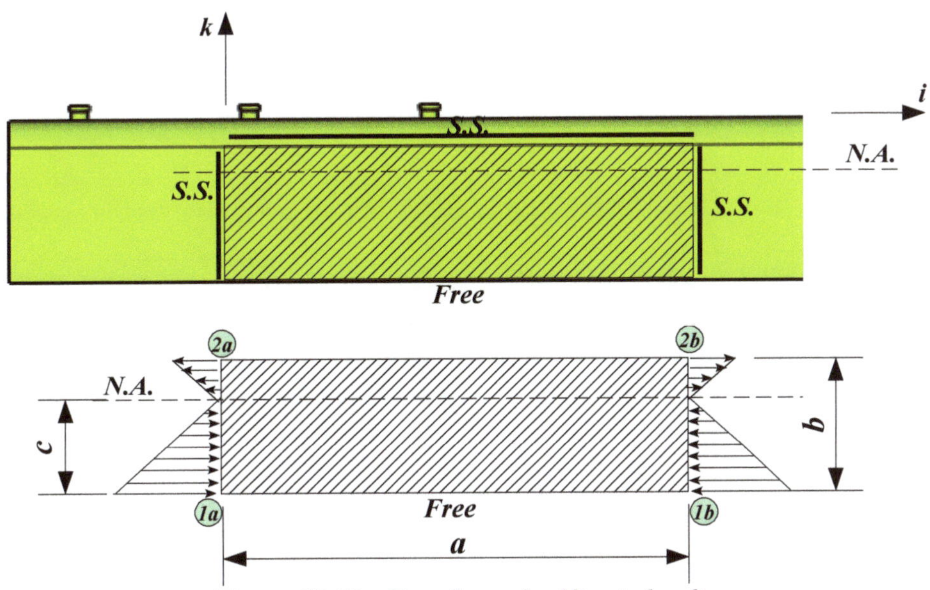

Figure 10-28 *– Free flange buckling in bending.*

For conservatism, the flange is considered simply supported at all sides except at the free edge. Using *Figure B-1* for flat plate buckling in bending, the bending stress allowable is calculated as:

$$F_{b\ cr} = \frac{k_b\,\eta_c\,\pi^2\,E}{12\left(1-\mu^2\right)}\left(\frac{t}{b}\right)^2$$

From *Figure 10-28*, the length *"a"* is arbitrarily shown since it is not required in this analysis. Since the beam is cantilevered, the bending stress at point *1a* is different (higher) than the one at point *1b*. The same for *2a* and *2b*. The maximum (at points *1a* and *2a*) are used for conservatism.

From *Table 10-8* we know that the total stress at point *1* is –20,870 *psi* and that it is 6,353 *psi* at point *2* located 1.5 *in.* away. For this analysis, we are interested to know the location where the stress is zero. That is *"c"* on *Figure 10-28*. By doing a linear interpolation between points *1 & 2* we get c = 1.163 *in.*

In this case, b = 1.25 *in.* and c = 1.163 *in.* Thus b/c = 1.25 / 1.163 = 1.07. Hence from *Figure B-1* the value for k_b read is: 0.50. The secondary Young's modulus in compression "E_{c_2}" is used from *Table 10-7.*

From the above equation:

$$\frac{F_{b\,cr}}{\eta_c} = \frac{k_b\,\pi^2\,E}{12\left(1-\mu^2\right)}\left(\frac{t}{b}\right)^2$$

$$\frac{F_{b\,cr}}{\eta_c} = \frac{(0.54)\,\pi^2\,9.5*10^6}{12\left(1-0.33^2\right)}\left(\frac{0.050}{1.25}\right)^2 = 7{,}576\ psi$$

For
$k_b = 0.54$
$b = 1.25\ in$
$t = 0.050\ in$
$\mu = 0.33$
$E_{c\_2} = 9.5*10^6\ psi$

Since this critical stress is relatively low, the true buckling stress is the same since it is in the elastic portion of the stress-strain curve. The inelastic correction faction "η_c" is 1.0 in this case.

Margin of safety

The stress in compression is much higher at point *1* than is the tension stress at point *2*. Hence, knowing that the allowable in tension "F_{TU}" is much higher than the allowable in compression "F_{bcr}", there is no need to check the tension side.

Thus the MS is calculated as:

$$MS = \frac{F_{b\,cr}}{f_{b1}} - 1 = \frac{7{,}576}{|-20{,}870|} - 1 = -0.64$$

The purpose of this analysis is to merely demonstrate the weakness in this design. Therefore, I didn't consider the impact of the torsional shear as it would have made the *MS* even more negative.

Recommendation: The Vertical Post should be flipped such that the free flange is in tension and that the compression stress is located at the strong beam corners.

10.4.8 SHEAR BUCKLING

Due to the side load "P_j", a constant shear exists in the web of the Vertical Post. It is assumed that the Post flanges don't pick-up any of that shear so that only the web carries the shear in the *i-j* plane. In *Figure 10-29*, the shear panel is conservatively assumed simply-supported on all four sides.

For $P_j = V = 17.4$ *lbs*, the shear stress "f_s" is calculated as: $f_s = V/A_{web} = 17.4 / (3.1 * 0.05) = 112$ *psi*.

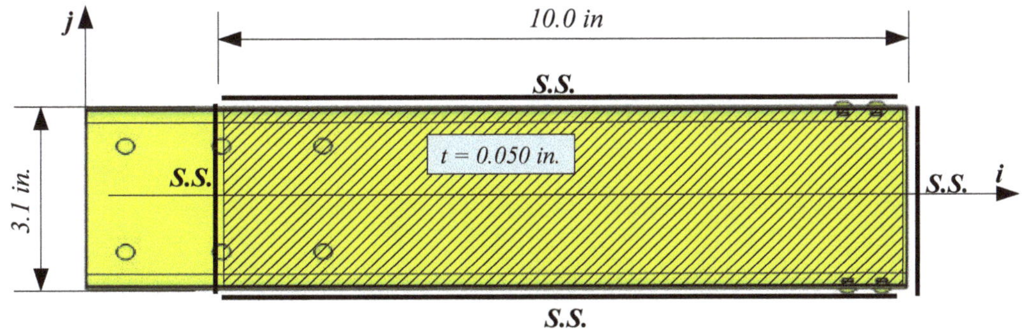

Figure 10-29 *– Web shear buckling.*

> In this design, the Vertical Post web is framed by the two flanges on the two long sides and also at the group of six fasteners at the bottom but there isn't really anything to frame it at the upper side. Thus this analysis would be somewhat un-conservative, or better yet, the design should be changed to incorporate some framing.
>
> The bottom edge is considered simply supported because of the six rivets. However, the fastener pitch is much larger than *6D* and, therefore, prone to inter-rivet buckling. Here the shear flow is so low that there is no risk of that happening.

Using (ref. 20.) for flat plate buckling in shear, the shear stress allowable is calculated as:

$$F_{S\,cr} = \frac{k_S\,\eta_S\,\pi^2\,E}{12\left(1-\mu^2\right)}\left(\frac{t}{b}\right)^2$$

From *Figure 10-29*, the dimension "*a*" is 10.0 *in*. and the dimension "*b*" is 3.1 *in*. Note that for shear buckling, the value for "*b*" is always the smallest of the two dimensions.

Thus $a/b = 10.0 / 3.1 = 3.2$. Hence from *Figure A-2*, the value for k_S read is: 5.8. The secondary Young's modulus in tension "E_2" is used from *Table 10-7*.

For
$k_s = 5.8$
$a/b = 3.2$
$t = 0.050$ in
$\mu = 0.33$
$E_2 = 9.5*10^6$ psi

From the above equation:

$$\frac{F_{s\,cr}}{\eta_s} = \frac{k_s\,\pi^2\,E}{12\left(1-\mu^2\right)}\left(\frac{t}{b}\right)^2$$

$$\frac{F_{s\,cr}}{\eta_s} = \frac{(5.8)\,\pi^2\,9.5*10^6}{12\left(1-0.33^2\right)}\left(\frac{0.050}{3.1}\right)^2 = 13{,}230\ psi$$

Since this critical stress is relatively low, the true shear buckling stress is the same since it is in the elastic portion of the stress-strain curve. The inelastic correction factor "η_c" is 1.0 in this case.

As for the free-flange buckling, because of the pure Aluminum cladding on the beam material, the secondary elastic modulus is used "E_2". However, the one for tension is used for conservatism since it is slightly lower than the one for compression.

A quick approach to see if the critical stress is in the elastic region of the stress-strain curve is to use (ref. 10, Fig. 11.2.5). But since in our case F_{scr} is way below the yield value, the 2024-T42 curve would be exactly the same in this region as the 2024-T3.

Margin of safety

Thus the MS in shear buckling is calculated as:

$$MS = \frac{F_{s\,cr}}{f_s} - 1 = \frac{13{,}230}{112} - 1 = \ >1.0$$

10.4.9 SHEAR STRENGTH

Using the *F.B.D.* shown on *Figure 10-21*, the torque "*T*" going though the Post at the shear center "*SC*" is 229 *in-lb*. This is the torque applied at the centroid "T_o" plus the impact of the transverse shear "V_X" applied at a distance "*d*" between the centroid and the shear center.

$$T = T_O + V_X * d = 209 + 17.4 * 1.159 = 229 \ in.-lb$$

For
T_o = 209 in-lb
d = 1.159 in
V_X = 17.4 lbs

The maximum shear stress due to torsion is calculated for a cantilevered Channels beam having one end free to twist and warp and the other end free to warp but not twist. Section properties are from *Figure 10-24* and *Figure 10-25*. *Figure 10-30* shows a model of the beam to evaluate the maximum shear stress due to torsion. The shear stress is located everywhere on the surface of the section.

$$G = \frac{E}{2(1+\mu)}$$

$$G = \frac{10.0\,E^6}{2(1+0.33)} = 3.76\,E^6 \ psi$$

For
T = 229 in-lb
K = 2.50E-4 in^4
G = 3.76E^6 psi
Hence
θ' = 0.2436

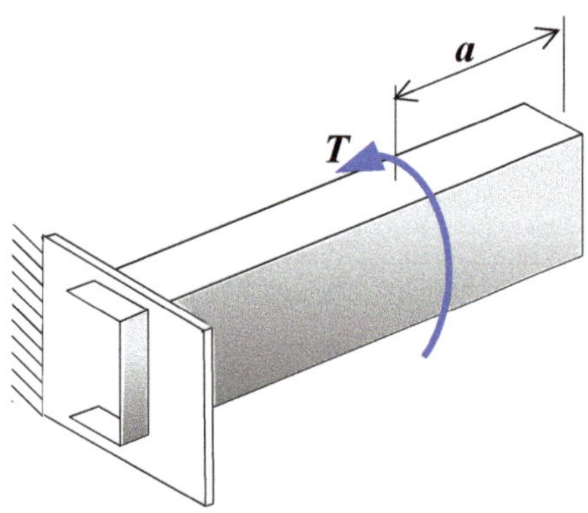

$$Max\ \tau_1 = t\ G\theta'$$
$$(at\ the\ surface\ everywhere)$$

where for $a = 0$

$$\theta' = \frac{T}{KG} = \frac{229}{2.50\,E^{-4}\ 3.76\,E^6} = 0.2436$$

so

$$Max\ \tau_1 = (0.050)\ 3.76\,E^6(0.2436) = 45{,}797 \ psi$$

Figure 10-30 *– Cantilevered beam with one end free to twist and warp and one end free to warp but not twist.*

The total shear stress "f_{s_Total}" is then calculated by summing up the torsional stress "$Max\ \tau_1$" and the transverse shear stress "f_s" calculated on page *87*.

Hence

$$f_{s\ Total} = Max\ \tau_1 + f_s = 45{,}797 + 112 = 45{,}909\ psi$$

Margin of safety

Thus for a shear allowable "F_{SU}" of 35 *ksi*, the MS in shear is calculated as:

$$MS = \frac{F_{SU}}{f_{s\ Total}} - 1 = \frac{35{,}000}{45{,}909} - 1 = -0.24$$

The shear stress due to torsion calculated above is for a fully developed torsion stress. That is at a distance sufficiently far from the restrained (clamped) end. So that stress is maximum at the free end and would get somewhat smaller as we get closer to the clamped end since the torque would be reacted by differential shear in the vertical flanges.

Recommendation: The Vertical Post should be thicker or, preferably, be a closed section in order to reduce the shear stress due to torsion.

10.4.10 OUT-OF-PLANE BENDING STRESS

Because the Post is quite flexible/soft in torsion (way more flexible than the Frame anyway), it is assumed that all the torque in the Post is reacted at point *(C)* by only the top two rivets from the group of six. Therefore, from *Figure 10-48*, fasteners ***1** & **6*** are the two effective rivets while fastener no. ***6*** is the most critical one in tension. The tension load in that fastener is determined as being 335 *lbs*. See *Table **10-15***.

From *Figure **10-31***, it can be seen that the portion of the web located between rivet no.***6*** and the Post flange acts as an Angle clip. This approach may be a bit too conservative but highlights the fact that the Post's web is solicited in bending as shown in *Figure **10-32***.

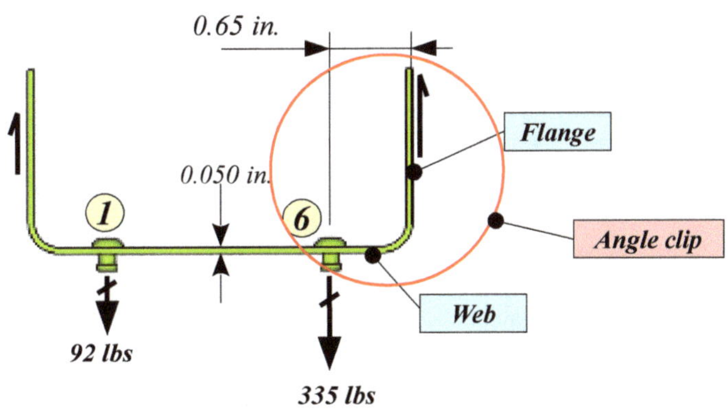

Figure 10-31 *– F.B.D. of rivets 1 & 6.*

Figure 10-32 shows an approximation of a curve for formed *2024* and *7075* Aluminum Alloy Angle Clips, the only information we need is:

- The Clip thickness; $t = 0.050$ in.
- The distance from the load to the fastener; $e = 0.65$ in.

From *Figure 10-32* we get: P_{all} = **105 *lbs*** ultimate.

Since the 105 *lbs* allowable is much lower than the 335 *lbs* applied tension load in rivet no. *6*, the Post is found inadequate and a re-design is required.

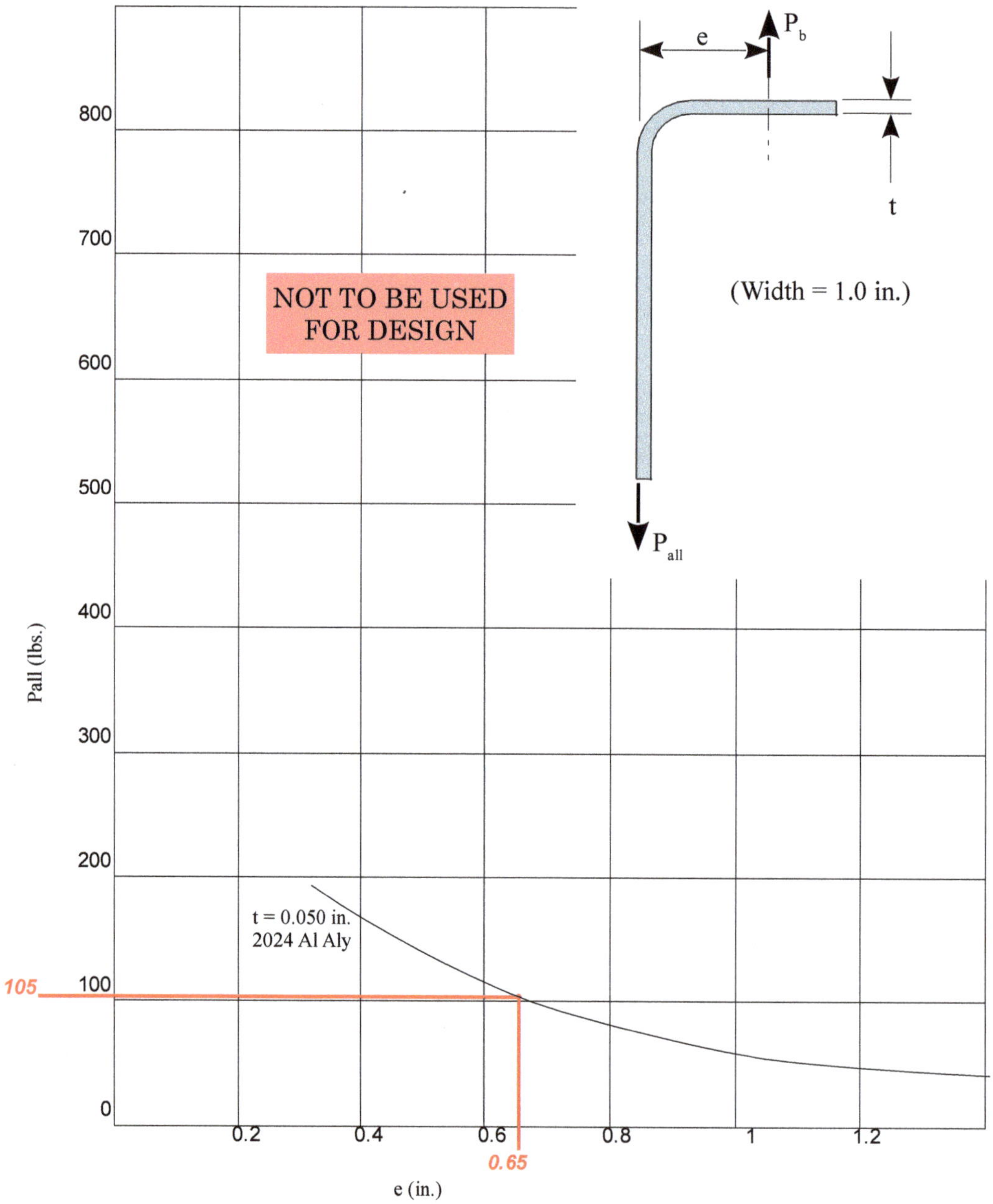

***Figure 10-32** – Angle Clip allowable approximation.*

10.5 FRAME

The Frame reacts all the loads coming from this structure. For the purpose of the exercise, the Frame is not considered as part of the airframe. Therefore, only the loads from this structure are considered. It is made out of *Al Aly 7050-T7451* machined Plate 1.0 *in*. thick. Its connections at each end are not looked at. *Figure 10-33* shows the dimensions of the Frame.

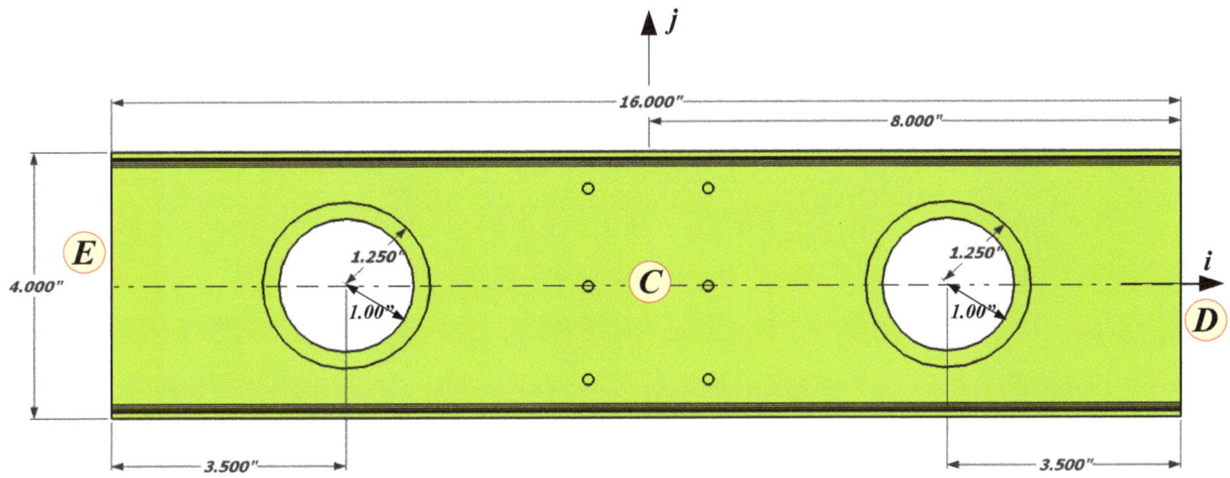

Figure 10-33 *– Frame dimensions.*

If the Frame had been part of the airframe (primary structure), that part would have been analyzed not only for the loads coming from this structure but also by taking into account (superimposing) the internal airframe loads for the same load case.

10.5.1 LOADS

As shown on *Figure 10-34* the loads at point *(C)* are applied at the center of the rivet group. The loads are then reacted at ends *(D)* and *(E)* which are considered simply supported. As specified in section *10.1*, the Frame is assumed fastened to a skin along its bottom edge where the 17.4 *lbs* load is reacted. Also, the torque M_j of 222 *in-lb* is reacted only by the top Frame cap at each end (14 *lbs*) in and out of the page direction. All the loads are at the section's centroid.

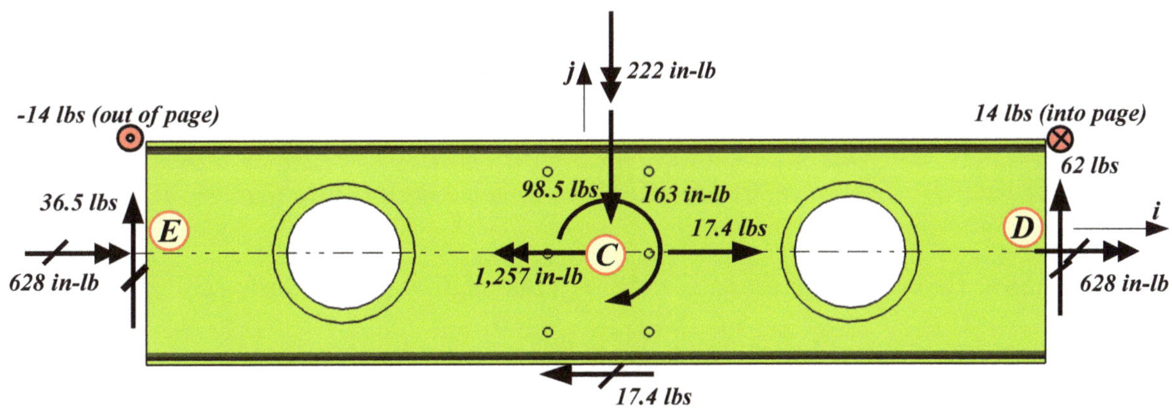

Figure 10-34 *– FBD in i-j plane at section's center of gravity (centroid).*

Shear and moment diagram

The shear and moment diagrams of the Frame can be seen in *Figure 10-35* and *Figure 10-37*. The Frame is considered simply supported at both ends but is considered fixed-fixed in torsion along the "*i*" axis. The applied loads are located at the fastener group centroid.

Plane i-k:

From *Figure 10-4* and *Figure 10-38*, the distance "*d*" in "*k*" direction between the load application point *(A)* and the section's centroid is: d_k = (12.551 + 0.212) = 12.76 *in*. Therefore, the torque "M_i" applied at point *(C)* along the section's c.g. is $T = P_j * d_k = 98.5 * 12.76 = 1{,}257$ *in-lb*. Similarly, the moment "M_j" applied at point *(C)* along the section's c.g. is $M_j = P_i * d_k = 17.4 * 12.76 = 222$ *in-lb*.

Since the loads are applied right in the middle of the Frame, then the internal loads are thus symmetrical from the Frame's center. *Figure 10-35* shows the shear and moment diagram of the Frame in the *i-k* plane.

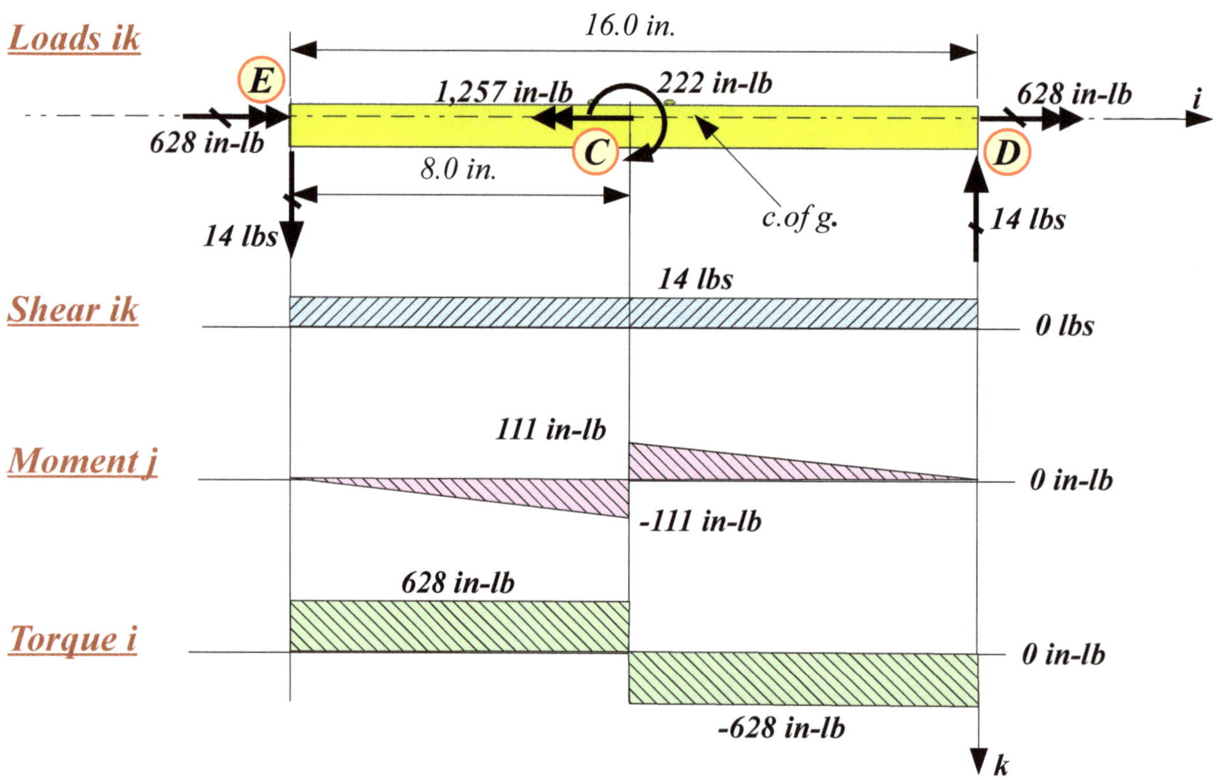

***Figure 10-35** – Frame Shear and moment diagram in plane i-k.*

However, as explained on page *96*, the torque in the Frame is to be taken as a couple between the Frame upper cap and the skin. An *FBD* as well as the shear and moment diagram for the Upper cap only is shown on *Figure 10-36*.

It is also assumed that all the M_j moment (222 *in-lb*) is applied to the upper cap only. The reason being that the vertical Post introduces its torque into the Frame. Therefore, if the Post was very stiff compared to the Frame, most of the torque would be reacted by the Frame lower cap as it is almost infinitely rigid because it is in the plane of the skin. But because the Post is quite weak in torsion (due to its open section), it won't be stiff enough to carry much torque all the way down to the Frame lower cap. So most of the torque is reacted by the Frame upper cap.

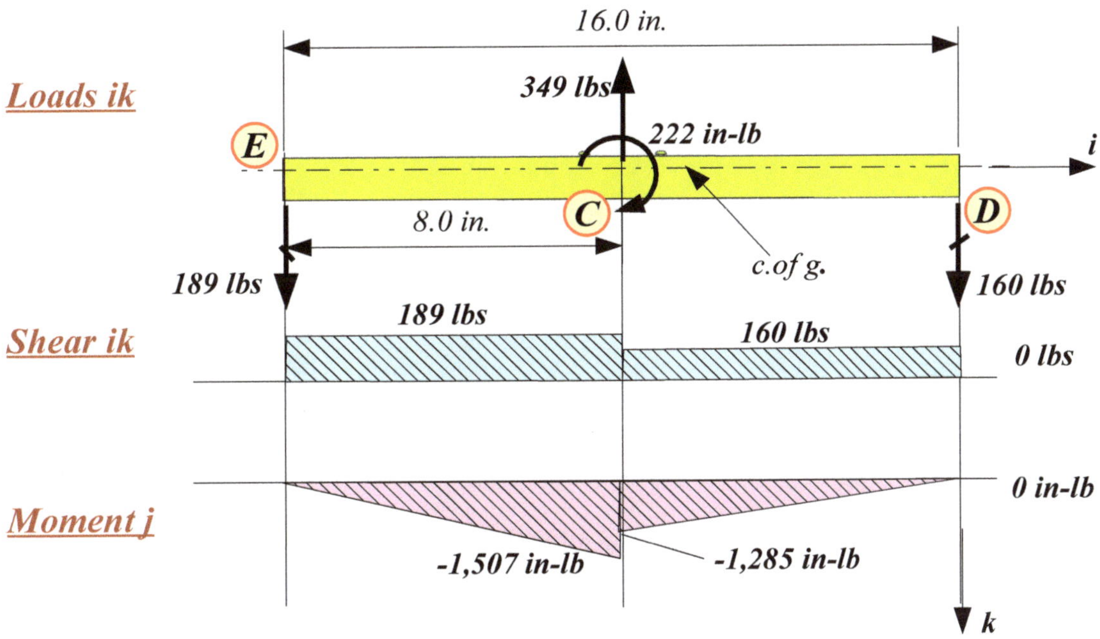

Figure 10-36 *– Frame Upper cap Shear and moment diagram in plane i-k.*

Plane i-j:

The torque is, of course, the same in this plane. The applied moment "M_k" and the axial load "P_i" are obtained from *Figure 10-23*. *Figure 10-37* shows the shear and moment diagram of the Frame in the *i-j* plane.

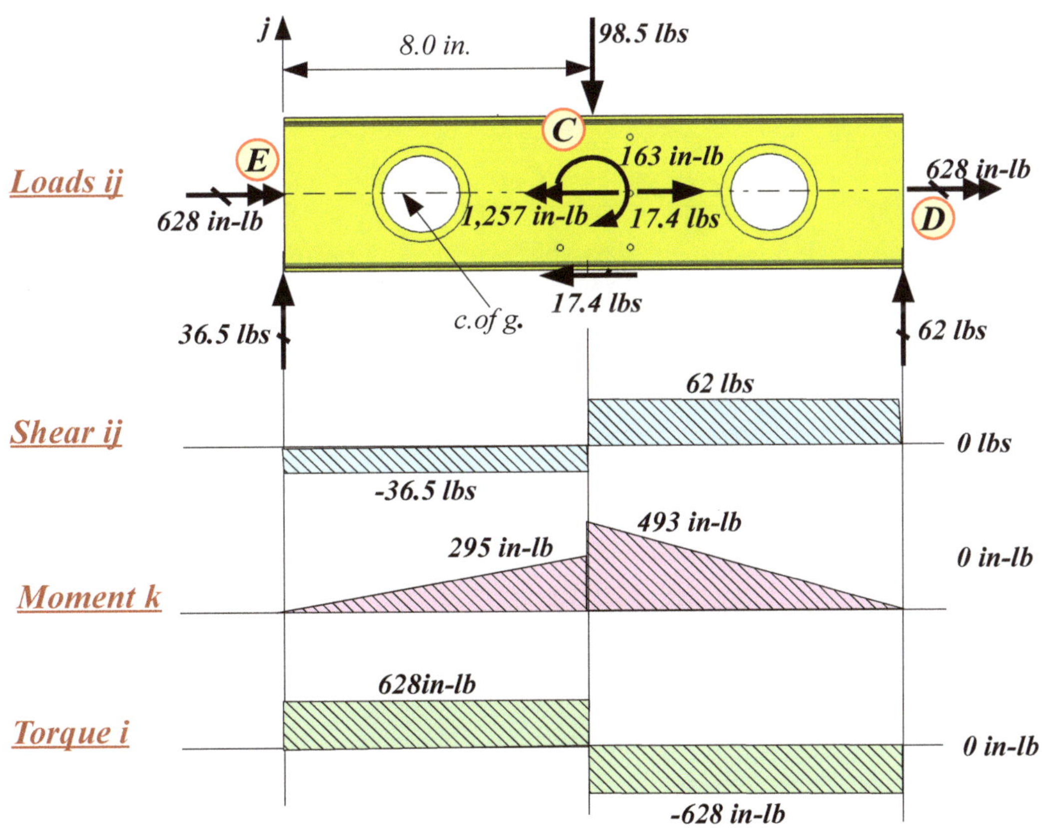

Figure 10-37 *– Frame Shear and moment diagram in plane i-j.*

10.5.2 ANALYSES

There is no axial load in the Frame. Thus, the following checks have to be performed:

- Bending
- Free-flange buckling
- Shear buckling of web
- Bending + Shear strength
- Deflections (See Section *10.7)*

10.5.3 MATERIAL ALLOWABLES

The actual and latest material allowables for all metallic materials can be obtained from the latest (ref. 8.). The **Frame**, is machined out of *Al Aly **7050-T7451*** plate between 1.00 *in.* and 2.50 *in.* thick. *Table **10-9*** shows fictitious values for that material.

Table 10-9 *- Al Aly 7050-T7451 Plate 2.50 in. fictitious material allowables.*

Aluminum Alloy 7050-T7451 Plate
Fictitious Values, NOT FOR DESIGN

	2.50" [ksi]
F_{tu}	75
F_{ty}	65
F_{cy}	60
F_{su}	40
F_{bru} (e/D = 2.0)	105
F_{bry} (e/D = 2.0)	85
E, 10^3	10
G, 10^3	3.8
μ	0.33
n	20

10.5.4 SECTION PROPERTIES

The Frame has a constant cross-section. Therefore, only one section needs to be calculated. The section properties are obtained from a *3D CAD* drawing software. See *Figure 10-38*.

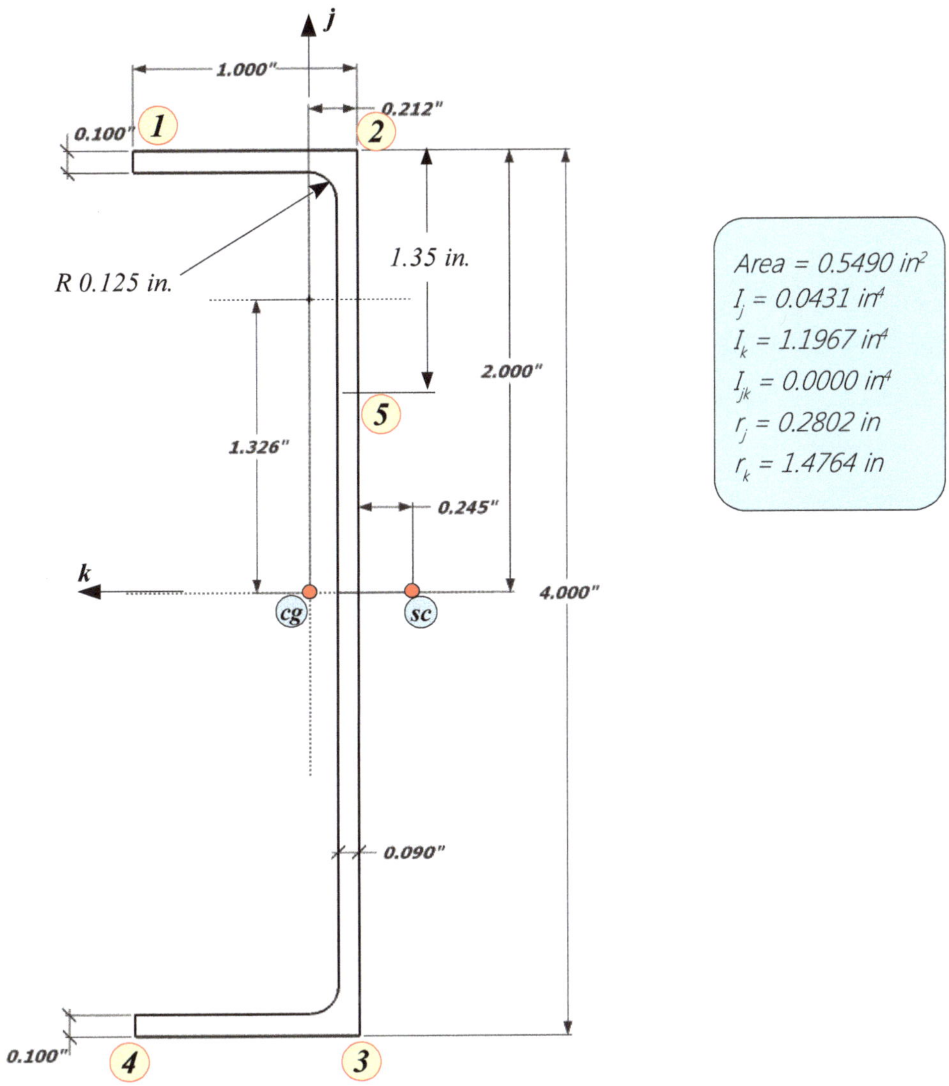

Figure 10-38 *– Frame section properties.*

However, the position of the shear center *"SC"* is calculated for Channels. The calculations are shown in *Figure 10-39*. Therefore, the distance from the face of the web to the shear center is: $e - t/2 = 0.290 - 0.090/2 = 0.245\ in$. The torsional stiffness *"K"* and the warping constant *"C_w"* of the section are calculated below.

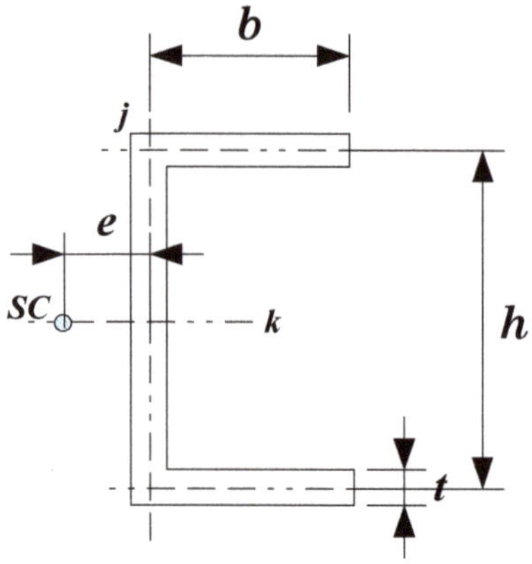

$e = b^2 h^2 t / 4I_k$

For

$b = 0.955 in$

$h = 3.90 in$

$t = 0.100 in$

$I_k = 1.1967 in^4$

Hence

$e = 0.290 in$

$$K = \frac{t^3}{3}(h + 2b)$$

$$C_w = \frac{h^2 b^3 t}{12} \frac{2h + 3b}{h + 6b}$$

And

$K = 1.937E^{-3} in^4$

$C_w = 1.2179 in^6$

Figure 10-39 *– Frame shear center position.*

As for the Horizontal beam and the Vertical post, I have calculated the position of the shear center as well as the section's properties in torsion *(K, C_w)* but those won't be required since the Frame is not free to twist due to its attachment to the skin.

Upper Cap section properties

Because it is assumed that the Frame is fastened to a skin along its bottom flange, it is therefore partially restrained from twisting. The applied torque is thus reacted as a couple between the Frame upper cap and the skin, which doesn't move. See *Figure 10-41* for the upper cap section properties and *Figure 10-40* to see how the torque in the Frame is reacted by the Upper cap and the skin.

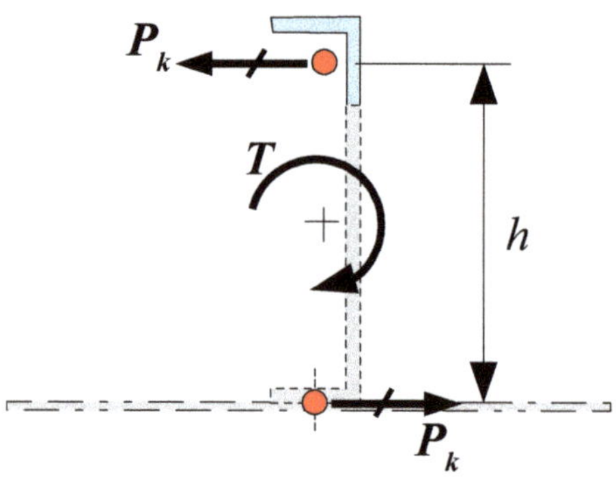

$h = 3.60 in$

$T = 1,257 in\text{-}lb$

$P_k = T / h = 349 lbs$

Figure 10-40 *– Couple reacting Torque in Frame.*

An effective web width of *15t* (or 15*0.090 = 1.35 *in.*) is assumed in the upper cap section.

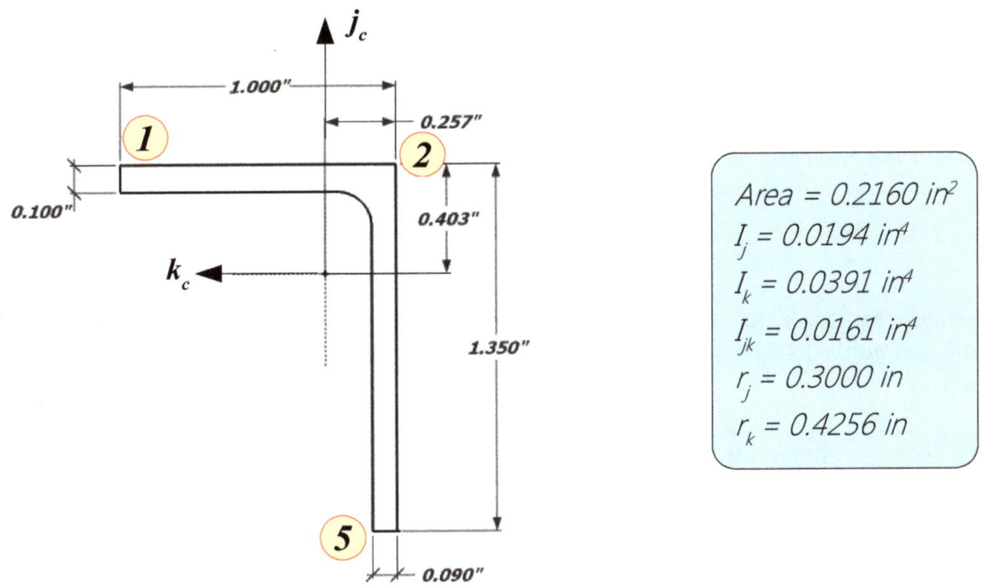

Figure 10-41 *– Frame upper cap section properties.*

10.5.5 STRESS ANALYSIS

Bending stress

The bending stress is calculated at the location where the bending moment is the largest which is right at the center. It is calculated at the points as shown on *Figure 10-38* and *Figure 10-41*. There will be two different bending stress calculations to consider.

1. Due to the moment M_k *(full section)*. (ref. *Figure 10-37)*
2. Due to the moment M_j *(Upper cap section only)*. (ref. *Figure 10-36)*

The resulting stresses obtained at points *1, 2* and *5* will be superimposed from the two calculations in order to get the total bending stress. Points *3 & 4* will be only influenced by the stress from M_k.

One can notice that the maximum bending moment M_k occurs just slightly to the right of the Frame center while the maximum bending moment M_j occurs just slightly to the left of the Frame center. The two maximum values are conservatively used in the total stress calculation.

The shear and moment diagrams are idealized and are usually more severe than in reality due principally to the width of the Vertical Post that load-up the Frame. Therefore those peak bending moments are already conservative and I could have used the value for M_k of 295 *in-lb* instead of 493 *in-lb* to combine with the maximum M_j value of 1,507 *in-lb*. But, since I wanted to stick to the assumed moment distribution model, then I conservatively decided to use the most severe value for M_k since there is only a very fine line separating the values of 493 and 295.

Due to M_k:

The Frame full section as shown in *Figure 10-38* is used. However, for conservatism, no effective skin is considered.

The results of the bending stress calculated at all stress points are shown in *Table 10-10*.

> Since the full frame section is symmetrical ($I_{jk} = 0$), no complex bending needs to be used. But for the Upper cap angle section, $I_{jk} \neq 0$. However, although I chose to consider the upper cap as being a beam by itself, in reality it is still attached to the rest of the Frame. Therefore, it is stabilized (prevented to twist). Hence, as for the full frame section, no complex bending needs to be used.
>
> Also, the skin is ignored in the section properties mainly because, by considering an effective skin working with the Frame section, that would require that the loads from the Frame be added to the skin internal loads for the skin analysis.
>
> The effective skin is, of course, considered for the fuselage frames but for secondary structures (as in this example) it is preferable to leave the skin alone and not take advantage of it.

Using the shear and moment diagram on *Figure 10-37,* the bending stress is thus calculated as:

$$f_b = \frac{M_k(j)}{I_k}$$

From *Figure 10-37* and *Figure 10-38,* the calculation example at point *1* is:

$$f_{b1} = \frac{-M_k(j)}{I_k}$$

$$f_{b1} = -\frac{(493)(2.0)}{1.1967}$$

$$f_{b1} = -824\ psi$$

> For
> $M_k = 493$ in-lb
> $I_k = 1.1967$ in^4
> $j = 2.0$ in

Bending stresses at all corners:

Table 10-10 *- Bending stresses at corners due to M_k.*

Points	j [in]	k [in]	f_b [psi]
1	2	0.788	**-824**
2	2	-0.212	-824
3	-2	-0.212	824
4	-2	0.788	824
5	0.65	0.788	-268

Due to M_j:

Now, the bending stresses will be calculated by considering the Upper cap only. The Frame upper cap section as shown in *Figure 10-41* is used.

Using the shear and moment diagram on *Figure 10-36*, the bending stress is thus calculated as:

$$f_b = \frac{M_j (k_c)}{I_j}$$

From *Figure 10-36* and *Figure 10-41*, the calculation example at point *1* is:

$$f_{b1} = \frac{M_j (k_c)}{I_j}$$
$$f_{b1} = \frac{(-1{,}507)(0.743)}{0.0194}$$
$$f_{b1} = -57{,}717 \ psi$$

For
$M_j = -1{,}507$ in-lb
$I_j = 0.0194$ in^4
$k_c = 1.0 - 0.257 = 0.743$ in

Bending stresses at all corners:

Table 10-11 *- Bending stresses at corners due to* M_j.

Points	j [in]	k [in]	f_b [psi]
1	0.403	0.743	**-57,717**
2	0.403	-0.257	19,964
5	-0.947	-0.257	19,964

Total bending stress:

The stress due to the moment M_k applied to the full Frame section is added to the stress due to the moment M_j applied to the Frame Upper cap section to get the total stress at each critical stress point.

Table 10-12 *- Total Bending stresses.*

Points	$f_{b\,(Mk)}$ [psi]	$f_{b\,(Mj)}$ [psi]	$f_{b\,Total}$ [psi]
1	-824	-57717	**-58,540**
2	-824	19964	19,140
3	824	0	824
4	824	0	824
5	-268	19964	19,696

As seen above, the most critical location is at point *1* where the compression stress is maximum at the free edge. Hence, a free flange buckling analysis will be performed.

10.5.6 FREE FLANGE BUCKLING

The Frame upper flange has a fairly high compression stress at the free flange (point *1*). Because of the constant bending stress between *point 1* and *point 2* due to the M_k moment, then the location where the stress is zero (neutral axis) does **not** correspond to the section's center of gravity. Therefore, considering the flange to start from the free edge (point *1*) and stop at the corner (point *2*), as shown on *Figure 10-41*, the variation of the stress goes from compression at the free edge and some tension stress at the corner.

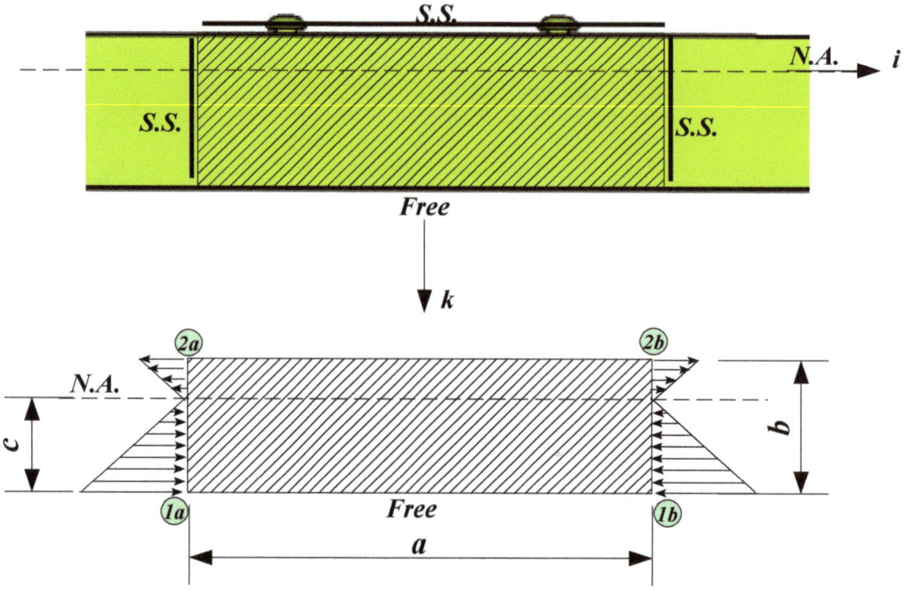

Figure 10-42 *– Free flange buckling in bending.*

For conservatism, the flange is considered simply supported at all sides except at the free edge. The bending stress allowable for flat plate buckling (ref. 20.) in bending is calculated as:

$$F_{b\ cr} = \frac{k_b\ \eta_c\ \pi^2\ E}{12\left(1-\mu^2\right)}\left(\frac{t}{b}\right)^2$$

From *Figure 10-42*, the length *"a"* is arbitrarily shown since it is not required in this analysis. Since the bending moment along the Frame varies along its length, the bending stress at point *1a* is different (higher) than the one at point *1b*. The same for *2a* and *2b*. The maximum (at points *1a* and *2a*) are used for conservatism.

From *Table 10-12* we know that the total stress at point *1* is –58,540 *psi* and that it is 19,140 *psi* at point *2* located 1.0 *in*. away. For this analysis, we are interested to know the location where the stress is zero. That is *"c"* on *Figure 10-42*. By doing a linear interpolation between points *1 & 2* we get c = 0.754 *in*. In this case, b = 1.0 *in*. and c = 0.754 *in*. Thus b/c = 1.0/ 0.754 = 1.33. Hence from *Figure B-1*, the value for k_b read is: 0.62. The Young's modulus in compression *"E"* is used from *Table 10-9*. From the above equation:

$$\frac{F_{b\ cr}}{\eta_c} = \frac{k_b\ \pi^2\ E}{12\left(1-\mu^2\right)}\left(\frac{t}{b}\right)^2$$

$$\frac{F_{b\ cr}}{\eta_c} = \frac{(0.62)\ \pi^2\ 10*10^6}{12\left(1-0.33^2\right)}\left(\frac{0.100}{1.0}\right)^2 = 57{,}224\ psi$$

For
k_b *= 0.62*
b = 1.00 in
t = 0.100 in
μ *= 0.33*
*E = 10.0*10*^{6} *psi*

This is quite high, therefore plasticity occurs and the plasticity reduction factor in compression "η_c" must be calculated.

Based on (ref. 20.), the Plastic Reduction Factor for panels, or columns, in compression; "η_c" is evaluated as follows:

$$\eta_c = \frac{E_t}{E} = \frac{1}{1+\left(\frac{0.002\ E\ n}{F_{cy}}\right)\left(\frac{F_{c,cr}}{F_{cy}}\right)^{n-1}}$$

The determination of this value is done by an iterative process. This is easily done in a spreadsheet for example using the *Goal Seek* function. The steps are as follows:

1. Assume a value for the critical bending compression stress; $F_{c\_cr\_assumed}$.
2. Use this value into the η_c equation.
3. With that η_c value, calculate a new value for F_{c_cr} from the buckling equation.
4. Calculate the difference between $F_{c\ cr\_assumed}$ and $F_{c\ cr}$.
5. Perform a *Goal Seek* by making the cell containing the difference to merge to 0 by changing $F_{c\ cr\_assumed}$.

For
$F_{cy} = 60{,}000\ psi$
$k_b = 0.62$
$b = 1.00\ in$
$t = 0.100\ in$
$\mu = 0.33$
$E_c = 10.0 \times 10^6\ psi$
$n = 20$

In our case we are looking for a compression buckling allowable due to bending. This is exactly the same process and same formulas, the only difference being that we're using the buckling factor "k_b" instead of "k_c" in the buckling equation.

I then started by assuming a value of 50,000 *psi* for $F_{c\ cr\_assumed}$ and went through the above process. I then got a value of **50,289 *psi*** for $F_{c\ cr}$ with a value for the plastic reduction factor of 0.811. This allowable is about 7 *ksi* lower than without considering plasticity. It then shows the importance of considering the reduction in the modulus due to the plasticity effect.

Margin of safety

The stress in compression is much higher at point *1* than is the tension stress at point *2*. Hence, knowing that the allowable in tension "F_{TU}" is much higher than the allowable in compression "F_{bcr}", there is no need to check the tension side.

Thus the MS is calculated as:

$$MS = \frac{F_{b\,cr}}{f_{b1}} - 1 = \frac{50{,}289}{|-58{,}540|} - 1 = -0.14$$

The analysis here is quite conservative. It could be re-done by considering the width of the fastener pattern from the Post instead of considering a point load. With this new assumption, maybe we could obtain a positive *MS*.

Recommendation: The fastener pattern connecting the Post to the Frame should be as wide as possible. Also, the Vertical Post should be stiffer so that more of its torque is reacted down at the skin. Ideally, the Frame should be flipped so that its free flanges are in the opposite direction and, therefore, in tension.

10.5.7 SHEAR BUCKLING

The Frame web is checked for shear buckling. As shown on *Figure 10-43*, there are three panels in the web, two identical large ones *(1 & 3)* and one small one *(2)*. Their boundaries are determined by the upper and lower flanges and by the lines of rivets. It is assumed that there would be shear clips at both ends of the Frame. The panels are assumed simply supported on all sides. Since the web's thickness is constant for all the panels, and since panels *1 & 3* are exactly the same size then only the largest panel having the highest shear flow need to be analyzed. That is panel *3*. See *Figure 10-44* for the dimensions of the critical panel.

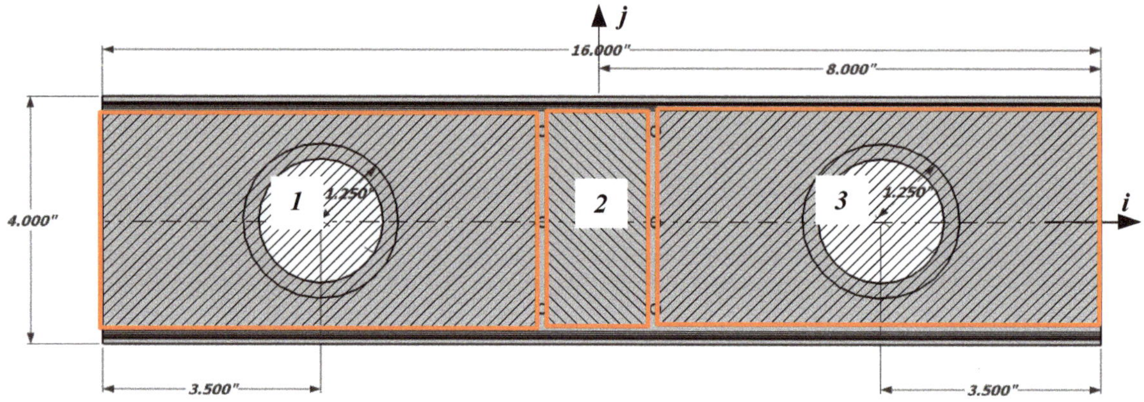

Figure 10-43 – *Frame web panels.*

From *Figure 10-37*, the maximum transverse shear going through the Frame web is $V_j = 62$ *lbs*. Since the web height is $h = 3.9$ *in*. and the thickness is $t = 0.090$ *in*, then the average shear stress is calculated as:

$f_s = V_j / A_{web} = 62 / (3.9 * 0.090) = \mathbf{177}$ ***psi***. The shear flow is; $q = V_j / h = 62 / 3.9 = 16$ *lb/in*.

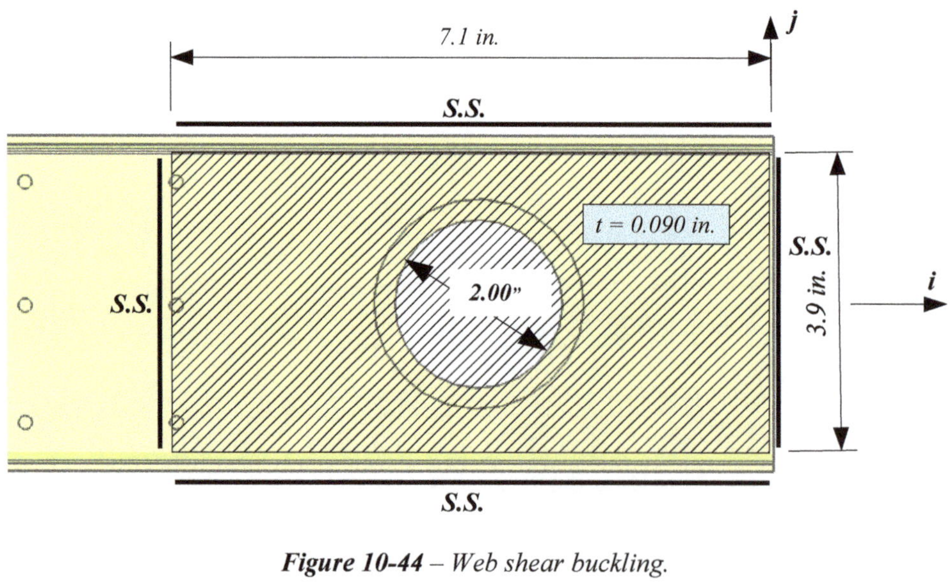

Figure 10-44 – *Web shear buckling.*

This panel is relatively thick and can be considered as shear buckling resistant. That is that no buckling will occur at ultimate load.

Panel without a hole

Since there is a lightening hole in the panel, the shear buckling will be calculated in two steps. First, the shear buckling coefficient "k_S" is obtained, then, we get the knock down due to the hole. From (ref. 20.), for flat plate buckling in shear, the shear stress allowable is calculated as:

$$F_{S\,cr} = \frac{k_S\, \eta_S\, \pi^2 E}{12\left(1-\mu^2\right)}\left(\frac{t}{b}\right)^2$$

From *Figure 10-44*, the dimension "a" is 7.1 *in.* and the dimension "b" is 3.9 *in*. Note that for shear buckling, the value for "b" is always the smallest of the two dimensions. Thus a/b = 7.1 / 3.9 = 1.8. Hence from *Figure A-2*, the value for k_S read is: 6.7. The primary Young's modulus in tension "E" is used from *Table 10-9*. From the above equation:

$$\frac{F_{s\,cr}}{\eta_s} = \frac{k_s\, \pi^2 E}{12\left(1-\mu^2\right)}\left(\frac{t}{b}\right)^2$$

$$\frac{F_{s\,cr}}{\eta_s} = \frac{(6.7)\,\pi^2\, 10*10^6}{12\left(1-0.33^2\right)}\left(\frac{0.090}{3.9}\right)^2 = 32{,}932 \ psi$$

For
K_s = 6.7
b = 3.9 in
a/b = 1.8
t = 0.090 in
μ = 0.33
E = 10.0*10^6 psi

Panel with hole

For this analysis, the hole is assumed un-reinforced and perfectly centered in a square panel. First the ratio D/b is calculated, that is 2.0 /3.9 = 0.51. Then the knock down ratio k/k_o can be approximated using the following equation.

$$k/k_o = 0.96 - \frac{D}{b} = 0.96 - 0.51 = 0.45$$

So using the shear buckling coefficient obtained previously (k_o = *6.7*), the value for k having a hole is thus 0.45 * 6.7 = 3.01.

Hence recalculating the shear buckling allowable we get:

$$\frac{F_{s\,cr}}{\eta_s} = \frac{(3.01)\,\pi^2\, 10.0*10^6}{12\left(1-0.33^2\right)}\left(\frac{0.090}{3.9}\right)^2 = 14{,}795 \ psi$$

For
D = 2.0
k = 3.01
b = 3.9 in
t = 0.090 in

As expected, this latter shear buckling allowable is much lower than the one calculated without the hole. Since this critical stress is relatively low compared to the yield shear stress allowable, the inelastic correction faction "η_s" is 1.0 in this case.

Margin of safety

Thus the MS in shear buckling is calculated as:

$$MS = \frac{F_{s\,cr}}{f_s} - 1 = \frac{14{,}795}{177} - 1 = >1.0$$

10.5.8 SHEAR STRESS

The transverse shear in the "k" direction is calculated at the neutral axis, where the shear stress is maximum, using the formula:

$$f_{s\,max} = \frac{VQ}{I\,t}$$

From *Figure 10-36*, the maximum shear load "V" is 189 *lbs*. The first moment of inertia "Q" of the upper cap flange hatched area from the j_C axis shown on *Figure 10-45* is calculated as follows:

$$Q = A * \bar{k} = 0.743 * 0.100 * 0.372 = 0.02764\, in.^3$$

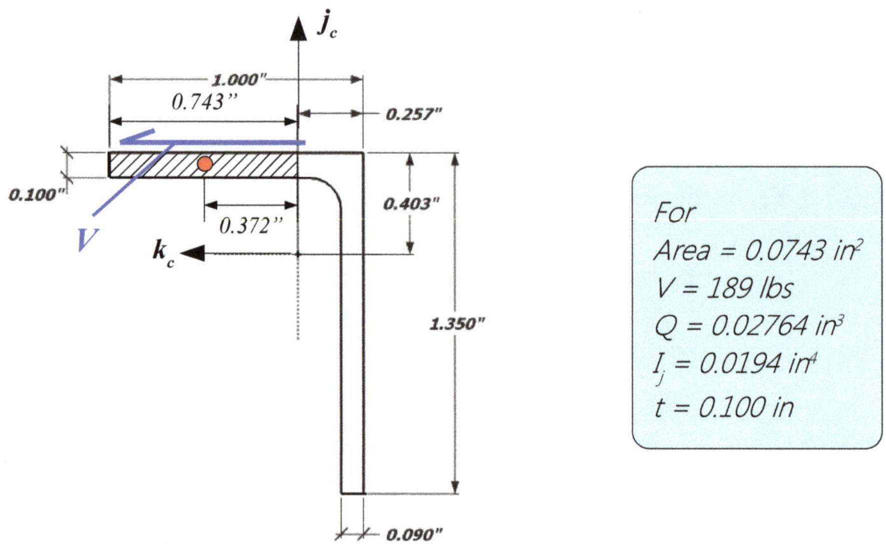

Figure 10-45 *– Upper cap first moment of Inertia dimensions.*

Then the maximum shear stress is:

$$f_{s\,max} = \frac{189 * 0.02764}{0.0194 * 0.10} = 2{,}693\, psi$$

To calculate the maximum shear stress as shown above is quite conservative when we're using ultimate loads as the average shear is normally used. In this case the average shear stress would be $fs_{avg} = V/A_f = 189 / (1.00*0.10) = 1{,}890\, psi$.

The reason is that, at ultimate loads, it is understood that some local yielding is to occur but no failure. Thus, as the section yields due to the applied ultimate load, the stress gets more evenly distributed across the section. Therefore the shear stress tends to get closer to the average through the flange.

Note that to calculate the average shear stress, only the flange is used as the effective area to react the shear load as it is in the same direction as the applied load. The web wouldn't be considered as it is out of plane.

Bending + Shear strength

The bending stress calculated at *point 1* from *Table 10-12* is combined with the maximum shear stress calculate above in order to get the strength MS. The following interaction equation is normally accepted when combining an axial (or bending) stress with a shear stress at yield or ultimate:

$$R_b^2 + R_S^2 = 1$$

Thus the MS is calculated as:

$$MS = \frac{1}{\sqrt{R_b^2 + R_S^2}} - 1$$

Where R_b and R_S are respectively the bending and shear stress ratios calculated as follows:

$$R_b = \frac{f_b}{F_{cy}} = \frac{|-58{,}540|}{60{,}000} = 0.976$$

$$R_s = \frac{f_s}{F_{su}} = \frac{2{,}693}{40{,}000} = 0.067$$

F_{cy} is used in this case as the allowable since the maximum stress is in compression. It is a common practice to use F_{cy} whenever an ultimate stress in compression exists in relatively thin parts as this one and most aircraft structures anyway.

The shear stress ratio is quite small and therefore could have been neglected without much impact on the *MS*.

Hence the MS is:

$$MS = \frac{1}{\sqrt{0.976^2 + 0.067^2}} - 1 = 0.02$$

As shown on page ***101***, the problem is not the strength but the local flange stability. Therefore some changes are required anyway since the stress level will never reach the strength level. The flange will buckle locally causing the section to basically lose its original geometry leading to a reduction of its section properties as the flange will be rendered useless.

Recommendation: Same as the comments for the Free Flange buckling.

"One cannot learn while talking."

10.6 CONNECTIONS

The connections include everything that attaches one part to another like rivets, bolts, welds or adhesives. In our example there are three locations where connections are used, namely points *(A)*, *(B)* and *(C)*. They are all comprised of *BJ5 (MS20470AD5)* rivets. We will look at each point to calculate a margin of safety for each of them.

10.6.1 POINT B

Rivets *MS20470AD5* shear:

Extracted from *Figure 10-12*, *Figure 10-46* shows the loads reacted by the group of fasteners at point *(B)*.

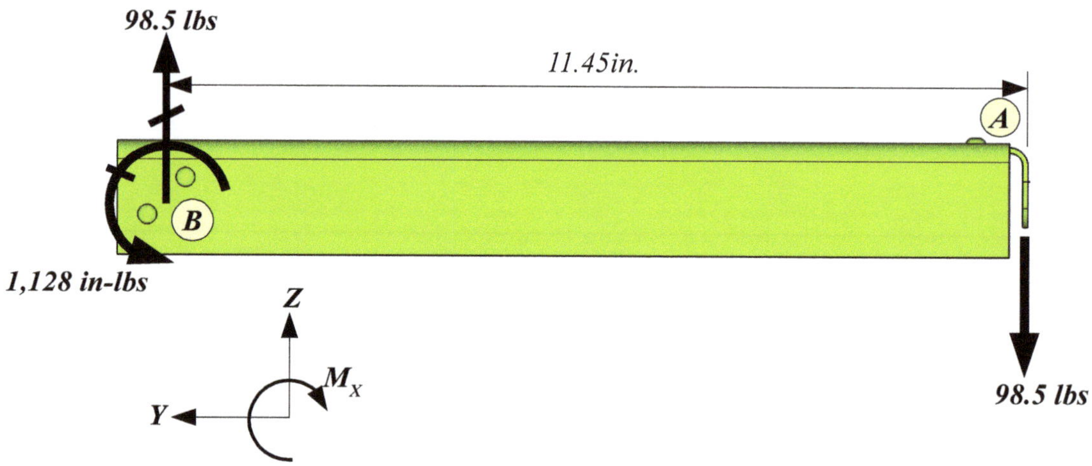

Figure 10-46 *– Close-up at rivets from point B with Free Body Diagram.*

The applied force *"P_Z"* of 98.5 *lbs* will cause a moment at point *(B)* of $M_X = P_Z \times l = 98.5 \times 11.45$ *in.* = 1,128 *in-lb.*, It is assumed that each pair of rivets react half the moment M_x. See *Figure 10-12* to see the two pairs of rivets. Assuming a fastener pitch *"s"* of *6D* (6 × 0.156 *in.* = 0.936 *in.*) and that the fasteners are about at a 45 degree angle, then the load in the rivets due to the moment are:

$F_M = |M_X / (2s)| = |\ 1{,}128 / (2 \times 0.936)| = 603\ lb$. The vertical shear force reacted by the rivets corresponds to $F_Z = 98.5 / 4 = 25\ lbs$ per rivet. That load is thus combined vectorially as shown on *Figure 10-47*.

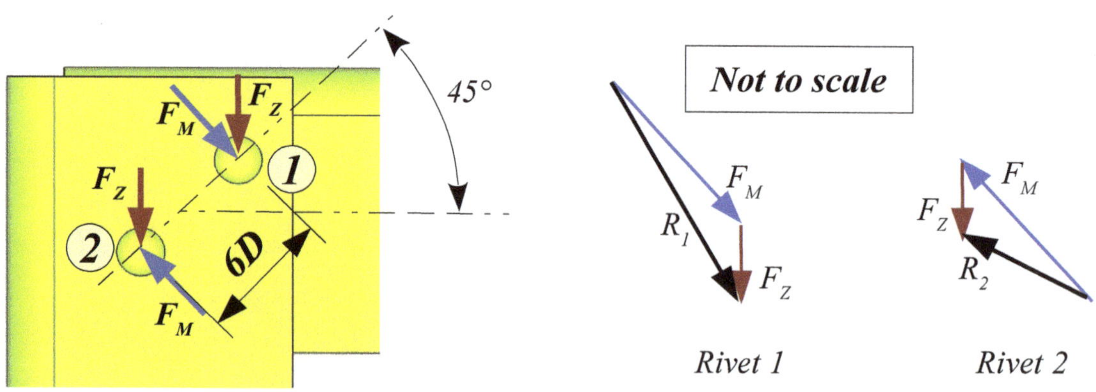

Figure 10-47 *– Applied Force vectors on rivets.*

From the figure above, one can see that rivet (*1*) is more critical than rivet (*2*). Hence, the resulting force ($\boldsymbol{R_1}$) in rivet (*1*) is thus:

$$R_1 = \sqrt{(F_M \cos(45) + F_Z)^2 + (F_M \sin(45))^2}$$

$$R_1 = \sqrt{(603\cos(45) + 25)^2 + (603\sin(45))^2} = 621\ lbs$$

A Fitting Factor "*FF*" of 1.15 (ref. ***25. FAR §25.625***) is used as well as a Shear Correction Factor "*SF*" of 0.996 for a BJ5 in single shear in 0.050" thick from *Table **10-3***. The shear allowable "P_{SU}" of a BJ5 (MS20470AD5) rivet is 596 *lbs* (ref. *Table **10-2***).

The Margin of Safety for shear is thus calculated as follows:

$$MS = \frac{SF \times P_{su}}{FF \times R_1} - 1 = \frac{0.996 \times 596}{1.15 \times 621} - 1 = -0.17$$

<u>Bearing:</u>

Since both the beam and the post have the same material and same thickness, then only one of the two needs to be checked. Otherwise, the bearing in all parts would need to be analyzed and an *MS* would be calculated for the most critical one.

The bearing ultimate allowable load "P_{bru}" in *Aluminum **2024-T42** Clad* 0.050 *in.* thick is calculated:

$$P_{bru} = F_{bru} \times D \times t$$

For: F_{bru} = 115 *ksi*, (ref. *Table **10-5***), e/D = 2.0, D = 5/16" and t = 0.050" hence:

$$P_{bru} = 115{,}000 \times 0.156 \times 0.050 = 897\ lbs$$

The Margin of Safety for ultimate bearing is thus calculated as follows:

$$MS = \frac{P_{bru}}{R_1} - 1 = \frac{897}{621} - 1 = 0.44$$

Usually, only a check at ultimate is necessary. However, one must be careful for certain materials, like in this case for *Al Aly **2024-T42** Clad* where F_{bru} / F_{bry} = 115 / 70 = 1.64 > 1.5! (Ultimate load factor). Thus a Yield *MS* shall be calculated at limit load.

The bearing yield allowable load "P_{bry}" in *Aluminum Alloy **2024-T42** Clad* 0.050 *in.* thick is calculated:

For: F_{bry} = 70 *ksi*, (ref. *Table **10-5***), e/D = 2.0, D = 5/16" and t = 0.050" hence:

$$P_{bry} = 70{,}000 \times 0.156 \times 0.050 = 546\ lbs$$

The Margin of Safety for yield bearing is thus calculated as follows:

$$MS = \frac{1.50 \times P_{bry}}{R_1} - 1 = \frac{1.50 \times 546}{621} - 1 = 0.31$$

The yield bearing MS is thus more critical than the ultimate one and therefore drives the design.

Therefore, the fastener shear is more critical than the bearing in the parts. A design change is thus required just for the connections. Since the bearing *MS* is adequate but the shear *MS* is not, one could simply change the fastener type for one that has a higher shear strength and be done with it.

However, at the beginning we assumed a fastener pitch "*s*" of 6 times the fastener diameter but it could also be as low as *4D* since that normally those rivets may not be jig-located. That would correspond to a fastener load due to the moment *Mx* alone of $P_M = |M_X / (2s)| = |\ 1{,}128 / (2 \times 4 \times 0.156)| = 904$ *lbs*! This is way higher than the fastener shear allowable "P_{SU}" and possibly higher than the bearing allowable "P_{bru}" and thus would lead to a more negative *MS*.

Although Hi-Lite fasteners *(HST10-5)* having a single shear strength of about 2000 *lbs* could be used instead of the *BJ5*, providing there is sufficient space, I would rather add two more rivets *(BJ5)* on each side that would make a square rivet pattern.

Recommendation: Add two more MS20470AD5 rivets on each side of the Beam-Post assembly to create a square bolt pattern.

10.6.2 POINT C

Rivets *MS20470AD5* shear:

Extracted from *Figure 10-21*, *Figure 10-48* shows the loads applied to the group of fasteners at point *(C)*.

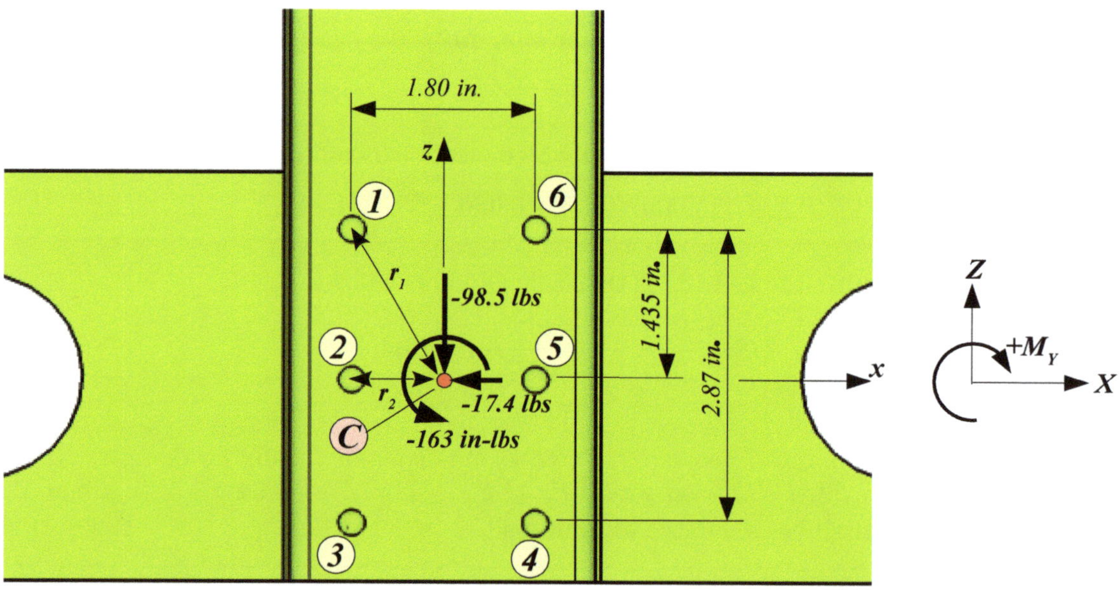

Figure 10-48 *– Close-up at rivets from point (C) with Free Body Diagram.*

Providing all the rivets are of the same type, it is fairly simple to calculate the load "*F*" in each rivet from the applied shear load only as it is the total load from each axis divided by the number of rivets "*n*". In our case, the contribution of that load for each rivet corresponds to:

$$F_{Xi} = \frac{P_X}{n} = \frac{-17.4}{6} = -3 \ lbs$$

$$F_{Zi} = \frac{P_Z}{n} = \frac{-98.5}{6} = -16.4 \ lbs$$

However, for the applied moment, in order to evaluate the load in each rivet, I will be using what we call a "***2D***" bolt group approach. In our case, the bolt group uses all the same rivet size and material and the thickness of the parts to be connected together is constant everywhere. Therefore, the centroid of the bolt pattern is right in the geometric center.

No explanations of the methodology is done here as I am only showing an example of calculations. From the theory, the force taken by each fastener from the applied moment is calculated as:

$$F_i = \frac{M \times r_i}{\sum_{i=1}^{n} r_i^2} = \frac{M\, r_i}{I}$$

Where "r_i" is the distance from the bolt group centroid and the center of the fastener "i" and "M" is the applied moment. In our case, the bolt group "Inertia; I" is calculated as:

$$I = \sum_{i=1}^{n} r_i^2 = r_1^2 + r_2^2 + r_3^2 + r_4^2 + r_5^2 + r_6^2$$

$$I = 4 \times 1.694^2 + 2 \times 0.90^2 = 13.0969\, in.^2$$

For
$r_1 = r_3 = r_4 = r_6 = 1.694\ in$
$r_2 = r_5 = 0.90\ in.$

Therefore, the loads taken by the fasteners due to the moment $M_Y = -163$ *in-lb* are shown in *Table 10-13*. As an example, for rivet no. *1*, its force (normal to the vector for r_1) due to the moment M_Y is:

$$F_i = \frac{M_Y r_1}{I} = \frac{-163 \times 1.694}{13.0969} = 21.1\ lbs$$

The components of this force along the "X" and "Z" axes are:

$$F_{X1} = -\frac{z}{r_1} \times F_1 = -\frac{1.435}{1.694} \times 21.1 = -17.9\, lbs$$

$$F_{Z1} = \frac{x}{r_1} \times F_1 = \frac{-0.900}{1.694} \times 21.1 = -11.2\, lbs$$

Then, all the forces are summed up to calculate the total load on rivet no. *1*:

$$F_{X1\,Total} = -2.9 - 17.9 = -20.8\, lbs$$

$$F_{Z1\,Total} = -16.4 - 11.2 = -27.6\, lbs$$

Hence the resultant force is:

$$F_{1\,Total} = \sqrt{(F_{X1Total})^2 + (F_{Z1Total})^2} = \sqrt{(-20.8)^2 + (-27.6)^2} = 34.5\, lbs$$

Table 10-13 - *Forces in rivets due to applied moment.*

bolt	x	z	r	r^2	F	F_X	F_Z
	[in]	[in]	[in]	[in²]	[lbs]	[lbs]	[lbs]
1	-0.9	1.435	1.694	2.86923	21.1	-17.9	-11.2
2	-0.9	0	0.900	0.81	11.2	0.0	-11.2
3	-0.9	-1.435	1.694	2.86923	21.1	17.9	-11.2
4	0.9	-1.435	1.694	2.86923	21.1	17.9	11.2
5	0.9	0	0.900	0.81	11.2	0.0	11.2
6	0.9	1.435	1.694	2.86923	21.1	-17.9	11.2
			I =	**13.0969**	**Σ =**	**0**	**0**

Then, as shown on *Figure 10-49,* the forces in each fastener from the shear loads are combined with the force from the moment. The results are shown in *Table 10-14*. The maximum shear load is 35 *lbs* at rivet no. *1*.

Table 10-14 - *Shear Forces in rivets*

	From V_X		From V_Z		From M_Y		Total		
bolt	F_X	F_Z	F_X	F_Z	F_X	F_Z	F_X	F_Z	F
	[lbs]	*[lbs]*	*[lbs]*	*[lbs]*	*[lbs]*	*[lbs]*	*[lbs]*	*[lbs]*	*[lbs]*
1	-2.9	0	0	-16.4	-17.9	-11.2	-20.8	-27.6	**34.5**
2	-2.9	0	0	-16.4	0.0	-11.2	-2.9	-27.6	27.8
3	-2.9	0	0	-16.4	17.9	-11.2	15.0	-27.6	31.4
4	-2.9	0	0	-16.4	17.9	11.2	15.0	-5.2	15.8
5	-2.9	0	0	-16.4	0.0	11.2	-2.9	-5.2	6.0
6	-2.9	0	0	-16.4	-17.9	11.2	-20.8	-5.2	21.4
$\Sigma =$	**-17.4**	**0**	**0**	**-98.5**	**0**	**0**	**-17.4**	**-98.5**	

The force vectors applied to rivets no. *1 & 2* are shown in *Figure 10-49* as an example.

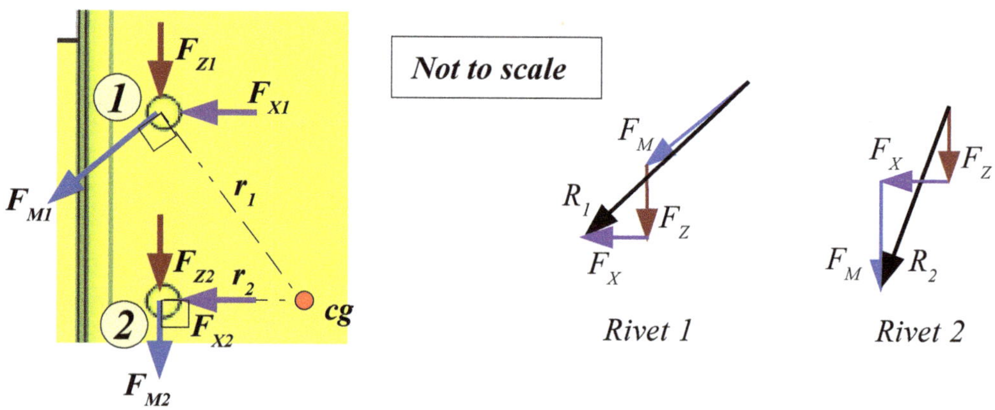

Figure 10-49 – *Force vectors on rivets 1 & 2.*

Before we can calculate an *MS* in shear for those six rivets, we have to calculate the tension force that some of them have due to the applied torque "M_Z" and the applied M_X" moment from the Post. That tension load will be combined to the shear and then a Margin of Safety will be calculated.

Tension forces in the rivets

As per *Figure 10-20* the M_X and M_Z moments reacted at the base of the Post (*point C*) are:

$M_Z = -219$ *in-lb*; that moment puts rivets *4, 5* and *6* in tension. However, it is assumed that only fastener *6* sees that tension load.

$M_X = 1{,}224$ *in-lb*; that moment puts rivets *1* and *6* in tension.

For conservatism, the moments are assumed to be reacted at the fasteners locations regardless whether the force is in tension or in compression. Of course, the compression forces are ignored as the parts are in intimate contact. See *Figure 10-50* for the *FBD* of the rivets for out-of-plane loads.

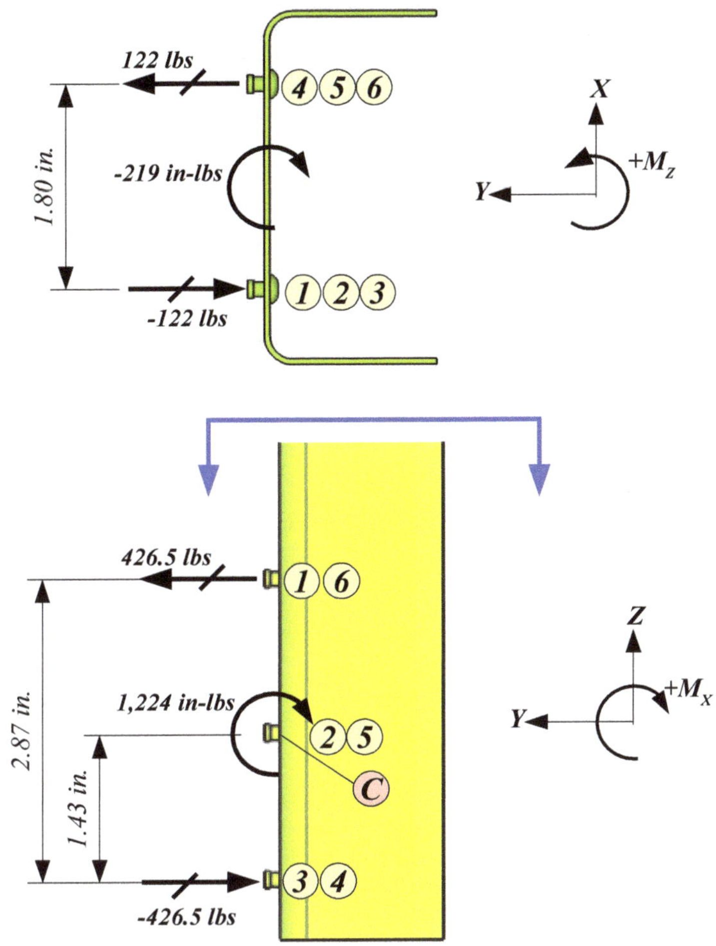

Figure 10-50 *– Out-of-plane rivet reactions.*

The total axial load in each fastener is shown in *Table 10-15*. The most critical fastener is no.6 with a tension load of 335 lbs. From *Table 10-14*, the shear load in rivet no. *6* is 21 *lbs*.

Table 10-15 *- Axial Forces in rivets.*

	From M_X	From M_Z	Total
Rivet	P_{axial}	P_{axial}	P_{axial}
	[lbs]	*[lbs]*	*[lbs]*
1	213.2	-121.7	92
2	0.0	0.0	0
3	-213.2	0.0	-213
4	-213.2	0.0	-213
5	0.0	0.0	0
6	213.2	121.7	**335**
$\sum =$	**0**	**0**	**0**

Because the rivet sees a combination of shear and tension, an interaction equation is required to calculate the Margin of Safety. This following equation is normally used for rivets.

$$R_s^2 + R_t^2 = 1$$

A Fitting Factor "*FF*" of 1.15 (ref. *25.. FAR §25.625*) is used as well as a single shear correction factor "*SF*" of 0.995 for a BJ5 in single shear in 0.050" thick material from *Table 10-3*. The shear allowable "P_{SU}" of a BJ5 (MS20470AD5) rivet is 596 *lbs* from *Table 10-2*.

Also from *Table 10-2,* the ultimate tension allowable of a MS20470AD5 solid rivet in a 0.050" thick sheet is given as; P_{tu} = 430 *lbs*.

The load ratios are thus:

$$R_s = \frac{P_s}{SF \times P_{su}} = \frac{21}{0.995 \times 596} = 0.035$$

$$R_t = \frac{P_t}{P_{tu}} = \frac{335}{430} = 0.779$$

Then from the interaction equation above, using a Fitting Factor *"FF"* of 1.15, the Margin of Safety for the MS20470AD5 no. *6* is thus calculated as follows:

$$MS = \frac{1}{FF\sqrt{R_s^2 + R_t^2}} - 1 = \frac{1}{1.15\sqrt{0.035^2 + 0.779^2}} - 1 = 0.11$$

Of course the other five rivets are *Passed by Comparison* to this one.

<u>Bearing:</u>

Since both the beam and the post have the same material and same thickness, then only one check needs to be done. Otherwise, the bearing in all parts would need to be analyzed and an MS would be calculated for the most critical one. Here, from *Table 10-14*, rivet no *1* is critical with 35 *lbs* of shear load.

The bearing ultimate allowable load "P_{bru}" in *Aluminum 2024-T42 Clad 0.050 in*. thick is calculated:

$$P_{bru} = F_{bru} \times D \times t$$

For: F_{bru} = 115 *ksi*, (ref. *Table 10-5*), $e/D = 2.0$, D = 5/16" and t = 0.050" hence:

$$P_{bru} = 115{,}000 \times 0.156 \times 0.050 = 897\,lbs$$

The Margin of Safety for ultimate bearing is thus calculated as follows:

$$MS = \frac{P_{bru}}{F_1} - 1 = \frac{897}{35} - 1 = >1.0$$

Therefore, the fastener shear is more critical than the bearing in the parts. This is to be expected for joints with ordinary solid rivets *(MS20470AD)*.

Usually, only a check at ultimate is necessary. However, one must be careful for certain materials, like in this case for Al Aly 2024-T42 Clad where $F_{bru} / F_{bry} = 115 / 70 = 1.64 > 1.5$! (Ultimate load factor). Thus a Yield *MS* should be calculated at limit load. However, judging from how high the Ult *MS* is, it would be a waste of time to calculate a Yield *MS;* especially since we know that a redesign is required!

Recommendations: The top fasteners no.*1* & *6* should be tension application fasteners like HI-LITEs HST11-5. Also, a radius block (or Doubler) should be added underneath those fasteners to reduce the out-of-plane bending stress in the Post while ribs should be added in the machined Frame to again reduce the out-of-plane bending stress in the Frame.

10.7 DEFORMATIONS

One aspect of stress analysis that is unfortunately often overlooked is: **deformations or displacements**. One could show a positive MS in strength but fails to check by how much the analyzed structure moves due to the applied load. For secondary structures, such as the one being analyzed here, it is generally accepted to limit the maximum deflections to 1/16" under *limit* loads. Designers usually leave about 1/4" of minimum gap between any two systems or between a system and the primary structure. So if two systems structures are next to each other and that they each move in opposite directions by 1/16" that would only leave 1/8" of gap. This isn't much.

In addition, if the deflection is a bit too large (even at 1/16") than somebody yanking on the structure could notice that the structure doesn't "feel" right. Similarly as with hand rails in houses where we tend not to trust those that move a bit too much.

For those reasons, I will check the rotational deflection of the Horizontal beam at the Clip end and see if it exceeds 0.100 *in.* under ultimate loads. That is about 1.5 * 0.063 *in.*

For this analysis it will be assumed that all the parts retain their original cross-section i.e. no buckling or warping is considered.

10.7.1 FRAME

It is assumed that the Frame is simply-supported at both ends and that the lower cap is restrained in the "*X*" and "*Y*" directions by the skin. Therefore, only the upper cap of the Frame can move laterally (in the "*Y*" direction). The total displacement will build-up from the Frame's ends all the way to the Clip lug where the load is applied.

<u>Upper cap lateral displacement</u>

Using the displacement equation for a beam simply supported at both ends having a point load applied mid-span, the following information is gathered:

The load applied at the center of the upper cap in the "*Y*" direction is 349 *lbs*, .ref. ***Figure 10-36,***

Span; $l = 16$ *in.*	ref. ***Figure 10-36***
Young's modulus; $E = 10.0*10^6$ *psi*	ref. ***Table 10-9***
Moment of Inertia; $I_Z = 0.0194$ in^4,	ref. ***Figure 10-40***

Since the ends are assumed simply supported, the displacement at the center is calculated as:

$$y = \frac{-Wl^3}{48EI} = \frac{-349*16^3}{48*10.0\,E^6*0.0194} = 0.154\,in.$$

This is already higher than the permissible maximum displacement and right there, the current concept would be subject to a redesign. But for the purpose of the demonstration, I'll go ahead and calculate the total deformation at the load application point anyway. This will then be compared to the result of a small detailed FEM later.

As shown on ***Figure 10-51*** and calculated next page, this deflection corresponds to an angular deformation "θ" of the Frame of 0.038 *rad* (2.2 *deg.*). The load application point (*A*) will thus move from a rigid body motion by $dy = 0.430$ *in.* and $dz = -0.490$ *in.*

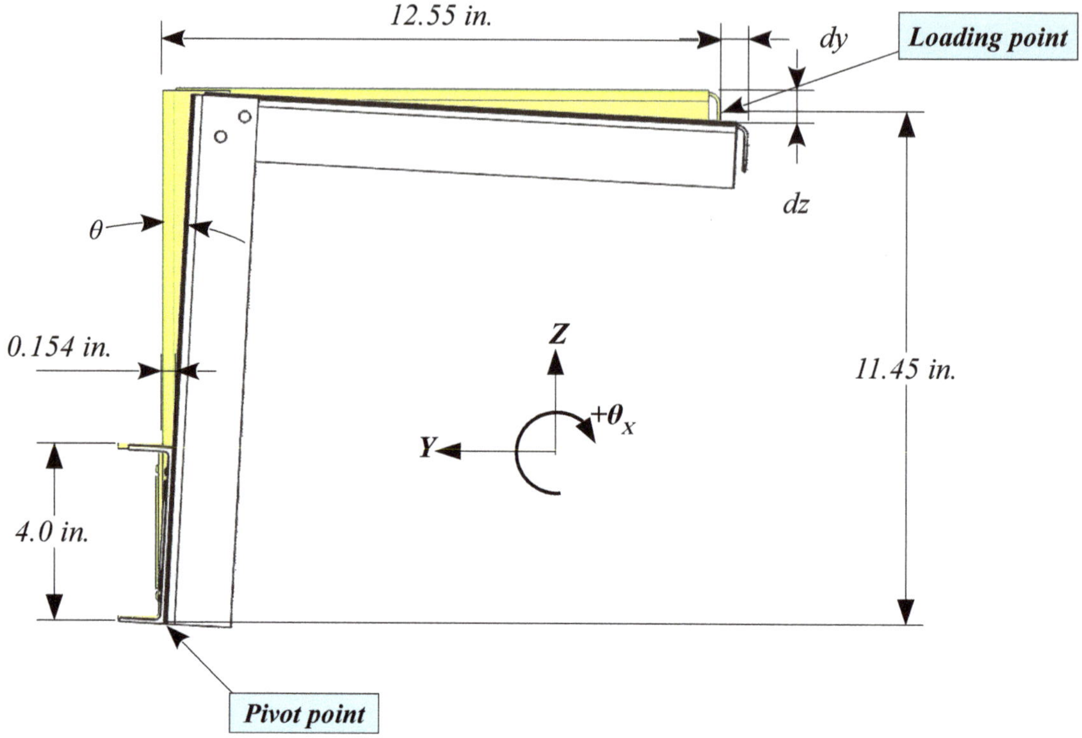

Figure 10-51 *– Rigid body rotation due to Frame upper cap displacement.*

The angular deformation "θ" is calculated as follows:

$$\theta = atan\left(\frac{y}{h}\right)$$

Where $y = 0.154$ *in*, $h = 4.0$ *in*. Therefore,

$$\theta = atan\left(\frac{0.154}{4.0}\right) = 0.038\ rad.$$

Hence, the new location of point "*A*" is calculated as follows:

$$y' = y\cos\theta - z\sin\theta$$
$$z' = z\cos\theta + y\sin\theta$$

Where $y = -12.55$" and $z = 11.45$", therefore:

$$y' = -12.55\cos(0.038) - 11.45\sin(0.038) = 12.98$$
$$z' = 11.45\cos(0.038) + (-12.55)\sin(0.038) = 10.96$$

So

$$dy = 12.98 - 12.55 = 0.430\ in.$$
$$dz = 10.96 - 11.45 = -0.490\ in.$$

This corresponds to a total displacement; d of:

$$d = \sqrt{dy^2 + dz^2} = \sqrt{0.430^2 + (-0.490)^2} = 0.652\ in.$$

This is way too flexible!

10.7.2 VERTICAL POST

As you'll see in a moment, the Vertical Post is way more flexible in torsion than it is in bending. However, for the demonstration, both displacements are going to be calculated.

Bending in the "*X*" direction

The Post is assumed fixed at point "*C*", corresponding to the center of the six rivets group, and free at point "*B*" where the shear load P_X is applied. The Post's maximum deflection in the "*X*" axis at the free end is calculated using the following equation:

$$y = \frac{W\,l^3}{3\,E\,I}$$

Where:

$W = P_X = 17.4$ *lbs.*	ref. *Figure **10-21***
Length; $l = 9.37$ *in.*	ref. *Figure **10-21***
Young's modulus; $E = 10.0{*}10^6$ *psi*	ref. *Table **10-5***
Moment of Inertia; $I_k = I_Y = 0.4373$ in^4	ref. *Figure **10-24***

Therefore;

$$x = \frac{P_X\,l^3}{3\,E\,I_Y} = \frac{17.4 \times 9.37^3}{3 \times 10.0 * 10^6 \times 0.4373} = 0.001\,in.$$

Bending in the "*Y*" direction

Again, the Post is assumed fixed at point (*C*), corresponding to the center of the six rivets group, and free at point (*B*) where the moment M_X is applied. The Post's maximum deflection in the "*Y*" axis at the free end is calculated using the following equation:

$$y = \frac{M_o\,l^2}{2\,E\,I}$$

Where:

$M_o = 1{,}184$ *in-lb.*	ref. *Figure **10-21***
Length; $l = 9.37$ *in.*	ref. *Figure **10-21***
Young's modulus; $E = 10.0{*}10^6$ *psi*	ref. *Table **10-5***
Moment of Inertia; $I_j = I_X = 0.0659$ in^4	ref. *Figure **10-24***

Therefore;

$$y = \frac{1{,}184 \times 9.37^2}{2 \times 10.0 * 10^6 \times 0.0659} = 0.079\,in.$$

While the deformation along the "*X*" axis is negligible, this not the case along the "*Y*" axis where it is already pretty large for this member only.

Twist

For a beam free to twist and warp at the top end (point *B*) and free to warp but not twist at the bottom end (point *C*, but at the first row of rivets), the following equations and parameters are used.

For the torque applied completely at the top end *(a = 0)*, the angular rotational displacement at the top is calculated as:

$$\theta = \frac{T_o l}{C_w E \beta^2}$$

Where:

$$\beta = \sqrt{\frac{KG}{C_w E}}$$

The torque applied at the shear center of the Post is: T_o = 229 *in-lb*. ref. ***Figure 10-20***

Length; l = 9.37–1.435 = 7.935 *in*.	ref. ***Figure 10-21*** and ***Figure 10-48***.
Young's modulus; E = 10.0*10^6 *psi*	ref. ***Table 10-5***
Warping coefficient; C_w = 1.0088 *in*^{4},	ref. ***Figure 10-25***
Torsional stiffness constant; K = 2.5*10^{-4} *in*^{4}	ref. ***Figure 10-30***
Shear modulus; G = 3.76*10^6 *psi*	ref. ***Figure 10-30***

Therefore;

$$\beta = \sqrt{\frac{KG}{C_w E}} = \sqrt{\frac{2.5*10^{-4} \times 3.76*10^6}{1.0088 \times 10.0*10^6}} = 0.009653$$

and

$$\theta = \frac{T_o l}{C_w E \beta^2} = \frac{229 \times 7.935}{1.0088 \times 10.0*10^6 \times 0.009653^2} = 1.933 rad$$

The angular displacement *(1.933 radians)* calculated above corresponds to more than 90 degrees! That means that the Horizontal Beam for instance would be oriented along the "*X*" axis instead of along the "*Y*" axis under load.

At that point it becomes obvious that this concept is inadequate and that further analysis to calculate the total deformation of the system would be a waste of time.

As one can see from the above results, it is quite often faster and more efficient to start by calculating the deformation of a system before getting into the stress analysis as it may right away tell us if the structure is sufficiently stiff.

Recommendations: TheFrame upper cap needs to be re-inforced. The Vertical Post and the Horizontal beam need to be stiffer in bending and much stiffer in torsion .

10.8 SUMMARY OF RECOMMENDATIONS

Clip:

- Relocated on top of the Horizontal Beam,
- The fasteners should be as far apart as possible and should be Hi-Lites 5/32".

Horizontal beam:

- Must be stiffer in bending and in torsion,
- If keeping a similar cross-section ("CEE" section) the beam should be flipped to have it's free flanges pointing upward,
- A shear clip (or two) most be added at the connection to the Vertical Post,
- Add more fasteners at the connection between the Horizontal beam and the Post.

Vertical Post:

- Must be stiffer in bending and in torsion,
- If keeping a similar cross-section ("CEE" section) the beam should be flipped to have it's free flanges pointing in the opposite direction,
- Preferably, a closed section should be used for much greater torsional stiffness,
- The top two fasteners *(1 & 6)* should be Hi-Lites 5/32",
- Add a radius block nested inside the Post at the two top fasteners locations.

Frame:

- Must be stiffer in bending at the Upper cap,
- Integral ribs should be added at the six fasteners bolt group location,
- The Frame should be flipped so that the flanges are pointing in the opposite direction.

> *"Perfection is attained not when there is nothing else to add but when there is nothing else to remove"*
>
> (Antoine de Saint-Exupéry)

11. IMPROVED QUIZ STRUCTURE

From the results of the Quiz design, shown in the previous chapter, it is clear that a re-design is necessary. This chapter shows the complete analysis of the improved structure. There is an infinite number of ways that something can be designed. The following configuration is but one way and might not be the best design but should do the job just fine. The beauty in starting from scratch is that we can do pretty much whatever we want providing the given constraints are respected. See *Figure 11-1* for the first iteration of the improved design.

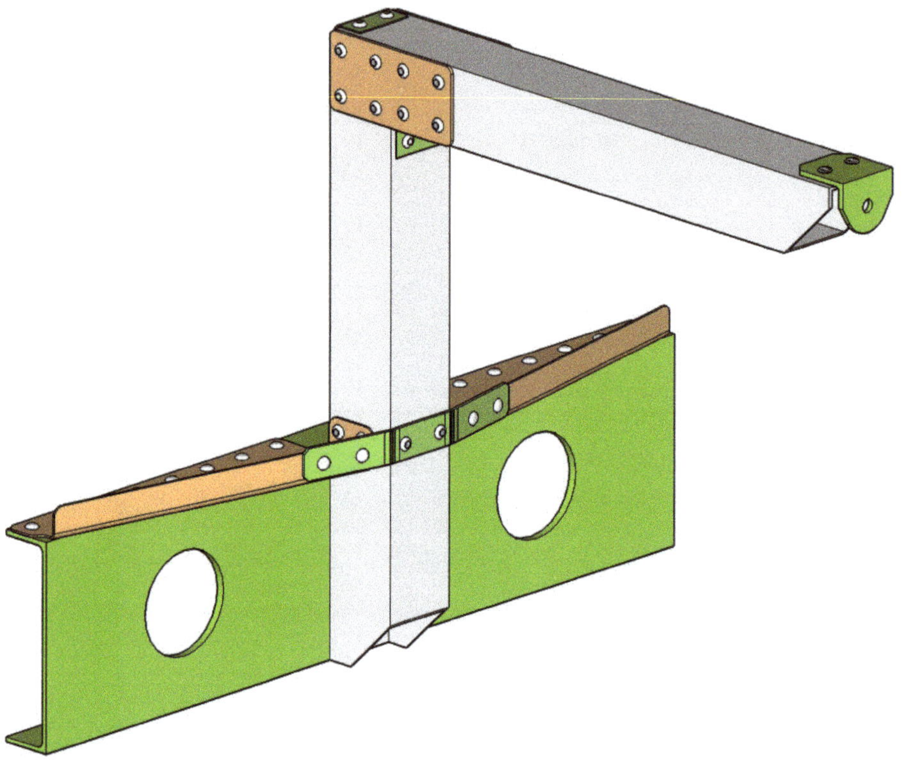

Figure 11-1 *– Improved structure-first iteration.*

Usually, the earlier in the design phase the configuration of a structure can be frozen and the less trouble there will be later as the systems groups start routing their pipes, tubes, electronic boxes, cables and wires.

It is vital to understand exactly what the constraints and the design criteria are as this will dictate how our design will look like. For instance, in the original design above, it was clear that the reason that the applied load was cantilevered off the Frame was to provide clearance to something. Providing this was not the case, then the design could have been simpler and much stiffer by using, for example, an arrangement of struts as shown in *Figure 11-2*. In this case, the three struts provide equilibrium in all 3 axes and yet allow the airframe to "deform" as it needs to.

This sketch is of course very simplistic as, ideally, the three struts should meet at the exact load application point in order to prevent the struts from reacting some bending moments. The top strut would connect to something like an Intercostal spanning between two fuselage frames (for example) and the two other struts would connect to small corner fittings at the intersection of the fuselage frames, the Frame (shown) and the skin.

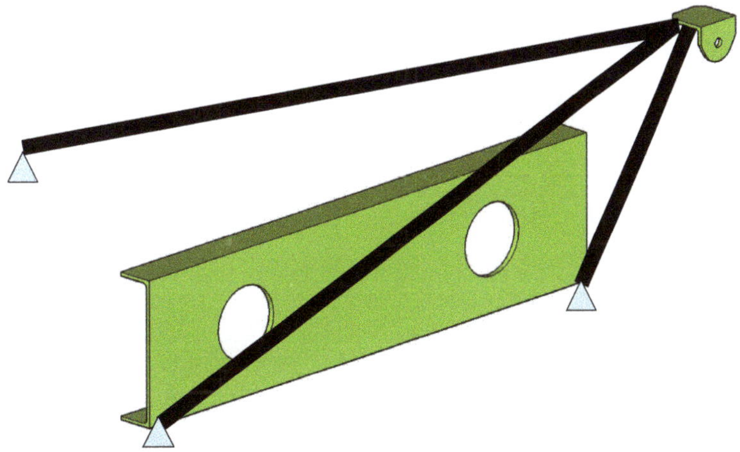

***Figure 11-2** – Schematic structure using struts.*

However, here I will stick to simply improving the current concept in order to make it work while remaining pretty much within the same volume of space. For the first design iteration, I shot from the hip by going with what I thought would work in terms of cross sections and thicknesses. Recommendations on the necessary changes will be made during the analysis process in order to get the final design.

Before I go ahead and spend days in analyzing the new structure in detail and then risk that I come to realize at the very end that it is still not adequate, I would like to get a *warm and fuzzy feeling* beforehand that the new structure is sufficiently stiff. Although I know for a fact that this new design is much stiffer than the previous one, is it going to meet the 0.100" maximum displacement target? What could I do that wouldn't require me too much time and yet be accurate enough? The answer: a quick and crude *FEM*! Here I'll simply use BAR elements and see what displacement I get under the applied load. *Figure 11-3* shows the *FEM* along with the applied load and the boundary conditions.

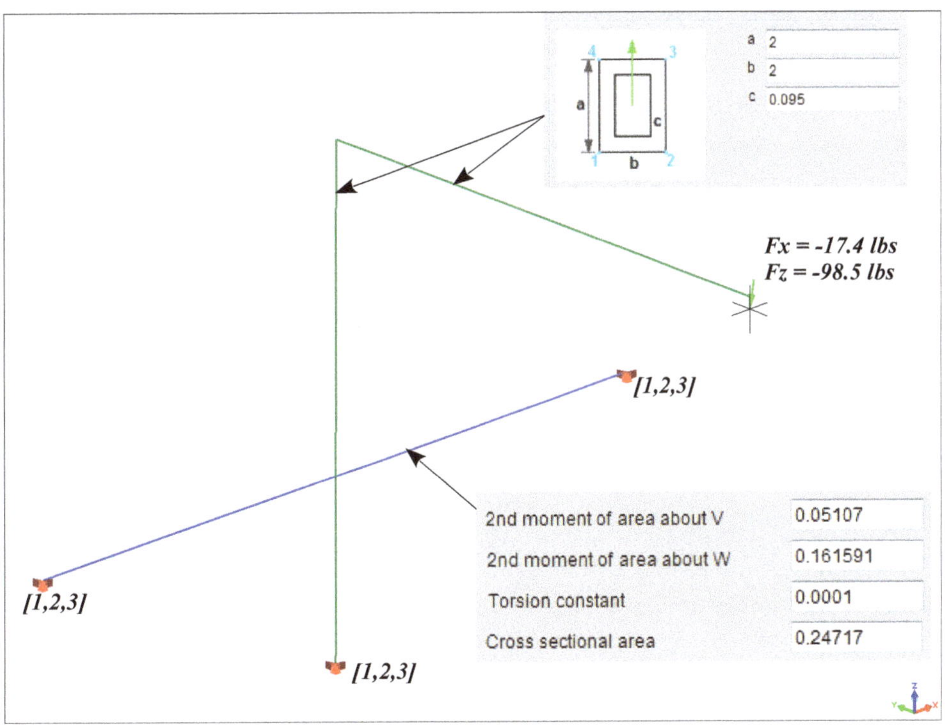

***Figure 11-3** – Crude FEM.*

The two dozens BAR elements have been given the section properties of the first trial where the Horizontal Beam and the Vertical Post were made out of Aluminum Alloy extruded square tube of 1.5" x 1.5" x 0.065". For the Frame, only the upper cap was modeled and was given the section properties of the Upper cap combined with the 0.040" thick Gussets.

The boundary conditions at the three nodes are considered pinned in all three axes [1,2,3].

After a few iterations, that took less than an hour, in order to meet the maximum displacement permissible, I came up with a cross section for the Horizontal Beam and the Post of 2.0" x 2.0" x 0.095" and the Gusset thickness had to be increased to 0.050" thick.

Note that the cross section properties used for the Frame upper cap are taken from *Figure 11-39* corresponding to the section at mid-span.

The deformed shape of the structure is shown in *Figure 11-4* where the maximum displacement obtained is 0.099 *in.* at the location where the load is applied.

Figure 11-4 *– FEM deformed shape.*

The assumption to pin both ends of the Frame is conservative since that, even if Gussets were added at the ends to connect to something like the airframe, there WILL be some sort of flexibility. Therefore, I ran the *FEM* by assuming the two ends of the Frame blocked in translation and around the "*Z*" axis (1,2,3,6) and found a maximum displacement of 0.055 *in.*

In reality, it'll be somewhere in between but at least I know that I'm pretty sure of meeting the maximum displacement criteria of 0.100 *in.*

I also ran the FEM in order to get the first modes. The lowest frequency found was 82 *Hz*. This is high enough as we're always trying to get the natural frequencies of secondary structures above the windmilling frequency. That frequency varies but it usually is between 20 and 30 *Hz*.

> In the case of a turbo fan airplane where, during flight, an engine would be shut off due to a broken off fan blade, the engine's large fan would still be rotating due to the incoming air. The fan is said to be wind-milling under that condition. Because one (or more) fan blades are gone, the fan is unbalanced and will cause very high vibrations through-out the airplane.
>
> In order to avoid the aircraft from destroying itself due to that excitation, the structures are thus designed to have their natural frequencies as far away as possible from the wind-milling frequency so that they do not get into resonance.

Using the cross-sections and thicknesses from the results of the above iterations, the latest design is shown in *Figure 11-5*.

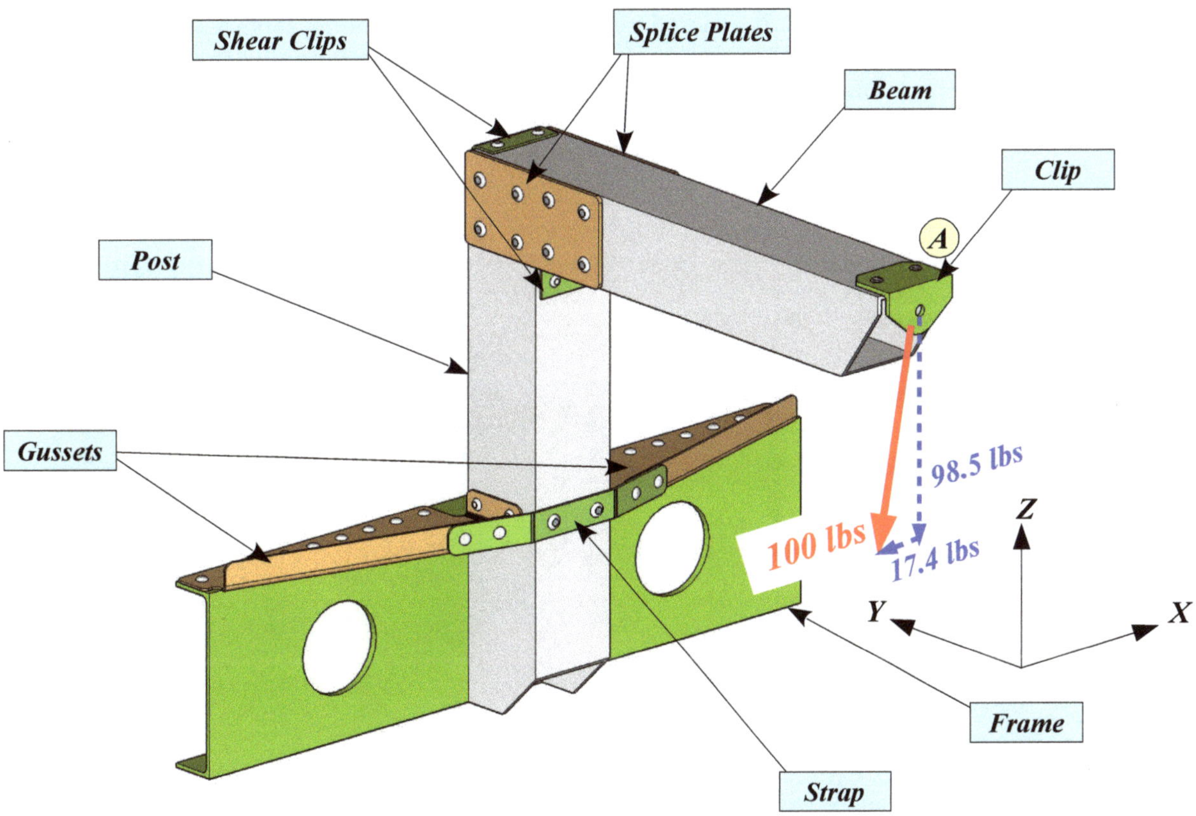

Figure 11-5 *– Improved structure-second iteration.*

The assembly is analyzed starting from the Clip down to the Frame.

11.1 CLIP

I changed the Clip to use an extrusion Angle instead of a bent sheet metal clip. This is mainly to increase its stiffness. It is installed on top of the Beam so that the tension load in the fasteners is reduced. Also, the load application point will be closer to the Beam's centroid so very little torque will be seen by the Beam. I now use two 5/32" Hi-Lites instead of the former BJ5 rivets in order to provide a good clamp-up.

The clip is now made from an Angle extrusion of 1.25" x 1.25" x 0.10" made out of *Al Aly 2024-T3511*. It is connected to the horizontal beam with *qty (2) HST10-5* Hi-Lite *6Al-4V Titanium* Pins and *HST79-5* Hi-Lite Aluminum Collars. *Figure 11-6* shows the *F.B.D.* of the Clip.

11.1.1 LOADS

To analyze the clip, I will assume that the Hi-Lites are pinned and thus can't react any moment locally (even if they can). Thus the moment "M_X" due to "P_Z" times the horizontal distance to the edge of the Horizontal Beam (about 0.175") is reacted by a couple between the fasteners and the end of the Beam "$R_{Z1,2}$". Also, due to the load "P_X", a torque moment "M_Y" will be reacted by the couple "R_{Z1}" and "R_{Z2}". The vectors of the reactions forces are shown as expected.

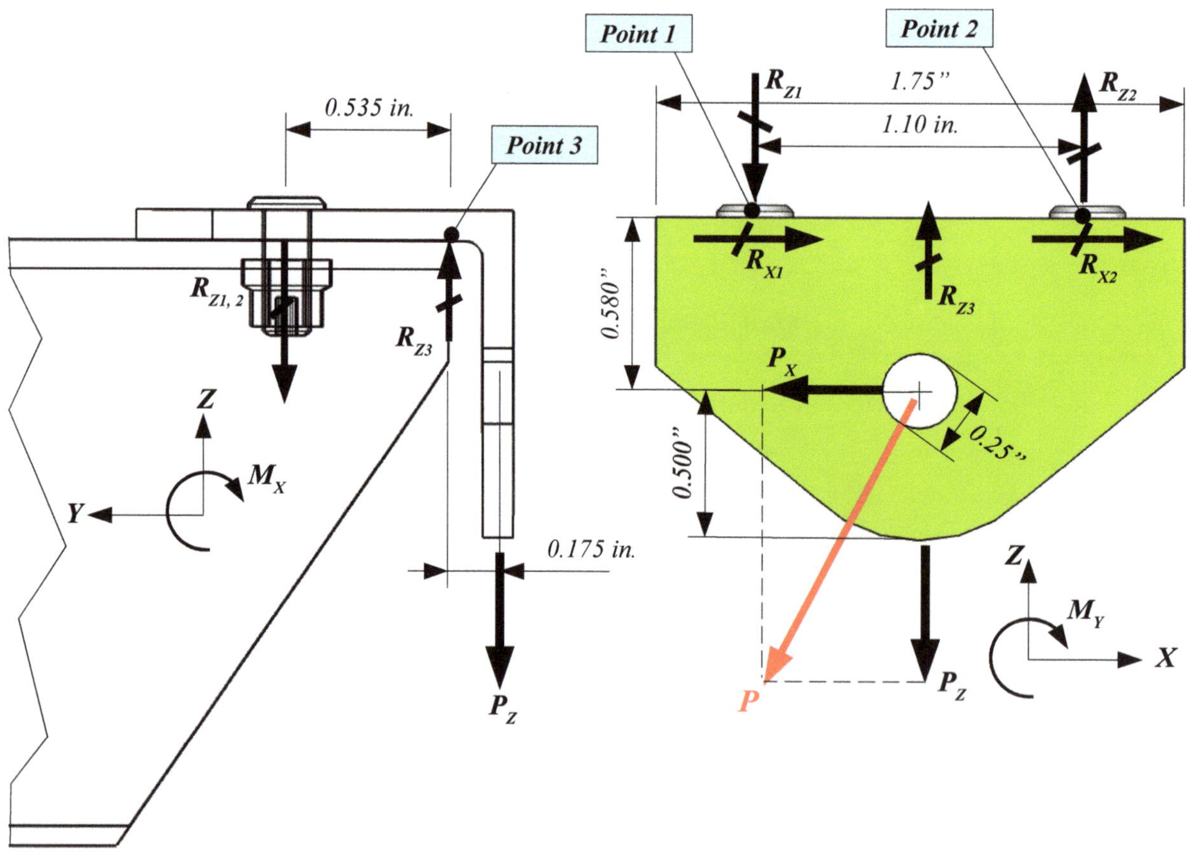

***Figure 11-6** – Clip dimensions and FBD.*

The reactions in the *YZ* plane are calculated by doing the sum of the moments at point *3* and the sum of the forces in the *Z* direction as follows:

$$\sum M_{X\,3} = P_Z\,(0.175) - R_{Z1,2}\,(0.535) = 0$$

$$R_{Z1,2} = \frac{(98.5)\,(0.175)}{0.535} = 32\ lbs$$

$$\sum F_Z = -R_{Z1,2} - P_Z + R_{Z3} = 0$$

$$R_{Z\,3} = 32 + 98.5 = 131\ lbs$$

The reactions in the *XZ* plane are calculated by doing the sum of the moments at point ***1*** and the sum of the forces in the *Z* direction as follows.

$$\sum M_{Y\,1} = (P_Z - R_{Z\,3})\left(\frac{1.10}{2}\right) + P_X\,(0.580) - R_{Z\,2}\,(1.10) = 0$$

$$R_{Z\,2} = \frac{(98.5 - 131)\,(0.55) + (17.4)(0.580)}{1.10} = -7\ lbs$$

$$\sum F_Z = -R_{Z1} + R_{Z2} - P_Z + R_{Z3} = 0$$

$$R_{Z\,1} = -7 - 98.5 + 131 = 25.5\ lbs$$

As a verification, $-R_{Z1} + R_{Z2} = -25.5 + (-7) = -32.5$ lbs (*downward*) which equals to $R_{Z1,2}$ calculated earlier (33 *lbs* pointing down) for the *YZ* plane.

Except for R_{Z2}, we can notice that the results of all the reaction load values calculated are positive, which means that the drawn force orientation vectors are indeed as assumed. Sometimes I prefer doing that than, as in section *9.1*, draw all force vectors in the positive global coordinate system. One simply has to stick to the assumed directions when calculating the sum of the forces and moments.

In this particular case, I preferred drawing the forces in the directions that I thought they were going to be. The main reason for it was that, at point 3, there is no fastener. The reaction force can only exist there when there is a contact. Otherwise, if the applied load P_Z were to be in the opposite direction (Up instead of down), then the Clip would tend to pivot around the line of rivets (points *1* and *2*) and no contact would occur at point *3*. There would be a contact between the edge of the Clip and the Horizontal Beam instead which would lead to a totally different *FBD*.

11.1.2 MATERIAL ALLOWABLES

The actual material allowables for the Clip can be obtained from the latest edition of (ref. 8.) for *Al Aly 2024-T3511 Extrusion* 0.100 *in.* thick. For the purpose of the demonstration, *Table 11-1* shows fictitious values.

Table 11-1 *- Al Aly 2024-T3511 Extrusion <0.250in. fictitious material allowables*

Aluminum Alloy 2024-T3511 Extrusion
Fictitious Values, NOT FOR DESIGN

	≤0.250" [ksi]
F_{tu}	55
F_{ty}	40
F_{cy}	35
F_{su}	30
F_{bru} $(e/D = 2.0)$	110
F_{bry} $(e/D = 2.0)$	75
E, 10^3	10
G, 10^3	3.76
μ	0.33
n	20

11.1.3 SECTION PROPERTIES

The section at point *3* of *Figure 11-6*, where the bending moment is maximum, is shown in *Figure 11-7*.

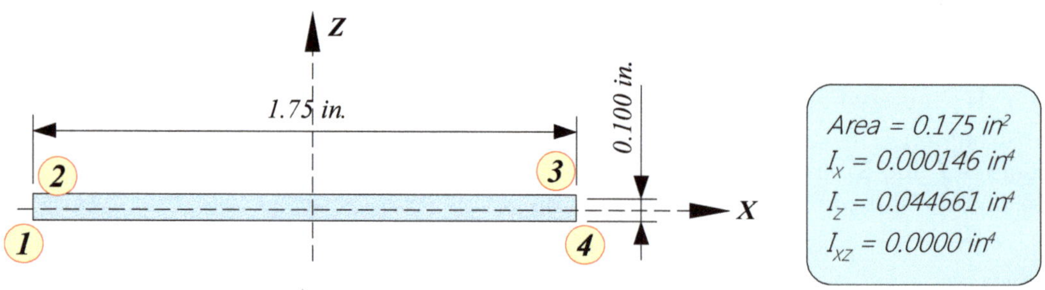

Figure 11-7 *– Clip Section along Beam edge (point 3).*

11.1.4 STRESS ANALYSIS

Bending

From *Figure 11-6*, the maximum bending moments occur along the edge of the Horizontal Beam (points *3*) and are calculated as: $M_X = P_Z\,(d)$, $M_Y = P_X\,(h)$ and $M_Z = P_X\,(d)$. For $d = 0.175\ in$, $h = 0.580 - 0.10 / 2 = 0.53\ in.$, $P_X = 14.7\ lbs$ and $P_Z = 98.5\ lbs$, then we obtain:

$$M_X = 98.5(0.175) = 17.2\ in.-lb$$
$$M_Y = T = 17.4(0.53) = 9.2\ in.-lb$$
$$M_Z = 17.4(0.175 + 0.535) = 12.4\ in.-lb \text{ (at the fastener line)}$$

The bending stresses are calculated at the four corners of the section. The maximum value for M_Z calculated at the fasteners is conservatively used in the bending stress calculation performed at the location where the maximum value for M_X exists i.e. at point *3*.

$$f_b = \frac{M_X(z)}{I_X} + \frac{-M_Z(x)}{I_Z}$$

Calculation example at stress point *1*:

$$f_{b1} = \frac{M_X(z_1)}{I_X} + \frac{-M_Z(x_1)}{I_Z}$$
$$f_{b1} = \frac{(17.2)(-0.050)}{0.000146} + \frac{-(-12.4)(-0.875)}{0.044661}$$
$$f_{b1} = -5{,}890 - 243 = -6{,}133\ psi$$

For
$M_X = 17.2$ *in-lb*
$M_Z = -12.4$ *in-lb*
$I_X = 0.000146\ in^4$
$I_Z = 0.044661\ in^4$
$z_1 = -0.050\ in$
$x_1 = -0.875\ in$

Bending stresses at all corners:

Table 11-2 *- Bending stresses at corners.*

Points	x [in]	z [in]	f_b [psi]
1	-0.875	-0.05	**-6,133**
2	-0.875	0.05	5,647
3	0.875	0.05	6,133
4	0.875	-0.05	-5,647

As seen above, the most critical location is at point *1* where the compression stress is maximum.

In addition to increasing the thickness slightly, having no holes at the maximum bending location and, installing the Clip at the top of the Horizontal Beam drastically reduced the bending stress.

Shear

The transverse shear is maximum at point *3*. It exists in both the "X" and the "Z" directions and are calculated as follows:

For $V_X = P_X = 17.4\ lbs$, $V_Z = P_Z = 98.5\ lbs$ and, $A = 0.175\ in^2$

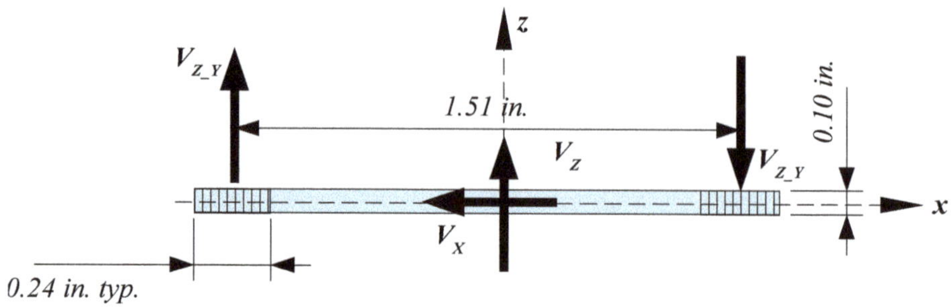

Figure 11-8 *– Shear forces at Clip Section at point 3*

A torque exists in the Clip "M_Y" due to the vertical offset between the load "P_X" application point and the shear center of the section which, in this case, corresponds to the center of gravity. Because of that torsion, one would be tempted to evaluate the torsional shear stress by using the equation:

$$\tau = \frac{T}{J} t$$

However, that would be too conservative as the torsional shear stress doesn't have time to fully develop due to the short distance between the free end (where the load is applied) and the restrained end (at the rivets *1 & 2*). A distance of about **four** to **six** times the section's width (or largest dimension) is needed to justify using the above equation.

What happens is that, for short distances (below four times the width), the torque is mainly reacted as differential transverse shear. See *Figure 11-8* where the differential shear load is shown as "V_{Z_Y}" assumed applied at the center of the most outboard material section area "A_S" of 0.240 * 0.100 = 0.024 in^2. That shear load is calculated as:

$$V_{Z-Y} = \frac{T}{arm} = \frac{9.2}{1.51} = 6.1\ lbs$$

Therefore, the transverse shear stresses are calculated as follows:

$$f_{sX} = \frac{V_X}{A} = \frac{17.4}{0.175} = 100\ psi$$

$$f_{sZ} = \frac{V_Z}{A} + \frac{V_{Z-Y}}{A_S} = \frac{98.5}{0.175} + \frac{9.2}{0.024} = 563 + 383 = 946\ psi$$

The vertical shear stress is calculated at the location where V_Z and $V_{Z\text{-}Y}$ are in the same direction (points *3* near the Fwd edge) as this will be the maximum shear stress. The effective width of 0.24" for the calculation of A_S was conservatively assumed as the material left between a rivet hole and the edge of the part.

Since the shear stresses calculated are orthogonal to each other, they are combined as follows:

$$f_{S-TOTAL} = \sqrt{f_{SX}^2 + f_{SZ}^2} = \sqrt{100^2 + 946^2} = 981\ psi$$

The resulting shear is low but yet must be combined to the bending stress.

11.1.5 COMBINED STRESSES

The bending stress is combined with the shear stress in order to get the MS. The following interaction equation is normally accepted when combining an axial (or bending) stress with a shear stress at yield or ultimate:

$$R_b^2 + R_S^2 = 1$$

Thus the MS is calculated as:

$$MS = \frac{1}{\sqrt{R_b^2 + R_S^2}} - 1$$

Where R_b and R_S are respectively the bending and shear stress ratios calculated as follows using the material allowables from *Table 11-1*:

$$R_b = \frac{f_b}{F_{tu}} = \frac{|-6{,}133|}{55{,}000} = 0.112$$

$$R_s = \frac{f_s}{F_{su}} = \frac{981}{30{,}000} = 0.033$$

"B" Basis allowables are normally used as the allowable for secondary structures like this one. It is a common practice to use *Ftu* even when there is a compression stress knowing that *Fcu* is normally a bit larger. However, that is only applicable to very stable cross sections (when no risk of buckling or warping exists). In which case, *Fcy* would have to be the allowable, just like for the crippling analysis.

The shear stress ratio is quite small and therefore could have been neglected without much impact on the *MS*.

Hence the MS is:

$$MS = \frac{1}{\sqrt{0.112^2 + 0.033^2}} - 1 => 1.0$$

A tension Clip analysis is not relevant here as the distance between the loaded Lug and the edge of the Beam (the unsupported length) is very short. In addition, the fasteners are not directly acting in tension anymore.

11.1.6 LUG

At first glance, the designer did a good job for the Lug. The hole is relatively large (0.25 *in*. dia.) and at 0.50 *inch* of edge distance, it is far enough from the edge at two times the bolt diameter *(2D)*. In this case, there is no bushing but if there was, we would have to consider the hole size without the bushing.

Since the loaded hole is at *2D* from the edge, we could simply do a bearing check and call it a day. Let's do it below.

The bearing load allowable "P_{bru}" is calculated as:

For:
$t = 0.100\ in$
$e = 0.50\ in$
$D = 0.25\ in$
$e/D = 2$
$F_{bru} = 110\ ksi$

$P_{bru} = F_{bru} * D * t = 110{,}000 * 0.25 * 0.100$

$P_{bru} = 2{,}750\ lbs$

This is way higher than the 100 *lbs* ultimate applied load. Therefore the lug is *Passed by engineering judgment*. But for comparison, let's do the complete lug analysis.

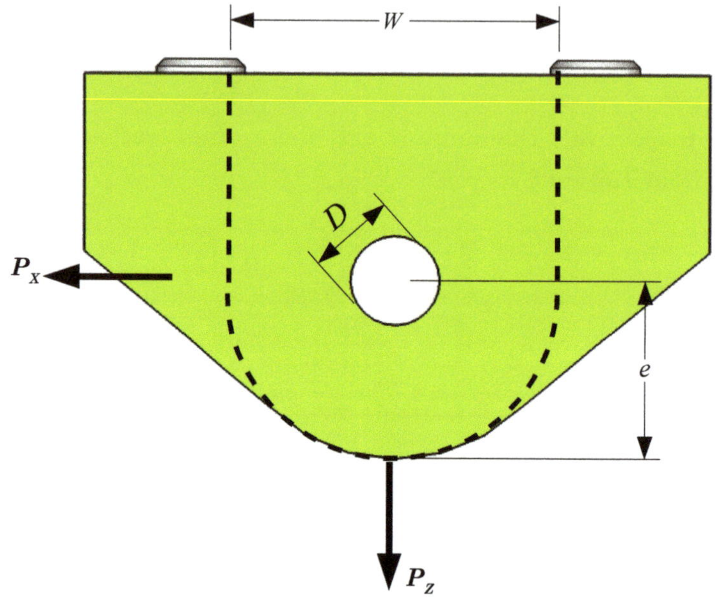

Figure 11-9 *– Angle Clip Lug geometry.*

Lug axial net section strength

For conservatism and to simplify the analysis, the lug geometry is assumed as the dotted line shown in *Figure 11-9*.

I'd rather have some additional conservatism in lugs (as in joints in general) and an *MS* of 0.25 minimum than trying to save weight in those locations. The joints are typically areas of concern in durability and DTA. There are plenty of other places to optimize the weight than at lugs.

Since I'm considering the lug as a joint, I would also use a 1.15 Fitting Factor. Make sure to verify with your boss, with your company's Design Criteria Manual, and your DER to get their approval before using (or not) any factor. One important thing, make sure you use the same approach throughout the stress report. If this leads to negative *MS* in some location during the sizing phase, consider re-designing.

The axial tension allowable "P_u" is calculated as:

$$P_u = K_t\, F_{tu}\, A_t$$

The tension efficiency factor "K_t" is obtained from *Figure B-6* using curve *1*. For $W/D = 1.0 / 0.25 = 4$, we find a K_t of 0.9.

The net tension area "A_t" is calculated as:

$$A_t = (W - D)t$$

Thus A_t = (1.0 – 0.25) 0.100 = 0.075 in^2

For F_{tu} = 55 *ksi*, ref. *Table 11-1,* therefore, the tension allowable is:

$$P_u = (0.9) * 55{,}000 * 0.075 = 3{,}712\ lbs$$

Shear out bearing Strength

As for the axial net section check, to be consistent, the lug geometry is assumed as the dotted line shown on *Figure 11-9.*

Ultimate:

The shear out-bearing ultimate allowable "P_{bru}" is calculated as:

$$P_{bru} = K_{bru}\ F_{tu}\ A_{br}$$

The shear-bearing efficiency factor "K_{bru}" for axially loaded lugs is obtained from *Figure B-7.* For D/t = 0.25 / 0.100 = 2.5 and e/D = 2, we find a value for K_{bru} of 1.95.

The bearing area "A_{br}" is calculated as:

$$A_{br} = D\ \ t$$

Thus A_{br} = 0.25 * 0.100 = 0.025 in^2

For F_{tu} = 55 *ksi*, ref. *Table 11-1,* therefore, the shear out-bearing allowable is:

$$P_{bru} = (1.95) * 55{,}000 * 0.025 = 2{,}681\ lbs$$

Note that this value should be very close to the value calculated on page 128. In our case, there is a slight difference because the allowables are fictitious and thus not the actual test based results.

Yield:

The shear bearing yield allowable "P_{bry}" is calculated as:

$$P_{bry} = K_{bry}\ F_{ty}\ A_{br}$$

The shear-bearing yield efficiency factor "K_{bry}" for axially loaded lugs is obtained from *Figure B-8.* For t/D = 0.100 / 0.25 = 0.4 and e/D = 2, we find a value for K_{bry} of 1.55.

For A_{br} = 0.025 in^2, F_{ty} = 40 *ksi*, ref. *Table 11-1.*

Therefore, the shear bearing yield allowable is:

$$P_{bry} = (1.55) * 40{,}000 * 0.025 = 1{,}550\ lbs$$

Since P_{bru} / P_{bry} = 2,681 / 1,550 = 1.73 and thus > 1.5, then a Yield MS needs to be calculated since it is more critical than the ultimate MS.

Lug Strength under transverse loading

Using the same lug geometry as the dotted line shown on *Figure 11-9* for consistency and conservatism.

Ultimate:

The transverse ultimate allowable "P_{tu}" is calculated as:

$$P_{tu} = K_{tu}\, F_{tu}\, A_{br}$$

The tension efficiency factor "K_{tu}" for transversely loaded lugs is obtained from *Figure B-9* curve *2* for 2024-T3 plate ≤ 0.5 in. For A_{av} / A_{br} = 0.03986 / 0.025 = 1.59 (using the maximum value of 1.4 in the graph), we find a value for K_{tu} of 1.1.

The average area "A_{av}" is calculated using *Figure 11-10* as:

$$A_{av} = \frac{6}{\dfrac{3}{A_1} + \dfrac{1}{A_2} + \dfrac{1}{A_3} + \dfrac{1}{A_4}}$$

Where, in our case;

$$A_2 = A_3 = \frac{(W-D)t}{2} = \frac{(1.0-0.25)(0.100)}{2} = 0.0375\ in.^2$$

$$A_1 = A_4 = \frac{(W-D\cos(45))t}{2} = \frac{(1.0-0.25\cos(45))(0.100)}{2} = 0.04116\ in.^2$$

Therefore;

$$A_{av} = \frac{6}{\dfrac{3}{0.04116} + \dfrac{1}{0.0375} + \dfrac{1}{0.0375} + \dfrac{1}{0.04116}} = 0.03986\ in.^2$$

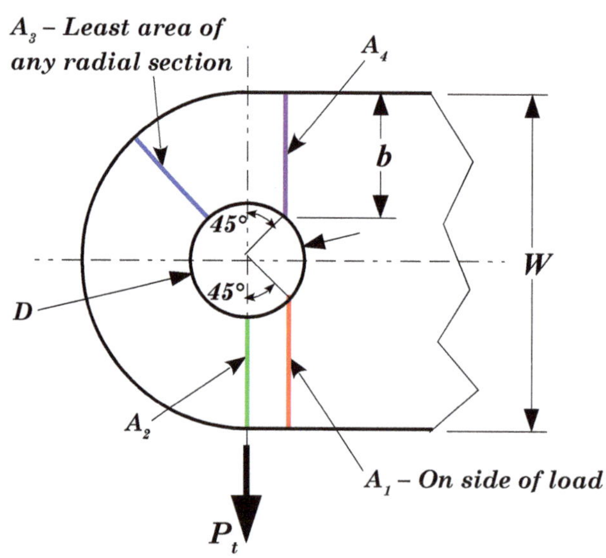

Figure 11-10 – Lug section areas for transverse loading.

For A_{br} = 0.025 in^2, F_{tu} = 55 *ksi* and K_{tu} = 1.1.

Therefore, the transverse ultimate allowable is:

$$P_{tu} = (1.1) * 55{,}000 * 0.025 = 1{,}512\ lbs$$

Yield:

The transverse yield allowable "P_{ty}" is calculated as:

$$P_{ty} = K_{ty}\,F_{ty}\,A_{br}$$

The tension efficiency factor "K_{ty}" for transversely loaded lugs is obtained from *Figure B-9* curve *1* for all Aluminum at yield. For A_{av} / A_{br} = 0.03986 / 0.025 = 1.59 (using the maximum value of 1.4 in the graph), we find a value for K_{ty} of 1.28.

For A_{br} = 0.025 in^2, $F_{ty\,(LT)}$ = 40 ksi and K_{ty} = 1.28.

Therefore, the transverse yield allowable is:

$$P_{ty} = (1.28) * 40{,}000 * 0.025 = 1{,}280\ lbs$$

Since P_{tu} / P_{ty} = 1,512 / 1,280 = 1.18 and thus < 1.5, then no Yield MS would need to be calculated. However, we'll do it anyway just for the demonstration.

Margin of safety

Since we have a combination of axial and transverse loading *(oblique loading)*, the following interaction equation is used at yield and ultimate:

$$R_a^{1.6} + R_{tr}^{1.6} = 1$$

Thus the MS is calculated as:

$$MS = \frac{1}{\left(R_a^{1.6} + R_{tr}^{1.6}\right)^{0.625}} - 1$$

Where R_a and R_{tr} are respectively the axial and transverse load ratios calculated as follows;

For Ultimate:

$$R_a = \frac{P_a}{Min(P_u\ ;\ P_{bru})} = \frac{98.5}{Min(3{,}712\ ;\ 2{,}681)} = 0.037$$

$$R_{tr} = \frac{P_t}{P_{tu}} = \frac{17.4}{1{,}512} = 0.012$$

Hence the Ult. MS is:

$$MS = \frac{1}{\left(0.037^{1.6} + 0.012^{1.6}\right)^{0.625}} - 1 => 1.00$$

For Yield:

$$R_a = \frac{P_a}{1.5\ P_{bry}} = \frac{98.5}{1.5 \times 1{,}550} = 0.042$$

$$R_{tr} = \frac{P_t}{1.5\ P_{ty}} = \frac{17.4}{1.5 \times 1{,}280} = 0.009$$

Hence the Yield MS is:

$$MS = \frac{1}{\left(0.042^{1.6} + 0.009^{1.6}\right)^{0.625}} - 1 => 1.00$$

The ultimate and yield *MS* calculated are both larger than 1.0. Therefore, that translates to a design which is more than two times stronger than required. So at that point, there is no need to show the actual MS.

A statement such as: *Passed by Inspection* or *Passed by Engineering Judgment* could also have been used instead of calculating the actual *MS*.

11.1.7 COMMENT

The Clip is thus found adequate strength-wise. The deformation will be calculated in section *11.11*.

"The most efficient people are often the most insecure."

11.2 HORIZONTAL BEAM

Regarding the Horizontal Beam, I was debating whether I should be using a bent sheet metal channel ("U" section) as before but flip it to have the free flanges pointing up or, replace it by the same square tube extrusion as for the Vertical Post.

There is no problem in using the sheet metal part providing that the load application point at the Clip is perfectly aligned with the shear center of the section. Although this could be done, I was trying to imagine a mechanic that would be installing this system and how, by possibly leaning onto the Channel during the installation, he/she would probably feel that it is kind of flimsy especially in torsion. Therefore, in order to avoid the part being damaged at installation and, especially, try to prevent mechanics from bitching about engineers, I decided that the Horizontal Beam would be made out of the same stock as the Vertical Post.

This was also confirmed during the small FEM (*Figure 11-3*) iterations in order to achieve the desired maximum displacement.

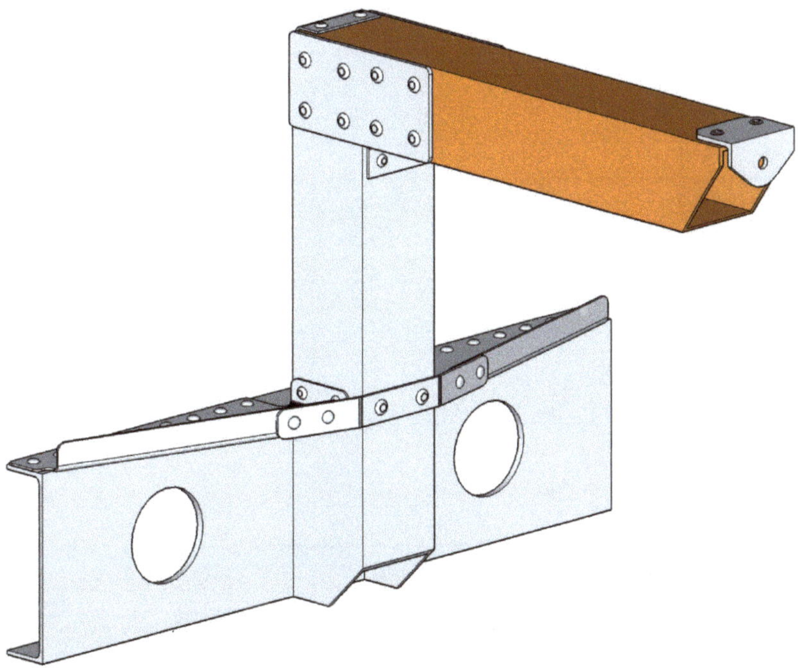

***Figure 11-11** – Horizontal beam highlighted.*

Using the same square tube extrusion has other advantages. Not only does this provide for a very stiff beam in bending and torsion but it also shows consistency in the design and in the manufacturing methods. See *Figure 11-11* for an isometric view of the structure with the new Horizontal Beam highlighted.

The Horizontal Beam transfers the relatively large vertical load and the small fore-aft load to the vertical post. It is made out of an *Aluminum Alloy 2024-T3511* square tube extrusion of 2.0" x 2.0" x 0.095". It is connected to the Vertical Post through two Splice plates and two shear Clips. The fasteners are Aluminum blind rivets *NAS1398D4* common to the clips and *NAS1398D5* common to the Splices.

11.2.1 LOADS

The following assumptions are made regarding the load paths:

- The two Splice Plates react all of the M_X, M_Y and M_Z moments.
- The vertical shear P_Z and the M_X moment are reacted at the centroid of the fasteners to the right end of the Splices.
- The two Shear Clips react the fore-aft load P_X. They are also used to stabilize the section.

> The M_Y and M_Z moments could have been assumed reacted as a couple at the two small shear clips. However, because they are thinner than the Splice plates and because they use a lesser number of smaller fasteners, the Shear Clips were considered a more flexible load path and was thus ignored. Hence, M_Z is considered reacted by the Splice Plates only as a couple.

Figure 11-12 shows the Horizontal beam Free Body Diagram.

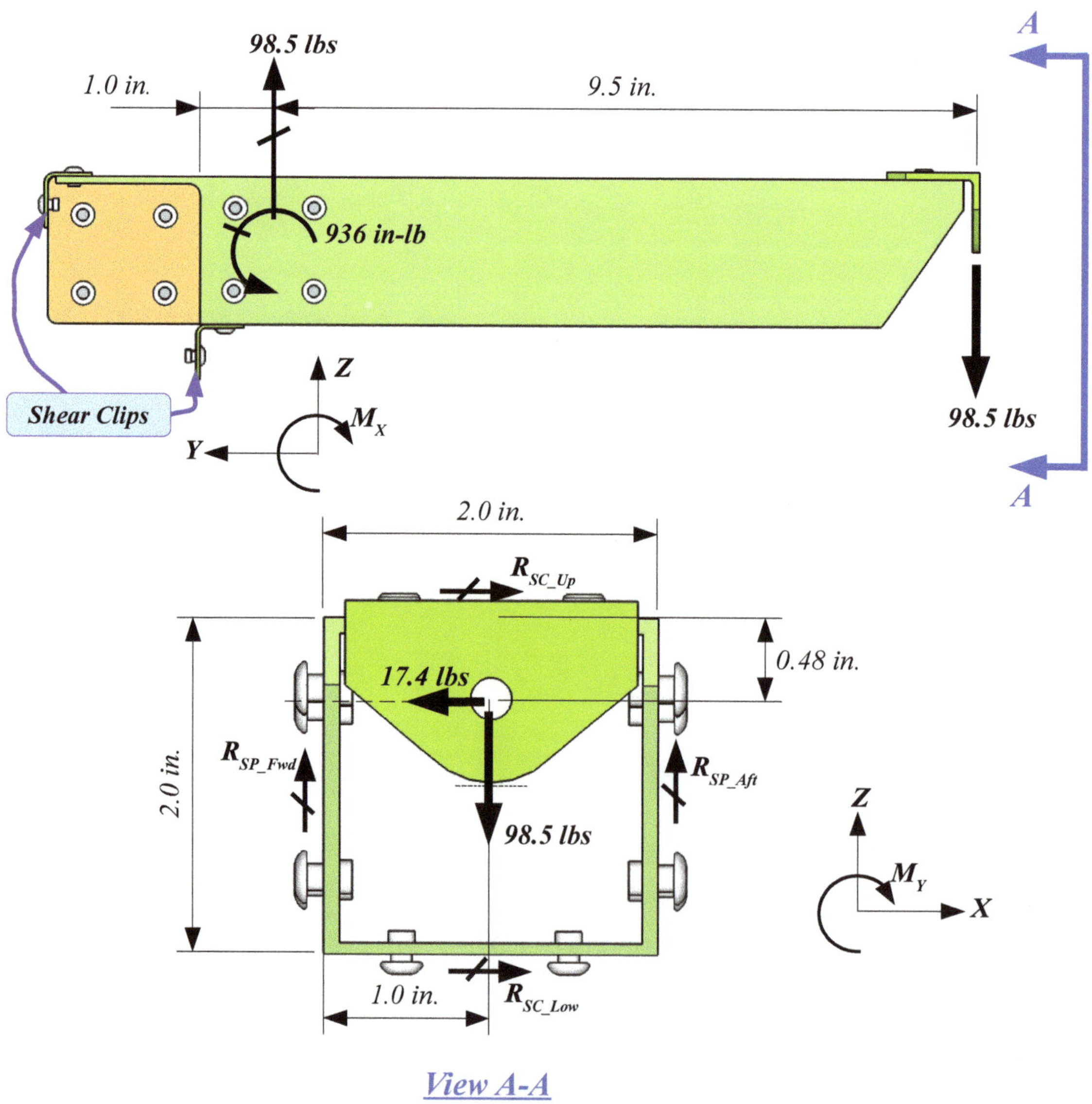

Figure 11-12 *– Horizontal beam dimensions and FBD.*

Reactions in the shear Clips

As assumed previously, the shear Clips only react the Fore-Aft *"P_X"* load. Since all moments are taken by the Splice Plates, the Shear Clips are simply left with the Fore-Aft load that they share evenly because they are located at the same distance from the Beam's centroid. Therefore, R_{SC_Up} and R_{SC_Low} are calculated as follows:

$$R_{SC_{Up}} = R_{SC_{Low}} = \frac{P_X}{2} = \frac{17.4}{2} = 8.7\ lbs$$

Reactions in the Splice Plates

From the previous assumptions, the Splice Plates react the vertical *"P_Z"* load and all the moments. However, in the following analysis, I wish to know the vertical shear load distribution between the Forward and the Aft Splices. In this case, only the torsional "M_Y" moment matters. Therefore, R_{SP_Fwd} and R_{SP_Aft} are calculated as follows:

Torque in the Beam:

Since the load application point is offset *"e"* from the tube's shear center (centroid), a torsional moment "M_Y" will be reacted by the Beam. The calculated torque is thus:

$$M_Y = P_X(e) = -17.4(1.0-0.48) = -9\ in.-lb$$

The *FBD* thus looks like in *Figure 11-13*.

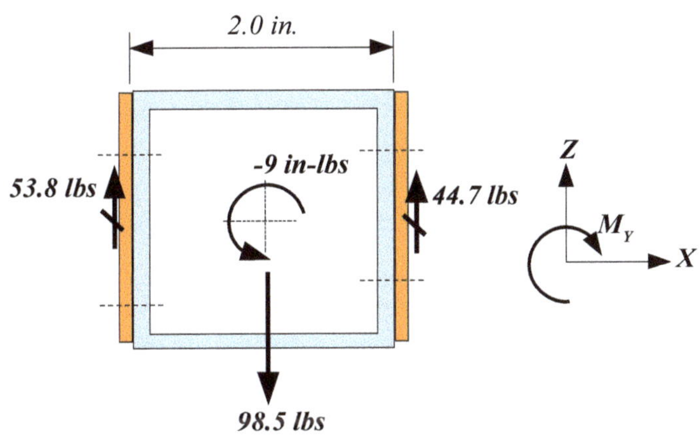

Figure 11-13 *– Splice Plates vertical reactions.*

Where;

$$R_{SP_{Aft}} = \frac{P_Z}{2} + \frac{M_Y}{2} = \frac{98.5}{2} + \frac{-9}{2} = 44.7\ lbs$$

and;

$$R_{SP_{Fwd}} = \frac{P_Z}{2} - \frac{M_Y}{2} = \frac{98.5}{2} - \frac{(-9)}{2} = 53.8\ lbs$$

Shear and moment diagram

The shear and moment diagrams of the Horizontal beam can be seen in *Figure 11-14* and *Figure 11-15*. The beam is cantilevered. The applied load is located at the free end *(lug of the Clip)* and the "clamped" end is considered located at the center of gravity of the rivet group. The moment at that location is thus calculated as P (*l*).

Plane i-k:

For $P_k = 98.5$ *lbs* and $l = 9.5$ *in*.

$$M_j = P_k\,(l) = -98.5\,(9.5) = -936\ in.-lb$$

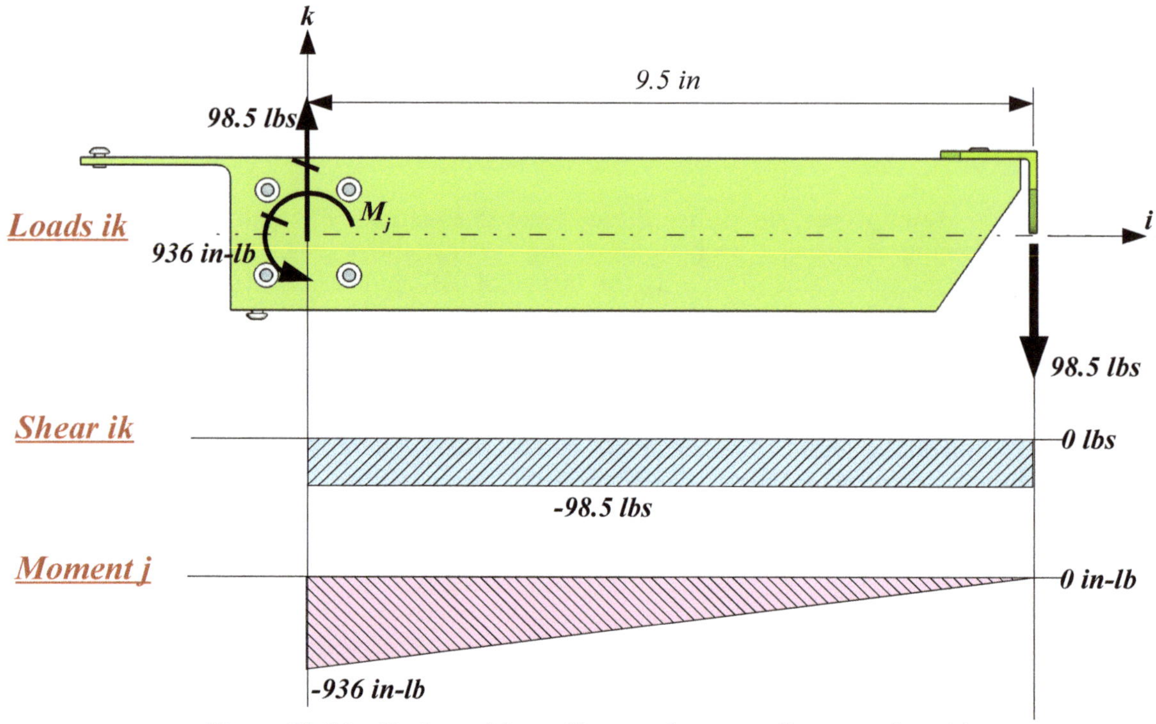

Figure 11-14 – *Horizontal beam Shear and moment diagram, plane i-k.*

Of course, in reality, the exact shear and moment curves are a bit more complex at the fasteners location. But the simplification shown is sufficient for the analysis performed.

Plane i-j:

In this plane, it's a bit more tricky since the inner and the outer Shear Clips, from *Figure 11-15* located at sections *A* and *B* respectively, are not located at the same cross section as the location where the Splice Plates react the "M_k" moment (*section C*). The internal bending moments at those three locations are calculated below.

The M_{j_A} moment at section *A* is 0, at section *B* it is: $M_{j\_B} = 9 * 2.0\ in. = 18\ in\text{-}lb$, and the moment M_j just to the left of section *C* is: $M_{j\_C\text{-}} = (9 * 2.0\ in.) + (17.4 * 1.0\ in.) = 35\ in\text{-}lb$.

For $P_j = -17.4$ *lbs* and $l = 9.5$ *in*., the moment M_j just to the right of section *C* is:

$$M_{j_C} = P_k\,(l) = -17.4\,(9.5) = -165\ in.-lb$$

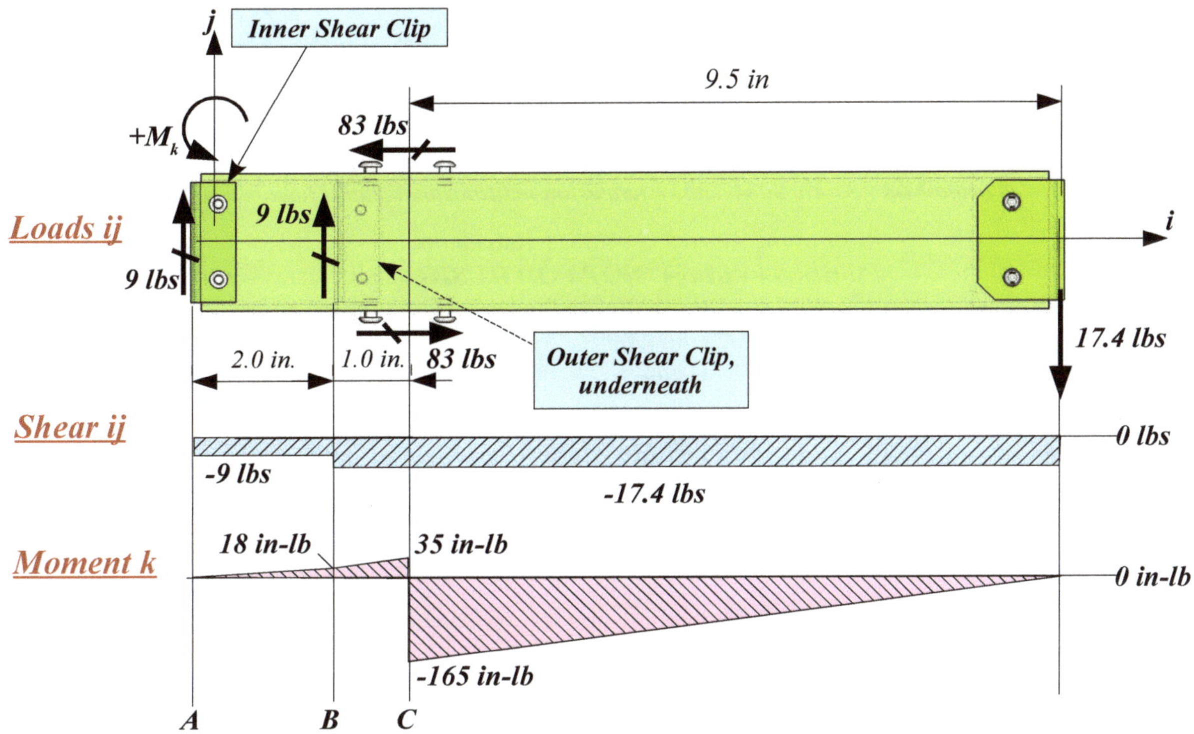

***Figure 11-15** – Horizontal beam Shear and moment diagram, plane i-j.*

11.2.2 ANALYSES

There is no axial load in the beam. Just shear and bending. Therefore, the following checks have to be performed:

- Crippling
- Shear buckling
- Combined stresses (Bending + Shear) strength
- Deflections (See section ***11.11)***
- Fasteners (See section ***11.10.3)***

11.2.3 MATERIAL ALLOWABLES

The actual material allowables for the Horizontal Beam can be obtained from the latest edition of (ref. 8.) for *Al Aly 2024-T3511 Extrusion* 0.095 *in.* thick. For the purpose of the demonstration, *Table 11-3* shows fictitious values.

Table 11-3 *- Al Aly 2024-T3511 Extrusion fictitious material allowables.*

Aluminum Alloy 2024-T3511 Extrusion
Fictitious Values, NOT FOR DESIGN

	≤0.250" [ksi]
F_{tu}	55
F_{ty}	40
F_{cy}	35
F_{su}	30
F_{bru} $(e/D = 2.0)$	110
F_{bry} $(e/D = 2.0)$	75
E, 10^3	10
G, 10^3	3.76
μ	0.33
n	20

11.2.4 SECTION PROPERTIES

The Horizontal beam has a constant cross-section. Therefore, only one section needs to be calculated. The section properties are obtained from a *3D* drawing software. See *Figure 11-16.*

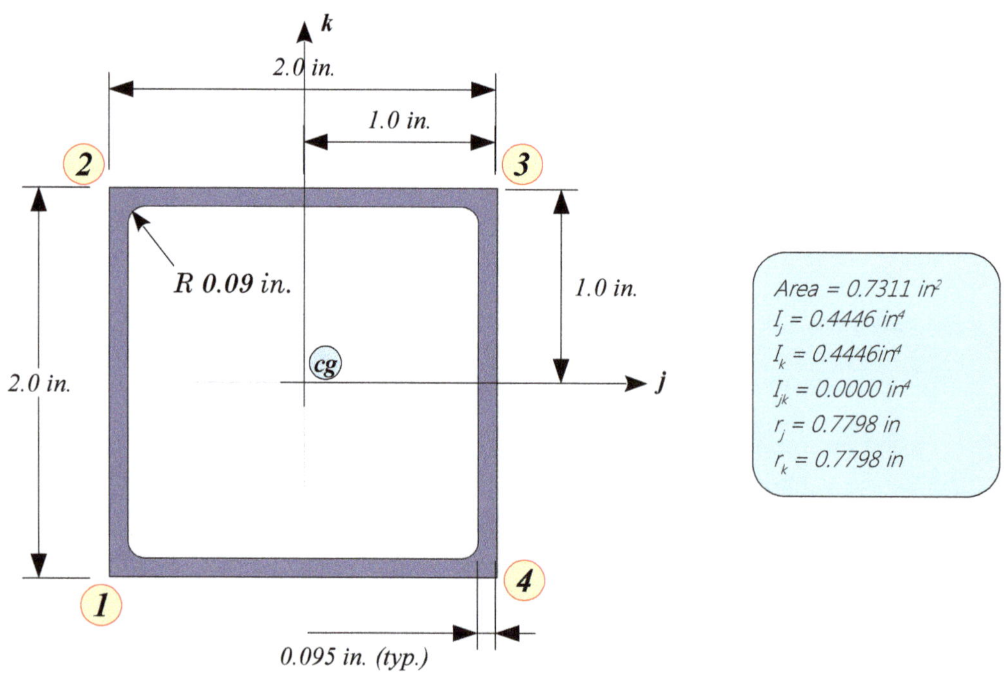

Figure 11-16 *– Horizontal beam section properties.*

Since this cross section has two planes of symmetry, the shear center "*SC*" corresponds to the centroid "*cg*". The torsional stiffness "*K*" of the section is calculated for a thin-walled hollow rectangle as follows:

$$K = \frac{2tt_1(a-t)^2(b-t_1)^2}{at + bt_1 - t^2 - t_1^2}$$

Then,

$$K = \frac{2(0.095)(0.095)(2-0.095)^2(2-0.095)^2}{2(0.095) + 2(0.095) - 0.095^2 - 0.095^2} = 0.3603\ in.^4$$

> *For*
> *$t = t_1 = 0.095$ in.*
> *$a = b = 2$ in.*
> *then*
> *$K = 0.3603\ in^4$*

11.2.5 STRESS ANALYSIS

The bending stress is calculated at the location where the bending moment is the largest. That is at the center of the rivet group common to the Splice Plates as shown on *Figure 11-14*. It is calculated at the four corners (points *1* through *4*) as shown on *Figure 11-16*. The results of the bending stress calculated at each of the four points are shown in *Table 11-4*.

> Since the section is symmetrical ($I_{jk} = 0$), no complex bending needs to be used.

The bending stress is thus calculated as:

$$f_b = -\frac{M_j(k)}{I_j} + \frac{-M_k(j)}{I_k}$$

From the shear and moment diagrams shown in *Figure 11-14* and *Figure 11-15* and, from the section properties shown in *Figure 11-16*, the bending stresses are calculated at the most critical location; section "*C*".

Calculation example at point *1*:

$$f_{b1} = -\frac{M_j(k)}{I_j} + \frac{-M_k(j)}{I_k}$$

$$f_{b1} = -\frac{(-936)(-1.0)}{0.4446} + \frac{-(-165)(-1.0)}{0.4446}$$

$$f_{b1} = -2{,}476\ psi$$

> *For*
> *$M_j = -936$ in-lb*
> *$M_k = -165$ in-lb*
> *$I_j = 0.4446\ in^4$*
> *$I_k = 0.4446\ in^4$*
> *$j = -1.0$ in*
> *$k = -1.0$ in*

The bending stresses at all corners (stress points) are:

Table 11-4 *- Bending stresses at corners.*

Points	*j* [in]	*k* [in]	f_b [psi]
1	-1.0	-1.0	**-2,476**
2	-1.0	1.0	1,734
3	1.0	1.0	2,476
4	1.0	-1.0	-1,734

As seen above, the most critical location is at point *1* where the compression stress is maximum at the corner. Hence, a crippling analysis will be performed.

11.2.6 CRIPPLING

Because there is no free flange in this section, this section is a lot sturdier than the previous opened section. The corner at point *1* is the most critical one since it is the one with the highest compression stress. The crippling analysis is performed by considering the effective flange width as half the tube's width. The edges of the flanges are then assumed simply supported (no free edges). See *Figure 11-17* for the effective segments used in the analysis.

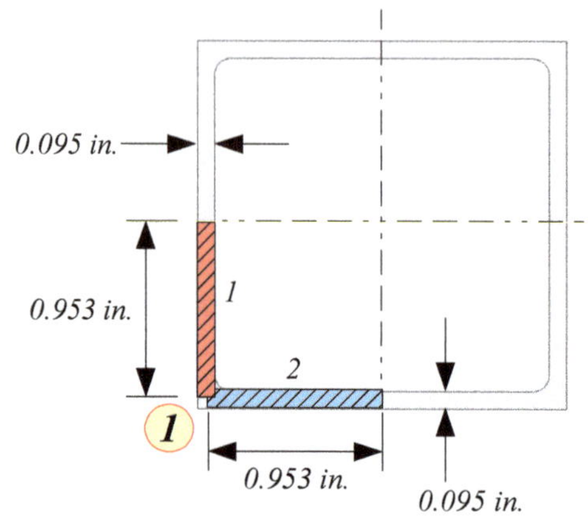

Figure 11-17 – *Effective segments for crippling at corner 1.*

For the segment's width, usually the smaller between half the web's length and 15t is used. In this case, the width of each segment is determined as half the section's width as it is shorter than 15 * (0.095) = 1.425 *in.*

The crippling allowable for each segment is calculated as shown below. See *Figure 11-18*. Note that the crippling allowable "F_{CC}" of a segment can never be higher than "F_{CY}". So the maximum ratio value for F_{CCn} / F_{cy} is 1.0. The value of the abscissa is calculated as:

$$abscissa=\sqrt{F_{CY}/E}(b/t)=\sqrt{35{,}000/10.0\,E^{6}}(0.953/0.095)=0.59$$

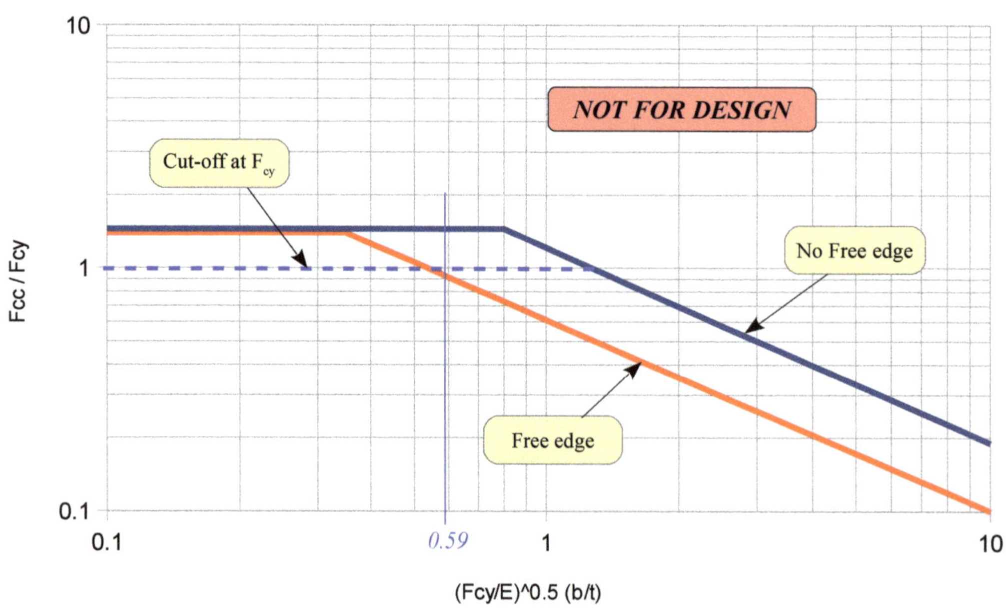

Figure 11-18 – *Crippling curves.*

Table 11-5 *- Crippling analysis at corner 1.*

	Free edge	b_n	t_n	b_nt_n	b_n / t_n	$(F_{cy} / E)^{0.5}$	$(F_{cy} / E)^{0.5}(b_n/t_n)$	F_{CCn} / F_{cy}	F_{CCn}	$t_nb_nF_{CCn}$
Segment		[in]	[in]	[in²]					[ksi]	[lbs]
1	0	0.953	0.095	0.091	10.032	0.0592	0.5935	1.00	35	3169
2	0	0.953	0.095	0.091	10.032	0.0592	0.5935	1.00	35	3169
			$\sum b_nt_n =$	0.1811				$\sum b_nt_nF_{ccn} =$		6337

thus:

$$F_{CC} = \frac{\sum b_n t_n F_{CCn}}{\sum b_n t_n} = \frac{6{,}337}{0.1811} = 35{,}000 \ psi$$

Because the ratio b/t is relatively low, the crippling stress allowable for each segment of the corner is equal to the compression yield allowable "F_{CY}". The larger that ratio is and the lower the crippling allowable.

Margin of safety

The bending stress in compression is much higher at point ***1*** than is the tension stress at point ***2***. Hence, knowing that the allowable in tension "F_{TU}" is much higher than the allowable in compression "F_{CC}", there is no need to check the tension side.

Thus the MS is calculated as:

$$MS = \frac{F_{CC}}{f_{b1}} - 1 = \frac{35{,}000}{|-2{,}476|} - 1 = >1.0$$

11.2.7 SHEAR BUCKLING

Due to the down load "P_K", a constant shear exists in the two vertical webs of the Horizontal beam. In addition, the torque applied to the beam also creates a shear in all four webs of the tube. Due to the direction of the applied torque, the forward web is the one that sees the maximum shear load. On *Figure 11-19*, the shear web is conservatively assumed simply-supported on all four sides.

Due to $P_K = V = 98.5\ lbs$, the shear stress "f_S" is calculated as: $f_S = V/A_{web}$ = 98.5 / (2 * 2.0 * 0.095) = 259 *psi*.

Due to the torque, the average shear stress "f_{ST}" around the section is:

$$Average\ f_{ST} = \frac{T}{2t(a-t)(b-t)}$$

Then

$$Average\ f_{ST} = \frac{9}{2(0.095)(2.0-0.095)(2.0-0.095)} = 13\ psi$$

For
t = 0.095 in.
a = b = 2 in.
T = -9 in-lb

Therefore, the total shear stress ($f_{S\ Total}$) in the forward tube web is:

$$f_{S\ Total} = f_S + f_{ST} = 259 + 13 = 272\ psi$$

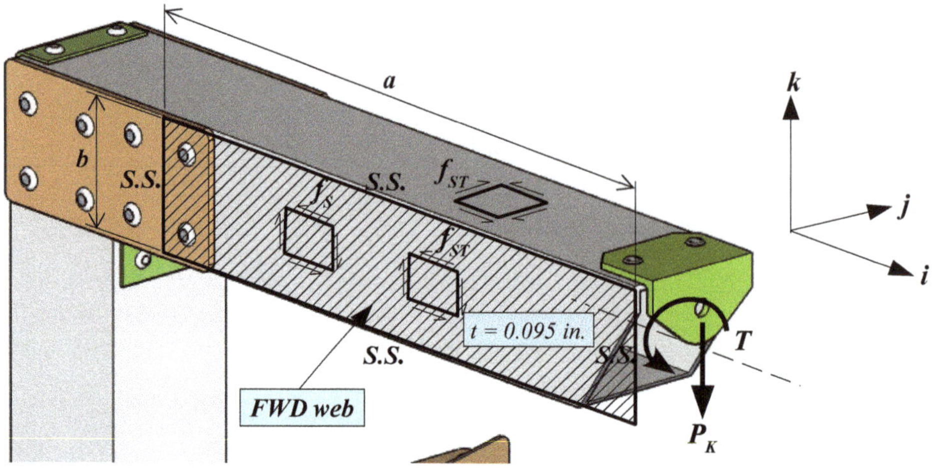

***Figure 11-19** – Web shear buckling.*

The forward web is conservatively assumed simply supported on all four sides. However, as can be noticed near the load application Clip, the web is unsupported. This edge is free! Because the applied loads are very small, it would be fine but normally all shear webs should be framed. This design wouldn't be acceptable if the applied load had been higher and/or if the web thickness had been thin (< 0.063").

This end of the beam should be capped to stabilize the beam section and in order to introduce the vertical shear load directly into the vertical webs, as it should be.

Using (ref. *20.)* for flat plate buckling in shear, the shear stress allowable is calculated as:

$$F_{S\,cr} = \frac{k_S\,\eta_S\,\pi^2\,E}{12\left(1-\mu^2\right)}\left(\frac{t}{b}\right)^2$$

From *Figure 11-19*, the dimension "***a***" is 9.3 *in.* and the dimension "***b***" is 2.0 *in.* Note that for shear buckling, the value for "***b***" is always the smallest of the two dimensions. The web is considered restrained (simply-supported) in the middle of the rivet group at the Splice end.

Thus a/b = 9.3 / 2.0 = 4.7. Hence from *Figure A-2*, the value for k_S read is: 5.5. The primary Young's modulus in tension *"E"* is used from *Table 11-3.*

From the above equation:

$$\frac{F_{s\,cr}}{\eta_s} = \frac{k_s\,\pi^2\,E}{12\left(1-\mu^2\right)}\left(\frac{t}{b}\right)^2$$

$$\frac{F_{s\,cr}}{\eta_s} = \frac{(5.5)\,\pi^2\,10*10^6}{12\left(1-0.33^2\right)}\left(\frac{0.095}{2.0}\right)^2 = 114{,}536\ psi$$

For
k_S = 5.5
b = 2.0
t = 0.095 in
μ = 0.33
*E = 10.0*10⁶ psi*

This is way higher than "F_{SU}". Therefore, a plasticity correction factor for shear "η_S" needs to be calculated using the following equation:

$$\eta_S = \frac{1}{1+\left(\frac{0.00346\,G}{F_{sy}}\right)\left(\frac{F_{scr}}{F_{sy}}\right)^{n-1}}$$

This equation is solved by iteration where a first value for "$F_{s\,cr}$" is assumed in the above equation until it is equal to the value obtained from the equation:

$$F_{S\,cr} = \frac{k_S\,\eta_S\,\pi^2\,E}{12\left(1-\mu^2\right)}\left(\frac{t}{b}\right)^2$$

For this analysis, we need the yield shear stress allowable "F_{sy}". This value is almost never provided in the references and, therefore, can be calculated using the *distortion-energy (Von Mises)* theory as shown below:

$$F_{SY} = \frac{1}{\sqrt{3}}F_{TY} = 0.577\,F_{TY}$$

Thus, from *Table 11-3*, for $F_{TY} = 40$ *ksi*;

$$F_{SY} = 0.577 \times 40 = 23.08\ ksi$$

For the iteration process, I use the *Goal Seek* function from a popular spreadsheet software.

The iteration process is described below:

1. Assume a value for F_{scr}'; 20,000 *psi* in our case *(could be anything!)*.
2. Calculate the plasticity factor "η_S" from the assumed value of F_{scr}'; 0.96425.
3. Calculate the critical shear buckling stress allowable "F_{scr}"; 110,442 *psi*.
4. Calculate the difference between F_{scr}' and F_{scr}; −90,442 *psi*.
5. Perform *Goal Seek* by applying a target of 0 to the cell containing the difference ($F_{scr}' - F_{scr}$) and by varying the cell that contains the first assumed value for F_{scr}'.
6. Once the *Goal Seek* operation is successful, the values for F_{scr}' and F_{scr} are equal and represent the actual critical shear buckling stress allowable.

Figure 11-20 shows a screen grab of the spreadsheet containing the initial values to perform the *Goal Seek* operation as well as the final results. The plasticity factor obtained is **0.22186** and the resulting critical shear buckling stress allowable is **25,411** *psi*.

Margin of safety

Thus the MS in shear buckling is calculated as:

$$MS = \frac{F_{scr}}{f_{S_{Total}}} - 1 = \frac{25{,}411}{272} - 1 = >1.0$$

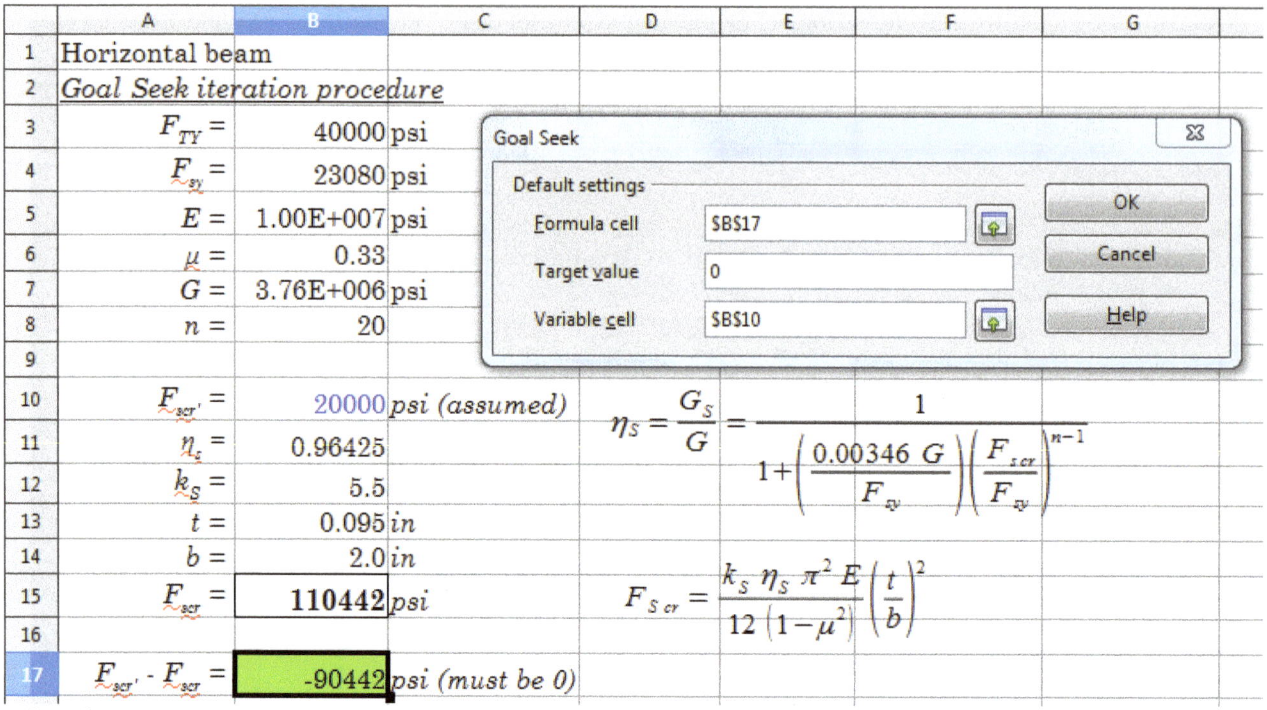

Results:

$F_{scr'}$ =	25411	*psi (assumed)*
η_s =	0.22186	
k_S =	5.5	
t =	0.095	*in*
b =	2.0	*in*
F_{scr} =	**25411**	*psi*
$F_{scr'} - F_{scr}$ =	0	*psi (must be 0)*

Figure 11-20 *– Goal Seek iteration procedure.*

11.2.8 COMBINED STRESSES

Shear

As calculated on page *141,* the total shear stress is 272 *psi*. This is way lower than the shear stress allowable "F_{SU}" of 30,000 *psi*. Therefore, the ultimate shear stress Ratio "R_{su}" is very small:

$$R_{SU} = 272 / 30{,}000 = 0.009$$

Combined bending + shear

The maximum stress due to the bending is combined with the maximum stress due to shear. This is conservative as both maximums don't occur at the exact same point. The MS is calculated for point *1* at section *C*.

The MS for the combined bending and shear is calculated as:

$$MS = \frac{1}{\sqrt{R_b^2 + R_S^2}} - 1$$

where:

$R_S = 0.009$, from previous page,

$R_b = |f_{b1}| \ / \ F_{CY}$, so for $f_{b1} = -2{,}476$ *psi* (ref. *Table 11-4)* and $F_{CY} = 35{,}000$ *psi* (ref. *Table 11-3*),
then $R_b = |-2{,}476| \ / \ 35{,}000 = 0.071$

then:

$$MS = \frac{1}{\sqrt{0.071^2 + 0.009^2}} - 1 = > 1.0$$

11.3 VERTICAL POST

The Vertical Post sees a relatively high torque. So the previous very soft opened section was replaced by a square tube extrusion of 2.0" x 2.0" x 0.095" made in *Aluminum Alloy 2024-T3511*. There is no need to use more expensive alloys like 7075 since we're mainly looking for a stiffness increase and not necessarily for a strength increase. Since the Young's modulus of all Aluminum alloys is pretty much the same, the cheaper 6061-T6 alloy could also probably be used if we wanted. The new vertical Post is shown highlighted in *Figure 11-21*.

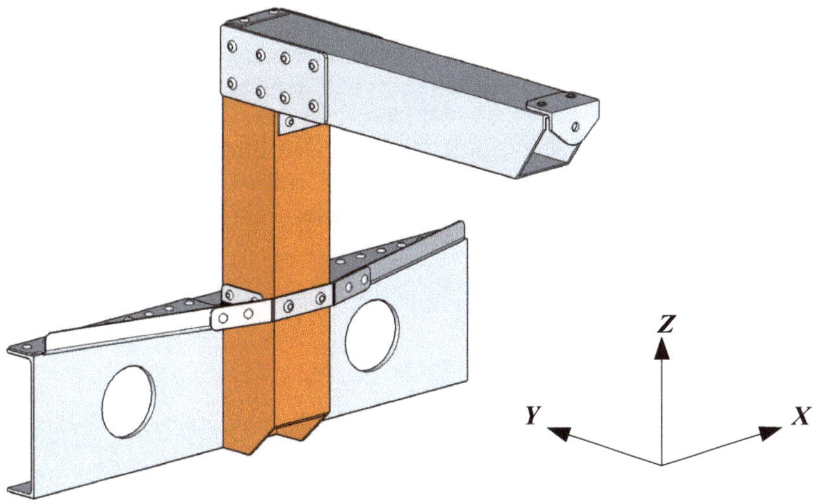

***Figure 11-21** – Vertical Post highlighted.*

11.3.1 LOADS

Referring to *Figure 11-5*, the applied loads from the Beam to the Post are assumed as follows:

- The bending moments "M_X", "M_Y" and "M_Z" and the shear loads (P_Y and P_Z) are introduced into the top of the Post via the two Splices,
- The shear "P_X" is introduced via the two small shear clips.

At the bottom, the Post's loads are assumed reacted as follows:

- The Fore-Aft load "P_X", the vertical load "P_Z" and the moment "M_Y" are reacted through the group of six fasteners common to the Frame.

- The moment "M_X" is reacted as a couple between the Strap and the two solid rivets at the bottom of the group of six.
- The lateral load "P_Y", if there was any, would be reacted by the fasteners common to the two Gussets.
- The moment "M_Z" is reacted by: the Strap, the two Gussets and the top row of fasteners from the group of six.

Because of the lack of accessibility due to the closed section, blind fasteners are used everywhere except for the two bottom fasteners from the group of six where solid rivets are used.

Knowing that the axial load in the bottom two fasteners from the group of six is in compression, I then decided to use solid rivets (MS20470AD5) instead of Aluminum blind rivets (NAS1398D5). In reality, the compression load transferred between the bottom of the Post and the Frame is going to be simply by contact along the bottom edge of the Post.

Sometimes (when we're not sure about the loading conditions), it is not uncommon to assume that the applied loads are completely reversible. In that case, the loads in the two bottom rivets would be in tension. That is why we usually assume the loads to be reacted at the center-line of the rivets so that the same *F.B.D.* can be used and, also for conservatism.

Providing that were the case, I would probably have Hi-Lites installed instead of the BJ5's.

Figure 11-22 shows the *F.B.D.* of the Vertical Post in the "*i-k*" and "*i-j*" planes.

As shown on *Figure 11-22(a)* for the "*i-k*" plane, the loads at end *(B)* are applied at the center of the rivet group. The loads are then reacted at end *(C)* at the center of the rivet group. *Figure 11-22(b)* shows the loads in the "*i-j*" plane. The calculations of the reactions follow for the shear and moment diagrams.

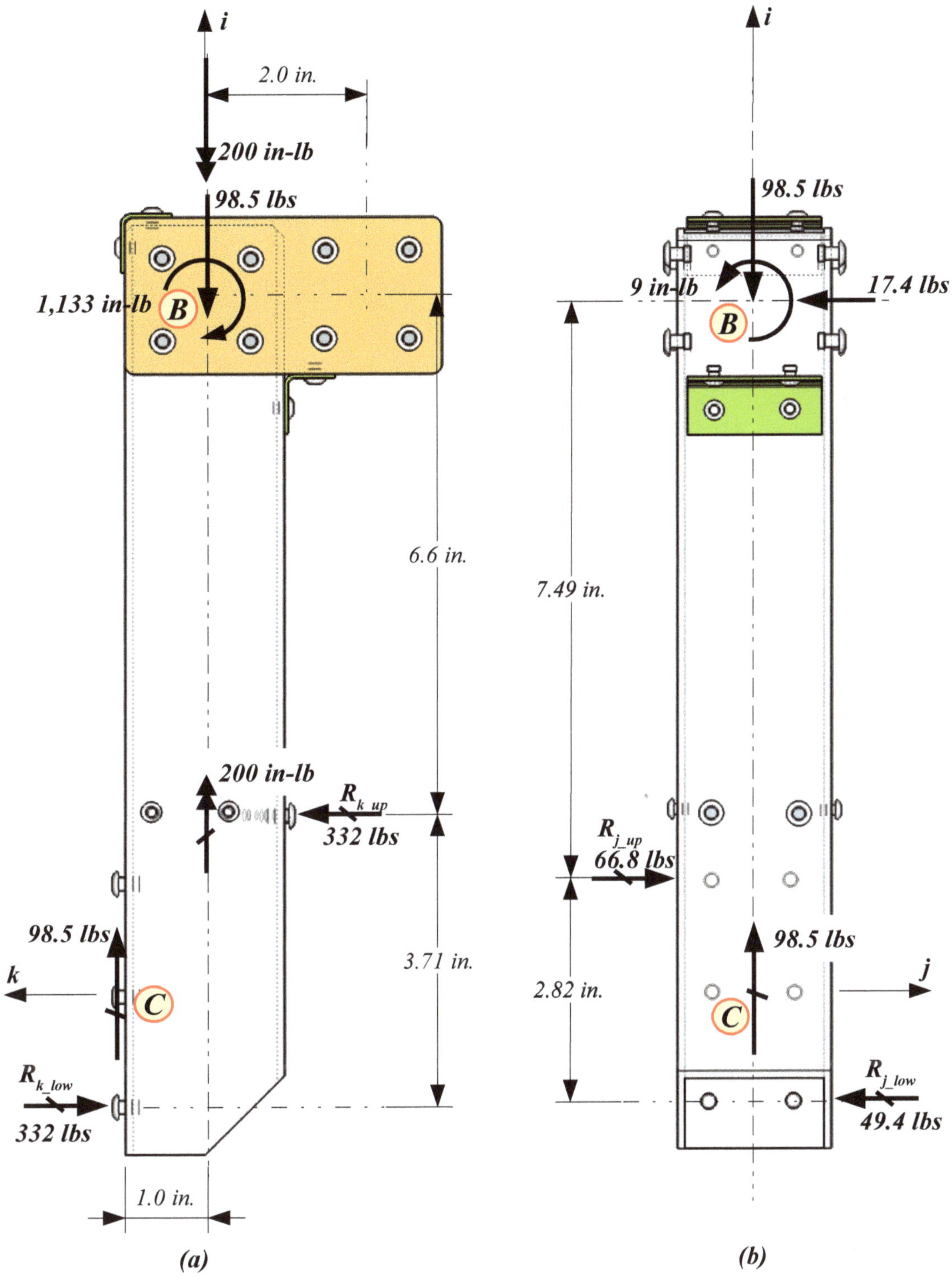

Figure 11-22 – *F.B.D of Post in i-k and i-j planes.*

<u>Shear and moment diagram</u>

The shear and moment diagrams of the Vertical Post can be seen in *Figure 11-23* and *Figure 11-24*. The beam is cantilevered off the Frame. The applied loads are located at the top end.

Plane i-k:

From *Figure 11-12* and *Figure 11-22(a)*, the moment arm "l" between the application point *(A)* and point *(B)* at the rivet group center is 11.5 *in*. For $P_i = -98.5\ lbs$ and $P_j = -17.4\ lbs$, then the applied moments are calculated as:

$$M_j = -P_i(l) = -98.5(11.5) = 1{,}133\ in.-lb$$
$$M_i = -P_j(l) = -17.4(11.5) = 200\ in.-lb$$

and the reactions are:

$$R_{k\ up} = R_{k\ low} = \frac{M_j + P_i}{3.71} = \frac{1{,}133 + 98.5}{3.71} = 332\,lbs$$

$$R_{j\ up} = \frac{M_k + P_j(7.49 + 2.82)}{2.82} = \frac{9 + 17.4(10.31)}{2.82} = 67\,lbs$$
$$R_{j\ low} = R_{j\ up} - P_j = 67 - 17.4 = 49.4\,lbs$$

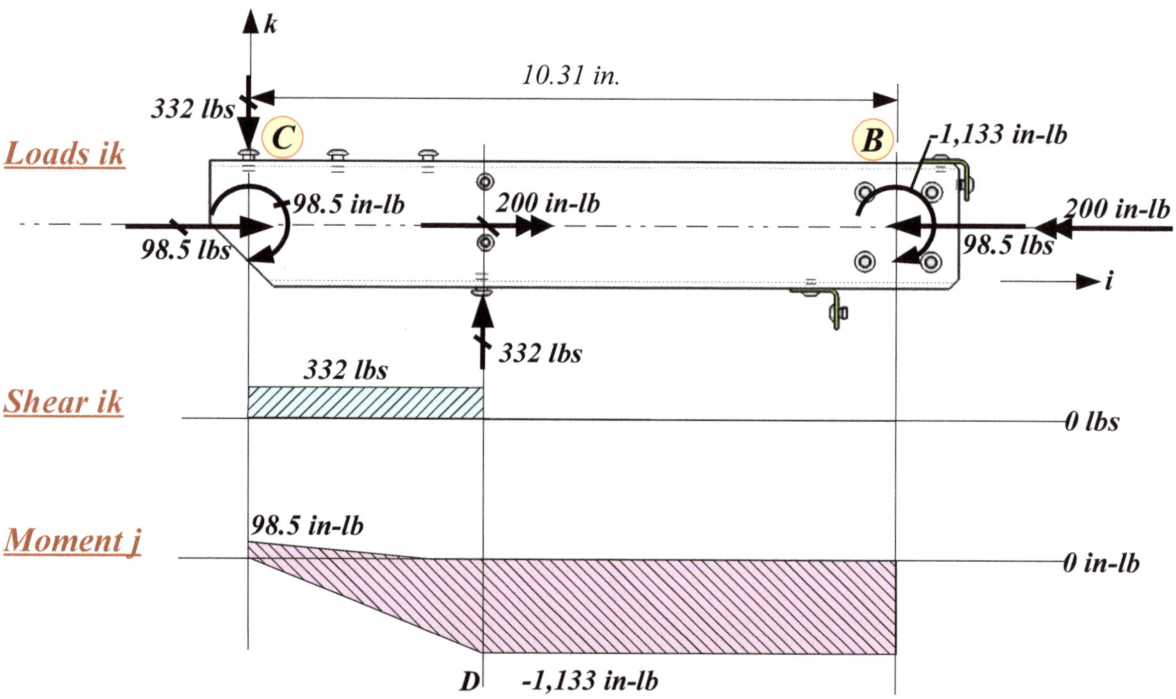

Figure 11-23 – *Vertical Post Shear and moment diagram in plane i-k.*

The moment of 98.5 *in-lb* at the bottom end (left on the above Figure) is due to the fact that the axial load "P_i" of 98.5 *lbs* is reacted one inch away at the faying face between the post and the frame where the group of six fasteners are. **The Shear and Moment diagrams must always follow the centroid of the beam so that those internal loads can readily be used in the stress analysis.**

Plane i-j:

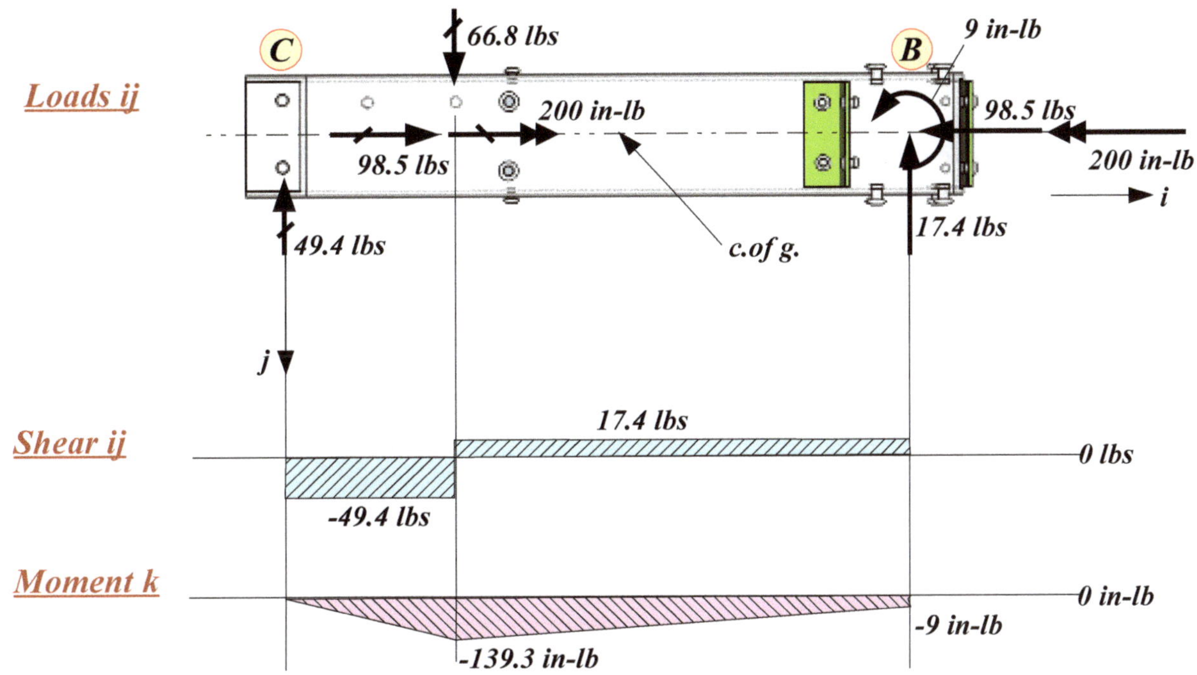

Figure 11-24 *– Vertical Post Shear and moment diagram in plane i-j.*

11.3.2 ANALYSES

There is an axial compression load and internal moments in the Post. Thus, the following checks have to be performed:

- Beam - column
- Crippling
- Shear buckling
- Combined stresses
- Deflections (See section ***10.7)***
- Fasteners (See section ***10.6.2)***

11.3.3 MATERIAL ALLOWABLES

The material allowables for the Vertical Post is the same as for the Horizontal Beam. It is made out of *Al Aly* ***2024-T3511 Extrusion***, 0.095 in. thick. Fictitious allowables are shown in *Table* ***11-6***.

Table 11-6 *- Al Aly 2024-T3511 Extrusion fictitious material allowables.*

Aluminum Alloy 2024-T3511 Extrusion
Fictitious Values, NOT FOR DESIGN

	≤0.250" [ksi]
F_{tu}	55
F_{ty}	40
F_{cy}	35
F_{su}	30
F_{bru} (e/D = 2.0)	110
F_{bry} (e/D = 2.0)	75
E, 10^3	10
G, 10^3	3.76
μ	0.33
n	20

11.3.4 SECTION PROPERTIES

The Vertical Post has a constant cross-section. Therefore, only one section needs to be calculated. The section properties are obtained from a 3D drawing software. See *Figure* ***11-25***.

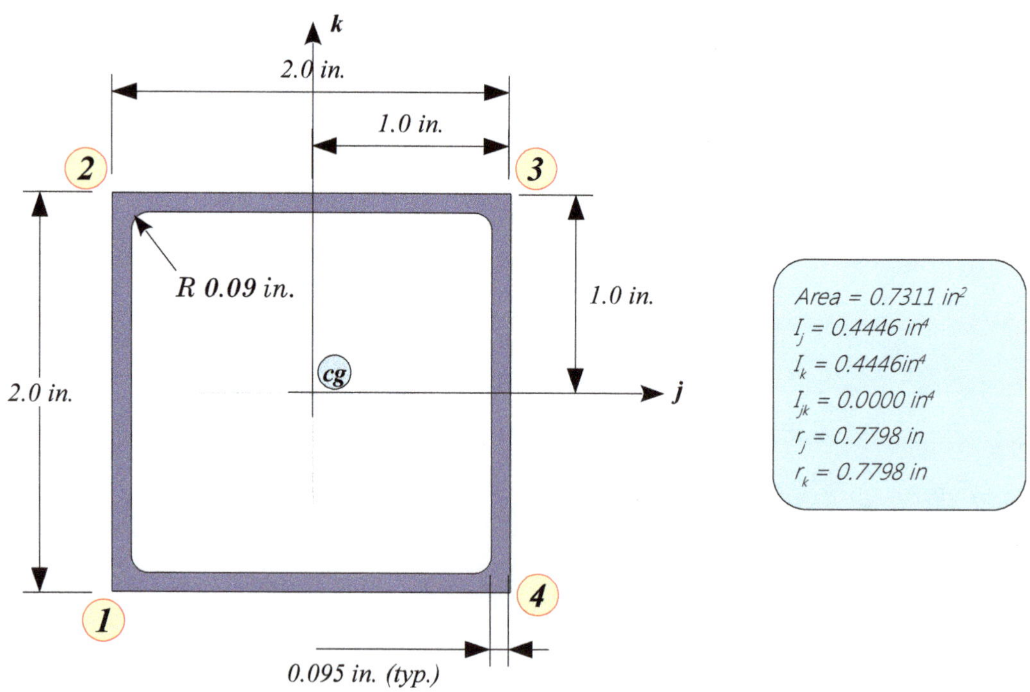

Figure 11-25 – *Vertical Post section properties.*

Since this cross section has two planes of symmetry, the shear center "*SC*" corresponds to the centroid "*cg*". The torsional stiffness "*K*" of the section is calculated for a thin-walled hollow rectangle as follows:

$$K = \frac{2t \cdot t_1 (a-t)^2 (b-t_1)^2}{at + bt_1 - t^2 - t_1^2}$$

Then,

$$K = \frac{2(0.095)(0.095)(2-0.095)^2(2-0.095)^2}{2(0.095)+2(0.095)-0.095^2-0.095^2} = 0.3603\ in.^4$$

For
$t = t_1 = 0.095\ in.$
$a = b = 2\ in.$
then
$K = 0.3603\ in^4$

11.3.5 COLUMN BUCKLING

The Vertical Post is considered clamped at its base (point (*C*)) and Free where the load is applied (point (*B*)). It is submitted to a 98.5 *lbs* ultimate compression load. The Euler column buckling allowable load is calculated as:

$$P_E = \frac{\pi^2 (EI)}{(L')^2}$$

where L' is the effective length calculated as:

$$L' = \frac{L}{\sqrt{c}}$$

Where "*L*" is the column length and "*c*" is the End Fixity Coefficient. In the case shown in *Figure 11-26,* the end fixity coefficient "*c*" for a free-fixed column is obtained from *Figure B-2*. "*L*" is determined as the distance between the center of the rivet group at the top to the center of the rivet group at the bottom.

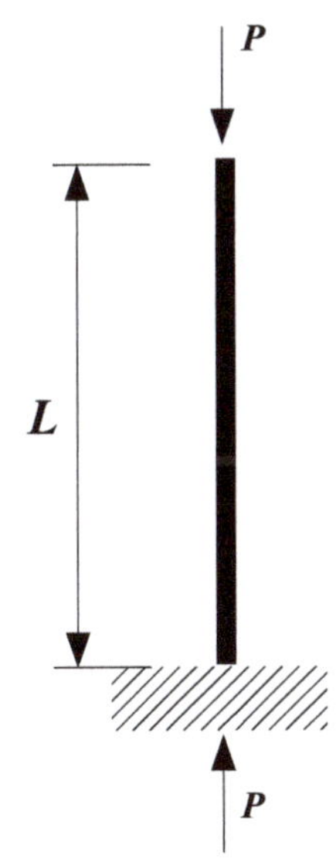

Uniform column, axially loaded, one end fixed, one end free

For
$L = 9.0\ in.$
$c = 0.25$
$\frac{1}{\sqrt{c}} = 2$
thus
$L' = 2L = 2 \times 9.0 = 18.0\ in.$

Figure 11-26 *– Vertical Post column model*

Hence, for $E = E_C = 10 * 10^6\ psi$, $I = min(I_j, I_k) = 0.4446\ in^4$, then:

$$P_E = \frac{\pi^2\ 10*10^6(0.4446)}{(18.0)^2} = 135{,}433\ lbs$$

This corresponds to a stress of: $F_E = P_E / A = 135{,}433 / 0.7311 = 185{,}245\ psi$. This is way above the yield stress allowable ($F_{cy} = 35\ ksi$) and is therefore unrealistic.

Before anything else is calculated, the general stability of a part must be checked. The Euler buckling is quite high since the column is relatively short with a value for $L'/\rho = 18.0 / 0.7798 = 23$. The column thus doesn't buckle.

Now, to obtain a more realistic compression load allowable "P_C", one must use the actual elastic modulus at the point of the allowable load. That is the tangent modulus "E_T".

This can be done by using the actual stress-strain curve of the material. In order to do that we will be using the Ramberg-Osgood methodology that uses a parameter "shape factor" value "n", the stress value at 70% of E "$F_{0.7}$" and the elastic modulus "E". See (ref. *18., 20.*) for detailed explanations on that methodology.

As for the $F_{0.7}$ value, a simple way to determine it is to use the actual stress-strain curve graphs provided in (ref. *8.*) approximated here in *Figure 11-27* for the purpose of the demonstration. In this case, I used the same graph as for the "n" value above and, using the tangent modulus curve in compression, got the value at 70% of the primary E_C value.

For $E_C = 10.0 * 10^6$ *psi*, then $E_{0.7} = 70\% * E_C = 7.0 * 10^6$ *psi.*, we then find a value for $F_{0.7}$ of 40.5 *ksi*.

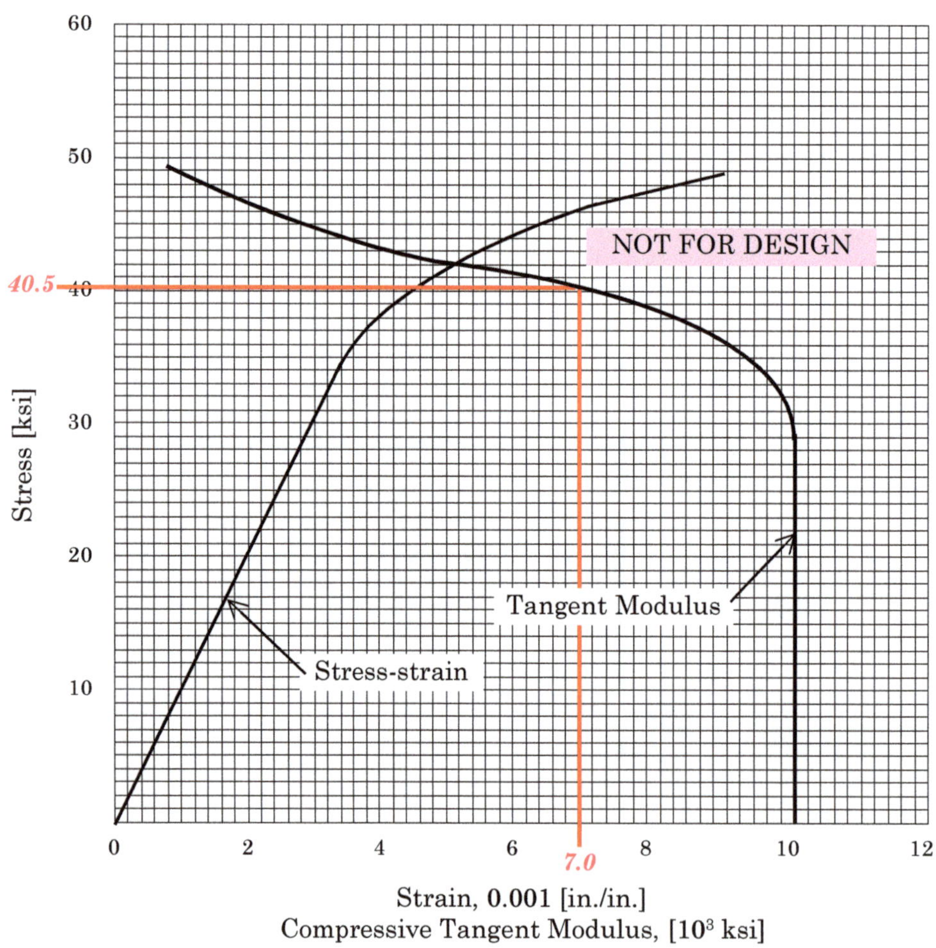

Figure 11-27 – *Approximation of Stress-strain and Tangent Modulus curves for Al Aly 2024-T3511.*

Now that we have the value for "$F_{0.7}$", the allowable compression load "P_C" can be calculated as follows. Using the *three-parameter* equation for the tangent modulus from (ref. 17), the compression stress allowable "F_C" is obtained by iteration. The *three-parameter* equation is:

$$\frac{E_t}{E} = \frac{1}{1 + \frac{3}{7} n \left(\frac{F_C}{F_{0.7}} \right)^{n-1}}$$

using the Euler equation:

$$F_C = \frac{\pi^2 E_t}{\left(\frac{L'}{\rho} \right)^2}$$

introducing the three-parameter equation into the Euler equation we get:

$$F_C = \frac{\pi^2}{\left(\frac{L'}{\rho} \right)^2} \frac{E}{1 + \frac{3}{7} n \left(\frac{F_C}{F_{0.7}} \right)^{n-1}}$$

Solving by iteration for "F_C" using a *Goal seek* we get:

$$F_C = \frac{\pi^2}{(23)^2} \frac{10*10^6}{1 + \frac{3}{7}(20)\left(\frac{38,789}{40,500} \right)^{20-1}} = 38,789 \ psi$$

For
$L' = 18.0$ in
$\rho = 0.7798$ in
$L'/\rho = 23$
$F_{0.7} = 40.5$ ksi
$n = 20$
$E = E_C = 10*10^6$ psi

The allowable compression load is thus, $P_C = F_C * A = 38,789 * 0.7311 = 28,358$ *lbs*.

This is still way higher than the applied 100 *lbs* or so. However, the bending will have to be taken into account to get the MS. Especially, because of the secondary bending effect, an "amplified" bending moment will be calculated using the approximate method as shown below. M_o is the primary bending moment while M is the "amplified" bending moment.

The "amplifying" factor is calculated as:

$$\frac{M}{M_o} = \frac{1}{1 - \frac{P}{P_{CR}}}$$

For $P = 98.5$ *lbs* and $P_{CR} = 28,358$ *lbs*, hence:

$$\frac{M}{M_o} = \frac{1}{1 - \frac{98.5}{28,358}} = 1.004$$

Therefore, because the applied axial load is relatively low, the maximum bending moment is amplified by less than 1%. This factor will be applied to both M_j and M_k.

Thus the amplified values become:

- $M_j = 1.004 * -1,133 = -1,138$ *in-lb*
- $M_k = 1.004 * -139.3 = -140$ *in-lb*

11.3.6 STRESS ANALYSIS

The bending stress is calculated at the location where the bending moment is the largest. That is at the section ***D*** as shown in *Figure 11-23*. It is calculated at the four corners (points *1* through *4*) as shown on *Figure 11-25*. The results of the bending stress calculated at each of the four points are shown in *Table 11-7*. For conservatism, the maximum moment "M_k" is used even though this maximum is not located exactly at the same section.

The axial and bending stresses are thus calculated as:

$$f_i = \frac{P_i}{A} - \frac{M_j(k)}{I_j} + \frac{-M_k(j)}{I_k}$$

From the shear and moment diagrams shown in *Figure 11-23* and *Figure 11-24* and, from the section properties shown in *Figure 11-25,* the bending stresses are calculated at the most critical location; section ***D***.

For
P_i = -98.5 lbs
M_j = -1,138 in-lb
M_k = -140 in-lb
A = 0.7311 in^2
I_j = 0.4446 in^4
I_k = 0.4446 in^4
j = -1.0 in
k = -1.0 in

Calculation example at point *1*:

$$f_1 = \frac{P_i}{A} - \frac{M_j(k)}{I_j} + \frac{-M_k(j)}{I_k}$$

$$f_1 = -\frac{98.5}{0.7311} - \frac{(-1{,}138)(-1.0)}{0.4446} + \frac{-(-140)(-1.0)}{0.4446}$$

$$f_1 = -135 - 2{,}559 - 315 = -3{,}009\ psi$$

Stresses at all corners:

Table 11-7 *- Stresses at corners.*

Points	j [in]	k [in]	f_b [psi]	f_a [psi]	f_{Total} [psi]
1	-1.0	-1.0	**-2,874**	-135	**-3,009**
2	-1.0	1.0	2,245	-135	2,110
3	1.0	1.0	2,874	-135	2,740
4	1.0	-1.0	-2,245	-135	-2,379

As seen above, the most critical location is at point *1* where the compression stress is maximum at the corner. Hence, a crippling analysis will be performed at that point.

11.3.7 CRIPPLING

Since the cross-section of the Vertical Post is exactly the same as the one for the Horizontal Beam, it thus has the same crippling allowable of F_{CC}= 35,000 *psi* calculated on page ***140***.

Thus the MS in crippling is calculated as:

$$MS = \frac{F_{CC}}{f_1} - 1 = \frac{35{,}000}{|-3{,}009|} - 1 = >1.0$$

11.3.8 SHEAR BUCKLING

The Vertical Post sees some transverse shear across its two axes (*j* & *k*) and also some torsion. From the Shear and Moment diagrams shown on *Figure 11-23* and *Figure 11-24*, the maximum transverse shear occurs from section ***D*** down to the bottom. At the same time, the torque is maximum from the top down to section ***D***. Therefore, there are two locations (panels *A* & *B*) to check for the shear buckling. All webs are assumed simply-supported on all sides. See *Figure 11-28*.

Panel *(A)*:

The transverse shear in the "*i-k*" plane is 332 *lbs* while it is 49 *lbs* in the "*i-j*" plane. Since the Post's section is symmetrical, the webs in the "*i-k*" plane from panel ***(A)*** will be checked hence passing the webs in the "*i-j*" plane by comparison.

Due to V_{i_k} = 332 *lbs*, the shear stress "f_s" is calculated as: $f_s = V/A_{web}$ = 332/(2* 2.0 * 0.095) = 874 *psi*.

Panel *(B)*:

Due to V_{i_j} = 49 *lbs*, the shear stress "f_S" is calculated as: $f_S = V/A_{web}$ = 49 /(2 * 2.0 * 0.095) = 129 *psi*.

Due to the torque, the average shear stress "f_{ST}" around the section is:

$$Average\ f_{ST} = \frac{T}{2t(a-t)(b-t)}$$

For
t = 0.095 in.
$a = b = 2$ in.
T = 200 in-lb

Then

$$Average\ f_{ST} = \frac{200}{2(0.095)(2.0-0.095)(2.0-0.095)} = 290\ psi$$

Therefore, the total shear stress "$f_{S\ Total}$" in panel ***(B)*** is:

$$f_{S_{Total}} = f_S + f_{ST} = 129 + 290 = 419\ psi$$

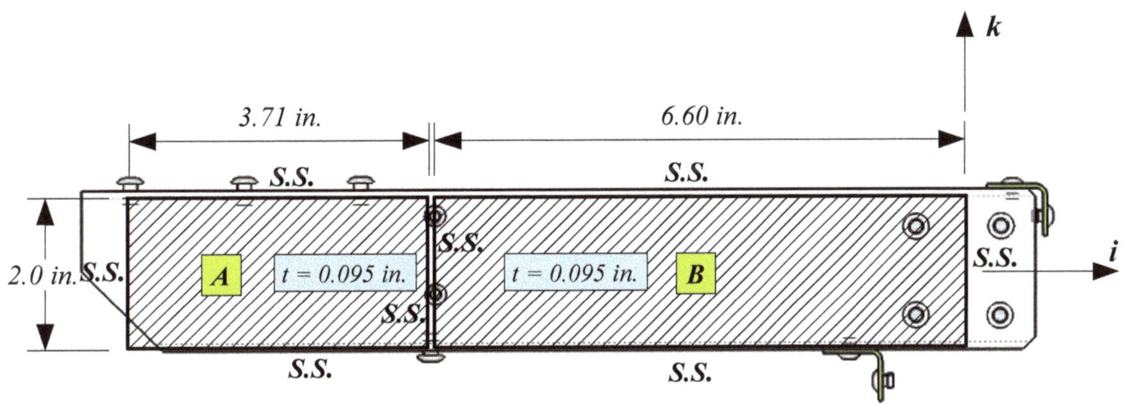

Figure 11-28 – *Webs shear buckling.*

The two panels are conservatively assumed simply supported on all four sides. However, as can be noticed at the bottom of panel *(A)*, the tube is opened and the web is unsupported. This edge is free! Because the applied loads are relatively small, it would be fine but normally all shear webs should be framed. This design wouldn't be acceptable if the applied load had been higher and/or if the web thickness had been thin (< 0.063"). This end of the beam should be capped to stabilize the beam section.

Using (ref. 20.) for flat plates in shear buckling, the shear stress allowable for each panel is calculated as:

$$F_{S\,cr} = \frac{k_S\,\eta_S\,\pi^2\,E}{12\left(1-\mu^2\right)}\left(\frac{t}{b}\right)^2$$

From page *143*, the yield shear allowable F_{sy} is 23.08 *ksi*. The material properties are also the same as for the Horizontal Beam.

Using the same procedure as described in section *11.2.7*, the critical shear buckling stress is calculated by iteration. See *Table 11-8* for the resulting critical shear stress and the associated Margins of Safety.

Table 11-8 *- Critical shear buckling stress for panels A & B.*

	a	*b*	*a/b*	k_S	*t*	η_s	F_{scr}	f_s	*MS*
Panel	*[in]*	*[in]*			*[in]*		*[psi]*	*[psi]*	
A	3.71	2.0	1.86	6.6	0.095	0.17342	**25825**	874	**>1.0**
B	6.6	2.0	3.3	5.7	0.095	0.19903	**25596**	419	**>1.0**

Margin of safety

From the above results, panel *(A)* is the most critical one even though the MS is very large. Thus the MS in shear buckling for the Vertical Post is calculated as:

$$MS = \frac{F_{s\,cr}}{f_s} - 1 = \frac{25{,}825}{874} - 1 = \;>1.0$$

11.3.9 COMBINED STRESSES

The most critical location is at section ***D***. See *Figure 11-23* and *Figure 11-24*. The maximum stresses due to the axial force and bending moments were calculated and summarized in *Table 11-7*. However, because a shear stress also exists at that location, both axial (+bending) and shear must be combined together in order to calculate the MS.

Shear

The maximum shear stress from either sides of section ***D*** will be used. As calculated on page *155*, the shear stress is 874 *psi*. This is way lower than the shear stress allowable "F_{SU}" of 30,000 *psi*. Therefore, the shear stress Ratio "R_s" is very small:

$$R_S = 874 \,/\, 30{,}000 = 0.029$$

Axial

From *Table 11-7*, the maximum stress is at point *1* with $f_1 = -3{,}009$ *psi*. The stress ratio *Ra* is calculated using the compression yield stress "F_{CY}" as the allowable.

$$Ra = 3{,}009 \,/\, 35{,}000 = 0.086$$

Combined axial + shear

The maximum stress due to the axial and bending is combined with the maximum stress due to shear. The MS is calculated for point *1* at section ***D*** as:

$$MS = \frac{1}{\sqrt{R_a^2 + R_S^2}} - 1$$

then:

$$MS = \frac{1}{\sqrt{0.086^2+0.029^2}} - 1 = >1.0$$

You may wonder why I used F_{cy} as the allowable to calculate the ultimate axial stress ratio R_a. Well because the maximum stress is in compression and the main mode of failure is crippling. When the maximum stress is in compression, I've seen people use F_{tu} (F_{cu}), but since the allowable crippling stress is limited to F_{cy}, any stress higher than this value (for thin sections, as usually used on aircraft) will cause the part to collapse thus never reaching F_{tu} no matter what. As always, this hypothesis is to be confirmed by your superiors before using it.

11.4 FRAME ASSEMBLY

The Frame was improved by adding vertical ribs on either sides of the group of six fasteners. In order to stiffen the upper cap in the lateral direction, gussets were added. A strap is used to provide continuity to the Gussets' outer flange. *Figure 11-29* shows the location of all these new features.

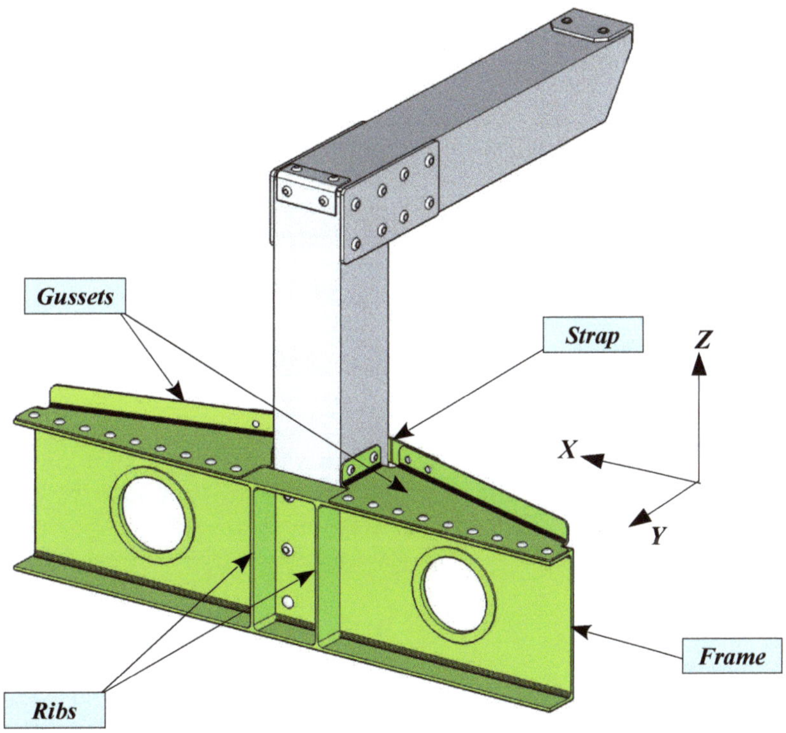

Figure 11-29 – *Improved Frame.*

Since the lower cap is assumed attached to the fuselage skin, then only the upper cap needs to be reinforced. In addition to making the upper cap stiffer in bending I wanted to prevent the fasteners in the group of six to be working in tension. In order to do that, I made sure that the Vertical Post was fastened all around at the upper cap so that the torque would be reacted adequately in shear and that this torque was transferred as a couple at each end of the Frame.

As can be seen from *Figure 11-29*, now all fasteners are used in shear, none is in tension. The Strap provides continuity between the two Gussets that, together with the Frame, work as a continuous beam.

For the Gussets to work with the machined Frame, it must not buckle and the pitch of the rivets must be between 4 to 6 D.

If that concept were installed in an actual aircraft I would certainly add gussets at each end of the Frame on the upper cap to transfer the lateral load efficiently to the adjacent structure.

The Frame assembly comprises of the machined Frame, the two Gussets and the Strap reacts all the loads coming from this structure. For the purpose of the exercise, the Frame assembly is not considered as part of the airframe. However, only the loads from this structure is considered.

The Frame is made out of *Al Aly 7050-T7451* machined Plate 1.0 *in.* thick, the Gussets are made out of *Al Aly 2024-T42 Clad* sheet 0.050 *in.* thick and, the Strap is made out of *Al Aly 2024-T3 Clad* sheet 0.063 *in.* thick. For this example, the connections to any structure at each end *(E & D)* are not looked at. *Figure 11-30* shows the dimensions of the Frame Assy.

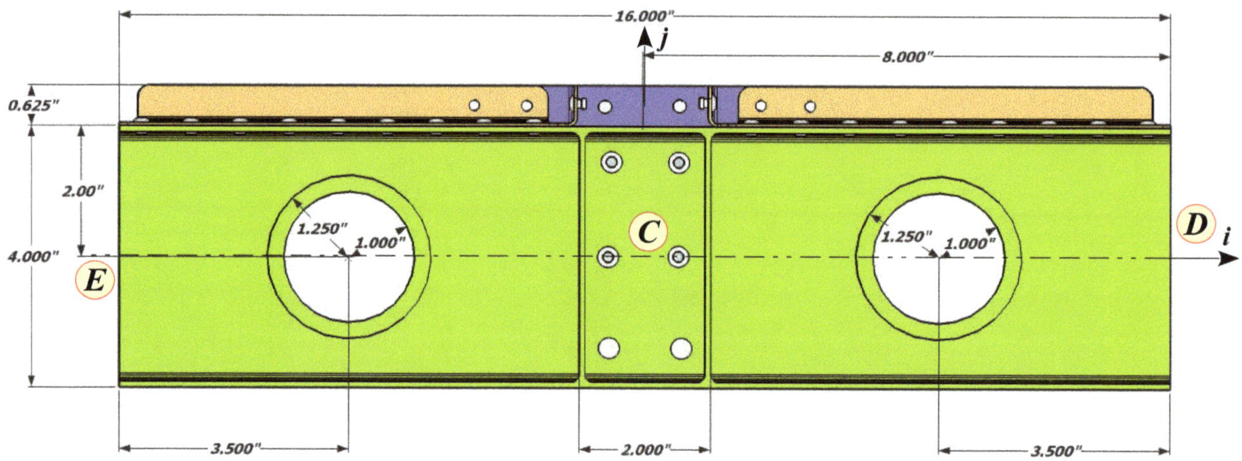

***Figure 11-30** – Improved Frame assembly dimensions.*

If the Frame had been part of the airframe (primary structure), that part would have been analyzed not only for the loads coming from this structure but also by taking into account (superimposing) the internal airframe loads for the same load case.

11.4.1 LOADS

In order to stiffen the Frame upper cap in the lateral direction "*Y*", two Gussets and a Strap were added. This arrangement drastically increased the upper cap's moment of Inertia around the "*Z*" axis. The load paths are assumed as shown in *Figure 11-31* where the machined Frame upper cap takes some axial compression loads and the Strap takes axial tension loads.

This add-on was made so that the total displacement at the load application point (*point A*) would be within the requirement. Although the Strap is efficient in tension, I most likely wouldn't consider this load path providing the Strap was loaded in compression. However, the Gussets and Strap are NOT considered efficient for the moment around the "*Y*" axis because the Strap would be in compression. Therefore, only the machined Frame is considered in the analysis except for the upper cap analysis due to the lateral load.

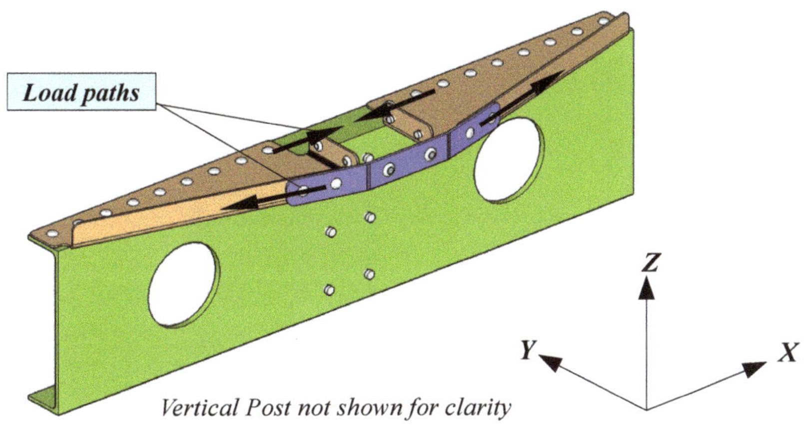

Figure 11-31 *– Load paths in Frame Assembly upper cap.*

As shown on *Figure 11-32* the loads at point (*C*) are applied at the center of the rivet group. The loads are then reacted at ends (*D*) and (*E*) which are considered simply supported. As specified in section ***10.1***, the Frame is assumed fastened to a skin along its bottom edge where the 17.4 *lbs* load is reacted. All the loads are at the centroid of the machined Frame section.

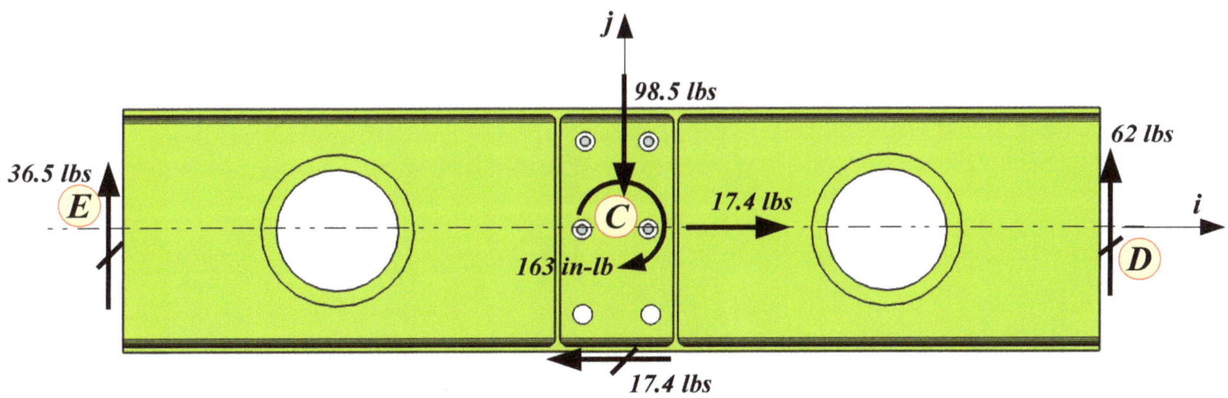

Figure 11-32 *– F.B.D. in i-j plane at section's center of gravity (centroid).*

Figure 11-33 shows the *FBD* of the Frame Assy in the "***i-k***" plane. The loads are applied at the center of the Vertical Post. The applied torque "M_j" is reacted at each end of the Frame while the "P_i" load is assumed reacted at the centroid of the Frame upper cap at section *C-C* from ***Figure 11-41***.

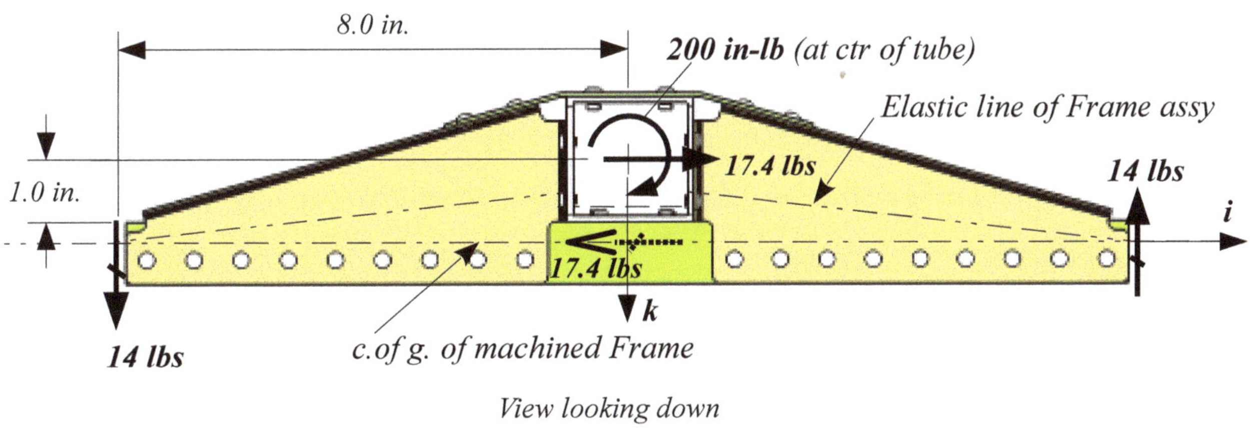

Figure 11-33 *– F.B.D. in i-k plane.*

Shear and moment diagram

The shear and moment diagrams of the Frame in the *"i-j"* plane can be seen in *Figure 11-34.*

Plane i-j:

The applied moment "M_k" and the axial load "P_i" are obtained from *Figure 11-32. Figure 11-34* shows the shear and moment diagram of the Frame in the *"i-j"* plane.

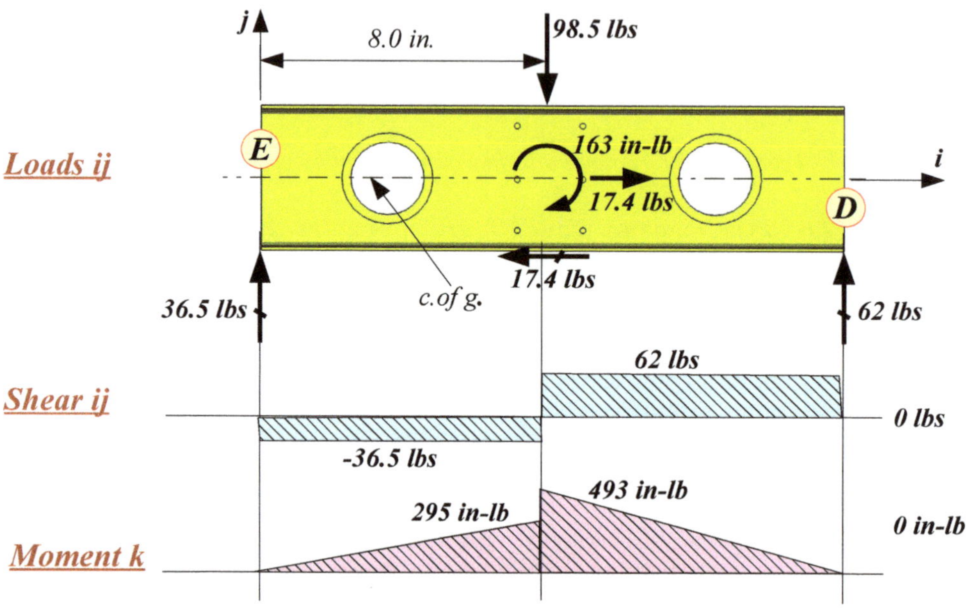

***Figure 11-34** – Frame Shear and moment diagram in plane i-j.*

Plane i-k:

As explained earlier, only the Frame Assy upper cap is considered in the *"i-k"* plane. A *FBD* as well as the shear and moment diagram for the Upper cap only is shown on *Figure 11-36*. From this new design where the Gussets effectively transfer the Post's torque into the Frame, therefore the torque "M_i" and "M_j" moments are considered applied to the upper cap only. Based on *Figure 11-22, Figure 11-35* shows how the torque "M_i" is reacted by the Frame upper and lower caps.

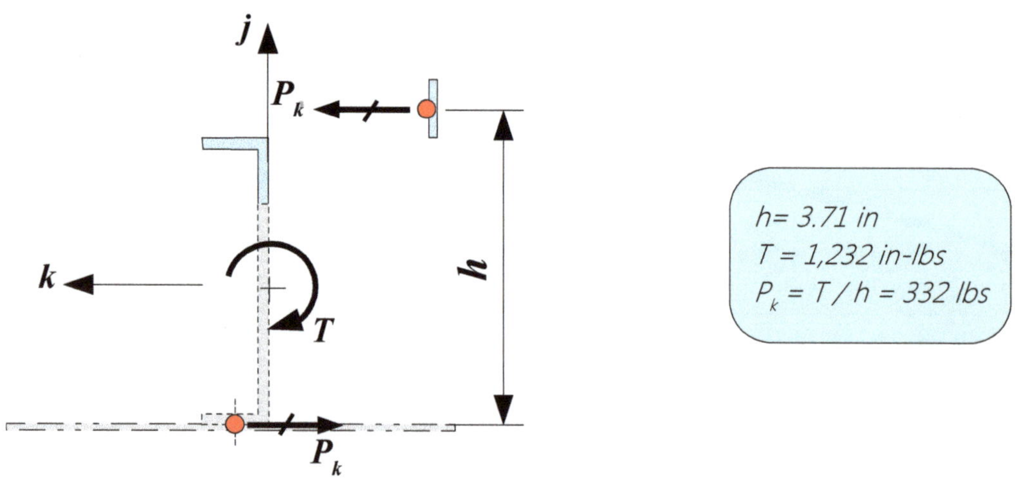

***Figure 11-35** – Couple reacting Torque in Frame upper & lower caps.*

Where the torque at the Frame web face (where the vertical load of 98.5 *lbs* is reacted) is calculated as: 1,133 + 98.5 (1.0) = 1,232 *in-lb*. The reacted lateral load in the upper cap is thus: $P_k = T / h$ = 1,232 / 3.71 = 332 *lbs*. The "M_j" moment is calculated at the Frame centroid at a distance *d* of (1.00 + 0.212) = 1.212 *in*. from the center of the Vertical Post. Therefore, M_j = 200 + 17.4 (1.212) = 222 *in-lb*.

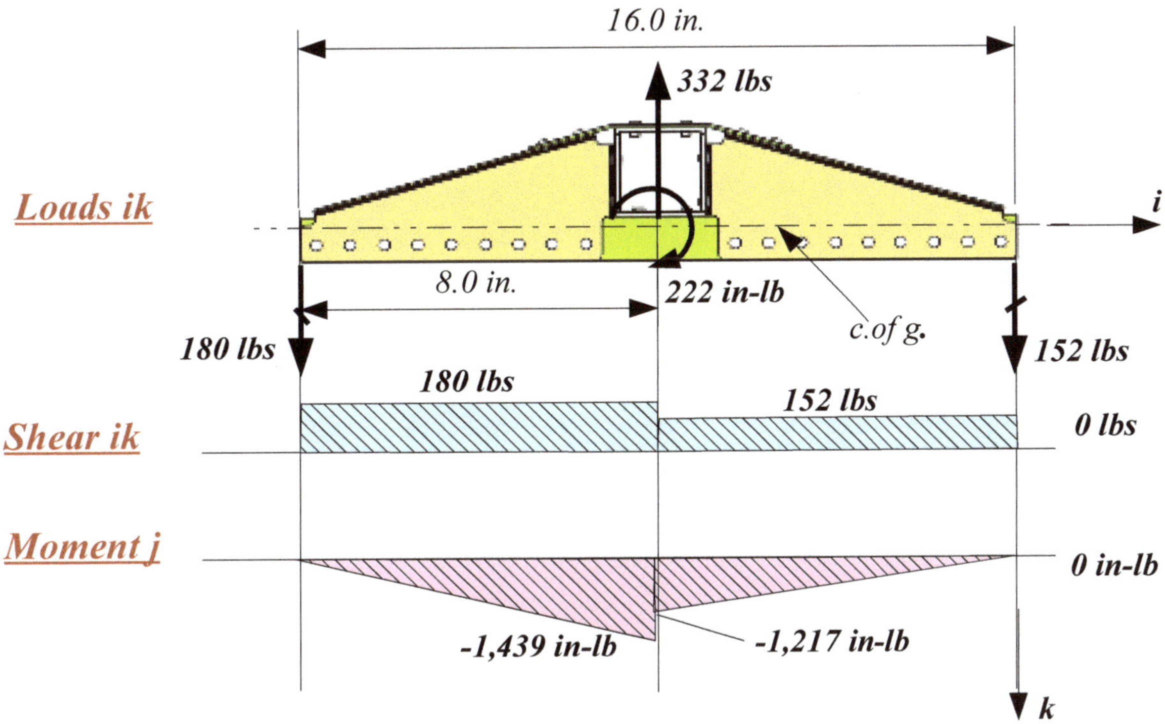

Figure 11-36 – *Frame Upper cap Shear and moment diagram in plane i-k.*

11.4.2 ANALYSES

There is no axial load in the Frame. Thus, the following checks have to be performed:

- Tension in Strap
- Inter-rivet buckling
- Free-flange buckling
- Shear buckling of webs
- Combined stresses
- Deflections (See section ***11.11***)
- Fasteners (See section ***11.10.4***)

11.4.3 MATERIAL ALLOWABLES

The **Frame** is made out of *Al Aly 7050-T7451* Plate, 1.0 *in.* thick. Fictitious material allowables are shown in *Table 11-9.*

Table 11-9 *- Al Aly 7050-T7451 Plate 1.0 in. fictitious material allowables.*

Aluminum Alloy 7050-T7451 Plate
Fictitious Values, NOT FOR DESIGN

	1.00"
	[ksi]
F_{tu}	75
F_{ty}	65
F_{cy}	60
F_{su}	40
F_{bru} $(e/D = 2.0)$	105
F_{bry} $(e/D = 2.0)$	85
$E, 10^3$	10
$G, 10^3$	3.8
μ	0.33
n	20

The **Strap** is made out of *Al Aly 2024-T3 Clad* sheet, 0.063 *in.* thick. Fictitious material allowables are shown in *Table 11-10.*

Table 11-10 *- Al Aly 2024-T3 Clad sheet 0.063 in. fictitious material allowables.*

Aluminum Alloy 2024-T3 Clad sheet
Fictitious Values, NOT FOR DESIGN

	0.063"
	[ksi]
F_{tu}	60
F_{ty}	45
F_{cy}	40
F_{su}	35
F_{bru} $(e/D = 2.0)$	120
F_{bry} $(e/D = 2.0)$	85
$E, 10^3$	10
$G, 10^3$	3.76
μ	0.33
n	20

The **Gussets** are made out of *Al Aly 2024-T42 Clad* sheet, 0.050 *in.* thick. Fictitious material allowables are shown in *Table 11-11.*

Table 11-11 *- Al Aly 2024-T42 Clad sheet 0.050 in. fictitious material allowables.*

Aluminum Alloy 2024-T42 Clad sheet
Fictitious Values, NOT FOR DESIGN

	0.050"
	[ksi]
F_{tu}	55
F_{ty}	35
F_{cy}	37
F_{su}	35
F_{bru} (e/D = 2.0)	115
F_{bry} (e/D = 2.0)	70
$E, 10^3$	
Primary	10
Secondary	9.5
μ	0.33
n	20

11.4.4 SECTION PROPERTIES

The Frame (without the Gussets and the Strap) has a constant cross-section. Therefore, only one section needs to be calculated. The section properties are obtained from a *3D* drawing software. See *Figure 11-37*.

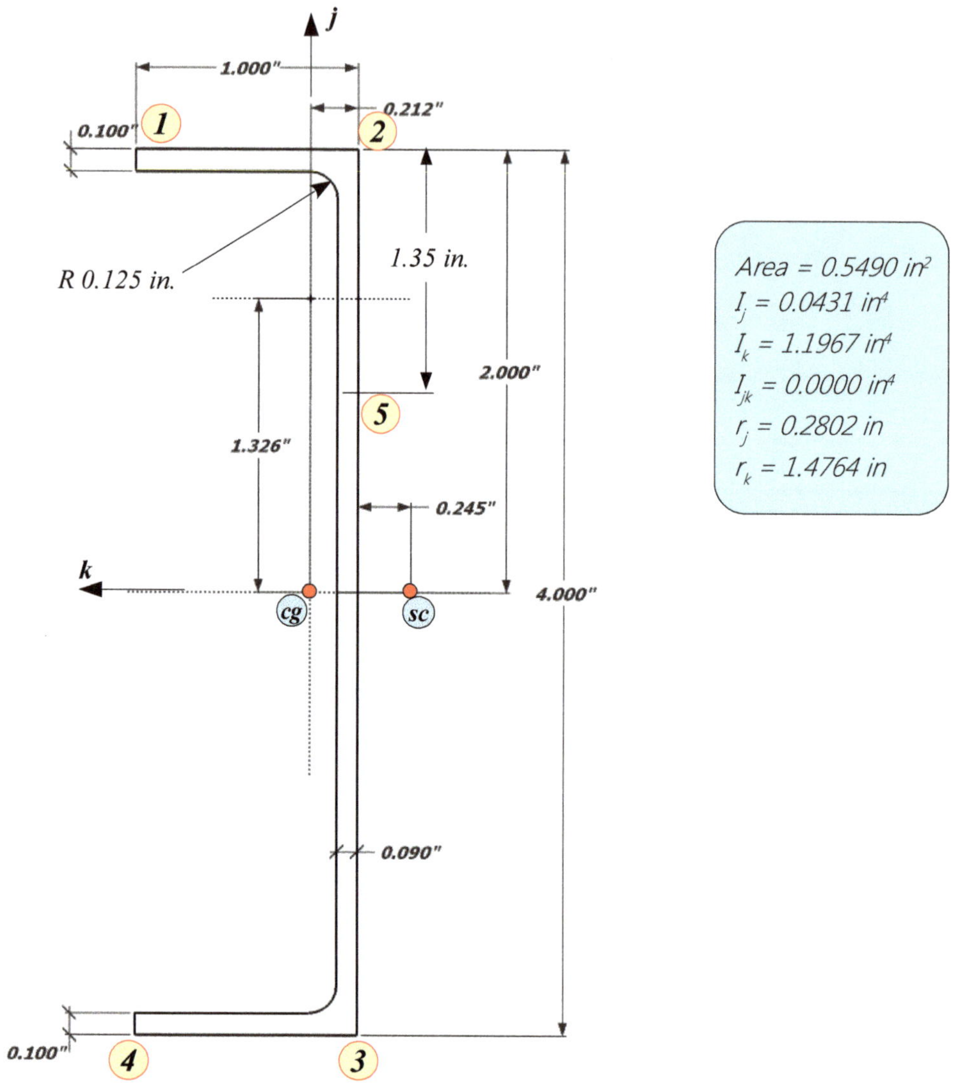

Figure 11-37 *– Frame section properties.*

Upper Cap section properties

Because it is assumed that the Frame is fastened to a skin along its bottom flange, it is therefore partially restrained from twisting. The applied torque is thus reacted as a couple between the Frame upper cap and the skin, which doesn't move. See ***Figure 11-40*** for the upper cap section properties and ***Figure 11-35*** to see how the torque in the Frame is reacted by the Upper cap and the skin.

An effective web width of *15t* (or 15*0.090 = 1.35 *in.*) is assumed in the upper cap section.

Because the Frame upper cap assembly section is not constant, its elastic line is not parallel to the upper cap's span. This line should follow the centroid of all the cross section along the Frame's length. *Figure 11-38* shows the approximated elastic line of the upper cap. Note that all of the cross sections used in the analysis shall be normal to that elastic line and that the internal load at each cross section shall be transformed to its centroid.

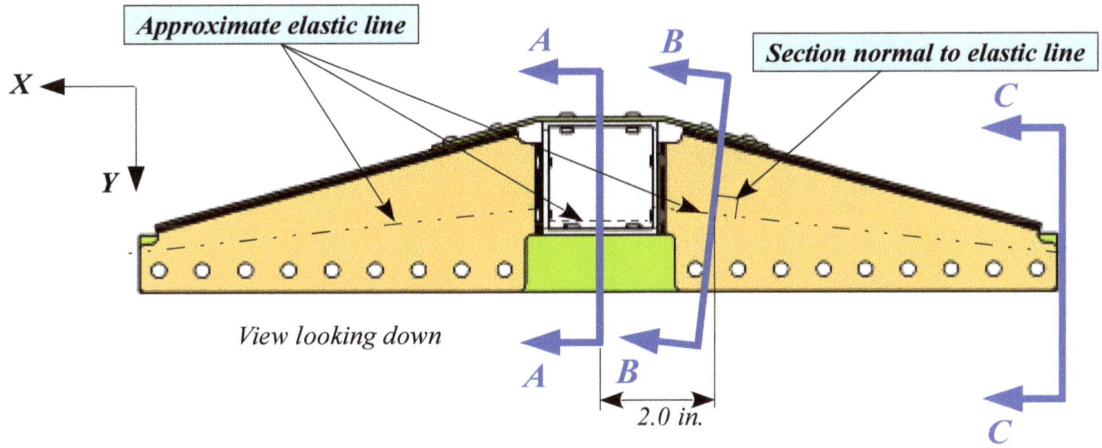

Figure 11-38 *– Critical sections along the approximated elastic line.*

Section ***A-A*** comprises of the machined Frame upper cap as well as the Strap. See *Figure 11-39.*

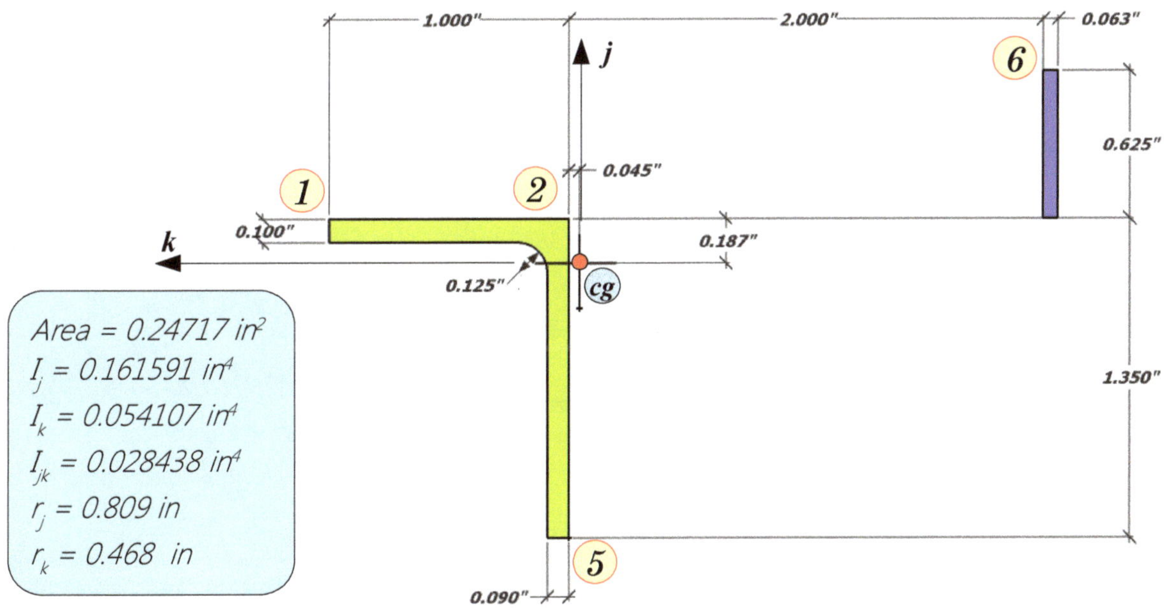

Figure 11-39 *– Frame upper cap section A-A properties.*

Section ***B-B*** is taken normal to the elastic line and through the point that goes in-between the two rivets connecting the Strap to the Gusset was chosen. At that location, it is assumed that the Strap has dumped all of its load into the Gusset. See *Figure 11-40.*

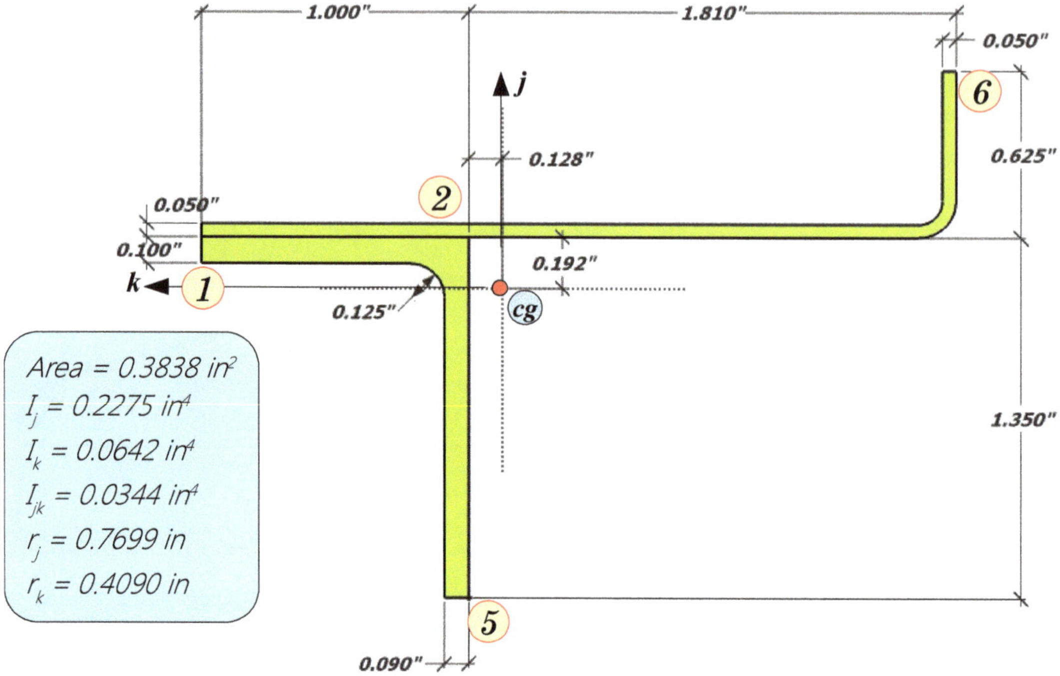

Figure 11-40 *– Frame upper cap section B-B properties.*

Section ***C-C*** is simply the machined Frame upper cap without the Gusset. See *Figure* ***11-41***.

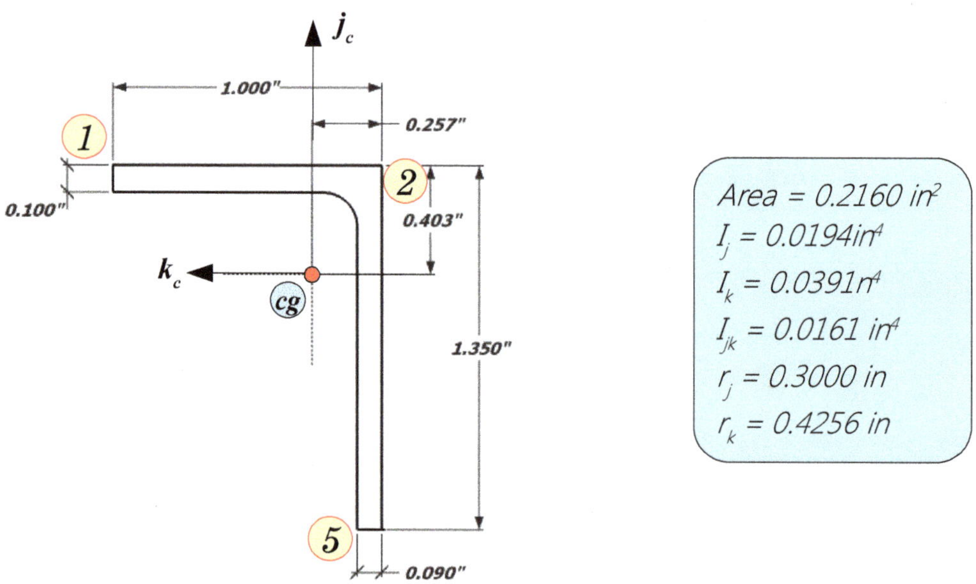

Figure 11-41 *– Frame upper cap section C-C properties.*

11.4.5 STRESS ANALYSIS

The bending stress is calculated at the location where the bending moment is the largest which is right at the center. It is calculated at the points as shown on *Figure 11-37* to *Figure 11-41*. There will be two different bending stress calculations to consider.

1. Due to the moment "M_k" (full machined Frame section).
2. Due to the moment "M_j" (Upper cap sections only).

The resulting stresses obtained at points *1, 2* and *5* will be superimposed from the two calculations in order to get the total bending stress. Points *3 & 4* will be only influenced by the stress from "M_k".

One can notice that the maximum bending moment "M_k" occurs just slightly to the right of the Frame center while the maximum bending moment "M_j" occurs just slightly to the left of the Frame center. The two maximum values are conservatively used in the total stress calculation.

The shear and moment diagrams are idealized and are usually more severe than in reality due principally to the width of the Vertical Post that load-up the Frame. Therefore those peak bending moments are already conservative and I could have used the value for "M_k" of 295 *in-lb* instead of 493 *in-lb* to combine with the maximum "M_j" value of 1,439 *in-lb*. But, since I wanted to stick to the assumed moment distribution model, then I conservatively decided to use the most severe value for "M_k" since there is only a very fine line separating the values of 493 and 295 *in-lb*.

11.4.5.1 Section A-A

<u>Due to M_k:</u>

The Frame full section as shown in *Figure 11-37* is used. However, for conservatism, no effective skin is considered.

The results of the bending stress calculated at 5 points are shown in *Table 11-12*.

Since the full frame section is symmetrical ($I_{jk} = 0$), no complex bending needs to be used. But for the Upper cap angle section, $I_{jk} \neq 0$. However, although I chose to consider the upper cap as being a beam by itself, in reality it is still attached to the rest of the Frame. Therefore, it is stabilized (prevented to twist). Hence, as for the full frame section, no complex bending needs to be used.

Also, the skin is ignored in the section properties mainly because, by considering an effective skin working with the Frame section, that would require that the loads from the Frame be added to the skin internal loads for the skin analysis. The effective skin is, of course, considered for the fuselage frames. However, for secondary structures (as in this example) it is preferable to leave the skin alone and not take advantage of it.

Using the shear and moment diagram on *Figure 11-34*, the bending stress is thus calculated as:

$$f_b = \frac{M_k(j)}{I_k}$$

From *Figure 11-34* and *Figure 11-37*, the calculation example at point *1* of section *A-A* is shown:

$$f_{b1} = \frac{-M_k(j)}{I_k}$$

$$f_{b1} = \frac{-(493)(2.0)}{1.1967}$$

$$f_{b1} = -824\ psi$$

For
M_k = 493 in-lb.
I_k = 1.1967 in^4
j = 2.0 in

The bending stresses due to "M_k" at the Frame center at all corners of section *A-A* are listed in *Table 11-12*.

Table 11-12 *- Bending stresses at corners of section A-A due to* M_k.

Points	j [in]	k [in]	f_b [psi]
1	2	0.788	-824
2	2	-0.212	-824
3	-2	-0.212	824
4	-2	0.788	824
5	0.65	0.788	-268

Due to M_j:

Now, the bending stresses will be calculated by considering the Upper cap only. The Frame upper cap section as shown in *Figure 11-39* to *Figure 11-41* are used.

Using the shear and moment diagram on *Figure 11-36*, the bending stress is thus calculated as:

$$f_b = \frac{M_j(k_c)}{I_j}$$

From *Figure 11-39*, the stress in the Strap will be calculated. However, since the moment diagram shown in *Figure 11-36* was calculated at the Frame centroid, the moment "M_j" must be transformed to be re-calculated at the centroid of section *A-A*. From *Figure 11-37* and *Figure 11-39*, the distance between the two centroids is; *d* = 0.212 + 0.045 = 0.257 *in*.

The transformed internal moment at the centroid of section *A-A* "$M_{j\,A\text{-}A}$" is thus; $M_{j\,A\text{-}A} = M_j - P_X * d$ = 1,439 – 17.4 * 0.257 = 1,435 *in-lb*. The calculation example for point *6* is shown:

$$f_{b6} = \frac{M_j(k_6)}{I_j}$$

$$f_{b6} = \frac{(-1{,}435)(-2.018)}{0.161591}$$

$$f_{b6} = 17{,}921\ psi$$

For
M_j = -1,435 in-lb.
I_j = 0.161591 in^4
k_6 = 2.0+0.063-0.045=2.018 in

The bending stresses due to "M_j" at all corners of section *A-A* are listed in *Table 11-13*.

Table 11-13 *- Bending stresses at corners of section A-A due to* M_j.

Points	j [in]	k [in]	f_b [psi]
1	0.187	1.045	-9,280
2	0.187	0.045	-400
5	-1.163	0.045	-400
6	0.812	-2.018	**17,921**

Total bending stress:

The stress due to the moment "M_k" applied to the full machined Frame section is added to the stress due to the moment "M_j" applied to the Frame Upper cap section to get the total stress at each critical point. *Table 11-14* shows a summary of the total stress at all stress points of section *A-A*.

Table 11-14 *- Total Bending stresses at section A-A.*

Points	$f_{b\,(Mk)}$ [psi]	$f_{b\,(Mj)}$ [psi]	$f_{b\,Total}$ [psi]
1	-824	-9280	**-10,104**
2	-824	-400	-1,224
3	824	0	824
4	824	0	824
5	-268	-400	-667
6	0	17921	17,921

As seen above, the most critical locations are at points *1* and *6* where the maximum compression stress and maximum tension stress exist. Since point *1* is in compression, a free flange buckling analysis will be performed. As for point *6*, the Strap will be checked in net tension.

11.4.5.2 Section B-B

Due to M_k:

The Frame full section as shown in *Figure 11-37* is used. However, for conservatism, no effective skin is considered. The results of the bending stress calculated at 5 points are shown in *Table 11-15*.

Using the shear and moment diagram on *Figure 11-34*, the bending stress is thus calculated as:

$$f_b = \frac{M_k(j)}{I_k}$$

Where M_k at section *B-B* (located 2.0 inches away from the center) is calculated as: 295 – 36.5 * 2.0 = 222 *in-lb*. From *Figure 11-34* and *Figure 11-37*, the calculation example at point *1* of section *B-B* is shown:

$$f_{b1} = \frac{-M_k(j_1)}{I_k}$$

$$f_{b1} = -\frac{(222)(2.0)}{1.1967}$$

$$f_{b1} = -371\ psi$$

For
M_k = 222 in-lb.
I_k = 1.1967 in^4
j_1 = 2.0 in

The bending stresses due to "M_k" at all corners of the Frame center of section *B-B* are listed in *Table 11-15*.

Table 11-15 *- Bending stresses at corners of section B-B due to M_k.*

Points	*j* [in]	*k* [in]	f_b [psi]
1	2	0.788	-371
2	2	-0.212	-371
3	-2	-0.212	371
4	-2	0.788	371
5	0.65	0.788	-121

Due to M_j:

From *Figure 11-38*, one can see that section *B-B* is located 2.0 inches away from the Frame center. Using *Figure 11-36*, the maximum bending moment at that location is thus; M_j = 1,439 – 180 * 2.0 = 1,079 *in-lb*.

However, since the moment diagram shown in *Figure 11-36* was calculated at the Frame centroid, the moment M_j must be transformed to be re-calculated at the centroid of section *B-B*. From *Figure 11-37* and *Figure 11-40*, the distance between the two centroids is; d = 0.212 + 0.128 = 0.34 *in*.

The transformed internal moment "$M_{j\,B\text{-}B}$" at the centroid of section *B-B* is thus; $M_{j\,B\text{-}B} = M_j - P_X * d$ = 1,079 – 17.4 * 0.34= 1,073 *in-lb*. The calculation example at point *6* of section *B-B* is shown:

$$f_{b6} = \frac{M_j\,(k_6)}{I_j}$$

$$f_{b6} = \frac{(-1{,}073)\,(-1.682)}{0.2275}$$

$$f_{b6} = 7{,}933\ psi$$

For
M_j = -1,073 in-lb.
I_j = 0.2275 in^4
k_6 = 0.128-1.81=-1.682 in

The bending stresses due to "M_j" at all corners of section *B-B* are listed in *Table 11-16.*

Table 11-16 *- Bending stresses at corners of section B-B due to M_j.*

Points	*j* [in]	*k* [in]	f_b [psi]
1	0.242	1.128	-5,320
2	0.192	0.128	-604
5	-1.158	0.128	-604
6	0.817	-1.682	7,933

Total bending stress:

The stress due to the moment "M_k" applied to the full machined Frame section is added to the stress due to the moment "M_j" applied to the Frame Upper cap section to get the total stress at each critical point of section *B-B*. *Table 11-17* shows a summary of the total bending stresses for section *B-B*.

Table 11-17 *- Total Bending stresses at section B-B.*

Points	$f_{b\,(Mk)}$ [psi]	$f_{b\,(Mj)}$ [psi]	$f_{b\,Total}$ [psi]
1	-371	-5320	**-5,691**
2	-371	-604	-975
3	371	0	371
4	371	0	371
5	-121	-604	-724
6	0	7933	**7,933**

As seen above, the stresses are lower than for section *A-A*. Therefore section *B-B* is not critical.

11.4.5.3 Section C-C

Since the bending moments are nil at section *C-C*, this section is therefore not critical.

11.5 STRAP

The Strap is fastened to the Vertical Post and to the free flanges of the Gussets. It is made out of *Al Aly 2024-T3 Clad* sheet 0.063" thick. From *Table 11-14*, the maximum stress at point ***6*** (on the aft side) corresponds to the stress in the Strap and is 17,921 *psi*. Using the "M_j" moment at the fwd side, the stress obtained is $f = -1{,}217\ (-2.018) / 0.161591 = 15{,}198\ psi$.

Since the Strap has a cross section area of: $A = 0.625 * 0.063 = 0.039375\ in^2$, the Strap maximal internal tension load (aft side) is thus: $P_{Strap} = f_{6\,Aft} * A = 17{,}921 * 0.039375 = 706\ lbs$, while the tension load at the other end (fwd side) is: $P_{Strap} = f_{6\,Fwd} * A = 15{,}198 * 0.039375 = 599\ lbs$.

Figure 11-42 shows the Free Body Diagram of the Strap by calculating the load in the "*X*" direction from the stresses at point ***6*** on either sides of the center-line. The load taken by the two fasteners common to the Vertical post is: $P_{post} = 706 - 599 = 107\ lbs$.

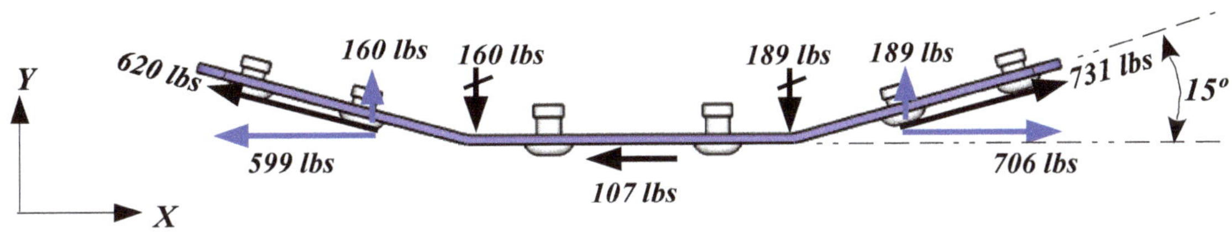

Figure 11-42 *– FBD of Strap.*

11.5.1 NET TENSION

The Strap is in pure tension so it is checked in net tension at the critical location. See *Figure 11-42* and *Figure 11-43* for the Free Body Diagram of the Strap. The reaction loads are calculated as:

$$P_{Strap}' = \frac{P_{Strap}}{\cos(15)} = \frac{706}{\cos(15)} = 731\ lbs$$

$$R = P_{Strap}' \sin(15) = 731 \sin(15) = 189\ lbs$$

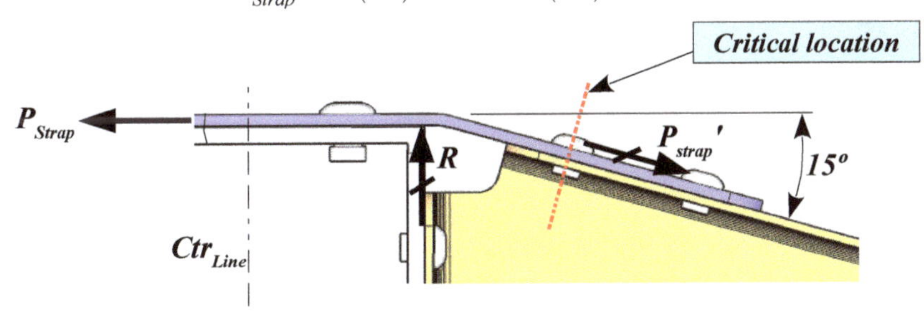

Figure 11-43 *– Strap FBD.*

Since the rivets at the critical location are 5/32" in diameter, the net tension area; A_{net} is:

$$A_{net} = A - (D * t) = 0.039375 - (0.156 * 0.063) = 0.0295\ in.^2$$

The net tension stress "ft_{net}" is therefore:

$$ft_{net} = \frac{P_{strap}'}{A_{net}} = \frac{731}{0.0295} = 24{,}780\ psi$$

Margin of safety

For the MS calculation, the net tension stress at the critical location is compared to F_{tu} from *Table 11-10*. Thus the net tension MS is calculated as:

$$MS = \frac{F_{tu}}{ft_{net}} - 1 = \frac{60{,}000}{24{,}780} - 1 = >1.0$$

11.6 GUSSETS

The two Gussets are fastened to the Frame upper flange using *MS20470AD4* solid rivets. They are used to increase the stiffness of the Frame in the lateral direction and also, to transfer efficiently the torque from the Vertical Post to the Frame. They are made out of *Al Aly 2024-T42 Clad* sheet 0.050" thick. See *Figure 11-44*.

At section *B-B*, from *Table 11-17*, the stress at point ***6*** corresponds to the maximum tension stress in the Gusset's free flange and is 7,933 *psi* while the stress at point ***1*** corresponds to the maximum compression stress in the Gusset and is −5,691 *psi*.

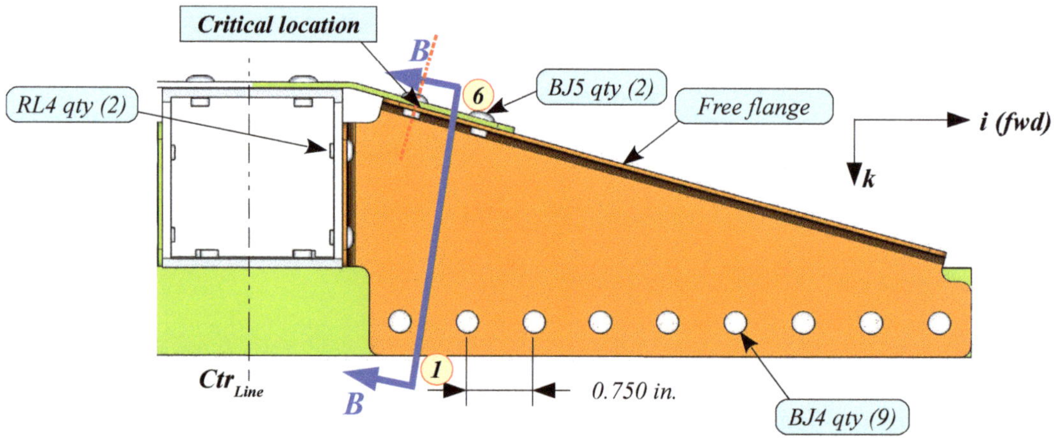

Figure 11-44 *– Fwd Gusset installation (symmetrical fwd-aft)*

11.6.1 NET TENSION

The free flange is checked for net tension at the connection with the Strap. The original free flange area is considered as; $A = h * t = 0.625 * 0.050 = 0.03125\ in^2$. Since the rivets at the critical location are 5/32" in diameter, the net tension area; A_{net} is:

$$A_{net} = (h - D) * t = (0.625 - 0.156) * 0.050 = 0.02345\, in.^2$$

The net tension stress "ft_{net}" is therefore:

$$ft_{net} = \frac{A}{A_{net}} f_1 = \frac{0.03125}{0.02345} 7{,}933 = 10{,}572\, psi$$

Margin of safety

For the MS calculation, the net tension stress at the critical location is compared to F_{tu} from *Table 11-11*. Thus the net tension MS is calculated as:

$$MS = \frac{F_{tu}}{ft_{net}} - 1 = \frac{55{,}000}{10{,}572} - 1 = >1.0$$

11.6.2 INTER-RIVET BUCKLING

The web of the Gusset is fastened to the Frame using BJ4 rivets and is considered working with the Frame upper cap as shown in the cross section **B-B** in *Figure 11-40*. However, since the edge at point ***1*** is in compression and that the web is thin (unstable) compared to the Frame upper cap, one must ensure that the web is not subjected to inter-rivet buckling. The pitch "***p***" of the rivet is 6D (or maximum) thus; $p = 6\ (0.125) = 0.75\ in$. The stress at the edge (point ***1***) is conservatively assumed constant across the web. The inter-rivet buckling allowable is calculated as:

$$F_{ir} = \frac{\pi^2 E_t}{\left(\frac{p}{\sqrt{C}}\frac{\sqrt{12}}{t}\right)^2}$$

In the above equation, the value for the thickness "***t***" has to be reduced since we're using some Clad material. Using the cladding reduction factor "***λ***" of 0.9 for sheets having a thickness < 0.064" (see ref. 20.), the effective thickness is thus calculated as; $t' = \lambda t = 0.9 * 0.050 = 0.045\ in.$

> From what I've seen, the equation from (ref. 2.) for the inter-rivet buckling analysis is slightly more conservative than others. The method for evaluating the effective thickness of Clad material is explained in (ref. 20) and is a bit cumbersome to use. I prefer to simply ignore the clad material so for a 10% total clad thickness, I only consider the remaining 90%.

The end fixity coefficient "***C***" for *Brazier* (protruding) type fasteners is ***3*** and the tangent elastic modulus "E_t" is found by iteration using a *Goal Seek* tool. A plasticity correction factor for compression "η_c" is used to multiply the Young's modulus "***E***" and is calculated as follows:

$$\eta_c = \frac{E_t}{E} = \frac{1}{1+\left(\frac{0.002\ E\ n}{F_{cy}}\right)\left(\frac{F_{c,cr}}{F_{cy}}\right)^{n-1}}$$

Therefore, using the same procedure as described in *section 11.2.7*, the critical inter-rivet buckling stress is calculated by iteration by first assuming a value for $F_{c\,cr}'$ (F_{ir}' in this case).

A first value of 30,000 *psi* was assumed for ***Fir'*** in the above equation, and from the iterative process, the value for "η_c" of 0.37414 was obtained leading to to a value for E_t of 0.37414 * 10,000,000 *psi* = $3.74*10^6$ *psi*. The final inter-rivet buckling allowable is thus;

> For
> $F_{cy} = 37{,}000\ psi$
> $n = 20$
> $E = 10.0*10^6\ psi$
> $p = 0.75\ in$
> $t = t' = 0.045\ in$
> $E_t = 3.74*10^6\ psi$

$$F_{ir} = \frac{\pi^2 3.74\times10^6}{\left(\frac{0.75}{\sqrt{3}}\frac{\sqrt{12}}{(0.045)}\right)^2} = 33{,}539\ psi$$

Margin of safety

With the stress at point ***1*** of -5,691 *psi*, the MS in inter-rivet buckling is calculated as:

$$MS = \frac{F_{ir}}{|f_{c1}|} - 1 = \frac{33{,}539}{5{,}691} - 1 = >1.0$$

11.6.3 SHEAR BUCKLING OF WEB

The web of the Gusset is checked to ensure it doesn't buckle under the shear load. *Figure 11-45* shows the Gusset web to be analyzed.

From *Figure 11-36*, the maximum internal shear load "*V*" is 180 *lbs*. The shear flow "*q*" is calculated as:

$$q = \frac{V}{h} = \frac{180}{2.50} = 72\ lb/in.$$

Since the shear stress "f_S" is q/t then $f_S = 72 / 0.050 = 1{,}440\ psi$.

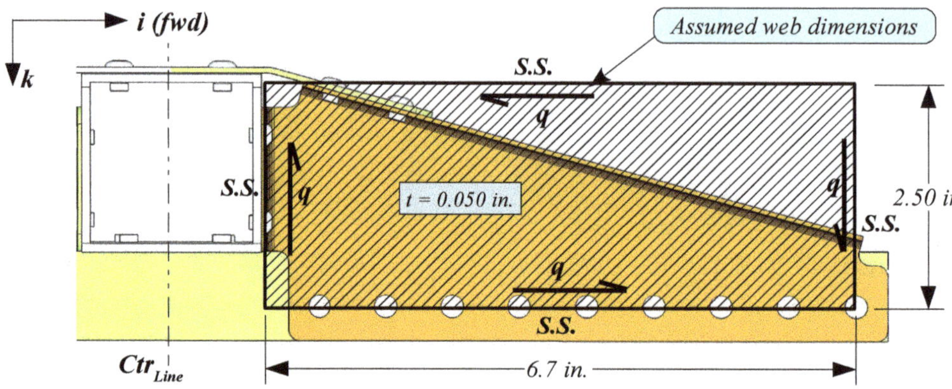

Figure 11-45 *– Web shear buckling.*

> The shear flow "*q*" is calculated here since it will be required later for the fasteners analysis. The web was conservatively assumed rectangular with its height constant at 2.50". However, to be more realistic, I could have used the average height of both ends as the height of the rectangle.

Using (ref. 20.) for flat plate buckling in shear, the shear stress allowable is calculated as:

$$F_{S\,cr} = \frac{k_S\,\eta_S\,\pi^2\,E}{12\left(1-\mu^2\right)}\left(\frac{t}{b}\right)^2$$

From *Figure 11-45*, the dimension "***a***" is 6.7 *in.* and the dimension "***b***" is 2.5 *in.* Note that for shear buckling, the value for "***b***" is always the smallest of the two dimensions. The web is considered simply-supported on all four sides.

Thus for ***a/b*** = 6.7 / 2.5 = 2.7, hence from *Figure A-2*, the value for k_S reads: 5.8.

From the above equation:

$$\frac{F_{s\,cr}}{\eta_s} = \frac{k_s\,\pi^2\,E}{12\left(1-\mu^2\right)}\left(\frac{t}{b}\right)^2$$

$$\frac{F_{s\,cr}}{\eta_s} = \frac{(5.8)\,\pi^2\,10*10^6}{12\left(1-0.33^2\right)}\left(\frac{0.045}{2.5}\right)^2 = 17{,}344\ psi$$

> *For*
> $k_s = 5.8$
> $b = 2.5\ in$
> $t = 0.045\ in$
> $\mu = 0.33$
> $E = 10.0*10^6\ psi$

The above value (assuming the critical shear stress to be in the elastic portion of the stress-strain curve; "η_S" = *1)* happens to be pretty close to the yield shear stress allowable calculated below. Therefore, the real critical shear stress will be calculated by the iterative process using the same procedure as described in section *11.2.7*. Once a tool is done for that procedure in a spreadsheet then it doesn't require too much effort to obtain the more accurate buckling allowable.

For this analysis, we need the yield shear stress allowable "F_{sy}". This value is almost never provided in the references and, therefore, can be calculated using the *distortion-energy (Von Mises)* theory as shown below:

$$F_{SY} = \frac{1}{\sqrt{3}} F_{TY} = 0.577\, F_{TY}$$

Thus for $F_{TY} = 35$ *ksi*;

$$F_{SY} = 0.577 \times 35 = 20.2\ ksi$$

Different approaches are available to evaluate the yield shear stress allowable; F_{SY} for any material. In our case, it's from fictitious material properties of *Al Aly 2024-T42 Clad*.

1) From the maximum shear theory (*Tresca*); F_{SY} = 0.5 F_{TY} = 0.50 * 35 *ksi* = 17.5 *ksi*, *2)* Ref. 10. *Eq. 11.2.8* gives F_{SY} = ***0.55*** F_{CY}. This corresponds to F_{SY} = 0.55* 37 *ksi* = 20.35 *ksi*. *3)* From the distortion-energy theory (*Von Mises)*; $F_{SY} = 1/\sqrt{3}\, F_{TY}$; $F_{SY} = 0.577\, F_{TY} = 0.577 * 35 =$ 20.2 *ksi*.

The (ref. 8.) also offers another equation for calculating F_{SY} using the values of F_{TY}, F_{CY}, F_{SU} and F_{TU} according to the grain directions of the material. Therefore, it's a matter of what approach is preferred at your stress office for the material in use.

Please note that fictitious values are used and will also lead to a fictitious value for F_{SY}. So it could be possible that, using real allowable values, the value for F_{SY} calculated using the different equations could be more or less conservative. Comparison between methods is always a good thing!

Therefore, using the same procedure as described in section *11.2.7*, the critical shear buckling stress is calculated by iteration by first assuming a value for $F_{S\,cr}'$. A plasticity correction factor for shear "η_S" is used to multiply the Young's modulus "***E***" and is calculated as follows:

$$\eta_S = \frac{1}{1+\left(\frac{0.00346\, G}{F_{sy}}\right)\left(\frac{F_{scr}}{F_{sy}}\right)^{n-1}}$$

This equation is solved by iteration until $F_{s\,cr}'$ is equal to the value obtained from the equation:

$$F_{S\,cr} = \frac{k_S\, \eta_S\, \pi^2 E}{12\left(1-\mu^2\right)}\left(\frac{t}{b}\right)^2$$

The same reduced thickness (t = 0.045") due to the Clad material as in *section 11.6.2* is used. A first value of 20,000 *psi* was assumed for F_{scr}, in the above equation, and from the iterative process, the value for "η_s" of 0.97745 was obtained. The final inter-rivet buckling allowable is thus;

$$F_{s\,cr} = \frac{5.8(0.97745)\,\pi^2\,10.0*10^6}{12\,(1-0.33^2)}\left(\frac{0.045}{2.5}\right)^2 = 16{,}953\ psi$$

For
$k_s = 5.8$
$F_{sy} = 20{,}200\ psi$
$n = 20$
$E = 10.0*10^6\ psi$
$t = t' = 0.045\ in$
$b = 2.5\ in.$

Margin of safety

Thus the MS in shear buckling is calculated as:

$$MS = \frac{F_{s\,cr}}{f_s} - 1 = \frac{16{,}953}{1{,}440} - 1 = >1.0$$

Using *Tresca* to determine F_{sy}, the value for the critical shear buckling stress F_{scr} would be 15,754 *psi*, or about 1.2 *ksi* lower than the value using the *Von Mises* criteria. However, since the *MS* is high, it wouldn't have any impact on it.

11.7 FRAME

The Frame is made out of *Al Aly 7050-T7451* machined Plate 1.0 *in.* thick. Since the Gussets are fastened to the machined Frame and are considered as working together (the Gussets don't buckle), the stresses in the Frame were calculated in section *11.4.5* and are used in the following analyses.

11.7.1 FREE FLANGE BUCKLING

Using the stresses at section ***A-A*** from *Table 11-14*, a free-flange buckling analysis is performed since the Frame upper flange has a fairly high compression stress at the free flange (point ***1***). Therefore, considering the flange to start from the free edge (point ***1***) and stop at the corner (point ***2***), as shown on *Figure 11-46*, the variation of the stress goes from a relatively high compression stress at the free edge to a smaller compression stress at the corner.

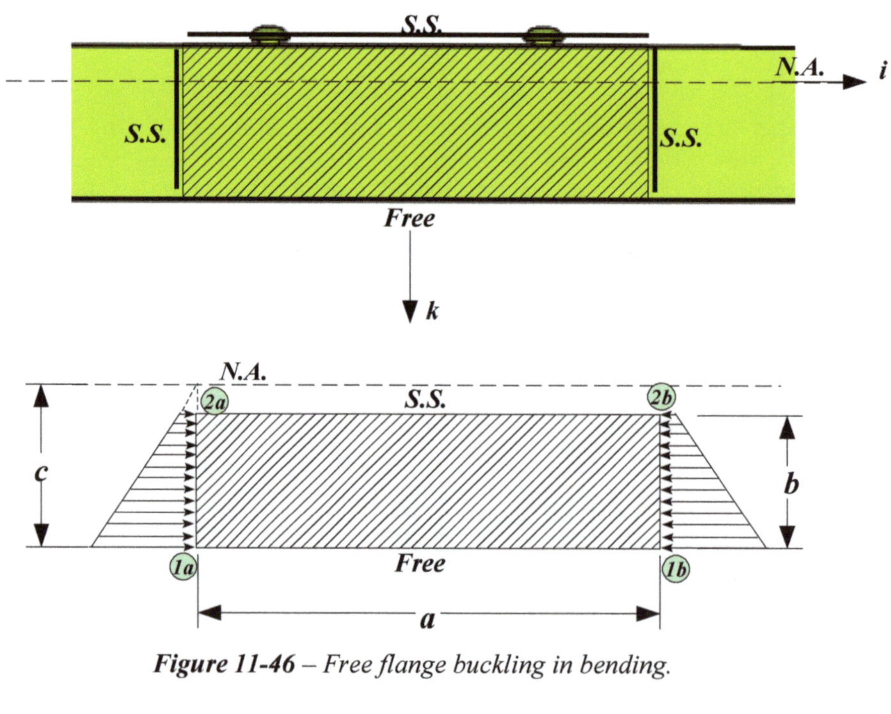

Figure 11-46 – *Free flange buckling in bending.*

For conservatism, the flange is considered simply supported at all sides except at the free edge. Using *Figure B-1* for flat plate buckling in bending, the bending stress allowable is calculated as:

$$F_{b\,cr} = \frac{k_b\,\eta_c\,\pi^2\,E}{12\left(1-\mu^2\right)}\left(\frac{t}{b}\right)^2$$

From *Figure 11-46,* the length "*a*" is arbitrarily shown since it is not required in this analysis. Since the stress varies along the Frame, the bending stress at point ***1a*** is different (higher) than the one at point ***1b***. The same for ***2a*** and ***2b***. The maximum (at points ***1a*** and ***2a***) are used for conservatism.

From *Table 11-14* we know that the total stress at point ***1*** is -10,104 *psi* and that it is −1,224 *psi* at point ***2*** located 1.0 *in.* away. For this analysis, we are interested to know the location where the stress is zero. That is *"c"* on *Figure 11-45*. By doing a linear extrapolation from points *1 & 2* we get *c* = 1.138 *in*.

In this case, $b = 1.0$ *in*. and $c = 1.138$ *in*. Thus $b/c = 1.0/\ 1.138 = 0.88$. Hence from *Figure B-1*, the value for k_b read is: 0.52. The Young's modulus from *Table 11-9* is used.

From the above equation:

$$\frac{F_{b\,cr}}{\eta_c} = \frac{k_b\,\pi^2\,E}{12\left(1-\mu^2\right)}\left(\frac{t}{b}\right)^2$$

$$\frac{F_{b\,cr}}{\eta_c} = \frac{(0.52)\,\pi^2\,10*10^6}{12\left(1-0.33^2\right)}\left(\frac{0.100}{1.0}\right)^2 = 47{,}995\ psi$$

For
$k_b = 0.52$
$b = 1.00$ in
$t = 0.100$ in
$\mu = 0.33$
$E = 10.0*10^6$ psi

This value is still not too far from F_{cy} (60 *ksi*). Thus, I'll use the iteration process to determine the actual critical stress in free flange buckling.

Based on (ref. 20.), a plasticity correction factor for compression "η_c" is used to multiply the Young's modulus "***E***" and is calculated as follows:

$$\eta_c = \frac{E_t}{E} = \frac{1}{1+\left(\frac{0.002\ E\ n}{F_{cy}}\right)\left(\frac{F_{c,cr}}{F_{cy}}\right)^{n-1}}$$

Therefore, using the same procedure as described in section *11.2.7*, the critical buckling bending stress is calculated by iteration by first assuming a value for $F_{c\,cr}'$.

A first value of 30,000 *psi* was assumed for $F_{c\,cr}'$ in the above equation, and from the iterative process, the value for "η_c"of 0.91894 was obtained. The final critical bending buckling stress allowable is thus;

$$F_{b\,cr} = \frac{0.52\,(0.91894)\,\pi^2\,10.0*10^6}{12\left(1-0.33^2\right)}\left(\frac{0.100}{1.0}\right)^2 = 47{,}785\ psi$$

For
$F_{cy} = 60{,}000$ psi
$n = 20$
$E = 10.0*10^6$ psi
$t = 0.10$ in
$b = 1.0$ in

Margin of safety

Thus the MS in free flange buckling is calculated as:

$$MS = \frac{F_{b\,cr}}{f_{b1}} - 1 = \frac{47{,}785}{|-10{,}104|} - 1 = \ >1.0$$

11.7.2 CRIPPLING

Since the corner of the Frame upper cap (point ***2***) is in compression, a crippling check is performed. However, this check is somewhat redundant since that, as shown is section *11.7.1*, the flange doesn't buckle. The flange buckling would be the first step to happen before the corner cripples therefore no crippling is expected.

But for the purpose of the demonstration, and since it is common to systematically perform a crippling check as soon as there is compression in a corner, the crippling stress allowable "F_{CC}" will be calculated below. This value is also required when the section analyzed is in compression and unsupported over a fairly long distance, as in the case of a column, where a *Johnson-Euler* analysis is necessary.

The crippling analysis is performed by considering the effective web width as *15t* minus half the flange thickness. The edge of the flange is considered free while the edge of the web is assumed simply supported (no free edge). See *Figure 11-47* for the effective segments used in the analysis.

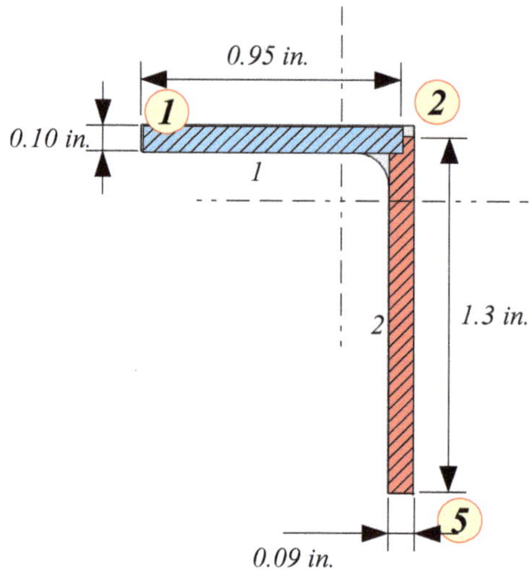

Figure 11-47 – *Effective segments for crippling at corner 2.*

> The same web width (15t) is used as assumed in the Frame upper cap section. Note that the Gusset web is NOT considered in the crippling analysis since it relies on the stability of the Frame upper cap.
>
> The purpose of the crippling analysis is to determine the loading capacity of a corner made of relatively thin material. The corner is the last line of defense before the structure fails in compression. Any web or skin attached to that corner will most likely buckle hence dumping their load into the corner.

The crippling allowable for each segment is calculated as shown below. Note that the crippling allowable "F_{CC}" of a segment can never be higher than F_{CY}. So the maximum value for F_{CCn} / F_{cy} is 1.0. The value at the abscissa of *Figure 11-48* is calculated as:

$$abscissa = \sqrt{F_{CY}/E}\,(b/t)$$

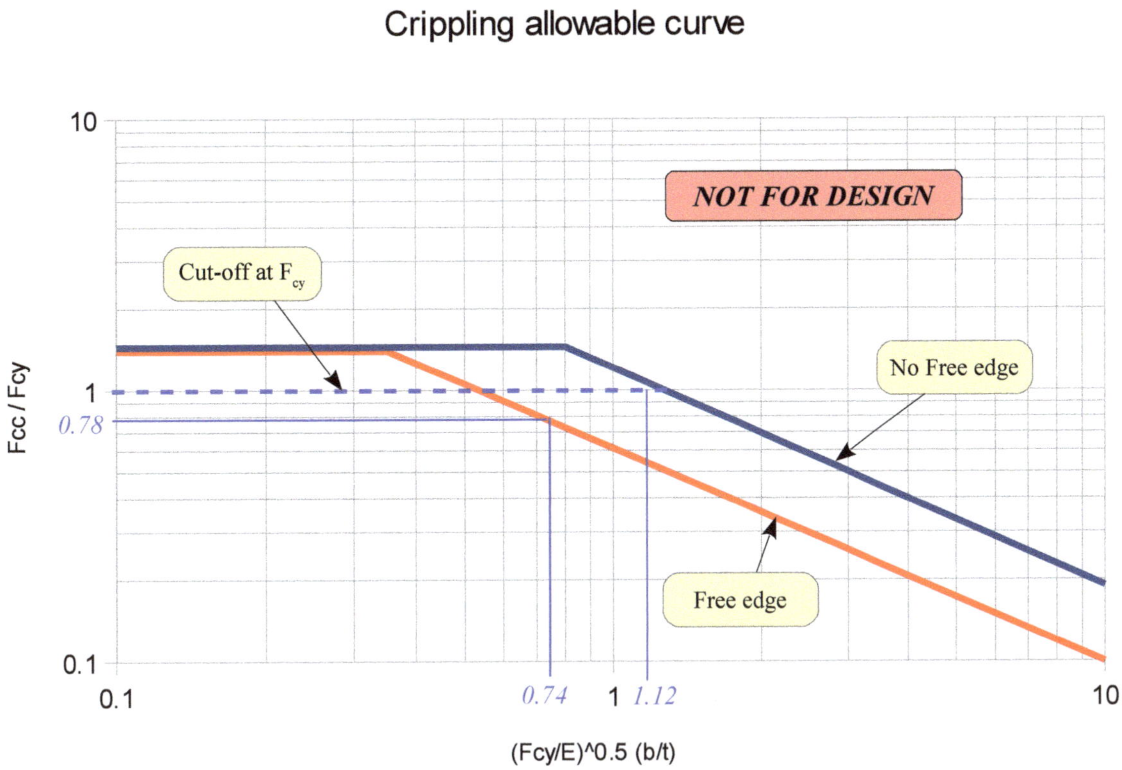

Figure 11-48 *– Typical Crippling curves.*

So for $E = 10.0{*}10^6\ psi$, $F_{cy} = 60{,}000\ psi$, the crippling allowable of each segment is calculated in *Table 11-18.*

Table 11-18 *- Crippling analysis at corner 2.*

	Free edge	b_n	t_n	$b_n t_n$	b_n / t_n	$(F_{cy} / E)^{0.5}$	$(F_{cy} / E)^{0.5}(b_n/t_n)$	F_{CCn} / F_{cy}	F_{CCn}	$t_n b_n F_{CCn}$
Segment		*[in]*	*[in]*	*[in²]*					*[ksi]*	*[lbs]*
1	1	0.95	0.1	0.095	9.500	0.0775	0.7359	0.78	46.8	4446
2	0	1.3	0.09	0.117	14.444	0.0775	1.1189	1.00	60	7020
			$\sum b_n t_n =$	0.2120					$\sum b_n t_n F_{ccn} =$	11466

thus:

$$F_{CC} = \frac{\sum b_n t_n F_{CCn}}{\sum b_n t_n} = \frac{11{,}466}{0.212} = 54{,}085\ psi$$

From *Table 11-14,* the stresses at points *1, 2* and *5* are −10,104, −1,224 and −667 *psi* respectively. An average stress could be calculated from those three points and compared to F_{CC} but, to be conservative, I will simply use the maximum compression stress at point *1.*

Margin of safety

Thus the MS is calculated as:

$$MS = \frac{F_{CC}}{f_1} - 1 = \frac{54{,}085}{|-10{,}104|} - 1 = >1.0$$

11.7.3 SHEAR BUCKLING OF WEBS

The Frame web is checked for shear buckling. As shown on *Figure 11-49*, there are three panels in the web, two identical large ones *(1 & 3)* and one small one *(2)*. Their boundaries are determined by the upper and lower flanges and by the vertical ribs and the lines of rivets. It is assumed that there would be shear clips at both ends of the Frame. The panels are assumed simply supported on all sides. Since the web's thickness is constant for all the panels, and since panels *1 & 3* are exactly the same size then only the largest panel having the highest shear flow need to be analyzed. That is panel *3*. See *Figure 11-50* for the dimensions of the critical panel.

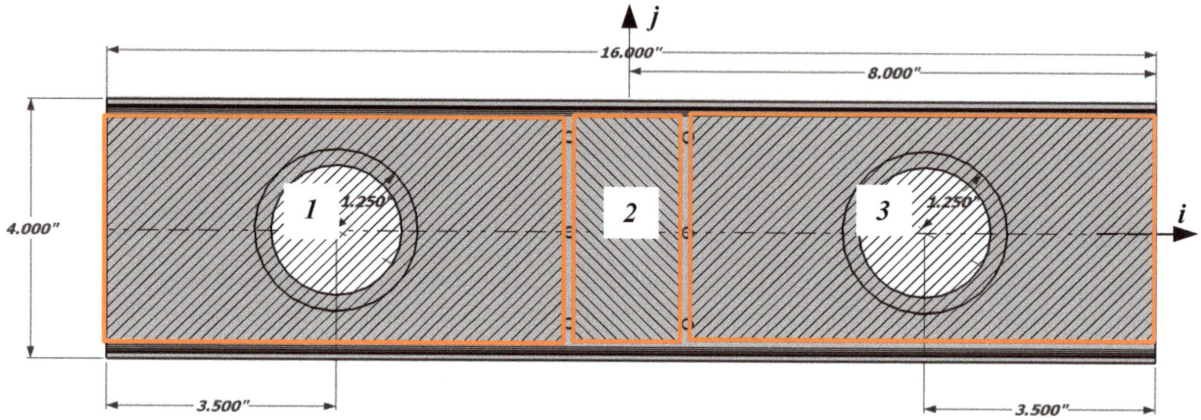

Figure 11-49 *– Frame web panels.*

From *Figure 11-34*, the maximum transverse shear going through the Frame web is V_j = 62 *lbs*. Since the web height is h = 3.9 *in*. and the thickness is t = 0.090 *in*, then the average shear stress is calculated as:

$f_s = V_j/A_{web}$ = 62 / (3.9 * 0.090) = **177 *psi***. The shear flow is; $q = V_j / h$ = 62 / 3.9 = 16 *lb/in*.

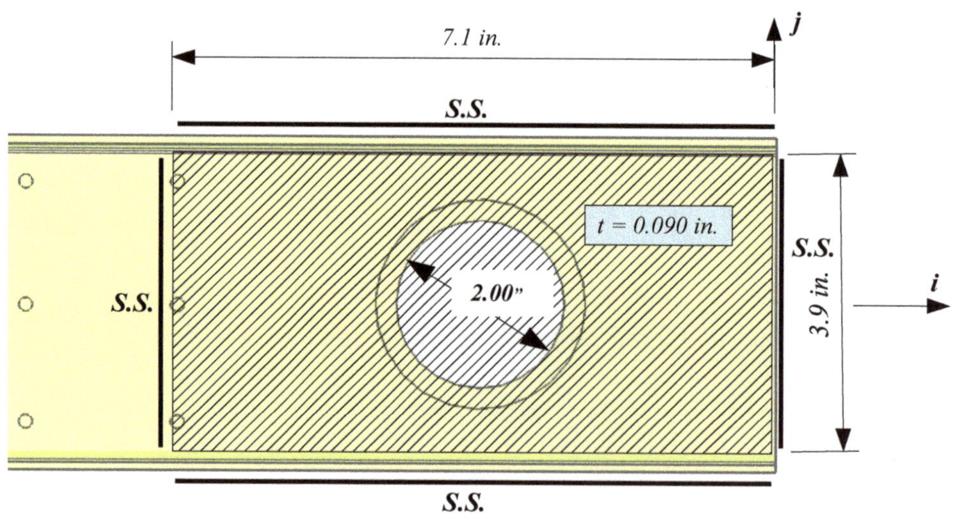

Figure 11-50 *– Web shear buckling.*

This panel is relatively thick and can be considered as a shear buckling resistant. That is such that no buckling is permitted at ultimate load.

11.7.3.1 Panel without a hole

Since there is a lightening hole in the panel, the shear buckling will be calculated in two steps. First, the shear buckling coefficient "k_S" is obtained, then, we get the knock down due to the hole. Using (ref. 20.) for flat plate buckling in shear, the shear stress allowable is calculated as:

$$F_{S\,cr} = \frac{k_S\,\eta_S\,\pi^2\,E}{12\left(1-\mu^2\right)}\left(\frac{t}{b}\right)^2$$

From *Figure 11-50*, the dimension "*a*" is 7.1 *in*. and the dimension "*b*" is 3.9 *in*. Note that for shear buckling, the value for "*b*" is always the smallest of the two dimensions. Thus $a/b = 7.1 / 3.9 = 1.8$. Hence from *Figure A-2*,, the value for k_S read is: 6.7. The primary Young's modulus in tension "*E*" is used from *Table 11-9*. From the above equation:

For
$k_s = 6.7$
$b = 3.9$ in
$a/b = 1.8$
$t = 0.090$ in
$\mu = 0.33$
$E = 10.0*10^6$ psi

$$\frac{F_{s\,cr}}{\eta_s} = \frac{k_s\,\pi^2\,E}{12\left(1-\mu^2\right)}\left(\frac{t}{b}\right)^2$$

$$\frac{F_{s\,cr}}{\eta_s} = \frac{(6.7)\,\pi^2\,10*10^6}{12\left(1-0.33^2\right)}\left(\frac{0.090}{3.9}\right)^2 = 32{,}932\ psi$$

11.7.3.2 Panel with hole

For this analysis, the hole is assumed non-reinforced and perfectly centered in a square panel. First the ratio D/b is calculated, that is 2.0 /3.9 = 0.51. Then the knock down ratio k/k_o can be approximated using the following equation.

$$k/k_o = 0.96 - \frac{D}{b} = 0.96 - 0.51 = 0.45$$

So using the shear buckling coefficient obtained previously (k_0 = *6.7*), the value for *k* having a hole is thus 0.45 * 6.7 = 3.01.

For
$D = 2.0$
$k = 3.01$
$b = 3.9$ in
$t = 0.090$ in

Hence recalculating the shear buckling allowable we get:

$$\frac{F_{s\,cr}}{\eta_s} = \frac{(3.01)\,\pi^2\,10.0*10^6}{12\left(1-0.33^2\right)}\left(\frac{0.090}{3.9}\right)^2 = 14{,}795\ psi$$

As expected, this latter shear buckling allowable is much lower than the one calculated without the hole. Since this critical stress is relatively low compared to the yield shear stress allowable, the inelastic correction faction "η_s" is 1.0 in this case.

Margin of safety

Thus the MS in shear buckling is calculated as:

$$MS = \frac{F_{s\,cr}}{f_s} - 1 = \frac{14{,}795}{177} - 1 = >1.0$$

11.7.4 SHEAR STRESS

The transverse shear in the "***k***" direction is calculated at the neutral axis, where the shear stress is maximum, using the formula:

$$f_{s\,max} = \frac{VQ}{I\,t}$$

From *Figure 11-36*, the maximum shear load "*V*" is 180 *lbs*. The first moment of inertia "*Q*" of the upper cap flange hatched area from the j_c axis shown on *Figure 11-51* is calculated as follows:

$$Q = A * \bar{k} = 0.743 * 0.100 * 0.372 = 0.02764\, in.^3$$

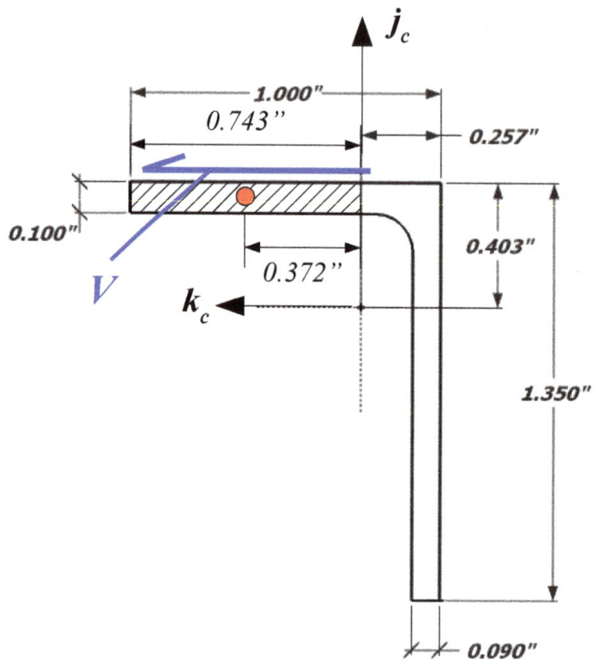

Figure 11-51 – *Upper cap first moment of Inertia dimensions.*

Then

$$f_{s\,max} = \frac{180 * 0.02764}{0.0194 * 0.10} = 2{,}565\ psi$$

For
Area = 0.0743 in^2
V = 180 lbs
Q = 0.02764 in^3
I_j = 0.0194 in^4
t = 0.100 in

To calculate the maximum shear stress as shown above is quite conservative when we're using ultimate loads as the average shear is normally used. In this case the average shear stress would be $fs_{avg} = V/A_f = 180 / (1.00*0.10) = 1{,}800\ psi$.

The reason is that, at ultimate loads, it is understood that some local yielding is to occur but no failure. Thus, as the section yields due to the applied ultimate load, the stress gets more evenly distributed across the section. Therefore the shear stress tends to get closer to the average through the flange.

Note that to calculate the average shear stress, only the flange is used as the effective area to react the shear load as it is in the same direction as the applied load. The web wouldn't be considered as it is out of plane.

11.7.5 COMBINED STRESSES

The bending stress calculated at point *1* from *Table 11-13* is combined with the maximum shear stress calculated above in order to get the strength MS. The following interaction equation is normally accepted when combining an axial (or bending) stress with a shear stress at yield or ultimate.

$$R_b^2 + R_S^2 = 1$$

Thus the MS is calculated as:

$$MS = \frac{1}{\sqrt{R_b^2 + R_S^2}} - 1$$

Where R_b and R_S are respectively the bending and shear stress ratios calculated as follows:

$$R_b = \frac{f_b}{F_{cy}} = \frac{|-10{,}104|}{60{,}000} = 0.168$$

$$R_s = \frac{f_s}{F_{su}} = \frac{2{,}565}{40{,}000} = 0.064$$

F_{cy} is used in this case as the allowable since the maximum stress is in compression. It is a common practice to use F_{cy} whenever an ultimate stress in compression exists in relatively thin parts as this one and most aircraft structures anyway.

The shear stress ratio is quite small and therefore could have been neglected without any impact on the *MS*.

Hence the MS is:

$$MS = \frac{1}{\sqrt{0.168^2 + 0.064^2}} - 1 => 1.0$$

11.7.6 COMMENT

The impact of adding the Gussets and the Strap to the machined Frame stiffened the structure greatly and, as a nice secondary effect, reduced the stresses significantly such that the Margins of Safety are now quite high in general.

11.8 SHEAR CLIPS

The two Shear Clips connect the Horizontal beam to the Vertical Post. They are made out of *Al Aly 2024-T3 Clad* 0.050 *in.* thick.

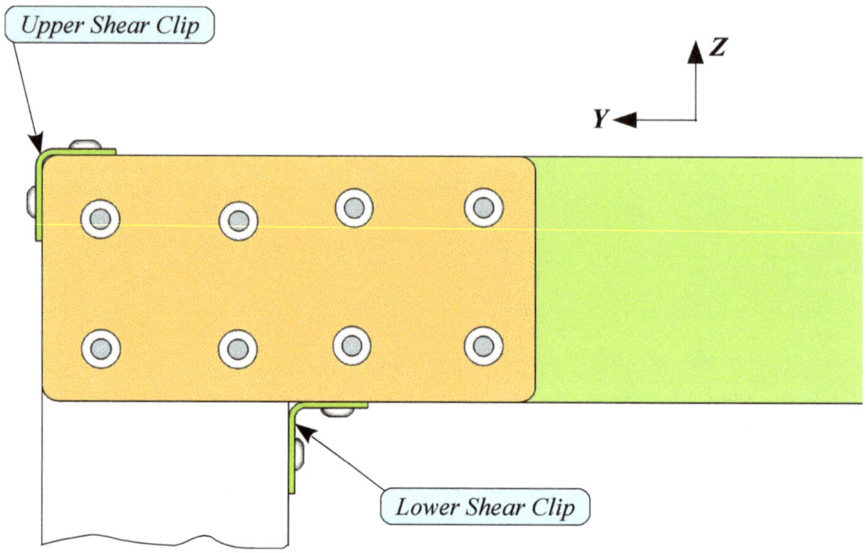

Figure 11-52 *– Shear Clips location.*

According to the assumptions made at the beginning of section *11.2.1*, the Shear Clips only take the load "$\boldsymbol{P_X}$". There are two Shear Clips, as shown in *Figure 11-52*. Conservatively, only the Lower Shear Clip is considered to transfer the $\boldsymbol{P_X}$ load. From page *135*, the load reacted by the two Shear Clips is 17.4 *lbs*. See *Figure 11-53* for the Shear Clip loading.

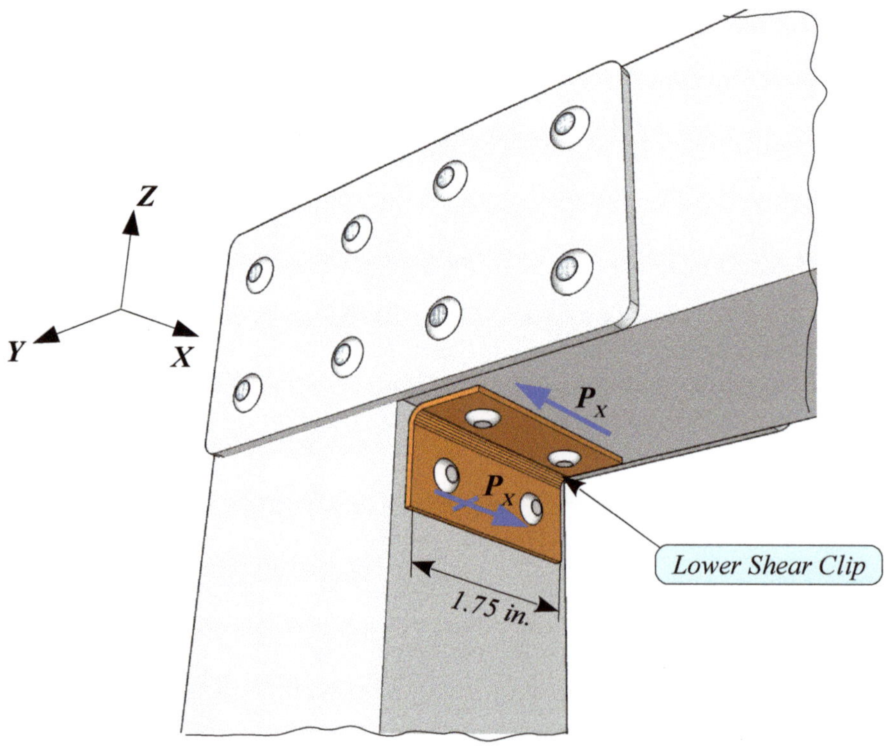

Figure 11-53 *– Lower Shear Clip loading.*

11.8.1 LOADS

The Free Body Diagram of the Shear Clip is shown in *Figure 11-54.* As in all shear clips, the shear load is considered applied at the faying surface of the flange and the shear clip is considered cantilevered off of the two (or more) rivets.

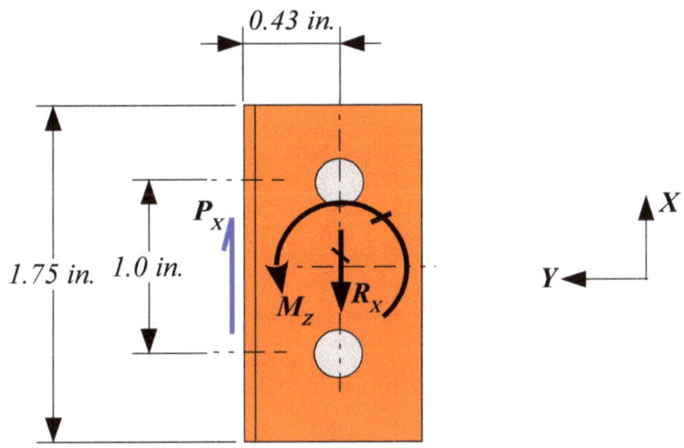

Figure 11-54 *– Shear Clip FBD.*

The shear clip is analyzed at the most critical location. That is through the fasteners where the bending moment is maximum and the cross section minimum. Therefore, for P_X = 17.4 *lbs* then M_Z = 17.4 * 0.43 = 7.5 *in-lb*.

In this example, there are only two rivets. This is a minimum to have equilibrium. If there were three rivets, all three would assume to share the shear load evenly while only the two rivets the furthest apart from each other would be reacting the moment. If four rivets or more are used, a 2D bolt group analysis is necessary.

11.8.2 MATERIAL ALLOWABLES

The Shear Clips are made out of *Al Aly 2024-T3 Clad* sheet, 0.050 *in*. thick. Fictitious material allowables are shown in *Table 11-19.*

Table 11-19 - *Al Aly 2024-T3 Clad sheet 0.050 in. fictitious material allowables.*

Aluminum Alloy 2024-T3 Clad sheet
Fictitious Values, NOT FOR DESIGN

	0.050" [ksi]
F_{tu}	60
F_{ty}	45
F_{cy}	40
F_{su}	35
F_{bru} (e/D = 2.0)	120
F_{bry} (e/D = 2.0)	85
E, 10^3	10
G, 10^3	3.76
μ	0.33
n	20

11.8.3 SECTION PROPERTIES

The cross section dimensions and properties across the fasteners are shown in *Figure 11-55.*

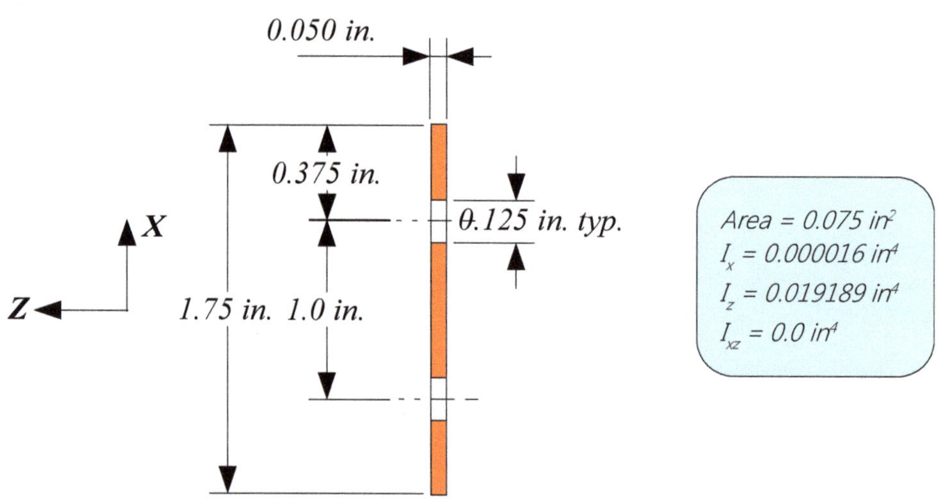

Figure 11-55 – *Shear Clip cross section.*

11.8.4 STRESS ANALYSIS

Bending stress

$$f_b = \frac{M_Z c}{I_{zz}} = \frac{7.5\,(0.875)}{0.019189} = 342\ psi$$

Shear stress

$$f_s = \frac{P_X}{A} = \frac{17.4}{0.075} = 232\ psi$$

Combined stresses

The bending and shear stresses are combined according to the equation:

$$R_b^2 + R_S^2 = 1$$

Thus the MS is calculated as:

$$MS = \frac{1}{\sqrt{R_b^2 + R_S^2}} - 1$$

Where R_b and R_S are respectively the bending and shear stress ratios calculated as follows:

$$R_b = \frac{f_b}{F_{cy}} = \frac{|342|}{40{,}000} = 0.009$$

$$R_s = \frac{f_s}{F_{su}} = \frac{232}{35{,}000} = 0.007$$

F_{cy} is used in this case as the allowable since the maximum stress is in compression. It is a common practice to use F_{cy} whenever an ultimate stress in compression exists in relatively thin parts as this one and most aircraft structures anyway.

Hence the MS is:

$$MS = \frac{1}{\sqrt{0.009^2 + 0.007^2}} - 1 => 1.0$$

Of course, the *MS* is quite high since the applied load is very small. One could have possibly used a thinner material for the shear clips but it is also desirable (for production) to use the same sheet gauge thickness for many parts.

Something to be careful is the distance between the flange and the row of fasteners (0.43 *in.* here). In the case that this distance is large, not only the bending moment at the fasteners will be large but there is a risk of free flange buckling of the clip. This distance should be as small as possible.

11.9 SPLICES

The two Splice Plates are used to transfer all the loads *(except for P_X)* from the Horizontal beam to the Vertical Post. There are made out of *Al Aly 2024-T3 Clad* 0.100 *in.* thick.

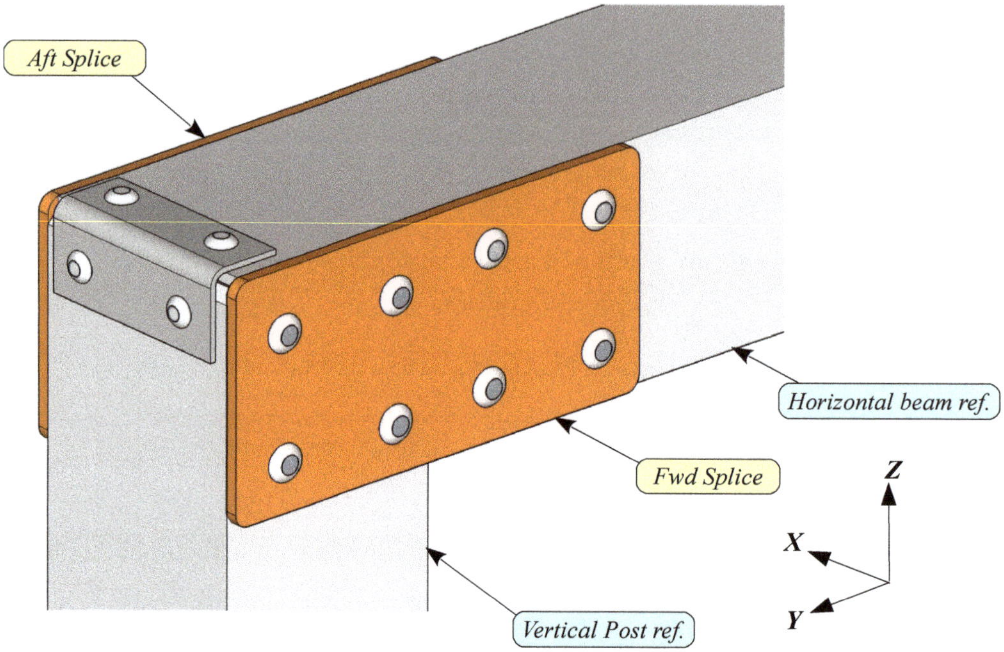

Figure 11-56 *– Splice Plates location.*

*Figure **11-56*** shows the location of the two Splice Plates. Splices are very common in aircraft structures and yet, they are often neglected or "forgotten" in the analysis. It doesn't matter if the longeron or the spar are well analyzed if the splices are done wrong. Again, joints are as important (if not more) and should be as strong or stronger than the connected members.

Joining two tubes is a bit tricky. In our example, the joint is never as strong/stiff as the square tubes. To make it as strong one would have to weld the two members together or use a casting or a machined part that would form an elbow into which both members would be inserted and fastened.

Welds are seldom used in aircraft as is weakens the base material due to the heat affected zones. Unless it is possible to heat-treat the assembly after it is welded, welded aluminum is too weak and therefore avoided. It is also more difficult to repair compared to a fastened part.

11.9.1 LOADS

The internal loads in each of the two Splices are calculated and since the two Splices are identical, only the most critical one is analyzed passing the other one by comparison. The *FBD's* are made using *Figure 11-13* to *Figure 11-15*. It is assumed that the moment "M_j" (or M_X) of 936 *in-lb* is shared equally between the two Splice Plates. It is assumed that the loads are applied and reacted at the bolt group centroids. See *Figure 11-57* for the aft Splice Plate Free Body Diagram and *Figure 11-58* for the fwd Splice Plate Free Body Diagram.

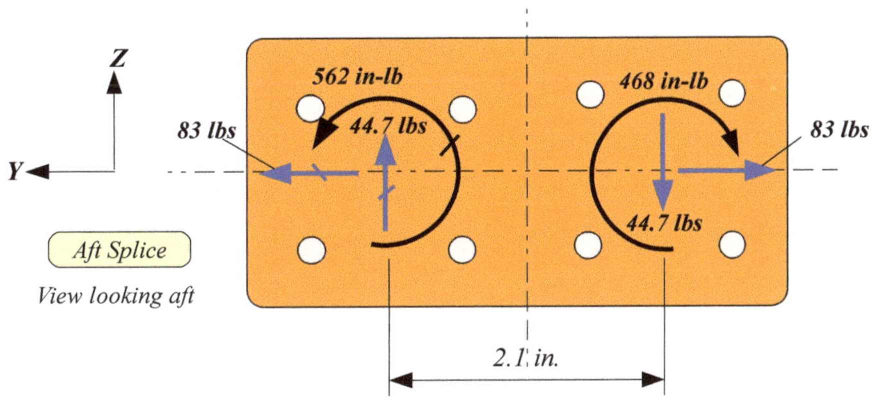

Figure 11-57 *– Aft Splice Plate FBD.*

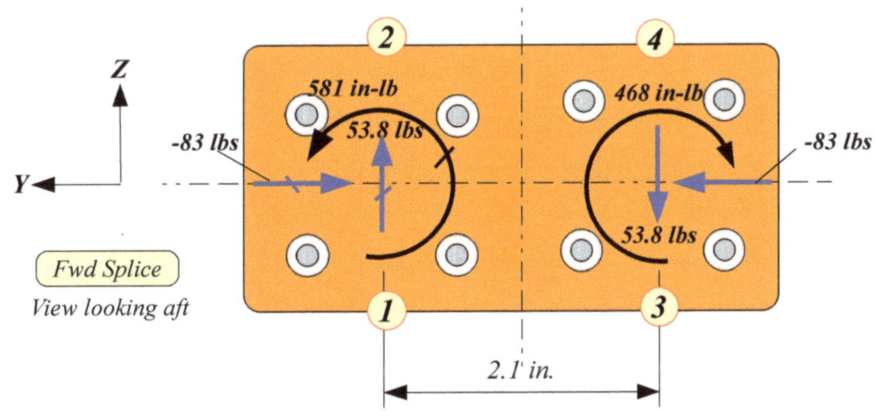

Figure 11-58 *– Fwd Splice Plate FBD.*

The reacting "M_X" moments for the two Splice Plates are calculated as follows:

$$M_{X\,Aft} = P_{Z\,Aft}(l) + M_{Xo} = 44.7(2.1) + \frac{936}{2} = 562\ in.-lb$$

$$M_{X\,Fwd} = P_{Z\,Fwd}(l) + M_{Xo} = 53.8(2.1) + \frac{936}{2} = 581\ in.-lb$$

From the two above figures, one can see that the Fwd Splice Plate is more critical since the shear and the moment are higher and since the axial load is in compression.

11.9.2 MATERIAL ALLOWABLES

The Splice Plates are made out of *Al Aly 2024-T3 Clad* sheet, 0.10 *in*. thick. Fictitious material allowables are shown in *Table 11-20*.

Table 11-20 *- Al Aly 2024-T3 Clad sheet 0.10 in. fictitious material allowables.*

Aluminum Alloy 2024-T3 Clad sheet
Fictitious Values, NOT FOR DESIGN

	0.100" [ksi]
F_{tu}	60
F_{ty}	45
F_{cy}	40
F_{su}	35
F_{bru} $(e/D = 2.0)$	120
F_{bry} $(e/D = 2.0)$	85
$E, 10^3$	10
$G, 10^3$	3.76
μ	0.33
n	20

11.9.3 SECTION PROPERTIES

The cross section dimensions and properties of one Splice Plate across points *1* & *2* are shown in *Figure 11-59*.

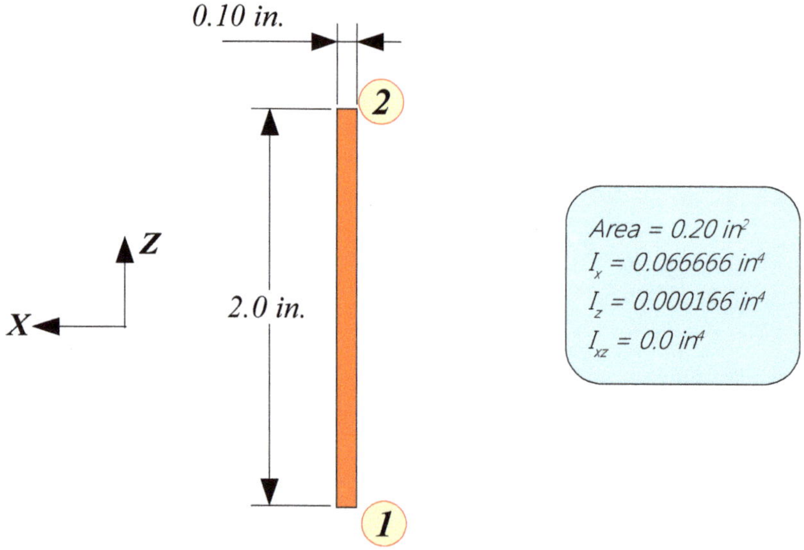

Figure 11-59 *– Splice Plate cross section.*

11.9.4 STRESS ANALYSIS

Axial stress

$$f_a = \frac{P_Y}{A} = \frac{-83}{0.20} = -415\ psi$$

Bending stress

$$f_{b1} = \frac{M_X c_1}{I_{xx}} = \frac{581\,(-1.0)}{0.066666} = -8{,}715\ psi$$

Shear stress

$$f_s = \frac{P_Z}{A} = \frac{53.8}{0.20} = 269\ psi$$

The Splice plate will be checked for free flange buckling due to the compression stress at the free edge.

11.9.5 FREE FLANGE BUCKLING

Using the axial and bending stresses calculated in section *11.9.4,* a free-flange buckling analysis is performed since the Splice's lower edge has a fairly high compression stress at point ***1***. Therefore, considering the unsupported edge to start from point ***1*** and stop at point ***3***, as shown on *Figure 11-60*, the variation of the stress is relatively small and thus the stress will be conservatively assumed constant between the two points.

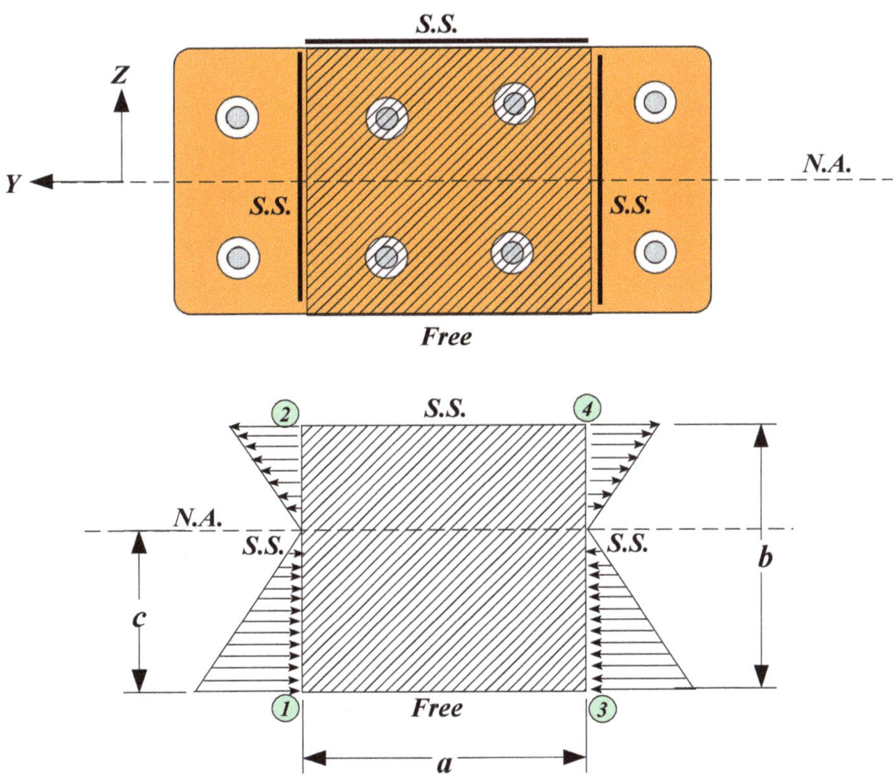

Figure 11-60 – *Free flange buckling in axial + bending.*

For conservatism, the Splice is considered simply supported at all sides except at the free edge. Using (ref. 20.) for flat plate buckling in bending, the bending stress allowable is calculated as:

$$F_{b\,cr} = \frac{k_b\,\eta_c\,\pi^2\,E}{12\left(1-\mu^2\right)}\left(\frac{t}{b}\right)^2$$

From *Figure 11-60*, the length "*a*" is 2.1 *in.* but it is not required in this analysis. Since the stress varies along the Frame, the bending stress at point ***1*** is different (higher) than the one at point ***3***. The same for ***2*** and ***4***. The maximum (at points ***1*** and ***2***) are used for conservatism.

From section ***11.9.4***, the total stress at point ***1*** is the sum of the axial and the bending stresses. At point ***1***, that is; f_{a1} = –8,715 – 415 = –9,130 *psi* and at point ***2***, it is f_{a2} = 8,715 – 415 = 8,300 *psi*. For this analysis, we are interested to know the location where the stress is zero. That is "*c*" on *Figure 11-60*. By doing a linear interpolation between points ***1*** & ***2*** we get c = 1.048 *in*.

In this case, b = 2.0 *in*. and c = 1.048 *in*. Thus b/c = 2.0 / 1.048 = 1.9. Hence from *Figure B-1*, the value for k_b read is: 0.77. The Young's modulus from *Table 11-9* is used.

From the above equation:

$$\frac{F_{b\,cr}}{\eta_c} = \frac{k_b\,\pi^2\,E}{12\left(1-\mu^2\right)}\left(\frac{t}{b}\right)^2$$

$$\frac{F_{b\,cr}}{\eta_c} = \frac{(0.77)\,\pi^2\,10*10^6}{12\left(1-0.33^2\right)}\left(\frac{0.100}{2.0}\right)^2 = 17{,}767\ psi$$

For
k_b = 0.77
b = 2.00 in
t = 0.100 in
μ = 0.33
E= 10.0*10⁶ psi

This value is far from F_{cy} (40 *ksi*) and thus it is well within the elastic (linear) portion of the stress-strain curve. Therefore, one can easily determine that "η_c" is *1* in this case.

Margin of safety

Thus the MS in free flange buckling is calculated as:

$$MS = \frac{F_{b\,cr}}{f_{b1}} - 1 = \frac{17{,}767}{\left|-9{,}130\right|} - 1 = 0.94$$

11.9.6 COMBINED STRESSES

The total longitudinal stress at point ***1*** is the sum of the axial and the bending stresses. This longitudinal stress is then combined with the shear stress according to the equation:

$$R_a^2 + R_S^2 = 1$$

Thus the MS is calculated as:

$$MS = \frac{1}{\sqrt{R_a^2 + R_S^2}} - 1$$

Where R_a and R_S are respectively the longitudinal/axial and shear stress ratios calculated as follows using the stresses from section ***11.9.4***:

For

$$f_{a\,total} = f_a + f_b = -415 - 8{,}715 = -9{,}130\ psi$$

then

$$R_a = \frac{f_{a\ total}}{F_{cy}} = \frac{|-9{,}130|}{40{,}000} = 0.228$$

$$R_s = \frac{f_s}{F_{su}} = \frac{269}{35{,}000} = 0.008$$

F_{cy} is used in this case as the allowable since the maximum stress is in compression.

Hence the MS is:

$$MS = \frac{1}{\sqrt{0.228^2 + 0.008^2}} - 1 = >1.0$$

11.10 CONNECTIONS

The connections include everything that attaches one part to another like rivets, bolts, welds or adhesives. In our example, there are four locations where connections are used, namely points *(A)*, *(B)*, *(C)* and common to the Frame assembly . A variety of fastener types and sizes are used. We will look at each location to calculate a margin of safety for each of them.

See *Figure 11-61* and *Figure 11-62* for details views of the fasteners callouts.

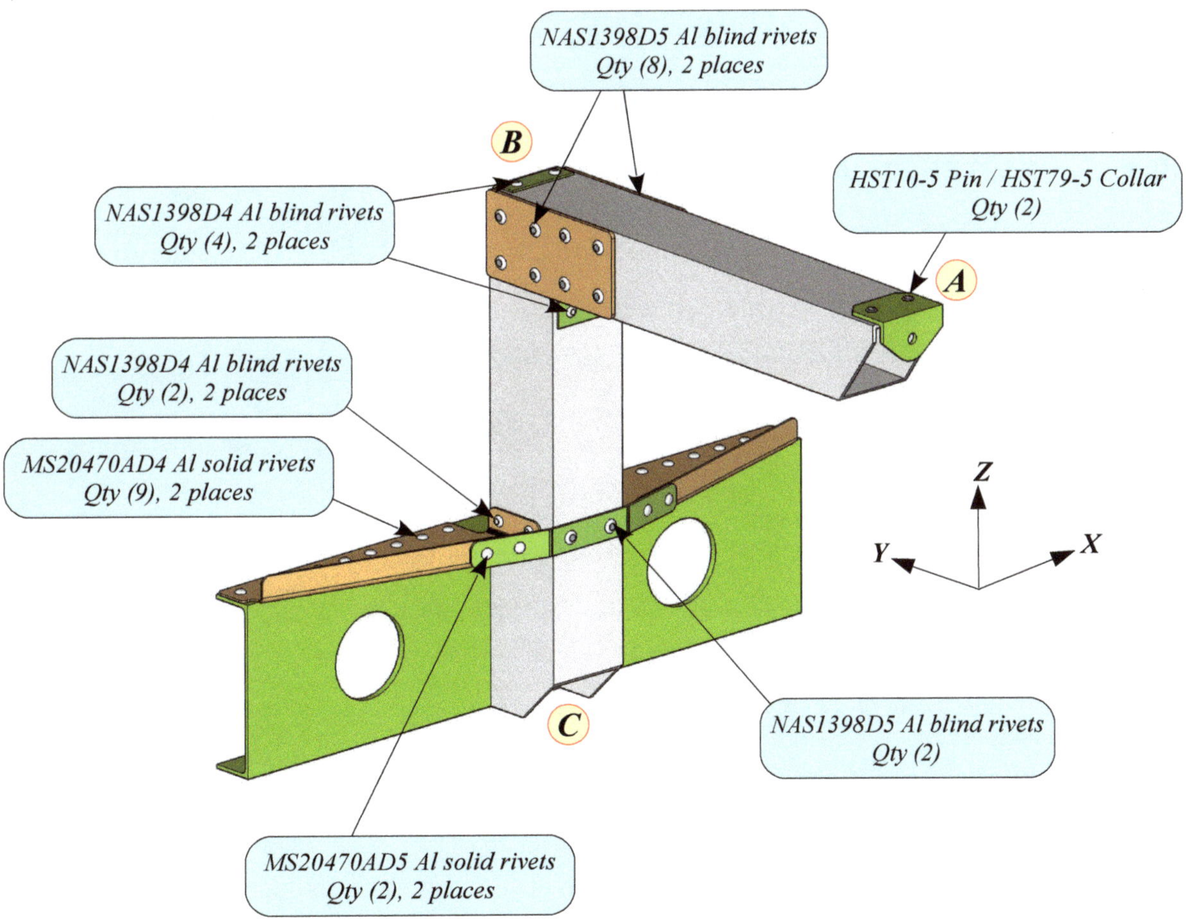

Figure 11-61 *– Fasteners callouts, view looking right.*

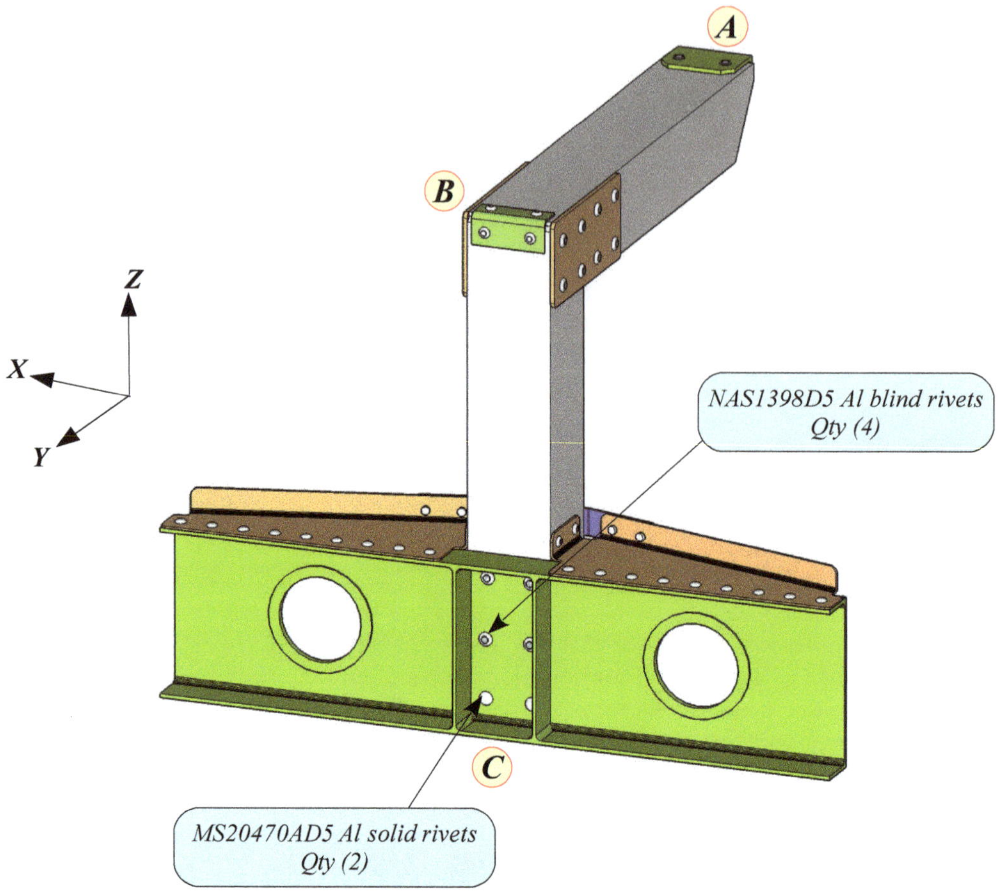

Figure 11-62 *– Fasteners callouts, view looking left.*

11.10.1 FASTENERS ALLOWABLES

Table 11-21 shows the ultimate shear allowables for 1/8" and 5/32" *MS20470AD* solid rivets.

Table 11-21 *- MS20470AD Solid rivets ultimate shear allowables.*

	MS20470AD Al Aly 2117-T3 $F_{su} = 30$ ksi	
	4	5
Dia. [in.]	0.1285	0.159
P_{su} *[lbs]*	**388**	**596**

Table 11-22 shows the ultimate shear allowables for 1/8" and 5/32" *NAS1398D* blind rivets from (ref. 26.) for Cherry Lock A Blind rivet.

Table 11-22 *- NAS1398D Blind rivets ultimate shear allowables.*

	NAS1398D Al Aly 2017-T4 $F_{su} = 38$ ksi	
	4	5
Dia. [in.]	0.130	0.162
P_{su} *[lbs]*	**494**	**755**

Table 11-23 shows the ultimate shear and tension allowables for 5/32" Hi-Lites *HST10-5* Titanium Pins and *HST79-5* Aluminum Collars from (ref. 24.).

Table 11-23 *- HST10, HST11 & HST79 Hi-Lites ultimate shear and tension allowables.*

	P_{tu} [lbs]	P_{su} [lbs]
Pin HST10-5	1,940	2,005
Pin HST11-5	1,650	2,005
Collar HST79-5	1,400	N/A

11.10.2 END A

This is the connection between the Clip and the Horizontal beam. Here, because of the way the fasteners are loaded (in tension), Hi-Lites are installed. The fasteners are *qty (2) HST10-5* Hi-Lite 6Al-4V Titanium Pins and *HST79-5* Hi-Lite Aluminum Collars.

I chose aluminum Collars, even though they are described as to be used in shear applications, since the loads are low. For high tension loads, Titanium or A286 CRES Collars are advisable as the tension allowable of the Collar shall be equal or higher than the tension allowable of the Pin.

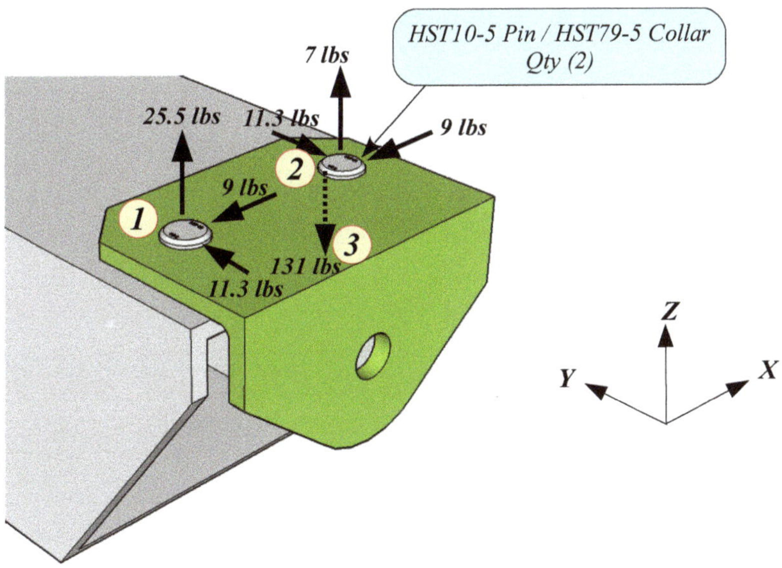

Figure 11-63 *– Fasteners applied loads.*

From section *11.1.1* and section *11.1.4*, the loads applied to the fasteners are summarized in *Figure 11-63*. The tension load is maximum for fastener ***1*** with P_t = 25.5 *lbs* and the shear force is the vector sum of F_X = 9 *lbs* and F_Y = 11.3 *lbs*, so *F* = 14.5 *lbs*. The MS in the Pin is calculated from an interaction equation for shear and tension while the MS for the Collar is simply calculated by comparing the tension load to the tension allowable of the Collar.

At point *3* there is no fastener. It is simply a contact point between the Clip and the Horizontal beam. The location of the point was assumed at the center of the Clip to simplify the calculations. However, because that contact point is out-of-plane of the Horizontal beam web, thus more flexible than at the corners, the actual contact points would more likely be at the ends of the Clip near the beam's corners.

The only analysis that could be done for the contact points would be to evaluate the width of the contacting area by assuming that the contact stress is at F_{cy}. Then, the contact force (R_3 in our case) would be moved to the center of that contact width and the *FBD* would then be recalculated based on the new position of R_3. This is an iterative process and is to be done to make sure that the contact area isn't too large as this would be an indication that this loading assumption is incorrect and something has to be changed.

Pin (HST10-5)

The interaction equation for bolts subjected to shear and tension loads is:

$$R_s^3 + R_t^2 = 1$$

See another example of that method on my blog regarding fasteners:
https://aircraftstress.com/2019/01/19/aircraft-fastener-analysis-part-2/

Where the load ratios are:

$$R_s = \frac{P_s}{P_{su}} = \frac{14.5}{1{,}940} = 0.007$$

$$R_t = \frac{P_t}{P_{tu}} = \frac{26}{2{,}005} = 0.013$$

Then from the interaction equation above and using a Fitting factor of 1.15, the Margin of Safety for the *HST10-5* no. *1* is thus calculated as follows:

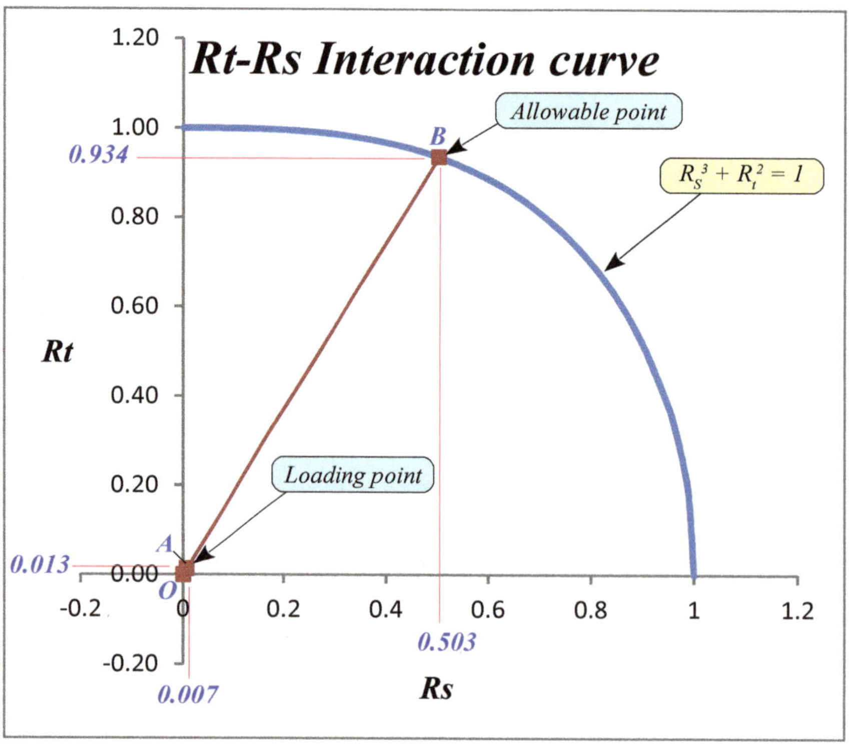

where

$$MS = \frac{\overline{OB}}{FF\,\overline{OA}} - 1 = \frac{R_{S\,All}}{FF\,R_S} - 1 = \frac{0.503}{1.15*0.007} - 1 = >1.0$$

or

$$MS = \frac{R_{t\,All}}{FF\,R_t} - 1 = \frac{0.934}{1.15*0.013} - 1 = >1.0$$

<u>*Collar (HST79-5)*</u>

Then, the tension force at fastener ***1*** of 26 *lbs* is compared to the tension allowable of the Collar. From *Table **11-23***, that is 1,400 *lbs*.

The Margin of Safety for the ***HST79-5*** Collar is thus calculated as follows using a Fitting Factor of 1.15:

$$MS = \frac{P_{tu}}{FF\,P_t} - 1 = \frac{1{,}400}{1.15 \times 26} - 1 = >1.0$$

11.10.3 END B

This is the connection between the Horizontal beam and the Vertical Post. There are two types of joints at that location. One uses shear clips and the other uses splice plates as shown in *Figure **11-64***.

Extracted from *Figure **11-58***, *Figure **11-64*** shows the loads applied to the group of fasteners at end (***B***). As stated earlier, all of the load in the *"X"* direction is assumed taken by the lower Shear Clip. Hence the upper Shear Clip sees no load at all and is thus passed by comparison to the lower Shear Clip.

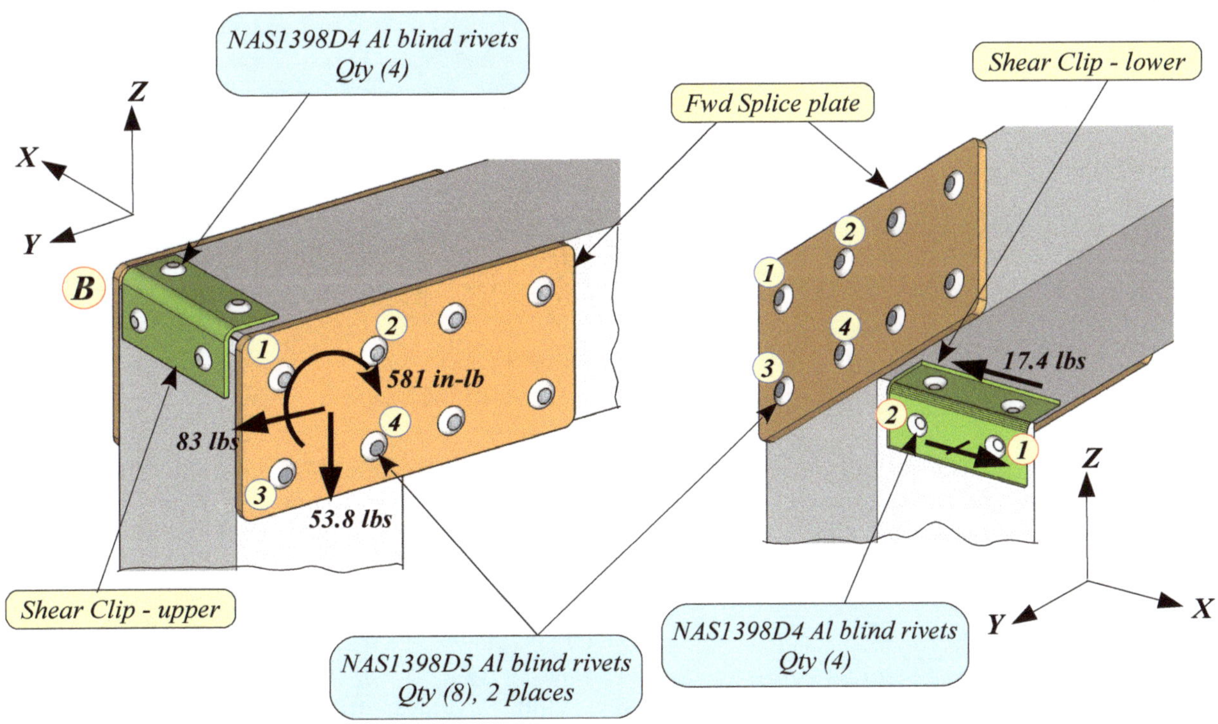

Figure 11-64 *– Close-up at rivets from point B with Applied loads.*

11.10.3.1 Splice Plate rivets

The rivet pattern on either sides of the joint is square and all four fasteners are the same type and same size (*NAS1398D5*). Therefore the bolt group centroid is located exactly at the geometric center and all the fasteners are at the same distance *"r"* from the centroid. *Figure 11-65* shows the applied force vectors at each rivet caused by the "M_X" moment (581 *in-lb*), the "P_Z" (53.8 *lbs*) and the "P_Y" (83 *lbs*) forces. The forces are calculated as follows:

Due to the "M_X" moment:

$$F_M = \frac{M}{n\,r} = \frac{581}{4\,(0.765)} = 190\ lbs$$

The orthogonal components of "F_M" are thus (example for rivet no. ***1***, where $\theta = 135$ *deg.*):

$$F_{Y1} = F_M \sin(\theta) = 190\sin(135) = 134\ lbs$$
$$F_{Z1} = F_M \cos(\theta) = 190\cos(135) = -134\ lbs$$

Due to P_Y:

$$F_Y = \frac{P_Y}{n} = \frac{83}{4} = 21\ lbs$$

Due to P_Z:

$$F_Z = \frac{P_Z}{n} = \frac{-54}{4} = -14\ lbs$$

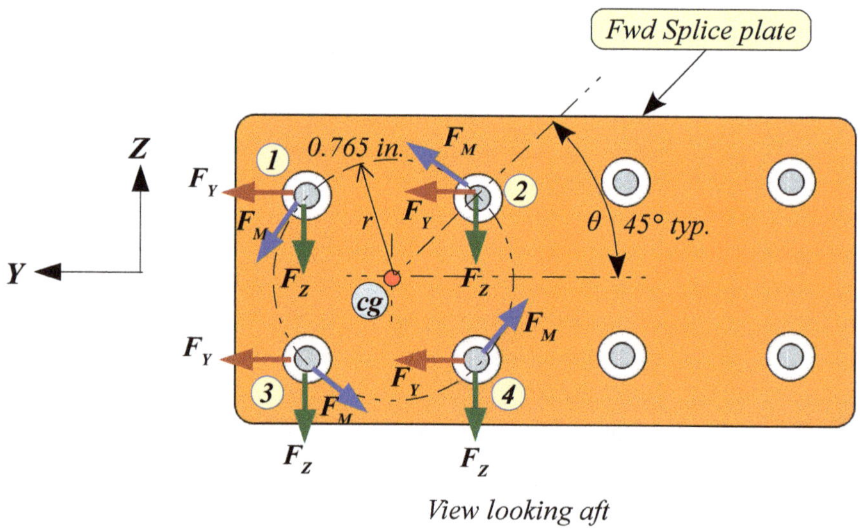

Figure 11-65 *– Applied Force vectors on rivets.*

See *Table 11-24* for the distributed forces in all four rivets due to the "M_X" moment. The resulting total forces in the rivets by combining vectorially all the rivets forces are shown in *Table 11-25*. From *Table 11-25*, rivet (***1***) is the most critical rivet with 214 *lbs* in shear.

Table 11-24 *- Fastener shear forces in Splice Plate due to M_X.*

bolt	y	z	r	F	θ	F_Y	F_Z
	[in]	[in]	[in]	[lbs]	[deg.]	[lbs]	[lbs]
1	0.541	0.541	0.765	190	135	134.2	-134.2
2	-0.541	0.541	0.765	190	45	134.2	134.2
3	0.541	-0.541	0.765	190	225	-134.2	-134.2
4	-0.541	-0.541	0.765	190	315	-134.2	134.2

Table 11-25 *- Total Fastener shear forces in Splice Plate.*

	From P_Y		From P_Z		From M_X		Total		
bolt	F_Y	F_Z	F_Y	F_Z	F_Y	F_Z	F_Y	F_Z	F
	[lbs]	[lbs]	[lbs]	[lbs]	[lbs]	[lbs]	[lbs]	[lbs]	[lbs]
1	21	0	0	-14	134	-134	155	-148	**214**
2	21	0	0	-14	134	134	155	121	196
3	21	0	0	-14	-134	-134	-113	-148	186
4	21	0	0	-14	-134	134	-113	121	166
$\sum =$	**83**	**0**	**0**	**-54**	**0**	**0**	**83**	**-54**	

Rivet shear:

A Fitting Factor "*FF*" of 1.15 (ref. 25. *FAR §25.625*) is used. From *Table 11-22*, the shear allowable "P_{SU}" of an *ACC5 (NAS1398D5)* rivet is 755 *lbs*. No shear factor *"SF"* is required since the Splice Plate and the Vertical Post are relatively thick.

The Margin of Safety for shear is thus calculated as follows:

$$MS = \frac{P_{su}}{FF \times F_1} - 1 = \frac{755}{1.15 \times 214} - 1 => 1.0$$

The Fitting Factor is used here since the Clip can be considered a fitting and that the fastener loads are not determined simply by multiplying a shear flow by a fastener pitch.

The use of a Fitting Factor should be clearly stated in the Design Criteria document for the program you're working on and your boss and the DER should agree with that as well.

Bearing:

This joint is shear critical since we're using relatively weak fasteners and therefore we expect the *MS* in bearing to be even higher than the one calculated above for shear. However, for the purpose of the demonstration (and as is normally done), a bearing *MS* will be calculated.

$$P_{bru} = F_{bru} \times D \times t$$

The bearing ultimate allowable load "P_{bru}" in the Splice Plate made out of *Aluminum 2024-T3 Clad* 0.100 *in*. thick is calculated:

For: F_{bru} = 120 *ksi* (ref. *Table 11-20)*, e/D = 2.0. And for D = 0.162" and t = 0.100" hence:

$$P_{bru} = 120{,}000 \times 0.162 \times 0.100 = 1{,}944 \; lbs$$

The bearing ultimate allowable load "P_{bru}" in the Vertical Post made out of *Aluminum 2024-T3511*, 0.095 *in*. thick is calculated:

For: F_{bru} = 110 *ksi* (ref. *Table 11-6)*, e/D = 2.0. And for D = 0.162" and t = 0.095" hence:

$$P_{bru} = 110{,}000 \times 0.162 \times 0.095 = 1{,}693 \; lbs$$

Margin of Safety

From the above calculations, the ultimate bearing allowable is lower for the Vertical Post with P_{bru} = 1,693 *lbs*. Therefore, the Ult. *MS* in bearing is calculated as follows:

$$MS = \frac{P_{bru}}{F_1} - 1 = \frac{1{,}693}{214} - 1 => 1.0$$

This *MS* also covers the other side of the Splice Joint and the Aft Splice Plate as their loads are lower. They are thus *Passed by Comparison.*

> For both materials involved, the ratio F_{bru} / F_{bry} is smaller than 1.5. Therefore, no bearing check at limit (yield) is necessary.

11.10.3.2 Shear Clip rivets

There are two shear clips that are used to transfer the load in the "X" direction, one upper and one lower as shown in *Figure 11-64.* They are connected to the Horizontal Beam and the Vertical Post with *NAS1398D4* rivets. As stated previously, only the lower one is considered effective to transfer the load "P_X" (17.4 *lbs*). *Figure 11-66* shows the applied forces at the rivets.

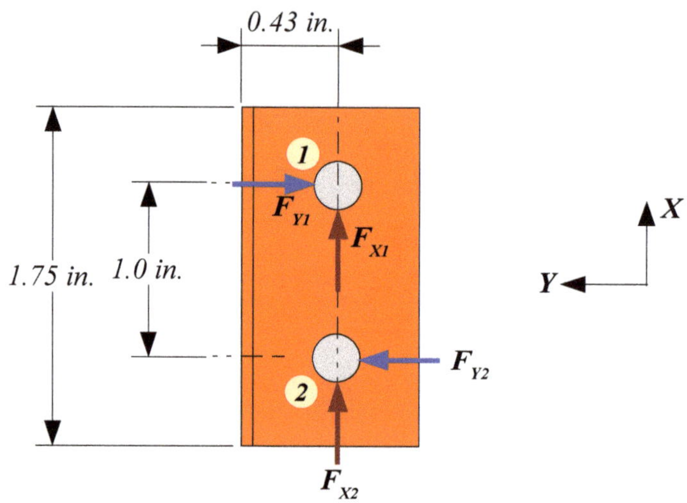

Figure 11-66 *– Applied forces on the Shear Clip rivets.*

From the loads at section *11.8.1*, the forces in the rivets are calculated as follows:

$$F_{X1} = F_{X2} = \frac{P_X}{n} = \frac{17.4}{2} = 9\ lbs$$

$$F_{Y1} = -F_{Y2} = -\frac{M_Z}{pitch} = -\frac{7.5}{1.0} = -7.5\ lbs$$

Combining the vectors, the resulting rivet load "*F*" is calculated as:

$$F = \sqrt{F_{Xi}^2 + F_{Zi}^2} = \sqrt{9^2 + 7.5^2} = 12\ lbs$$

<u>Rivet shear:</u>

A Fitting Factor "*FF*" of 1.15 (ref. 25. *FAR §25.625*) is used. From *Table 11-22*, the shear allowable "P_{SU}" of an ACC4 (NAS1398D4) rivet is 494 *lbs*. No shear factor "*SF*" is required since the Horizontal beam, the Vertical Post and the shear clip are relatively thick. (ref. *Table 10-3)*. The Margin of Safety for shear is thus calculated as follows:

$$MS = \frac{P_{su}}{FF \times F} - 1 = \frac{494}{1.15 \times 12} - 1 = >1.0$$

<u>Bearing:</u>

This joint is shear critical since we're using relatively weak fasteners and therefore we expect the *MS* in bearing to be even higher than the one calculated above for shear. However, for the purpose of the demonstration, a bearing *MS* will be calculated.

$$P_{bru} = F_{bru} \times D \times t$$

The bearing ultimate allowable load "P_{bru}" in the Shear Clip made out of ***Aluminum 2024-T3 Clad*** 0.050 *in*. thick is calculated:

For: F_{bru} = 120 *ksi* (ref. ***Figure 10-32***), e/D = 2.0. And for D = 0.130" and t = 0.050" hence:

$$P_{bru} = 120{,}000 \times 0.130 \times 0.050 = 780\,lbs$$

The bearing ultimate allowable load "P_{bru}" in the Vertical Post and the Horizontal Beam made out of ***Aluminum 2024-T3511***, 0.095 *in*. thick is calculated:

For: F_{bru} = 110 *ksi* (ref. ***Table 11-6***), e/D = 2.0. And for D = 0.130" and t = 0.095" hence:

$$P_{bru} = 110{,}000 \times 0.130 \times 0.095 = 1{,}358\,lbs$$

<u>Margin of Safety</u>

From the above calculations, the ultimate bearing allowable is lower for the Shear Clip with P_{bru} = 780 *lbs*. Therefore, the Ult. *MS* in bearing is calculated as follows:

$$MS = \frac{P_{bru}}{F} - 1 = \frac{780}{12} - 1 = > 1.0$$

For both materials involved, the ratio F_{bru} / F_{bry} is smaller than 1.5. Therefore, no bearing check at limit (yield) is necessary. Although no *FF* is used, some knock-down factor is used sometimes for the bearing *MS*. I have seen 0.80 be used as a coefficient to P_{bru}.

11.10.4 END C

Extracted from *Figure 11-22*, *Figure 11-67* shows the loads applied to the group of fasteners at end (*C*). For this joint analysis, to simplify the analysis in a conservative way, only the rivets located the farthest from the bolt group centroid are considered in taking the "M_Y" moment and the "P_X" load while all six rivets share the "P_Z" load.

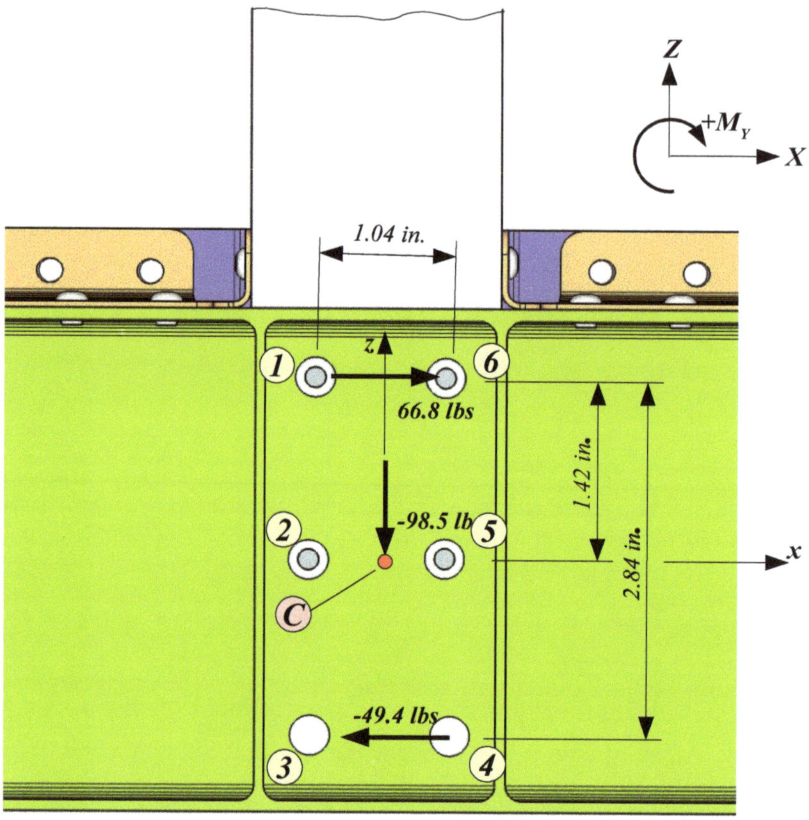

Figure 11-67 *– Close-up at rivets from end (C) with Free Body Diagram.*

Since all the rivets are of the same size and all made out of Aluminum, the bolt group centroid coincides with the geometric center of the six rivets. However, because of the simplified assumption, the bolts' centroid isn't necessary. The force in each rivet "***F***" is simply calculated by dividing the load applied to a group of rivets by the number of rivets "***n***" in that group. In our case, the contribution of that load for each rivet corresponds to:

For all six rivets:

$$F_{Zi} = \frac{P_Z}{n} = -\frac{98.5}{6} = -16.4\ lbs$$

For rivets *1* & *6*:

$$F_{X1,6} = \frac{P_{X1,6}}{n} = \frac{66.8}{2} = 33.4\ lbs$$

For rivets *3* & *4*:

$$F_{X3,4} = \frac{P_{X3,4}}{n} = -\frac{49.4}{2} = -24.7\ lbs$$

By adding the forces vectorially in each rivet we get:

For rivets *1 & 6:*

$$F_1 = F_6 = \sqrt{(F_{X1,6})^2 + (F_{Zi})^2} = \sqrt{(33.4)^2 + (-16.4)^2} = 37.2\ lbs$$

For rivets *2 & 5*:

$$F_2 = F_5 = |F_{Zi}| = 16.4\ lbs$$

For rivets *3 & 4*:

$$F_3 = F_4 = \sqrt{(F_{X3,4})^2 + (F_{Zi})^2} = \sqrt{(-24.7)^2 + (-16.4)^2} = 29.6\ lbs$$

Margins of safety in shear :

Since there are two types of fasteners, two *MS* will be calculated.

Blind rivets *NAS1398D5* (fasteners *1, 2, 5 & 6*):

The highest load in any of those four rivet is from rivets *1 & 6* with 37.2 *lbs*. A Fitting Factor "*FF*" of 1.15 (ref. 25. ***FAR §25.625***) is used however no Shear correction Factor "*SF*" is necessary as both the Vertical Post and the machined Frame are relatively thick.. From *Table **11-22***, the shear allowable "P_{SU}" of an ACC5 (NAS1398D5) rivet is 755 *lbs*.

$$MS = \frac{P_{SU}}{FF \times F_{1,6}} - 1 = \frac{755}{1.15 \times 37.2} - 1 => 1.0.$$

Solid rivets *MS20470AD5* (fasteners *3 & 4*):

The load in rivets *3 & 4* is 29.6 *lbs*. A Fitting Factor "*FF*" of 1.15 (ref. 25. ***FAR §25.625***) is used however no Shear correction Factor "*SF*" is necessary as both the Vertical Post and the machined Frame are relatively thick.. From *Table **11-21***, the shear allowable "P_{SU}" of a BJ5 (MS20470AD5) rivet is 596 *lbs*.

$$MS = \frac{P_{SU}}{FF \times F_{3,4}} - 1 = \frac{596}{1.15 \times 29.6} - 1 => 1.0.$$

Bearing:

This joint is shear critical since we're using relatively weak fasteners and therefore we expect the *MS* in bearing to be even higher than the one calculated above for shear. However, for the purpose of the demonstration (and as is normally done), a bearing *MS* will be calculated.

$$P_{bru} = F_{bru} \times D \times t$$

The ultimate bearing allowable load "P_{bru}" in the machined Frame made out of *Al Aly **7050-T7451*** machined Plate 1.0 *in*. thick is calculated using the smallest rivet diameter between the two types:

For: $F_{bru} = 105$ *ksi* (ref. *Table **11-9***), $e/D = 2.0$. And for $D = 0.159$" and $t = 0.090$" hence:

$$P_{bru} = 105{,}000 \times 0.159 \times 0.090 = 1{,}502\ lbs$$

The ultimate bearing allowable load "P_{bru}" in the Vertical Post made out of *Aluminum **2024-T3511***, 0.095 *in*. thick is calculated using the smallest rivet diameter between the two types:

For: $F_{bru} = 110$ *ksi* (ref. *Table **11-6***), $e/D = 2.0$. And for $D = 0.159$" and $t = 0.095$" hence:

$$P_{bru} = 110{,}000 \times 0.159 \times 0.095 = 1{,}661\ lbs$$

Margin of Safety in bearing

From the above calculations, the ultimate bearing allowable is lower for the machined Frame with P_{bru} = 1,502 *lbs*. The maximum rivet load occurs for rivets *1 & 6* with 37.2 *lbs*. Therefore, the Ult. *MS* in bearing is calculated as follows:

$$MS = \frac{P_{bru}}{F_{1,6}} - 1 = \frac{1{,}502}{37.2} - 1 = >1.0$$

For both materials involved, the ratio F_{bru} / F_{bry} is smaller than 1.5. Therefore, no bearing check at limit (yield) is necessary.

11.10.5 STRAP ATTACHMENTS

The Strap, made out of *Al Aly 2024-T3 Clad* sheet 0.063" thick, is fastened to the Vertical Post with *qty (2) NAS1398D5* and to the two Gussets with *qty (2) MS20470AD5* on either sides. *Figure 11-68* shows the loads in the fasteners common to the Strap are extracted from *Figure 11-42*.

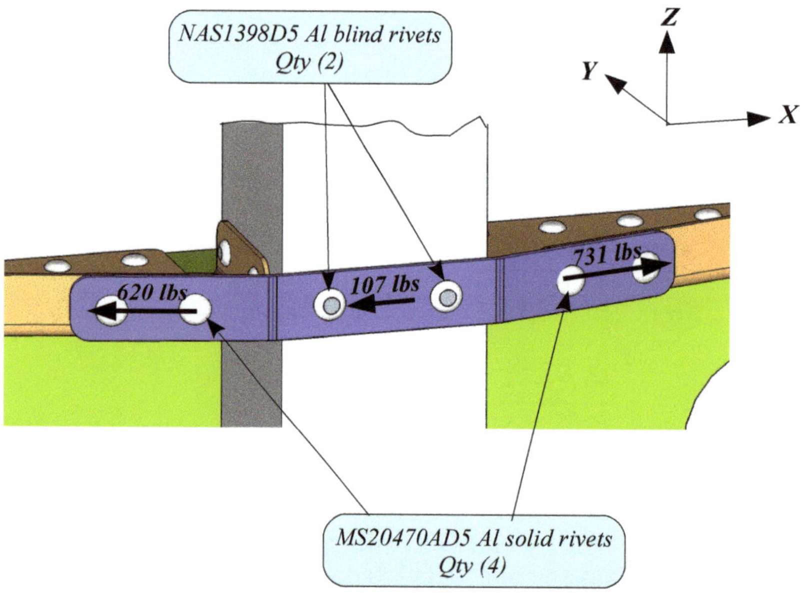

***Figure 11-68** – Fastener loads in Strap.*

Rivet shear:

MS20470AD5

The maximum load in the *MS20470AD5* rivets is 731 *lbs* and is considered distributed equally to the two rivets. Thus, the maximum load per rivet is; $F = 731 / 2 = 366$ *lbs*.

A Fitting Factor "*FF*" of 1.15 (ref. 25. *FAR §25.625*) is used. From *Table 11-21,* the shear allowable "P_{SU}" of a BJ5 *(MS20470AD5)* rivet is 596 *lbs*. A shear factor "*SF*" is required since the Gusset is only 0.050 *in*. thick, so from (ref. *Table 10-3)* for single shear rivets: *SF* = 0.996.

The Margin of Safety for shear is thus calculated as follows:

$$MS = \frac{SF \times P_{su}}{FF \times F} - 1 = \frac{0.996 \times 596}{1.15 \times 366} - 1 = 0.41$$

NAS1398D5

The maximum load in the *NAS1398D5* blind rivets is 107 *lbs* and is considered distributed equally to the two rivets. Thus, the maximum load per rivet is; $F = 107 / 2 = 54$ *lbs*.

A Fitting Factor "*FF*" of 1.15 (ref. 25. *FAR §25.625*) is used. From *Table 11-22*, the shear allowable "P_{SU}" of an ACC5 *(NAS1398D5)* rivet is 755 *lbs*. From *Table 10-3*, for a thickness of 0.063 *in*. (thickness of the Strap), the shear factor "*SF*" in single shear is 1.0. Hence, the *SF* is not needed.

The Margin of Safety for shear is thus calculated as follows:

$$MS = \frac{P_{su}}{FF \times F} - 1 = \frac{755}{1.15 \times 54} - 1 = > 1.0$$

Bearing:

This joint is shear critical since we're using relatively weak fasteners and therefore we expect the *MS* in bearing to be even higher than the one calculated above for shear. However, a bearing *MS* will be calculated.

$$P_{bru} = F_{bru} \times D \times t$$

Strap

The ultimate bearing allowable load "P_{bru}" in the Strap made out of *Al Aly **2024-T3** Clad* sheet 0.063" thick is calculated:

For: $F_{bru} = 120$ *ksi* (ref. *Table 11-10)*, $e/D = 2.0$. And for $D = 0.159"$ and $t = 0.063"$ hence:

$$P_{bru} = 120{,}000 \times 0.159 \times 0.063 = 1{,}202\ lbs$$

Post

The ultimate bearing allowable load "P_{bru}" in the Vertical Post made out of *Aluminum **2024-T3511***, 0.095 *in*. thick is calculated:

For: $F_{bru} = 110$ *ksi* (ref. *Table 11-6)*, $e/D = 2.0$. And for $D = 0.162"$ and $t = 0.095"$ hence:

$$P_{bru} = 110{,}000 \times 0.162 \times 0.095 = 1{,}693\ lbs$$

Gusset

The ultimate bearing allowable load "P_{bru}" in the Gusset made out of *Al Aly **2024-T42** Clad* sheet 0.050" thick is calculated:

For: $F_{bru} = 115$ *ksi* (ref. *Table 11-11)*, $e/D = 2.0$. And for $D = 0.159"$ and $t = 0.050"$ hence:

$$P_{bru} = 115{,}000 \times 0.159 \times 0.050 = 914\ lbs$$

Margin of Safety in bearing

From the above calculations, the ultimate bearing allowable is the lowest for the Gussets with $P_{bru} =$ 914 *lbs*. Since the Gussets are common to the most highly loaded rivets at 366 *lbs*, the Ult. *MS* in bearing is calculated as follows:

$$MS = \frac{P_{bru}}{F} - 1 = \frac{914}{366} - 1 = > 1.0$$

The Gusset is made out of *Al Aly 2024-T42 Clad* material and, from *Table 11-11,* the ratio F_{bru} / F_{bry} is 115 / 70 = 1.64. Because that ratio is greater than 1.5, a bearing check at yield is required.

So for: $F_{bry} = 70$ *ksi* (ref. *Table 11-11*), $e/D = 2.0$. And for $D = 0.159$" and $t = 0.050$" hence:

$$P_{bry} = 70{,}000 \times 0.159 \times 0.050 = 556\ lbs$$

The Yield. *MS* in bearing is calculated as follows:

$$MS = \frac{P_{bry}}{F_{ult}/1.5} - 1 = \frac{556}{366/1.5} - 1 = >1.0$$

11.10.6 GUSSETS ATTACHMENTS

The two Gussets are made out of *Al Aly 2024-T42 Clad* sheet 0.050" thick. They are fastened to the Vertical Post with *qty (2) NAS1398D4* and to the upper flange of the machined Frame with *qty (9) MS20470AD4*, on either sides. *Figure 11-69* shows the shear flow "*q*" in the fasteners of the Gusset. The shear flow is obtained from *Figure 11-45* for $q = 72\ lb/in$.

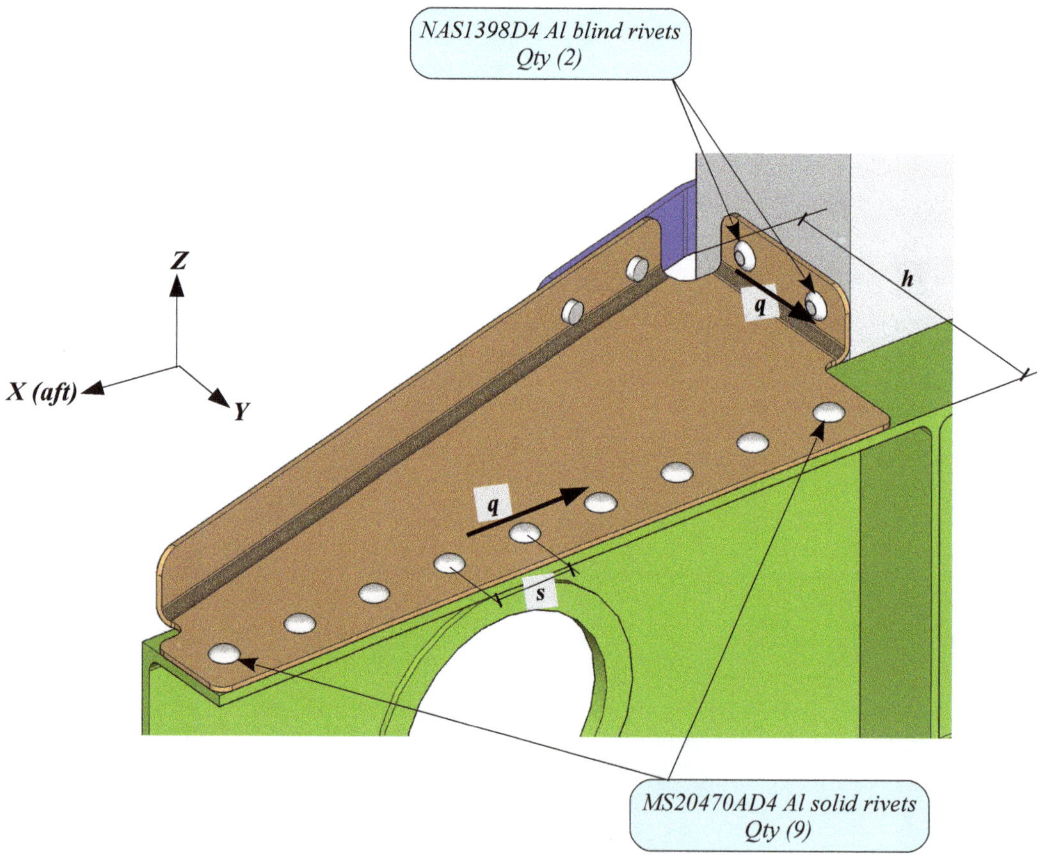

Figure 11-69 *– Fastener loads in aft Gusset.*

In the shear & moment diagrams shown in *Figure 11-36*, the Frame assembly was treated as a constant section beam having a constant transverse shear of 180 *lbs*. In reality, the Gusset forms a triangle that could be modeled with three struts. In which case, there would be no shear flow as the load in each of the three struts would be constant between each end. This analysis is therefore conservative by considering a shear flow.

Rivet shear:

MS20470AD4

The load in each of the ***MS20470AD4*** rivets is $F = q * s$ where $\boldsymbol{q}$ is 72 *lb/in* and $\boldsymbol{s}$ is 0.75 *in*. Thus $F = 72 * 0.75 = 54$ *lbs*. Refer to ***Figure 11-69***.

A Fitting Factor "*FF*" of 1.15 (ref. 25. ***FAR §25.625***) is used. From ***Table 11-21***, the shear allowable "P_{SU}" of a BJ4 ***(MS20470AD4)*** rivet is 388 *lbs*. A shear factor "*SF*" is required since the Gusset is only 0.050 *in*. thick, so from ***Table 10-3*** for single shear rivets: $SF = 0.996$.

The Margin of Safety for shear in the ***MS20470AD4*** is thus calculated as follows:

$$MS = \frac{SF \times P_{su}}{FF \times F} - 1 = \frac{0.996 \times 388}{1.15 \times 54} - 1 = > 1.0$$

Calculating the load in the fasteners by multiplying the shear flow by the pitch can only be done accurately when there is a fair quantity of fasteners (I'd say: >4) and that they fill all of the joint row. Below that amount, or in the case where there are large gaps between some fasteners along an edge or at the end of a row, it is preferable to calculate the fastener load by multiplying the shear flow by the total edge length of a joint and divide by the number of fasteners.

In both cases, it is assumed that all the rivets in the joint analyzed are the same and that they share the load evenly. In the case that one or more fasteners are larger than most, then it is to be expected that those larger ones are going to "attract" more load to them while off-loading slightly the smaller ones.

In which case, one can create different groups of fasteners in the same joint by their type and size to evaluate the load. The load distribution can be done according to the bearing areas of each fastener group or, by the stiffness of the fasteners. This choice depends whether the joint is bearing or shear critical and is also up to the analyst.

NAS1398D4

Referring to ***Figure 11-69***, since there are only two fasteners along that joint, the load in the ***NAS1398D4*** blind rivets is calculated as follows: $F = q * h / n$ where $\boldsymbol{q}$ is 72 *lb/in*, $\boldsymbol{h}$ is 3.0 *in*. and $\boldsymbol{n}$ is *2*. Thus, the load per rivet is; $F = 72 * 3.0 / 2 = 108$ *lbs*.

A Fitting Factor "*FF*" of 1.15 (ref. 25. ***FAR §25.625***) is used. From ***Table 11-22***, the shear allowable "P_{SU}" of an ACC4 ***(NAS1398D4)*** rivet is 494 *lbs*. A shear factor "*SF*" is required since the Gusset is only 0.050 *in*. thick, so from ***Table 10-3*** for single shear rivets: $SF = 0.996$.

The Margin of Safety for shear in the ***NAS1398D4*** is thus calculated as follows:

$$MS = \frac{SF \times P_{su}}{FF \times F} - 1 = \frac{0.996 \times 494}{1.15 \times 108} - 1 = > 1.0$$

Bearing:

This joint is shear critical since we're using relatively weak fasteners and therefore we expect the *MS* in bearing to be even higher than the one calculated above for shear. However, a bearing *MS* will be calculated anyhow.

$$P_{bru} = F_{bru} \times D \times t$$

Frame

The ultimate bearing allowable load "P_{bru}" in the Frame made out of *Al Aly* ***7050-T7451*** machined Plate 1.0 *in.* thick is calculated:

For: F_{bru} = 105 *ksi* (ref. *Table* ***11-9***), e/D = 2.0. And for D = 0.128" and t = 0.100" hence:

$$P_{bru} = 105{,}000 \times 0.128 \times 0.100 = 1{,}344\ lbs$$

Post

The ultimate bearing allowable load "P_{bru}" in the Vertical Post made out of *Al Aly* ***2024-T3511***, 0.095 *in.* thick is calculated:

For: F_{bru} = 110 *ksi* (ref. *Table* ***11-6***), e/D = 2.0. And for D = 0.130" and t = 0.095" hence:

$$P_{bru} = 110{,}000 \times 0.130 \times 0.095 = 1{,}358\ lbs$$

Gusset

The ultimate bearing allowable load "P_{bru}" in the Gusset made out of *Al Aly* ***2024-T42*** *Clad* sheet 0.050" thick is calculated:

For: F_{bru} = 115 *ksi* (ref. *Table* ***11-11***), e/D = 2.0. And for D = 0.128" and t = 0.050" hence:

$$P_{bru} = 115{,}000 \times 0.128 \times 0.050 = 736\ lbs$$

Margin of Safety in bearing

From the above calculations, the ultimate bearing allowable is the lowest for the Gussets with P_{bru} = 736 *lbs*. Since the highest load occur for the joint between the Gusset and the Vertical Post with F = 108 *lbs* and since the diameters are pretty much the same, the Ult. *MS* in bearing is calculated as follows:

$$MS = \frac{P_{bru}}{F} - 1 = \frac{736}{108} - 1 = >1.0$$

The Gusset is made out of *Al Aly* ***2024-T42*** *Clad* material and, from *Table* ***11-11***, the ratio F_{bru} / F_{bry} is 115 / 70 = 1.64. Because that ratio is greater than 1.5, a bearing check at yield is required.

So for: F_{bry} = 70 *ksi* (ref. *Table* ***11-11***), e/D = 2.0. And for D = 0.128" and t = 0.050" hence:

$$P_{bry} = 70{,}000 \times 0.128 \times 0.050 = 448\ lbs$$

The Yield. *MS* in bearing is calculated as follows:

$$MS = \frac{P_{bry}}{F_{ult}/1.5} - 1 = \frac{448}{108/1.5} - 1 = >1.0$$

11.11 DEFORMATIONS

As was done in section *10.7,* the total deformation at the load application point "*A*" is going to be calculated starting from the Frame since it is connected to the airframe. This hand analysis will serve as a verification for the results obtained from the small crude FEM shown in *Figure 11-4.*

11.11.1 FRAME ASSEMBLY

It is assumed that the Frame is simply-supported at both ends and that the lower cap is restrained in the "*X*" and "*Y*" directions by the skin. Therefore, only the upper cap of the Frame assembly can move laterally (in the "*Y*" direction). The total displacement will build-up from the Frame's ends all the way to the Clip lug where the load is applied.

Upper cap lateral displacement

Using the displacement equation for a beam simply supported at both ends having a point load applied mid-span, the following information is gathered:

The load applied at the center of the upper cap in the "*Y*" direction is 332 *lbs*. ref. *Figure 11-36.*

Span; $l = 16$ *in*. ref. *Figure 11-36.*

Young's modulus; $E = 10.0*10^6$ *psi* ref. *Table 11-9.*

Moment of Inertia; $I_Z = 0.161591$ in^4, ref. *Figure 11-39.*

Since the ends are assumed simply supported, the displacement at the center is calculated as:

$$y = \frac{-Wl^3}{48EI} = \frac{-332*16^3}{48*10.0\,E^6*0.161591} = 0.017\,in.$$

The angular deformation (θ), shown in *Figure 11-70*, is calculated as follows:

$$\theta = atan\left(\frac{y}{h}\right)$$

Where $y = 0.017$ *in*, $h = 4.0$ *in*. Therefore,

$$\theta = atan\left(\frac{0.017}{4.0}\right) = 0.00425\ rad.$$

Hence, the new location of point "*A*" is calculated as follows:

$$y' = y\cos\theta - z\sin\theta$$
$$z' = z\cos\theta + y\sin\theta$$

Where $y = -12.55$ *in* and $z = 11.45$ *in*, therefore:

$$y' = -12.55\cos(0.00425) - 11.45\sin(0.00425) = 12.599\ in.$$
$$z' = 11.45\cos(0.00425) + (-12.55)\sin(0.00425) = 11.397\ in.$$

So

$$dy = 12.599 - 12.55 = 0.049\ in.$$
$$dz = 11.397 - 11.45 = -0.053\ in.$$

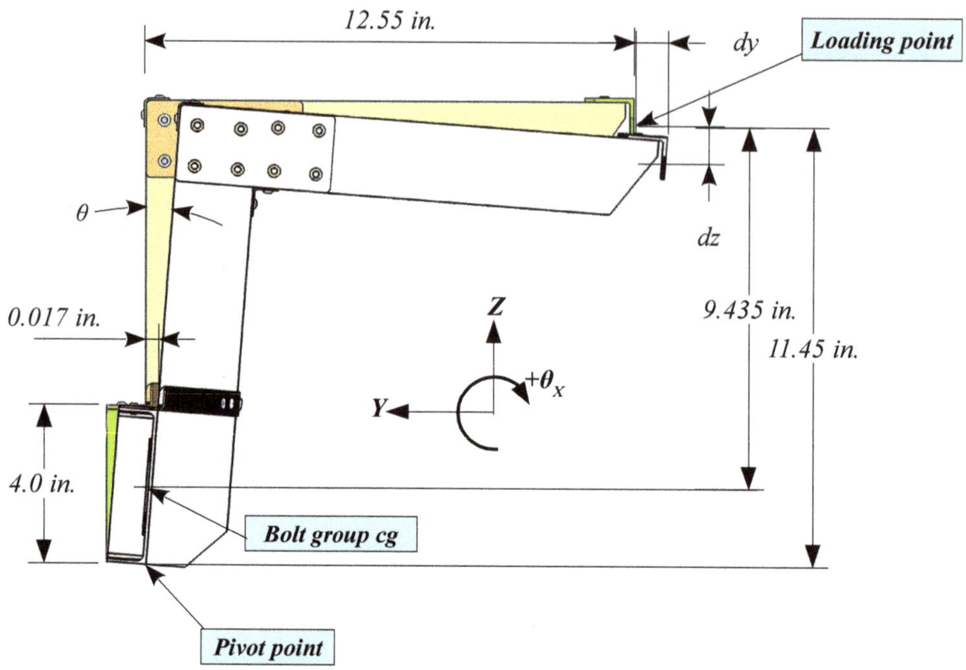

***Figure 11-70** – Rigid body rotation due to Frame upper cap displacement.*

As shown in *Figure 11-70*, this deflection corresponds to an angular deformation "θ" of the Frame of 0.00425 *rad* (0.243 *deg.*). The load application point *(A)* will thus move from a rigid body motion by dy = 0.049 *in.* and dz = −0.053 *in.*

This corresponds to a total displacement; d of:

$$d = \sqrt{dy^2 + dz^2} = \sqrt{0.049^2 + (-0.053)^2} = 0.072 \ in.$$

To this rigid body motion, the deformation of the individual parts shall be added in order to obtain the total displacement.

11.11.2 VERTICAL POST

The Vertical Post is now quite stiffer in torsion and in bending so we should expect very small displacements.

<u>Bending in the "X" direction</u>

The Post is assumed fixed at point *(C)*, corresponding to the center of the six rivets group, and free at point *(B)* where the shear load "P_X" is applied. The Post's maximum deflection in the "X" axis is calculated using the following equation:

$$x = \frac{W \, l^3}{3 \, E \, I}$$

Where:

$W = P_X = 17.4 \ lbs.$	ref. *Figure 11-5*
Length; $l = 9.435 \ in.$	ref. *Figure 11-70*
Young's modulus; $E = 10.0 * 10^6 \ psi$	ref. *Table 11-6*
Moment of Inertia; $I_k = I_Y = 0.4446 \ in^4$	ref. *Figure 11-25*

Therefore;

$$x = \frac{P_X l^3}{3EI_Y} = \frac{17.4 \times 9.435^3}{3 \times 10.0 * 10^6 \times 0.4446} = 0.0011\ in.$$

Bending in the "Y" direction

Again, the Post is assumed fixed at point *(C)*, corresponding to the center of the six rivets group, and free at point *(B)* where the moment "M_X" is applied. The Post's maximum deflection in the "Y" axis is calculated using the following equation:

$$y = \frac{M_o l^2}{2EI}$$

Where:

$M_o = 1{,}133$ *in-lb.*	ref. *Figure 11-22*
Length; $l = 9.435$ *in.*	ref. *Figure 11-70*
Young's modulus; $E = 10.0*10^6$ *psi*	ref. *Table 11-6*
Moment of Inertia; $I_j = I_X = 0.4446$ in^4	ref. *Figure 11-25*

Therefore;

$$y = \frac{1{,}133 \times 9.435^2}{2 \times 10.0 * 10^6 \times 0.4446} = 0.0113\ in.$$

End rotation

The rotation angle at the top is calculated as:

$$\theta = \frac{M_o l}{EI}$$

Where:

$M_o = 1{,}133$ *in-lb.*	ref. *Figure 11-22*
Length; $l = 9.435$ *in.*	ref. *Figure 11-70*
Young's modulus; $E = 10.0*10^6$ *psi*	ref. *Table 11-6*
Moment of Inertia; $I_j = I_X = 0.4446$ in^4	ref. *Figure 11-25*

Therefore;

$$\theta = \frac{1{,}133 \times 9.435}{10.0 * 10^6 \times 0.4446} = 0.0024\ rad.$$

The angular rotation around the "X" axis (0.0024 *radian*) calculated above corresponds to 0.137 *degree*.

The resulting rigid body displacement at the load application point (11.5 *in.* away) due to the above rotation is calculated as:

$$dy = l(1-\cos\theta) = 11.5(1-\cos(0.137)) \approx 0$$
$$dz = l(\sin\theta) = 11.5(\sin(0.137)) = -0.0275\ in.$$

Twist

For the torque applied completely at the top end (a = 0) and restrained at its base, the angular rotational displacement of the Post at the top is calculated as:

$$\theta = \frac{TL}{KG}$$

Where:

The torque applied at the center of the Post is; T= 200 *in-lb*. ref. *Figure 11-22*
Length; L = 6.60 *in*. ref. *Figure 11-22*
Young's modulus; $E = 10.0*10^6$ *psi* ref. *Table 11-6*
Torsional stiffness constant; $K = 0.3603$ in^4 ref. section *11.3.4*
Shear modulus; $G = 3.76*10^6$ *psi* , ref. *Table 11-6*

Therefore;

$$\theta = \frac{T\,L}{K\,G} = \frac{200\times 6.60}{0.3603\times 3.8*10^6} = 0.000964\ rad$$

The angular displacement (0.000964 *radian*) calculated above corresponds to 0.055 *degree*.

The resulting rigid body displacement at the load application point due to the above rotation is calculated as:

$$dy = l(1-\cos\theta) = 6.60(1-\cos(0.055)) \approx 0$$
$$dx = l(\sin\theta) = 6.60(\sin(0.055)) = -0.0063$$

11.11.3 HORIZONTAL BEAM

The Horizontal Beam is also quite stiffer in torsion and in bending so we should expect very small displacements.

Bending in the "X" direction

The Beam is assumed fixed at point *(B)*, corresponding to the center of the six rivets group, and free at point *(A)* where the shear load "P_X" is applied. The Beam's maximum deflection in the "X" axis is calculated using the following equation:

$$x = \frac{W\,l^3}{3\,E\,I}$$

Where:

$W = P_X = 17.4$ *lbs*. ref. *Figure 11-5*
Length; l = 11.5 *in*. ref. *Figure 11-12*
Young's modulus; $E = 10.0*10^6$ *psi* ref. *Table 11-6*
Moment of Inertia; $I_k = I_Y = 0.4446$ in^4 ref. *Figure 11-25*

Therefore;

$$x = \frac{P_X\,l^3}{3\,E\,I_Y} = \frac{17.4\times 11.5^3}{3\times 10.0*10^6\times 0.4446} = 0.00198\ in.$$

Bending in the "Z" direction

Again, the Beam is assumed fixed at point *(B)*, corresponding to the center of the Vertical Post, and free at point *(A)* where the force "P_Z" is applied. The Beam's maximum deflection in the "Z" axis is calculated using the following equation:

$$z = \frac{W\,l^3}{3\,E\,I_X}$$

Where:

$W = P_Z = 98.5\ lbs.$ ref. *Figure 11-5*
Length; $l = 11.5\ in.$ ref. *Figure 11-12*
Young's modulus; $E = 10.0*10^6\ psi$ ref. *Table 11-6*
Moment of Inertia; $I_j = I_X = 0.4446\ in^4$ ref. *Figure 11-25*

Therefore;

$$z = \frac{98.5 \times 11.5^3}{3 \times 10.0 * 10^6 \times 0.4446} = 0.0112\ in.$$

End rotation

The rotation angle at the loaded end is calculated as:

$$\theta = \frac{W\,l^2}{2\,E\,I}$$

Where:

$W = P_Z = 98.5\ lbs.$ ref. *Figure 11-5*
Length; $l = 11.5\ in.$ ref. *Figure 11-12*
Young's modulus; $E = 10.0*10^6\ psi$ ref. *Table 11-6*
Moment of Inertia; $I_j = I_X = 0.4446\ in^4$ ref. *Figure 11-25*

Therefore;

$$\theta = \frac{98.5 \times 11.5}{2 \times 10.0 * 10^6 \times 0.4446} \approx 0\,rad.$$

The angular rotation around the "*X*" axis is thus negligible. Hence, for the same parameters but a lower load in "*X*" at 17.4 *lbs*, the rotation angle around the "*Z*" axis is also 0.

Twist

For the torque applied completely at the load application point *(A)* ($a = 0$) and assumed restrained at the center of the Vertical Post, the angular rotational displacement of the Horizontal beam at the Clip hole is calculated as:

$$\theta = \frac{T\,L}{K\,G}$$

Where:

The torque applied at the center of the Beam is; $T = 9\ in\text{-}lb.$ ref. *Figure 11-13*
Length; $L = 11.5\ in.$ ref. *Figure 11-12*
Young's modulus; $E = 10.0*10^6\ psi$ ref. *Table 11-6*
Torsional stiffness constant; $K = 0.3603\ in^4$ ref. section *11.3.4*
Shear modulus; $G = 3.76*10^6\ psi$ ref. *Table 11-6*

Therefore;

$$\theta = \frac{T\,L}{K\,G} = \frac{9 \times 11.5}{0.3603 \times 3.76 * 10^6} \approx 0$$

The deformation due to the contribution of the torque is thus negligible.

11.11.4 COMBINED DISPLACEMENTS

All of the deformations are summed up to get the total deformation at the load application point. *Table 11-26* shows the contribution of each member and the total deformation.

Table 11-26 *- Total displacement at load application point.*

	Local deformation			Rotation Angle			Deformation at pt. A		
	dx [in]	dy [in]	dz [in]	θx [rad]	θy [rad]	θz [rad]	dx [in]	dy [in]	dz [in]
Frame upper cap	0	-0.017	0	0.00425	0	0	0	-0.049	-0.053
Vertical Post	-0.0011	-0.0113	0	0.0024	0	0.000964	-0.00630	0	-0.0275
Horizontal Beam	-0.00198	0	-0.0112	0	0	0	-0.00198	0	-0.0112
Total							**-0.00828**	**-0.049**	**-0.0917**

Resulting displacement: **0.104** *in.*

This is comparable to the value obtained from the crude FEM of 0.099 *in.* as shown in ***Figure 11-4*** within **5%**! However, the hand calculated value slightly exceeds the 0.100 *in.* requirement but since we assumed all boundary conditions to be pinned and that in reality there will be some kind of rotational restraint, then we can call it a day with this concept.

In cases when the result fails to meet the requirements by having a slightly negative *MS* or by exceeding deflection limits slightly as demonstrated above, always add an "EXCUSE ME NOTE" to justify why you failed to meet the requirements. This note can be, as the example above, to reassure the reader that the analysis was a bit too conservative so that in reality, the requirements will be met. NEVER SHOW A NEGATIVE MARGIN OF SAFETY.

The excuse me note can NEVER be something like: "*because my boss says it's fine*" or "*because I don't have time to redo the analysis*" or "*because the designer doesn't want to change the design*"...

Reassure the reader *(the certifying agency representative (DER))* that the design **IS** good and meets the requirements, don't give excuses!

11.12 CONCLUSION

From the results of the analysis done in section *11*, the improved design meets all of the requirements and is thus acceptable as it shows only positive MS*'s* and that the total displacement under the ultimate load is the critical criteria and is within the limits. Of course, other concepts could have been proposed that could have been stiffer/stronger, lighter or cheaper but the one analyzed wasn't drastically modifying the concept and appeared simpler to manufacture.

Studying many concepts in order to pick the best one is time consuming and, quite often, time is a luxury that most programs don't have, on the commercial aerospace side anyway. So a concept that had been used on other previous programs (whether or not it was good or had been optimized) is sometimes "imposed" to the stress analyst. A program like "*six sigma*", which is based on a decision matrix approach, is a great tool where the best solution can be determined by a group of people coming from all the different departments involved (engineering, manufacturing, in-service support, procurement, methods, etc). Normally that works pretty well... providing the V.P. doesn't show-up at the last minute and imposes a solution that had already been eliminated...

To my opinion (and from experience), even if a room would be full of people from all the different departments in a company and that some of them may have good ideas, the stress engineer is still the one that will have to do most of the work in order to evaluate every single proposed concept. That is why it is essential that the stress analyst be capable to quickly reject ideas that don't make sense and be able to propose something that does work simply by looking at them or doing quick calcs. That capability comes with relevant experiences, not time.

> *"20% of the workers do 80% of the workload."*

12. BOUNDARY CONDITIONS

Struts work only (mostly) in axial loads; either tension or compression. Therefore, they're hard to beat regarding stiffness, simplicity, compactness and weight. So, if we could, that would certainly be the first choice in most design concepts. As soon as some bending or torsion appear in some of the members from a certain design, then the complexity of the design increases and the weight goes up. Simply look at the first airplanes to see that all the structure was fabricated with struts and cables.

In the dawn of aviation, because all aircraft structures used struts and cables, I suspect that stress engineers didn't use the Euler-Bernoulli beam theory ($\sigma = My/I$) and simply used $\sigma = P/A$. In those days, up to probably when CAD systems became quite common, the designer and the stress analyst were the same person. Therefore, they could only design things that they could analyze.

Remember that next time you have difficulty analyzing something, the design could probably be simplified/improved.

The advantage of using arrays of struts can also be a disadvantage if not designed well. Especially because of their very high stiffness, one must take every precaution not to "bridge" the primary structure. What this means is that, by installing an array of struts (or any type of structure in fact) to support something (especially as a secondary structure), the airframe must still be allowed to deform freely without being constrained by the struts.

The way to do that is by avoiding boundary conditions redundancy. As you know, the fuselage and the wing deflect due to flight loads and pressurization. If, for example, a very stiff secondary structure were to be attached to a fuselage at multiple points with the same boundary conditions at all points in all three axes (*X.Y,Z* or *1,2,3* and maybe *4,5,6*), then when the airplane gets pressurized, the fuselage would try to stretch but the secondary structure, because it would be very stiff, would try to refrain it from stretching. This would lead to premature cracking of the fuselage and/or the secondary structure or simply the failure of the weakest link.

"You can do all the mistakes you want as long as you catch them before drawing signatures."

That is why that, when we install a very stiff box for example, we try to restrain it with only three attachments if possible instead of the typical four. See *Figure 12-1* and *Figure 12-2* for a typical avionic box attachment.

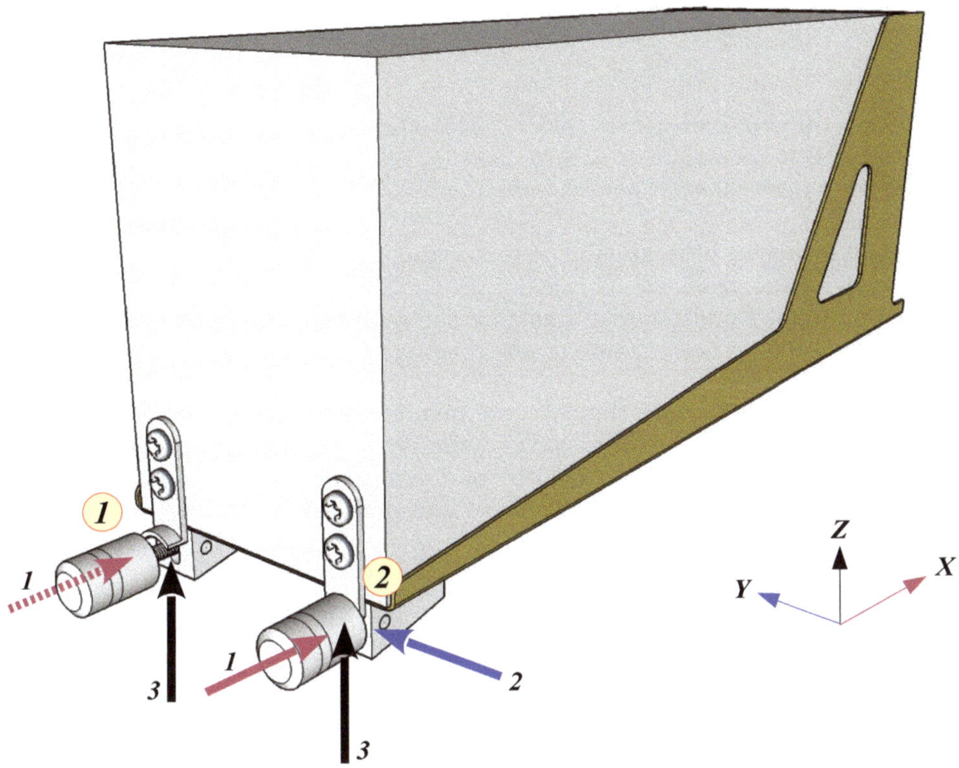

Figure 12-1 – *Typical front boundary conditions for an Avionic box.*

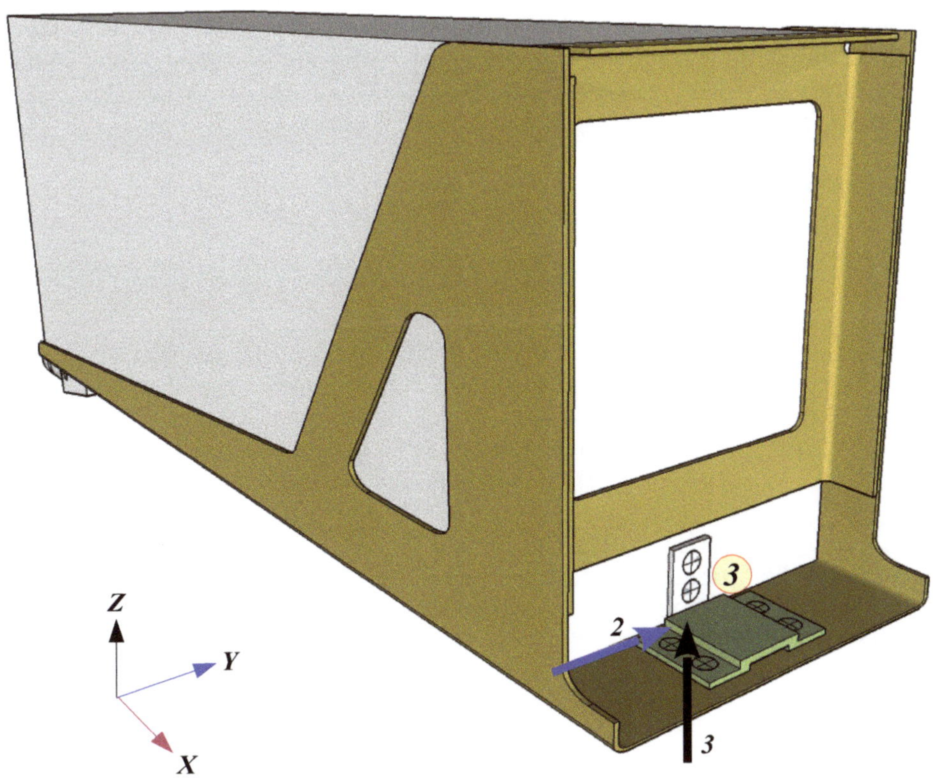

Figure 12-2 – *Typical rear boundary conditions for an Avionic box.*

From the above Avionic box example, point ***2*** is restrained in *X,Y,Z (1,2,3)*, point ***3*** in *Y, Z (2,3)* and finally, point ***1*** in *X, Z (1,3)*. If a forth point should exists (or more), then it would have to be blocked in *Z* only *(3)*.

Here point ***1*** has a redundant constraint in *X*. That's why I showed it in dashes in *Figure 12-1* as it wouldn't be necessary. But because the Avionic box is quite narrow, we can certainly assume that points ***1*** and ***2*** share the same load in the *X* direction.

The system is thus found to be statically determinate and the reaction loads at each point can easily be calculated by static analysis. This exercise can of course easily be done using a *FEM* by setting the proper Boundary Conditions and getting the resulting reaction loads at each attachment point.

A friend of mine was telling me once that, during hiring interviews, he would ask the candidate to calculate by hand the reaction loads from a very simple box submitted to a *9G FWD* emergency landing condition. Although all of the candidates had graduated from engineering schools and had (supposedly) experience, only about half could solve the problem correctly from which of course he would make his pick.

Determining the correct boundary conditions from which to derive the Free Body Diagram is crucial in getting the correct reaction loads at the attachment points and, in the case of my friend's company, getting a job!

12.1 BACK-UP STRUCTURE STIFFNESS

Parts that we analyze are always attached to some other part or structure that we call the *back-up structure*. When calculating the reaction forces at the fasteners of a given part, we usually look at that part by itself when doing its *Free Body Diagram*. In doing so, we generally assume the back-up structure to be infinitely rigid and that is usually fine. However, this is rarely the case! There are two modes of loading for a fastener:

1. Shear loading,
2. Axial loading.

Shear loading

At ultimate loads, for a group of fasteners acting on the same plane, we usually assume the applied load to be evenly distributed among all the fasteners. That is providing that the fasteners are all of the same type and size and that the thickness of each of the two connected parts is constant at all fasteners' locations. When different fasteners having different materials, diameters or lengths (due to different material thicknesses) are used, then the distribution of the load among them will vary.

The fasteners having the material with the largest Young's modulus, the largest diameter or the largest bearing area will be “attracting” more load as their stiffness in shear will be higher. One could visualize the shear stiffness of a fastener as a spring normal to the fastener's axis. The larger the fastener, the stronger the spring is.

Axial loading

Because it is more efficient, we always try to have fasteners to work in shear and avoid as much as we can fasteners to work in tension. A lot of times though, fasteners have to react tension loads. Although we sometimes assume fasteners to react compression loads when doing a *FBD*, the compression force being transferred between two parts is done by contact and not by the fastener.

There is nothing wrong in having fasteners to work in tension. The problem is more regarding the parts that are connected through that fastener as those parts will most likely be loaded in the out-of -plane direction. Normal loads (out-of-plane) cause local bending and relatively large displacements. As demonstrated below, the displacement of a bar in tension is a lot less than for the same bar in bending for the same load and length.

Considering a square bar shown in *Figure 12-3* having one end fixed and the other end free, the axial displacement "d_a" at the free end from a tension force is mathematically compared to the normal displacement "d_n" at the free end also.

Where

$$d_a = \frac{PL}{AE} \qquad d_n = \frac{PL^3}{3EI}$$

and

$$A = b\,h \qquad I = \frac{b\,h^3}{12}$$

Therefore, the ratio of the normal displacement over the axial displacement is:

$$\frac{d_n}{d_a} = \frac{PL^3}{3EI} \cdot \frac{AE}{PL} = \frac{4L^2}{h^2}$$

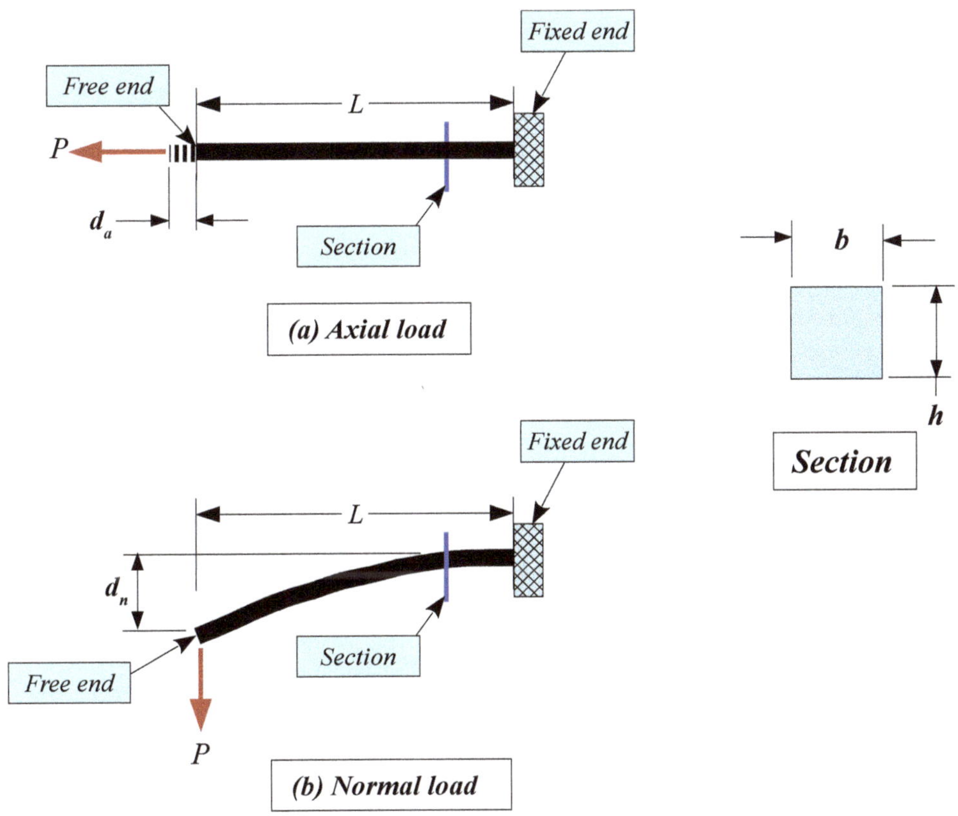

Figure 12-3 *– Displacement comparison (a) axial loading, (b) normal loading.*

In general, the "bar" thickness is the thickness of a flange and is in the range of 0.050 *in.* to 0.125 *in.* while the length of the bar is the fastener pitch and is in the range of 0.80 *in.* to 1.25 *in.* So, for the sake of the study, using a thickness of 0.10 *in.* and a length of 1.0 *in.* we get a d_n / d_a ratio of:

$$\frac{d_n}{d_a} = 4\frac{(1)^2}{(0.1)^2} = 400$$

Since the stiffness is inversely proportional to the displacement, the out-of-plane stiffness of a component is much lower than in shear by hundreds of times.

As a real example, *Figure 12-4* shows two configurations of a fitting (which is analyzed in section *14.14*) fastened to a back-up structure.

On the left side of the figure *(a)*, fasteners *1 & 2* connecting the fitting are positioned on either sides of an integral rib in the Intercostal. The Intercostal flange thickness is also the same for fasteners *1 & 2*. The integral rib greatly helps in reducing the out-of-plane displacement and the bending stress in the Intercostal flange.

On the right side of the figure *(b)*, the fitting had been moved such that fasteners *1 & 2* are now both positioned on the same side of the integral rib of the Intercostal. Fastener *1* is now quite far from the stiffening rib and is also positioned where the Intercostal flange is thinner. Therefore, not only that the shear stiffness of fastener *1* is reduced a bit due to the reduction of the bearing area (thickness) but more importantly, its normal stiffness is drastically reduced from not being positioned in a corner anymore. So, since the stiffness of fastener *2* is the same as before (as fastener *1* was) then, fastener *1* is now much softer than fastener *2* and will thus attract less loads.

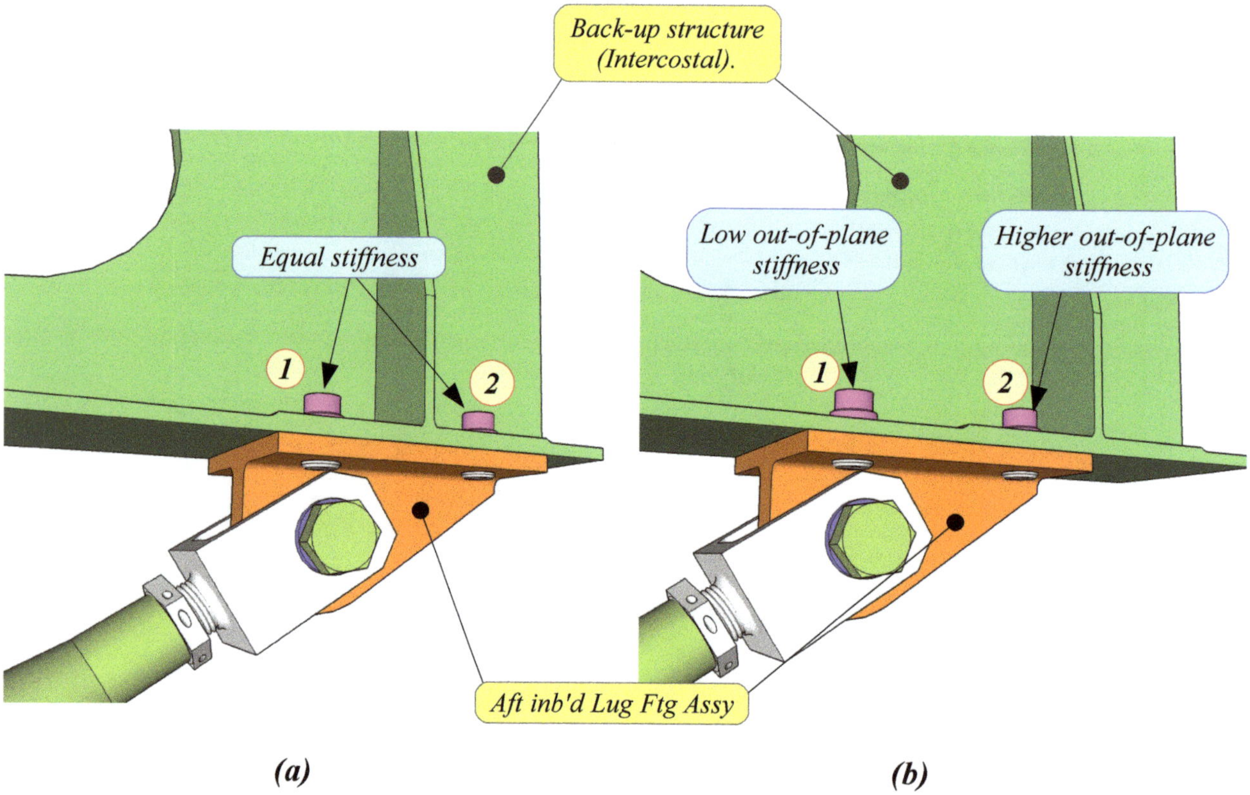

Figure 12-4 *– Connections Out-of-plane stiffness comparison, (a) equal stiffness, (b) higher stiffness near a rib.*

This condition (*b*) is not desirable as, even if we're analyzing for ultimate loads and that a load path would certainly find its way, premature fatigue cracks could occur.

In the above example of *Figure 12-4(a)*, a load distribution between fasteners *1 & 2* of 50%-50% would be considered. However, for the configuration shown in *Figure 12-4*(*b*) (in the case that we couldn't get the design to be changed...), one would probably want to assume a load distribution of something like 25% for fastener *1* and 75% for fastener *2*. As one *DER* would say: "*No reasonable offer will be refused!*"

PART III

Airframe analysis examples

In the following chapters, I've tried to provide some typical analyses from different aircraft components which could be used as the basis for analyzing any other component in an airframe. I am covering the fuselage, the wing, and some systems of a fictitious commercial airplane. Each analysis example is complete and uses pretty much all or most of the calculation equations/tools that are available from different references.

In large aircraft manufacturing companies, they segregate each part of an aircraft into groups. Therefore the different stress groups are typically: Fuselage, wing, empennage, doors and systems. The systems group comprising of primary and secondary structures like: flight controls, landing gear and seats are considered primary structures, while the avionic racks, water and waste, cargo, galleys and over-head bins for example are considered secondary structures.

All those aircraft components deal with primary structures. A primary structure is a structure that is absolutely necessary for the aircraft to fly. On the other hand, secondary structures are not essential even though they could be very important.

I wish I could show analysis examples of every part of an aircraft, but that would require me many more years and many volumes of this book. As you'll notice, the analysis approach is pretty much the same for all structures and to show examples of every part of an aircraft would be somewhat redundant.

In this book I've decided to show two complete analysis examples:

1. Monument installation analysis in chapter ***14***
2. Wing Spar detail analysis in chapter ***15***

These might not seem like a lot but, as you'll see, they're detailed and complete and take-up most of the pages in this book. That is about 280 pages for the monument installation analysis and 100 pages for the analysis of a small portion of the wing rear spar. That should give you an idea on how much work is required to substantiate both of those *"drawings"*.

The monument installation analysis, detailed in chapter ***14,*** involves analyzing its back-up structure which is mainly the floor structure while the wing spar analysis, detailed in chapter ***15***, involves analyzing part of the wing rear spar at the location of a cutout using internal loads extracted from a *FEM*.

But, before starting with the Monument Installation analysis, since this involves the checking of the fuselage floor structure, I'll start first by demonstrating the analysis on part of the fuselage, detailed in chapter ***13,*** as this will be necessary as a prerequisite.

Let's start!

"Running solves nothing, we have to start at the right time."

13. FUSELAGE

The fuselage is a large tube —for pressurized airplanes— that starts with the cockpit in the front and finishes with the Tail cone at the aft end. The empennage, containing the elevators and the rudder, is usually not considered as part of the fuselage but is taken care of by its own group.

In sub-chapters *13.1* and *13.2*, I will be demonstrating static analysis examples of the floor Crossbeams and the Gussets that connect the Crossbeams to the Seat track rails.

13.1 FUSELAGE CROSSBEAMS

The Crossbeams support all the vertical loads from the floor panels and transfer it to the fuselage Frames as vertical shear. Also, because they are connected to the Frames at each end, they also get axial tension loads and end moments as the fuselage gets pressurized and expands radially. See *Figure 13-1* for a typical Crossbeam installation.

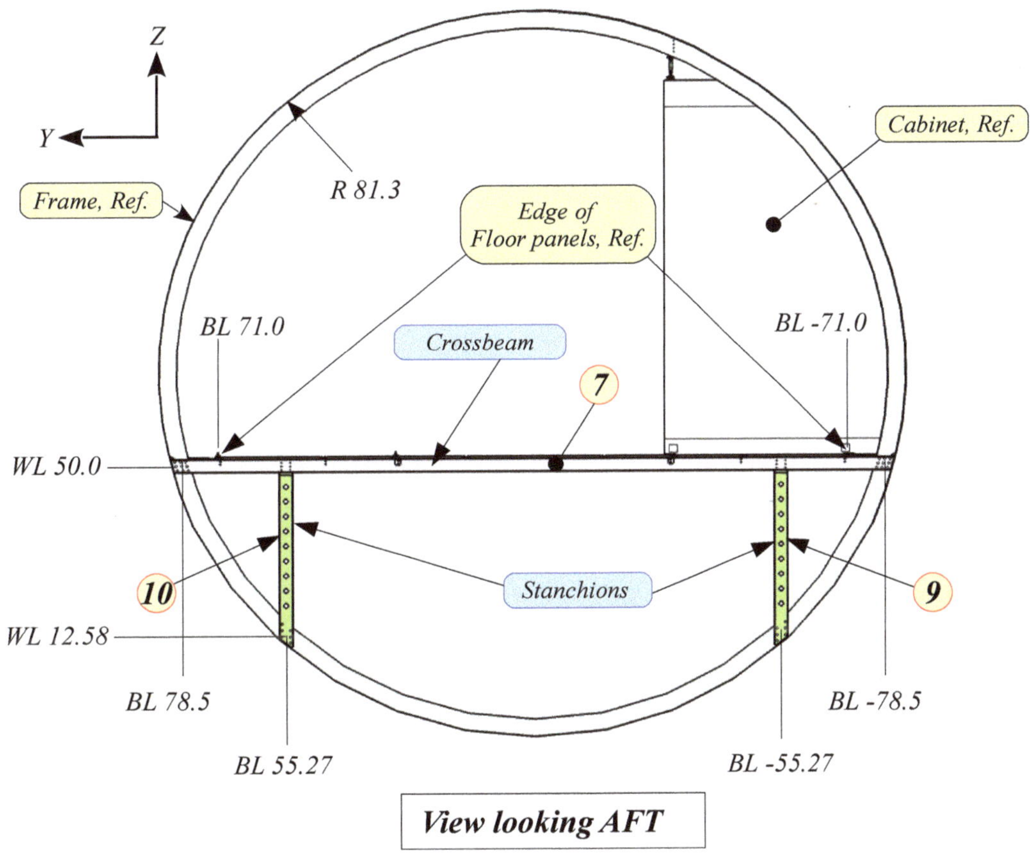

Figure 13-1 – *Crossbeam installation.*

The Crossbeams are machined out of *Al Aly 7050-T7451* plate 2.50 *in*. thick. They are connected to the Frames at each end and to the Stanchions using *MS20470AD6* solid rivets.

13.1.1 SUMMARY OF MARGINS OF SAFETY

Table 13-1 shows the summary of minimum Margins of Safety for all parts analyzed in the Crossbeam installation.

Table 13-1 *- Crossbeam install. Summary of minimum Margins of Safety*

Nomenclature	*Description*	*Material*	*Mode of failure*	*Load case*	*MS*	*Page*
Crossbeam instl.	Lower cap crippling	Al Aly 7050-T7451	Crippling	6g Down	0.07	239
	Lower cap instability	Al Aly 7050-T7451	Column buckling	3g Up	0.55	250
	Free Flange stability	Al Aly 7050-T7451	Bending buckling	6g Down	0.26	258
	Web stability	Al Aly 7050-T7451	Shear buckling	6g Down	>1.0	260
	Connection to Stanchion	MS20470AD6	Fastener shear	6g Down	0.59	262
	Connection to Frame	MS20470AD6	Fastener shear	6g Down	0.13	265
Gusset instl.	Stress through section A-A	Al Aly 2024-T3 Clad	Bending + axial + shear	3g Up	>1.0	273
	Lower cap instability	Al Aly 2024-T3 Clad	Column buckling	3g Up	>1.0	280
	Connection to crossbeam	MS20470AD5	Fastener shear	3g Up	>1.0	282

13.1.2 LOADS

For the purpose of the demonstration, I have assumed an evenly distributed floor longitudinal *(along the X axis)* running load "w_X" of 30 *lb/in.* and that the most critical condition was the emergency landing ***6g Down*** from the (ref. 25. ***FAR §25.561***). Then, the ultimate floor loading applied to the *Crossbeam* is calculated below:

For a Frame pitch "***p***" of 18 *in.*, the total ultimate vertical load "P_Z" applied to a Crossbeam is:

$$P_Z = N_Z w_X \cdot p = 6 \cdot 30 \cdot 18 = 3{,}240\ lbs$$

For a Floor panels total width "l_{fp}" of 142 *in.*, the distributed vertical load "w_Z" on the Crossbeam is:

$$w_Z = \frac{P_Z}{l_{fp}} = \frac{3{,}240}{142} = 22.8\ lbs/in.$$

The structure comprising of the Crossbeam, the Stanchions and the Frame along with the applied and the assumed reacted loads from the *6g Down LC* are shown in *Figure 13-2*. The applied floor loads are normally reacted in shear from the whole fuselage. It is shown here as vertical tension in the Frames for simplicity.

The reaction loads due to an emergency loading condition are not easy to determine. For example, in a *6g Down LC* one could assume that the aircraft is hitting the ground flat and that the reaction loads are applied as a pressure on the belly of the aircraft. The idea behind analyzing the structure for an emergency *LC* is to ensure that the structure, although expected to deform extensively and yield, will not break.

In our example, we simply want to check the Crossbeam and its connections to ensure the floor won't collapse for the safety of the passengers without worrying too much about where the load goes afterwards.

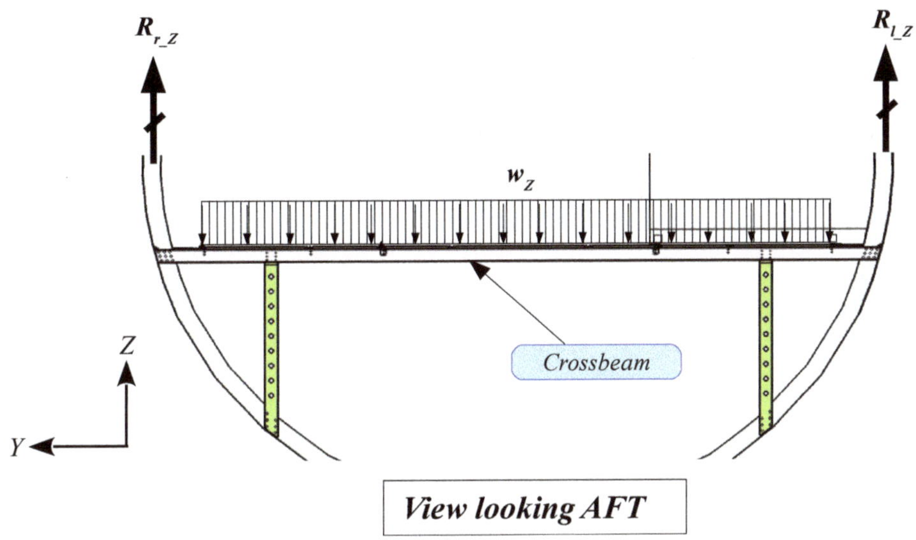

Figure 13-2 *– Crossbeam FBD.*

By isolating the Crossbeam from the rest of the structure, its dimensions and *FBD* look like as shown in *Figure 13-3*. For this exercise, the bending stiffness (moment of Inertia) of the Crossbeam is assumed comparable to the bending stiffness of the Frame, a 50% rotational stiffness is assumed at each end of the Crossbeam. I also considered the Stanchions to be somewhat more flexible in bending then the Crossbeam so I've assumed the Crossbeam to be simply supported at the Stanchions locations.

In reality, the moment of Inertia of the Frame including the effective skin is about twice the one of the Crossbeam, so an end-fixity lower than 50% at the Crossbeam ends is expected.

To calculate the end moments of the Crossbeam for a 50% end fixity, I calculate the average end-moment from the two following conditions for the *6g Down LC*:

1. End-moments by assuming the Crossbeam simply supported at the Frames (0%) stiffness.
2. End-moments by assuming the Crossbeam fixed at the Frames (100%) stiffness.

Both *FBD's* and, shear and moments diagrams are shown in *Figure 13-4* for condition *1* and *Figure 13-5* for condition *2*.

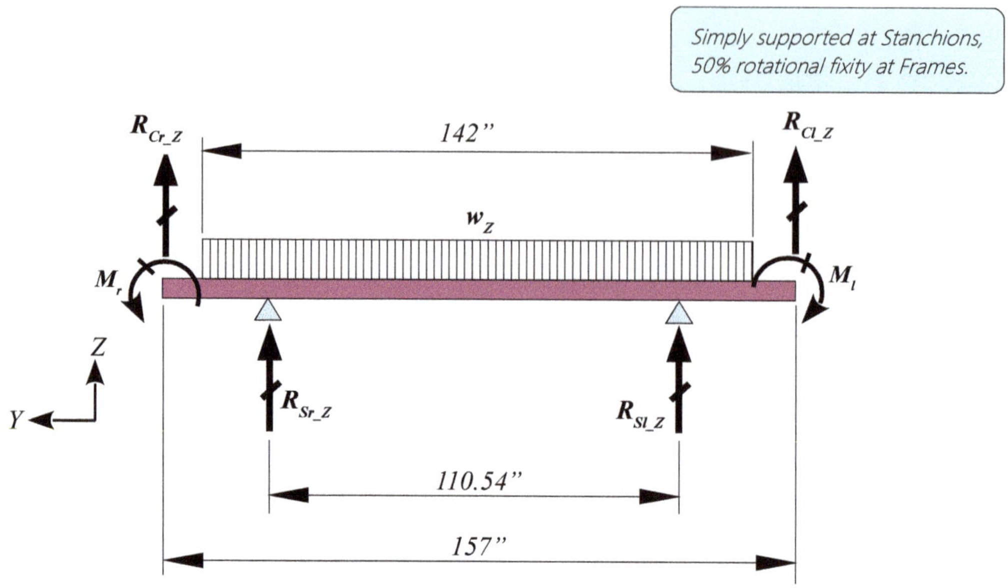

Figure 13-3 *– Crossbeam FBD.*

To calculate the internal loads and the reactions of the Crossbeams, I used a small load FEM since the Crossbeam is statically indeterminate. A hand calculation method called: "*The moment distribution method"* as described in (ref. 2.§*A.11)* could have also been used as a double check.

Of course, the end-moment for condition *1* is 0, while the end-moment for condition *2* is 9,461 *in-lb*. Therefore, the end-moment for a 50% end fixity is:

$$M_{50} = \frac{M_0 + M_{100}}{2} = \frac{0 + 9{,}461}{2} = 4{,}731 \ in.-lb$$

The corresponding shear and moment diagram is then shown in *Figure 13-6.*

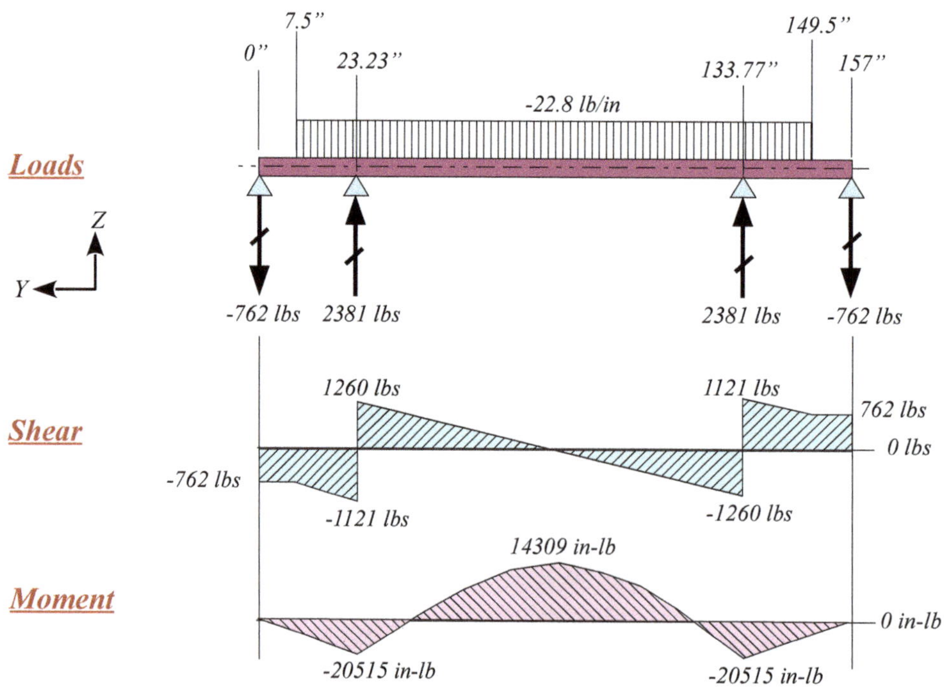

Figure 13-4 *– Crossbeam FBD, Simply supported at ends.*

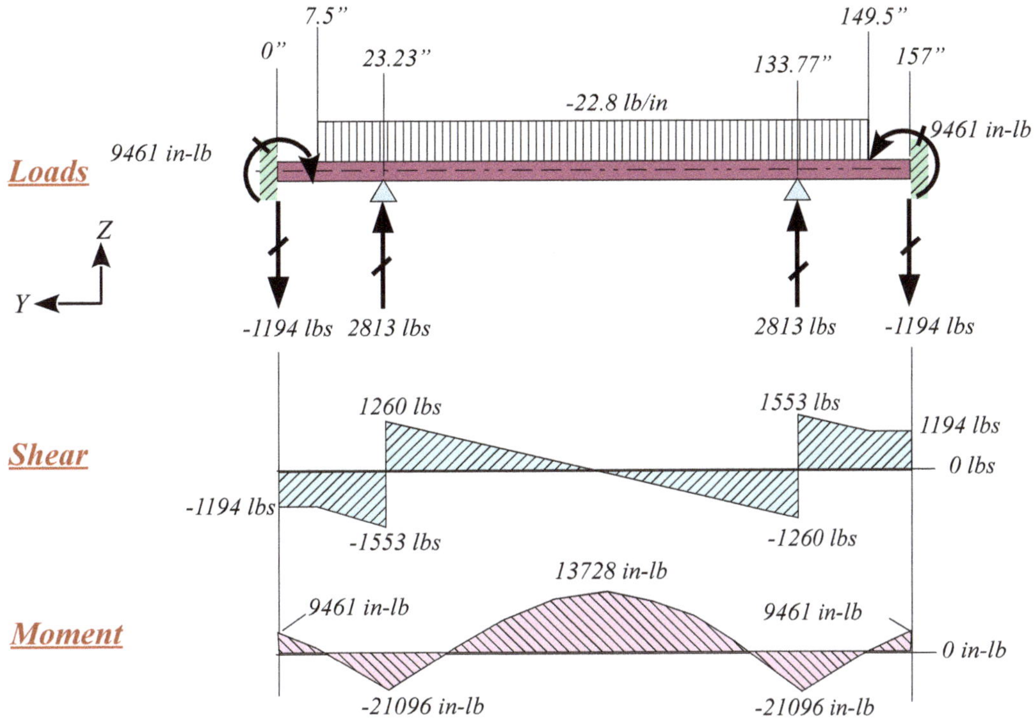

Figure 13-5 *– Crossbeam FBD, with Fixed ends.*

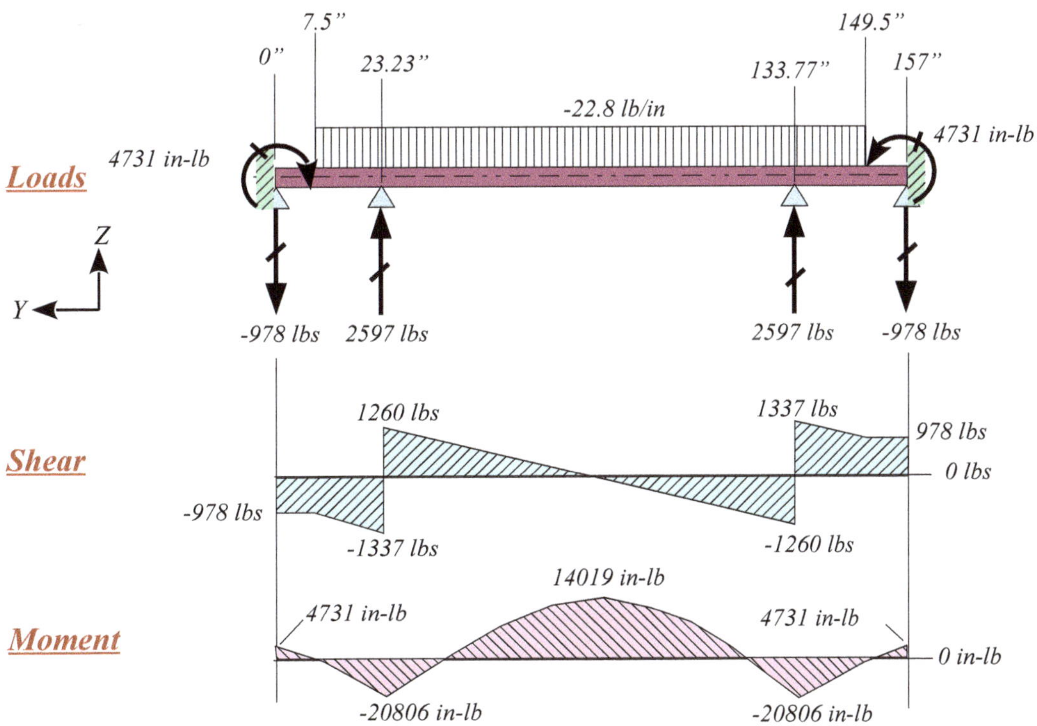

Figure 13-6 *– Crossbeam FBD, with 50% end fixity.*

As one can see, there isn't much difference in the maximum bending moment between all three end-fixity assumptions. However, you may want to go with the 50% end-fixity assumption all the way or, simply use the maximum shear and moment between all three. Just make sure that your choice is clearly stated and that everybody (boss, *DER*, company policy) is in agreement with it.

Comparison to a simple *FEM*.

In order to double-check my previous loads calculations and to verify the 50% end-fixity assumption I made, I constructed a small *FEM* by simply using *BAR* elements having the correct section properties for the Frame, the Crossbeam and the Stanchions. The following are the boundary conditions "*BC*":

- All the nodes are blocked in the out-of-plane direction (*X*).
- To be consistent with my *FBD*, I released the rotational restraint of the Stanchions at the Crossbeam attachments.
- The vertical reactions in "*Z*" are done at the Frame where the *BAR* elements are the most vertical.
- One node in the middle of the Crossbeam is blocked in the lateral "*Y*" direction to prevent rigid body movements.

Figure 13-7 shows the *FEM* section properties, Boundary Conditions and the floor loading under *6g Down LC* while *Figure 13-8* shows the maximum bending moments around the "*X*" axis.

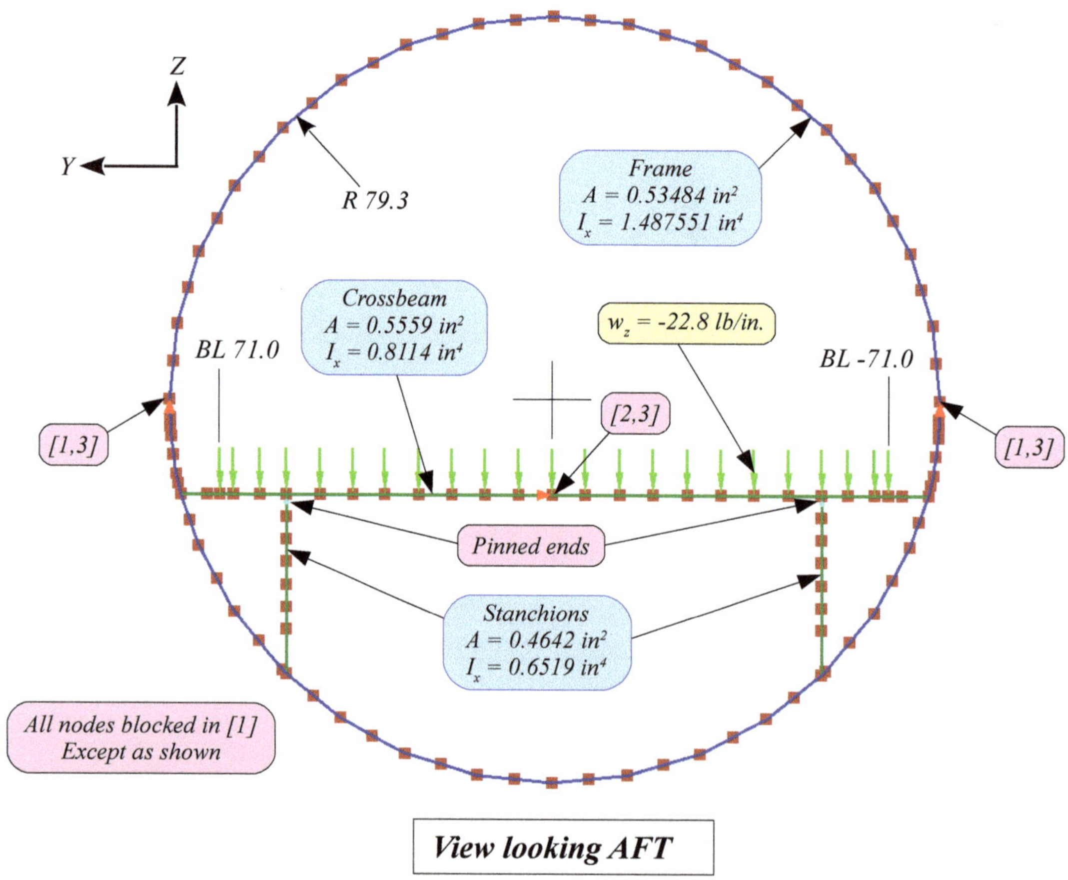

Figure 13-7 *– Fuselage section FEM BC and 6g Down ultimate floor loading.*

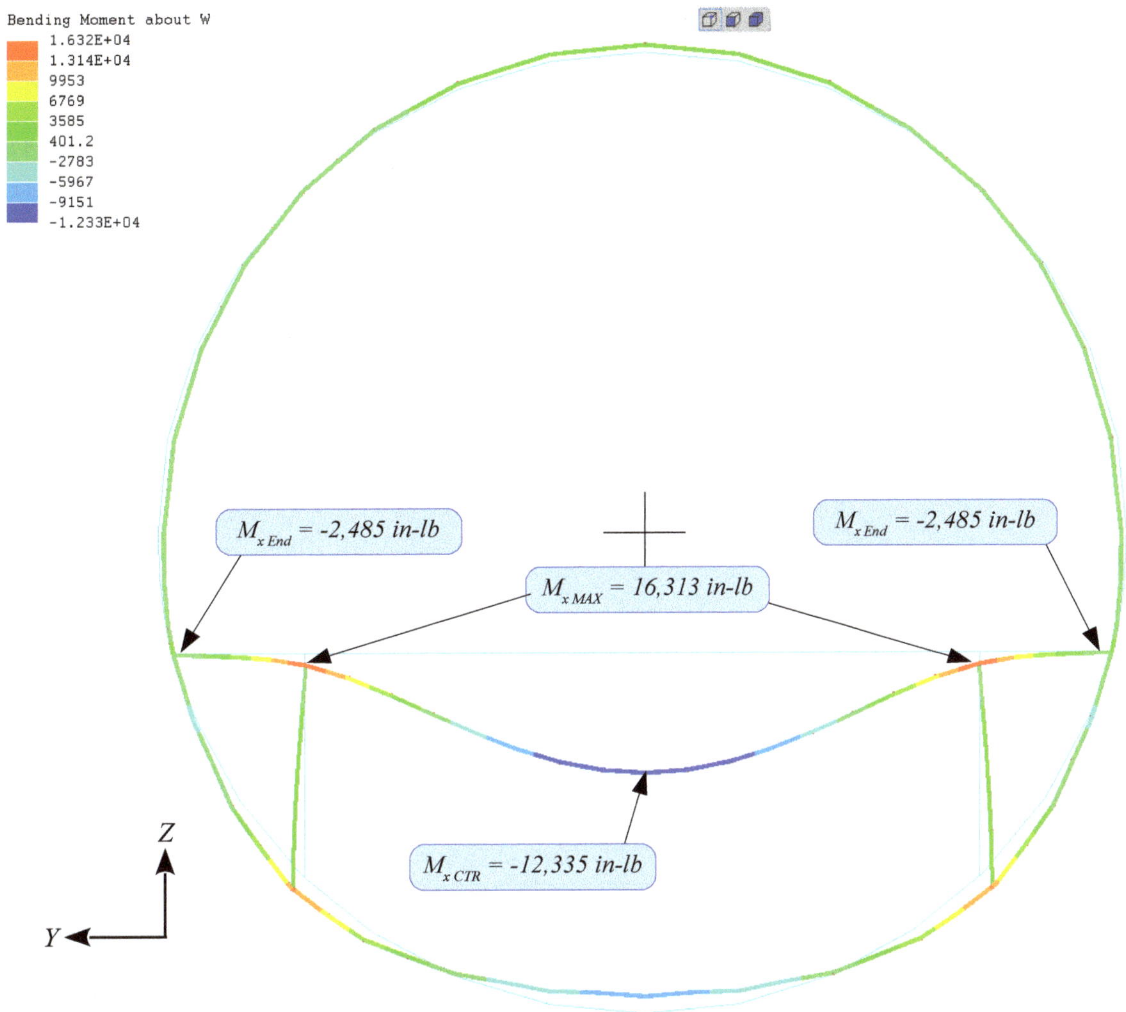

Figure 13-8 *– Fuselage section FEA bending moment results, 6g Down.*

From *Figure 13-8*, the end moment "M_{FEA}" obtained is −2,485 *in-lb*. To be consistent with the coordinate system convention I'm using, the moment is actually +2,485 *in-lb;* which produces compression in the Crossbeam upper cap. Comparing this moment to the one for an end-fixity of 100% (9,461 *in-lb*) from *Figure 13-5*, we can then calculate the end-fixity percentage as follows:

$$\%_{Fixity} = \frac{M_{FEA}}{M_{100}} \cdot 100 = \frac{2{,}485}{9{,}461} \cdot 100 = 26\%$$

From this result, I now know that the 50% end-fixity was conservative for the connections between the Crossbeam and the Frame but not for the Crossbeam itself. However, as noticed earlier, the end-fixity doesn't seem to affect the maximum bending moment in the Crossbeam much.

Therefore, for conservatism (and since one can't be certain about the actual fixity unless a very detailed FEM or a static test is performed), the Crossbeam and its connections will be analyzed by using the highest shear load and bending moment between the *simply supported* beam and the *fixed-fixed* beam configurations.

13.1.3 MATERIAL ALLOWABLES

The actual material allowables for all materials can be obtained from the latest edition of (ref. 8.). The **Crossbeams** and the **Stanchions**, are made out of *Al Aly 7050-T7451* plate between 1.00 *in.* and 2.50 *in.* thick. *Table 13-2* shows fictitious values.

Table 13-2 *- Al Aly 7050-T7451 Plate 2.50 in. fictitious material allowables.*

Aluminum Alloy 7050-T7451 Plate
Fictitious Values, NOT FOR DESIGN

	2.50" [ksi]
F_{tu}	75
F_{ty}	65
F_{cy}	60
F_{su}	40
F_{bru} $(e/D = 2.0)$	105
F_{bry} $(e/D = 2.0)$	85
$E, 10^3$	10
$G, 10^3$	3.8
μ	0.33
n	20

13.1.4 FASTENERS ALLOWABLES

Table 13-3 shows the ultimate shear allowables "P_{SU}" for 1/8", 5/32" and 3/16" *MS20470AD* solid rivets. Note that this data is fictitious, it is thus required to use the latest allowables possible from either (ref. 8.) or from the test results of the company you work for in your analysis.

Table 13-3 *- MS20470AD Solid rivets ultimate shear allowables.*

	MS20470AD Al Aly 2117-T3 $F_{su} = 30$ ksi		
	4	**5**	**6**
Dia. [in.]	0.1285	0.159	0.191
P_{su} *[lbs]*	**388**	**596**	**862**

Table 13-4 *- Shear correction factor "SF".*

	Diameter of rivet [in]			
Thickness [in]	3/32"	1/8"	5/32"	3/16"
0.025	0.972	–	–	–
0.032	1.000	0.964	–	–
0.036		0.980	–	–
0.040		0.996	0.964	–
0.045		1.000	0.98	–
0.050			0.996	0.972
0.063			1.000	1.000

13.1.5 SECTION PROPERTIES

The Crossbeam has a constant section all along its span except at the attachment to the Frame. Therefore two cross-sections exist. The floor panels are not considered effective in working with the Crossbeams. *Figure 13-9* shows the locations of both cross-sections while *Figure 13-10* and *Figure 13-11* show the dimensions and section properties for sections *A-A* and *B-B* respectively.

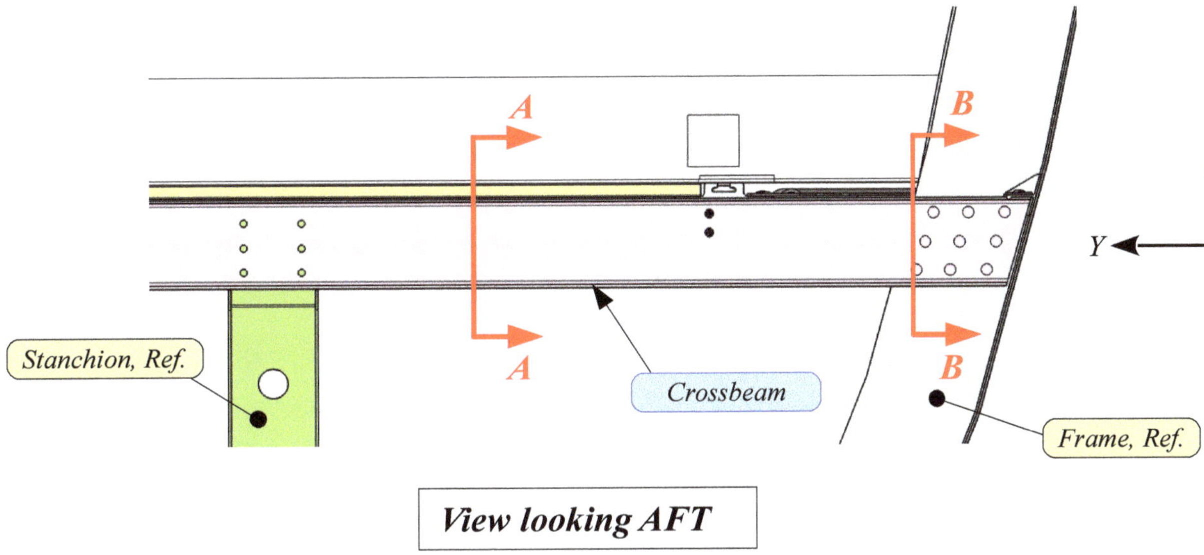

Figure 13-9 – *Crossbeam cross sections locations.*

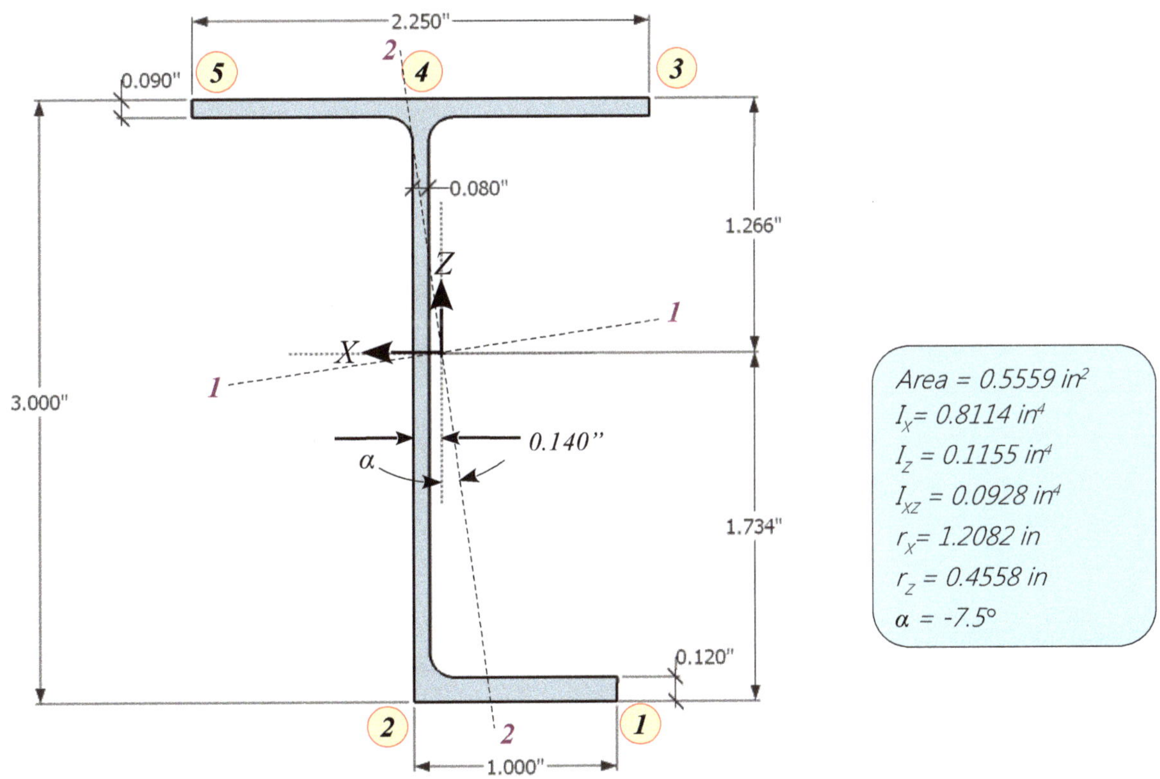

Figure 13-10 – *Crossbeam section A-A.*

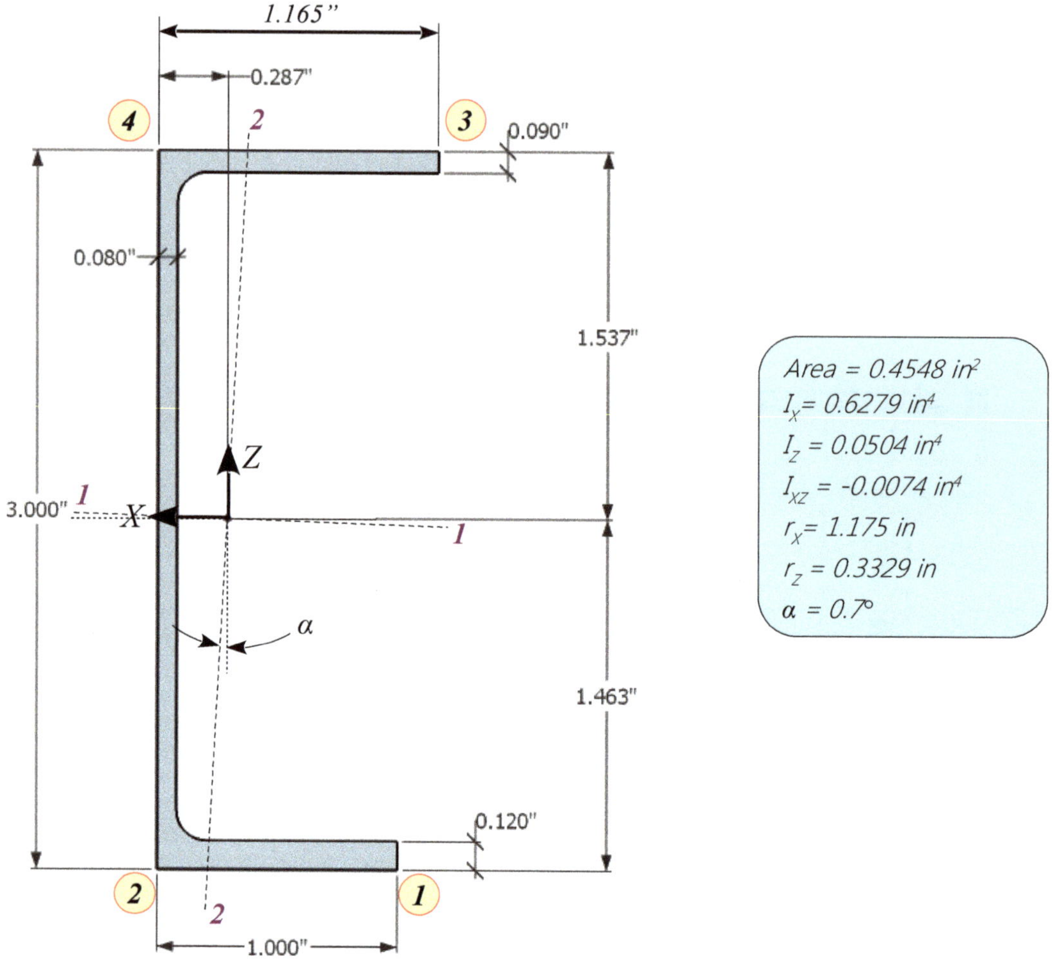

Figure 13-11 *– Crossbeam section B-B.*

13.1.6 CRIPPLING AT SECTION A-A

The crippling stress allowable "F_{CC}" will be calculated below for the section *A-A*. The section is divided into segments as shown in *Figure 13-12*. The length of segments *2* and *3* are calculated as *15t* or 15 (0.080) = 1.20 *in*.

All segments have one edge free except segments *2* and *3*.

The crippling allowable for each segment is calculated as shown below. Note that the crippling allowable "F_{CC}" of a segment can never be higher than F_{CY}. So the maximum value for F_{CCn} / F_{cy} is 1.0. The value at the abscissa of *Figure 13-13* is calculated as:

$$abscissa = \sqrt{F_{CY}/E}\left(b/t\right)$$

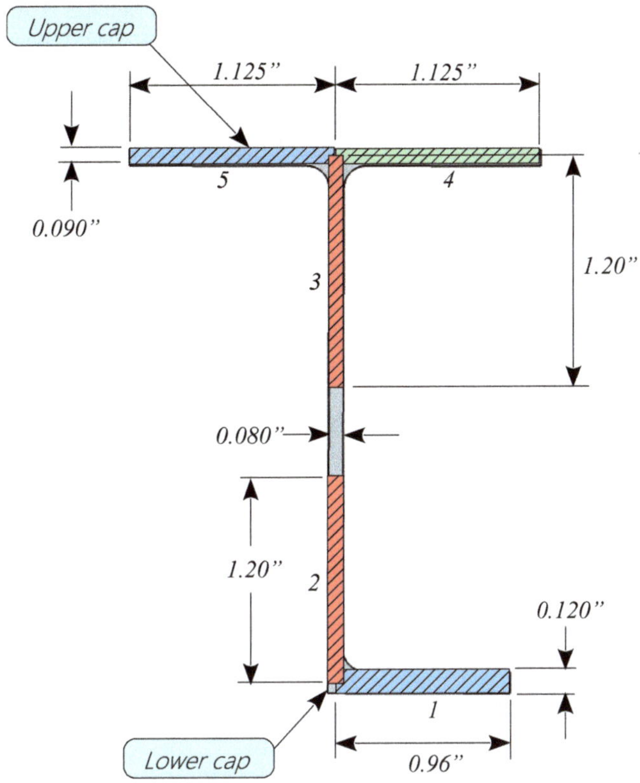

Figure 13-12 – *Effective segments for crippling at section A-A.*

See *Figure 13-13* for typical crippling allowable curves.

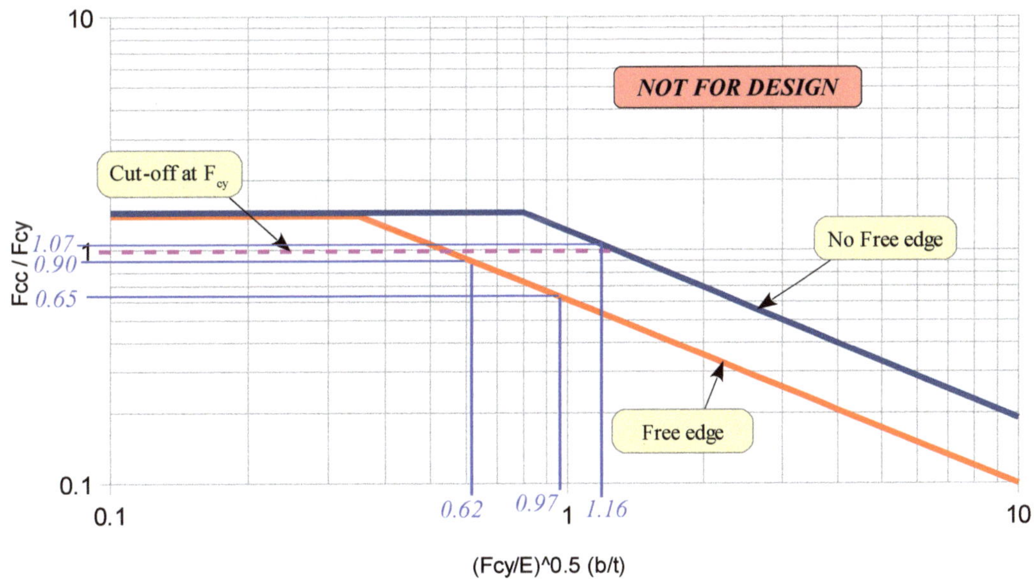

Figure 13-13 – *Typical Crippling curves.*

So for $E = 10.0*10^6$ *psi*, F_{cy} = 60,000 *psi* and from the dimensions shown in *Figure 13-12*, the crippling allowable of each segment is calculated in *Table 13-5*.

***Table 13-5** - Crippling analysis for the Crossbeam Section A-A.*

Segment	Free edge	b_n	t_n	$b_n t_n$	b_n / t_n	$(F_{cy} / E)^{0.5}$	$(F_{cy} / E)^{0.5}(b_n/t_n)$	F_{CCn} / F_{cy}	F_{CCn}	$t_n b_n F_{CCn}$
		[in]	[in]	[in²]					[ksi]	[lbs]
1	1	0.96	0.12	0.115	8.000	0.0775	0.62	0.90	54	6221
2	0	1.2	0.08	0.096	15.000	0.0775	1.16	1.00	60	5760
3	0	1.2	0.08	0.096	15.000	0.0775	1.16	1.00	60	5760
4	1	1.125	0.09	0.101	12.500	0.0775	0.97	0.65	39	3949
5	1	1.125	0.09	0.101	12.500	0.0775	0.97	0.65	39	3949
			$\Sigma b_n t_n =$	0.50970					$\Sigma b_n t_n F_{ccn} =$	25638

Thus the crippling allowable stress for the whole section *A-A* is:

$$F_{CC} = \frac{\sum b_n t_n F_{CCn}}{\sum b_n t_n} = \frac{25{,}638}{0.5097} = 50{,}301\ psi$$

13.1.7 CRIPPLING AT SECTION B-B

The crippling stress allowable "F_{CC}" will be calculated below for the section *B-B*. The section is divided into segments as shown in *Figure 13-14*. The length of segments *2* and *3* are calculated as *15t* or 15* (0.080") = 1.20 *in*. Segments *1* and *4* have one edge free except segments *2* and *3*.

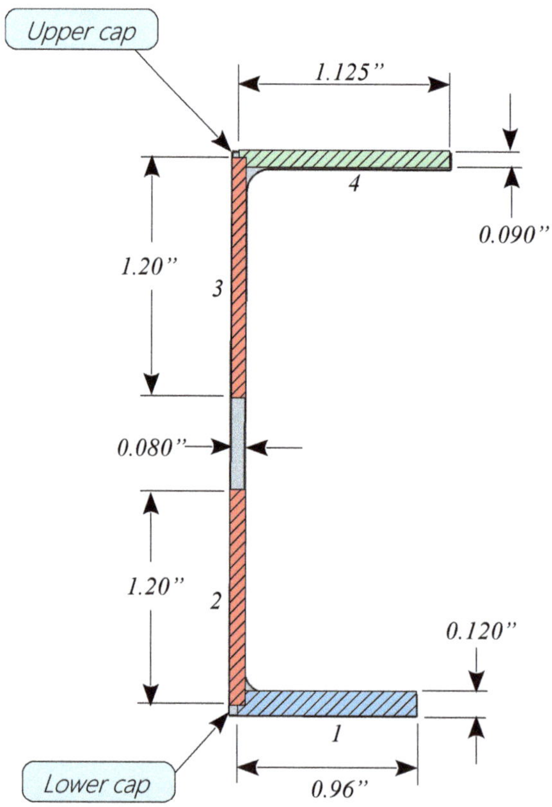

***Figure 13-14** – Effective segments for crippling at the Section B-B.*

The crippling allowable for each segment is calculated as shown below. Note that the crippling allowable "F_{CC}" of a segment can never be higher than F_{CY}. So the maximum value for F_{CCn} / F_{cy} is 1.0. The value at the abscissa of *Figure 13-13* is calculated as:

$$abscissa = \sqrt{F_{CY}/E}\,(b/t)$$

See *Figure 13-13* for typical crippling allowable curves.

So for $E = 10.0*10^6$ *psi*, F_{cy} = 60,000 *psi* and from the dimensions shown in *Figure 13-14*, the crippling allowable of each segment is calculated in *Table 13-6*.

Table 13-6 *- Crippling analysis for the Crossbeam Section B-B.*

Segment	Free edge	b_n	t_n	$b_n t_n$	b_n / t_n	$(F_{cy} / E)^{0.5}$	$(F_{cy} / E)^{0.5}(b_n/t_n)$	F_{CCn} / F_{cy}	F_{CCn}	$t_n b_n F_{CCn}$
		[in]	[in]	[in²]					[ksi]	[lbs]
1	1	0.96	0.12	0.115	8.000	0.0775	0.62	0.90	54	6221
2	0	1.2	0.08	0.096	15.000	0.0775	1.16	1.00	60	5760
3	0	1.2	0.08	0.096	15.000	0.0775	1.16	1.00	60	5760
4	1	1.125	0.09	0.101	12.500	0.0775	0.97	0.65	39	3949
			$\sum b_n t_n =$	0.40845					$\sum b_n t_n F_{ccn} =$	21690

Thus, the crippling allowable stress for section *B-B* is:

$$F_{CC} = \frac{\sum b_n t_n F_{CCn}}{\sum b_n t_n} = \frac{21{,}690}{0.40845} = 53{,}102 \ psi$$

13.1.8 CRIPPLING AT SECTION A-A LOWER CAP

The crippling stress allowable "F_{CC}" will be calculated below for the lower cap of section *A-A*. The section is divided into segments as shown in *Figure 13-15*. The length of segment *2* is calculated as *15t* or 15* (0.080") = 1.20 *in*. Segment *1* has one edge free while segment *2* has none.

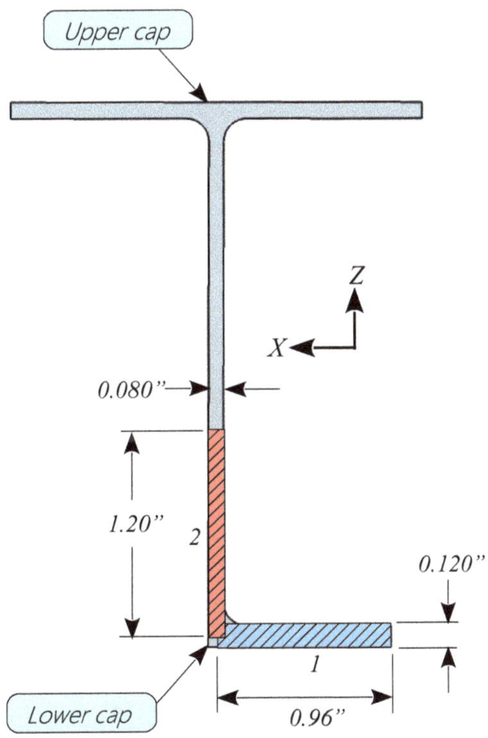

Figure 13-15 *– Effective segments for crippling at the lower cap of Section A-A.*

The crippling allowable for each segment is calculated as shown below. Note that the crippling allowable "F_{CC}" of a segment can never be higher than F_{CY}. So the maximum value for F_{CCn} / F_{cy} is 1.0. The value at the abscissa of *Figure 13-13* is calculated as:

$$abscissa = \sqrt{F_{CY}/E}\,(b/t)$$

See *Figure 13-13* for typical crippling allowable curves.

So for $E = 10.0*10^6$ *psi*, $F_{cy} = 60{,}000$ *psi* and from the dimensions shown in *Figure 13-15*, the crippling allowable of each segment is calculated in *Table 13-7*.

Table 13-7 *- Crippling analysis for the Crossbeam lower cap of Section A-A.*

Segment	*Free edge*	b_n [in]	t_n [in]	$b_n t_n$ [in²]	b_n / t_n	$(F_{cy} / E)^{0.5}$	$(F_{cy} / E)^{0.5}(b_n/t_n)$	F_{CCn} / F_{cy}	F_{CCn} [ksi]	$t_n b_n F_{CCn}$ [lbs]
1	1	0.96	0.12	0.115	8.000	0.0775	0.62	0.90	54	6221
2	0	1.2	0.08	0.096	15.000	0.0775	1.16	1.00	60	5760
			$\sum b_n t_n =$	0.21120				$\sum b_n t_n F_{ccn} =$		11981

Thus, the crippling allowable stress for the lower cap of section *A-A* is:

$$F_{CC} = \frac{\sum b_n t_n F_{CCn}}{\sum b_n t_n} = \frac{11{,}981}{0.2112} = 56{,}727 \ psi$$

13.1.9 CRIPPLING AT SECTION A-A UPPER CAP

The crippling stress allowable *"Fcc"* will be calculated below for the upper cap of section *A-A*. The section is divided into segments as shown in *Figure 13-16*. The length of segment *3* is calculated as *15t* or 15* (0.080") = 1.20 *in*. Segments *4* and *5* have one edge free while segment *3* has none.

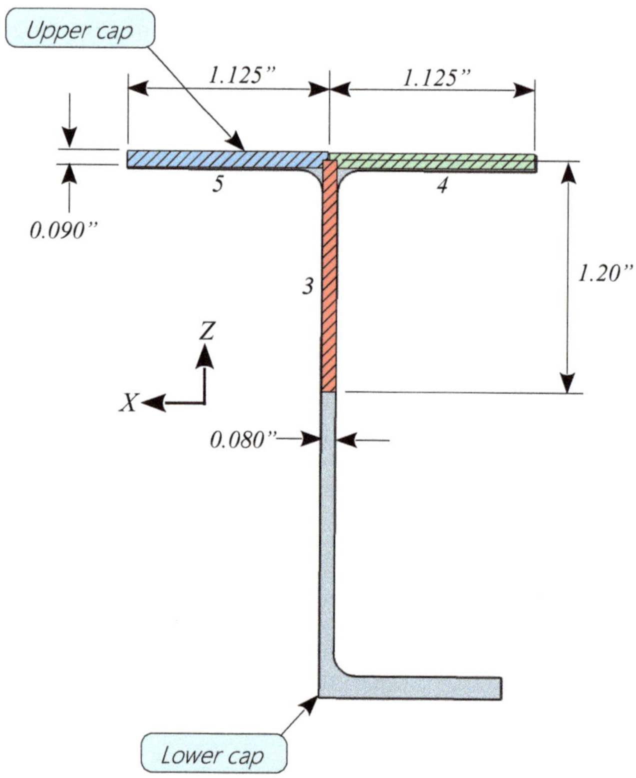

Figure 13-16 *– Effective segments for crippling at the upper cap of Section A-A.*

The crippling allowable for each segment is calculated as shown below. Note that the crippling allowable "F_{CC}" of a segment can never be higher than F_{CY}. So the maximum value for F_{CCn} / F_{cy} is 1.0. The value at the abscissa of *Figure 13-13* is calculated as:

$$abscissa = \sqrt{F_{CY}/E}\,(b/t)$$

See *Figure 13-13* for typical crippling allowable curves. So for $E = 10.0*10^6$ *psi*, $F_{cy} = 60{,}000$ *psi* and from the dimensions shown in *Figure 13-16*, the crippling allowable of each segment is calculated in *Table 13-8*.

Table 13-8 *- Crippling analysis for the Crossbeam upper cap of Section A-A.*

Segment	*Free edge*	b_n	t_n	$b_n t_n$	b_n / t_n	$(F_{cy} / E)^{0.5}$	$(F_{cy} / E)^{0.5}(b_n/t_n)$	F_{CCn} / F_{cy}	F_{CCn}	$t_n b_n F_{CCn}$
		[in]	*[in]*	*[in²]*					*[ksi]*	*[lbs]*
3	0	1.2	0.08	0.096	15.000	0.0775	1.16	1.00	60	5760
4	1	1.125	0.09	0.101	12.500	0.0775	0.97	0.65	39	3949
5	1	1.125	0.09	0.101	12.500	0.0775	0.97	0.65	39	3949
			$\Sigma b_n t_n =$	0.29850					$\Sigma b_n t_n F_{ccn} =$	13658

Thus, the crippling allowable stress for the upper cap of section *A-A* is:

$$F_{CC} = \frac{\sum b_n t_n F_{CCn}}{\sum b_n t_n} = \frac{13{,}658}{0.2985} = 45{,}754 \ psi$$

13.1.10 STRESS ANALYSIS

There is no axial load in the Crossbeam. Thus, the following checks have to be performed:

- Bending stress
- Crippling
- Free flange buckling
- Column stability of lower cap
- Shear buckling of webs
- Connections (See section *13.1.14)*

13.1.10.1 Section A-A

From my above assumption, the Crossbeam is simply modeled as a beam reacting the vertical distributed ultimate load. Hence, from that model, no axial load exists, only bending and shear.

Bending

The maximum bending moment occurs for the *6g Down LC* at the Stanchion attachment for the fixed-fixed condition *(Figure 13-5)* with $M_X = -21{,}096$ *in-lb* while, it is 14,309 *in-lb* at mid span *(Figure 13-4)* for the simply-supported condition. Complex bending is used as the Crossbeam is not considered stabilized laterally. The complex bending stress is thus calculated as:

$$f_{by} = -\frac{(M_X I_Z + M_Z I_{XZ})z + (M_Z I_X + M_X I_{XZ})x}{I_X I_Z - I_{XZ}^2}$$

Calculation example at point ***3*** of section *A-A* due to the *6g Down LC* at the Stanchions (posts):

For
$M_X = -21{,}096$ *in-lb* $\quad I_X = 0.8114\ in^4 \quad x_3 = -1.025\ in$
$M_Z = 0$ *in-lb* $\quad I_Z = 0.1155\ in^4 \quad I_{XZ} = 0.0928\ in^4 \quad z_3 = 1.266\ in$

$$f_{by3} = -\frac{(-21{,}096 \cdot 0.1155 + 0 \cdot 0.0928)(1.266) + (0 \cdot 0.8114 + (-21{,}096)0.0928)(-1.025)}{0.8114 \cdot 0.1155 - 0.0928^2} = 59{,}825\ psi$$

Table 13-9 shows the bending stresses in the Crossbeam at the Stanchions (*BL ±55.27*) at all corners for the *6g Down LC*.

Table 13-9 *- Bending stresses at corners of section A-A due to* M_X *at Stanchions due to 6g Down.*

Points	x [in]	z [in]	f_b [psi]
1	-0.860	-1.734	-29,862
2	0.140	-1.734	-52,866
3	-1.025	1.266	59,825
4	0.140	1.266	33,026
5	1.225	1.266	8,067

Table 13-10 shows the bending stresses in the Crossbeam at mid-span (*BL 0.0*) at all corners.

Table 13-10 *- Bending stresses at corners of section A-A due to* M_X *at Mid-span due to 6g Down.*

Points	x [in]	z [in]	f_b [psi]
1	-0.860	-1.734	20,255
2	0.140	-1.734	35,858
3	-1.025	1.266	-40,578
4	0.140	1.266	-22,401
5	1.225	1.266	-5,472

Lower cap Margin of safety

The maximum bending stress in compression obtained above at the Stanchions for section *A-A* is compared to the crippling stress allowable of F_{CC} = 50,301 *psi* calculated in section *13.1.6*. *Table 13-9* shows that the maximum compression stress occurs at point *2* with a stress value of −52,866 *psi*. At first, one would think that we have a negative MS since that stress value is higher than the crippling allowable. In order to assess whether it is actually the case or not, we have to take the following into consideration:

- The crippling allowable was calculated from the whole cross section. In a case like this, where there is no axial compression load, the crippling allowable of the lower cap only (*or the one in compression*) could have been calculated using segments *1* and *2* only from *Figure 13-12*. By doing so, we would have obtained a much higher F_{CC} of 56,727 *psi*.
- The supports at the stanchions were considered being simple points while the Stanchions have a 3.0 *in.* width. Therefore, having taken that into consideration, the peak stress would have been somewhat a little less.
- Complex bending was used as if no lateral support existed. The Stanchions being fastened to the Crossbeam exactly at the location where the maximum bending moment occurs, one could expect the Stanchions to provide some lateral restraint which would be forcing the Crossbeam to bend straight down with little or no twist. Using the simpler bending stress equation (*Mc/I*) one would get a maximum bending stress of −45,083 *psi*.
- The critical load case is an ultimate *Emergency Landing* condition and some material yielding leading to some load redistribution is expected to occur. Thus, one would expect the Crossbeam lower cap to see a constant load hence a constant stress across its section. Knowing that, one could calculate the average compression stress between points *1* and *2* and compare it to the whole section (*or more accurately to the lower cap section*) crippling allowable.

The above information tells me that by using the maximum stress at point *2* only and the crippling allowable from the whole section (*where I know that the upper cap flanges are long and thin and contribute in lowering the whole section crippling allowable*) may be too conservative. From the many options available above (*and after consulting with the most experienced stressman in the room...*), I decided to be a little more accurate yet remaining conservative in my approach. Therefore, I chose to keep the maximum stress at point *2* but compare it to the lower cap crippling allowable calculated in section *13.1.8*.

The MS in crippling for the lower cap of section *A-A* is calculated below.

Upper cap Margin of safety

$$MS = \frac{F_{CC}}{f_{b2}} - 1 = \frac{56{,}727}{52{,}866} - 1 = 0.07$$

Crossbeam Lower cap Section A-A, Ultimate	***0.07***
6g Down	*Crippling*
Al Aly 7050-T7451	

How much "fat" (conservatism) you can remove to get a positive *MS* may depends on the company policy, the *DER*, some test results, or whether the drawing release is due on Friday... One thing for sure, it is always wise to be as transparent as possible in your decision and consult your stress group lead about it. Obviously, you also have to be comfortable with the decision. If you believe a redesign is necessary than ask for it.

Similarly as for the lower cap, the upper cap of section *A-A* is checked for crippling. The maximum bending stress in compression obtained above at mid-span for section *A-A* is compared to the crippling stress allowable of F_{CC} = 45,754 *psi* calculated in section *13.1.9*. *Table 13-10* shows that the maximum compression stress occurs at point *3* with a stress value of −40,758 *psi*.

The MS in crippling for the upper cap of section *A-A* is calculated below.

$$MS = \frac{F_{CC}}{f_{b3}} - 1 = \frac{45{,}754}{40{,}758} - 1 = 0.12$$

Thus, section *A-A* is thus more critical at the Stanchions than at mid-span.

Free flange buckling

Since the bending moment is around the longitudinal "*X*" axis, the flanges that make up the upper/lower caps will be loaded together simultaneously. In a case where the flanges get loaded in compression and that the free flanges were to buckle, the integrity of the structure wouldn't necessarily be compromised as the section will still retain most of its bending capability. Crippling of the caps being the last line of defense.

The free-flange buckling will be performed for the demonstration for both the upper and the lower caps. See section *13.1.12* for those analyses.

Tension stress

From *Table 13-9*, the maximum tension stress is obtained at point *3* with 59,825 *psi*. That stress is compared to F_{TU} = 75 *ksi* from *Table 13-2* and the MS is calculated below.

$$MS = \frac{F_{TU}}{f_{b3}} - 1 = \frac{75{,}000}{59{,}825} - 1 = 0.25$$

Shear stress

The shear stress is calculated through *Section A-A* by assuming the shear to be taken by the vertical web only. Therefore, the web shear area is: A_S = (3.0)*(0.080) = 0.24 in^2. The maximum transverse shear load "V_Z" is 1,553 *lbs,* from *Figure 13-5* for the fixed-fixed condition, just outboard of the Stanchion attachment. The average ultimate shear stress is calculated as:

$$f_s = \frac{V_Z}{A_S} = \frac{1{,}553}{0.24} = 6{,}471\ psi$$

This shear stress will be used later to check for the web shear buckling.

13.1.10.2 Section B-B

Section *B-B* is right where one side of the Crossbeam upper cap is trimmed so that the Crossbeam web can contact the web of the Frame for the connection. From my above assumption, the Crossbeam is simply modeled as a beam reacting the vertical distributed ultimate load. Hence, from that model, no axial load exists, only bending and shear.

Bending

The end moment is used to analyze the bending stresses for the fixed-fixed condition *(Figure 13-5)* with M_X = 9,461 *in-lb.* Because the Crossbeam is right next to the Frame which provides some lateral stability, it wouldn't be necessary to use the complex bending equation. However, since the section is almost symmetric, there isn't much difference whether we use the complex bending or not. To be consistent, I'll simply use it again. The complex bending stress is thus calculated as:

$$f_{by} = -\frac{(M_X I_Z + M_Z I_{XZ})z + (M_Z I_X + M_X I_{XZ})x}{I_X I_Z - I_{XZ}^2}$$

Calculation example at point ***4*** of section *B-B* due to the ***6g*** *Down LC* at the Frame:

For

M_X = 9,461 in-lb	I_X = 0.6279 in^4	x_4 = 0.287 in	z_4 = 1.537 in
M_Z = 0 in-lb	I_Z = 0.0504 in^4	I_{XZ}= -0.0074 in^4	

$$f_{by4} = -\frac{(9{,}461 \cdot 0.0504 + 0 \cdot (-0.0074))(1.537) + (0 \cdot 0.6279 + 9{,}461 \cdot 0.0928)(0.287)}{0.6279 \cdot 0.0504 - (-0.0074)^2} = -23{,}835\ psi$$

Table 13-11 shows the bending stresses in the Crossbeam through section *B-B* at the Frame (*BL ±76.75*) at all corners.

Table 13-11 - *Bending stresses at corners of section B-B due to M_X at the Frame.*

Points	x [in]	z [in]	f_b [psi]
1	-0.713	-1.463	23,662
2	0.287	-1.463	21,446
3	-0.878	1.537	-21,253
4	0.287	1.537	-23,835

Margin of safety

The maximum bending stress in compression obtained above for section *B-B* is compared to the crippling stress allowable of F_{CC} = 53,102 *psi* calculated in section *13.1.7*. The MS in crippling is calculated below.

$$MS = \frac{F_{CC}}{|f_{b4}|} - 1 = \frac{53{,}102}{|-23{,}835|} - 1 = 1.22$$

Shear stress

The shear stress is calculated through section *B-B* by assuming the shear to be taken by the vertical web only. Therefore, the web shear area is: $A_S = (3.0)*(0.080) = 0.24\ in^2$. The transverse shear load "V_Z" is 1,194 *lbs,* from *Figure 13-5* for the fixed-fixed condition, at section *B-B*. The average ultimate shear stress is calculated as:

$$f_s = \frac{V_Z}{A_S} = \frac{1{,}194}{0.24} = 4{,}975\ psi$$

Since the shear stress calculated above is well below the ultimate shear allowable of F_{SU} = 40 *ksi* and no shear buckling can occur due to the existing group of fasteners for the connection to the Frame, hence the MS is greater than 1.0 in shear.

13.1.11 COLUMN STABILITY OF LOWER CAP

Because Seat track rails and floor panels are fastened to the upper cap of the Crossbeams, any compression load in the upper cap couldn't lead to a stability problem. The upper cap is prevented from moving sideways, it can only move up and down from bending around the "*X*" axis, as intended. For the *−6g Down LC*, the upper cap was in compression and the lower cap was in tension so no stability problem could occur.

On the other hand, in the case of another load condition, for example the emergency landing case N_Z = *+3.0g Up*, the Crossbeam lower cap would be in compression. This could lead to a column stability problem as the lower cap is not laterally stabilized as well as the upper cap.

Using *Figure 13-4*, since the internal loads of the Crossbeam due to vertical loads can simply be calculated by using a linear ratio, the new shear and moment diagram for a simply supported Crossbeam under the *+3.0g Up LC* is shown in *Figure 13-17* using the ratio: 3/−6 = ***−0.5***.

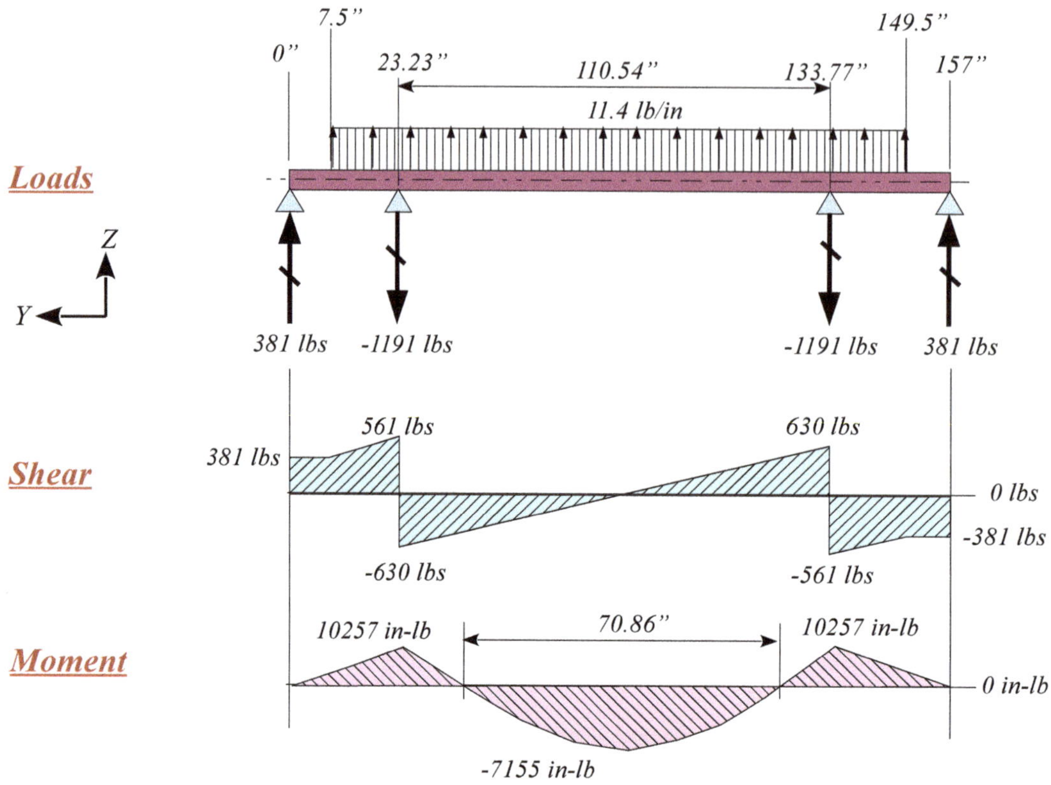

Figure 13-17 – *Crossbeam FBD 3g Up, Simply supported at ends.*

13.1.11.1 Lower cap without lateral supports.

There are recipes "equations" to evaluate the lateral stability of beams but, in this case, I like to conservatively consider the lower cap as a column of its own that is stabilized in the vertical direction. See *Figure 13-20* for the dimensions and the section properties of the lower cap. The lower cap is restrained vertically but it is conservatively assumed disconnected from the rest of the beam therefore free to move laterally.

As shown in *Figure 13-17*, the portion where the lower cap is in compression is where the bending moment is negative and that is smaller than the distance between the two stanchions. In other circumstances, the beam could have been assumed pinned at the point where the bending moment is 0. But here, the lower cap would tend to move laterally from the center-line all the way to the Stanchions; the only actual lateral supports. The column length is then 110.54 *in*. As shown in *Figure 13-18*, since it is symmetrical, and to help with the analysis, the column is considered split in half with one end guided in rotation around *"Z"* at the center-line and, one end pinned at the Stanchion.

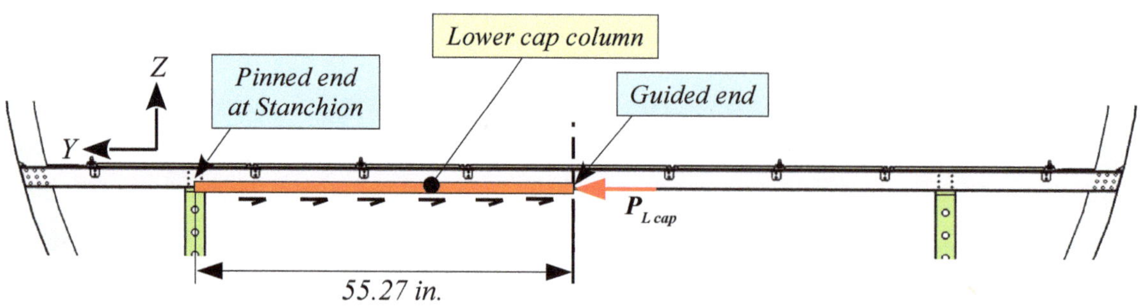

Figure 13-18 – *Crossbeam lower cap column assumption.*

Now, I couldn't find any end-fixity coefficient for pinned-guided columns. However, this is exactly the same as a free-fixed column, for which an end-fixity coefficient exists. *Figure 13-19* shows the pinned-guided *(a)* and the free-fixed *(b)* idealized boundary conditions that have the exact same behavior since the reaction in "*X*" was moved from the left end, in the figure, to the right end.

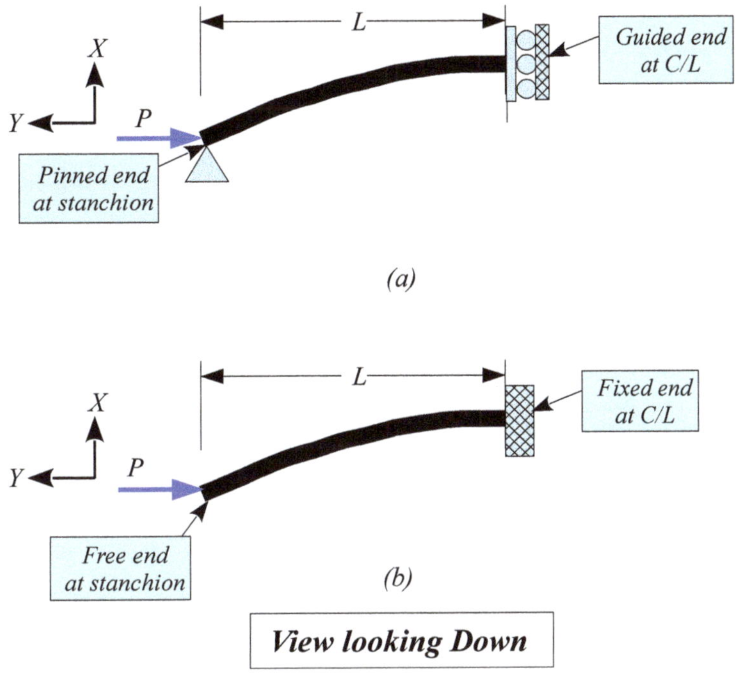

Figure 13-19 *– Crossbeam lower cap column B.C.'s*

Figure 13-21 shows the dimensions and the section properties of the Crossbeam lower cap.

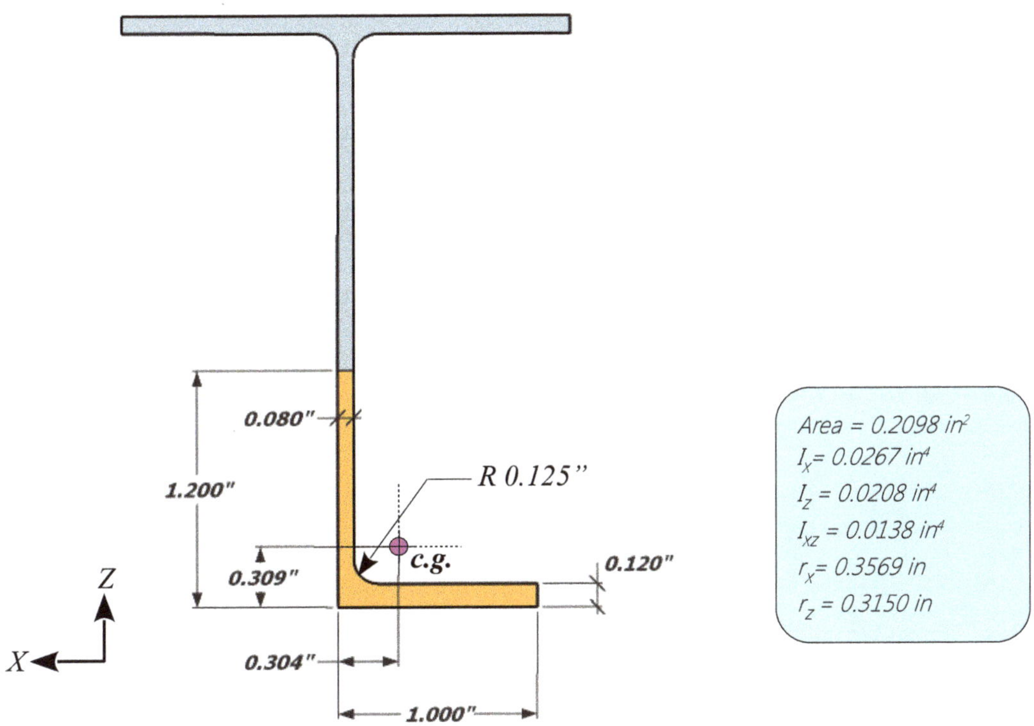

Figure 13-20 *– Crossbeam lower cap dimensions and section properties.*

Free-Fixed column

For this assumption, the column is therefore idealized as shown in *Figure 13-19 (b)* and *Figure 13-21*. From *Figure B-2* for a uniform column distributed axis load, one end fixed, one end free, the end-fixity coefficient is thus $C = 0.794$. Therefore, the effective column length "L'" is calculated as follows.

Hence, for $L = 55.27$ *inches*;

$$L' = \frac{L}{\sqrt{c}} = \frac{55.27}{\sqrt{0.794}} = 62.0 \ in.$$

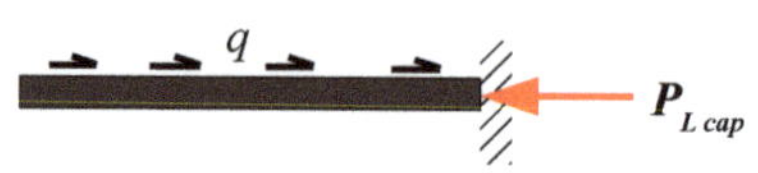

***Figure 13-21** – Seat track rail lower cap idealized column.*

The critical Euler column buckling load is calculated as:

$$P_{CR} = \frac{\pi^2 (E_t I_Z)}{(L')^2}$$

then

$$P_{CR} = \frac{\pi^2 \, 10*10^6 (0.0208)}{(62.0)^2} = 534 \ lbs$$

> For
> $E_t = E = 10.0*10^6$ *psi*
> $I_z = 0.0208$ *in*^{4}
> $L' = 62.0$ *in.*
> $A = 0.2098$ *in*^{2}

The critical compression stress is thus:

$$F_{CR} = \frac{P_{CR}}{A} = \frac{534}{0.2098} = 2{,}545 \ psi$$

Bending stress

The maximum stress in the lower cap from *Table 13-10* occurred at point ***2*** with 35,858 *psi* for the *6g Down LC*. Using the ***-0.5*** ratio, the stress for the *3g Up LC* is calculated as: $f_b = (-0.5)\ 35{,}858 = -17{,}929$ *psi*. This is much higher than the compression allowable calculated above. Therefore, it is clear to me that additional lateral supports are necessary!

> As a double check, I did a Lateral Stability check based on (ref. 10. §10.5 case 8). I used the approach for the equal flanged I-beam (A) by assuming the beam simply supported at the Stanchions and a uniformly distributed load on it. This very conservative approach neglected the fact that the upper cap of the beam was completely restrained laterally and that only the lower cap could move. With that assumption I got an *MS* of −0.12.
>
> Then as a second double *(triple)* check, I assumed the beam to be laterally supported at its center only (ref. 10. §10.5 case 6). I know this was un-conservative for the lower cap but it was still very conservative for the upper cap. As a result, this gave me an *MS* of +0.15.
>
> Therefore, to the light of the above double checks, the actual *MS* would be somewhere in between and most likely be very close to 0. Too close for comfort anyway...

Small FEM verification

In order to make sure about the above result, I did a simple FEM of the whole Crossbeam lower cap (***110.54*** *in.* long) pinned at both ends and having a symmetrical (from the center) distributed unit load. As shown in ***Figure*** *13-22,* I obtained a critical compression load of 498 *lbs* for the first buckling mode. As one can notice, the FEM shows the twisting effect of the *"L"* shaped cap which is, of course, ignored in the hand calcs and contributes to lower the allowable. Nevertheless, this is within 7% of the hand calculated number! Therefore I have a good confidence about the column idealization.

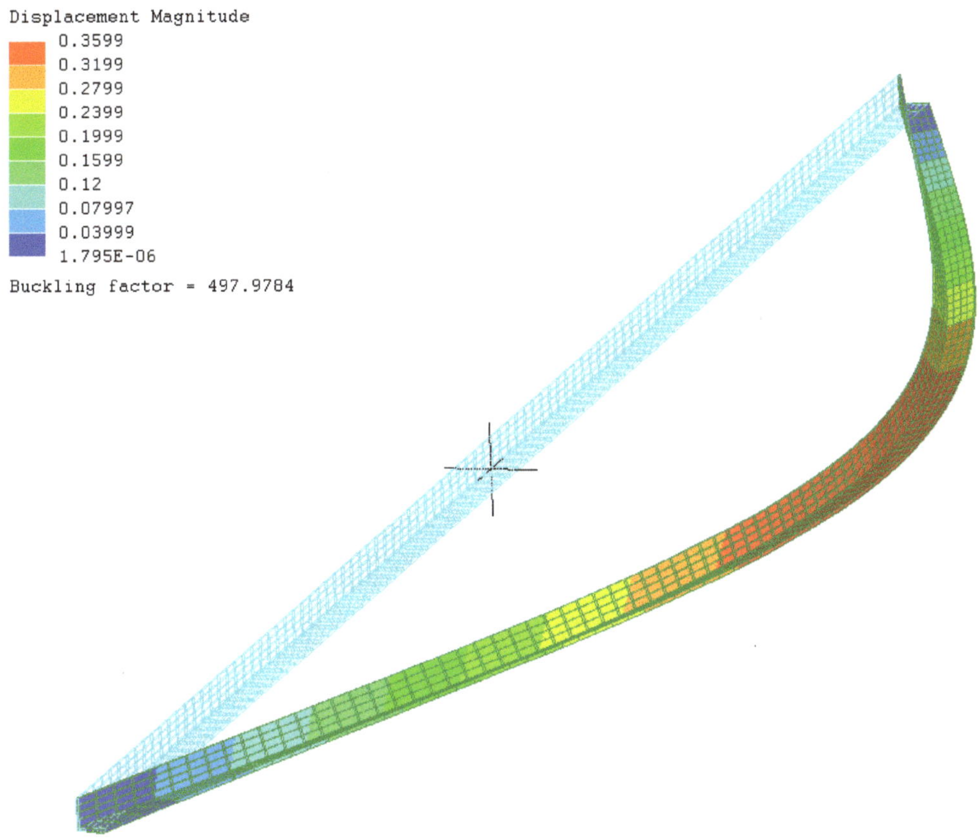

Figure 13-22 – *Seat track rail lower cap idealized column.*

Therefore, I came to the conclusion that the Crossbeam lower cap needed to be stabilized laterally with some additional structure located at some of the Seat track rails. The lower cap could also be "*beefed-up*" by making it thicker, and/or by adding a flange to make it an inverted "*TEE*" cap instead of an "*L*". This is where different solutions need to be analyzed to determine which one is best.

Sometimes, the best solution is not the lightest or the prettiest but it is the one that requires the least compromise and, generally, the one chosen is the fastest and the cheapest to implement.

Recommendation: Add lateral supports to Crossbeam lower cap at Seat track rails.

New Lateral supports

There are many ways one can add lateral supports to the Crossbeam lower cap. *Figure 13-23* shows many possible means of stabilizing the lower cap. Each of them is described below:

1. **Longitudinal stringers:** They connect all Crossbeams lower caps together up to a location where an Intercostal or any structural member connected to the Seat track rails exists. Due to fatigue and Damage tolerance, we tend to avoid having fasteners in the lower cap flange; especially when the lower cap gets in tension every flight.
2. **Built-up Intercostal:** This structure is fastened to the ribs (*panel breakers*) of the Crossbeam and is connected to the Seat track rail. This also stiffens the Seat track rail substantially.
3. **Gussets:** The Gussets are connected to the Crossbeam ribs and to the Seat track rails.
4. **New lower cap:** The Crossbeams are redesigned to have stronger lower caps possibly avoiding the need for lateral supports. This requires more analyses as it is a trial and error process.
5. **Intercostal:** In locations where no seat track rail exists, an Intercostal is installed. It is connected to the crossbeam ribs and to the floor panels directly.
6. **Diagonal braces:** Also called *St-Andrew's cross*, this arrangement connects the lower cap to the adjacent Crossbeam's upper cap. Prone to vibrations and seldom used.

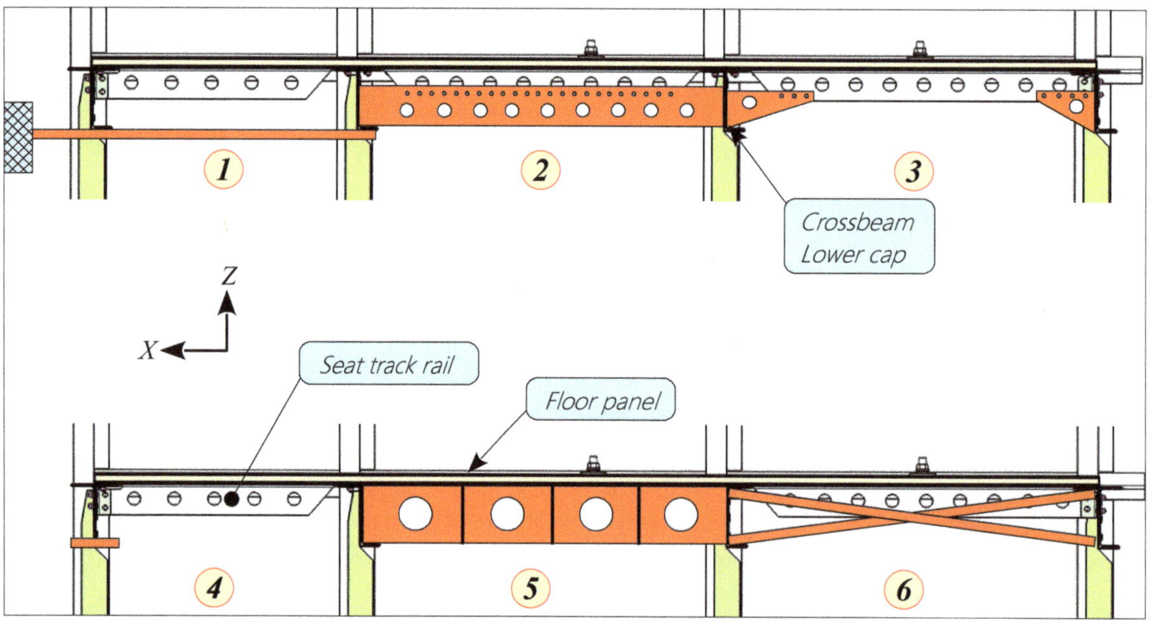

Figure 13-23 *– Crossbeam lower cap stabilizing features.*

13.1.11.2 Re-analyzing the Lower cap with lateral supports.

In our example, I will use the configuration no. *3* with the Gussets. (*any configuration would have done the same for this analysis*). Those Gussets will be located at the most inboard Seat track rails located at *BL* ±15.27. The Crossbeam span between those two Seat track rails is the longest with 30.54 *in*. Since this is the location where the largest bending moment exists (and the largest compression stress in the lower cap), the analysis of this span only is sufficient and covers the other Crossbeam spans.

For this analysis I am conservatively assuming that the column is pinned at *BL* ±15.27 and that it is subjected to a constant compression load "*P*" (no shear flow).

The internal load in the lower cap is conservatively calculated by assuming the maximum stress to be across the whole lower cap section. Hence, taking the maximum compression stress (−17,929 *psi*) and multiplying it by the lower cap cross section area (0.2098 in^2 from *Figure 13-20*) one gets an internal compression force "*P*" of −3,762 *lbs*.

Pinned-pinned column

For this assumption, the column is therefore idealized as shown in *Figure 13-24*. For a uniform column having end loads, both ends pinned, the end-fixity coefficient is thus $C = 1.0$. Therefore, the effective column length "L'" is calculated as follows:

Hence, for $L = 30.54$ *inches*;

$$L' = \frac{L}{\sqrt{c}} = \frac{30.54}{\sqrt{1.0}} = 30.54 \ in.$$

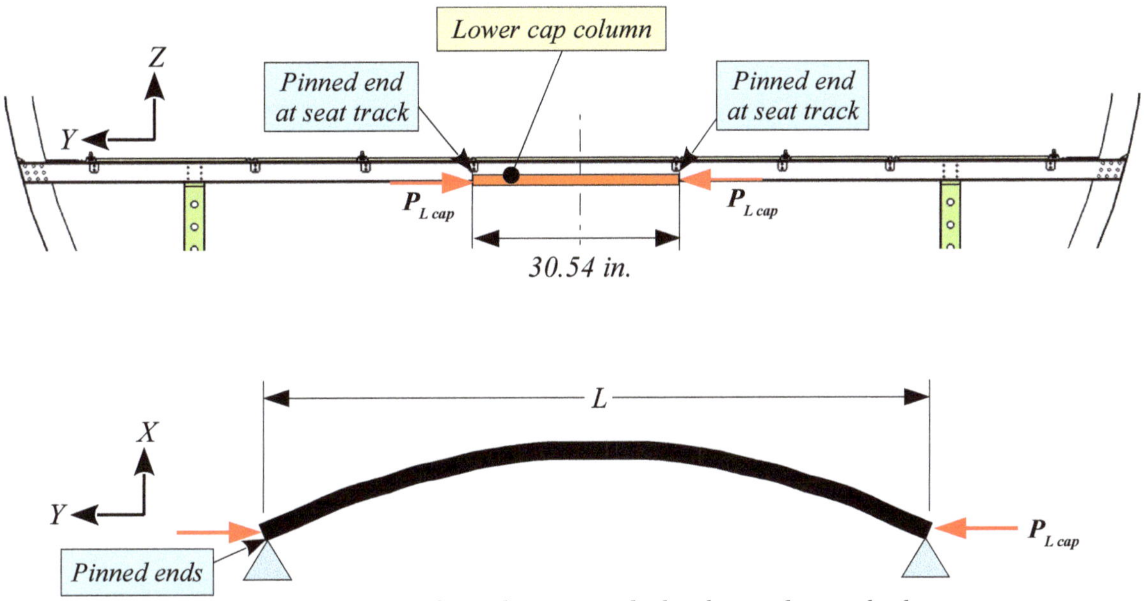

Figure 13-24 – *Crossbeam lower cap idealized pinned-pinned column.*

The critical Euler column buckling load is calculated as:

$$P_{CR} = \frac{\pi^2 (E_t I_Z)}{(L')^2}$$

then

$$P_{CR} = \frac{\pi^2 \ 10*10^6 (0.0208)}{(30.54)^2} = 2{,}201 \ lbs$$

For
$E_t = E = 10.0*10^6$ psi
$I_z = 0.0208 \ in^4$
$L' = 30.54$ in.
$A = 0.2098 \ in^2$

This critical compression load is well below the lower internal load of −3,762 *lbs*. However, assuming the column to have a constant load was overly conservative since the internal load varies due to the shear flow "*q*". Therefore, I decided to be closer to the "truth" while keeping some conservatism. So I made the following boundary conditions idealization for the lower cap.

Fixed-Free column with lateral spring

For this assumption, the column is therefore idealized as shown in *Figure 13-27*. Since the shear flow "*q*" is relatively small, I'll conservatively ignore it and I'll assume that the maximum load is constant from one end of the column to the other. However, I'll take into account the lateral spring that exists since the lower cap is attached to the upper cap which is fastened to the floor panels.

As shown in *Figure 13-25*, the upper cap flanges are thus considered fixed and the Crossbeam web, considered as a vertical beam, is free at the lower cap's *c.g.* where the spring is considered. For conservatism, the part of the web making up the beam is only assumed to be 1.0 *inch* wide.

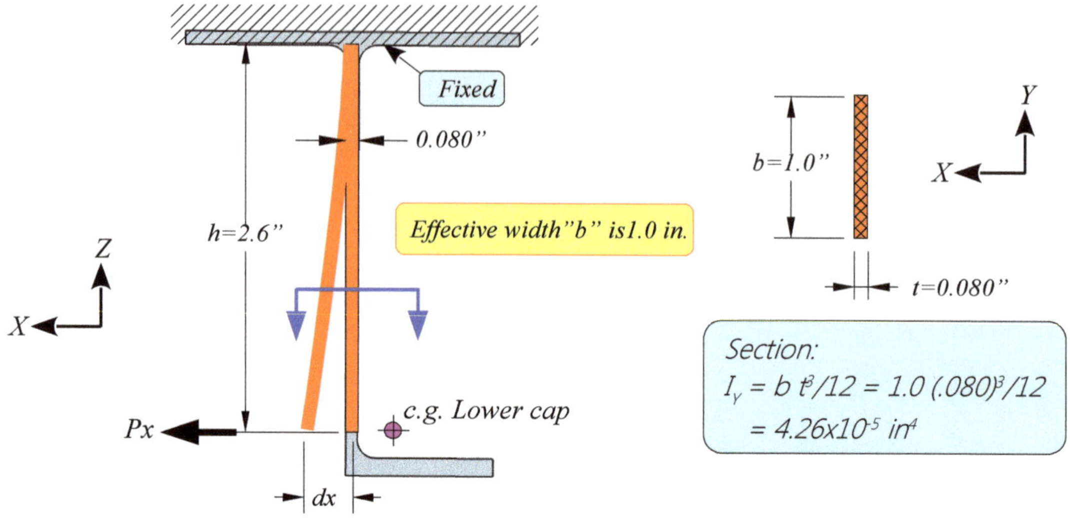

***Figure 13-25** – Lateral stiffness (spring) at lower cap.*

In addition to this idealization, instead of conservatively assuming that the extreme fiber bending stress is constant through the cap and calculating the internal cap load from it, the Crossbeam is considered as having an upper and a lower cap with very small cross-section areas. Thus, the upper and lower caps internal loads are calculated as a couple by dividing the bending moment "*M*" by the vertical distance between the centroid of the upper and lower caps "*h*". *Figure 13-26* shows the distance between the upper and the lower caps. The couple force "*P*" is calculated as:

$$P = \frac{M}{h} = \frac{-7{,}155}{2.466} = -2{,}902\ lbs$$

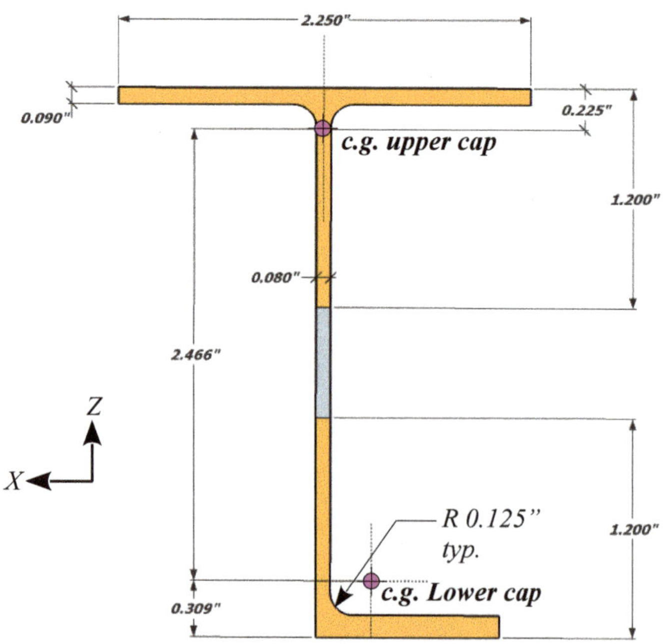

***Figure 13-26** – Crossbeam distance between upper and lower caps.*

At the aircraft center-line where, from *Figure 13-17* for the *3g Up LC*, the maximum bending moment is −7,155 *in-lb*, the cap load is thus calculated as $P_{CL} = -7{,}155 \,/\, 2.466 = -2{,}902$ *lbs*. Then, considering half the span (from the center-line to the seat track support), the column is idealized as guided having a lateral spring at the center due to symmetry and simply-supported at the Seat track rail as shown in *Figure 13-27*.

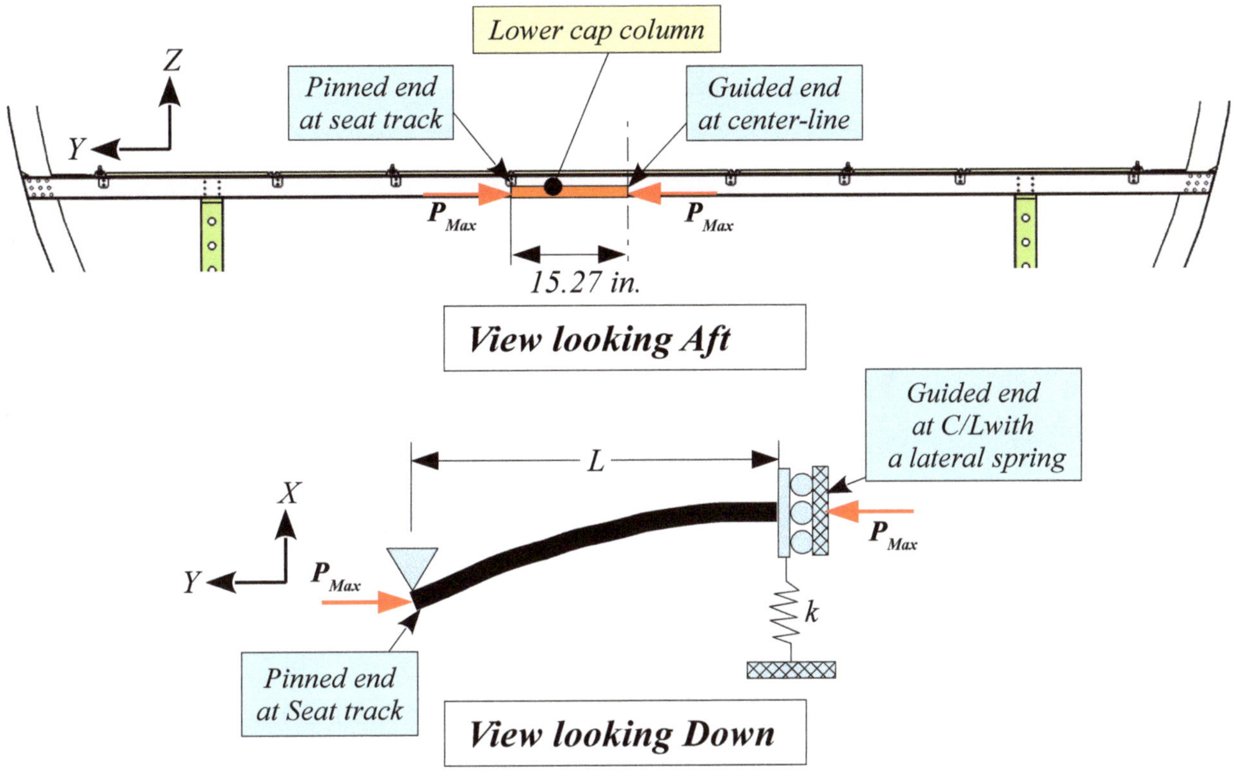

Figure 13-27 – *Crossbeam lower cap idealized Free-Fixed column.*

Figure 13-28 shows the equivalent column idealization to use in the analysis.

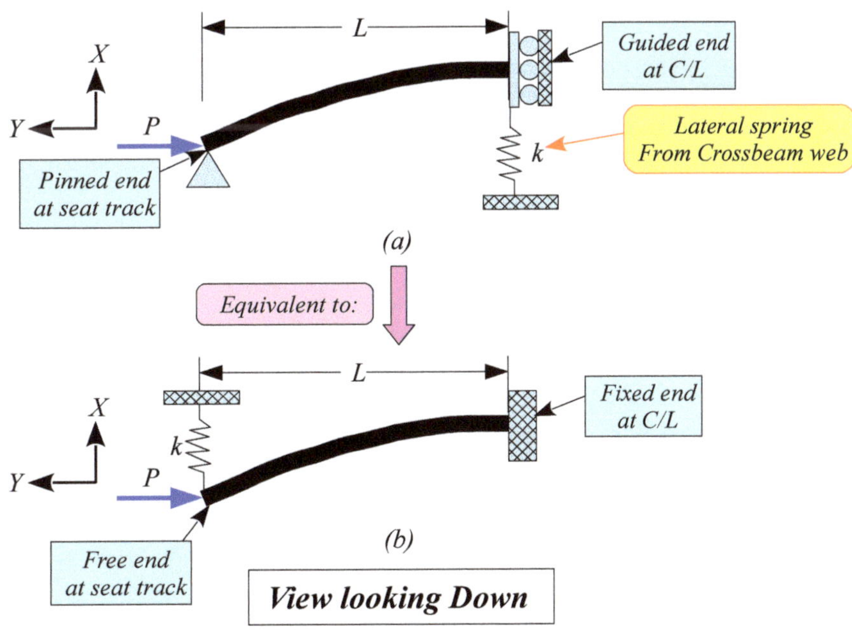

Figure 13-28 – *Crossbeam lower cap idealized Free-Fixed column.*

For the above column idealization, the critical compression load "P_{CR}" is calculated as follows:

$$P_{cr} = \beta^2 \frac{EI}{L^2}$$

where β is solved by iteration using the following equation:

$$\frac{\gamma(\beta - \sin\beta) - \beta^3}{\gamma(1 - \cos\beta)} = \tan\beta$$

and where γ is calculated as:

$$\gamma = \frac{kL^3}{EI}$$

The lateral spring stiffness "k" is calculated as:

$$k = \frac{P_X}{d_X} = \frac{P_X}{P_X h^3 / 3EI_{web}} = \frac{3EI_{web}}{h^3} = \frac{(3)10\times10^6(4.26\times10^{-5})}{2.6^3} = 189\ lbs/in.$$

Hence we obtain:

$$\gamma = \frac{189 \cdot 15.27^3}{10*10^6\ (0.0208)} = 3.2$$

and solving by iteration:

$$\beta = 2.246\ rad.$$

For
$E_t = E = 10.0*10^6$ psi
$I_z = 0.0208$ in^4
$L = 15.27$ in.
$\beta = 2.246$ rad.
$A = 0.2098$ in^2

Finally, the critical Euler column buckling load is calculated as:

$$P_{CR} = 2.246^2 \frac{10\times10^6(0.0208)}{(15.27)^2} = 4{,}499\ lbs$$

thus the critical compression stress is:

$$F_{CR} = \frac{P_{CR}}{A} = \frac{4{,}499}{0.2098} = 21{,}446\ psi$$

The stress is sufficiently below the compression Yield stress "F_{CY}" so no plasticity occurs.

Lower cap Margin of Safety

The MS in column buckling for the Crossbeam lower cap of section *A-A* is calculated below.

$$MS = \frac{P_{CR}}{P_{Max}} - 1 = \frac{4{,}499}{2{,}902} - 1 = 0.55$$

Crossbeam Lower cap Section A-A, Ultimate	***0.55***
3g Up *Al Aly 7050-T7451*	*Column*

Since I had made a lot of assumptions in this analysis, I wanted to get some assurance. So I decided to do a small FEM of the complete Crossbeam in buckling to see what sort of allowable I would get. See section ***13.1.11.3***.

13.1.11.3 FEA column buckling double check

A general buckling of the Crossbeam in the case of a *3g Up* Emergency landing would mean the loss of the Crossbeam's strength and stiffness. When buckled, the Crossbeam would then be acting as a cable pulling to two sides of the fuselage towards each other. That doesn't sound too good for the passengers fastened to the failed Crossbeam. So, to ensure that the above calculations are correct, a small FEM was done to test the validity of the assumptions of the idealized column.

The small FEM for the Crossbeam was made using *QUAD8* elements. A row of upper cap nodes were blocked in the *X* direction while the nodes at the ends of the beam and at the Stanchions were blocked in *X* and *Z*. One node was blocked in *Y* to prevent rigid body motion.

Two models were done. One without lateral supports (Gussets) at *BL ±15.27* (section *13.1.11.1*) and one with the supports (section *13.1.11.2*). Static and buckling runs were done for both models.

Figure 13-29 shows the FEM having no additional lateral supports. *Figure 13-30* shows the resulting stress and displacements, and *Figure 13-31* shows the first buckling mode for the no-lateral supports configuration.

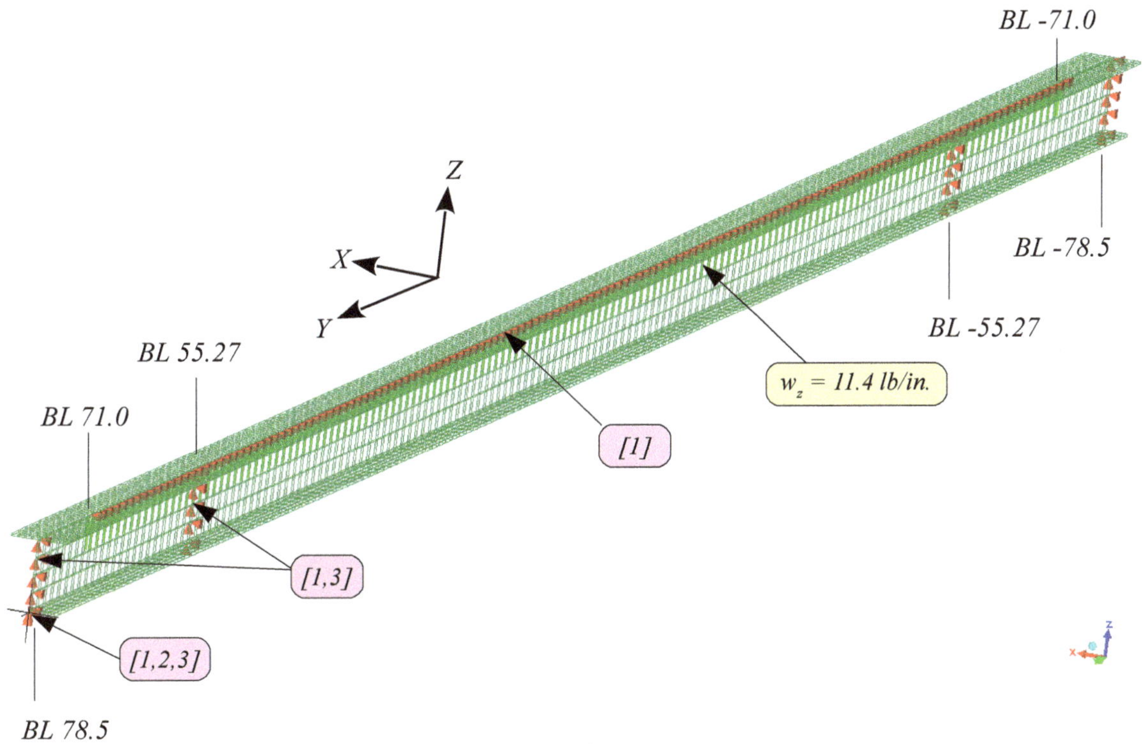

Figure 13-29 *– Crossbeam FEM 3g Up LC, no lateral supports.*

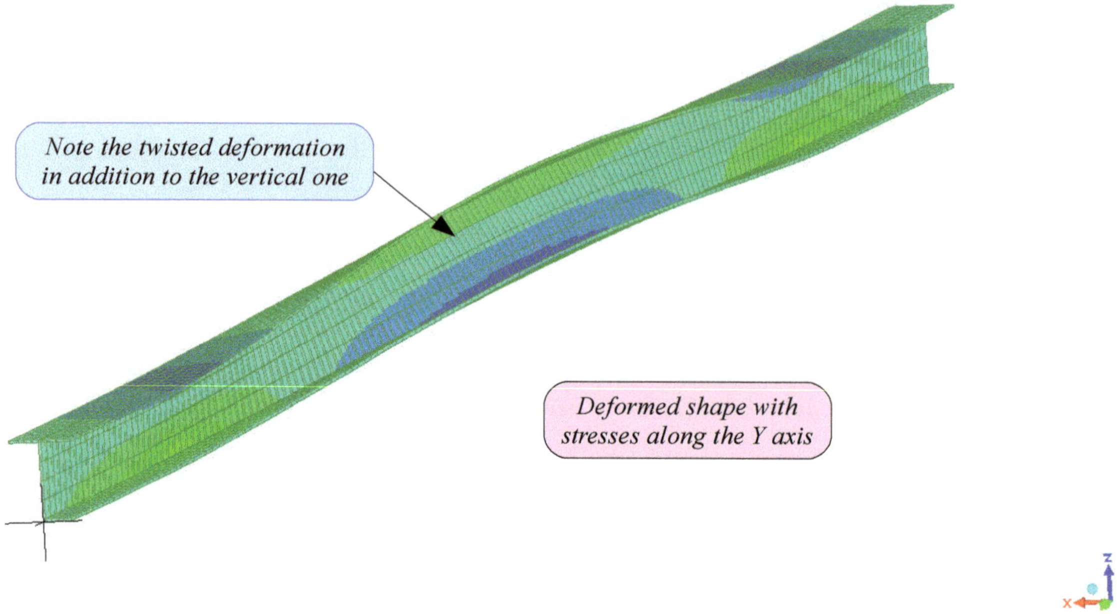

Figure 13-30 *– Crossbeam FEA static results 3g Up LC, no lateral supports.*

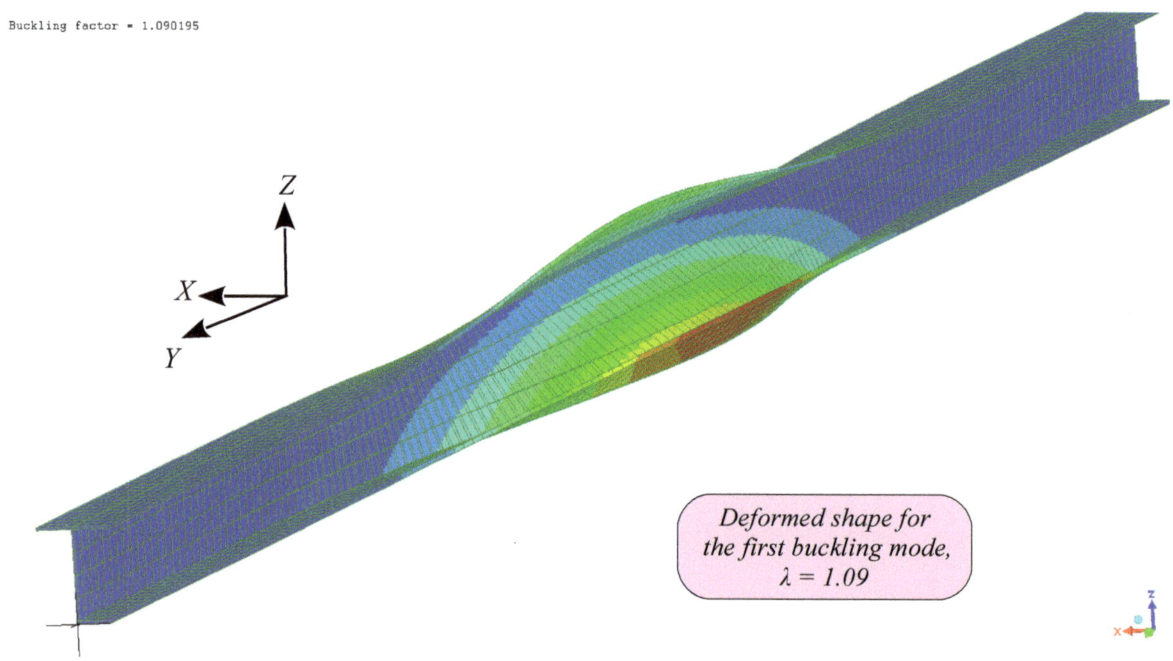

Figure 13-31 *– Crossbeam FEA buckling results 3g Up LC, no lateral supports.*

The results from *Figure 13-30* from the static run, although amplified in the picture, shows that the Crossbeam substantially twists under the perfectly vertical load applied at the center of the upper cap. This behavior hints us that the beam is probably lacking some lateral supports. This is confirmed by the buckling results shown in *Figure 13-31*.

Although the result shows a slight positive MS in buckling (MS = $\lambda - 1 = 0.09$), I find it way too low and, as specified earlier, lateral supports should be added at *BL ±15.27* as dictated by the negative MS obtained earlier. The hand calcs should ideally be a bit more conservative than the FEM.

Figure 13-32 shows the FEM with the lateral supports at *BL ±15.27*. *Figure 13-33* shows the resulting stress and displacements, and *Figure 13-34* shows the first buckling mode for that configuration.

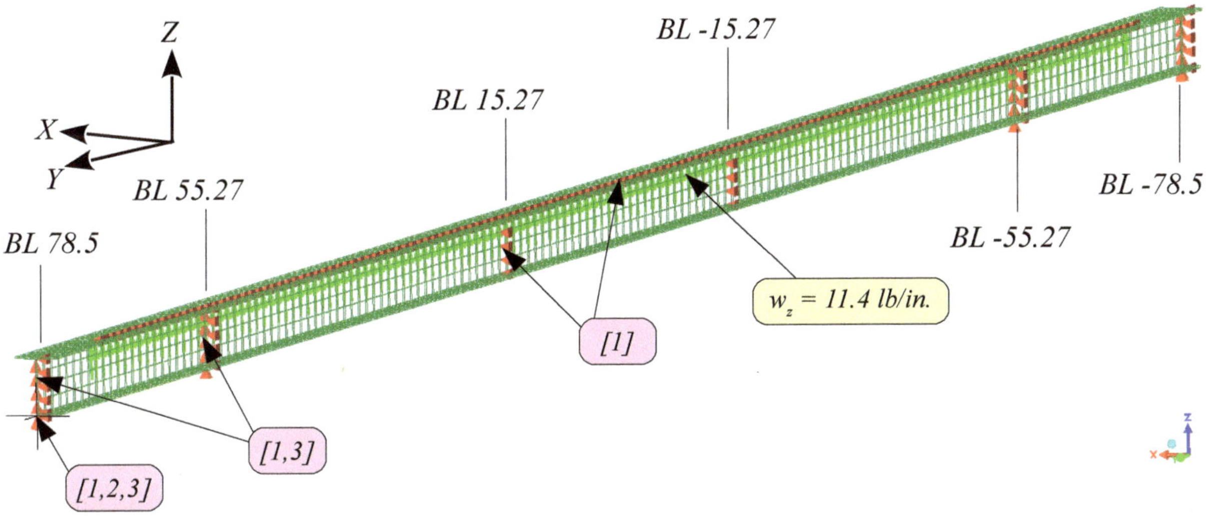

Figure 13-32 *– Crossbeam FEM 3g Up LC, with lateral supports at BL±15.27.*

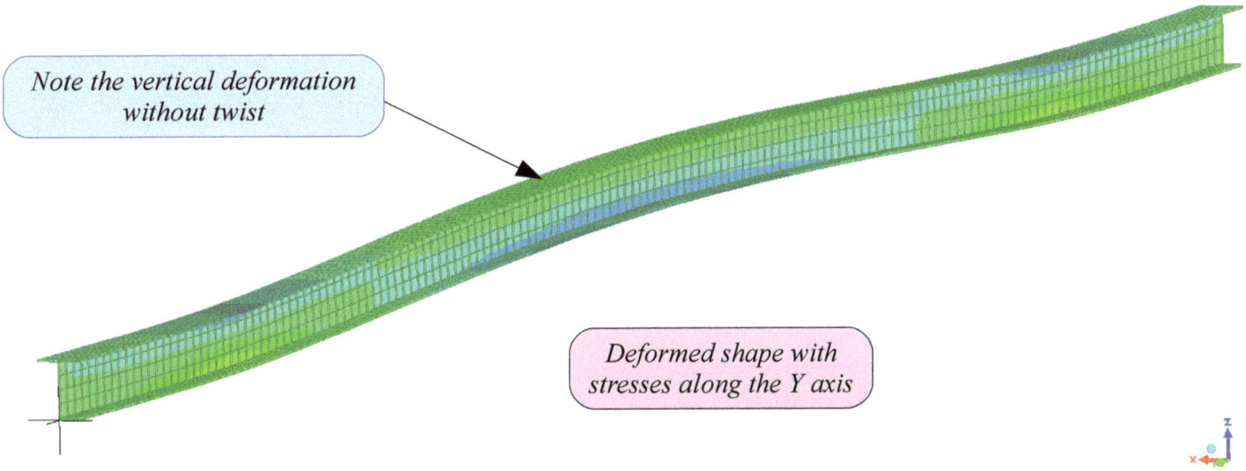

Figure 13-33 *– Crossbeam FEA static results 3g Up LC, with lateral supports at BL±15.27.*

The results from *Figure 13-33* from the static run shows that the Crossbeam doesn't twist anymore showing an almost purely vertical displacement. This is much better and desirable. The buckling behavior, as shown in *Figure 13-34*, also doesn't show much twist but only some free flange buckling as expected.

The result shows a good positive MS in general buckling (MS = $\lambda - 1 = 0.73$). This demonstrates that the lateral supports added at *BL ±15.27* definitely help and that the hand calculations done earlier are indeed conservative and provide similar results with a MS of 0.55.

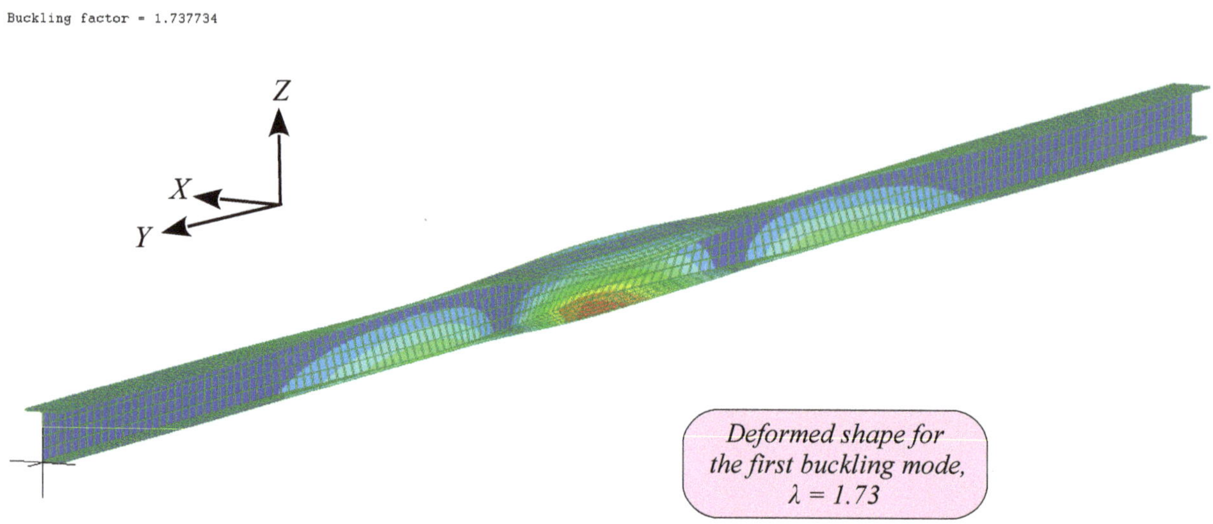

Figure 13-34 *– Crossbeam FEA buckling results 3g Up LC, with lateral supports at BL±15.27.*

Of course, the worst floor loading scenario was analyzed. Providing that the floor loading and/or the Crossbeam pitch were smaller in some areas, the additional Gussets may not be required at those places. Here I am simply assuming that the Gussets are being installed at every Crossbeam throughout the fuselage.

Some readers may argue that the need for additional lateral supports is not justified and that my approach is too conservative. Maybe I'm chicken, or maybe I don't like having a beam meant to react simple vertical loads to twist in addition to the normal bending.

Unnecessary twisting can lead to premature cracking while, in addition, a beam lacking lateral supports may be vibrating in flight. Just for fun, I did some *FEA* modal runs to compare both configurations.

The configuration without the lateral supports showed a first vibration mode at 1.08 *Hz* and a second at 2.56 *Hz* all in torsion! With the supports at *BL ±15.27*, the first vibration mode was at 2.78 *Hz* in the vertical direction while the second mode was in torsion at 4.54 *Hz*. A noticeable difference!

One may be analyzing a structure for stress, buckling and deflection but the designing criteria could very well be a target frequency to avoid, a fatigue life to meet or, an inspection interval requirement from the *DTA*. The static analysis is done mainly to size the parts and release drawings but, while doing that, it is good to keep in mind that other criteria have to be met.

For those reasons, (if I am given the time) I will optimize the part using quick tools to ensure that all criteria will be met. Otherwise, I'll try to be a bit more generous in the *MS* by having them not too close to 0.00; especially for flight conditions.

13.1.12 FREE FLANGE BUCKLING

There are two critical locations to check for the Free-Flange buckling: the first is the lower cap at the Stanchions (*BL ±55.27*) and the second, is the upper cap at mid-span (*BL 0.0*). From the stresses at section ***A-A*** calculated previously from *Table 13-9* and *Table 13-10*, free-flange buckling analyses are performed between points *1 & 2* and between points *3 & 4*. In both cases, the flange is assumed simply-supported at the web and, of course, free at the tip. See *Figure 13-35* for their locations.

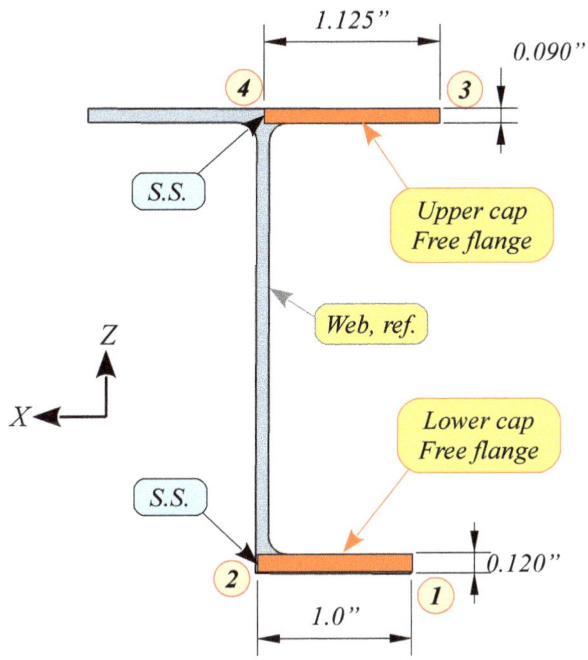

Figure 13-35 *– Critical Free flange locations on Crossbeam.*

13.1.12.1 Lower cap flange

Figure 13-36 shows the stress distribution of the lower cap free flange at the Stanchions. Because, the compression stress at the free flange is smaller than the one at the flange root, the Neutral Axis "*NA*" would be beyond the free edge. In this case, I prefer calculating the average compression stress and assuming that it is constant throughout the flange's width.

For conservatism, the flange is considered simply supported at all sides except at the free edge. Using (ref. 20.) for flat plate buckling in compression, the compression stress allowable is calculated as:

$$F_{C\ cr} = \frac{k_C\ \eta_c\ \pi^2\ E}{12\left(1-\mu^2\right)}\left(\frac{t}{b}\right)^2$$

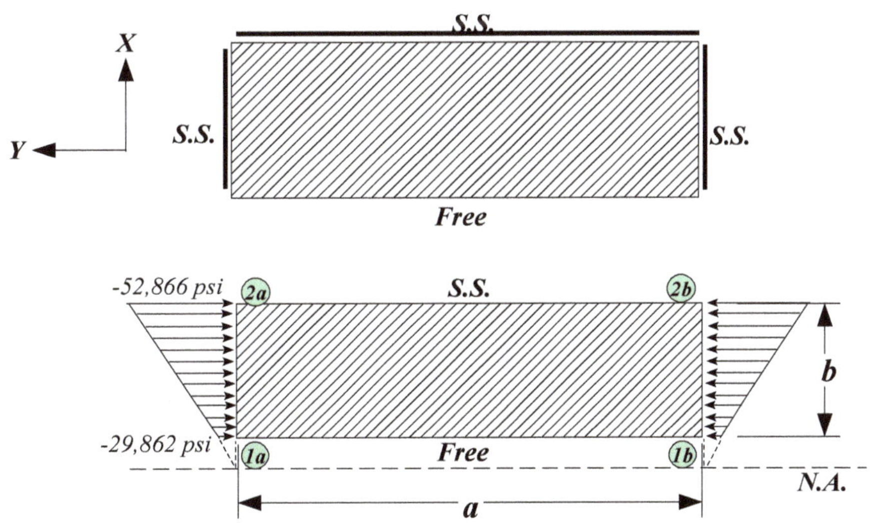

Figure 13-36 *– Lower cap Free flange buckling in compression.*

From *Figure 13-36*, the length "*a*" is quite large. Therefore, I will use the lowest asymptotic value for k_C. Hence from *Figure A-1*, the value for k_C read is: **0.43**. The Young's modulus from *Table 13-2* is used. Since the stress varies along the Crossbeam, the bending stress at point ***1a*** is different (higher) than the one at point ***1b***. The same for ***2a*** and ***2b***. The maximum (at points ***1a*** and ***2a***) are used for conservatism.

From the stresses in *Table 13-9*, the stress at point ***1*** is −29,862 *psi* and it is −52,866 *psi* at point ***2*** located 1.0 *in.* away. The average stress "f_{Avg}" is thus: −41,364 *psi.*

From the above compression stress buckling equation:

For
$k_C = 0.43$
$b = 1.0$ in
$t = 0.120$ in
$\mu = 0.33$
$E_C = 10.0*10^6$ psi

$$\frac{F_{C\ cr}}{\eta_c} = \frac{k_C\,\pi^2\,E}{12\left(1-\mu^2\right)}\left(\frac{t}{b}\right)^2$$

$$\frac{F_{C\ cr}}{\eta_c} = \frac{(0.43)\,\pi^2\,10*10^6}{12\left(1-0.33^2\right)}\left(\frac{0.120}{1.0}\right)^2 = 57{,}150\ psi$$

This value is very close to F_{CY}. Therefore, a plasticity reduction factor should be used.

A plasticity correction factor for compression "η_c" is used to multiply the Young's modulus "***E***" and is calculated as follows:

$$\eta_C = \frac{E_t}{E} = \frac{1}{1+\left(\frac{0.002\ E\ n}{F_{cy}}\right)\left(\frac{F_{c\,cr'}}{F_{cy}}\right)^{n-1}}$$

Therefore, using the same procedure as described in section ***10.5.6***, the critical buckling compression stress is calculated by iteration by first assuming a value for $F_{c\,cr}'$.

First, a value of 45,000 *psi* was assumed for $F_{c\,cr}'$ in the above equation, and from the iterative process, the value for "η_c" of 0.84453 was obtained. The final critical compression buckling stress allowable is thus;

For
$F_{cy} = 60{,}000$ psi
$n = 20$
$E = 10.0*10^6$ psi
$t = 0.120$ in
$b = 1.0$in

$$F_{C\ cr} = \frac{k_C\,\eta_c\,\pi^2\,E}{12\left(1-\mu^2\right)}\left(\frac{t}{b}\right)^2 = \frac{0.43\,(0.84453)\,\pi^2\,10*10^6}{12\left(1-0.33^2\right)}\left(\frac{0.120}{1.0}\right)^2 = 52{,}293\ psi$$

<u>Lower cap Free Flange Margin of safety</u>

The MS in Free flange buckling for the Crossbeam lower cap flange of section *A-A* is calculated below for the *6g Down LC*.

$$MS = \frac{F_{C\,CR}}{f_{Avg}} - 1 = \frac{52{,}293}{41{,}364} - 1 = 0.26$$

13.1.12.2 Upper cap flange

Figure 13-37 shows the stress distribution of the upper cap free flange at mid-span. For conservatism, the flange is considered simply supported at all sides except at the free edge. Using (ref. 20.) for flat plate buckling in bending, the bending stress allowable is calculated as:

$$F_{b\ cr} = \frac{k_b\,\eta_c\,\pi^2\,E}{12\left(1-\mu^2\right)}\left(\frac{t}{b}\right)^2$$

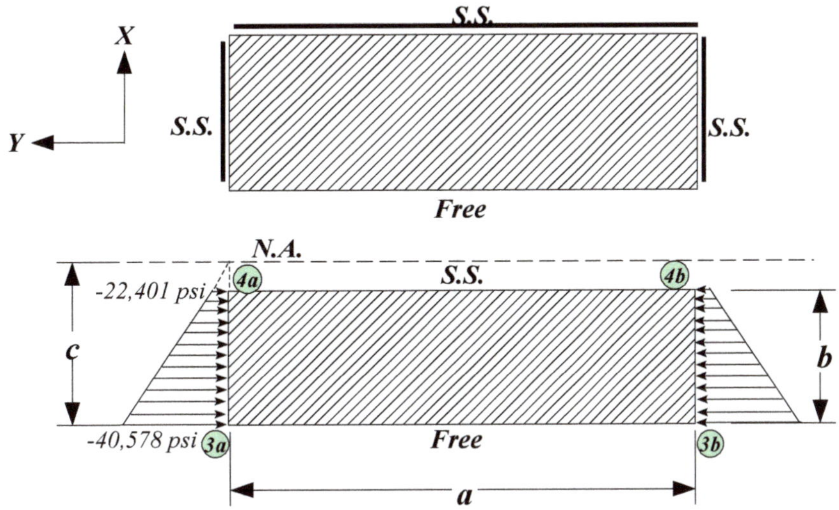

Figure 13-37 *– Upper cap Free flange buckling in compression.*

From *Figure 13-37*, the length "***a***" is arbitrarily shown since it is not required in this analysis. Since the stress varies along the Crossbeam, the bending stress at point ***3a*** is different (higher) than the one at point ***3b***. The same for ***4a*** and ***4b***. The maximum (at points ***3a*** and ***4a***) are used for conservatism.

From the stress calculations done previously, the stress at point ***3*** is −40,578 *psi* and it is −22,401*psi* at point ***4*** located 1.125 *in.* away. For this analysis, we are interested to know the location where the stress is zero. That is *"**c**"* on *Figure 13-37*. By doing a linear extrapolation from points ***3*** & ***4*** we get ***c*** = 2.511 *in.* In this case, b = 1.125 *in.* and c = 2.511 *in.* Thus b/c = 1.125 / 2.511 = 0.45. Hence from *Figure B-1*, the value for k_b read is: 0.46. The Young's modulus from *Table 13-2* is used.

From the above bending stress buckling equation:

$$\frac{F_{b\,cr}}{\eta_c} = \frac{k_b\,\pi^2\,E}{12\left(1-\mu^2\right)}\left(\frac{t}{b}\right)^2$$

$$\frac{F_{b\,cr}}{\eta_c} = \frac{(0.46)\,\pi^2\,10*10^6}{12\left(1-0.33^2\right)}\left(\frac{0.090}{1.125}\right)^2 = 27{,}172\ psi$$

For
k_b = 0.46
b = 1.125 in
t = 0.090 in
μ = 0.33
*E_c= 10.0*10^6 psi*

This value is lower than the compression stress at the free edge therefore one would expect a local buckling to occur. However, the assumption that all sides of the flange are simply supported (0% fixity) is quite conservative. In reality, it is somewhere between simply supported and fixed. It is a common practice to consider a 50% fixity along the edges for machined and extruded parts. This assumption was deemed reasonable as similar results were obtained from tests. Thus, another value for k_b by assuming the edge opposite to the free edge to be fixed is read from *Figure B-1* using the curve for fixed edge. The value obtained is 1.38 for a 100% fixity.

Calculating the average of the two k_b values, the value for a 50% fixity is:

$$k_{b\,50} = \frac{k_{b0}+k_{b\,100}}{2} = \frac{0.46+1.38}{2} = 0.92$$

The critical buckling stress for a 50% fixity free flange is:

$$\frac{F_{b\,cr}}{\eta_c}=\frac{k_b\,\pi^2\,E}{12\left(1-\mu^2\right)}\left(\frac{t}{b}\right)^2$$

$$\frac{F_{b\,cr}}{\eta_c}=\frac{(0.92)\,\pi^2\,10*10^6}{12\left(1-0.33^2\right)}\left(\frac{0.090}{1.125}\right)^2=54{,}345\ psi$$

For
k_b = 0.92
b = 1.125 in
t = 0.090 in
E_c= 10.0*10^6 psi

This new value is higher than the maximum compression stress at point ***3***, however, since it is close to F_{CY}, a plasticity correction factor for compression "η_c" is used to multiply the Young's modulus "***E***" and is calculated as follows:

$$\eta_C=\frac{E_t}{E}=\frac{1}{1+\left(\dfrac{0.002\ E\ n}{F_{cy}}\right)\left(\dfrac{F_{c\,cr'}}{F_{cy}}\right)^{n-1}}$$

Therefore, using the same procedure as described in section ***10.5.6***, the critical buckling bending stress is calculated by iteration by first assuming a value for $F_{b\,cr}'$.

A first value of 45,000 psi was assumed for $F_{b\,cr}'$ in the above equation, and from the iterative process, the value for "η_c" of 0.87533 was obtained. The final critical bending buckling stress allowable is thus;

For
F_{cy}= 60,000 psi
k_b = 0.92
n = 20
μ = 0.33
E = 10.0*10^6 psi
t = 0.090 in
b = 1.125in

$$F_{b\,cr}=\frac{k_b\,\eta_c\,\pi^2\,E}{12\left(1-\mu^2\right)}\left(\frac{t}{b}\right)^2=\frac{0.92\,(0.87533)\,\pi^2\,10*10^6}{12\left(1-0.33^2\right)}\left(\frac{0.090}{1.125}\right)^2=51{,}539\ psi$$

Upper cap Free Flange Margin of safety

The MS in Free flange buckling for the Crossbeam upper cap flange of section *A-A* is calculated below for the *6g Down LC*.

$$MS=\frac{F_{b\,cr}}{f_3}-1=\frac{51{,}539}{40{,}578}-1=0.27$$

Crossbeam Lower cap Ultimate	***0.26***
6g Down *Al Aly 7050-T7451*	*Free flange buckling*

13.1.13 SHEAR BUCKLING

The Crossbeam web is also checked for shear buckling. Although shear buckling is not a mode of failure, we need to verify that the web is non-buckling. As shown on *Figure 13-38*, there are six *(6)* web panels. The panel boundaries are determined by the upper and lower flanges and, the vertical integral ribs that act as panel breakers (dividers). Panels *D* and *E* are divided by the Stanchion. The panel is assumed simply supported on all sides.

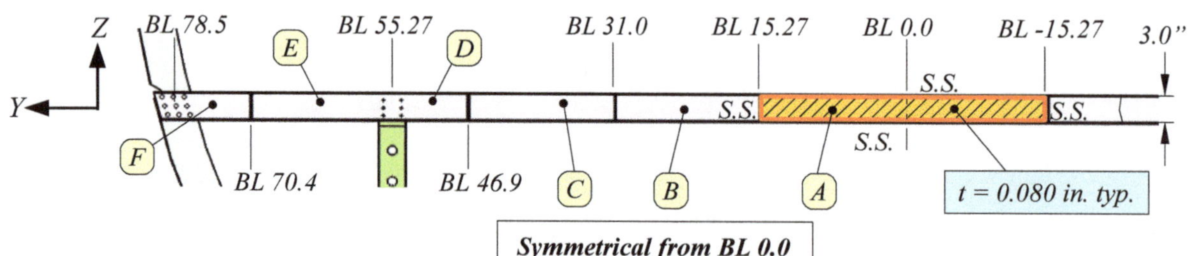

Figure 13-38 *– Crossbeam web panels.*

Each panel is checked using the most conservative transverse shear from *Figure 13-5*; that is the shear diagram for the Crossbeam having the fixed ends for the *6g Down LC*. The shear diagram is reproduced in *Figure 13-39* super-imposed on top of the actual Crossbeam drawing. The average shear load is used for each panel.

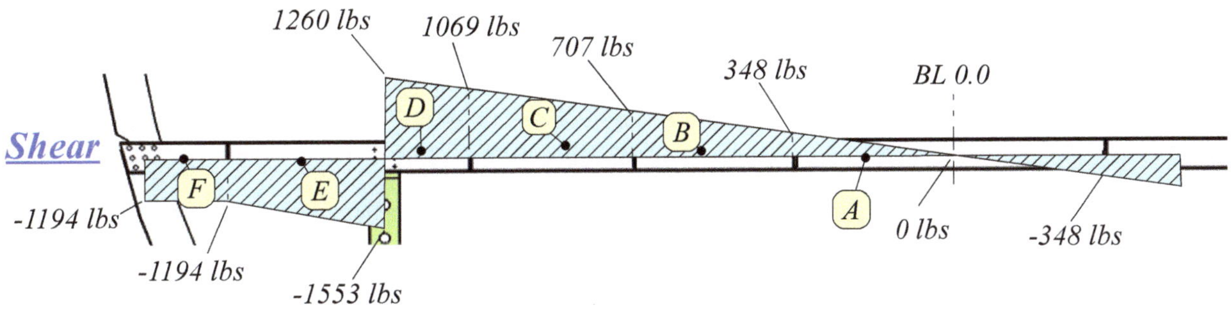

Figure 13-39 *– Crossbeam web panels transverse shear.*

Table 13-12 shows the shear buckling MS summary for the Crossbeam panels. From the results, all panels are shear buckling resistant for the ultimate *6g Down LC*.

Table 13-12 *- Crossbeam panels shear buckling MS summary.*

	a	b	a/b	k_S	t	η_s	F_{scr}	V_Z	f_s	MS
Panel	[in]	[in]			[in]		[psi]	[lbs]	[psi]	
A	30.54	3.0	10.2	5.35	0.080	0.88601	**31112**	348	1450	>1.0
B	15.73	3.0	5.2	5.35	0.080	0.88601	**31112**	528	2198	>1.0
C	15.9	3.0	5.3	5.35	0.080	0.88601	**31112**	888	3700	>1.0
D	8.37	3.0	2.8	5.8	0.080	0.83571	**31814**	1165	4852	>1.0
E	15.13	3.0	5.0	5.35	0.080	0.88601	**31112**	1374	5723	>1.0
F	8.1	3.0	2.7	5.8	0.080	0.83571	**31814**	1194	4975	>1.0

Refer to section ***14.6.10*** for an example of a web with a hole in a shear buckling analysis.

You might have noticed that no lightening holes were put into the web of the Crossbeam. In the real life, you (*or certainly someone from another group*) would probably want some holes, not only to make the Crossbeam a little lighter but mainly, to allow some systems to go through like electrical bundles, tubes, etc. Flanged lightening holes increase the shear buckling allowable so why not take advantage of it.

Analysis example for the most critical panel "***E***"

The MS from *Table 13-12* were all calculated following the same procedure. Panel ***E*** is found to be the most critical panel with an MS of 4.44. Using (ref. 20.) for flat plate buckling in shear having simply supported edges, the shear stress allowable is calculated as:

$$F_{S\,cr} = \frac{k_S\,\eta_S\,\pi^2\,E}{12\left(1-\mu^2\right)}\left(\frac{t}{b}\right)^2$$

From *Figure 13-38,* the dimension "*a*" is 70.4 – 55.27 = 15.13 *in.* and the dimension "*b*" is 3.0 *in.* Note that for shear buckling, the value for "*b*" is always the smallest of the two dimensions. Thus a/b = 15.13 / 3.0 = 5.0. Hence from *Figure A-2*, the value for k_S is: 5.35 (the asymptotic value). The primary Young's modulus in tension "*E*" is used from *Table 13-2*.

From the above equation:

$$\frac{F_{s\,cr}}{\eta_s} = \frac{k_s\,\pi^2\,E}{12\left(1-\mu^2\right)}\left(\frac{t}{b}\right)^2$$

$$\frac{F_{s\,cr}}{\eta_s} = \frac{(5.35)\,\pi^2\,10*10^6}{12\left(1-0.33^2\right)}\left(\frac{0.080}{3.0}\right)^2 = 35{,}114\ psi$$

For
F_{SU} = 40,000 psi
k_s = 5.35
b = 3.0 in
a/b = 5.0
t = 0.080 in
μ = 0.33
E = 10.0*10⁶ psi

Since the above value is very close to the shear strength of the material "F_{su}", the plastic reduction factor in shear "η_S" has to be evaluated as follows.

Therefore, using the same procedure as described in section *11.2.7*, the critical shear buckling stress is calculated by iteration by first assuming a value for $F_{S\,cr}'$. A plasticity correction factor for shear "η_S" is used to multiply the Young's modulus "***E***" and is calculated as follows:

$$\eta_S = \frac{1}{1+\left(\frac{0.00346\,G}{F_{sy}}\right)\left(\frac{F_{scr}}{F_{sy}}\right)^{n-1}}$$

For
F_{sy} = 33,000 psi
n = 20
G = 3.76*10⁶ psi

Where

$$F_{sy} = 0.55\,F_{cy} = (0.55)60{,}000 = 33{,}000\ psi$$

This equation is solved by iteration until $F_{s\,cr}'$ is equal to the value obtained from the equation:

$$F_{S\,cr} = \frac{k_S\,\eta_S\,\pi^2\,E}{12\left(1-\mu^2\right)}\left(\frac{t}{b}\right)^2$$

A first value of 20,000 *psi* was assumed for $F_{scr'}$ in the above equation, and from the iterative process, the value for "η_S"of 0.88601 was obtained. The final initial shear buckling allowable is thus;

$$F_{S\,cr} = \frac{k_S\,\eta_S\,\pi^2\,E}{12\left(1-\mu^2\right)}\left(\frac{t}{b}\right)^2 = \frac{5.35\,(0.88601)\,\pi^2\,10*10^6}{12\left(1-0.33^2\right)}\left(\frac{0.080}{3.0}\right)^2 = 31{,}112\ psi$$

For
k_s = 5.35
E = 10.0*10⁶ psi
t = 0.080 in
b = 3.0 in.

Margin of safety

The average transverse shear load at the location of panel "*E*" is; $V_{E\,avg}$ = (1,194 + 1,553) / 2 = 1,374 *lbs*. The shear stress is thus; $f_s = V_{E\,avg} / (bt)$ = 1,374 / (3.0 * 0.080) = 5,723 *psi*. Thus the MS in initial shear buckling of the Crossbeam web for the *6g Down LC* is calculated as:

$$MS = \frac{F_{s\,cr}}{f_s} - 1 = \frac{31{,}112}{5{,}723} - 1 = 4.43$$

Crossbeam web Ultimate	***> 1.0***
6g Down *Al Aly 7050-T7451*	*Shear buckling*

13.1.14 CONNECTIONS

The Crossbeams are connected to the Frame and to the Stanchions using *MS20470AD6* solid rivets. See *Figure 13-40* for the location of the connections. Only the left hand side "*LHS*" is shown since the *RHS* is the same as the structure is symmetrical.

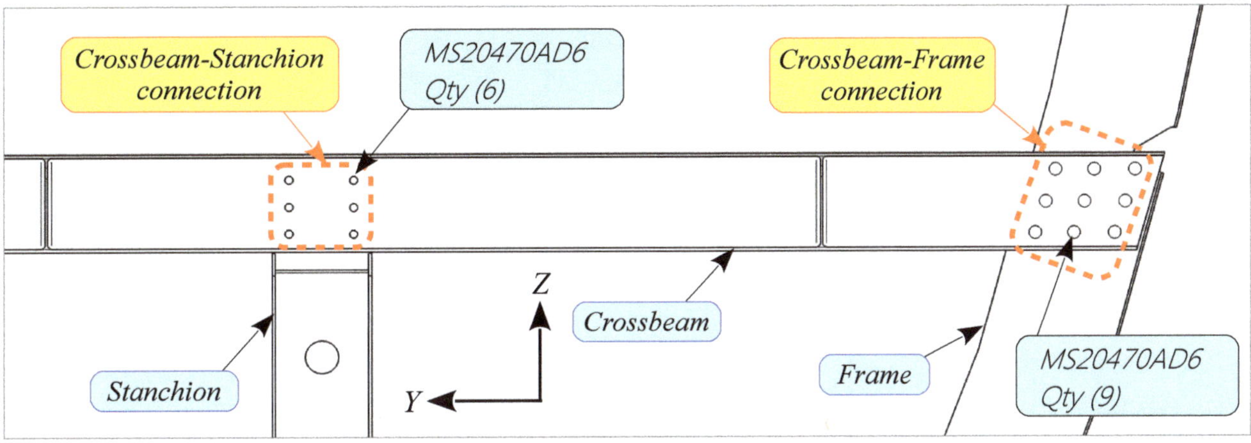

Figure 13-40 *– Location of Crossbeam connections to Frame and Stanchion.*

13.1.14.1 Crossbeam-Stanchion joint

That connection is comprised of qty*(6) MS20470AD6* rivets as shown in *Figure 13-41*. In order to be consistent with my original assumption of considering the Crossbeam simply-supported at the Stanchions, only vertical shear is being transferred in that joint.

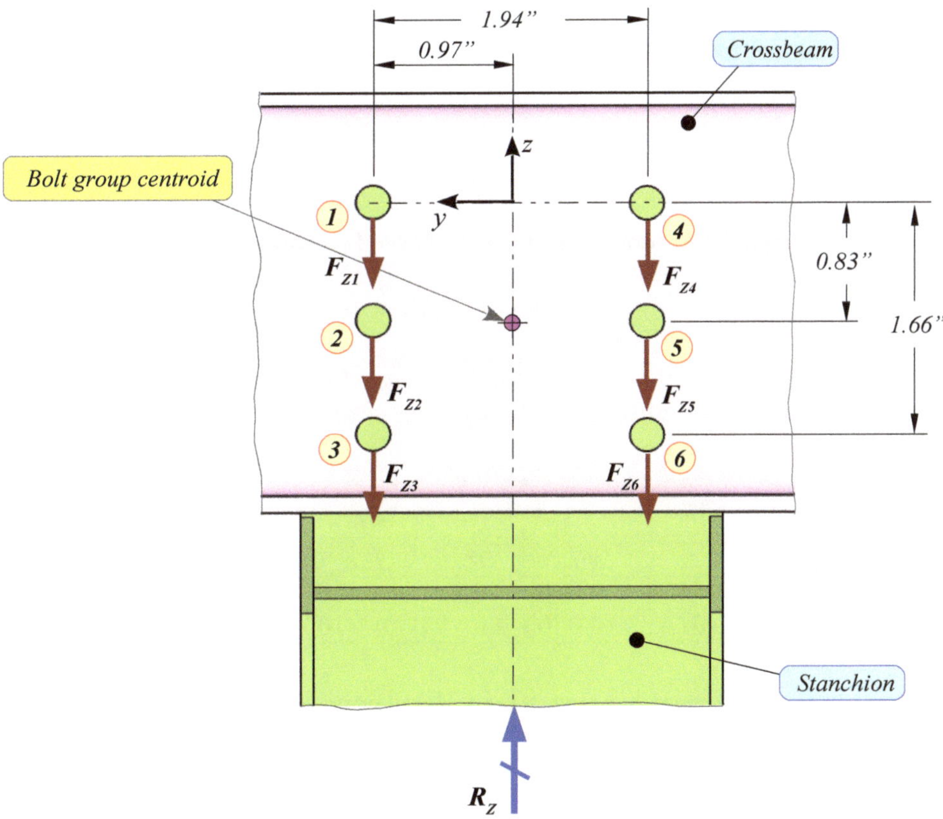

Figure 13-41 *– Crossbeam to Stanchion connection FBD.*

The highest reaction force at the Stanchion is obtained from *Figure 13-5* for the constraint assumption having *fixed ends* at each end of the Crossbeam. That force for the *6g Down LC* is 2,813 *lbs.*

Since there is no moment and since the Stanchion internal load is considered going right through its center and also through the centroid of the bolt group, the force in each rivet "F_{zi}" is all the same and is calculated as follows:

$$F_{Zi} = \frac{R_Z}{6} = \frac{2{,}813}{6} = 469\ lbs$$

Rivet shear:

The shear allowable "P_{SU}" for an *MS20470AD6 (BJ6)* is 862 *lbs* as shown in *Table 13-3*.

A Fitting Factor "*FF*" of 1.15 (ref. 25. *FAR §25.625*) is used. No shear factor "*SF*" is required since the Crossbeam and the Stanchion are relatively thick. (ref. *Table 13-4*). The Margin of Safety for shear is thus calculated as follows:

$$MS = \frac{P_{su}}{FF \times F} - 1 = \frac{862}{1.15 \times 469} - 1 = 0.59$$

Crossbeam to Stanchion, Ultimate	***0.59***
6g Down *MS20470AD6 in Al Aly 7050-T7451*	*Fastener shear*

In the case that *BJ5's* would have been used with a shear allowable of 596 *lbs*, an *MS* of 0.10 would have been obtained. This is a bit too tight for my taste for this kind of joint. Besides, the Stanchion and the Crossbeam webs are 0.080" thick; so in order to keep good proportions, 5/32" fasteners are a bit small for that material gauge. As a rule of thumb, the fastener diameter should be close to 3 times the smallest material thickness.

Bearing:

The bearing stress will be checked in the two parts involved in the joint. The bearing load allowable is calculated as follows:

$$P_{bru} = F_{bru} \cdot D \cdot t$$

Table 13-13 shows the summary of the resulting bearing *MS*. The Crossbeam and the Stanchions are made out of the same material and have the same thickness at the fasteners.

Table 13-13 - *Bearing stress MS table summary.*

Part	Material Al Aly	t [in]	D [in]	Fbru [ksi]	Pbru [lbs]	F [lbs]	MS bearing
Stanchion	7050-T7451	0.080	0.191	105	1604	469	2.42
Crossbeam	7050-T7451	0.080	0.191	105	1604	469	2.42

A calculation example for the lowest bearing stress *MS* is done below:

Both parts are equally critical in bearing. The bearing ultimate allowable load "P_{bru}" in the *Aluminum Alloy 7050-T7451* is calculated. For: F_{bru} = 105 *ksi* (ref. *Table 13-2*), $e/D = 2.0$. And for D = 0.191 *in.* and t = 0.080 *in.* hence:

$$P_{bru} = 105{,}000 \cdot 0.191 \cdot 0.080 = 1{,}604\,lbs$$

No Fitting Factor is used in bearing.

The Margin of Safety for the Crossbeam-Stanchion connections in bearing is:

$$MS = \frac{P_{bru}}{F} - 1 = \frac{1{,}604}{469} - 1 = 2.42$$

Since the *MS* in shear is lower than the one in bearing, the joint is thus shear critical. For the material involved, the ratio F_{bru} / F_{bry} is smaller than 1.5. Therefore, no bearing check at limit (yield) is necessary.

13.1.14.2 Crossbeam-Frame joint

That connection is comprised of *qty(9) MS20470AD6* rivets as shown in *Figure 13-42*.

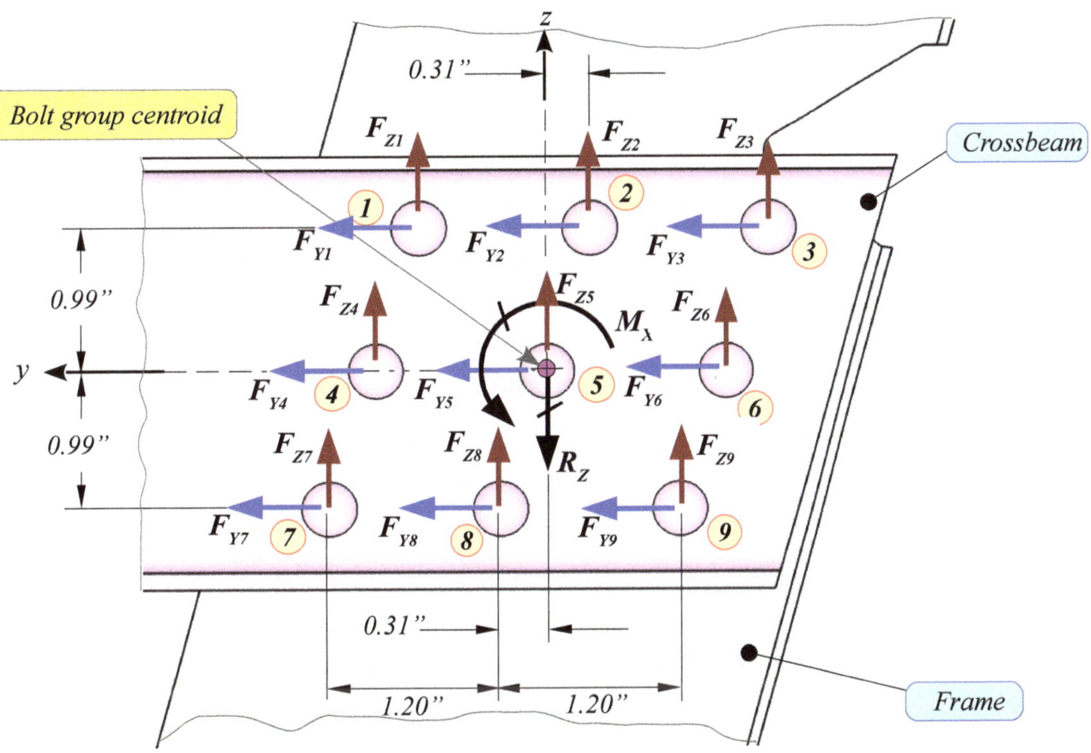

Figure 13-42 *– Crossbeam to Frame connection FBD.*

The highest reaction forces at the Frame are obtained from *Figure 13-5* for the constraint assumption having *fixed ends* at each end of the Crossbeam. The end-moment is 9,461 *in-lb* putting the upper cap in compression, and the vertical force is 1,194 *lbs* pulling the Crossbeam down, for the *6g Down LC* .

A "*2D*" bolt group analysis is performed to determine the force in each of the *9* rivets. It is calculated as previously explained in section *10.6.2*.

However, while performing that analysis on my spreadsheet, the results I obtained showed too high shear forces in the rivets for the *MS20470AD6*. Therefore, before taking the decision to change the fasteners for stronger ones, I questioned my original assumption that used the maximum end moment from the fixed-fixed Crossbeam configuration since it was very conservative. As shown in *Figure 13-8*, the resulting Crossbeam end-fixity from the small *FEA* was **26%**.

So, by consulting with my peers... I decided to consider a **50%** end-fixity as it is still conservative compared to the *FEA* result but not overly conservative. Hence, using the shear and moment diagrams from *Figure 13-6*, the end-moment is 4,731 *in-lb* putting the upper cap in compression, and the vertical force is 978 *lbs* pulling the Crossbeam down, for the *6g Down LC* . The resulting rivets shear forces due to the moment alone are shown in *Table 13-14*.

Of course, in a formal stress report, you may want to be consistent with the assumption you use. Therefore, in our example, I would have stated my intention to use the 50% end-fixity all along; for the connections at least.

Table 13-14 *- Forces in rivets due to applied moment.*

bolt	y	z	r	r^2	F	F_Y	F_Z
	[in]	[in]	[in]	[in²]	[lbs]	[lbs]	[lbs]
1	0.89	0.99	1.331	1.7722	417	-310	279
2	-0.31	0.99	1.037	1.0762	325	-310	-97
3	-1.51	0.99	1.806	3.2602	566	-310	-473
4	1.2	0	1.200	1.44	376	0	376
5	0	0	0.000	0	0	0	0
6	-1.2	0	1.200	1.44	376	0	-376
7	1.51	-0.99	1.806	3.2602	566	310	473
8	0.31	-0.99	1.037	1.0762	325	310	97
9	-0.89	-0.99	1.331	1.7722	417	310	-279
			I =	**15.0972**	**Σ =**	**0**	**0**

Then, *Table 13-15* shows the resulting rivet shear forces by combining the forces due to the applied moment "M_X" and the forces due to the applied vertical shear "V_Z". The maximum shear load obtained is 659 *lbs* at rivet no. *7*.

The rivet shear force from the applied vertical shear load is calculated as:

$$F_{Zi} = \frac{V_Z}{9} = \frac{978}{9} = 109\ lbs$$

Table 13-15 *- Shear Forces in rivets.*

	From V_Z		From M_X		Total		
bolt	F_Y	F_Z	F_Y	F_Z	F_Y	F_Z	F
	[lbs]	[lbs]	[lbs]	[lbs]	[lbs]	[lbs]	[lbs]
1	0	109	-310	279	-310	388	496
2	0	109	-310	-97	-310	12	310
3	0	109	-310	-473	-310	-365	479
4	0	109	0	376	0	485	485
5	0	109	0	0	0	109	109
6	0	109	0	-376	0	-267	267
7	0	109	310	473	310	582	**659**
8	0	109	310	97	310	206	372
9	0	109	310	-279	310	-170	354
Σ =	**0**	**978**	**0**	**0**	**0**	**978**	

<u>Rivet shear:</u>

The shear allowable "P_{SU}" for an *MS20470AD6 (BJ6)* is 862 *lbs* as shown in *Table 13-3*.

A Fitting Factor "*FF*" of 1.15 (ref. 25. *FAR §25.625*) is used. No shear factor "*SF*" is required since the Crossbeam and the Frame are relatively thick. (ref. *Table 13-4*). The Margin of Safety for shear is thus calculated as follows:

$$MS = \frac{P_{su}}{FF \times F} - 1 = \frac{862}{1.15 \times 659} - 1 = 0.13$$

Bearing:

The bearing stress will be checked in the two parts involved in the joint. The bearing load allowable is calculated as follows:

$$P_{bru} = F_{bru} \cdot D \cdot t$$

Table 13-16 shows the summary of the resulting bearing *MS*. The Crossbeam web is 0.080 *in.* thick and made out of *Al Aly 7050-T7451* while the Frame web is 0.060 *in.* thick and made out of *Al Aly 7075-T7351*. The bearing stress allowables are obtained from *Table 13-2* and *Table 14-20* respectively.

Table 13-16 *- Bearing stress MS table summary.*

Part	Material Al Aly	t [in]	D [in]	Fbru [ksi]	Pbru [lbs]	F [lbs]	MS bearing
Frame	7075-T7351	0.060	0.191	135	1547	659	1.35
Crossbeam	7050-T7451	0.080	0.191	105	1604	659	1.43

A calculation example for the lowest bearing stress *MS* is done below:

The joint is slightly more critical in the Frame. The bearing ultimate allowable load "P_{bru}" in the *Aluminum Alloy 7075-T7351* is calculated. For: F_{bru} = 135 *ksi, e/D* = 2.0. And for *D* = 0.191 *in.* and *t* = 0.060 *in.* hence:

$$P_{bru} = 135{,}000 \cdot 0.191 \cdot 0.060 = 1{,}547 lbs$$

No Fitting Factor is used in bearing.

$$MS = \frac{P_{bru}}{F} - 1 = \frac{1{,}547}{659} - 1 = 1.35$$

As the *MS* in shear is smaller than the *MS* in bearing, the joint is shear critical.

The Margin of Safety for the Crossbeam-Frame connection is:

Crossbeam to Frame, Ultimate	***0.13***
6g Down *MS20470AD6 in Al Aly 7050-T7451*	*Fastener shear*

13.1.15 COMMENT

Therefore, from the above results, the Crossbeam installation with the Gussets is adequate.

13.2 GUSSET ANALYSIS

The Gussets that were added at *BL ±15.27* are analyzed. They work as lateral stabilizers for the Crossbeams and, as shear clips to help transfer the vertical shear load from the Seat track rails to the Crossbeams. Due to the Seat track rail design, two types of lateral supports are provided as shown in *Figure 13-43*. This section only covers the Gusset type while the type having the direct connection between the Seat track rail and the Crossbeam is covered in section *13.1.14*.

The Gussets are made out of *Al Aly 2024-T3 Clad* 0.050 *in.* sheet and are fastened to the Crossbeam ribs and Seat track rail using *MS20470AD5* solid rivets.

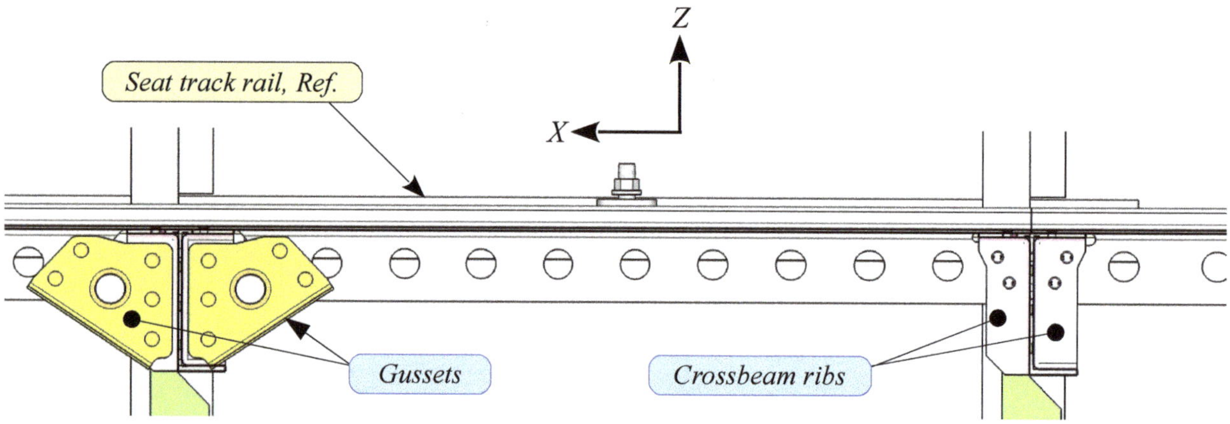

Figure 13-43 – *Two types of lateral supports at BL±15.27.*

The Crossbeams themselves were modified by adding vertical ribs that are used in the following ways:

1. Prevent the cross-section from warping at those locations.
2. Serve as panel breakers for the web to increase the buckling allowable.
3. Provide very good fasteners shear connections to the Seat track rails and the Gussets thus preventing having to drill holes into the lower cap flange; which is usually critical in durability.

Figure 13-44 shows the vertical ribs that are machined integral to the Crossbeams.

"Those who don't innovate, disappear."

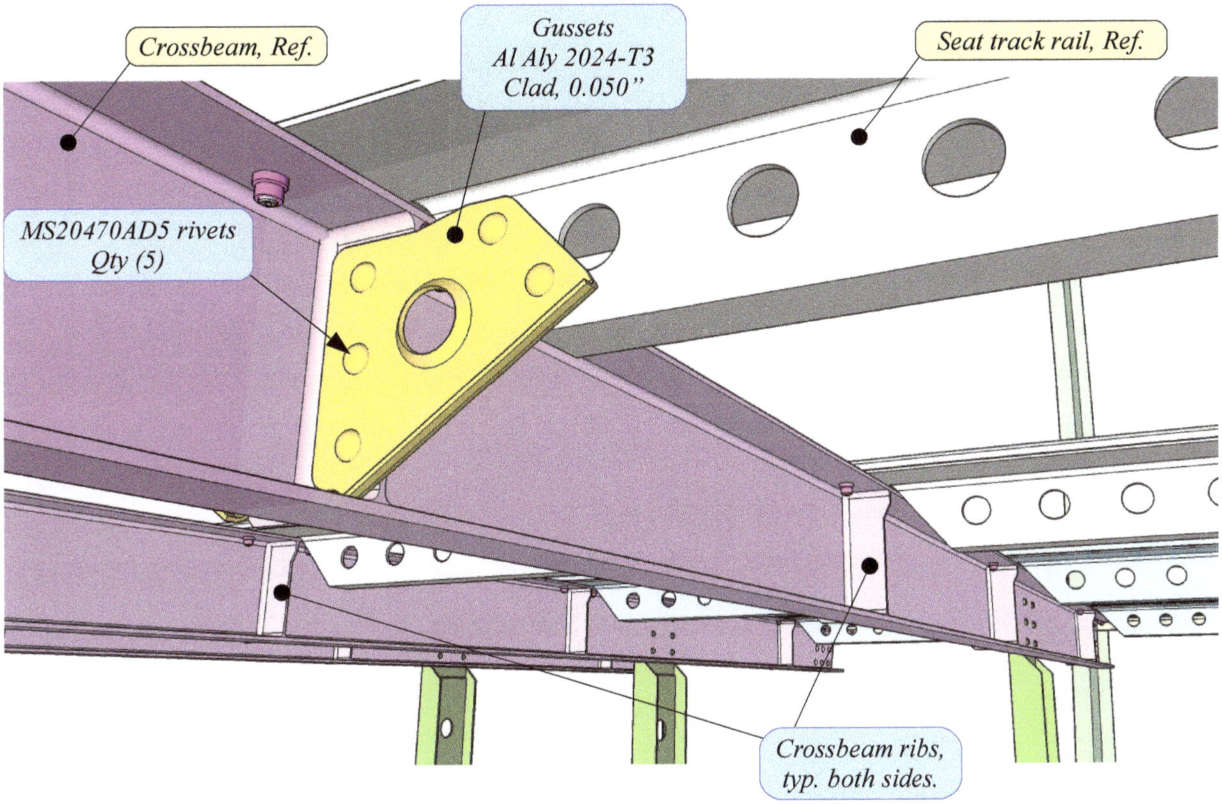

Figure 13-44 *– Typical Gusset installation at BL±15.27.*

13.2.1 LOADS

It is difficult to determine exactly what is the magnitude of the force required to prevent a beam (or a column) to buckle. Sometimes, just a finger located at the right place would be sufficient to prevent or delay the instability. It is however common to assume the stabilizing load (applied at a right angle to the column) to be a percentage of the internal load "*P*" in the member. It is usually around 5% to 10% of *P*.

In our example, the Crossbeam lower cap internal compression force at *BL±15.27* "$P_{15.27}$" as calculated on page *248* is −2,902 *lbs* for the *3g Up LC*. Conservatively, using 10% of that value as the lateral support force acting perpendicular to the Crossbeam, the force "P_{supp}" to be used to analyze the Gusset is:

$$P_{Supp} = 0.10\,P_{15.27} = (0.10)2{,}902 = 290\ lbs.$$

This force is assumed applied in the "*X*" direction at the *c.g.* of the Crossbeam lower cap and, since there are two Gussets for each location on a Crossbeam, half the load is assumed reacted by each Gusset. One Gusset will then be in compression while the other one will be in tension. See *Figure 13-45* for Gusset details and applied forces.

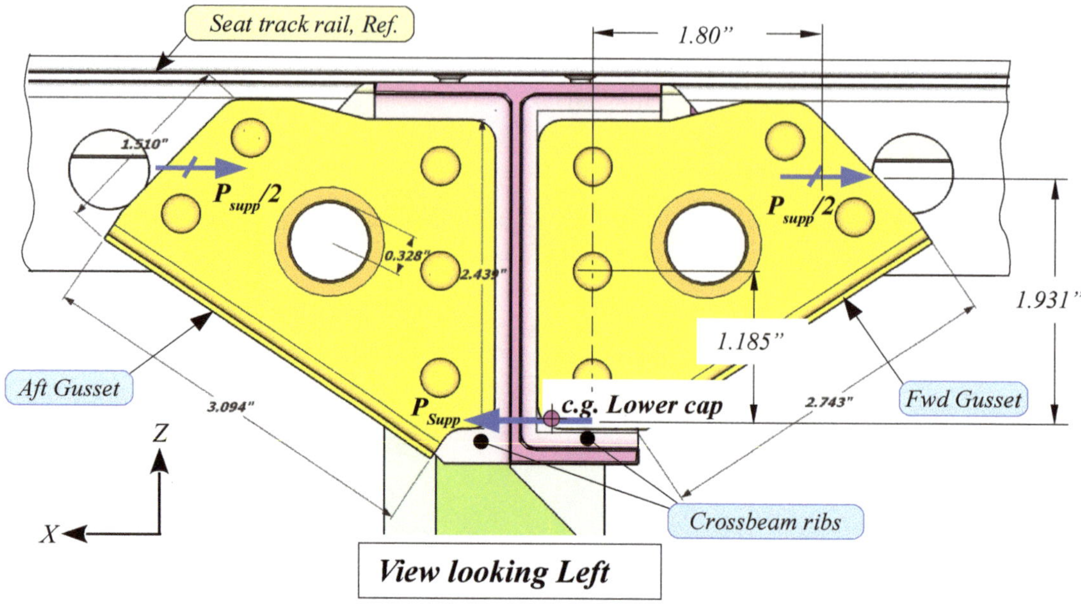

Figure 13-45 *– Gusset details and forces.*

From the small *FEM* with the configuration having lateral supports at *BL±15.27*, the reaction forces due to the static run at the nodes located where the Gussets would be was about 50 *lbs*. Our assumption using *10%P* is thus conservative.

In the following analysis, it is assumed that the Gussets are only loaded by the lateral load and no vertical shear load from the Seat track rail is considered.

Figure 13-46 shows the *FBD* of the Fwd Gusset due to the lateral force of P_{supp} / 2 = 290/2 = 145 *lbs*. The load is considered completely reversible of course. Both Aft and Fwd Gussets are the same except the Fwd one was trimmed at its lower end to clear the Crossbeam's lower flange.

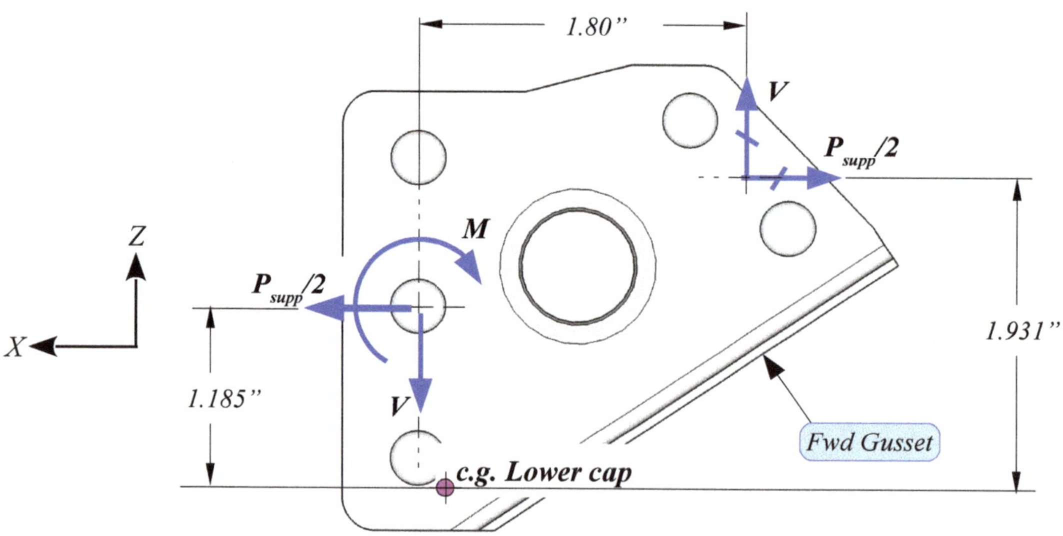

Figure 13-46 *– Gusset FBD.*

Since the connection to the Seat track rail has only two rivets (that are close to each other), this end has a much lower rotational stiffness than the 3 rivet connection to the Crossbeam. Therefore, the

Gusset is assumed pinned at the 2 rivet centroid and fixed at the 3 rivet centroid. The Gusset is basically idealized as a cantilevered beam. The internal forces are calculated as follows:

$$P_{Supp}/2 = 290/2 = 145\ lbs.$$

$$M_Y = \frac{P_{Supp}}{2} \cdot 1.185 = 145 \cdot 1.185 = 172\ in.-lb.$$

And from the sum of the moments at the middle rivet:

$$M = V \cdot 1.80 - \frac{P_{Supp}}{2} \cdot (1.931 - 1.185) = 1.80\text{V} - 145(0.746) = 172\ in.-lb$$

$$V = \frac{172 + 145(0.746)}{1.80} = 156\ lbs$$

To facilitate the analysis, the internal loads are transformed to meet the local coordinate system *[i,j,k]* as shown in *Figure 13-47*. For simplicity, the elastic axis is assumed to go through the centroids of each of the two rivet groups. The transformed internal forces are calculated below.

$$P_i = P_X \cos 22.5 + V_Z \sin 22.5 = 145 \cos 22.5 + 156 \sin 22.5 = 190\ lbs$$
$$V_k = -P_X \sin 22.5 + V_Z \cos 22.5 = -145 \sin 22.5 + 156 \cos 22.5 = 79\ lbs$$

As for the moment around the "*j*" axis, it is the same as "M_Y" but in the opposite direction, hence:

$$M_j = -M_Y = -172\ in.-lb.$$

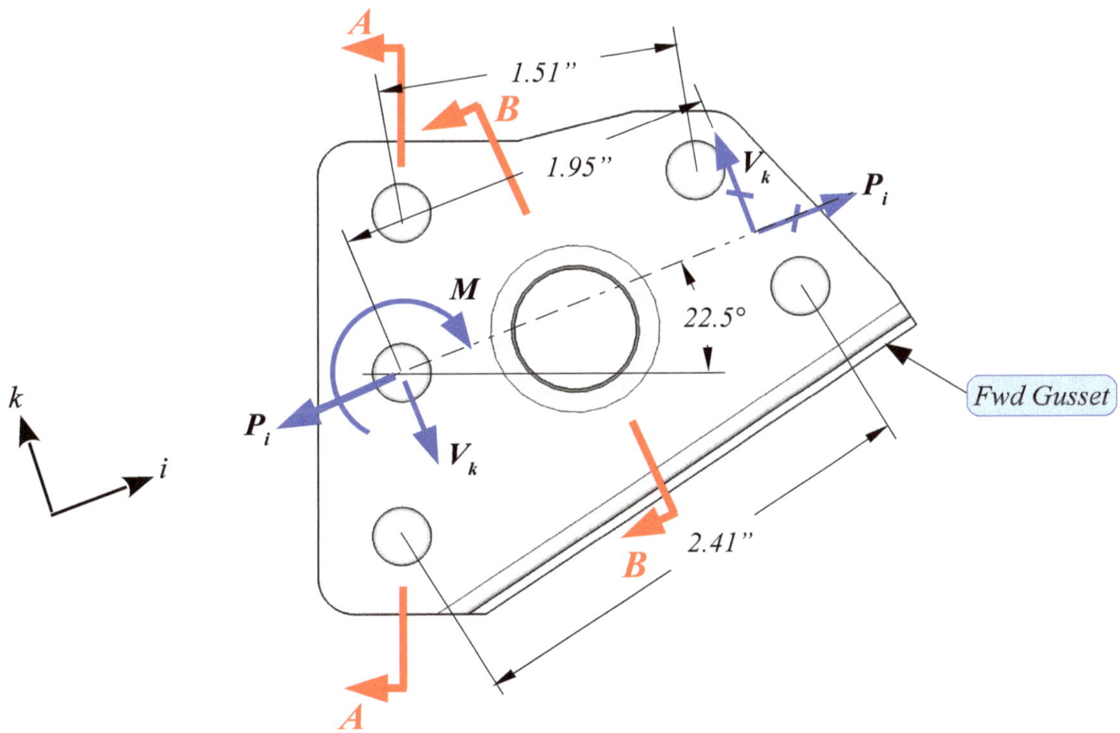

Figure 13-47 *– Gusset FBD in i-j-k coordinate system.*

Since the neutral axis of this relatively complex "beam" varies a lot and, also because the loads are low, it is assumed that the elastic axis coincides with the neutral axis. Therefore, no further transformation of the internal loads is necessary. *Figure 13-48* shows the shear and moment diagrams of the idealized cantilevered beam representing the Gusset.

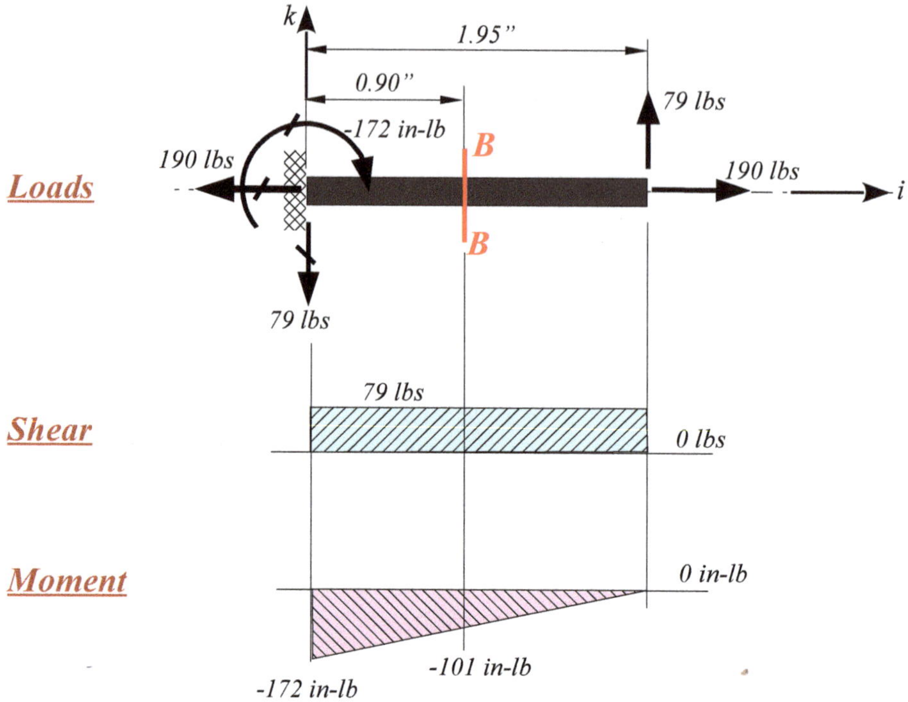

Figure 13-48 – *Gusset Shear & Moment diagrams in i-j-k coordinate system.*

13.2.2 MATERIAL ALLOWABLES

The actual material allowables for all materials can be obtained from the latest edition of (ref. 8.) The **Gussets**, are made out of *Al Aly 2024-T3 Clad* sheet 0.050 *in.* thick. *Table 13-17* shows fictitious values for this example.

Table 13-17 - *Al Aly 2024-T3 Clad sheet fictitious material allowables.*

Aluminum Alloy 2024-T3 Clad sheet
Fictitious Values, NOT FOR DESIGN

	0.050"
	[ksi]
F_{tu}	60
F_{ty}	45
F_{cy}	40
F_{su}	35
F_{bru} $(e/D = 2.0)$	120
F_{bry} $(e/D = 2.0)$	85
E, 10^3	10
G, 10^3	3.76
μ	0.33
n	20

13.2.3 FASTENERS ALLOWABLES

Solid rivets

Table 13-18 shows the fictitious ultimate shear allowables "P_{SU}" for 1/8", 5/32" and 3/16" *MS20470AD* solid rivets.

Table 13-18 *- MS20470AD Solid rivets ultimate shear allowables.*

	MS20470AD		
	Al Aly 2117-T3		
	F_{su} = 30 ksi		
	4	**5**	**6**
Dia. [in.]	0.1285	0.159	0.191
P_{su} *[lbs]*	**388**	**596**	**862**

Hi-Lites

Table 13-19 shows the ultimate single shear and tension allowables for 5/32" protruding head Hi-Lite *HST10-5* and flush head Hi-Lite *HST11-5* Titanium Pins and *HST79-5* Aluminum Collars from (ref. 24.).

Table 13-19 *- HST10, HST11 & HST79 Hi-Lites ultimate shear and tension allowables.*

	P_{tu} [lbs]	P_{su} [lbs]
Pin HST10-5	1,940	2,005
Pin HST11-5	1,650	2,005
Collar HST79-5	1,400	N/A

Table 13-20 Shows the shear correction factors *"SF"* for rivets in sheets of aluminum alloys.

Table 13-20 *- Shear correction factor "SF".*

	Diameter of rivet [in]			
Thickness [in]	*3/32"*	*1/8"*	*5/32"*	*3/16"*
0.025	0.972	–	–	–
0.032	1.000	0.964	–	–
0.036		0.980	–	–
0.040		0.996	0.964	–
0.045		1.000	0.98	–
0.050			0.996	0.972
0.063			1.000	1.000

13.2.4 SECTION PROPERTIES

The cross sections dimensions and properties for sections *A-A* and *B-B* are shown in *Figure 13-49* and *Figure 13-50* respectively. Section *A-A* is through the three rivets (at the fixed end) and section *B-B* is through the lightening hole perpendicular to the assumed elastic axis of the Gusset.

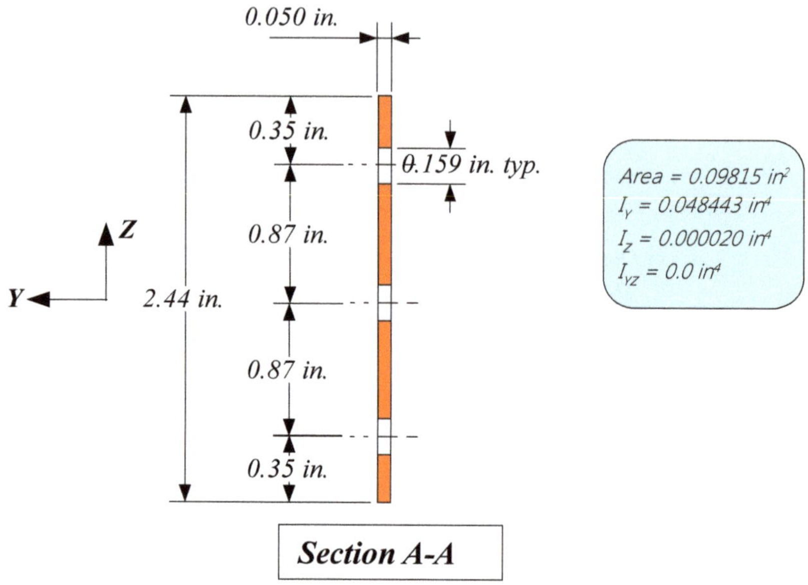

Figure 13-49 *– Gusset Section A-A cross section properties.*

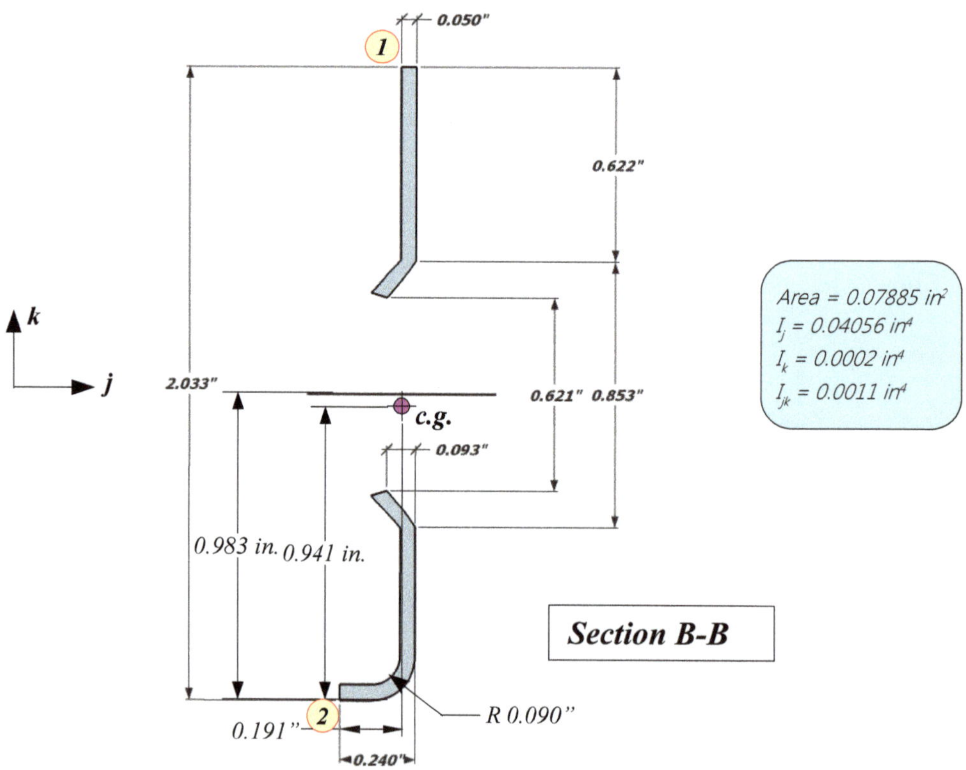

Figure 13-50 *– Gusset Section B-B cross section properties.*

13.2.5 STRESS ANALYSIS SECTION A-A

Using *Figure 13-48* and *Figure 13-49*, the bending stresses are calculated at the extreme fibers and, since it is symmetrical and that the Aft Gusset is also covered, is completely reversible.

<u>Axial stress</u>

$$f_a = \frac{P_i}{A} = \frac{190}{0.09815} = 1{,}936\ psi$$

<u>Bending stress</u>

$$f_b = \frac{M_j\, c}{I_Y} = \frac{172\,(1.22)}{0.048443} = \pm 4{,}332\, psi$$

<u>Shear stress</u>

$$f_s = \frac{V_k}{A} = \frac{79}{0.09815} = 805\ psi$$

<u>Combined stresses</u>

The axial, bending and shear stresses are combined according to the equation:

$$(R_a + R_b)^2 + R_S^2 = 1$$

Since there is no risk of instability at that location, the MS is calculated as:

$$MS = \frac{1}{\sqrt{(R_a + R_b)^2 + R_S^2}} - 1$$

Where, using *Table 13-17*, R_a, R_b and R_S are respectively the axial, bending and shear stress ratios calculated as follows:

$$R_a = \frac{f_a}{F_{cy}} = \frac{|1{,}936|}{40{,}000} = 0.048$$

$$R_b = \frac{f_b}{F_{cy}} = \frac{|4{,}332|}{40{,}000} = 0.108$$

$$R_s = \frac{f_s}{F_{su}} = \frac{805}{35{,}000} = 0.023$$

> F_{cy} is used in this case as the allowable since the maximum stress is in compression. It is a common practice to use F_{cy} whenever an ultimate stress in compression exists in relatively thin parts.

Hence the MS for section *A-A* of the Fwd Gusset is:

$$MS = \frac{1}{\sqrt{(0.048+0.108)^2 + 0.023^2}} - 1 = 5.34$$

Gusset, section A-A Ultimate	***>1.0***
3g Up *Al Aly 2024-T3 Clad*	*Axial + Bending + shear*

13.2.6 STRESS ANALYSIS SECTION B-B

Using *Figure 13-48* and *Figure 13-50*, the stresses are calculated at stress points *1* & *2*. In order to cover the Aft Gusset, the internal forces are considered completely reversible.

Axial stress

$$f_a = \frac{P_i}{A} = \frac{190}{0.07885} = 2{,}410\ psi$$

Bending stress at point *1*

$$f_{b1} = \frac{M_j\,k}{I_Y} = \frac{-101\,(2.033-0.941)}{0.04056} = -2{,}719\ psi$$

Bending stress at point *2*

$$f_{b2} = \frac{M_j\,k}{I_Y} = \frac{-101\,(-0.941)}{0.04056} = 2{,}343\ psi$$

Shear stress

For the shear stress, only the web area "A_S" is conservatively considered. Thus:

$$A_S = (2.033-0.621)0.050 = 0.0706\ in.^2$$

$$f_s = \frac{V_k}{A_S} = \frac{79}{0.0706} = 1{,}119\ psi$$

Combined stresses

No shear stress is combined with the axial and bending stresses as the shear is assumed taken by the web away from the extreme fibers.

Total stress at point *1*:

$$f_1 = f_a + f_{b1} = 2{,}410 + (-2{,}719) = -309\ psi$$

Total stress at point *2*:

$$f_2 = f_a + f_{b2} = 2{,}410 + 2{,}343 = 4{,}753\ psi$$

Since the internal forces are reversible, the stress at point *2* could be in compression. Therefore, the Gusset lower cap where point *2* is located will be checked in crippling and in column buckling. Point *1* sees almost no stress and hence is really not a concern neither in tension or in compression.

13.2.7 CRIPPLING AT SECTION B-B

The crippling stress allowable "F_{CC}" will be calculated below for the Gusset lower cap at section *B-B*.

The section is divided into segments as shown in *Figure 13-51*. The length of segments *2* stops at the edge of the lightening hole as the *45 deg.* flange is ignored as a restraint. All segments have one edge free.

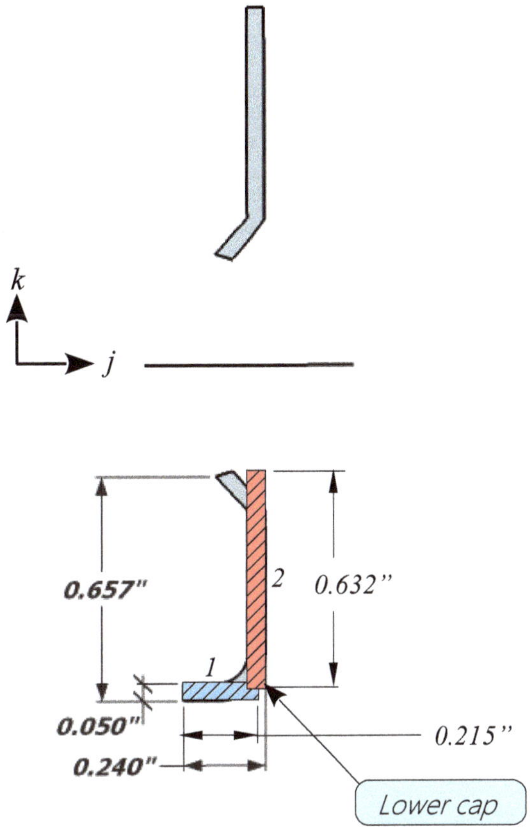

Figure 13-51 *– Effective segments for crippling at Gusset lower cap through section B-B.*

The crippling allowable for each segment is calculated as shown below. Note that the crippling allowable "F_{CC}" of a segment can never be higher than F_{CY}. So the maximum value for F_{CCn} / F_{cy} is 1.0. The value at the abscissa of *Figure 13-52* is calculated as:

$$abscissa = \sqrt{F_{CY}/E}\,(b/t)$$

The return flange "lip" appears to be a bit short for me. In order to provide a sufficient restraint to the web so that the lower cap can work as intended, it is essential that the lip be of a minimum size. Therefore, I am making sure that the lip is adequate before evaluating the crippling allowable stress.

The ratio of the length over thickness "***b/t***" for the lip has to be, according to (ref. 2. *§C7.9)*, at least 0.328 times the ratio for the web. Hence, from *Table 13-21,* the ***b/t*** ratio for segment *1* (lip) is 4.3 and the one for segment *2* (web) is 12.64. Thus, 0.328 * 12.64 = 4.15. Since, the lip ratio is larger than 4.15, then the lip is sufficiently large to provide an adequate edge support to the web.

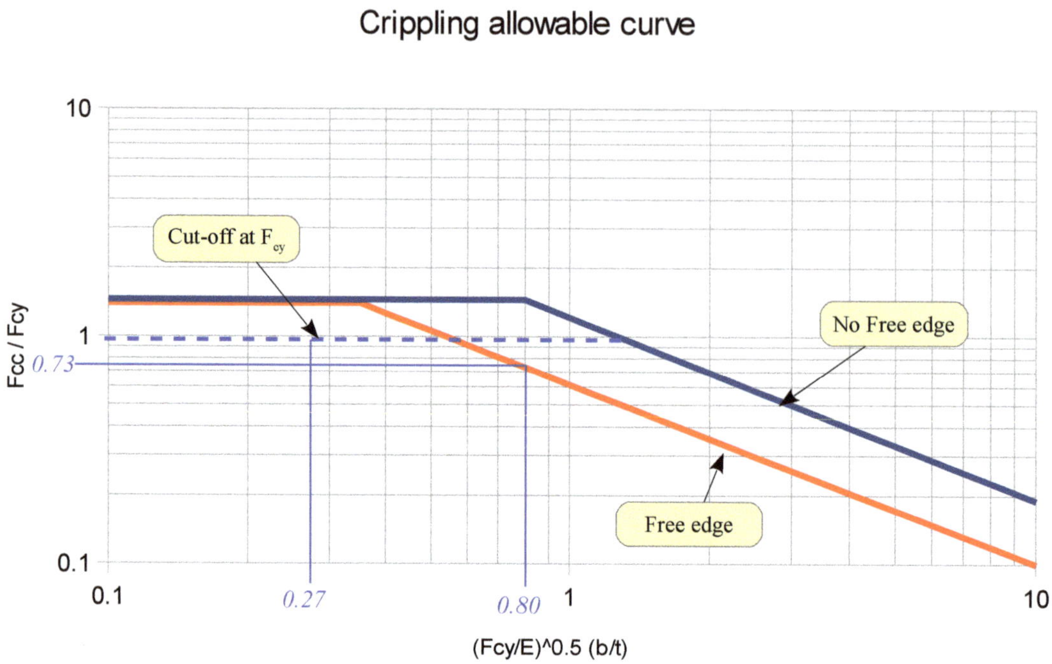

Figure 13-52 – *Typical Crippling curves.*

So for E = $10.0*10^6$ *psi*, F_{cy} = 40,000 *psi* and from the dimensions shown in *Figure 13-51*, the crippling allowable of each segment is calculated in *Table 13-21.*

Table 13-21 - *Crippling analysis for the Gusset lower cap.*

	Free edge	b_n	t_n	$b_n t_n$	b_n / t_n	$(F_{cy} / E)^{0.5}$	$(F_{cy} / E)^{0.5}(b_n/t_n)$	F_{CCn} / F_{cy}	F_{CCn}	$t_n b_n F_{CCn}$
Segment		[in]	[in]	[in²]					[ksi]	[lbs]
1	1	0.215	0.05	0.011	4.300	0.0632	0.27	1.00	40	430
2	1	0.632	0.05	0.032	12.640	0.0632	0.80	0.73	29.2	923
			$\sum b_n t_n =$	0.0424					$\sum b_n t_n F_{ccn} =$	1353

Thus the crippling allowable stress for the Gusset lower cap is:

$$F_{CC} = \frac{\sum b_n t_n F_{CCn}}{\sum b_n t_n} = \frac{1,353}{0.0424} = 31,941 \ psi$$

No MS is calculated for crippling as the crippling allowable is simply used later in the *Johnson-Euler* equation for the column buckling analysis; which is more critical.

13.2.8 COLUMN BUCKLING

Although the length of the Gusset lower cap is relatively short, I want to make sure that it is stable. *Figure 13-53* shows how the lower cap (highlighted) is idealized as a column as shown in *Figure 13-54.* It is assumed that half the axial load goes through the lower cap "$P_i/2$" and, since the Gusset is considered as a cantilevered beam, the reacting moment "M" at the center of the three rivets is considered as a couple "M/h" that is added as a compression load at the base of the lower cap column. The column is conservatively assumed pinned-pinned.

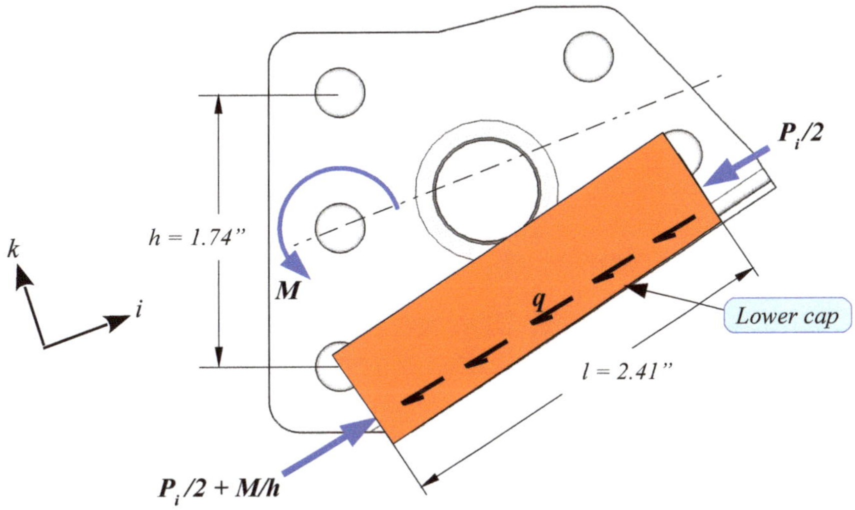

Figure 13-53 *– FBD on lower cap column.*

For $P_i = 158$ *lbs*, $M = 140$ *in-lb*, $l = 2.41$ *in.* and $h = 1.74$ *in.* then, since $ql = M/h$, the shear flow "q" is calculated as:

$$q = \frac{M}{hl} = \frac{140}{1.74 \cdot 2.41} = 33.4 \; lb/in.$$

Hence, the the column is loaded as shown below.

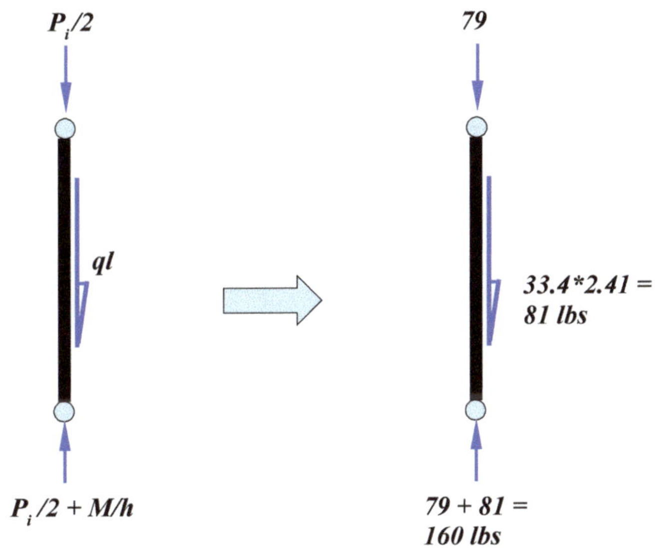

Figure 13-54 *– Idealized lower cap column.*

In order to determine the column buckling allowable, one needs to know the fixity coefficient *"C"* for the particular configuration. For a pinned-pinned uniform column having an applied end load and a shear flow, the ratio qL/P is evaluated as 33.4 (2.41)/ 160 = 0.50 and the *"C"* value is approximated as follows:

$$C = e^{(0.625\, qL/P)} = e^{(0.625 * 0.5)} = 1.37$$

Therefore, for $L = 2.41$ *in.*, the effective column length "L'" is calculated as:

$$L' = \frac{2.41}{\sqrt{1.37}} = 2.06 \; in.$$

See *Figure 13-55* for the section properties of the column.

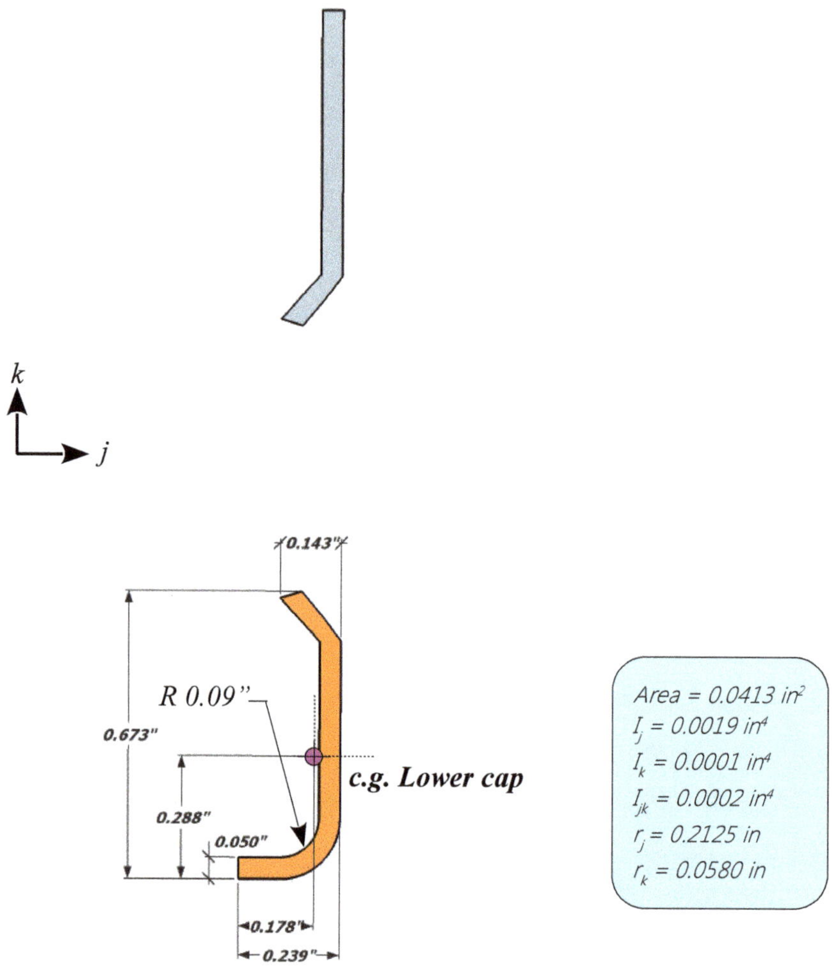

Figure 13-55 – *Gusset lower cap dimensions and section properties.*

Using the *Johnson-Euler* method, the critical column buckling stress is calculated as:

$$F_C = F_{CC} - \frac{F_{CC}^2}{4\pi^2 E_t}\left(\frac{L'}{\rho}\right)^2$$

For
$E_t = E = 10.0*10^6$ psi
$\rho_k = 0.0580$ in⁴
$L' = 2.06$ in.
$F_{CC} = 31{,}941$ psi

then

$$F_C = 31{,}941 - \frac{31{,}941^2}{4\pi^2 10*10^6}\left(\frac{2.06}{0.0580}\right)^2 = 28{,}681 \; psi$$

That value of F_C is sufficiently below F_{CY} that no plasticity should exist. However, just to make sure, I calculated the tangent modulus "E_t" using the *Ramberg-Osgood* equation from (ref. 18.).

$$\frac{E_t}{E} = \frac{1}{1 + \frac{3}{7}n\left(\frac{f}{F_{0.7}}\right)^{n-1}}$$

That ratio is found by iteration using a *goal seek* function as follows:

1. Knowing E and n, obtain the value for $F_{0.7}$.
2. Assume a value for $\boldsymbol{f}$ as equal to F_C calculated previously (28,681 *psi*).(*in fact, any value would do*).
3. Use the obtained value for "E_t" into the *Johnson-Euler* equation.
4. Calculate the difference between the initial value of $\boldsymbol{f}$ and the new value for $\boldsymbol{F_C}$; it must be 0.

Determination of $F_{0.7}$

To get the $F_{0.7}$ value, a simple way to determine it is to use the stress-strain curve graphs provided (among other sources) in (ref. 8.). A fictitious approximation is shown in *Figure 13-56* for the purpose of demonstration. Using the graph for the appropriate "*n*" value and the stress-strain curve in compression, I got the value at 70% of the primary E_C value.

For $E_C = 10.0 * 10^6$ *psi*, then $E_{0.7} = 70\% * E_C = 7.0 * 10^6$ *psi*. For a value of $n = 20$, I then found a value for $F_{0.7}$ of 37.0 *ksi*. See *Figure 13-56*.

Iteration results

From my results, I obtained the following values:

$$\frac{E_t}{E} = \eta = 0.98$$
$$E_t = (0.98)10{,}000{,}000 = 9{,}842{,}674\ psi$$
$$\frac{f}{F_{0.7}} = 0.718$$
$$F_C = 29{,}094\ psi$$

This value, although slightly more conservative, is very close to the value of 28,681 *psi* by not considering plasticity. The lowest number of the two will be used into the *MS* calculation below.

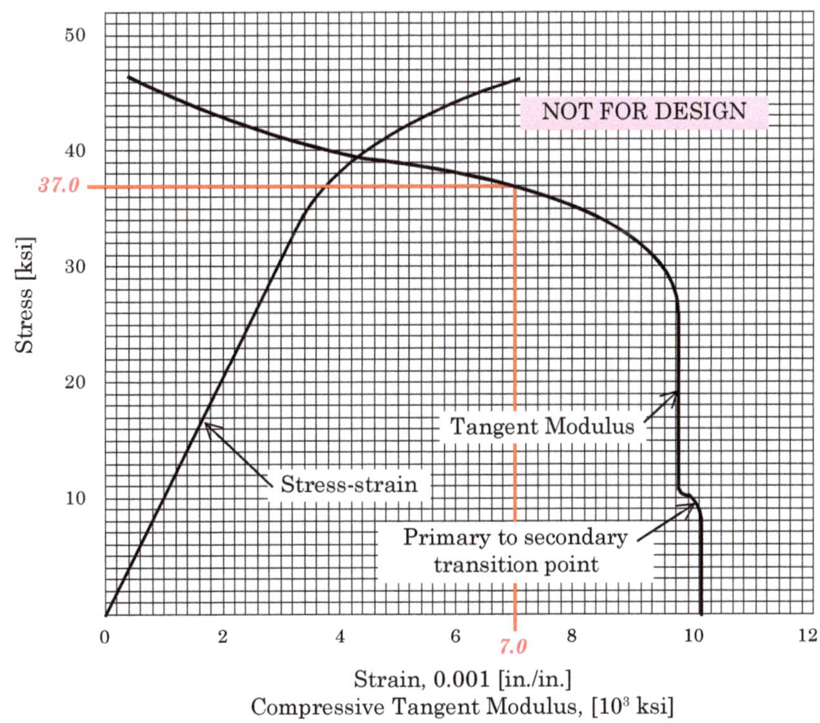

Figure 13-56 – *Approximation of Stress-strain and Tangent Modulus curves for Al Aly 2024-T3 Clad.*

<u>Margin of safety for column buckling</u>

Thus the column buckling MS is calculated using the maximum compression stress at point *2*:

$$MS = \frac{F_{CC}}{f_2} - 1 = \frac{29{,}094}{|-4{,}753|} - 1 = 5.12$$

Gusset Lower cap Section B-B, Ultimate	***>1.00***
3g Up *Al Aly 2024-T3 Clad*	*Column buckling*

I could have simplified the analysis (and be more conservative) by assuming a fixity coefficient "C" of 1 assuming the load to be equal at both ends of the column and still show a positive *MS*.

13.2.9 CONNECTIONS

<u>Gusset *MS20470AD5* solid rivets</u>

Three rivets connect the Gusset to the Crossbeam and two rivets connect it to the Seat track rail. *Figure 13-57* shows the applied forces at the rivet groups centroids from the 3*g Up* LC.

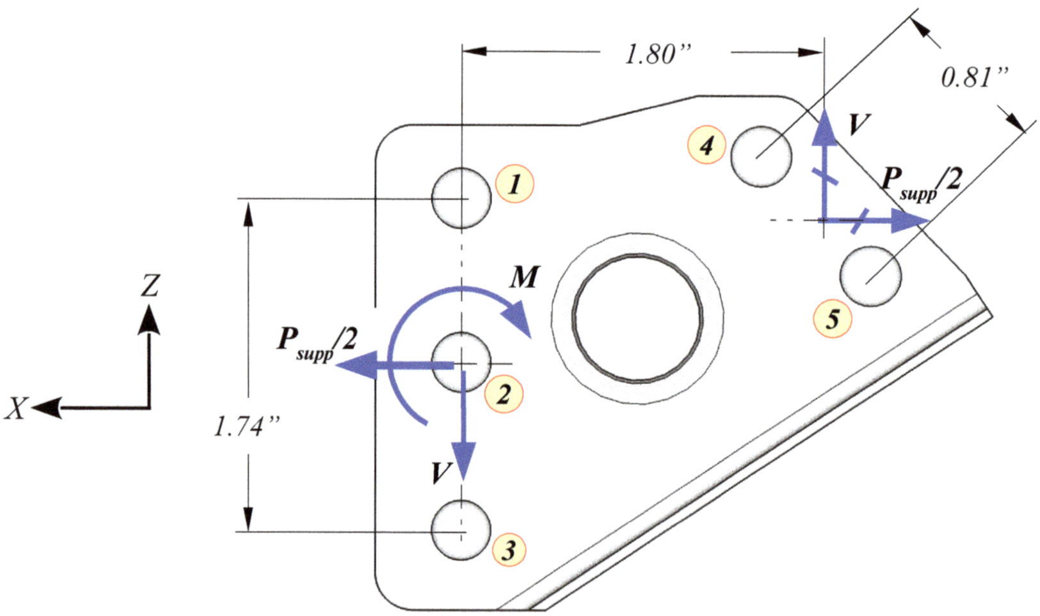

***Figure 13-57** – Applied forces on the Gusset rivets.*

As assumed earlier, the rivets (*1, 2 & 3*) common to the Crossbeam react a moment, axial and shear forces while the ones common to the Seat track rail (*4 & 5*) only react axial and shear forces. For $M = -172$ *in-lb*, $P_{supp}/2 = 145$ *lbs* and $V = 156$ *lbs*, the forces in the rivets are calculated as follows.

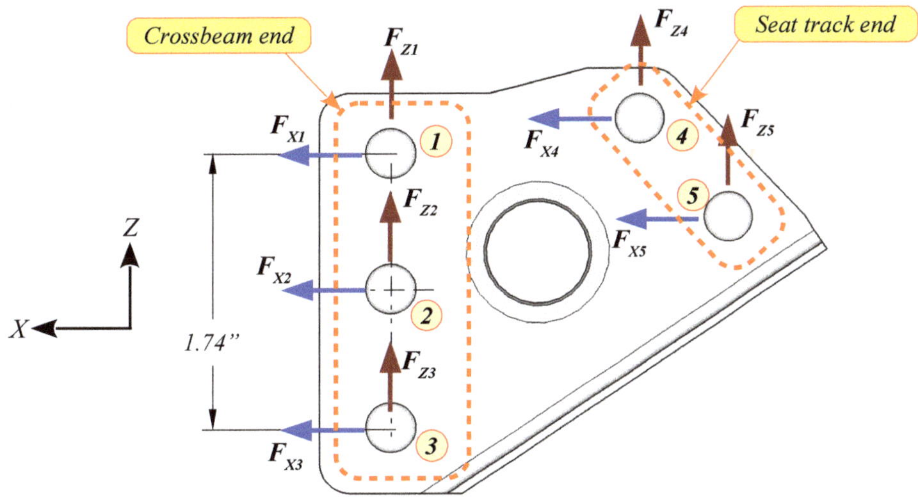

***Figure 13-58** – Individual reaction forces on the Gusset rivets.*

At the Crossbeam:

$$F_{Z1} = F_{Z2} = F_{Z3} = \frac{V}{n} = \frac{-156}{3} = -52 \ lbs$$

Due to $P_{supp}/2$:

$$F_{X1p} = F_{X2p} = F_{X3p} = \frac{P_{Supp}/2}{n} = \frac{145}{3} = 48.3 \ lbs$$

$$F_{X1M} = -F_{X3M} = \frac{M}{pitch} = \frac{-172}{1.74} = -99 \ lbs$$

Due to M:

Combining the vectors, the resulting rivet load *"F"* for rivets ***1, 2 & 3*** are calculated as:

$$F_1 = \sqrt{(F_{X1p} + F_{X1M})^2 + F_{Z1}^2} = \sqrt{(48.3 - 99)^2 + 52^2} = 72.6 \ lbs$$

$$F_2 = \sqrt{F_{X2p}^2 + F_{Z2}^2} = \sqrt{48.3^2 + 52^2} = 71 \ lbs$$

$$F_3 = \sqrt{(F_{X3p} + F_{X3M})^2 + F_{Z3}^2} = \sqrt{(48.3 + 99)^2 + 52^2} = 156 \ lbs$$

At the Seat track rail:

$$F_{Z4} = F_{Z5} = \frac{V}{n} = \frac{156}{2} = 78 \ lbs$$

$$F_{X4} = F_{X5} = \frac{P_{Supp}/2}{n} = \frac{-145}{2} = -72.5 \ lbs$$

Combining the vectors, the resulting rivet load *"F"* for rivets 4 & *5* are calculated as:

$$F_4 = \sqrt{F_{X4}^2 + F_{Z4}^2} = \sqrt{(-72.5)^2 + 78^2} = 106.5 \ lbs$$

$$F_5 = \sqrt{F_{X5}^2 + F_{Z5}^2} = \sqrt{(-72.5)^2 + 78^2} = 106.5 \ lbs$$

The rivet load summary is shown in *Table 13-22*.

***Table 13-22** - Gusset rivet forces table summary.*

	From $P_{supp}/2$		From V_Z		From M_Y		Total		
bolt	F_X	F_Z	F_X	F_Z	F_X	F_Z	F_X	F_Z	F
	[lbs]	[lbs]	[lbs]	[lbs]	[lbs]	[lbs]	[lbs]	[lbs]	[lbs]
1	48.3	0	0	52.0	-98.9	0	-50.5	52.0	72
2	48.3	0	0	52.0	0	0	48.3	52.0	71
3	48.3	0	0	52.0	98.9	0	147.2	52.0	**156**
4	72.5	0	0	78.0	0	0	72.5	78.0	106
5	72.5	0	0	78.0	0	0	72.5	78.0	106

Rivet shear:

The maximum resulting shear load value *"F"* calculated above, where rivet ***3*** sees the highest force with 127 *lbs,* is compared to the shear allowable *"P_{SU}"* for an *MS20470AD5* of 596 *lbs* from *Table 13-18.* A Fitting Factor *"FF"* of 1.15 (ref. 25. *FAR §25.625*) is used. Also, a shear factor *"SF"* of 0.996 is required since the Gusset is only 0.050 *in.* thick. (ref. *Table 13-20).* The Margin of Safety for shear is thus calculated as follows:

$$MS = \frac{SF \cdot P_{su}}{FF \cdot F} - 1 = \frac{(0.996) \cdot 596}{1.15 \cdot 156} - 1 = 2.30$$

Bearing:

The bearing stress will be checked in all of the three parts involved. The bearing load allowable is calculated as follows:

$$P_{bru} = F_{bru} \times D \times t$$

Table 13-23 shows the resulting bearing *MS.* No 1.15 Fitting Factor is used in bearing here.

***Table 13-23** - Bearing stress MS table summary.*

	Material	t	D	Fbru	Pbru	F	MS
Part	Al Aly	[in]	[in]	[ksi]	[lbs]	[lbs]	bearing
Gusset	2024-T3 Clad	0.050	0.159	120	954	156	5.11
Seat track	7075-T6	0.080	0.159	145	1844	156	10.82
Crossbeam	7050-T7451	0.070	0.159	105	1169	156	6.49

A calculation example for the lowest bearing stress MS is done below:

The most critical part is the Gusset. The bearing ultimate allowable load "P_{bru}" in the Gusset made out of *Aluminum Alloy 2024-T3 Clad* 0.050 *in.* thick is calculated:

For: F_{bru} = 120 *ksi* (ref. *Table 13-17), e/D* = 2.0. And for *D* = 0.159" and *t* = 0.050" hence:

$$P_{bru} = 120{,}000 \times 0.159 \times 0.050 = 954 \ lbs$$

As the *MS* in shear is smaller than the *MS* in bearing (shear critical), the *MS* for the joints shown below is in shear.

Margin of Safety for the Gusset connections

$$MS = \frac{P_{bru}}{F} - 1 = \frac{954}{156} - 1 = 5.11$$

Gusset, Ultimate	***>1.00***
3g Up *MS20470AD5 in Al Aly 2024-T3 Clad*	*Fastener shear*

13.2.10 COMMENT

The Gussets are adequate. The analysis of the forward Gussets also covers the aft Gusset by similarity.

Since the *MS* in shear is lower than the one in bearing, the joint is thus shear critical. For all materials involved, the ratio F_{bru} / F_{bry} is smaller than 1.5. Therefore, no bearing check at limit (yield) is necessary.

It would be interesting to see if a smaller material gauge (thickness) would work in order to save weight. Maybe 0.040"? Maybe going with a stronger material like 7075-T6, 0.032" thick? Some food for thought!

13.3 CONNECTIONS BETWEEN FRAME AND SKIN

A typical Frame to skin joint is shown in the example below. The fasteners attaching the skin to the Frame are loaded by the difference in shear flow from the skin panels on either sides of the Frame. The skin panels are assumed non-buckling. *Figure 13-59* shows a portion of a global FEM representing the fuselage along with the shear flow extracted for each QUAD element for a given load case.

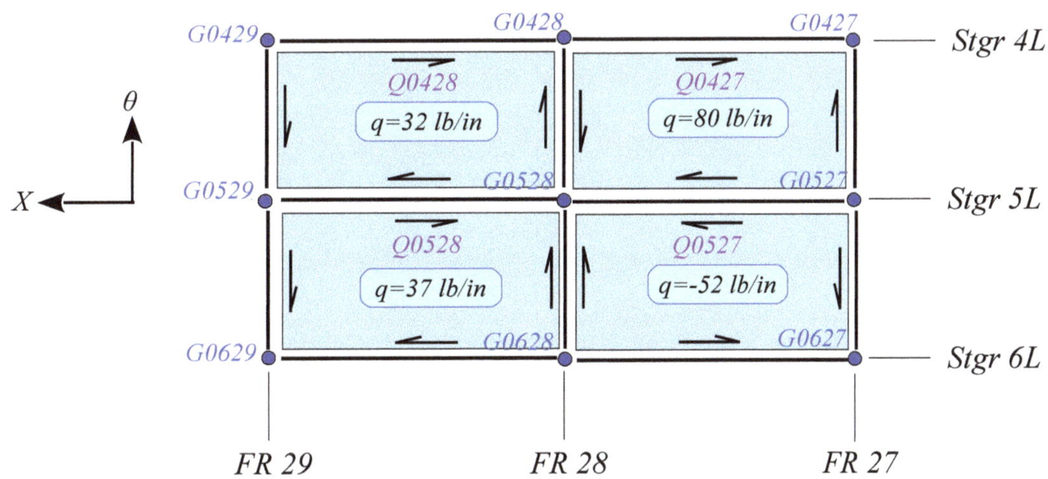

Figure 13-59 *– Shear flows in skin panels from the global FEM.*

The differences in shear flows "Δq" at *Frame 28* are calculated as:

Between *Stringer 4 Left* and *Stringer 5 Left*:

$$\Delta_{q1} = q_{Q0427} - q_{Q0428} = 80 - 32 = 48 \ lbs/in.$$

Between *Stringer 5 Left* and *Stringer 6 Left*:

$$\Delta_{q2} = q_{Q0527} - q_{Q0528} = -52 - 37 = -89 \ lbs/in.$$

Therefore, with a fastener pitch "s" of 1.20 *in.*, the maximum fastener shear load "P_S" between the skin and the Frame *28* occurs between *Grid 0528* and *Grid 0628* and is calculated as:

$$P_S = |\Delta_{q2}| \cdot s = 89 \cdot 1.20 = 107 \ lbs$$

The same approach is used to calculate the shear in the fasteners common to the Skin and the Stringers. In the above example, the "Δq" value is the highest at Stringer 5L between *Grid 0527* and *Grid 0528* with a maximum fastener load of:

$$P_S = |q_{Q0427} - q_{Q0527}| \cdot s = |80 - (-52)| \cdot 1.20 = 159\ lbs$$

One can notice, from the *Figure 13-59*, the shear flow arrows are pointing in the same direction when the adjacent QUADS have shear flows of opposite signs. Therefore the *delta q* is the greatest in those cases.

Note that this is true only if the QUAD's orientations (coordinate system) are the same. Be careful to verify the FEM to ensure this is the case. An FEM that is properly constructed should have all the elements' orientations in the same direction as well as their normal; but we can't never be too sure...

"Lockwire (safety wire) is very good at keeping the broken pieces of a bolt attached to the airframe."

14. MONUMENT INSTALLATION

An aircraft monument is not a memorial in a park but a structure *(usually considered non structural)* installed in the interior of an airplane. For example, galleys, cabinets, lavatories and trolley stowages are all monuments. On the other hand, seats are not considered monuments as they are structural and are considered differently since they have to meet different requirements.

Monuments are typically attached to the aircraft floor using either seat tracks fittings or other types of fittings and fasteners. They can be fastened to the floor only, providing they are not too high, or to the floor and the "ceiling" when their height is significant.

The monuments themselves are generally provided by vendors and are already certified. Hence, as stress analysts, our job consists of analyzing the attachments and the airframe back-up structure, not the monument itself. Note that the monuments have been certified providing they had the intended supports. Therefore, it is important to provide the same supports as recommended by the vendor. What that means is, all of the attachment fittings shall be used and as intended. For instance, if a monument attachment point is to take vertical load only, then use this point without trying to react loads in other directions. Also, make sure that the reaction loads you get at the attachment points do not exceed the vendor's specifications.

14.1 DESCRIPTION

The following example is a tall stowage Cabinet that is restrained to the floor by seat track fittings and to the airframe at the top via two struts. For the purpose of our example, the center of gravity of the Cabinet is located at *FS 515* and its maximum weight capacity is 375 *lbs* including the Cabinet's empty weight. The general arrangement of the Cabinet installation is shown in *Figure 14-1* to *Figure 14-4*.

Struts are used at the top of the cabinet to react the Fore-Aft loads only and, at the same time, allowing the cabinet to move vertically with the floor relative to the fuselage frames. This is very important as the fuselage will deform under the pressurization and other flight loads.

One could have used other means that would have done a similar thing like vertical pins that slide into a fitting. The struts are more appropriate when the monument is far from the fuselage skin while the pins are probably preferable when the distance is short.

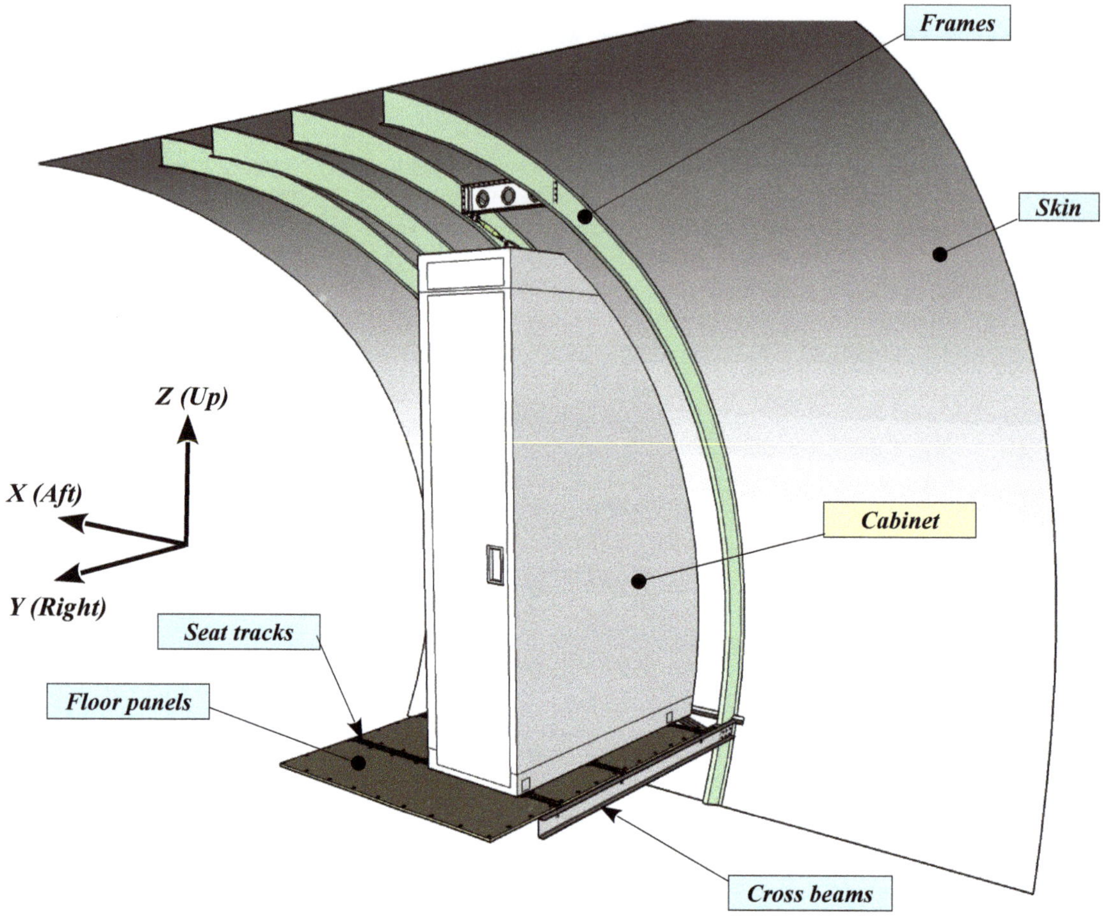

Figure 14-1 *– Cabinet installation arrangement, view looking Aft-Outb'd.*

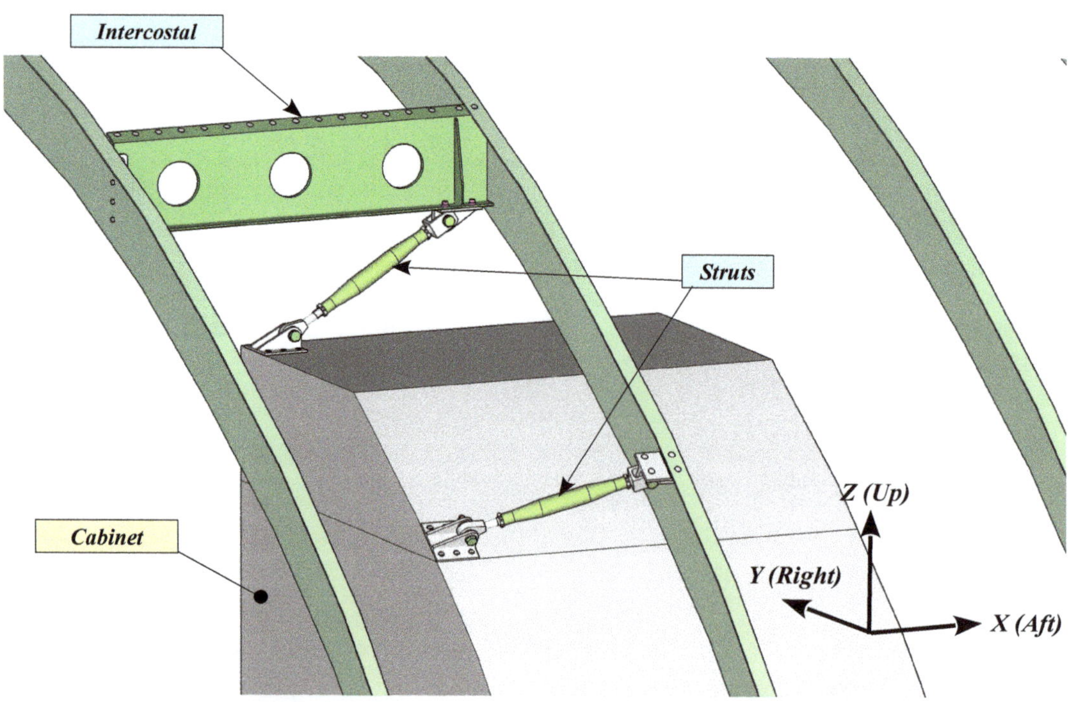

Figure 14-2 *– Cabinet installation upper Struts arrangement, view looking Inboard LHS.*

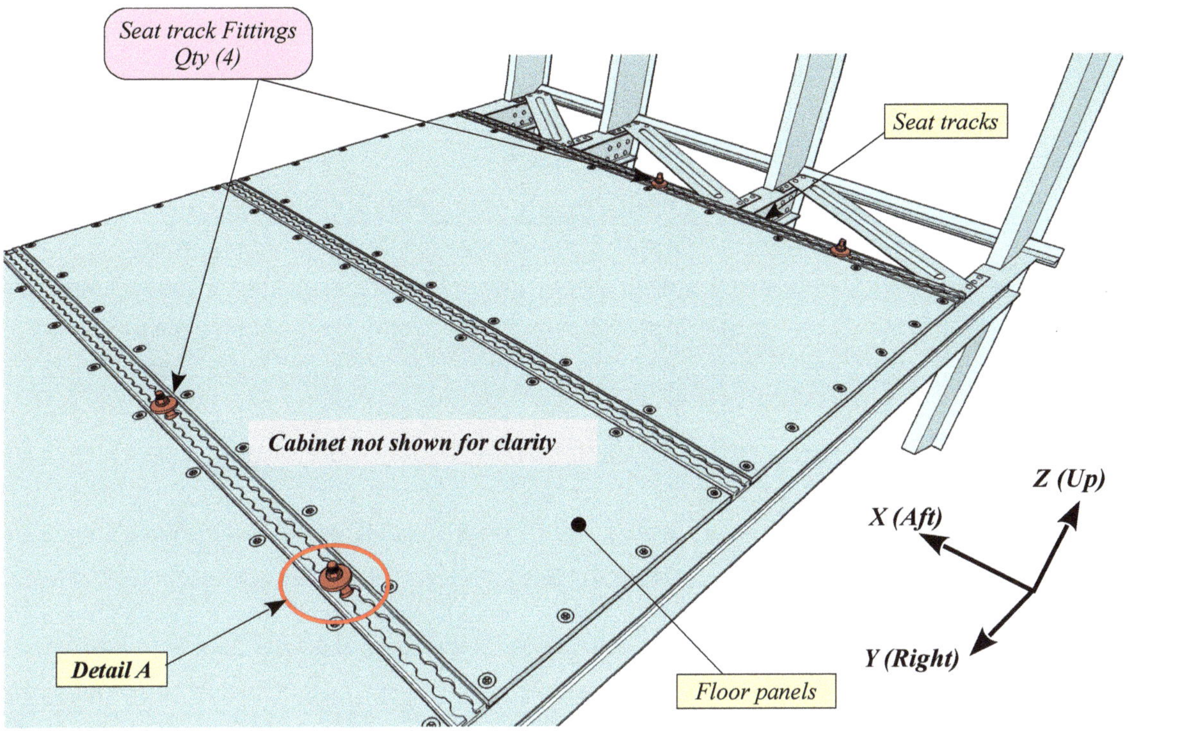

Figure 14-3 *– Cabinet floor attachments to seat tracks.*

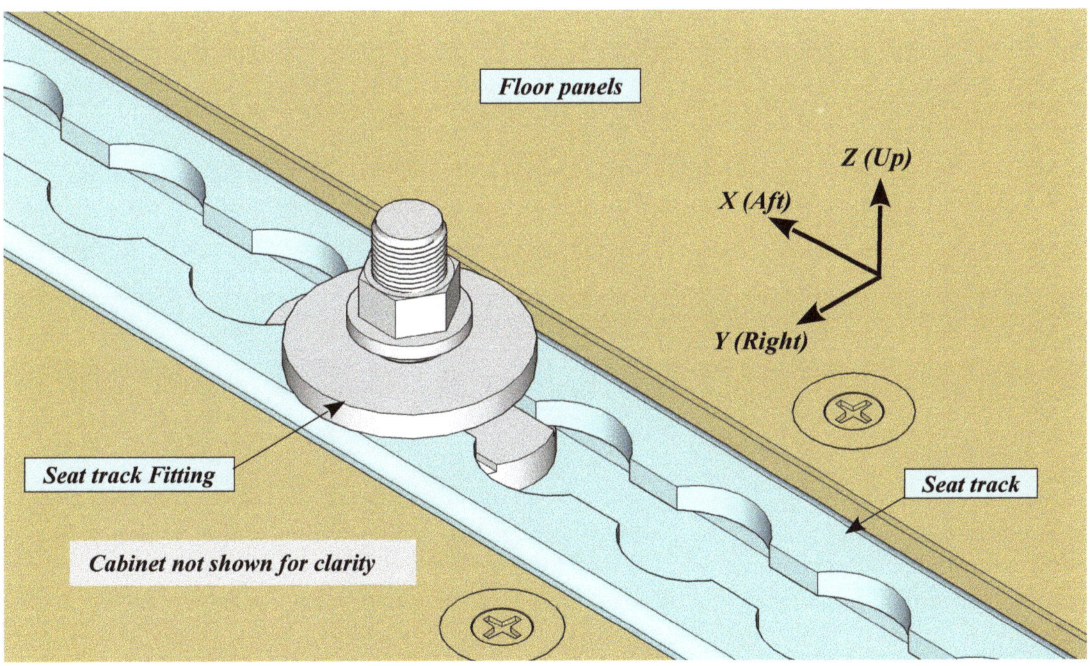

Figure 14-4 *– Detail A - Typical Cabinet floor attachment to seat tracks.*

The location of the Cabinet's center of gravity and its attachment points are shown in *Figure 14-5*. Points ***7*** and ***8*** are shown in order to calculate the strut angles.

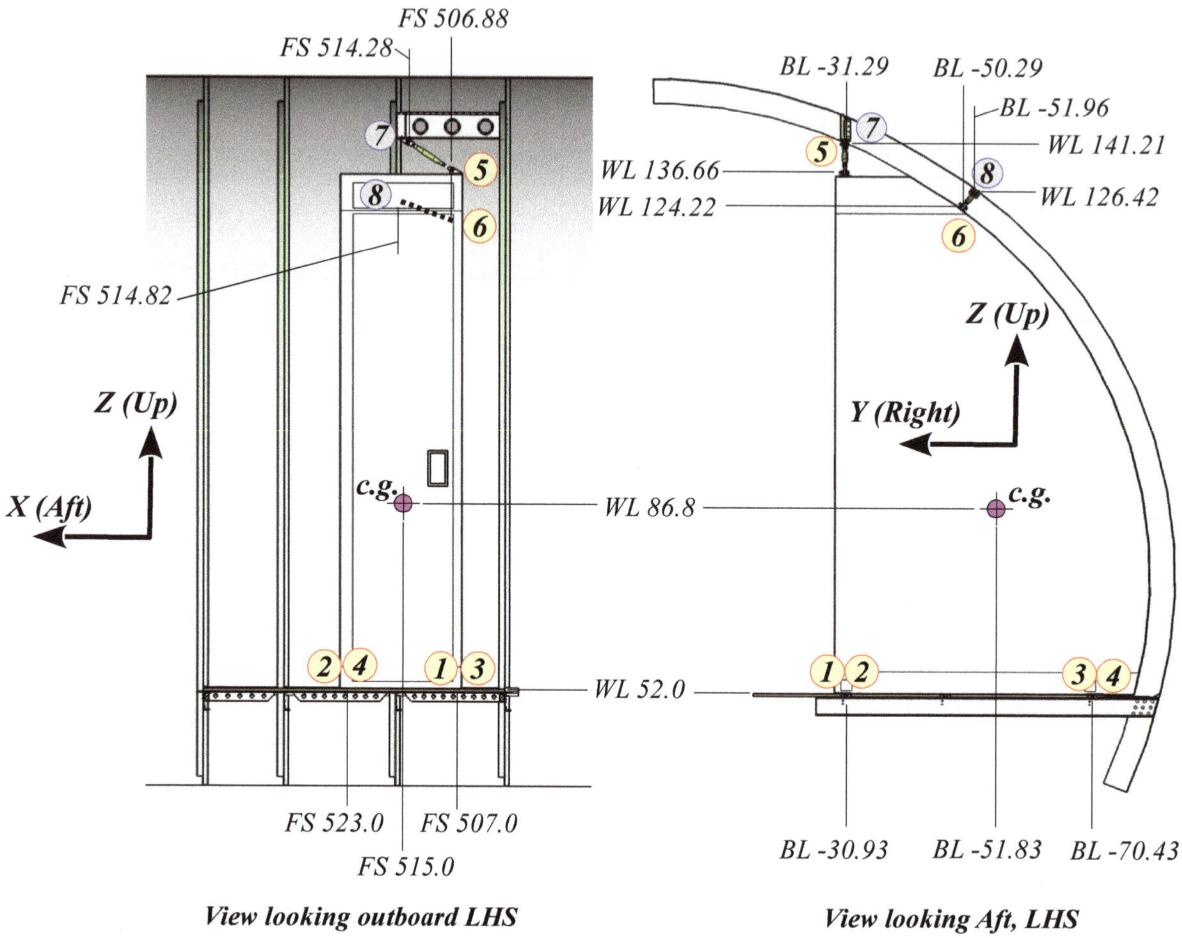

Figure 14-5 *– Center of gravity and reaction points locations of the Cabinet.*

Table 14-1 shows the attachments of the Cabinet in the aircraft fuselage coordinate system.

Table 14-1 *- Cabinet attachment points in fuselage Global Coordinate System.*

Points	*FS*	*BL*	*WL*
1	507	-30.93	52
2	523	-30.93	52
3	507	-70.43	52
4	523	-70.43	52
5	506.88	-31.29	136.66
6	506.88	-50.29	124.22
7	514.28	-31.29	141.21
8	514.82	-51.96	126.42
c. of g.	515	-51.83	86.8

14.2 SUMMARY OF MARGINS OF SAFETY

Table 14-2 shows the summary of minimum Margins of Safety for all parts analyzed in the monument installation.

Table 14-2 *- Summary of minimum Margins of Safety for the Monument installation.*

Nomenclature	Description	Material	Mode of failure	Load case	MS	Page
Seat track rail	Section A-A	Al Aly 7076-T6 Ext.	Compression + bending + shear	9g Fwd	0.10	332
	Section A-A	Al Aly 7076-T6 Ext.	Free flange buckling	3g Left	0.34	342
	Section B-B	Al Aly 7076-T6 Ext.	Axial + bending + shear	9g Fwd	0.13	344
	Tension clip	Al Aly 7076-T6 Ext.	Plastic bending + shear	9g Fwd	0.23	360
Seat track Ftg	Vendor part	CRES	Tension + shear	9g Fwd	>1.00	361
Shear Clip	Section through fasteners	Al Aly 2024-T3	Bending + shear	9g Fwd	0.05	364
Floor panel	Panel A	GR/EP Bonded panel	Shear buckling	9g Fwd	>1.00	371
	Panel A connection	GR/EP	Bearing	9g Fwd	0.69	383
Crash Strap	HST10-5 Pin	Al Aly 2024-T3 Clad	Bearing	9g Fwd	0.12	390
Crossbeam		Al Aly 7050-T7451	Bending + shear	6g Down	P.B.C.	394
Fwd Lug Ftgs	Section A-A	Al Aly 7050-T7451	Tension + bending + shear	9g Fwd	>1.0	407
	Tension clip	Al Aly 7050-T7451	Bending + shear	9g Fwd	0.38	421
Aft inb'd Lug Ftg	HST10-5 Pin	Al Aly 6061-T6	Shear & bearing	9g Fwd	>1.0	429
Aft outb'd Lug Ftg	Section B-B	Al Aly 7050-T7451	Combined stresses	9g Fwd	>1.0	439
	Bearing instl.	Al Aly 7050-T7451	Hoop stress	Interference	0.77	443
Strut Assy	Stability	Al Aly 7075-T62	Compression + bending	9g Fwd	>1.0	456
	Bolt	Alloy steel	Combined stresses	9g Fwd	0.71	463
Intercostal	Lower cap	Al Aly 7050-T7451	Column buckling	9g Fwd	>1.0	484
Shear clip	Section through fasteners	Al-Aly 2024-T3	Bending + shear	9g Fwd	>1.0	496

14.3 LOADS

The Cabinet being installed inside the aircraft, where passengers are located, has to meet (ref. 25. *FAR §25.561*) for the Emergency Landing Conditions. The acceleration factors (number of g's) are as follows:

- Nx = −9 (Fwd), +1.5 (Aft)
- Ny = ±3.0 (Left and right)
- Nz = −6 (Down), +3.0 (Up)

In addition, one must ensure that no ultimate flight load case causes a higher load factor than the ones shown above. In which case, the latter shall be used of course. See *Figure 14-6* for a fictitious graph of the load factors in the "*Z*" directions along the fuselage of the aircraft. From that example, at the Fuselage station "*FS*" where the center of gravity of the bin is located (*FS 515*), the "*Z* " load factor is higher in the upward direction (+N_Z) with a value of 3.15 and thus will be used.

Similar graphs would exist of course for the *X* and *Y* directions but are not shown since the approach is the same.

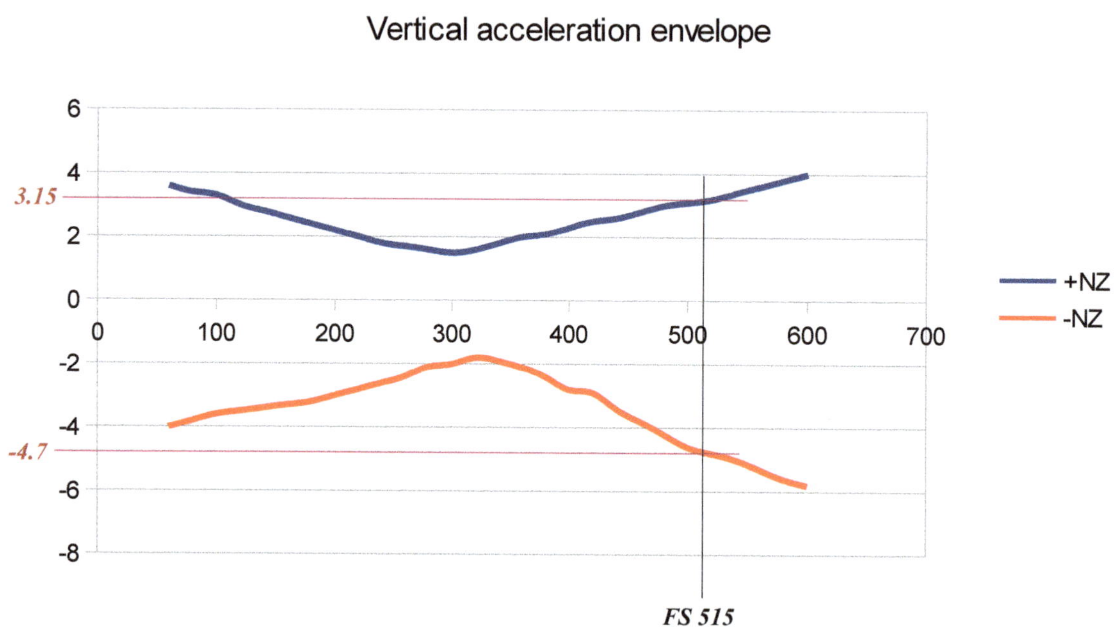

Figure 14-6 *– Flight loads vertical acceleration envelop distribution along the fuselage.*

The resulting load factors to be used in the analysis are thus:

- Nx = −9 (Fwd), +1.5 (Aft)
- Ny = ±3.0 (Left and right)
- Nz = −6 (Down), +3.15 (Up)

For a weight of 375 *lbs*, the forces to be applied at the c.g. of the cabinet are as shown in *Table 14-3.*

Table 14-3 - *Applied ultimate forces at Cabinet c. of g.*

		Ult. N	Force
			[lbs]
+NX	Aft	1.5	563
-NX	Fwd	-9.0	-3375
+NY	Right	3.0	1125
-NY	Left	-3.0	-1125
+NZ	Up	3.15	1181
-NZ	Down	-6.0	-2250

Free Body Diagrams

Since there are applied loads in all three axes, an *FBD* is done for each axis. To simplify the analysis, only one *FBD* is done per axis as the reactions loads at the attachment points can be factored linearly since the boundary conditions are the same for the positive and the negative directions of all axes. Hence, for the "*X*" axis, the *9g Fwd* load is used, for the "*Y*" axis, the *3g Right* is used and, for the "*Z*" axis, the *6g Down* is used. The Cabinet and the airframe structure are considered rigid.

14.3.1 *X* AXIS; 9G FWD

For this load case, as shown in *Figure 14-7*, at the floor, only the forward attachment points (***1 & 3***) are considered effective. Since there are two struts at the top that are not located at the same WL (points ***5 & 6***), the system is thus found to be statically indeterminate in the *X-Z* plane. To resolve this issue, one could be tempted to use a "***2D***" (or *3D*) bolt group approach to calculate the reaction loads but the results would be incorrect mainly due to the angled struts. Therefore, I will be using the "manual" approach by drawing *FBD*'s for each direction.

Table 14-4 shows the assumed Boundary Conditions (B.C.) at all the attachment points for the *9g Forward* Loading Condition. Note that, at the struts locations (points ***5 & 6***), the B.C. are assumed to be in the "*X*" direction only. However, because the struts are at an angle from the "*X*" axis, there can't be any restraint in that direction without some restraint in the other directions involved. The actual restrained points at the struts in all the three axes are points ***7 & 8*** common to the airframe.

Ideally, for this load case, we would want the struts to be along the "*X*" axis as much as possible therefore allowing the fuselage to "breathe" freely without introducing any "parasite" loads into the airframe. The Cabinet must be able to move vertically and laterally (at the struts) independently from the fuselage skin/frames.

Table 14-4 - *Boundary Conditions at all points for the 9g Fwd LC.*

Points	X	Y	Z	RX	RY	RZ	B.C.
	[in]	[in]	[in]	[deg.]	[deg.]	[deg.]	
1	0	0	0	–	–	–	[1,2,3]
2	–	–	–	–	–	–	[]
3	0	–	0	–	–	–	[1,3]
4	–	–	–	–	–	–	[]
5	0	–	–	–	–	–	[1]
6	0	–	–	–	–	–	[1]
7	0	0	0	–	–	–	[1,2,3]
8	0	0	0	–	–	–	[1,2,3]

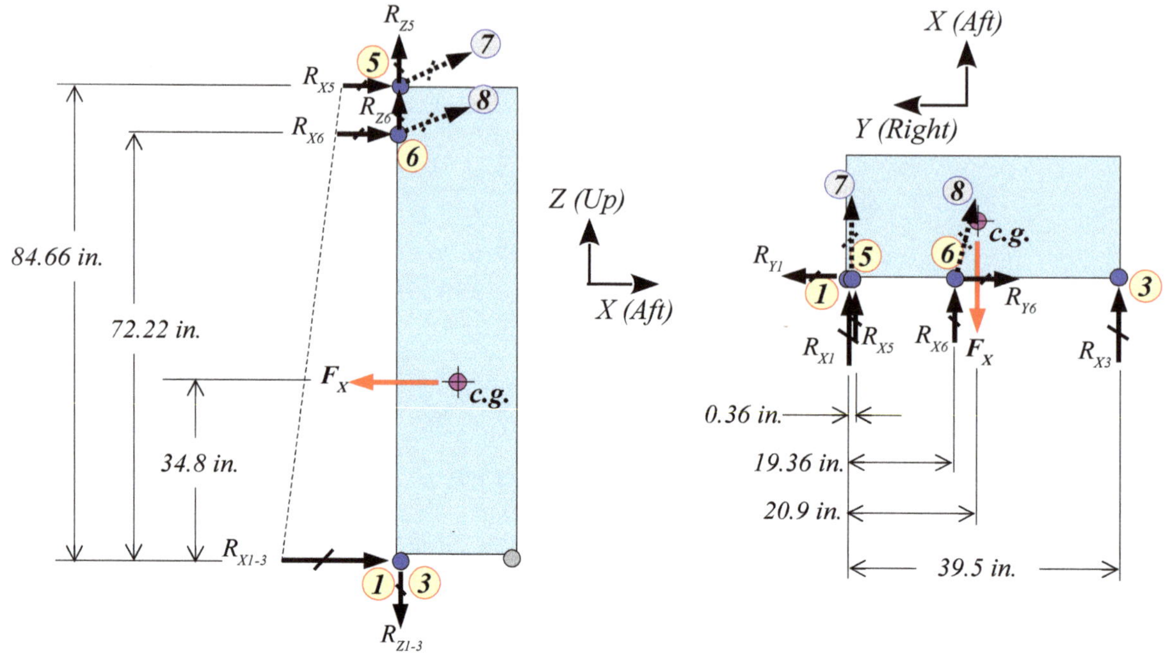

Figure 14-7 *– Cabinet FBD due to F_X in the X-Z and X-Y planes.*

For all the load cases, it is assumed that the reaction points have the same stiffness. As shown below, I'll group points (***1 & 3***) together and, because the Cabinet is assumed rigid, the reaction load at point ***6*** "$\boldsymbol{R_{X6}}$" is linearly interpolated from the loads at point ***1-3*** and point ***5***.

As mentioned earlier in section ***12*** for the boundary conditions, we do not want any secondary structure to carry primary loads from the aircraft primary structure. Therefore, when attaching a cabinet or a seat to the floor structure, only the aft or only the forward attachments will be designed to react the longitudinal loads (in "*X*"). It is done by either having different attachment fittings or by having a sufficient play in the attachments that we don't want it to carry any load.

For this load case, we have three unknowns: $\boldsymbol{R_{X1\text{-}3}}$, $\boldsymbol{R_{X5}}$ and $\boldsymbol{R_{X6}}$. Therefore we need three equations to solve this problem.

1) From the sum of the forces: $F_X = R_{X1-3} + R_{X5} + R_{X6}$

2) From the linear interpolation: $R_{X6} = \dfrac{R_{X5} - R_{X1-3}}{84.66} \times 72.22 + R_{X1-3}$

by re-arranging, we get: $0.147\, R_{X1-3} + 0.853\, R_{X5} - R_{X6} = 0$

3) From the sum of the moments at point ***1-3***: $R_{X5}(84.66) + R_{X6}(72.22) = F_X(34.8)$

Hence, for the *9g Fwd* load case, the applied force is: $F_X = -3{,}375$ *lbs*. The reaction loads matrix "*[F]*" is calculated by multiplying the inverse matrix "$[A^{-1}]$" by the "*[k]*" matrix. See *Table 14-5* for the results.

Table 14-5 - *Reaction forces for the 9g Fwd LC using matrices.*

Equations	R_{X1_3}	R_{X5}	R_{X6}	F_X
1)	1	1	1	3375
2)	0.147	0.853	-1	0
3)	0	84.66	72.22	117450

[A] (coefficients of R_{X1_3}, R_{X5}, R_{X6}); [k] (column F_X)

Inverse Matrix $[A]^{-1}$

0.98765	0.084	-0.01251
-0.07169	0.48767	0.00775
0.08404	-0.57167	0.00477

$[F]=[A]^{-1} \times [k]$

R_{X1_3} =	**1864**
R_{X5} =	**668**
R_{X6} =	**844**

Since the struts can only develop axial loads along their longitudinal axis and because they are at an angle from the "*X*" direction, components in the "*Z*" direction and in the "*Y*" direction (for point ***6*** only) exist.

The loads in the struts are thus calculated as follows.

Strut at point *5*

The strut shown in *Figure 14-9* is in the *X-Z* plane so since we know the force component in the "*X*" direction (R_{X5} = 668 *lbs*), the force component in the "*Z*" direction is calculated as follows:

$$R_{Z5} = \frac{4.55}{7.40} R_{X5} = \frac{4.55}{7.40} \times 668 = 411 \ lbs$$

Similarly, the axial load in the strut; P_5 is:

$$P_5 = \frac{8.69}{7.40} R_{X5} = \frac{8.69}{7.40} \times 668 = 784 \ lbs$$

The force vectors of the strut at point ***5*** is shown in *Figure 14-8.*

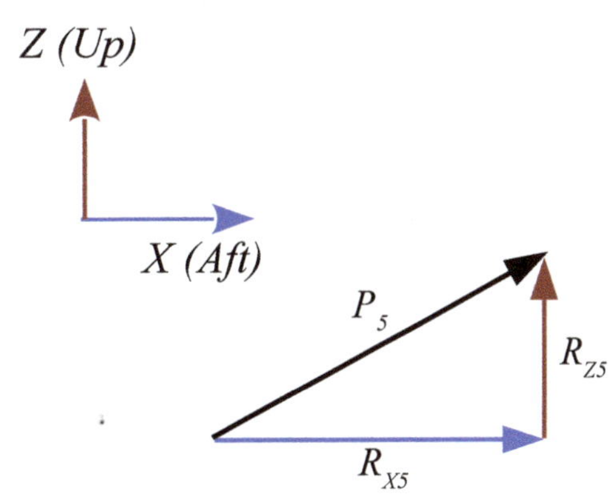

Figure 14-8 – *Force vectors of Strut at point 5.*

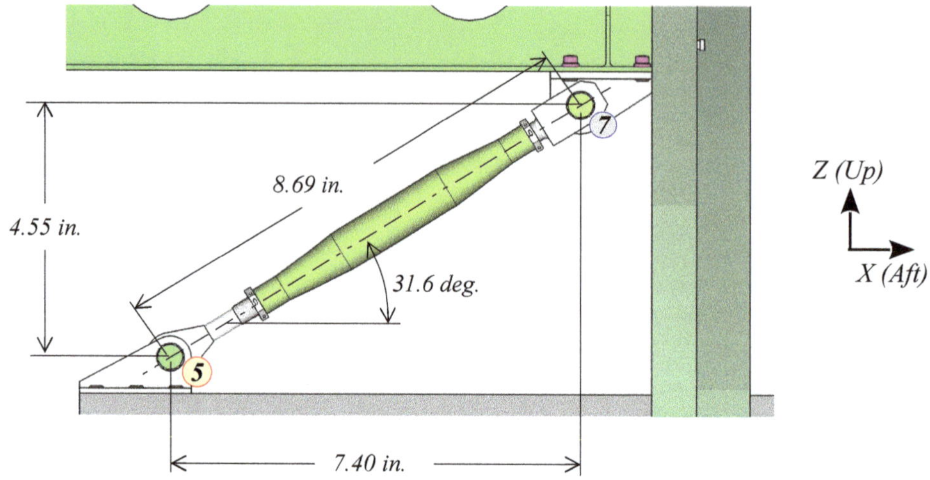

***Figure 14-9** – Geometry of Strut at point 5.*

Strut at point *6*

The strut shown in *Figure 14-11* is in a plane that follows the "*X*" axis. Therefore, since we know the force component in the "*X*" direction (R_{X6} = 844 *lbs*) the force component in the "*Z*" direction is calculated as follows:

$$R_{Z6} = \frac{2.20}{7.94} R_{X6} = \frac{2.20}{7.94} \times 844 = 234 \; lbs$$

the force component in the "*Y*" direction is calculated as follows:

$$R_{Y6} = \frac{1.67}{7.94} R_{X6} = \frac{1.67}{7.94} \times 844 = 178 \; lbs$$

Similarly, the axial load in the strut; P_6 is:

$$P_6 = \frac{8.40}{7.94} R_{X6} = \frac{8.40}{7.94} \times 844 = 893 \; lbs$$

The force vectors of the strut at point ***6*** is shown in *Figure **14-10***.

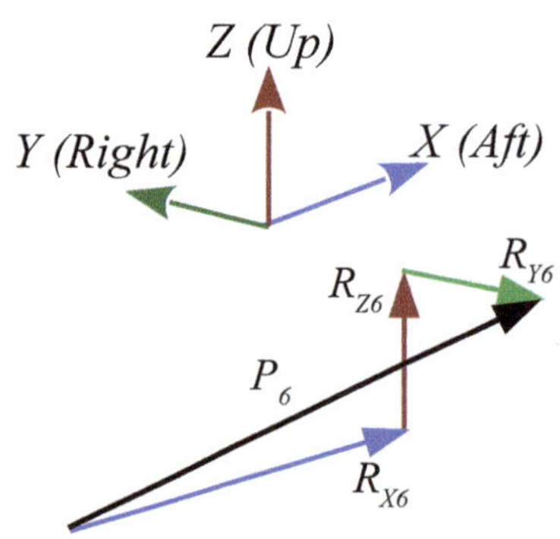

***Figure 14-10** – Force vectors of Strut at point 6.*

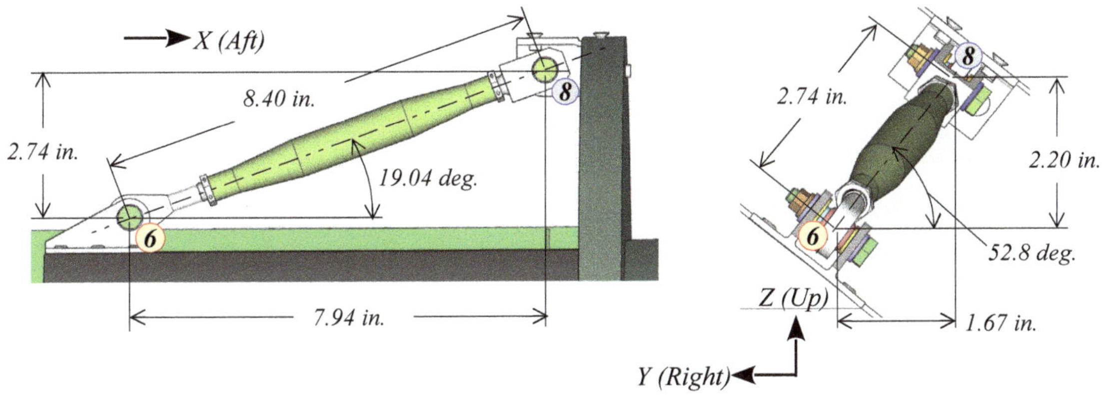

Figure 14-11 *– Geometry of Strut at point 6.*

Now the remaining forces that we have to determine are R_{X1}, R_{Y1}, R_{Z1}, R_{X3} and R_{Z3}. From *Table 14-5*, the reaction force in "*X*" for the combined points ***1 & 3*** "R_{X1_3}" is 1,864 *lbs*. From the right part of *Figure 14-7*, the force at point ***3*** "R_{X3}" is calculated by doing the sum of the forces at point ***1***.

$$M_Z+\text{: CCW} \qquad \sum M_{Z1} = R_{X5}(0.36)+R_{X6}(19.36)-R_{Y6}(0.12)-F_X(20.9)+R_{X3}(39.5)=0$$

hence

$$R_{X3} = \frac{F_X(20.9) - R_{X5}(0.36) - R_{X6}(19.36) + R_{Y6}(0.12)}{39.5}$$

$$= \frac{3375(20.9)-668(0.36)-844(19.36)+178(0.12)}{39.5} = 1{,}367\ lbs$$

From the sum of the forces in "*X*"; $R_{X1} = R_{X1\_3} - R_{X3} = 1{,}864 - 1{,}367 = 497\ lbs$,

and from the sum of the forces in "*Y*"; $R_{Y1} = R_{Y6} = 178\ lbs$.

From our initial boundary condition assumptions (*Table 10-3*), point ***1*** reacts loads in [1,2,3] while point ***3*** reacts loads in [1,3].

For R_{Z3}, the vertical force at point ***3*** is calculated by doing the sum of the forces at point ***1*** around the "*X*" axis. See *Figure 14-12* for the *FBD* in the *Y-Z* plane.

$$M_X+\text{: CW} \qquad \sum M_{X1} = -R_{Z5}(0.36) - R_{Z6}(19.36) + R_{Y6}(72.22)+R_{Z3}(39.5)=0$$

hence

$$R_{Z3} = \frac{R_{Z5}(0.36) + R_{Z6}(19.36) - R_{Y6}(72.22)}{39.5}$$

$$= \frac{411(0.36)+234(19.36)-178(72.22)}{39.5} = -207\ lbs$$

So, because of the negative sign, R_{Z3} is thus in the opposite direction then assumed on *Figure 14-12*. From the sum of the forces in "*Z*", the vertical reaction at point ***1*** is calculated;

$$\sum F_Z = R_{Z5}+R_{Z6}-R_{Z1}-R_{Z3}=0$$

$$R_{Z1} = R_{Z5}+R_{Z6}-R_{Z3} = 411+234-(-207) = 852\ lbs$$

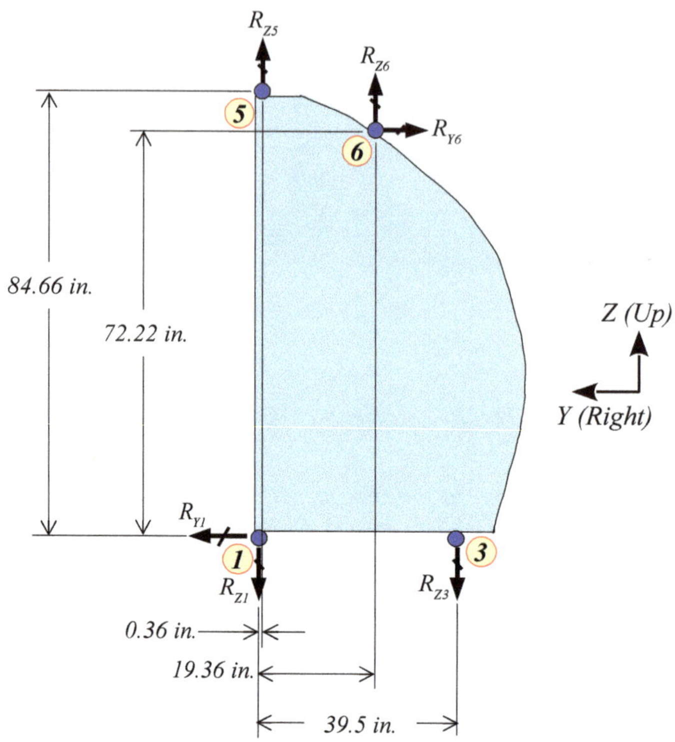

Figure 14-12 *– Cabinet FBD due to F_X in the Y-Z plane.*

Table 14-6 shows the summary of all the reaction forces in the GCS (Global Coordinate System) at all points for the ***9g Fwd*** load case.

Table 14-6 *- Summary of the Reaction forces for the 9g Fwd LC.*

Points	R_X	R_Y	R_Z
	[lbs]	[lbs]	[lbs]
1	497	178	-852
2	0	0	0
3	1367	0	207
4	0	0	0
5	668	0	411
6	844	-178	234
SUM:	3376	0	0

At this point you must be wondering: "*Why didn't he simply do a **FEM** instead of the long and cumbersome hand calculations and be done with it?*" Well, the short answer is: "*There shouldn't be any **FEM** described in the stress report that cannot be validated and correlated by static test.*" This is the message that I was getting from most DER's. However, it could be possible that, in some particular case, a FAA representative could accept a small FEM.

The long answer (which is my personal opinion by the way) is that any stress analyst, including the DER, should be able to replicate or edit the analysis after any number of years. If someone does a FEM using a certain software on a certain machine running a certain operating system, there is absolutely no guarantee that in twenty years time we will still be able to re-run it. As for hand calculations done on paper, anybody with a pocket calculator could edit the analysis regardless of the number of years that have passed.

Another reason is that stress reports are not only used for certification but they are also used by the MRB (Material Review Board) people. MRB engineers are constantly using stress reports to re-do analyses in cases where there is a deviation from the drawing. For example, if a part is slightly damaged during its assembly, MRB engineers will first look to salvage that part or come up with a disposition to repair it if they can. Therefore, they need to re-do the analysis of that part with the reduced section properties and see if the MS is still acceptable.

MRB's are usually under quite a bit of pressure to provide a disposition quickly for the discrepant part. Therefore, if they can simply mark-up the analysis in red kind-of-thing, then it's quick and easy. However, if they have to go back into a FEM and re-import the new modified solid from the CAD system and re-mesh it and run it (providing they still can do that! See comment above), one can easily see that it could take quite some time and be more difficult.

My opinion is that, if you have to tell the production manager that the discrepant part needs to be removed and replaced (scrapped!) then it's much preferable to let him know sooner than later. Therefore, the simpler the better.

Don't get me wrong! I don't hate *FEM's*. To the contrary I believe them to be absolutely necessary. But during my career I found that if an analysis can be done by hand then it is usually easier to "sell" to the DER. I usually use small FEM's when necessary to verify myself and to show the DER as a double check without necessarily including it into the stress report.

14.3.2 *X* AXIS; 1.5G AFT

For the *1.5g Aft* loading condition, it is assumed that the same *B.C.* as for the *9g Fwd* are applicable and that the reaction loads can be scaled linearly. Therefore, the reaction forces calculated previously, as shown in *Table 14-6*, are scaled by the factor 1.5 / −9 = −0.167. The results are shown in *Table 14-7*.

Table 14-7 *- Summary of the Reaction forces for the 1.5g Aft LC.*

Points	R_X *[lbs]*	R_Y *[lbs]*	R_Z *[lbs]*
1	-83	-30	142
2	0	0	0
3	-228	0	-35
4	0	0	0
5	-111	0	-69
6	-141	30	-39
SUM:	-563	0	0

Struts loads

The loads in the struts are thus obtained from the loads components at points ***5 & 6***.

$$P_5 = \sqrt{R_{X5}^2 + R_{Z5}^2} = \sqrt{(-111)^2 + (-69)^2} = 131\ lbs\,(compression)$$

and

$$P_6 = \sqrt{R_{X6}^2 + R_{Y6}^2 + R_{Z6}^2} = \sqrt{(-141)^2 + (30)^2 + (-39)^2} = 149\ lbs\,(compression)$$

14.3.3 *Y* AXIS; 3G RIGHT

For this load case, as shown in *Figure 14-13*, only the four attachment points at the floor (***1 to 4***) are considered effective. The two Struts at the top are considered completely ineffective laterally. *Table 14-8* shows the *BC* for the *3g Lateral LC*.

Table 14-8 *- Boundary Conditions at all points for the 3g Side LC.*

Points	X [in]	Y [in]	Z [in]	RX [deg.]	RY [deg.]	RZ [deg.]	B.C.
1	0	0	0	–	–	–	[1,2,3]
2	–	–	0	–	–	–	[3]
3	0	–	0	–	–	–	[1,3]
4	–	–	0	–	–	–	[3]

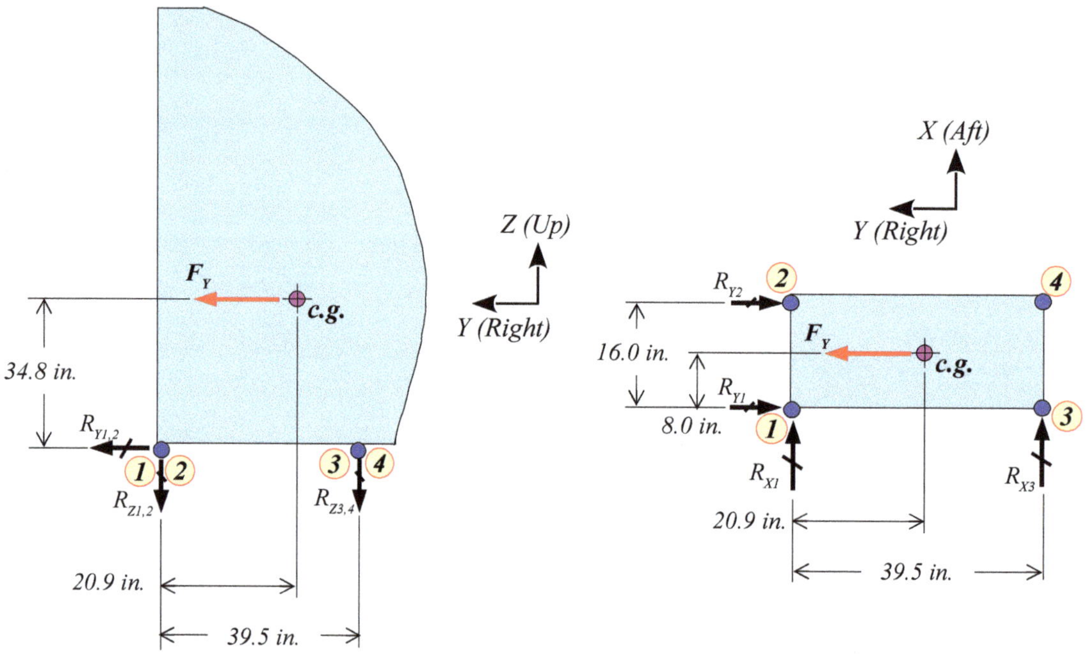

Figure 14-13 *– Cabinet FBD due to F_Y in the Y-Z and X-Y planes.*

For this load case, it is assumed that the lateral stiffness at the reaction point "***2***" is softer than the longitudinal stiffness at reaction point "***3***" and thus ineffective. The logic behind this is, for one, that point ***3*** is further away from point ***1*** (where the lateral load is reacted) than from point ***2***, and, for two, the seat track rails are designed to take higher loads in the longitudinal direction than laterally. Therefore one can see that for a unit rotation around point ***1***, point ***3*** will react a higher load than point ***2***, so point ***2*** is simply neglected in the lateral direction.

> Point ***1*** was chosen to react all of the lateral load but any point could have been chosen. The idea is to pick the one that is the stiffest. That would be the point that is the closest to a crossbeam for instance.

Hence, for the 3g *Right* load case, the applied force is: F_Y = 1,125 *lbs*. The reaction loads are calculated as follows. Since we determined that R_{Y2} was 0 for this load case, the forces at the four reaction points are statically determined. Therefore:

From the sum of the forces in "***Y***": $R_{Y1} = F_Y = 1{,}125\ lbs$

From the sum of the moments around the "*Z*" axis at point ***1***:

$$M_Z+: \text{CCW} \qquad \sum M_{Z1} = R_{X3}(39.5) - R_{Y2}(16.0) + F_Y(8) = 0$$

hence

$$R_{X3} = \frac{R_{Y2}(16.0) - F_Y(8.0)}{39.5}$$

$$= \frac{(0) - 1{,}125(8)}{39.5} = -228\ lbs$$

Because of the negative sign, R_{X3} is thus in the opposite direction then assumed in *Figure 14-13*.

From the sum of the forces in "*X*", the longitudinal reaction at point ***1*** is calculated;

$$\sum F_X = R_{X1} + R_{X3} = 0$$

$$R_{X1} = -R_{X3} = -(-228) = 228\ lbs$$

Because all of the fours points are assumed to have the same stiffness in the vertical direction, from the sum of the moments around the "*X*" axis at point ***1***:

$$M_X+: \text{CW} \qquad \sum M_{X1} = R_{Z3,4}(39.5) - F_Y(34.8) = 0$$

hence

$$R_{Z3,4} = \frac{F_Y(34.8)}{39.5}$$

$$= \frac{1{,}125(34.8)}{39.5} = 991\ lbs$$

From the sum of the forces in "*Z*", the vertical reaction at points ***1*** & ***2*** "$R_{Z1,2}$" is calculated;

$$\sum F_Z = R_{Z1,2} + R_{Z3,4} = 0$$

$$R_{Z1,2} = -R_{Z3,4} = -(991)\ lbs$$

Since the $R_{Z1,2}$ and $R_{Z3,4}$ forces are in theory located in the *Y-Z* plane going through the Cabinet's center of gravity, and since that plane is 8.0 *inches* aft of point ***1*** & ***3*** while point ***2*** & ***4*** are 16.0 *inches* aft, then the reactions at points ***1*** & ***2*** are half of $R_{Z1,2}$ and similarly, the reactions at points ***3*** & ***4*** are half of $R_{Z3,4}$. Therefore:

$$R_{Z1} = R_{Z2} = \frac{R_{Z1,2}}{2} = \frac{-991}{2} = -496\ lbs$$

$$R_{Z3} = R_{Z4} = \frac{R_{Z3,4}}{2} = \frac{991}{2} = 496\ lbs$$

Table 14-9 shows the summary of all the reaction forces in the *Global Coordinate System* at all four points for the ***3g Right*** load case.

Table 14-9 *- Summary of the Reaction forces for the 3g Right LC.*

Points	R_X [lbs]	R_Y [lbs]	R_Z [lbs]
1	228	-1125	496
2	0	0	496
3	-228	0	-496
4	0	0	-496
SUM:	0	-1125	0

14.3.4 *Y* AXIS; 3G LEFT

For the *3g Left* loading condition, it is assumed that the same *B.C.* as for the *3g Right* are applicable and that the reaction loads can be scaled linearly. Therefore, the reaction forces calculated previously, as shown in *Table 14-9*, are simply multiplied by -1. The results are shown in *Table 14-10*.

Table 14-10 *- Summary of the Reaction forces for the 3g Left LC.*

Points	R_X	R_Y	R_Z
	[lbs]	[lbs]	[lbs]
1	-228	1125	-496
2	0	0	-496
3	228	0	496
4	0	0	496
SUM:	0	1125	0

14.3.5 *Z* AXIS; 6G DOWN

For this load case, as shown in *Figure 14-14*, only the four attachment points at the floor (***1 to 4***) are considered effective. The two Struts at the top are considered completely ineffective vertically. *Table 14-11* shows the *BC* for the *6g Down LC*. All the four points are assumed to have the same vertical stiffness.

Table 14-11 *- Boundary Conditions at all points for the 6g Down LC.*

Points	X	Y	Z	RX	RY	RZ	B.C.
	[in]	[in]	[in]	[deg.]	[deg.]	[deg.]	
1	0	0	0	–	–	–	[1,2,3]
2	–	–	0	–	–	–	[3]
3	0	–	0	–	–	–	[1,3]
4	–	–	0	–	–	–	[3]

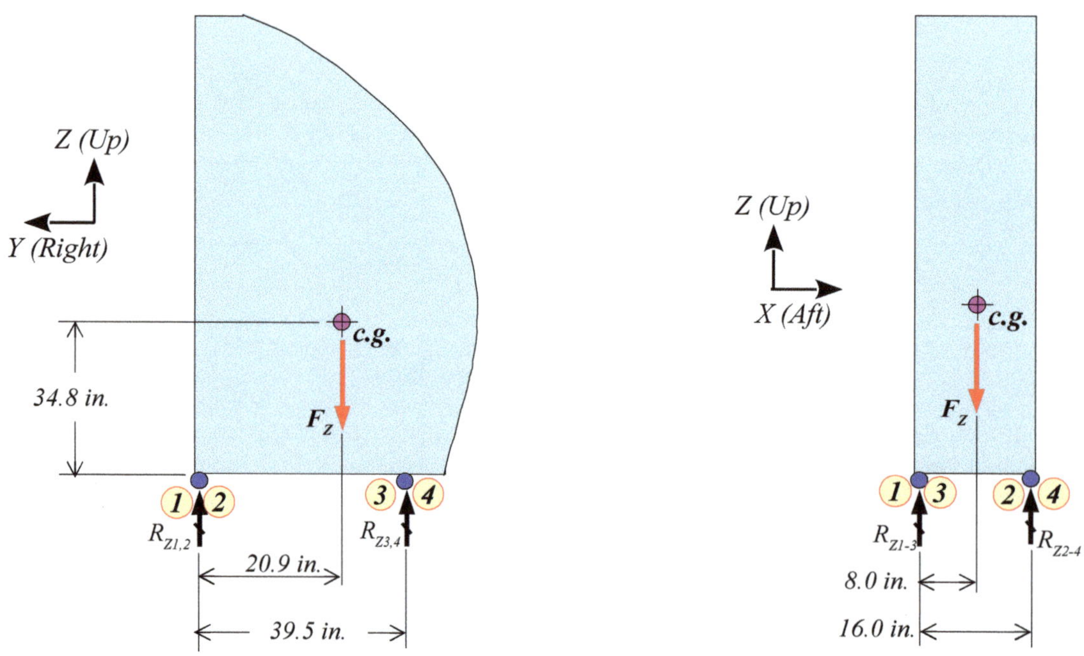

Figure 14-14 *– Cabinet FBD due to F_Z in the Y-Z and X-Z planes.*

Hence, for the *6g Down* load case, the applied force is: $F_Z = -2{,}250\ lbs$. The reaction forces at the four reaction points are statically determined. Therefore:

From the sum of the moments around the "*X*" axis at point ***1***:

$$M_X+: CW \qquad \sum M_{X1} = F_Z(20.9) - R_{Z3,4}(39.5) = 0$$

hence

$$R_{Z3,4} = \frac{-F_Z(20.9)}{39.5}$$

$$= \frac{-(-2{,}250)(20.9)}{39.5} = 1{,}191\ lbs$$

From the sum of the moments around the "*Y*" axis at point ***1***:

$$M_Y+: CW \qquad \sum M_{Y1} = -F_Z(8.0) - R_{Z2,4}(16.0) = 0$$

hence

$$R_{Z2,4} = \frac{-F_Z(8)}{16}$$

$$= \frac{-(-2{,}250)(8)}{16} = 1{,}125\ lbs$$

From the sum of the forces in "*Z*", the vertical reaction at points ***1&3*** "$R_{Z1,3}$" is calculated;

$$\sum F_Z = R_{Z1,3} + R_{Z2,4} + F_Z = 0$$

$$R_{Z1,3} = -F_Z - R_{Z2,4} = -(-2{,}250) - 1{,}125 = 1{,}125\ lbs$$

Again, since the $R_{Z1,3}$ and $R_{Z2,4}$ forces are equal, then the reactions at points ***1*** & ***3*** are half of $R_{Z1,3}$ and similarly, the reactions at points ***2*** & ***4*** are half of $R_{Z2,4}$. Therefore:

$$R_{Z1} = R_{Z3} = \frac{R_{Z1,3}}{2} = \frac{1{,}191}{2} = 596\ lbs$$

$$R_{Z2} = R_{Z4} = \frac{R_{Z2,4}}{2} = \frac{1{,}059}{2} = 530\ lbs$$

Table 14-12 shows the summary of all the reaction forces in the *Global Coordinate System* at all four points for the ***6g Down*** load case.

Table 14-12 *- Summary of the Reaction forces for the 6g Down LC.*

Points	R_X *[lbs]*	R_Y *[lbs]*	R_Z *[lbs]*
1	0	0	596
2	0	0	530
3	0	0	596
4	0	0	530
SUM:	0	0	2250

14.3.6 Z AXIS; 3.15G UP

For the *3.15g Up* loading condition, it is assumed that the same *B.C.* as for the *6g Down* are applicable and that the reaction loads can be scaled linearly. Therefore, the reaction forces calculated previously, as shown in *Table 14-12*, are multiplied by the factor: 3.15 / −6 = −0.525. The results are shown in *Table 14-13*.

Table 14-13 *- Summary of the Reaction forces for the 3.15g Up LC.*

Points	R_X [lbs]	R_Y [lbs]	R_Z [lbs]
1	0	0	-313
2	0	0	-278
3	0	0	-313
4	0	0	-278
SUM:	0	0	-1181

Table 14-14 Shows the summary of the maximum and minimum reaction forces at all points from all the ultimate load cases.

Table 14-14 *- Summary of the maximum and minimum Ult. Reaction forces at all points.*

Points	Max R_X		Min R_X		Max R_Y		Min R_Y		Max R_Z		Min R_Z	
	[lbs]	LC	[lbs]	LC	[lbs]	LC	[lbs]	LC	[lbs]	LC	[lbs]	LC
1	497	9g Fwd	-228	3g Left	1125	3g Left	-1125	3g Right	596	6g Down	-852	9g Fwd
2	0	–	0	–	0	–	0	–	530	6g Down	-496	3g Left
3	1367	9g Fwd	-228	3g Right	0	–	0	–	596	6g Down	-496	3g Right
4	0	–	0	–	0	–	0	–	530	6g Down	-496	3g Right
5	668	9g Fwd	-111	1.5g Aft	0	–	0	–	411	9g Fwd	-69	1.5g Aft
6	844	9g Fwd	-141	1.5g Aft	30	1.5g Aft	-178	9g Fwd	234	9g Fwd	-39	1.5g Aft

14.4 MATERIAL ALLOWABLES

Actual values of all metallic material allowables can be obtained from the latest edition of (ref. 8.).

The **Seat tracks rails** and **Stringers** are made out of *Al Aly **7075-T6** Extrusion* <0.25 *in.* thick. For the purpose of the demonstration, *Table **14-15*** shows fictitious values.

Table 14-15 *- Al Aly 7075-T6 Extrusion <0.250in. fictitious material allowables.*

*Aluminum Alloy **7075-T6** Ext*
Fictitious Values, NOT FOR DESIGN

	<0.250" [ksi]
F_{tu}	80
F_{ty}	70
F_{cy}	73
F_{su}	40
F_{bru} $(e/D = 2.0)$	145
F_{bry} $(e/D = 2.0)$	110
$E, 10^3$	10
$G, 10^3$	3.8
μ	0.33
n	20

The **Crossbeams**, **Lug Fittings** and **Intercostal** are made out of *Al Aly **7050-T7451*** plate between 1.00 *in.* and 2.50 *in.* thick. *Table **14-16*** shows fictitious values.

Table 14-16 *- Al Aly 7050-T7451 Plate 2.50 in. fictitious material allowables.*

*Aluminum Alloy **7050-T7451** Plate*
Fictitious Values, NOT FOR DESIGN

	2.50" [ksi]
F_{tu}	75
F_{ty}	65
F_{cy}	60
F_{su}	40
F_{bru} $(e/D = 2.0)$	105
F_{bry} $(e/D = 2.0)$	85
$E, 10^3$	10
$G, 10^3$	3.8
μ	0.33
n	20

The **Struts** are made out of *Al Aly 7075-T62* Tubes 0.75 *in. O.D.* x 0.035 *in.* wall. For the purpose of the demonstration, *Table 14-17* shows fictitious values.

Table 14-17 *- Al Aly 7075-T62 Tubing 0.75 x 0.035in. fictitious material allowables.*

Aluminum Alloy 7075-T62 Tubing
Fictitious Values, NOT FOR DESIGN

	0.035"
	[ksi]
F_{tu}	75
F_{ty}	65
F_{cy}	65
F_{su}	45
F_{bru} $(e/D = 2.0)$	125
F_{bry} $(e/D = 2.0)$	90
$E, 10^3$	10
μ	0.33
n	20

The **Rod Ends** are made out of *CRES 17-4PH* bar heat treated to *H1025*, 1.00 *in.* thick. For the purpose of the demonstration, *Table 14-18* shows fictitious values.

Table 14-18 *- CRES 17-4PH Bar H1025, 1.00 in. Bar fictitious material allowables.*

CRES 17-4PH, H1025 Bar
Fictitious Values, NOT FOR DESIGN

	1.00"
	[ksi]
F_{tu}	150
F_{ty}	140
F_{cy}	135
F_{su}	90
F_{bru} $(e/D = 2.0)$	325
F_{bry} $(e/D = 2.0)$	225
$E, 10^3$	30
$G, 10^3$	11.5
μ	0.3
n	21

The **Skin, Shear Clips** and **Crash-straps** are made out of *Al Aly 2024-T3 Clad* sheet. For the purpose of the demonstration, *Table 14-19* shows fictitious values.

Table 14-19 *- Al Aly 2024-T3 Clad sheet fictitious material allowables.*

Aluminum Alloy 2024-T3 Clad sheet
Fictitious Values, NOT FOR DESIGN

	0.100"
	[ksi]
F_{tu}	60
F_{ty}	45
F_{cy}	40
F_{su}	35
F_{bru} $(e/D = 2.0)$	120
F_{bry} $(e/D = 2.0)$	85
E, 10^3	10
G, 10^3	3.76
μ	0.33
n	20

The **Frames** are made out of *Al Aly 7075-T7351* Plate, 1.00 *in.* thick. For the purpose of the demonstration, *Table 14-20* shows fictitious values.

Table 14-20 *- Al Aly 7075-T7351, 1.00 in. Plate fictitious material allowables.*

Aluminum Alloy 7075-T7351 Plate
Fictitious Values, NOT FOR DESIGN

	1.00"
	[ksi]
F_{tu}	70
F_{ty}	60
F_{cy}	55
F_{su}	40
F_{bru} $(e/D = 2.0)$	135
F_{bry} $(e/D = 2.0)$	85
E, 10^3	10
G, 10^3	3.8
μ	0.33
n	20

The **New Shear Clips** are made out of *Al Aly* ***7075-T62*** *Clad* sheet. For the purpose of the demonstration, *Table* ***14-21*** shows fictitious values.

Table 14-21 *- Al Aly 7075-T62 Clad sheet fictitious material allowables.*

Aluminum Alloy 7075-T62 Clad sheet
Fictitious Values, NOT FOR DESIGN

	0.071" [ksi]
F_{tu}	75
F_{ty}	65
F_{cy}	62
F_{su}	42
F_{bru} (e/D = 2.0)	135
F_{bry} (e/D = 2.0)	100
E, 10^3	10
μ	0.33
n	20

The **Cabinet doubler** are made out of *Al Aly* ***6061-T6*** sheet. For the purpose of the demonstration, *Table* *14-22* shows fictitious values.

Table 14-22 *- Al Aly 6061-T6 sheet fictitious material allowables.*

Aluminum Alloy 6061-T6 sheet
Fictitious Values, NOT FOR DESIGN

	0.063" [ksi]
F_{tu}	40
F_{ty}	33
F_{cy}	33
F_{su}	25
F_{bru} (e/D = 2.0)	85
F_{bry} (e/D = 2.0)	55
E, 10^3	10
G, 10^3	3.8
μ	0.33
n	20

The faces of the **Floor panels** are made out of, *qty(2)* plies of *GR/EP* pre-preg fabric having a cured thickness of 0.0077 *in.* The plies are oriented as [(0)/(45)]. For the purpose of the demonstration, *Table 14-23* shows fictitious values for one typical ply.

***Table 14-23** - Carbon-Epoxy face plies fictitious material allowables.*

Carbon-Epoxy pre-preg fabric ply
Fictitious Values, NOT FOR DESIGN

	0.0077"
	[ksi]
E_{11}, 10^3	9.0
E_{22}, 10^3	9.0
G_{12}, 10^3	0.85
F_{1T}	95
F_{1C}	70
F_{2T}	85
F_{2C}	70
F_{12}	15
μ_{12}	0.04

For this layup (assumed quasi-isotropic), the equivalent Young's modulus "E_{eff}" and the equivalent Poisson coefficient "υ_{eff}" are calculated and the following values are respectively obtained: **6.67x10⁶** ***psi*** and **0.29**. The reader could refer for instance to (ref. 3., 11.) to familiarize himself with one of the methodologies.

The details of those calculations are not shown here as they are deemed not necessary to meet the intent of this book. The reader is invited to create a tool in a spreadsheet in order to perform those calculations easily and accurately.

The core of the **Floor panels** is made out of honeycomb core (similar to *Nomex® HRH10-3/16-3.0)*, 0.500 *in.* thick. This core has a cell size of 3/16 *in.* and a density of 3.0 *lb* per cubic foot. For the purpose of the demonstration, *Table 14-24* shows fictitious values for the core.

***Table 14-24** - Honeycomb core fictitious material allowables.*

Honeycomb Core - 3/16" - 3.0
Fictitious Values, NOT FOR DESIGN

	[ksi]
$E_C = E_{33}$, 10^3	20.0
$G_L = G_{13}$, 10^3	6.5
$G_W = G_{23}$, 10^3	3.4
μ	0.05

The **Rod Ends** comprise of a spherical bearing encased in a machined fitting made out of CRES, as shown in *Figure 14-15*. They allow for 1/4" bolts and have a 0.375" threaded end. For the purpose of the demonstration, their ultimate radial load allowable is 3,160 *lbs*. They are vendor parts therefore we only need to check that the applied loads do not exceed the allowable provided by the vendor.

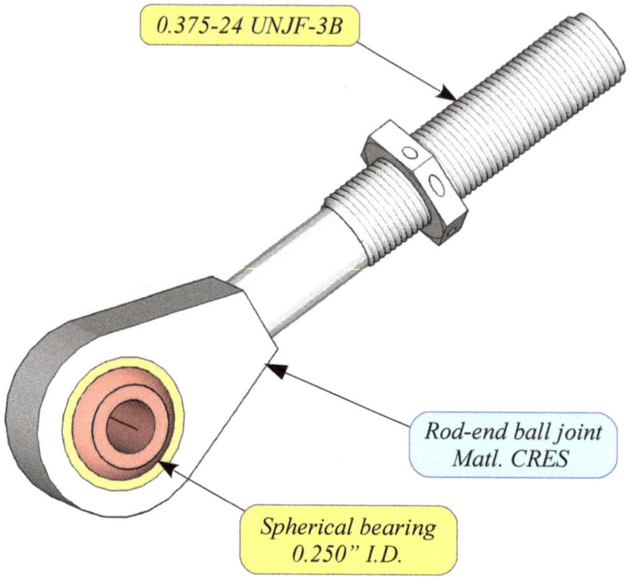

Figure 14-15 – *Rod-End from the Struts assemblies.*

The **Spherical Bearings** are installed in the upper strut lug fittings common to the airframe. See *Figure 14-16*. They are of the type *SAE-AS81820-4* and have an ultimate radial allowable load of P_{radial} = 5,500 *lbs* and an ultimate axial allowable load of P_{axial} = 400 *lbs*.

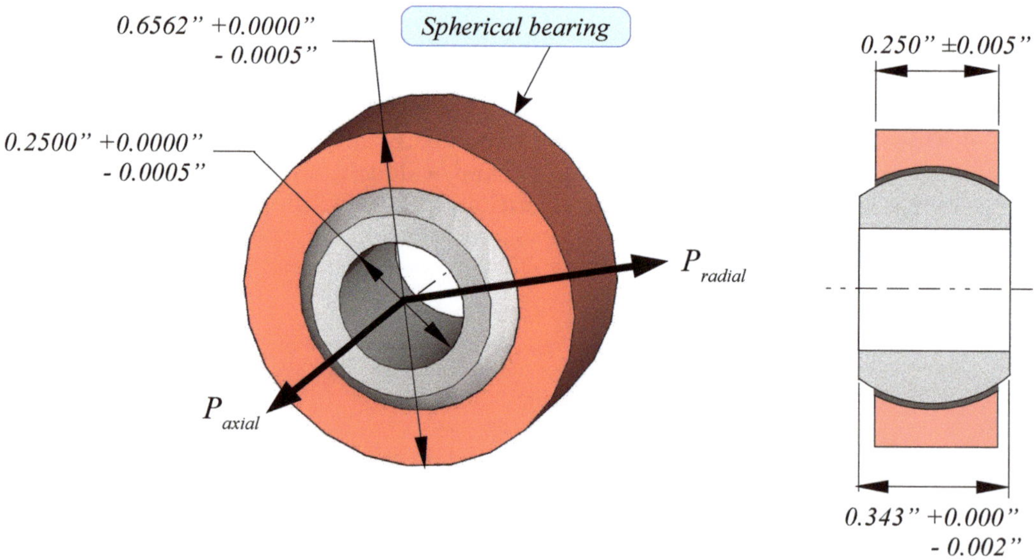

Figure 14-16 – *Spherical Bearing from the Lug Ftg assemblies.*

The **Seat track Fittings** are made out of CRES. For the purpose of the demonstration, their ultimate horizontal load allowable "P_H" is 5,000 *lbs* and their ultimate vertical load allowable "P_V" is 6,000 *lbs*. They are vendor parts therefore we only need to check that the applied loads do not exceed the allowable provided by the vendor. See *Figure 14-17* for a fictitious Seat track Ftg.

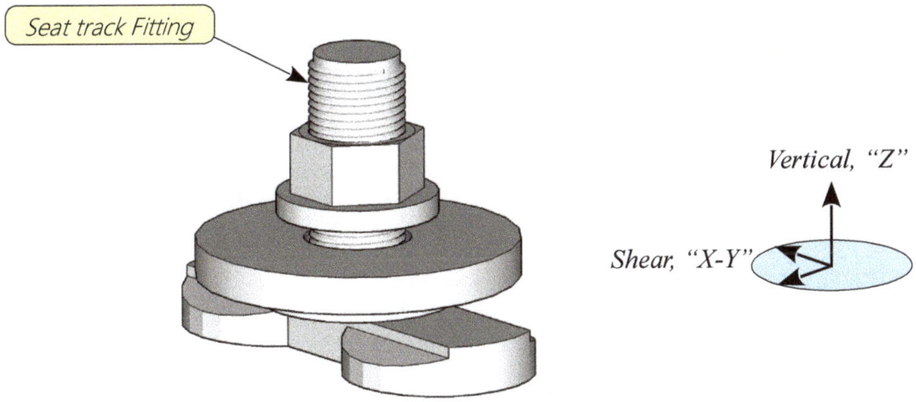

Figure 14-17 *– Seat Track Fittings.*

> *"There is no greater reward than to see the aircraft you worked on take-off ... and land..."*

14.5 FASTENERS ALLOWABLES

Solid rivets

Table 14-25 shows the ultimate shear allowables for *MS20470AD* protruding and *MS20426AD* flush solid rivets.

Table 14-25 - *MS20470AD Solid rivets ultimate shear allowables.*

	MS20470AD			*MS20426AD*		
	Al Aly 2117-T3			*Al Aly 2117-T3*		
	F_{su} = 30 ksi			*F_{su} = 30 ksi*		
	4	**5**	**6**	**4**	**5**	**6**
Dia. [in.]	0.1285	0.159	0.191	0.1285	0.159	0.191
P_{su} [lbs]	**388**	**596**	**862**	**388**	**575**	**795**
*P_{tu} [lbs] **	**375**	**430**	**520**	**230**	**360**	**480**

** in 0.050" 2024-T3 & 7075-T6 sheets.*

For flush rivets, it is imperative to get the joint allowables from (ref. 8.) (or actual tests) according to the material and the thickness as it could be lower than the shear strength. Otherwise, one could also use (ref. 7.) for flush solid rivets shear strength.

Hi-Lites

Table 14-26 shows the ultimate single shear and tension allowables for 5/32" protruding head Hi-Lite *HST10-5* and flush head Hi-Lite *HST11-5* Titanium Pins and *HST79-5* Aluminum Collars from (ref. 24.).

Table 14-26 - *HST10, HST11 & HST79 Hi-Lites ultimate shear and tension allowables.*

	P_{tu} *[lbs]*	P_{su} *[lbs]*
Pin HST10-5	1,940	2,005
Pin HST11-5	1,650	2,005
Collar HST79-5	1,400	N/A

Bolts

Hexagonal bolt ***NAS6204***, *0.2500-28.*, material is alloy steel 160 *ksi* tensile strength and 95 *ksi* shear strength.

Flanged bushings

The Flanged bushings are made out of alloy steel 160 *ksi* tensile strength and 95 *ksi* shear strength.

14.6 SEAT TRACKS

The Seat Tracks, for our example, are made out of *Aluminum Alloy 7075-T6* Extrusions. They run longitudinally and are used to attach the seats, monuments and any other accessories. They are located, in our example, at *BL -28, -44* and *-67.5* and are all the same part number. They are assumed simply supported at the Crossbeams located at *FS 498.5, 516.5, 534.5* and *547.5*. See *Figure 14-18* for a typical Seat Track arrangement.

Since Cabinets (or seats) can sometimes be moved longitudinally, the loads from the cabinet's attachment points are conservatively assumed at the worst location (mid-span) between the Crossbeams.

It is not uncommon to see the configuration of the interior of an airplane be revised at a later date. Therefore, it is wise for the analysis to assume that the attachment points are located at the most critical location. By doing that, in the case of a different configuration, no analysis will need to be re-done.

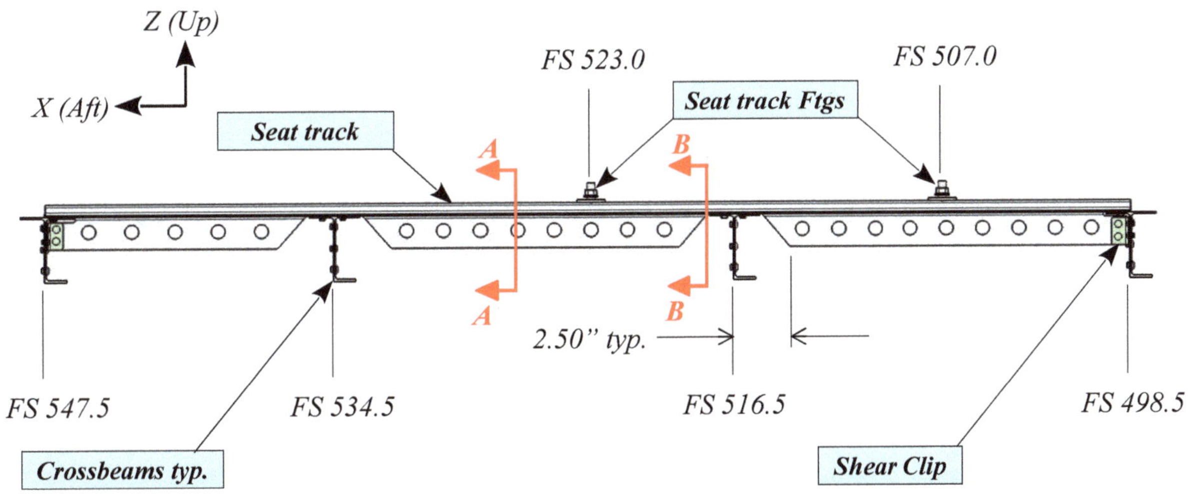

Figure 14-18 *– Typical Seat Track arrangement.*

14.6.1 SECTION PROPERTIES

There are two critical sections in the Seat Track rails: Sections *A-A* and *B-B*, see *Figure 14-18*. The first one is the typical section for the most part of the mid-span portion while, the second one, is at the Crossbeams. See *Figure 14-19* and *Figure 14-20* for the dimensions and the properties of section *A-A* and section *B-B* respectively.

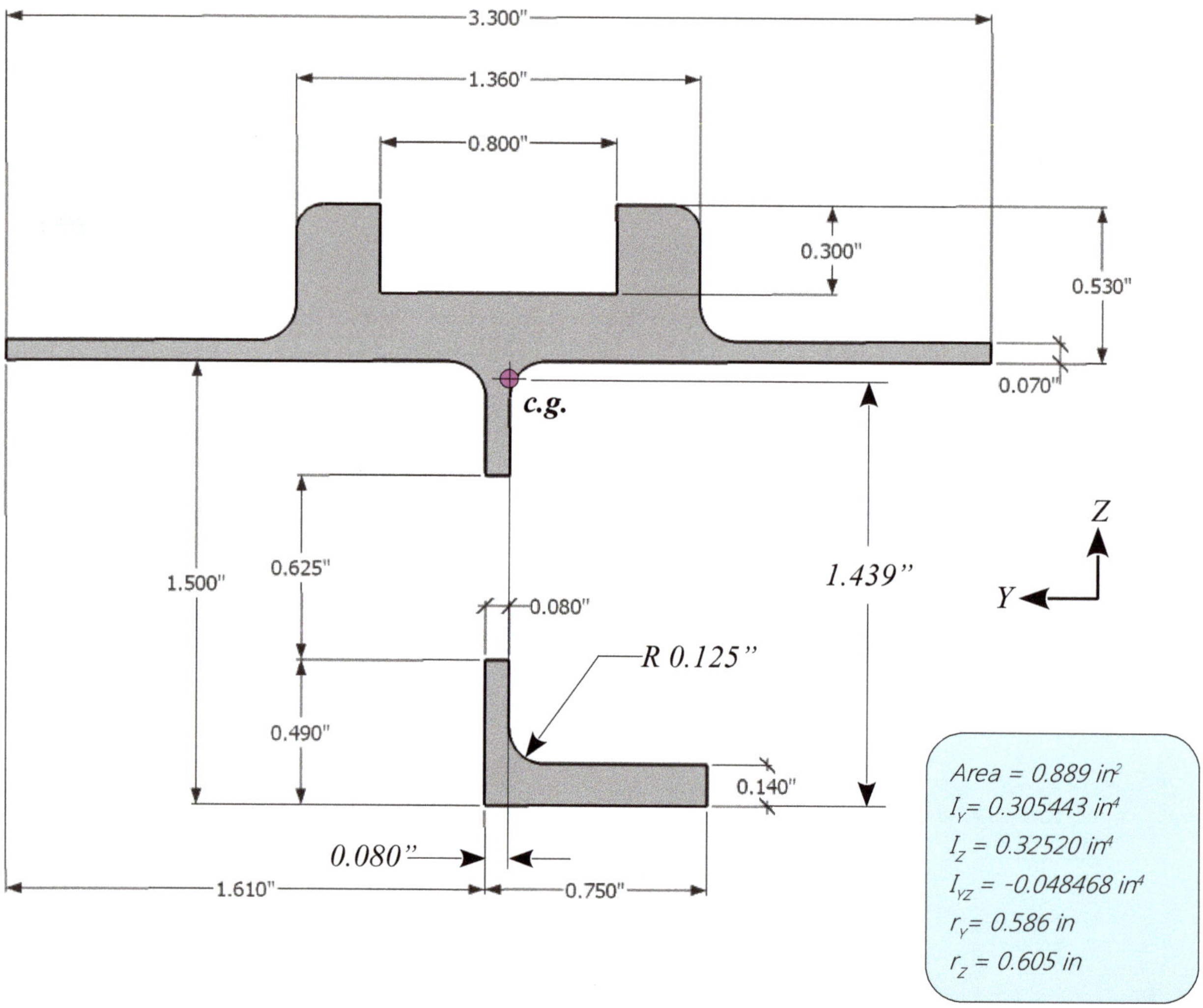

Figure 14-19 – *Seat Track rail Section A-A dimensions and properties.*

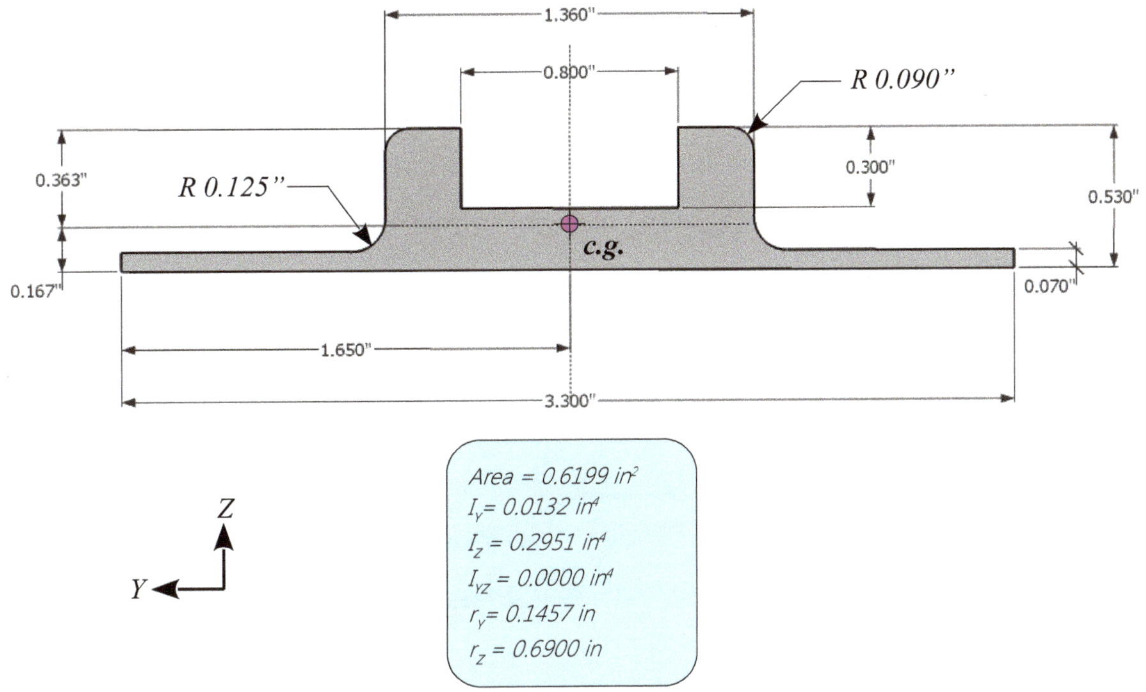

Figure 14-20 *– Seat Track rail Section B-B dimensions and properties.*

14.6.2 CRIPPLING AT SECTION A-A

Necessary for the *Johnson-Euler* analysis performed in section *14.6.5*, the crippling stress allowable *"F_{CC}"* will be calculated below for the section *A-A*.

The section is divided into segments as shown in *Figure 14-21*. The segments adjacent to the web lightening hole are considered free at the hole.

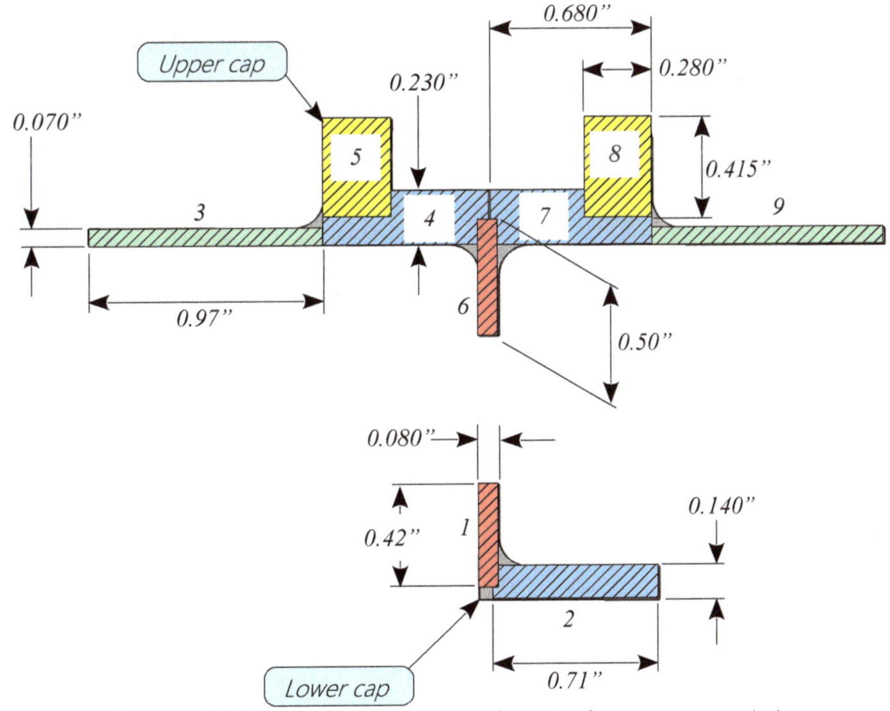

Figure 14-21 *– Effective segments for crippling at section A-A.*

The crippling allowable for each segment is calculated as shown below. Note that the crippling allowable "F_{CC}" of a segment can never be higher than F_{CY}. So the maximum value for F_{CCn} / F_{cy} is 1.0. The value at the abscissa is calculated as:

$$abscissa = \sqrt{F_{CY}/E}(b/t)$$

See *Figure 14-22* for typical crippling allowable curves.

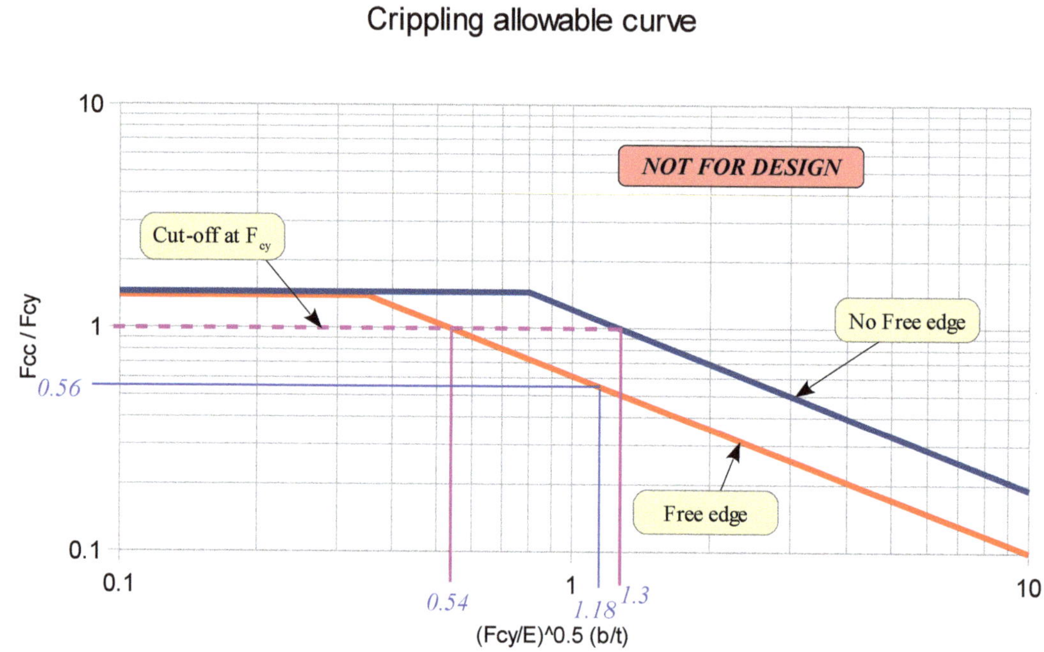

Figure 14-22 – *Typical Crippling curves.*

So for E = 10.0*10[6] *psi*, F_{cy} = 73,000 *psi* and from the dimensions shown in *Figure 14-21*, the crippling allowable of each segment is calculated in *Table 14-27*.

Table 14-27 - *Crippling analysis at Seat track rail Section A-A.*

Segment	Free edge	b_n [in]	t_n [in]	$b_n t_n$ [in²]	b_n / t_n	$(F_{cy} / E)^{0.5}$	$(F_{cy} / E)^{0.5}(b_n/t_n)$	F_{CCn} / F_{cy}	F_{CCn} [ksi]	$t_n b_n F_{CCn}$ [lbs]
1	1	0.42	0.08	0.034	5.250	0.0854	0.4486	1.00	73	2453
2	1	0.71	0.14	0.099	5.071	0.0854	0.4333	1.00	73	7256
3	1	0.97	0.07	0.068	13.857	0.0854	1.1840	0.56	40.88	2776
4	0	0.68	0.23	0.156	2.957	0.0854	0.2526	1.00	73	11417
5	1	0.415	0.28	0.116	1.482	0.0854	0.1266	1.00	73	8483
6	1	0.5	0.08	0.040	6.250	0.0854	0.5340	1.00	73	2920
7	0	0.68	0.23	0.156	2.957	0.0854	0.2526	1.00	73	11417
8	1	0.415	0.28	0.116	1.482	0.0854	0.1266	1.00	73	8483
9	1	0.97	0.07	0.068	13.857	0.0854	1.1840	0.56	40.88	2776
			$\sum b_n t_n$ =	0.85400				$\sum b_n t_n F_{ccn}$ =		57980

Thus the crippling allowable stress for section *A-A* is:

$$F_{CC} = \frac{\sum b_n t_n F_{CCn}}{\sum b_n t_n} = \frac{57{,}980}{0.854} = 67{,}892 \ psi$$

14.6.3 CRIPPLING AT SECTION B-B

Necessary for the *Johnson-Euler* analysis performed in section *14.6.5*, the crippling stress allowable "F_{CC}" will be calculated below for the section *B-B*. The section is divided into segments as shown in *Figure 14-23*.

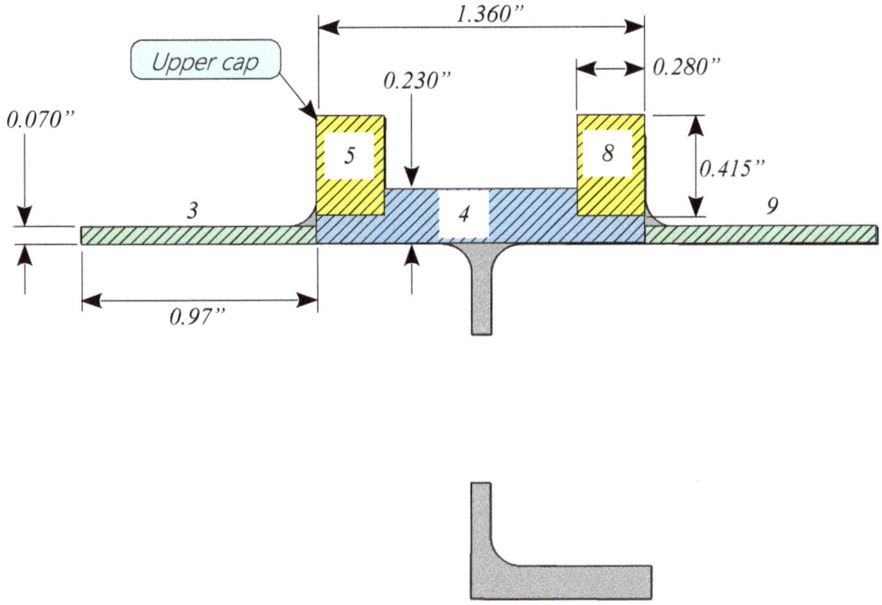

Figure 14-23 *– Effective segments for crippling at the Section B-B.*

The crippling allowable for each segment is calculated as shown below. Note that the crippling allowable "F_{CC}" of a segment can never be higher than F_{CY}. So the maximum value for F_{CCn} / F_{cy} is 1.0. The value at the abscissa is calculated as:

$$abscissa = \sqrt{F_{CY}/E}\,(b/t)$$

See *Figure 14-22* for typical crippling allowable curves.

So for $E = 10.0 \times 10^6$ *psi*, $F_{cy} = 73{,}000$ *psi* and from the dimensions shown in *Figure 14-23,* the crippling allowable of each segment is calculated in *Table 14-28.*

Table 14-28 *- Crippling analysis at Seat track rail Section B-B.*

Segment	Free edge	b_n [in]	t_n [in]	$b_n t_n$ [in²]	b_n / t_n	$(F_{cy} / E)^{0.5}$	$(F_{cy} / E)^{0.5}(b_n/t_n)$	F_{CCn} / F_{cy}	F_{CCn} [ksi]	$t_n b_n F_{CCn}$ [lbs]
3	1	0.97	0.07	0.068	13.857	0.0854	1.1840	0.56	40.88	2776
4	0	1.36	0.23	0.313	5.913	0.0854	0.5052	1.00	73	22834
5	1	0.415	0.28	0.116	1.482	0.0854	0.1266	1.00	73	8483
8	1	0.415	0.28	0.116	1.482	0.0854	0.1266	1.00	73	8483
9	1	0.97	0.07	0.068	13.857	0.0854	1.1840	0.56	40.88	2776
			$\sum b_n t_n =$	0.68100					$\sum b_n t_n F_{ccn} =$	45351

Thus, the crippling allowable stress for section *B-B* is:

$$F_{CC} = \frac{\sum b_n t_n F_{CCn}}{\sum b_n t_n} = \frac{45{,}351}{0.681} = 66{,}595\ psi$$

14.6.4 CRIPPLING OF LOWER CAP

Since the corner of the Seat track rail lower cap (bottom) is in compression due to bending, a crippling allowable calculation *"F_{CC}"* is performed solely for the lower cap of section *A-A*.

The crippling analysis is performed by considering the lower flange and part of the web up to the lightning hole. The edges of the flange and the web are considered free. See *Figure 14-24* for the effective segments used in the analysis.

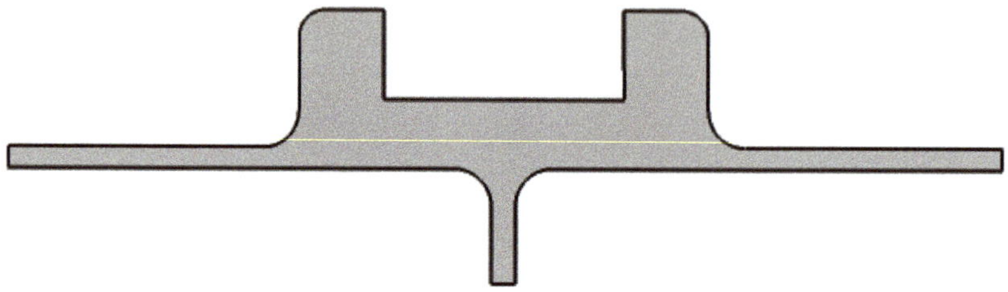

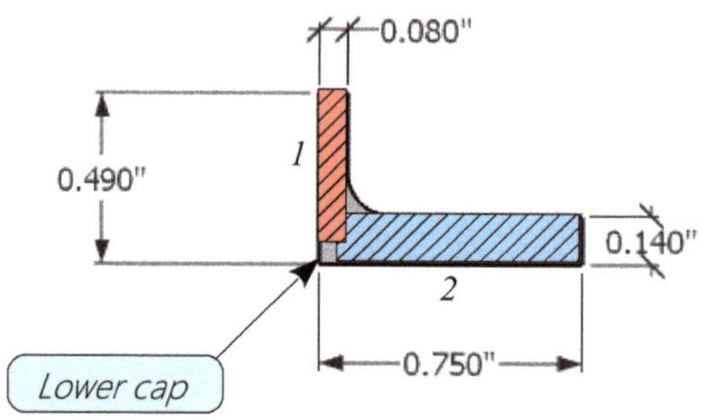

Figure 14-24 *– Effective segments for crippling at the lower cap.*

The crippling allowable for each segment is calculated as shown below. Note that the crippling allowable "F_{CC}" of a segment can never be higher than F_{CY}. So the maximum value for F_{CCn} / F_{cy} is 1.0. The value at the abscissa is calculated as:

$$abscissa = \sqrt{F_{CY} / E}(b/t)$$

See *Figure 14-25* for typical crippling allowable curves.

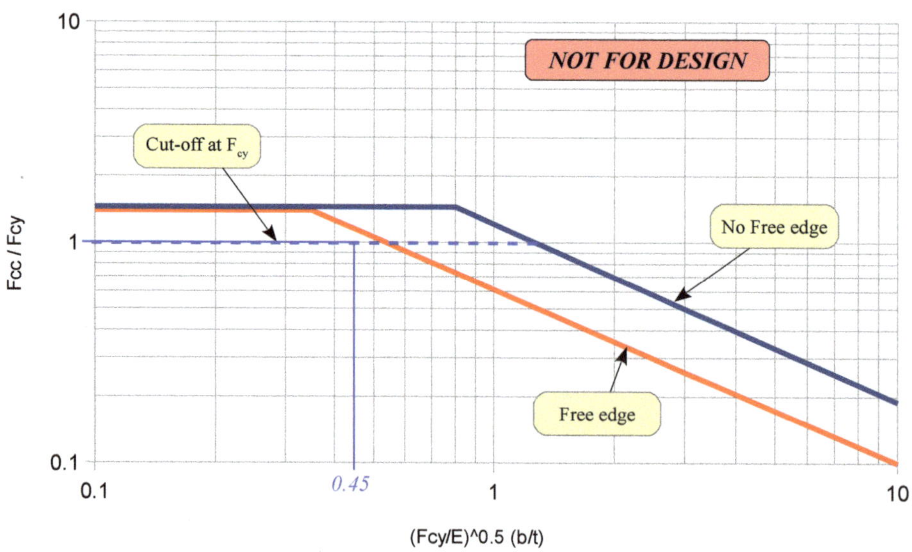

Figure 14-25 *– Typical Crippling curves.*

So for $E = 10.0*10^6$ *psi*, $F_{cy} = 73{,}000$ *psi* and from the dimensions shown in *Figure 14-24*, the crippling allowable of each segment is calculated in *Table 14-29*.

Table 14-29 *- Crippling analysis at Seat track rail lower cap.*

Segment	Free edge	b_n [in]	t_n [in]	$b_n t_n$ [in²]	b_n / t_n	$(F_{cy} / E)^{0.5}$	$(F_{cy} / E)^{0.5}(b_n/t_n)$	F_{CCn} / F_{cy}	F_{CCn} [ksi]	$t_n b_n F_{CCn}$ [lbs]
1	1	0.42	0.08	0.034	5.250	0.0854	0.4486	1.00	73	2453
2	1	0.71	0.14	0.099	5.071	0.0854	0.4333	1.00	73	7256
			$\sum b_n t_n =$	0.1330					$\sum b_n t_n F_{ccn} =$	9709

thus:

$$F_{CC} = \frac{\sum b_n t_n F_{CCn}}{\sum b_n t_n} = \frac{9{,}709}{0.133} = 73{,}000 \ psi$$

"Never burn any bridges."

14.6.5 COLUMN BUCKLING

The segment of the Seat Track rail is considered simply supported at the Crossbeams. It is considered stable as it is fastened to the floor panels, although, the floor panels are not taken into account in the section properties. It is a double stepped column since the second moment of inertia is different at the center than at the supports. It is submitted to a 1,367 *lbs* ultimate compression load. The Euler allowable load for a stepped column is calculated as:

$$P_E = \frac{c_2 \pi^2 (EI_2)}{(L')^2}$$

where L' is the effective length calculated as:

$$L' = \frac{L}{\sqrt{c}}$$

Where "L" is the column length and "c" is the End Fixity Coefficient. In the case shown in *Figure 14-26*, the End Fixity Coefficient "c" is obtained from *Figure B-2* and is equal to 1.0 for pinned-pinned columns. "L" is determined as the distance spanning between the Crossbeams. *Figure 14-26* shows the actual loading of the Seat track rail segment compared to the assumed loading. Since the reaction of the applied load "P" is discrete at the three floor panel fasteners, it can't really be considered as an evenly distributed shear flow. Thus, the loading is conservatively assumed as applied to each end.

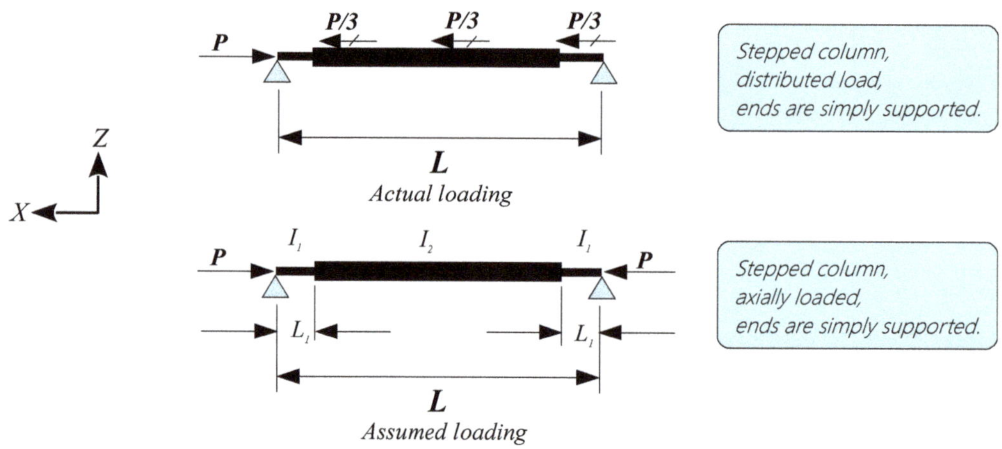

Figure 14-26 *– Seat track rail column model.*

Hence, for $E_1 = E_2 = 10 * 10^6$ *psi*, $I_1 = 0.0132$ in^4, $I_2 = 0.305443$ in^4, thus from *Figure B-3* for L_1 = 1.25 *in*. and L = 18.0 *in*. then $L_1 / L = 1.25 / 18 = 0.07$:

$$\sqrt{\frac{I_1}{I_2}} = \sqrt{\frac{0.0132}{0.305443}} = 0.21$$

we find the value for C_2 is: 0.89. Then:

$$P_E = \frac{0.89\pi^2\ 10*10^6(0.305443)}{(18.0)^2} = 82{,}808\ lbs$$

This corresponds to an internal axial stress through section *A-A* of: $F_E = P_E / A_{A\text{-}A}$ = 82,808 / 0.889 = 93,147 *psi* and, through section *B-B* of: $F_E = P_E / A_{B\text{-}B}$ = 82,808 / 0.6199 = 133,583 *psi*. These are way above the yield stress allowable (F_{cy} = 73 *ksi*) and is therefore unrealistic.

Now, to obtain a more realistic compression load allowable "P_C", one must use the actual elastic modulus at the point of the allowable load. That is the tangent modulus "E_T".

This can be done by using the actual stress-strain curve of the material. In order to do that we will be using the Ramberg-Osgood methodology that uses a parameter *(shape factor)* value "***n***", the stress value at 70% of E "$F_{0.7}$" and the elastic modulus "E". See (ref. 18.) for the theory.

The "n" value for the material used is fictitious and was assumed as 20 as shown in *Table 14-15*. As for the $F_{0.7}$ value, a simple way to determine it is to use the stress-strain curve graphs provided in (ref. 8.) approximated as a fictitious curve shown in *Figure 14-27* for the purpose of the demonstration. In this case, I used the same graph as for the "n" value above and, using the tangent modulus curve in compression, got the value at 70% of the primary "E_C" value.

For $E_C = 10.0 * 10^6\ psi$, then $E_{0.7} = 70\% * E_C = 7.0 * 10^6\ psi.$, we then find a value for $F_{0.7}$ of 66.0 *ksi*. See *Figure 14-27*.

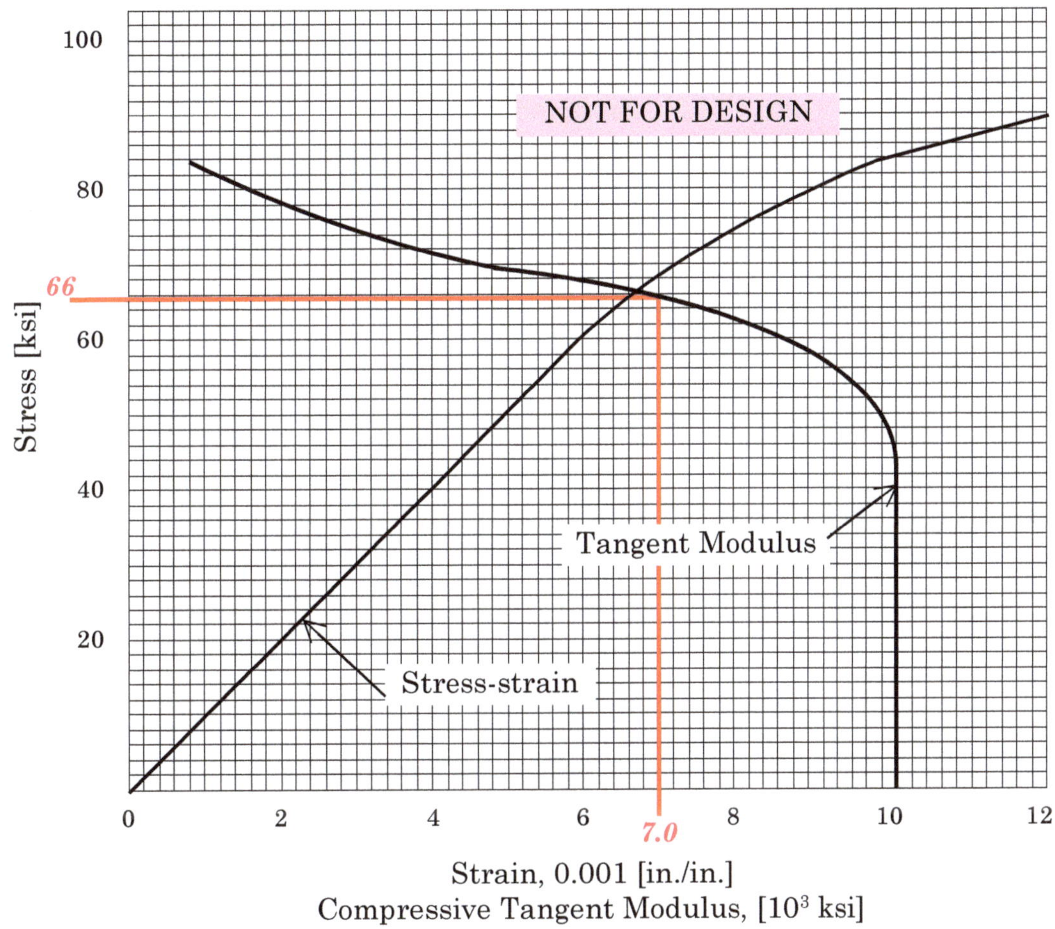

Figure 14-27 *– Fictitious Stress-strain and Tangent Modulus curves for Al Aly 7075-T6.*

Now that we have the value for "$F_{0.7}$", the allowable compression load "P_C" can be calculated as follows. Using the *three-parameter* equation for the tangent modulus from (ref. 18.), the compression stress allowable "F_C" is obtained by iteration.

The *three-parameter* equation is:

$$\frac{E_t}{E} = \frac{1}{1 + \frac{3}{7}\,n\left(\frac{F_C}{F_{0.7}}\right)^{n-1}}$$

Then, referring to *Figure 14-26* and using *Figure B-4* for columns having different second moments of inertia and modulus, the following modified Euler equation is used to calculate the critical compression load:

$$P_{CR} = B_2 \frac{E_{t2} I_2}{L^2}$$

As in all cases involving plasticity, the tangent modulus for each section is found by iteration by first assuming a value for P_{CR} until the same value is obtained from the above equation. *Table 14-30* shows a summary of the resulting calculations. As shown below, the calculation of the abscissa for *Figure B-4* was greater than 17 and was outside the graph.

$$abscissa = \frac{E_2 I_2}{E_1 I_1} = \frac{3{,}040{,}822}{25{,}264} = 120.4$$

The value for L_2/L to determine what curve to use is:

$$L_2 / L = 15.5/18 = 0.86$$

Then the *0.8* curve will be used for conservatism. Since the curve for $L_2/L = 0.8$ is almost straight, I decided to find the value "B_2" by linear extrapolation and found $B_2 = 4.2$.

For
$L' = L = 18.0$ in
$I_2 = 0.305443$ in^4
$B_2 = 4.2$
$E_{t2} = 9{,}955{,}449$ psi

The allowable compression load is thus:

$$P_{CR} = 4.2 \frac{9{,}955{,}449(0.305443)}{18^2} = 39{,}418 \ lbs$$

Table 14-30 *- Parameters for the inelastic column compression allowable.*

$P_{assumed} =$ 39418 *lbs*

$L' = L =$ 18 *in*

	Sect. B-B	Sect. A-A	
	Portion 1	Portion 2	
$A =$	0.6199	0.889	in^2
$I =$	0.0132	0.305443	in^4
$\rho =$	0.1459	0.5862	*in*
$L =$	1.25	15.5	*in*
$L'/\rho =$	123	31	
$f =$	63588	44340	*psi*
$f/F_{0.7} =$	0.963	0.672	
$E_t/E =$	0.19	1.00	
$E_t =$	1913906	9955449	*psi*
$E_t I =$	25264	3040822	$lb\text{-}in^2$

Although this last compression allowable is more accurate than the one that neglected the plasticity effect *(82,808 lbs)*, as you may have learned in maths classes, extrapolating is not recommended. Therefore, a more thorough and longer procedure will have to be used.

This method is for symmetrical columns having non-uniform cross-sections and having stresses in the inelastic zone of the stress-strain curve. It is an iterative procedure that requires the use of a spreadsheet.

The iteration process is described as follows:

1. Assume a value for ***P***; 40,000 *lbs* in our case *(it has to be a high number to ensure the stress is in the plastic range!)*. As a first iteration, the column lateral deformed shape is first assumed to follow a sine curve where: $y = \sin(\pi x / L)$. The deflection at the center is thus ***1.0***.
2. Plot the second moment of Inertia *"I"* of the column along its length. Use small length intervals. (I used dx = 0.25 *in.* for an 18 *inch* column).
3. Calculate the bending moment ***"M"*** by multiplying *P* by the lateral deformation "*y*" along the length of the column.
4. Divide the moments by the ***EI*** values to obtain the *M/EI* curve. Use E_t instead of *E* for stresses in the inelastic zone. (I use E_t all the time since $E_t = E$ in the elastic zone).
5. Obtain the slope curve by calculating the area under the *M/EI* curve for each *dx*, then, obtain the deflection curve by multiplying each area under the *M/EI* curve by the distance from its center of gravity to the center of the column. Correct the deflection curve to obtain a unit deflection at the center by multiplying the curve by the required factor.
6. Compare the plot of this deflection curve to the plot of the sine curve. Both curves should have a unit value at their center.
7. Providing the curves are very close to each other then stop. Otherwise, repeat steps 3 to 6 but every time calculating "***M***" using the latest deflection curve values "*y*" obtained from the last iteration instead of $y = \sin(\pi x / L)$.
8. Repeat until the new deflection curve almost overlaps the previous one.
9. Once you are happy with the accuracy of the deflection curve *(it took me 3 iterations in this example, see Figure 14-28)* then evaluate the critical compression load "***P***" by iteration. From the moment area principle, we know that the unit deflection "$y = 1$" is equal to half the area "*A*" under the *M/EI* curve times the distance between the centroid of that area to the column pinned end "$\bar{x}$", see *Figure 14-29*. The critical load is obtained when $A * \bar{x} = 1$. This part of the iteration is done using the *Goal Seek* operation.

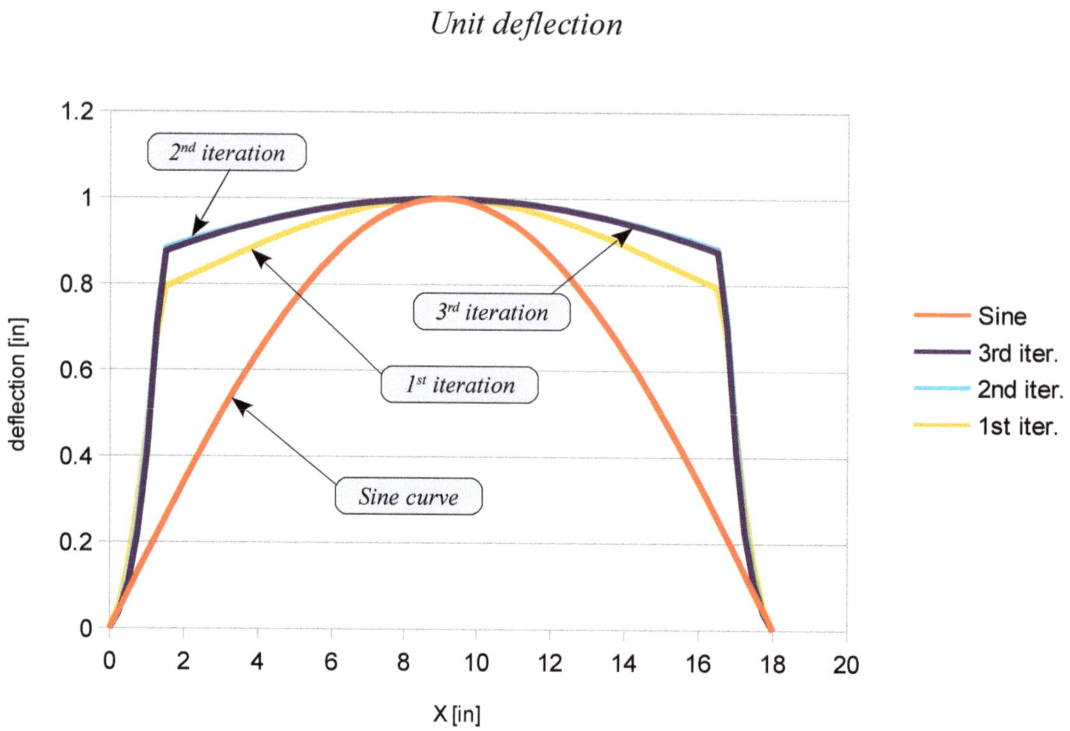

Figure 14-28 *– Column deformed shape for each iteration.*

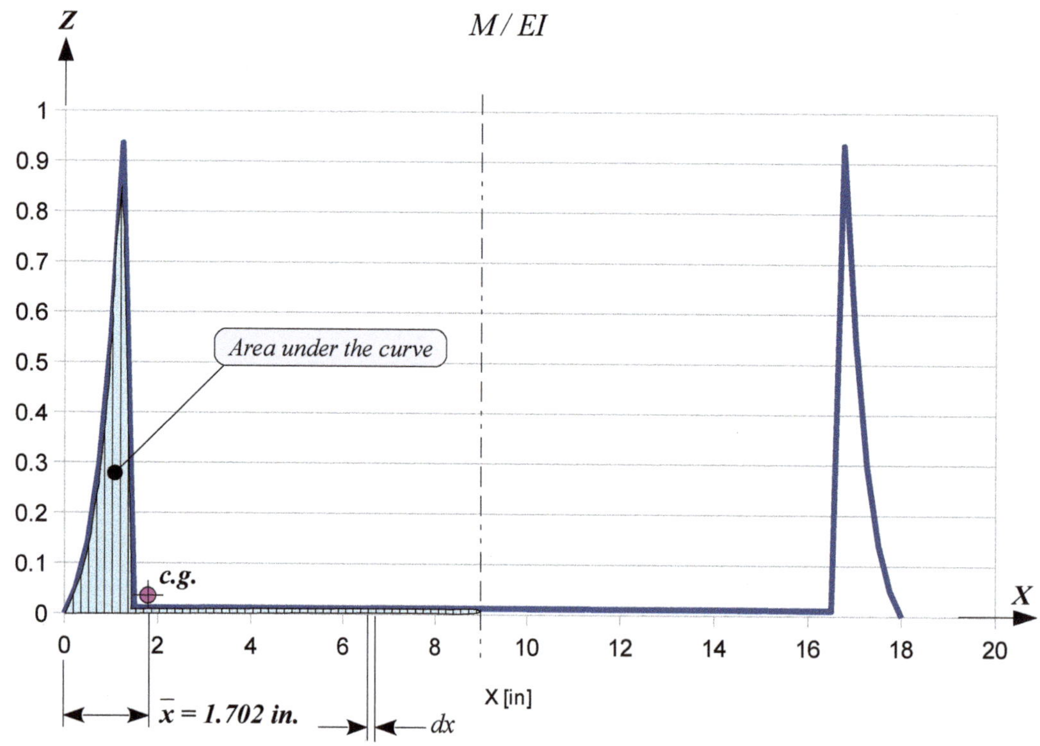

Figure 14-29 *– M/EI curve showing the centroid of the area.*

Figure 14-29 shows the final *M/EI* curve for our example. This was obtained after our 3rd iteration for the deflection curve as shown in *Figure 14-28.* Because the column is symmetrical, only half the area under the curve is considered. Note that the tangent modulus "E_t" is always used. Since we're doing this because some of the internal stresses are in the plastic zone and vary with the cross section area, the value for E_t also varies from point to point. Therefore, the *M/EI* curve will change with the applied load *"P"* and an iterative procedure is required to obtain the right location of the area centroid "$\bar{x}$".

Because $M = P*y$ and since the deflection $y = 1$, then $M = P$ so the *M/EI* curve is really the *P/EI* curve. The distance between point *O* and the *c. of g.* of the area *"A"* under half the curve is "$\bar{x}$" and is obtained by iteration using a *Goal Seek* function in a spreadsheet. Referring to *Figure 14-30*, the value for "$\bar{x}$" is calculated as follows:

$$\bar{x} = \frac{\sum A_i \times \frac{(x_i + x_{i+1})}{2}}{\sum A_i}$$

where

$$A_i = \frac{(z_i + z_{i+1})}{2} \times dx$$

and

$$A = \sum A_i$$

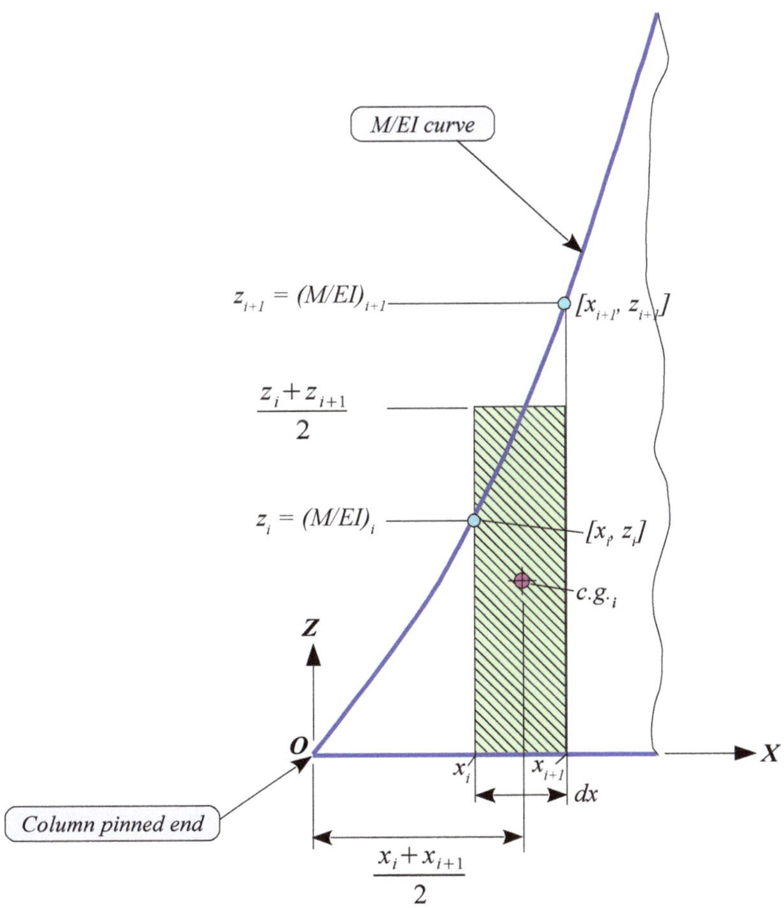

Figure 14-30 *– Close-up of M/EI curve using a numerical method of integration.*

Since the deflection of the column at its center is $y = A(\bar{x}) = 1$, the iterative procedure is as follows:

For the iteration process, P is varied until the difference between the result of "$A(\bar{x})$" and 1 is 0. For our example, the iteration procedure converged at "$\bar{x}$" = 1.702 *in.* and "P" = P_C = **39,032 *lbs*.**

This value is only about 1% lower than the value obtained previously using the simpler methodology using *Figure B-4*. At this point, the longer and more complex approach is certainly not worth the effort but was essential to confirm the result.

It is important to start with a high value for ***P*** to ensure that the stress is in the non-linear (plastic or non-elastic) zone. Otherwise, if the assumed applied load ***P*** is too low (*the stress is in the elastic range and* ***E*** *is constant*), varying ***P*** won't change the shape of the ***P/EI*** curve and the *Goal Seek* won't find convergence since $\bar{x}$ would be constant in that region.

Unstable cross-section

One could argue that the assumption to consider the Seat Track rail as a stable section in the first place was un-conservative. Of course, because the cross-section is made out of a relatively thin web and thin flanges, local buckling could occur before the column reaches its full compression capability. This would drastically reduce the allowable compression load. Therefore, as a double check, a column buckling analysis using the *Johnson-Euler* approach is done.

Johnson-Euler:

$$F_C = F_{CC} - \frac{F_{CC}^2}{4\pi^2 E_t}\left(\frac{L'}{\rho}\right)^2$$

This equation provides the allowable compression stress "F_C" of an unstable constant cross-section column. Since our column doesn't have a constant cross-section, the column will be considered uniform but the lowest crippling stress "F_{CC}" between sections *A-A* and *B-B* will be used conservatively along with the weighted average radius of gyration "ρ_{AVG}" and using the whole effective length. Refer to *Table 14-30* for values.

For:

$$\rho_{AVG} = \frac{2 \cdot L_1 \cdot \rho_1 + L_2 \cdot \rho_2}{L} = \frac{2 \cdot 1.25 \cdot 0.1459 + 15.5 \cdot 0.5862}{18} = 0.52485 \ in.$$

Hence:

$$F_C = 66{,}595 - \frac{66{,}595^2}{4\pi^2 10\times10^6}\left(\frac{18}{0.52485}\right)^2 = 53{,}382 \ psi$$

For
ρ_1 = 0.1459 in.
ρ_2 = 0.5862 in.
L_1 = 1.25 in.
L_2 = 15.5 in.
L = 18 in.
F_{CC} = 66,595 psi
E = 10x10^6 psi

This stress is a bit high and I expect the tangent *Young's* modulus "E_t" to be slightly lower than "E" thus leading to a lower compression stress allowable. By doing the same iterative approach as described on page *319*, the new value for *Fc* is calculated below by first assuming a value for "f" and, using a *goal seek*, iterating until $f = F_C$.

Thus for our "equivalent" uniform column, the allowable compression stress while considering plasticity is:

$$F_C = 66{,}595 - \frac{66{,}595^2}{4\,\pi^2 9.12\times10^6}\,(34.3)^2 = 52{,}111\ psi$$

For
$E = 10.0x10^6$ psi.
$F_{0.7} = 66{,}000$ psi.
$f/F_{0.7} = 0.79$
$E_t/E = 0.91$
$E_t = 9.12x10^6$ psi.

Calculating the weighted average cross section area "A_{AVG}":

$$A_{AVG} = \frac{2\,L_1\cdot A_{B-B} + L_2\cdot A_{A-A}}{L} = \frac{2(1.25)\cdot 0.6199 + 15.5\cdot 0.889}{18} = 0.851625\,in.^2$$

Thus, multiplying the average compression stress allowable "F_C" by the weighted average area, we get the compression load allowable "$P_{C\,AVG}$":

$$P_{C\,AVG} = F_C\cdot A_{AVG} = 52{,}111\cdot(0.851625) = 44{,}379\ lbs$$

One can observe that the critical compression load obtained from the *Johnson-Euler* equation is higher than the one obtained previously for the stable cross-section at 39,032 *lbs*. Therefore, I will be using the lowest critical compression load of the two methods for the MS calculation.

Compression load ratio

From the above analysis, the compression load allowable obtained was: P_{CR} = 39,032 *lbs*.

Therefore, the Seat Track rail is not subject to general instability (column buckling) since the maximum applied compression load of 1,367 *lbs* is much lower than the allowable. The compression load ratio "R_C" is thus:

$$R_C = \frac{P_X}{P_C} = \frac{1{,}367}{39{,}032} = 0.035$$

This value of R_C will be used later on along with the bending stress ratio "R_b" to calculate the MS.

To calculate the bending stress, because of the secondary bending effect, an "amplified" bending moment will be calculated using the approximate method as shown below. "M_o" is the primary bending moment while "M" is the "amplified" bending moment.

$$M = \frac{M_o}{1 - \frac{P}{P_{CR}}}$$

From the above equation, the "amplifying" factor is calculated as:

$$\frac{M}{M_o} = \frac{1}{1 - \frac{P}{P_{CR}}}$$

For P = 1,367 *lbs* and P_{CR} = 39,032 *lbs*, hence:

$$\frac{M}{M_o} = \frac{1}{1 - \frac{1{,}367}{39{,}032}} = 1.04$$

Therefore, because the applied axial load is relatively low, the maximum bending moment is amplified only by about 4%. This factor will be applied to the "M_Y" bending moment.

14.6.6 STRESS ANALYSIS SECTION A-A

Longitudinal load:

The critical loading for Seat track rails is vertical. Although it carries loads in the longitudinal direction "*X*" and in the lateral direction "*Y*", the vertical loading will cause most of the internal stress. Generally, there is not much longitudinal internal loading in the Seat tracks since the floor panels are fastened to the seat track rails and therefore all of the longitudinal load in the Seat track rails gets transferred by shear all the way to the sides of the fuselage where, ultimately, the fuselage skin will be reacting all the longitudinal loads.

From *Table 14-14*, the maximum longitudinal load is: F_X = 1,367 *lbs* for the *9g Fwd* load case, at *point 3*. As we know, the floor panels are fastened to the Seat Track rails so any longitudinal load from the rail gets dumped into the outboard floor panel (the one adjacent to the rail and closest to the fuselage skin) via fasteners pitched every 6 *inches*. For ultimate loads, we would generally assume that the longitudinal load from the Seat track rail is dumped over the totality of the floor panel fasteners.

However, for conservatism and because there are probably other monuments or seats, I will assume that the longitudinal load gets dumped into the floor panel by three fasteners. Then each fastener would see: 1,367 / 3 = 456 *lbs* of shear load. Of course, any load coming from other monuments or seats that are located inboard of this cabinet shall be added to the value calculated above.

Now, as for the Seat track rail itself, is the 1,367 *lbs* internal axial load a tension load or a compression load? Is the rail "pulling" on the floor panels or is it "pushing"? I guess that depends where the cabinet attachment loading up the rail is located. For instance, if the attachment point is near the rear end of the rail, then the internal axial load would definitely be in compression while, if the attachment is near the forward end, then it would definitely be in tension. In our case, *point 3* is near the forward end so we could assume that the load causes tension in the rail.

But since I'm chicken and since I suspect that, in the future, the operator of the aircraft may very well decide to move the cabinet aft causing the attachment point to be closer to the rear end, I'd rather assume the internal axial load to be in compression and check the Seat track rail for column stability.

For the column stability analysis, the most critical position for the attachment would be at a Crossbeam such that the column length would be maximum at the crossbeam pitch (18 *inches*). However, this axial load is to be combined to the bending stress caused by the applied vertical load and the most critical location for the vertical load to be applied is mid-span.

In addition, the longitudinal load is introduced into the Seat track rail at its top and therefore, because of the offset between the top and the section's center of gravity, it will cause a local bending moment around the "*Y*" axis that will have to be superimposed to the bending moment due to the vertical load.

Using section *A-A* as it covers most of the Seat track rail segment length, the longitudinal load is assumed applied at the center of the tabs that are 0.145 *in.* thick while the c. of g. is located 0.591 *in.* below the top. Thus the offset "*e*" is 0.591 – 0.145/2 = 0.519 *in.* The applied moment is therefore:

$$M_o = P_X e = 1{,}367(0.591) = 709 \; in.-lb$$

The reactions at the Crossbeams are thus:

$$R_{Z aft} = \frac{M_o}{l} = \frac{-709}{18} = -39\ lbs$$

$$R_{Z\ fwd} = \frac{M_o}{l} = \frac{709}{18} = 39\ lbs$$

Figure 14-31 shows the shear and moment diagrams resulting from the longitudinal load "P_X" of 1,367 *lbs* (ref. *Table 14-14 point 3)* applied mid-span.

As for the longitudinal compression load, I will be considering the whole span (18 *inches*) to determine the column buckling allowable and consider the vertical load to be introduced mid-span. This is a conservative assumption and will cover this configuration and any future ones.

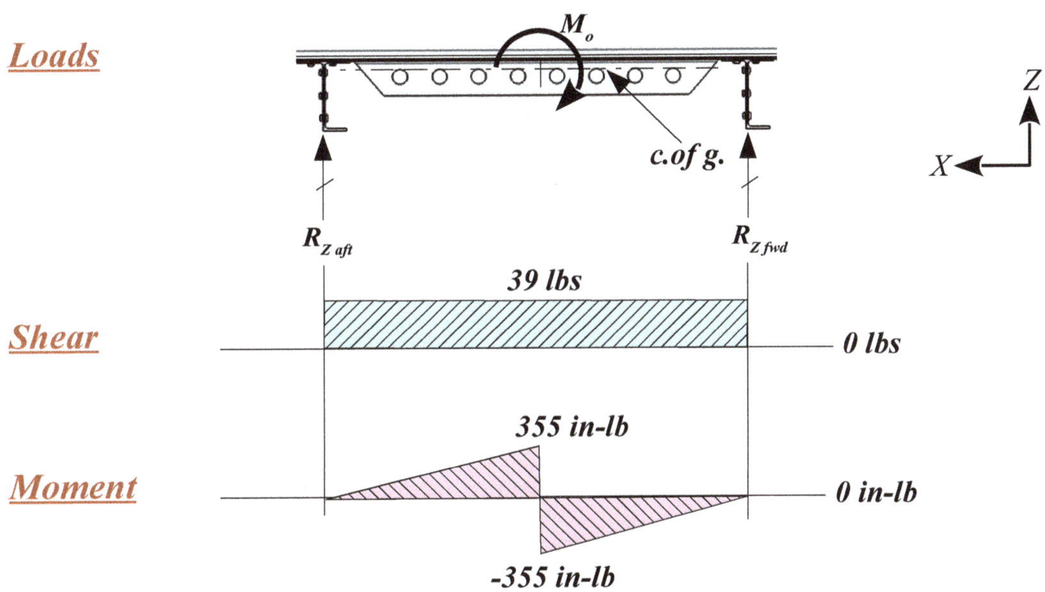

Figure 14-31 *– Seat track rail Shear & Moment diagrams due to P_X.*

Vertical load:

The longest span between the crossbeams is 18.0 *inches*. As a first check, it will be conservatively assumed that the attachments of the Cabinet are located at mid-span in order to cause the highest bending moment possible at section *A-A*. In that case, the seat track is assumed simply supported at the Crossbeams.

From *Table 14-14*, the maximum applied downward force is 596 *lbs* from the *6g Down* load case at points *1 & 3*, while the maximum applied upward force is 852 *lbs* from the *9g Fwd* load case at point *1*.

The maximum bending moment at mid-span due to a vertical force is calculated as:

$$M_{max} = \frac{W\,L}{4}$$

where $W = 852$ *lbs* and $L = 18.0$ *in.*, thus:

$$M_{max} = \frac{852 \times 18.0}{4} = 3{,}834\ in.-lb$$

This moment causes tension in the upper fibers of the Seat Track rail through section *A-A*. *Figure 14-32* shows the shear and moment diagram resulting from the vertical load "P_Z" of 852 *lbs* applied mid-span.

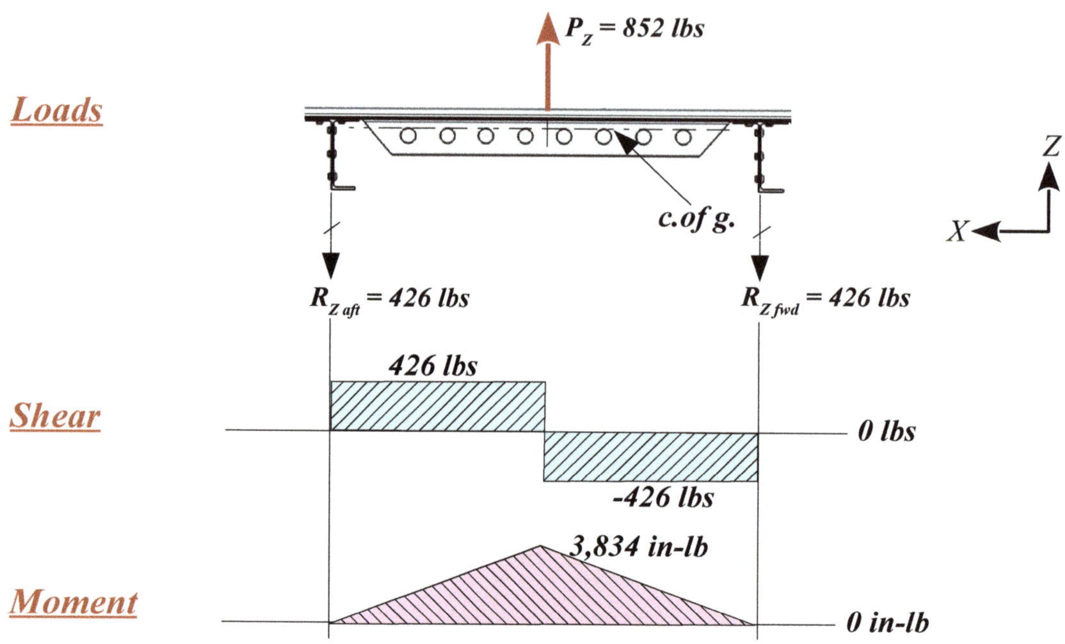

***Figure 14-32** – Shear and moment diagrams due to the vertical force "P_Z".*

Figure 14-33 shows the resulting shear and moment diagrams from combining the moment from the longitudinal load "M_o" of 709 *in-lb* with the vertical load "P_Z" of 852 *lbs,* both applied mid-span.

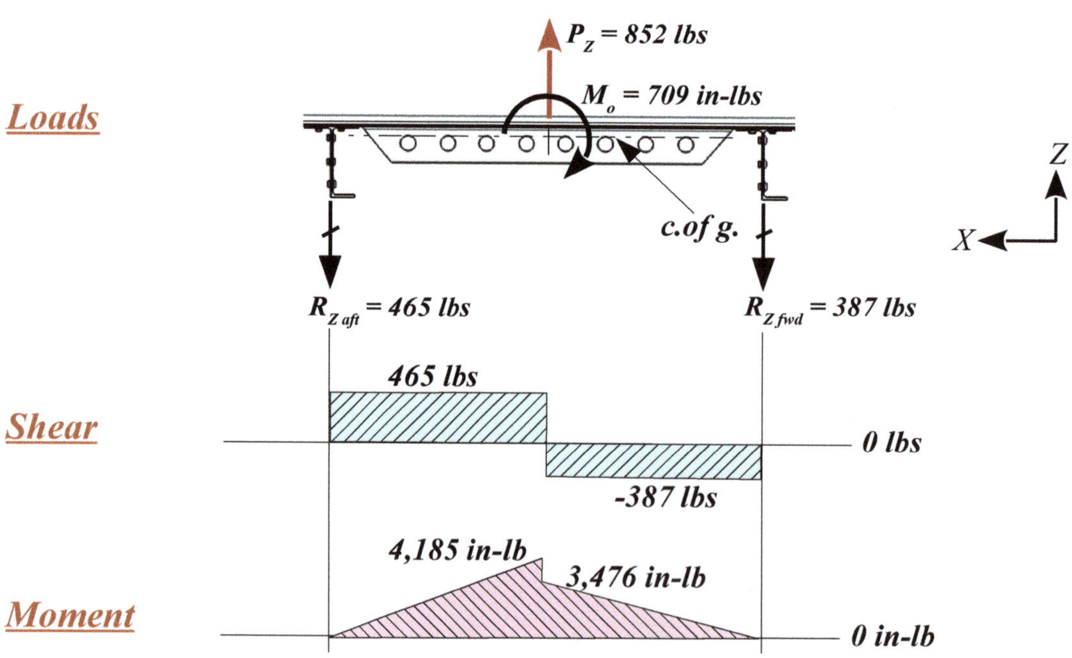

***Figure 14-33** – Shear and moment diagrams due to the combined forces "P_X" & "P_Z".*

Thus, using the calculated amplification factor from page *325,* the amplified bending moment value due to the secondary bending effect becomes: $M_Y = 1.04 * 4{,}185 = 4{,}353$ *in-lb.*

Axial stress

According to the *FAR's*, the emergency loading conditions are to be analyzed independently from each other. Therefore, one has to be "**load case consistent**" in the analysis. Having said that, I will only combine loads from the same load case but, to be conservative, I will use the maximum bending moment combined with the maximum axial load regardless of the attachment points. From *Table 14-14,* I will thus use the maximum axial load of 1,367 *lbs* at point *3* from the *9g Fwd* for the axial stress using the section properties from *Figure 14-20.*

$$f_a = \frac{P_X}{A} = \frac{-1{,}367}{0.889} = -1{,}538\,psi$$

Bending stress

The maximum bending moment from *Figure 14-33* amplified by the secondary bending moment factor leads to a bending moment of 4,353 *in-lbs.* The bending stress is calculated at the top and at the base.

$$f_{btop} = \frac{M_Y\, c_{top}}{I_{YY}} = \frac{4{,}353\,(0.591)}{0.305443} = 8{,}423\ psi$$

$$f_{bbottom} = \frac{M_Y\, c_{bottom}}{I_{YY}} = \frac{4{,}353\,(-1.439)}{0.305443} = -20{,}508\ psi$$

Total axial stress

$$f_{top} = f_a + f_{btop} = -1{,}538 + 8{,}423 = 6{,}885\ psi$$

$$f_{bottom} = f_a + f_{b\,bottom} = -1{,}538 + (-20{,}508) = -22{,}046\ psi$$

This stress could be compared to F_{CY} (or F_{CC} if lower) to get the bending stress ratio "R_b" but, since I'm concerned about the lateral stability of the Seat Track rail lower cap in compression, I will see first if the allowable stress from the lateral stability is lower than F_{CY} before calculating the bending stress ratio. See section *14.6.7* for the lateral stability analysis.

Shear stress

The shear stress is calculated through *Section A-A* by assuming the shear to be taken by the vertical web only. Therefore, the web shear net area is: A_S = (1.50−0.625) (0.080) = 0.070 in^2. From the shear diagram in *Figure 14-33*, the maximum shear is 465 *lbs.*

$$f_s = \frac{V_Z}{A_S} = \frac{465}{0.07} = 6{,}643\ psi$$

The shear stress calculated above is the average ultimate shear across the Seat track rail web. It is common to calculate the average shear for ultimate loads instead of the maximum shear stress that would normally occur at the neutral axis of the section for limit loads. Conservatively, this average shear stress will be combined to the bending stress in the MS calculation.

For that purpose, for F_{SU} = 40 *ksi* (ref. *Table 14-15*), the shear stress ratio "R_S" is calculated as:

$$R_S = \frac{f_S}{F_{SU}} = \frac{6{,}643}{40{,}000} = 0.166$$

14.6.7 LATERAL STABILITY OF LOWER CAP

The Seat track rail lower cap is in compression and is thus subject to lateral instability. We already checked the general stability of the rail but, since the lower cap is not laterally supported, I like to make sure that the lower cap is not tempted to move sideways as this would lead to the failure of the beam.

There are recipes to evaluate the lateral stability of beams but, in this case, I like to conservatively consider the lower cap as a column of its own that is stabilized in the vertical direction. See *Figure 14-36* for the dimensions and the section properties of the lower cap. The lower cap is restrained vertically but it is conservatively assumed disconnected from the rest of the beam therefore free to move laterally.

As one can notice from *Figure 14-34*, the ends of the column are free! So, to analyze it, since it is symmetrical, the column is considered split in half with one end fixed at the center-line and, one end free.

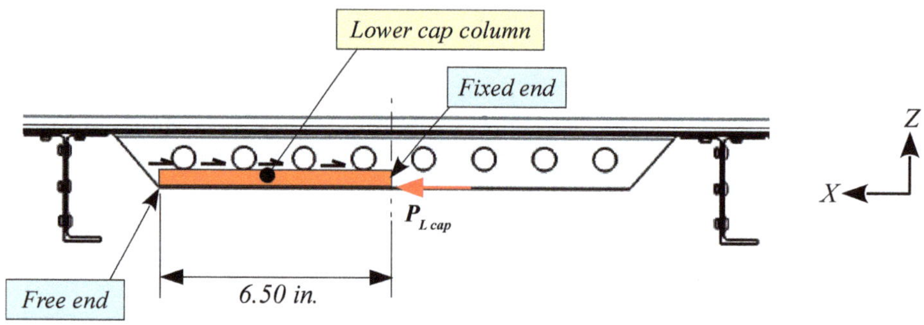

Figure 14-34 *– Seat track rail lower cap column assumption.*

The column is therefore idealized as shown in *Figure 14-35*. From *Figure B-2* for a uniform column distributed axis load, one end fixed, one end free, the end-fixity coefficient is thus C = 0.794. Therefore, the effective column length "***L'***" is calculated as:

$$L' = \frac{L}{\sqrt{c}}$$

Hence, for L = 6.50 *inches*;

$$L' = \frac{6.50}{\sqrt{0.794}} = 7.30\ in.$$

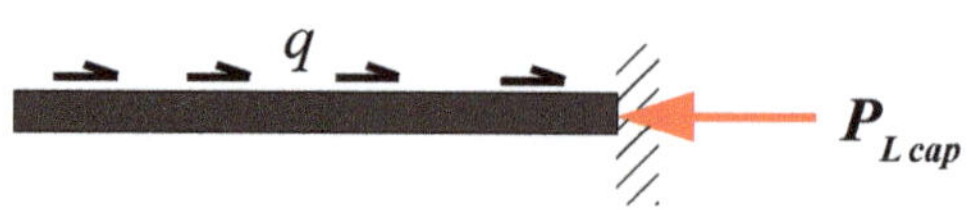

Figure 14-35 *– Seat track rail lower cap idealized column.*

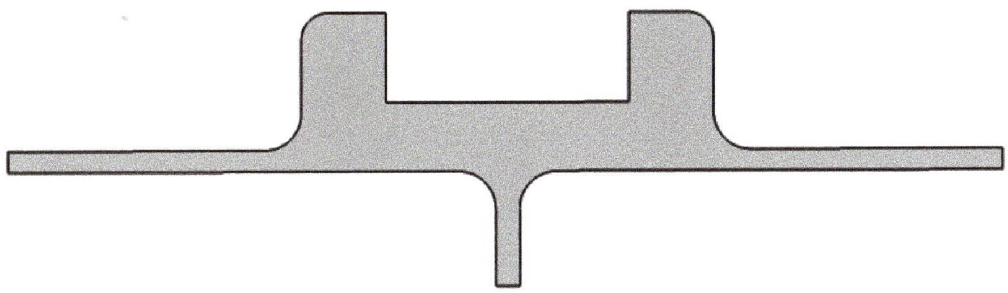

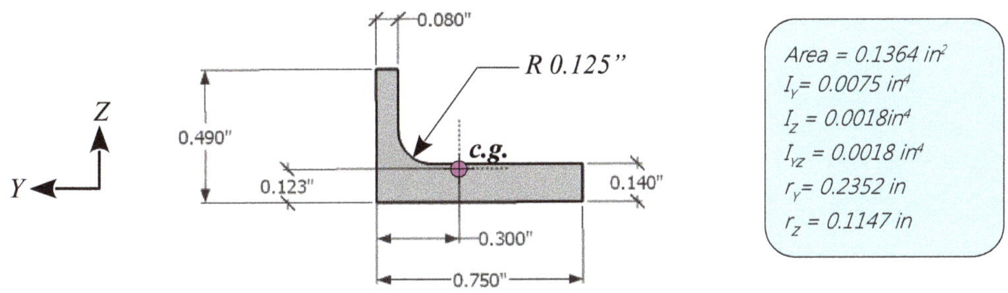

Figure 14-36 – *Seat track rail lower cap dimensions and section properties.*

The critical column buckling load is calculated as:

$$P_{CR} = \frac{\pi^2 (E_t I_Z)}{(L')^2}$$

For
$E_t = E = 10.0*10^6\ psi$
$I_Z = 0.0018\ in^4$
$L' = 7.30\ in.$

then

$$P_{CR} = \frac{\pi^2\ 10*10^6(0.0018)}{(7.30)^2} = 3{,}333\ lbs$$

The critical compression stress is thus:

$$F_{CR} = \frac{P_{CR}}{A} = \frac{3{,}333}{0.1364} = 24{,}435\ psi$$

Since this stress is well below the yield limit (F_{CY} = 73 *ksi*) and thus in the elastic zone, then $E_t = E$ and the critical column load allowable is good as is without the need to iterate. F_{CR} calculated above is also lower than the crippling stress allowable "F_{CC}" (73,000 *psi*) of the lower cap calculated in section *14.6.4* and thus will be used as the allowable bending stress in the bending stress ratio "R_b".

Lower cap internal load

In order to evaluate the internal load of the lower cap, I will conservatively assume the cross section to be made out of two concentrated unit areas having no moment of Inertia, one for the upper cap and one for the lower cap, located at their respective center of gravity. The internal moment "M_Y" is thus assumed to act as a couple, as shown on *Figure 14-37*. From page *329*, the maximum bending moment is 4,353 *in-lb*. Then, knowing that the sum of the forces has to be zero, $P_{U\ Cap} = -P_{L\ Cap} = P$, then the following equation for the sum of the moments at the Seat track rail neutral axis must be satisfied:

$$M_Y = P_{UCap}(0.209) + P_{LCap}(-1.316)$$
$$= P(0.209) + (-P)(-1.316) = 4{,}353\ in.-lb$$

Solving for P we get: $P_{U\ Cap}$ = 2,854 *lbs* and $P_{L\ Cap}$ = −2,854 *lbs*. This corresponds to an average compression stress due to bending in the lower cap of: f_b = −2,854 / 0.1364 = −20,924 *psi*.

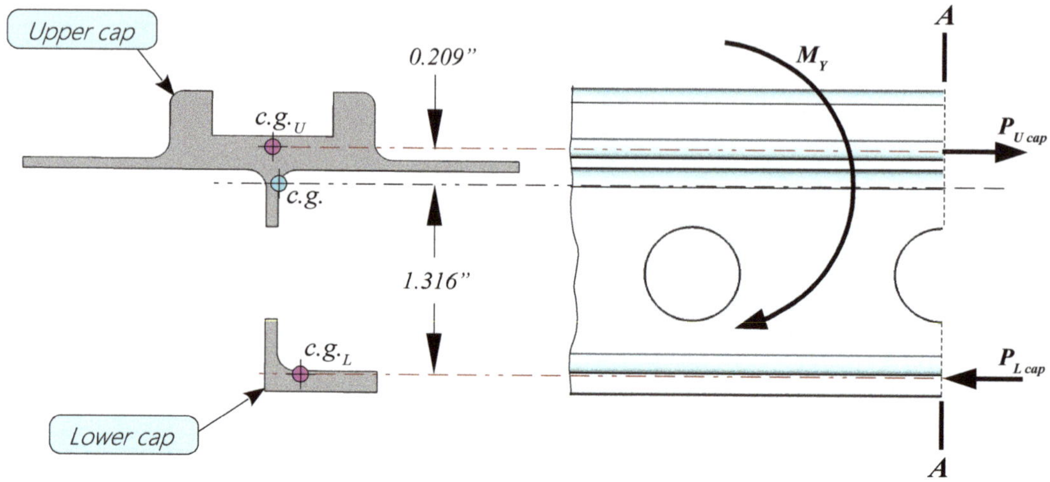

Figure 14-37 *– Seat track rail equivalent moment couple.*

Since this stress is really caused by the bending moment, a bending stress ratio "R_b" will be derived from it.

$$R_b = \frac{|f_b|}{F_{CR}} = \frac{|-20{,}924|}{24{,}435} = 0.856$$

The actual bending stresses across the lower cap calculated from Mc/I could have been used to evaluate the internal load but this is a bit more cumbersome to do although more accurate. This would have lead to a slightly lower stress ratio and a slightly higher *MS*.

Providing an applied internal load higher then the critical load of 3,333 *lbs* would have been obtained, I would have considered adding straps (for example) attaching the lower cap to the upper caps of the adjacent Seat track rails on either sides in order to provide lateral stability or, adding material to the lower cap.

<u>Margin of Safety for Section A-A</u>

Using the following interaction equation as the expression for the margin of safety for combined compression, bending and shear for a tube, hence the MS at section *A-A:*

$$MS = \frac{1}{R_C + \sqrt{R_b^2 + R_S^2}} - 1$$

For
$R_C = 0.035$
$R_b = 0.856$
$R_S = 0.166$

and then:

$$MS = \frac{1}{0.035 + \sqrt{0.856^2 + 0.166^2}} - 1 = 0.10$$

Seat track rail Section A-A Ultimate	***0.10***
9g Fwd *Al Aly 7075-T6 Ext.*	*Compression + bending + shear*

Lateral load:

For the side loading, the components that are reacting those loads in the "Y" direction are the Crossbeams. Then again, the floor panels will carry those loads in the Crossbeams by shear towards the front and the rear of the fuselage.

> To help visualizing the load paths in the fuselage, one can see the skin/stringers as a large beam in the shape of a tube (which it is pretty much) having a horizontal shear web in the middle that consists of the floor panels. Floor panels are thus essential in carrying fuselage loads and are normally required for flight.

The floor panels ultimately react the side loads from the Crossbeams that originally were coming from the Seat tracks. In theory, in the case where the side load is applied mi-span of the Seat track, one could argue that the floor panels attached to it would be taking that side load directly thus preventing any bending moment in the Seat track around the "Z" axis. However, for the analysis of the Seat track rails, the floor panels are considered softer compared to the Crossbeams and are thus conservatively ignored.

Because the second moment of Inertia "*Iz*" around the "*Z*" axis in the Seat track rail is pretty much the same at mid-span (section *A-A*) and at the Crossbeams (section *B-B*), and because the seat track rails are continuous, one could reasonably assumed the Seat track rails to be fixed at the Crossbeams. However, the Seat track rails will be conservatively assumed simply-supported at the Crossbeams while the side force will be assumed applied mid-span.

Using *section A-A* as it covers most of the Seat track rail segment length, the lateral load is assumed applied at mid-span. The maximum lateral load "P_Y" is 1,125 *lbs* from the most critical load case of *3g Right*. The maximum bending moment "M_Z" at mid-span due to a lateral force is calculated as:

$$M_{max} = \frac{W\,L}{4}$$

where $W = 1{,}125$ *lbs* and $L = 18.0$ *in.*, thus:

$$M_{max} = \frac{1{,}125 \times 18.0}{4} = 5{,}063\,in.-lb$$

This moment causes tension in the right fibers of the Seat Track rail through section *A-A*. *Figure 14-38* shows the shear and moment diagram resulting from the lateral load "P_Y" of 1,125 *lbs* (ref. *Table 14-14 point 1)* applied mid-span. The reactions at the Crossbeams are thus:

$$R_{Y\,aft} = \frac{P_Y}{2} = \frac{1{,}125}{2} = 563\;lbs$$

$$R_{Y\,fwd} = \frac{P_Y}{2} = \frac{1{,}125}{2} = 563\;lbs$$

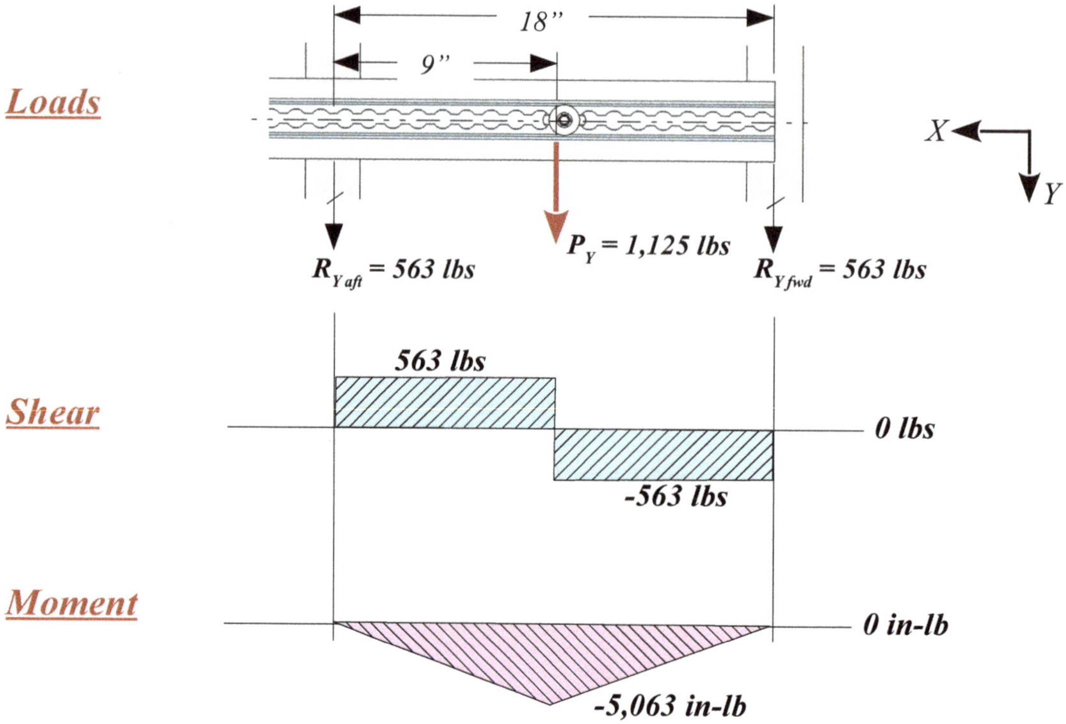

Figure 14-38 *– Seat track rail Shear & Moment diagrams due to P_Y.*

Bending stress

The maximum bending moment from *Figure 14-38* is 5,063 *in-lb*. The bending stress is calculated at the edges of the free flanges (points *1* through *4* from *Figure 14-39*) using the dimensions and properties as shown in *Figure 14-19*.

$$f_{b1} = -\frac{M_Z\, c_1}{I_{ZZ}} = -\frac{-5{,}063\,(1.69)}{0.32520} = 26{,}311\ psi$$

$$f_{b2} = -\frac{M_Z\, y_2}{I_{ZZ}} = -\frac{-5{,}063\,(-1.61)}{0.32520} = -25{,}066\ psi$$

$$f_{b3} = -\frac{M_Z\, y_3}{I_{ZZ}} = -\frac{-5{,}063\,(0.64)}{0.32520} = 9{,}964\ psi$$

$$f_{b4} = -\frac{M_Z\, y_4}{I_{ZZ}} = -\frac{-5{,}063\,(-0.72)}{0.32520} = -11{,}210\ psi$$

However, the applied load is reversible since the *3g Left* load case is simply the opposite of the *3g Right* load case. Therefore, the RHS free flange becomes the critical one. The tension stress is numerically about the same as the compression stress so it is not really a concern for me as it is much lower than F_{TU}. An MS could be calculated for it but it would simply show an MS of > ***1.0***. A statement like: "*Tension stress Passed by engineering judgment*" could be written providing the *DER* agrees to that.

Otherwise, show the MS. Therefore, my attention will go to the compression side since it is a free edge flange and it is subject to local instability.

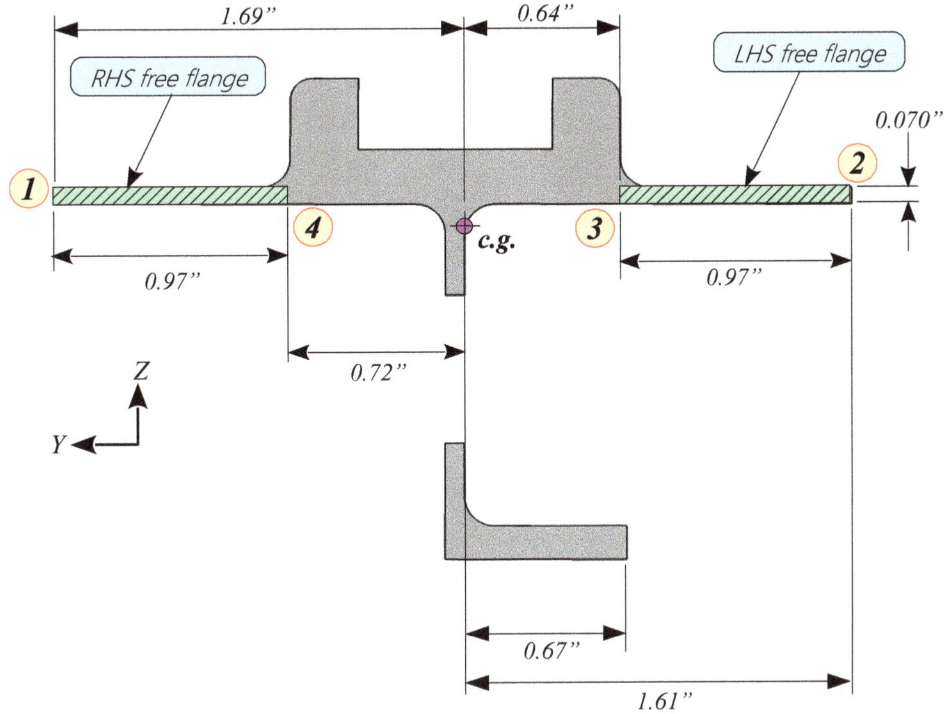

Figure 14-39 – *Critical stress locations at free flanges of Section A-A.*

Complex bending

Since section *A-A* has no axis of symmetry and because the Seat track rail is unsupported/non-stabilized in-between the Crossbeams, a complex bending analysis is performed as a comparison to the simpler *"Mc/I"* calculations done previously. The floor panels are ignored as always. See *Table 14-31* for the resulting bending stresses at each stress point from using the complex bending and the simple (beam considered stabilized) equation.

The very well known complex bending equation adapted to our coordinate system is:

$$f_{bx} = \frac{-(M_Z I_Y + M_Y I_{YZ})\,y + (M_Y I_Z + M_Z I_{YZ})\,z}{I_Y I_Z - I_{YZ}^2}$$

Figure 14-40 shows all of the stress points where the bending stresses are calculated along with the principal axes of the section. As an example, the complex bending stress calculation at the most critical location "point *2*" is shown below with the reversed bending moment M_Z = 5,063 *in-lb*.

$$f_{b2} = \frac{-(5{,}063(0.305443)+0)1.69 + (0+5{,}063(-0.048468))0.131}{0.305443 \times 0.3252 - (-0.048468)^2} = -27{,}278 \ psi$$

For
M_Z = 5,063 in-lb
M_Y = 0
I_Y = 0.305443 in^4
I_Z = 0.32520 in^4
I_{YZ} = -0.048468 in^4
y = 1.690 in.
z = 0.131 in.

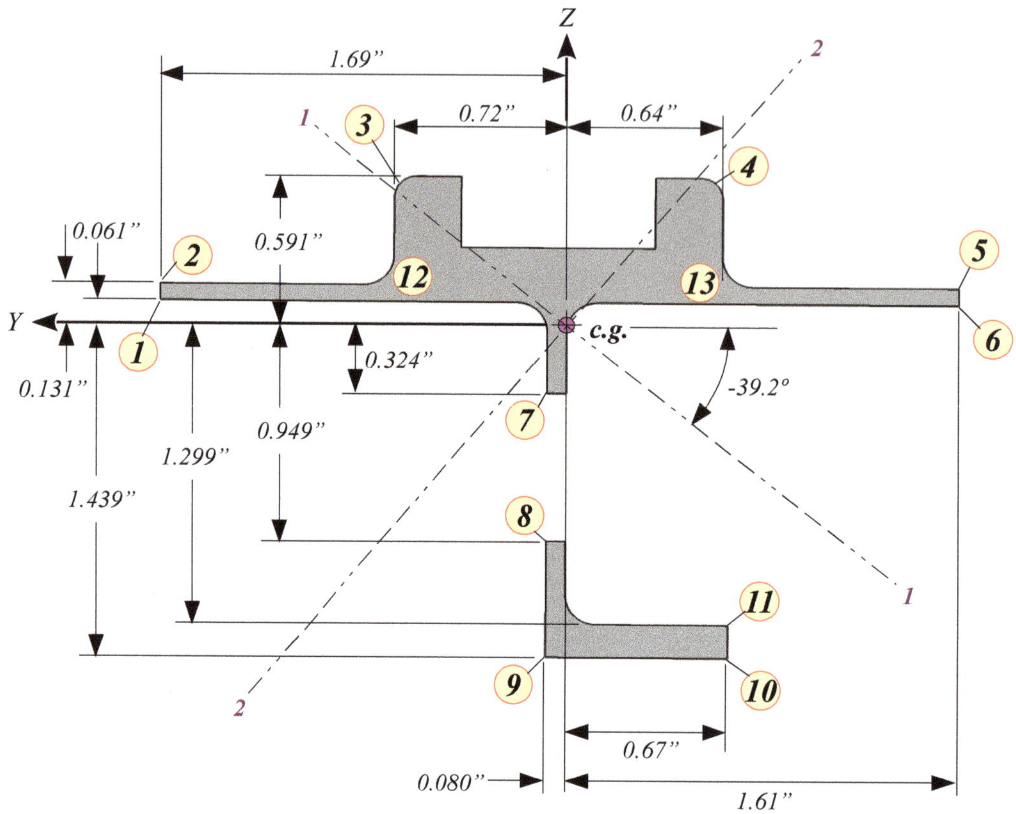

Figure 14-40 *– Section A-A dimensions and stress points for complex bending.*

Table 14-31 *- Bending stresses at points using simple and complex bending equations.*

Points	*y* [in]	*z* [in]	$f_{bx\_complex}$ [psi]	$f_{b=Mz(y)/Iz}$ [psi]
1	1.690	0.061	27,100	26,309
2	1.690	0.131	27,278	26,309
3	0.720	0.591	12,975	11,208
4	-0.640	0.591	-8,709	-9,963
5	-1.610	0.131	-25,339	-25,063
6	-1.610	0.061	-25,516	-25,063
7	0.080	-0.324	456	1,245
8	0.080	-0.949	-1,125	1,245
9	0.080	-1.439	-2,365	1,245
10	-0.670	-1.439	-14,324	-10,430
11	-0.670	-1.299	-13,969	-10,430
12	0.720	0.131	11,811	11,208
13	-0.640	0.061	-10,050	-9,963

The above results show the small difference in using the complex bending equation (in this case anyway). For the critical points (points *1, 2, 5 & 6*), shown in *Figure 14-40,* the difference is about *2-3%*. However, unless the beam is very well stabilized in torsion, the complex bending equation should be used as it is usually always more conservative.

14.6.8 FREE FLANGE BUCKLING

Using the stresses at section ***A-A*** calculated previously from *Table 14-31*, a free-flange buckling analysis is performed at the *RHS* free flange (point ***2***) by reversing the load. Therefore, considering the flange to start from the thicker area (point ***12***), as shown on *Figure 14-40* and *Figure 14-41*, the variation of the stress goes from a relatively high compression stress at the free edge to a smaller compression stress at the thicker base where the free flange is conservatively considered simply supported.

For conservatism, the flange is considered simply supported at all sides except at the free edge. Using (ref. 20.) for flat plate buckling in bending, the bending stress allowable is calculated as:

$$F_{b\ cr} = \frac{k_b\, \eta_c\, \pi^2\, E}{12\left(1-\mu^2\right)}\left(\frac{t}{b}\right)^2$$

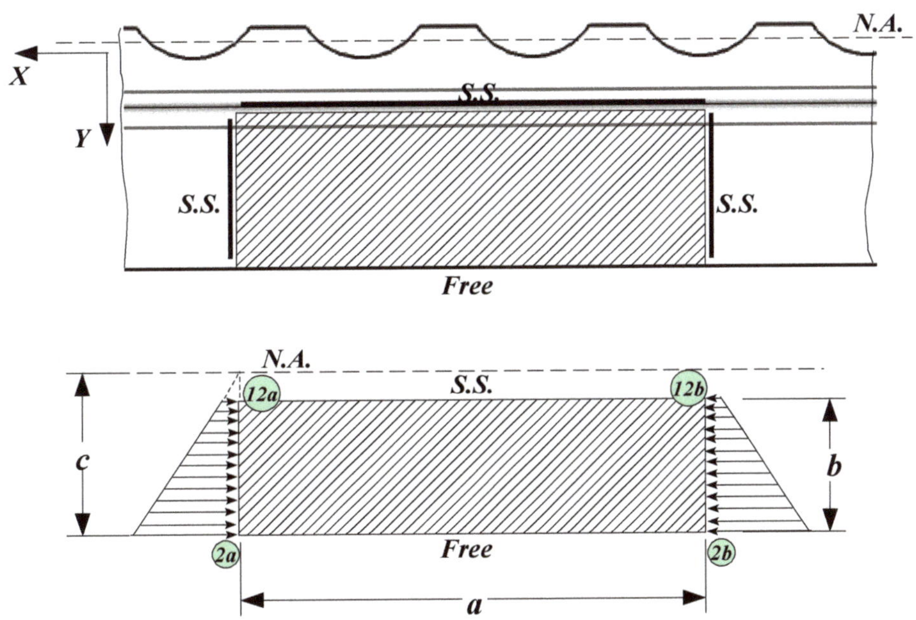

Figure 14-41 – *Free flange buckling in bending.*

From *Figure 14-41*, the length "***a***" is arbitrarily shown since it is not required in this analysis. Since the stress varies along the Seat track rail, the bending stress at point ***2a*** is different (higher) than the one at point ***2b***. The same for ***12a*** and ***12b***. The maximum (at points ***2a*** and ***12a***) are used for conservatism.

From the stress calculations done previously, the stress at point ***2*** is −27,278 *psi* and it is −11,811 *psi* at point ***12*** located 0.97 *in.* away. For this analysis, we are interested to know the location where the stress is zero. That is "***c***" on *Figure 14-41*. By doing a linear extrapolation from points ***2*** & ***12*** we get ***c*** = 1.71 *in.* In this case, b = 0.97 *in.* and c = 1.71 *in.* Thus b/c = 0.97/ 1.71 = 0.57. Hence from *Figure B-1*, the value for $\boldsymbol{k_b}$ read is: 0.44. The Young's modulus from *Table 14-15* is used. From the above bending stress buckling equation:

$$\frac{F_{b\ cr}}{\eta_c} = \frac{k_b\, \pi^2\, E}{12\left(1-\mu^2\right)}\left(\frac{t}{b}\right)^2$$

$$\frac{F_{b\ cr}}{\eta_c} = \frac{(0.44)\, \pi^2\, 10*10^6}{12\left(1-0.33^2\right)}\left(\frac{0.070}{0.97}\right)^2 = 21{,}149\ psi$$

For
$k_b = 0.44$
$b = 0.97\ in$
$t = 0.070\ in$
$E_c = 10.0*10^6\ psi$

This value is lower than the compression stress at the free edge therefore one would expect a local buckling to occur. However, the assumption that all sides of the flange are simply supported is quite conservative. In reality, it is somewhere between simply supported and fixed. It is a common practice to consider a 50% fixity along the edges for machined and extruded parts. This assumption was deemed reasonable as similar results were obtained from tests. Thus, another value for k_b by assuming the edge opposite to the free edge to be fixed is read from *Figure B-1* for $b/c = 0.57$. The value obtained is 1.3 for a 100% fixity.

Calculating the average of the two k_b, the value for a 50% fixity is:

$$k_{b\,50} = \frac{k_{b0} + k_{b\,100}}{2} = \frac{0.44 + 1.30}{2} = 0.87$$

The critical buckling stress for a 50% fixity free flange is:

$$\frac{F_{b\,cr}}{\eta_c} = \frac{k_b\,\pi^2\,E}{12\left(1-\mu^2\right)}\left(\frac{t}{b}\right)^2$$

$$\frac{F_{b\,cr}}{\eta_c} = \frac{(0.87)\,\pi^2\,10*10^6}{12\left(1-0.33^2\right)}\left(\frac{0.070}{0.97}\right)^2 = 41{,}818\ psi$$

For
$k_b = 0.87$
$b = 0.97$ in
$t = 0.070$ in
$\mu = 0.33$
$E_c = 10.0*10^6$ psi

A plasticity correction factor for compression "η_c" is used to multiply the Young's modulus "E" and is calculated using the following equation.

$$\eta_C = \frac{E_t}{E} = \frac{1}{1+\left(\dfrac{0.002\ E\ n}{F_{cy}}\right)\left(\dfrac{F_{c\,cr'}}{F_{cy}}\right)^{n-1}}$$

Therefore, using the same procedure as described in section ***10.5.6***, the critical buckling bending stress is calculated by iteration by first assuming a value for $F_{c\,cr}'$.

A first value of 45,000 *psi* was assumed for $F_{c\,cr}'$ in the above equation, and from the iterative process, the value for "η_c" of 0.99986 was obtained. The final critical bending buckling stress allowable is thus;

For
$F_{cy} = 73{,}000$ psi
$n = 20$
$E = 10.0*10^6$ psi
$t = 0.070$ in
$b = 0.97$in

$$F_{b\,cr} = \frac{k_b\,\eta_c\,\pi^2\,E}{12\left(1-\mu^2\right)}\left(\frac{t}{b}\right)^2 = \frac{0.87\,(0.99986)\,\pi^2\,10*10^6}{12\left(1-0.33^2\right)}\left(\frac{0.070}{0.97}\right)^2 = 41{,}812\ psi$$

<u>Bending stress ratio</u>

Since this stress at the edge of the free flange is really caused by the bending moment, a bending stress ratio "R_b" will be derived from it.

$$R_b = \frac{|f_b|}{F_{b\,cr}} = \frac{|-27{,}278|}{41{,}812} = 0.652$$

As one can see from the above calculations, the assumption regarding the panel's edge restraints has a big impact on the *MS*. Using a 50% fixity restraint is adequate in this case since the part is an extrusion. As additional conservatism, the peak bending stress was assumed constant across the whole length of the Seat track rail's free edge while it is obviously not. On top of that, the floor panels are fastened to that flange and will also help in stabilizing that free flange or, to the least, pick-up some of the applied side load.

Transverse Shear stress

The shear stress is calculated through section *A-A* by assuming the shear to be taken by the base only as shown in *Figure 14-42*. Therefore, the base shear area is: $A_S = 2\ (0.97 * 0.070) + 1.36 * 0.23 = 0.4486\ in^2$. From the shear diagram in *Figure 14-38,* the maximum transverse shear force "V_Y" is 563 *lbs.*

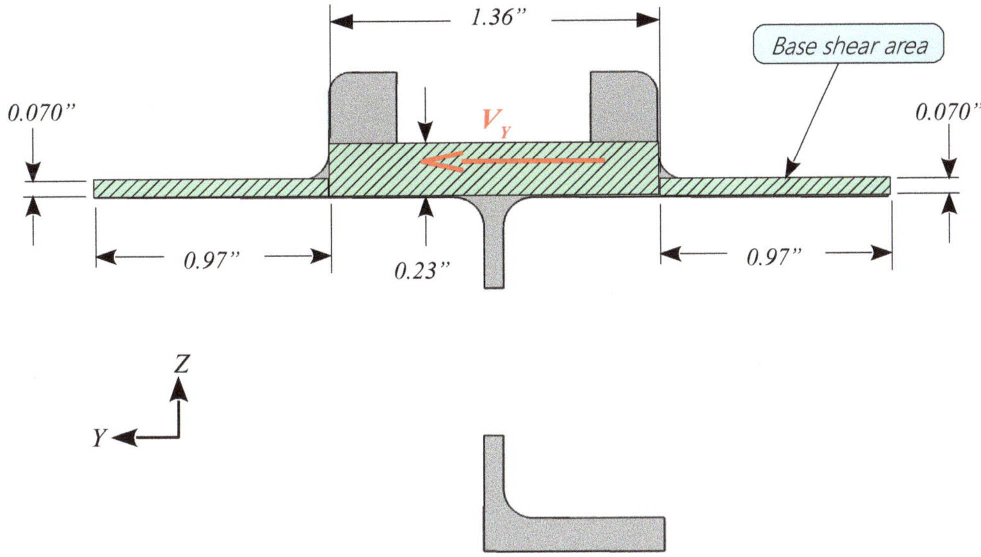

Figure 14-42 *– Seat track rail Base lateral shear area.*

Hence, the average shear stress "f_s" across the Seat track rail base is:

$$f_s = \frac{V_Y}{A_S} = \frac{563}{0.4486} = 1{,}255\ psi$$

The shear stress calculated above is the average ultimate shear stress across the Seat track rail base. It is common to calculate the average shear for ultimate loads instead of the maximum shear stress that would normally occur at the neutral axis of the section for limit loads using *VQ / It* when it is to be combined to the bending stress.

Torsion shear stress

As shown in *Figure 14-44*, the lateral load "P_Y" is applied at a point near the top of the section at a distance of 0.408 *in.* from the section's shear center "*SC*". For this section, the vertical position "*e*" of the shear center "*SC*" is calculated as an idealized wide flange beam with unequal flanges. The calculations of the *SC* position "*e*", the torsional stiffness "*K*" and the warping constant "C_w" of the section are shown in *Figure 14-43*.

By referring to *Figure 14-42*, the thickness of the upper flange "t_1" to be used in this idealization is calculated as the weighted average of the base thicknesses as follows:

$$t_1 = \frac{2 \times 0.97 \times 0.07 + 1.36 \times 0.23}{3.30} = 0.136\ in.$$

The applied torque "*T*" applied at mid-span is thus:

$$T = P_Y * d = 1{,}125 * 0.408 = 459\ in.-lb$$

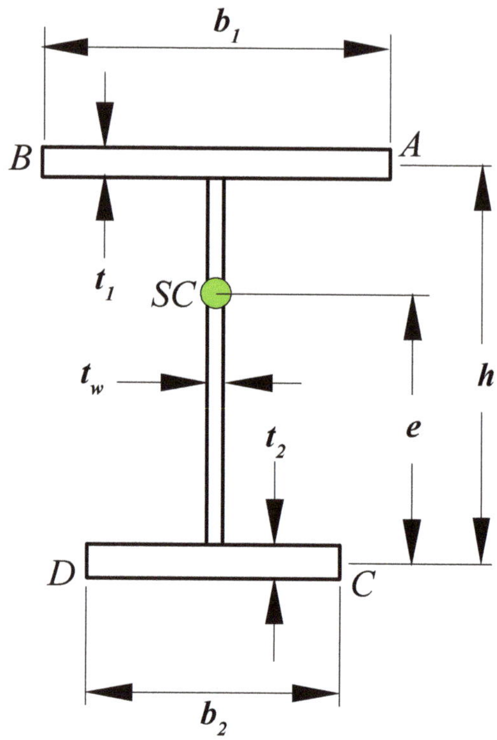

$$e = \frac{t_1 b_1^3 h}{t_1 b_1^3 + t_2 b_2^3}$$

$$K = \frac{1}{3}\left(t_1^3 b_1 + t_2^3 b_2 + t_w^3 h\right)$$

$$C_w = \frac{h^2 t_1 t_2 b_1^3 b_2^3}{12\left(t_1 b_1^3 + t_2 b_2^3\right)}$$

For
$b_1 = 3.3$
$t_1 = 0.136$ in.
$b_2 = 0.75$ in.
$t_2 = 0.14$ in.
$t_w = 0.08$ in.
$h = 1.498$ in.

Then
$e = 1.480$ in.
$K = 0.003705$ in^4
$C_w = 0.010912$ in^6

Figure 14-43 *– Idealized Seat track rail section A-A shear center position.*

The maximum shear stress due to torsion is calculated for an "*I*" beam having both ends free to warp but not twist. Section properties are from *Figure 14-43*. ***Figure 14-45*** shows a model of the beam to evaluate the maximum shear stress due to torsion. The maximum shear stress " *Max* τ_l" is located on the surface of the thickest section. Although an average thickness of the upper flange was used in determining the torsion parameters, I will used the actual maximum thickness of 0.23 *in.* in the shear stress calculation.

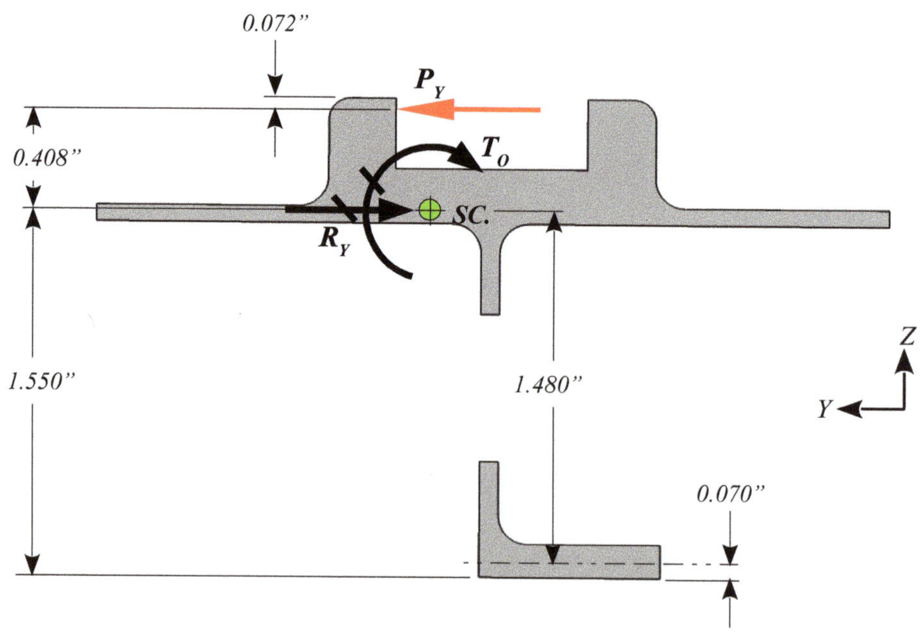

Figure 14-44 *– FBD at Seat track rail section A-A shear center.*

This shear stress is thus calculated at the location of the maximum thickness which is more at the center of the Seat track rail, away from the location of the maximum bending stress. Therefore, by combining the bending stress at the free edge of the flanges with the maximum shear is conservative.

Whatever assumptions you decide to chose for the analysis, always explain your line of thoughts to justify them. Tell a story! Describe how you see the load path, what you think happens in a certain area regarding the loads, the stresses, the strains and, the deformations. The DER (or your boss) might or might not agree with you on some of the assumptions but at least you will have demonstrated a rational for them. By reading your "story", that person may think: *–Oh! I see. I wouldn't have done it this way but it makes sense.*

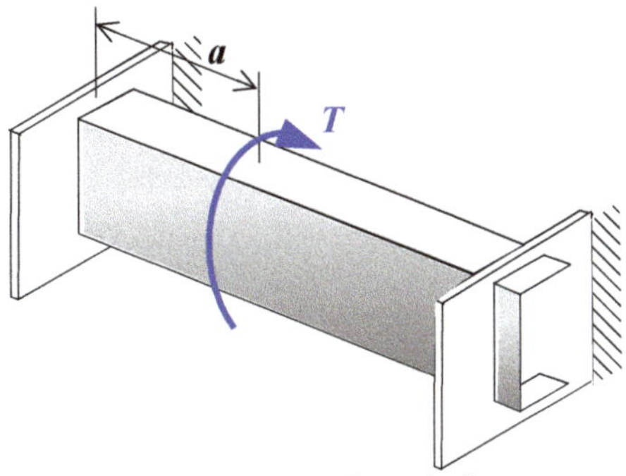

Figure 14-45 – *Beam in torsion with ends free to warp but not twist.*

The necessary parameters and the torsion shear stress occurring around the perimeter of the section are calculated as follows:

$$G = \frac{E}{2(1+\mu)}$$

$$\theta' = \frac{T_O}{2C_w E \beta^2}\left[1 - \frac{1}{\cosh(\beta l/2)}\right] \qquad G = \frac{10.0\,E^6}{2(1+0.33)} = 3.76E^6\ psi$$

$$\beta = \sqrt{\frac{KG}{C_w E}} = \sqrt{\frac{(0.003705)3.76\times10^6}{(0.01091)10.0\times10^6}} = 0.35727\,in.^{-1}$$

$$\theta' = \frac{459}{2(0.01091)\ 10.0\times10^6\,(0.35727)^2}\left[1 - \frac{1}{\cosh\left(\frac{(0.35727)18}{2}\right)}\right] = 0.01515\,in.^{-1}$$

$$Max\ \tau_1 = t_{max} G\,\theta' = (0.23)\ 3.76\times10^6\ (0.01515) = 13{,}103\ psi$$

For
T = 459 in-lb
$a = l/2$ = 9 in.
$K = 3.705\ E^{-3}\ in^4$
$C_w = 0.01091\ in^6$
$E = 10.0\ E^6$ psi
$G = 3.76\ E^6$ psi
$\beta = 0.35727\ in^{-1}$
l = 18 in
hence
$\theta' = 0.01515\ in^{-1}$

The torsion shear stress will add to the transverse shear in the top of the flange as they are in the same direction and subtract on the bottom since they are in the opposite directions. Therefore, the total shear stress "f_{S_Total}" is then calculated by summing up the torsional stress "$Max\ \tau_1$" and the transverse shear stress "f_S" calculated on page *339*.

Hence $$f_{S\ Total} = Max\ \tau_1 + f_s = 13{,}103 + 1{,}255 = 14{,}358\ psi$$

Conservatively, this shear stress will be combined to the bending stress in the *MS* calculation. For that purpose, for F_{SU} = 40 *ksi* (ref. ***Table 14-15***), the shear stress ratio "R_S" is calculated as:

$$R_S = \frac{f_S}{F_{SU}} = \frac{14{,}358}{40{,}000} = 0.359$$

<u>Margin of Safety for section *A-A* due to the *3g Left* Load case</u>

Using the following interaction as the expression for the margin of safety for combined bending and shear, hence the MS at section *A-A* due to the *3g Left* load case is:

$$MS = \frac{1}{\sqrt{R_b^2 + R_S^2}} - 1$$

For
$R_b = 0.652$
$R_s = 0.359$

and then:

$$MS = \frac{1}{\sqrt{0.652^2 + 0.359^2}} - 1 = 0.34$$

Seat track rail Section A-A Ultimate	***0.34***
3g Left *Al Aly 7075-T6 Ext.*	*Free flange buckling + shear*

A lot of times, an *MS* for the free flange buckling will be calculated separately while another one will be calculated for the shear. Quite often, local buckling doesn't mean a failure but, in our case, if the free flange were to buckle due to bending, that flange would be considered gone and the moment of inertia would then be greatly reduced leading to a possible complete failure of the beam.

Thus, since the bending stress occurs at the same location and at the same time as the shear stress, it made sense to combine both and obtain one single *MS*.

14.6.9 STRESS ANALYSIS SECTION B-B

In order to analyze the seat track for all loading configuration possible, a second check is performed by assuming the seat track to be fixed at the crossbeams. This maximizes the bending moment at each end through section *B-B*.

The maximum bending moment is obtained from a vertical force located at a third of the span and is calculated as:

$$M_{max} = -0.1481\,W\,L$$

where $W = 852$ *lbs* and $L = 18.0$ *in*, thus:

$$M_{max} = -0.1481(852 \times 18.0) = -2{,}271\ in.-lb$$

This moment causes compression in the upper fibers of the Seat Track rail through section *B-B*.

<u>Axial stress</u>

According to the *FAR's*, the emergency loading conditions are to be analyzed independently from each other. Therefore, one has to be "**load case consistent**" in the analysis. Having said that, I will only combine loads from the same load case but, to be conservative, I will use the maximum bending moment combined with the maximum axial load regardless of the attachment points of the seat track Fittings. From *Table 14-14*, I will thus use the maximum axial load of 1,367 *lbs* at point *3* from the *9g Fwd* for the axial stress using the section properties from *Figure 14-20.*

$$f_a = \frac{P_X}{A} = \frac{-1{,}367}{0.6199} = -2{,}205\ psi$$

Bending stress

To calculate the bending stress, I'll be using the bending moment calculated earlier (−2,271 *in-lb*) caused by the vertical force (852 *lbs*) assumed applied at a third of the span from the load case *9g Fwd* at point *1*. The bending stress is calculated at the top and at the base.

$$f_{btop} = \frac{M_Y\, c_{top}}{I_{YY}} = \frac{-2{,}271\,(0.363)}{0.0132} = -62{,}453\ psi$$

$$f_{bbase} = \frac{M_Y\, c_{base}}{I_{YY}} = \frac{-2{,}271\,(-0.167)}{0.0132} = 28{,}732\ psi$$

Shear stress

The shear stress is calculated through section *B-B* but by removing the base flanges as those won't be carrying much vertical shear due to their vertical flexibility. Therefore, the remaining shear area is: $A_S =$ 0.6199 – 1.94 (0.070) = 0.4841 in^2. In addition, the applied vertical force is conservatively assumed right at section *B-B* to maximize the shear stress in the section.

$$f_s = \frac{P_Z}{A_S} = \frac{852}{0.4841} = 1{,}760\ psi$$

Combined stresses

The axial, bending and shear stresses are combined according to the equation:

$$(R_a + R_b)^2 + R_S^2 = 1$$

In the above equation, the axial stress ratio "*Ra*" and the bending stress ratio "*Rb*" are separate even though the internal stresses in both cases are in the same direction. This is normally done when the allowables for the axial and the bending stresses are different. That occurs for instance when, for example, the plastic bending methodology is used. Or when the axial stress allowable is limited to the column buckling stress (instability).

In our case, the axial and the bending stresses are simply added up and a single axial stress ratio "R_a'" is calculated. Thus the MS is calculated as:

$$MS = \frac{1}{\sqrt{R_a'^2 + R_S^2}} - 1$$

Since there is no secondary bending effect and that buckling is not a concern at that location, the internal stresses are thus simply compared to the material allowables. Because the stress is the greatest at the top and because it is in compression, the MS is calculated at the top by comparing the internal compression stress to the crippling stress allowables "F_{CC}" of segments *5 & 8* as shown in *Table 14-28*. Hence, R_a' and R_S are respectively the axial and shear stress ratios calculated as follows:

$$R_a = \frac{f_a + f_{btop}}{F_{cy}} = \frac{|-2{,}205 + (-62{,}453)|}{73{,}000} = 0.886$$

$$R_s = \frac{f_s}{F_{su}} = \frac{1{,}760}{40{,}000} = 0.044$$

The lower crippling stress "F_{CC}" of the whole section *B-B* could have been used but would have been too conservative since it is reduced by the long and thin flanges (segments *3 & 9*) and since those flanges are located near the section's neutral axis, the bending stress in those flanges is quite small.

Hence the MS at section *B-B* is:

$$MS = \frac{1}{\sqrt{0.886^2 + 0.044^2}} - 1 = 0.13$$

Seat track rail Section B-B, Ultimate	***0.13***
9g Fwd *Al Aly 7075-T6 Ext.*	*Axial + bending + shear*

14.6.10 SHEAR BUCKLING

The Seat track rail web is also checked for shear buckling. Although shear buckling is not a mode of failure, we need to verify that the web is non-buckling. As shown on *Figure 14-46*, there is only one long and narrow panel that contains many 0.625 *in.* diameter lightning holes in the web. The panel boundaries are determined by the upper and lower flanges. The panel is assumed simply supported on all sides.

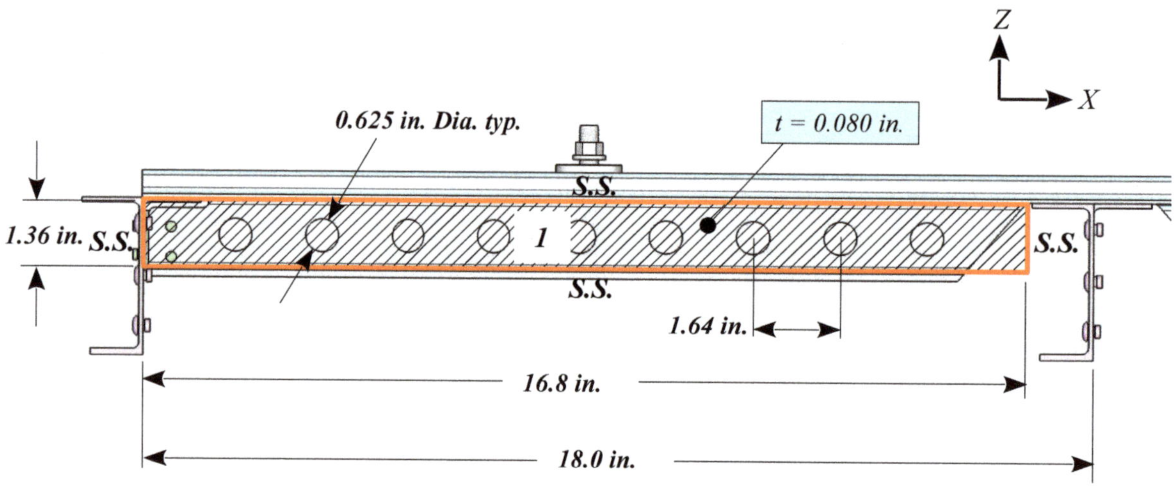

Figure 14-46 *– Seat track rail web panel.*

From *Figure 14-33*, the maximum transverse shear going through the Frame web is V_Z = 465 *lbs*. Since the web height is h = 1.36 *in*. and the thickness is t = 0.080 *in*, then the average shear stress is calculated as: $f_S = V_Z / A_{web}$ = 465 / (1.36 * 0.080) = 4,274 *psi*. The shear flow is thus; $q = V_Z / h$ = 465 / 1.36 = 342 *lb/in*.

This panel is relatively thick and can be considered as a shear buckling resistant. That is such that no buckling is permitted at ultimate load.

14.6.10.1 Panel without a hole

Since there are lightening holes in the panel, the shear buckling will be calculated in two steps. First, the shear buckling coefficient "k_S" is obtained, then, we get the knock down due to the holes. Using (ref. 20.) for flat plate buckling in shear, the shear stress allowable is calculated as:

$$F_{S\,cr} = \frac{k_S\, \eta_S\, \pi^2\, E}{12\left(1-\mu^2\right)}\left(\frac{t}{b}\right)^2$$

From *Figure 14-46*, the dimension "*a*" is 16.8 *in.* and the dimension "*b*" is 1.36 *in*. Note that for shear buckling, the value for "*b*" is always the smallest of the two dimensions. Thus a/b = 16.8 / 1.36 = 12.4. Hence from *Figure A-2*, since the a/b value is out of the graph, the asymptotic value for k_S read is: 5.35. The primary Young's modulus in tension "*E*" is used from *Table 14-15*. From the above equation:

$$\frac{F_{s\,cr}}{\eta_s} = \frac{k_s\,\pi^2\,E}{12\left(1-\mu^2\right)}\left(\frac{t}{b}\right)^2$$

$$\frac{F_{s\,cr}}{\eta_s} = \frac{(5.35)\,\pi^2\,10*10^6}{12\left(1-0.33^2\right)}\left(\frac{0.080}{1.36}\right)^2 = 170{,}862\ psi$$

For
k_s = 5.35
b = 1.36 in
a/b = 12.4
t = 0.080 in
μ = 0.33
E = 10.0*10⁶ psi

14.6.10.2 Panel with holes

For this analysis, the hole is assumed non-reinforced and perfectly centered in a square panel. First the ratio D/b is calculated, that is 0.625 /1.36 = 0.46. Then the knock down ratio k/k_o can be approximated using the following equation.

$$k/k_o = 0.96 - \frac{D}{b} = 0.96 - 0.46 = 0.50$$

So using the shear buckling coefficient obtained previously (k_o = *5.35*), the value for k having a hole is thus 0.50 * 5.35 = 2.67.

For
D = 0.625
a = 16.8 in
b = 1.36 in
t = 0.080 in

Hence recalculating the shear buckling allowable we get:

$$\frac{F_{s\,cr}}{\eta_s} = \frac{(2.67)\,\pi^2\,10.0*10^6}{12\left(1-0.33^2\right)}\left(\frac{0.080}{1.36}\right)^2 = 85{,}271\ psi$$

As expected, this latter shear buckling allowable is much lower than the one calculated without the hole. Since the above value is above the shear strength of the material "F_{su}", the plastic reduction factor in shear "η_S" has to be evaluated as follows.

Therefore, using the same procedure as described in section ***11.2.7***, the critical shear buckling stress is calculated by iteration by first assuming a value for $F_{S\,cr}'$. A plasticity correction factor for shear "η_S" is used to multiply the Young's modulus "***E***" and is calculated as follows:

$$\eta_S = \frac{1}{1+\left(\frac{0.00346\,G}{F_{sy}}\right)\left(\frac{F_{scr}}{F_{sy}}\right)^{n-1}}$$

For
F_{sy} = 40,150 psi
n = 20
G = 3.76*10⁶ psi

Where

$$F_{sy} = 0.55\,F_{cy} = (0.55)73{,}000 = 40{,}150\ psi$$

This equation is solved by iteration until $F_{s\,cr}'$ is equal to the value obtained from the equation:

$$F_{S\,cr} = \frac{k_S\,\eta_S\,\pi^2\,E}{12\left(1-\mu^2\right)}\left(\frac{t}{b}\right)^2$$

A first value of 20,000 *psi* was assumed for $F_{scr'}$ in the above equation, and from the iterative process, the value for "η_s"of 0.43286 was obtained. The final initial shear buckling allowable is thus;

$$F_{S\,cr} = \frac{2.67 \cdot (0.43286)\,\pi^2\; 10\times10^6}{12\left(1-0.33^2\right)}\left(\frac{0.080}{1.36}\right)^2 = 36{,}910\;psi$$

For
k_s = 2.67
*E = 10.0*10⁶ psi*
t = 0.080 in
b = 1.36 in.
η = 0.43286

Margin of safety

Thus the MS in initial shear buckling is calculated as:

$$MS = \frac{F_{s\,cr}}{f_s} - 1 = \frac{36{,}910}{4{,}274} - 1 = \;>1.0$$

Seat track rail Section A-A Ultimate	***> 1.0***
9g Fwd	*Shear buckling*
Al Aly 7075-T6 Ext.	

The *MS* calculated above is not a real Margin of Safety since, even if the web were to buckle, the Seat track rail would still be able to carry its load. The consequence of a buckled web would be that the web would get into the post-buckling phase and develop some diagonal tension. That is when the web acts like a bunch of cables in tension at an angle of more or less 45 degrees.

In that case, additional compression loads would be developed in the upper and lower caps of the beam that would have to be added to the already existing loads from the axial and bending loads.

This occurs mainly to very thin webs like in some locations of the fuselage skin where the skin is the web and, the Frames outer caps and Stringers frame the web. Refer to (ref. 19.) for details on diagonal tension analysis.

14.6.11 CONNECTIONS

HST11-5 Pins & HST79-5 Collars

The Seat track rails are fastened to the Crossbeams using ***100 deg.*** flush Titanium Hi-Lites 5/32". There are *qty(2)* fasteners at each end, otherwise, *qty(4)* fasteners are used at each Crossbeam connection. Since the Crossbeams cannot take any load in the fore-aft direction "*X*", the fasteners therefore can only transfer vertical up (tension) loads. However, that is true wherever there is no shear clip as the latter would be taking the vertical loads. Otherwise, all the fasteners transfer the lateral loads "*Y*" into the Crossbeams in shear.

Figure 14-47 shows the location of the fasteners connecting the Seat track rail to the Crossbeams. Only two positions are shown (*FS 498.5 & FS 516.5*) as the same arrangement is used at the other Crossbeams locations (*FS 534.5 & FS 547.5*).

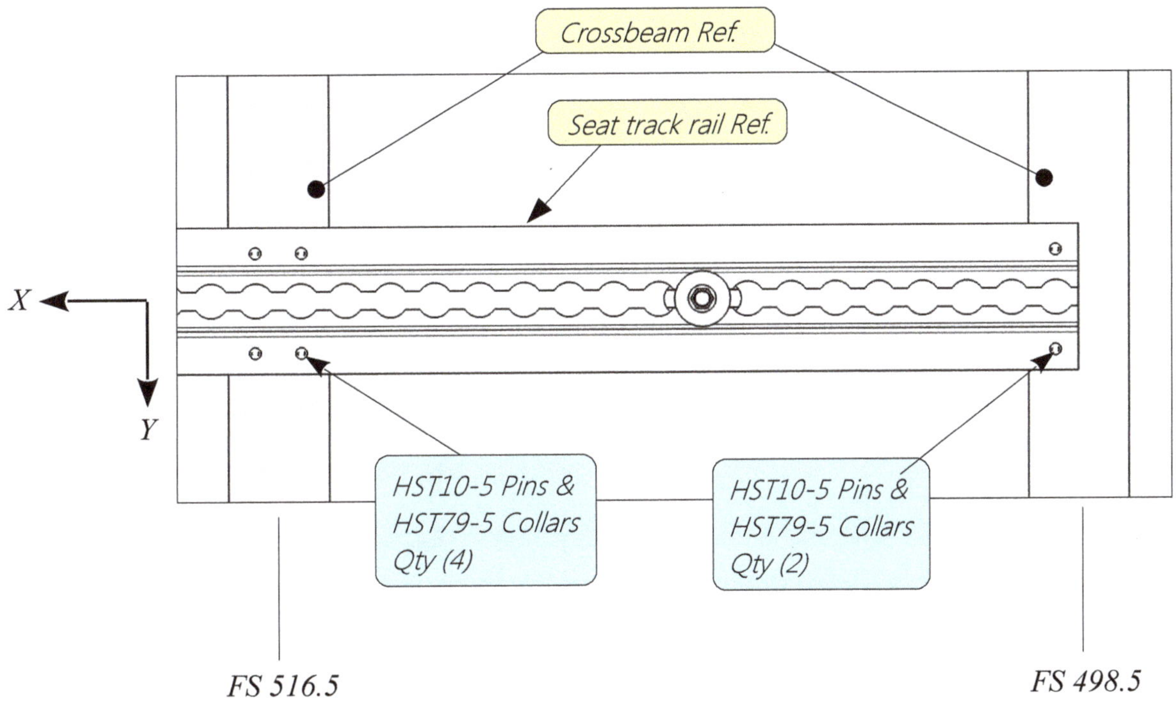

Figure 14-47 *– Locations of the fasteners connecting the Seat track rail to the Crossbeams.*

14.6.11.1 Connection at FS 498.5

For conservatism, the contribution of the Floor panels will be neglected and the total longitudinal and lateral loads in the Seat track rail will be assumed taken by the two end fasteners. The Seat track fitting is assumed located right at the Crossbeam to maximize the loads transferred through that joint.

See *Figure 14-48* for a detailed view of the connection at the Fwd end of the Seat track rail *FS 498.5*. This analysis also covers the connection at *FS 547.5*.

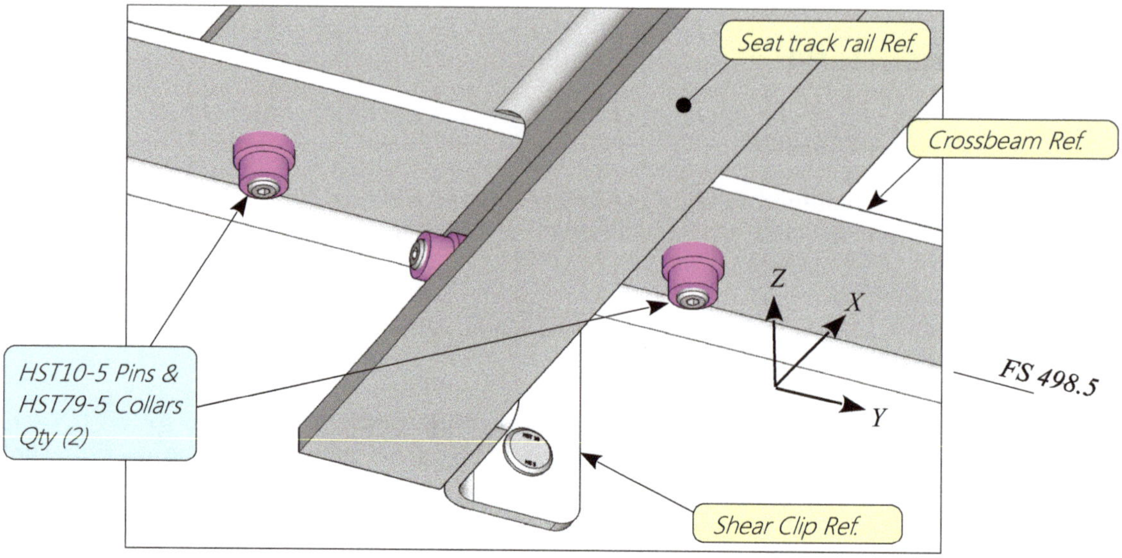

Figure 14-48 *– Detail view of the Seat track rail – Crossbeam connection at FS 498.5.*

9g FWD load case

Point 3:

From *Table 14-14*, the maximum longitudinal load *"F_X"* is 1,367 *lbs*, from the *9g Fwd* LC, occurs at point *3* while the corresponding vertical load *"F_Z"* at that same point from *Table 14-6* is 207 *lbs* acting down (in compression). Therefore, the fasteners only see shear in this case.

For the *9g Fwd* ultimate load case, I consider that the load is evenly distributed among the two *Hi-Lites*, so that each fastener sees 1,367 / 2 = 684 *lbs* of shear load.

With an allowable in single shear *"P_{SU}"* for the *HST11-5* Pins of 2,005 *lbs* from *Table 14-26*, and using a 1.15 *FF*, the *MS* for the *9g Fwd* LC is thus:

$$MS = \frac{P_{su}}{FF \times F} - 1 = \frac{2{,}005}{1.15 \times 684} - 1 = > 1.0$$

Point 1:

From *Table 14-14*, the maximum vertical load *"F_Z"* is 852 *lbs*, from the *9g Fwd* LC, occurs at *point 1* while the corresponding vertical load "*F_X*" at that same point from *Table 14-6* is 497 *lbs* acting forward and the corresponding lateral load "*F_Y*" at that same point from *Table 14-6* is 178 *lbs*. Therefore, the fasteners see a combination of shear and tension in this case.

The loads are assumed distributed evenly among the two fasteners. So each fastener sees the following forces:

- F_X = 497 / 2 = 249 *lbs*
- F_Y = 178 / 2 = 89 *lbs*
- F_Z = 852 / 2 = 426 *lbs*

Pin (HST11-5) fastener shear

Combining the shear force vectors, the resulting fastener shear load *"F"* is calculated as:

$$F = \sqrt{F_{Xi}^2 + F_{Yi}^2} = \sqrt{249^2 + 89^2} = 265\ lbs$$

With an allowable in single shear *"P_{SU}"* for the ***HST11-5*** Pins of 2,005 *lbs* from ***Table 14-26***, the shear load ratio *"R_S"* for the ***9g Fwd*** LC is thus:

$$R_S = \frac{F}{P_{su}} = \frac{265}{2{,}005} = 0.132$$

Fastener tension

The tension force in the fasteners is 426 *lbs*. With an allowable in tension *"P_{TU}"* for the ***HST11-5*** Pins of 1,650 *lbs* from ***Table 14-26***, the tension load ratio *"R_T"* for the ***9g Fwd*** LC is thus:

$$R_T = \frac{F_Z}{P_{tu}} = \frac{426}{1{,}650} = 0.258$$

The interaction equation for bolts subjected to shear and tension loads is:

$$R_s^3 + R_t^2 = 1$$

See another example of that method regarding fasteners in (ref. 27.).

Where the load ratios are:

$$R_s = 0.132 \qquad R_t = 0.258$$

Then from the interaction equation above and using a Fitting Factor of 1.15, the Margin of Safety for the ***9g Fwd*** LC for the ***HST11-5*** Pin is thus calculated as follows:

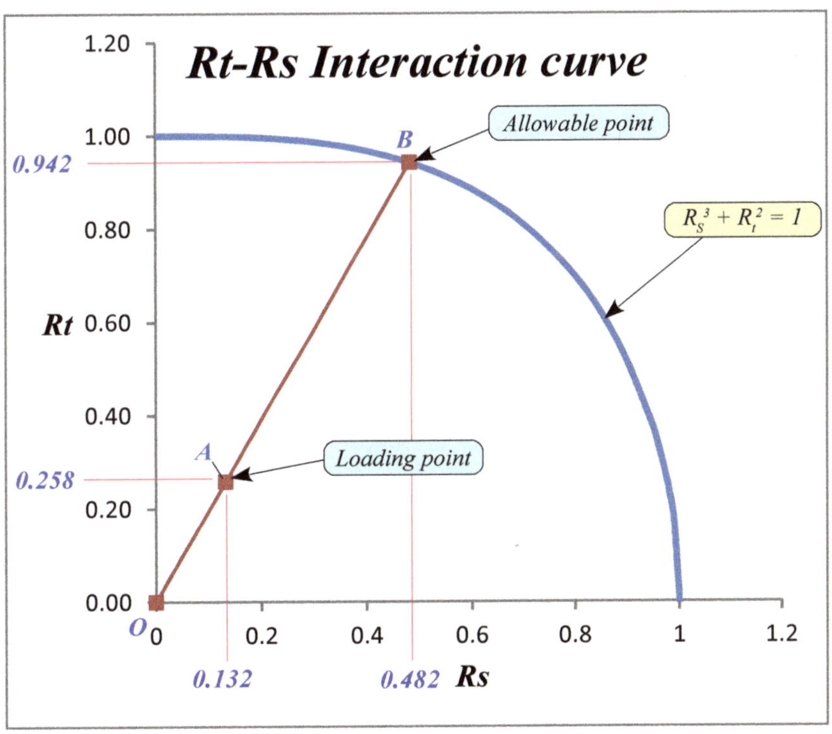

where

$$MS = \frac{\overline{OB}}{FF\,\overline{OA}} - 1 = \frac{R_{S\,All}}{FF\,R_S} - 1 = \frac{0.482}{1.15*0.132} - 1 = >1.0$$

or

$$MS = \frac{R_{t\,All}}{FF\,R_t} - 1 = \frac{0.942}{1.15*0.258} - 1 = >1.0$$

Collar (HST79-5)

Then, the tension force of 426 *lbs* is compared to the tension allowable of the Collar. From *Table 14-26*, that is 1,400 *lbs*. The Margin of Safety for the *HST79-5* Collar for the *9g Fwd* LC is thus calculated as follows using a Fitting Factor of 1.15:

$$MS = \frac{P_{tu}}{FF\,P_t} - 1 = \frac{1{,}400}{1.15\times426} - 1 = >1.0$$

3g SIDE load case

From *Table 14-14,* the maximum lateral load "F_Y" is 1,125 *lbs,* from the 3g *Left* LC, occurs at *point 1* while the corresponding longitudinal load "F_X" and vertical load "F_Z" at that same point from *Table 14-10* are respectively 228 *lbs* acting forward and 496 *lbs* acting up (in tension). Therefore, the fasteners see a combination of shear and tension in this case.

The loads are assumed distributed evenly among the two fasteners. So each fastener sees the following forces:

- $F_X = 228 / 2 = 114$ *lbs*
- $F_Y = 1{,}125 / 2 = 563$ *lbs*
- $F_Z = 496 / 2 = 248$ *lbs*

Pin (HST11-5)

Fastener shear:

Combining the shear force vectors, the resulting fastener shear load "F" is calculated as:

$$F = \sqrt{F_{X\,i}^2 + F_{Y\,i}^2} = \sqrt{114^2 + 563^2} = 575\ lbs$$

With an allowable in single shear "P_{SU}" for the *HST11-5* Pins of 2,005 *lbs* from *Table 14-26*, the shear load ratio "R_S" for the 3g *Left/Right* LC is thus:

$$R_S = \frac{F}{P_{su}} = \frac{575}{2{,}005} = 0.287$$

Fastener tension

The tension force in the fasteners is 248 *lbs*. With an allowable in tension "P_{TU}" for the *HST11-5* Pins of 1,650 *lbs* from *Table 14-26*, the tension load ratio "R_T" for the 3g *Left/Right* LC is thus:

$$R_T = \frac{F_Z}{P_{tu}} = \frac{248}{1{,}650} = 0.150$$

The interaction equation for bolts subjected to shear and tension loads is:

$$R_s^3 + R_t^2 = 1$$

See another example of that method regarding fasteners in (ref. 27.).

Where the load ratios are:

$$R_s = 0.287 \qquad R_t = 0.150$$

Then from the interaction equation above and using a Fitting Factor of 1.15, the Margin of Safety for the *3g Side* LC for the *HST11-5* Pin is thus calculated as follows:

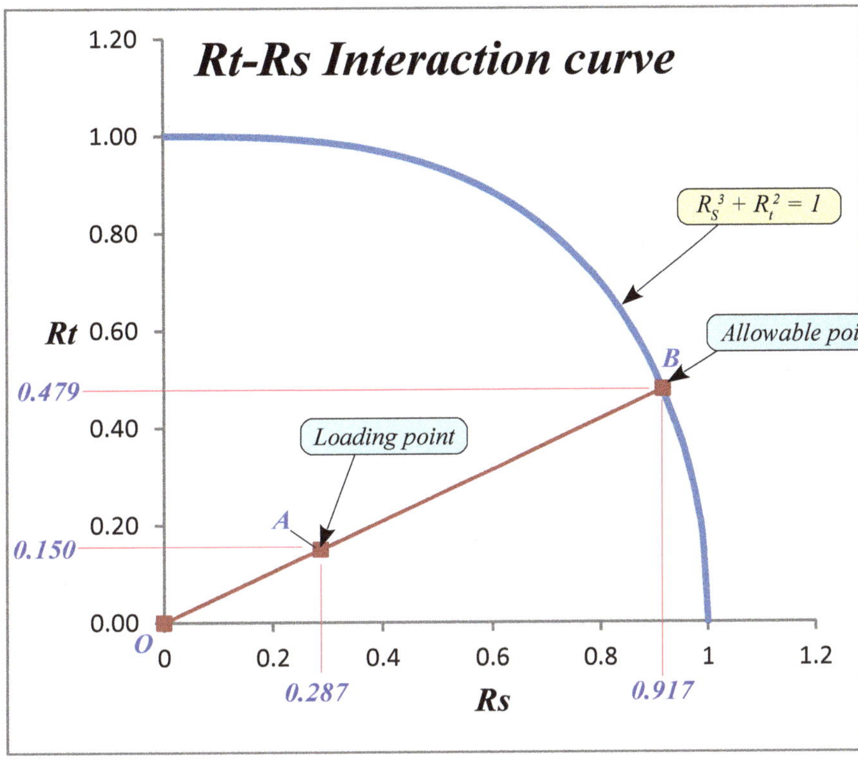

where

$$MS = \frac{\overline{OB}}{FF\,\overline{OA}} - 1 = \frac{R_{S\,All}}{FF\,R_S} - 1 = \frac{0.917}{1.15 * 0.287} - 1 = > 1.0$$

or

$$MS = \frac{R_{t\,All}}{FF\,R_t} - 1 = \frac{0.479}{1.15 * 0.150} - 1 = > 1.0$$

Collar (HST79-5)

Then, the tension force of 248 *lbs* is compared to the tension allowable of the Collar. From *Table 14-26*, that is 1,400 *lbs*.

The Margin of Safety for the *HST79-5* Collar for the *3g Side* LC is thus calculated as follows using a Fitting Factor of 1.15:

$$MS = \frac{P_{tu}}{FF\,P_t} - 1 = \frac{1{,}400}{1.15 \times 248} - 1 = > 1.0$$

6g DOWN / 3.15g UP load cases

From *Table 14-12*, the *6g Down* LC only provides "compression" in the fasteners. Thus the load is transferred to the Crossbeams by simple contact hence, the fasteners see no load. For the *3.15g Up* LC, as shown in *Table 14-13*, the maximum up load that put the fasteners into tension is 313 *lbs* at points *1* & *3*.

Since this tension load is lower than the 852 *lbs* tension load from the *9g Fwd* LC, the *3.15g Up* LC is enveloped by the former so no *MS* calculation is necessary here.

Margin of Safety for the Seat track rail connections

From the above calculations, the most critical load case is *9g Fwd* LC since it provides the lowest *MS*, however, the calculations for all of the load cases show *MS's* greater than 1.0.

Seat track rail connection Ultimate	***> 1.0***
9g Fwd	*Shear*
HST11-5 Pin	

14.6.12 TENSION CLIP

The Seat track rail is fastened to the Crossbeams with Hi-Lites. For the *9g Fwd* LC, a tension load of 852 *lbs* exists as the reacting force at point *1*. This tension force pulls onto the Crossbeam flanges. This causes the flanges to work as tension clips. Since the Seat track rail free flanges are thinner than the Crossbeam flanges, a tension clip analysis is performed on the Seat track rail side thus passing the Crossbeam flanges by comparison. See *Figure 14-49* for details.

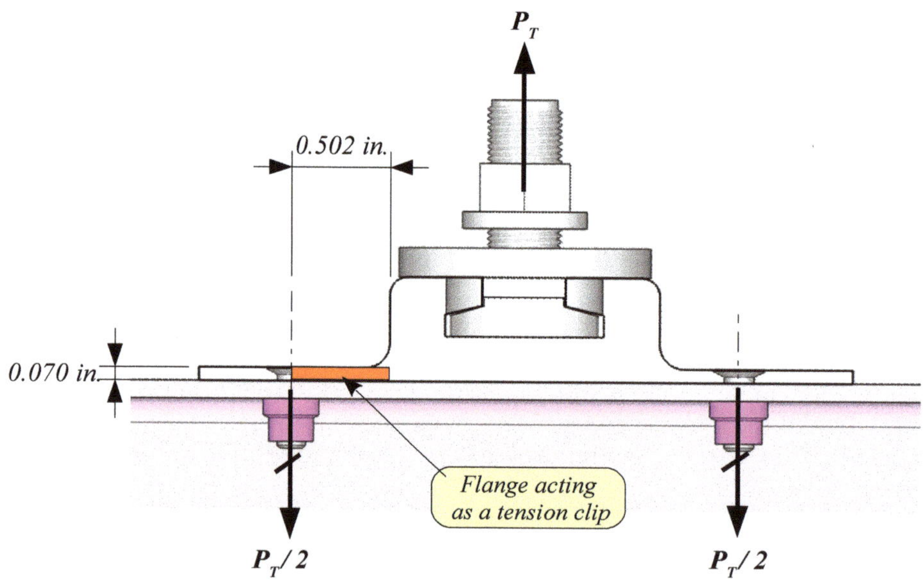

Figure 14-49 *– Tension load applied out of plane at the Seat track – Crossbeam connection.*

For the tension clip analysis, the Seat track rail flange is idealized as shown in *Figure 14-50*. The flange is assumed fixed at the thick end and guided at the center of the Hi-Lite fastener. Because the Hi-Lite is a bolt with a preload, it is fair to assume that it is clamped to the Crossbeam flange and therefore provide a local moment reaction. An effective flange width of 1.0 *inch* is considered.

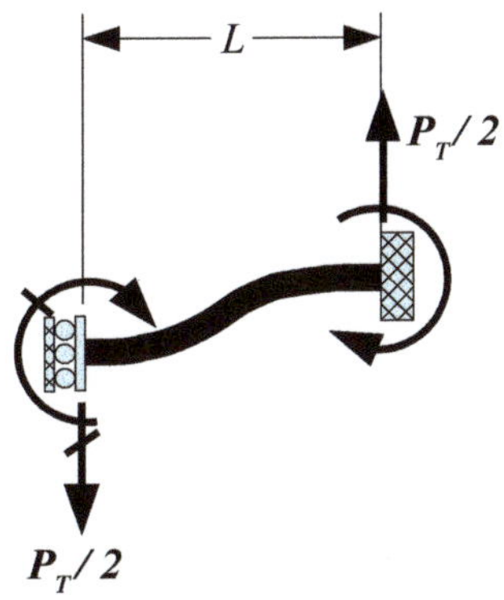

Figure 14-50 *– Idealized Seat track rail flange in out-of-plane bending.*

For a beam with the *Left end guided* and *right end fixed*, as shown in *Figure 14-50,* the maximum bending moment for $a = 0$ (load applied at left end) is:

$$M_{Max} = \frac{Wl}{2}$$

For
$W = P_T/2 = 852/2 = 426$ lbs.
$l = 0.502$ in.

then

$$M_{Max} = \frac{426 \times 0.502}{2} = 107\,in.-lb$$

Bending stress

For rectangular cross-section, the bending stress equation is simplified as follows:

$$f_b = \frac{6M}{bt^2} = \frac{6 \times 107}{1.0 \times 0.070^2} = 131{,}020\ psi$$

For
$M = 107$ in-lb
$b = 1.0$ in.
$t = 0.070$ in.

The above calculated bending stress is higher than the tensile strength of the material ($F_{TU} = 80\ ksi$) and one could say that it is not adequate. However, in the case of ultimate load cases, emergency landing cases in our case, it is common to use *plastic bending* to calculate the MS. At ultimate loads, it is understood that the structure WILL yield but it shall not fail. To determine the ultimate plastic bending moment capability of the section, the *Cozzone* procedure is used.

The fictitious plastic bending stress allowable "F_{pb}" is calculated as follows:

$$F_{pb} = f_m + f_o(k-1)$$

Where $f_m = F_{TU}$, $f_O = 74{,}734\ psi$ and $k = 1.5$ for rectangular sections.

Note that the value for fo was determined using the *Cozzone* methodology with a maximum fictitious ultimate elongation "ε" for that material at 10% *(for F_{TU})*. See also (ref. 2. *Figures C3.9* to *C3.22)* for fo stress-strain curves for the most common aircraft materials.

Hence, the plastic bending stress allowable "F_{pb}" is:

$$F_{pb} = 80{,}000 + 74{,}734(1.5-1) = 117{,}367\ psi$$

One can notice that the plastic bending stress allowable is lower than the bending stress. That means that the section definitely fails in bending.

However, very conservative assumptions were made to determine the applied tension load at the tension clip. For instance, the 852 *lbs* tension load "P_T" is considered applied at the end of the Seat track rail where a shear clip exists. The shear clip would normally take most of the vertical load.

Otherwise, P_T would be applied at the other Crossbeam locations where *qty (4)* Hi-Lites would share the load instead of two.

In the case where "P_T" is applied at any of the middle Crossbeams, the applied load "W" is distributed to the four fasteners according to their distance from the Seat track Fitting. See *Figure 14-51* for dimensions.

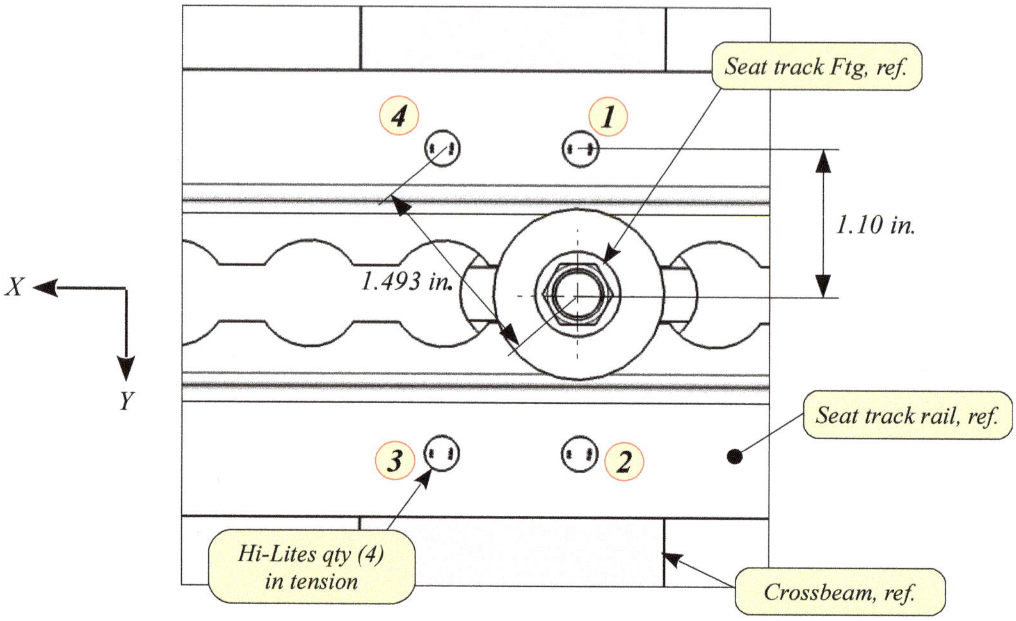

Figure 14-51 *– Seat track rail connection arrangement with qty(4) Hi-Lites.*

Since the deflection of a *guided-fixed* beam is according to its length cube, then by assuming that all fasteners have the same deflection, the load ratio between fasteners ***1&2*** "W_{12}" and ***3&4*** "W_{34}"is

$$\delta = \frac{W_{12} L_{12}^3}{12\,EI} = \frac{W_{34} L_{34}^3}{12\,EI}$$

calculated as follows:

rearranging:

$$\frac{W_{12}}{W_{34}} = \frac{L_{34}^3}{L_{12}^3} = \frac{1.493^3}{1.10^3} = 2.50$$

and since:

$$W_{12} + W_{34} = P_T = 852\ lbs$$

then:

$$W_{12} = \frac{2.50}{3.50} P_T = 0.714\, P_T = 0.714 \times 852\, lbs = 609\, lbs$$

Therefore, fasteners *1* and *2* would see a tension load of 609 / 2 = 305 *lbs*.

My first guess was that, since they are closer to the Fitting, fasteners *1 & 2* were going to take about 2/3 of the Fitting load. As you'll see later, I wouldn't have been too far off but the above calculation provides a much better rational than a simply thrown out number.

When the Seat track fitting is located at the end of the Seat track rail, as can be seen in *Figure 14-52*, and that the Shear clip connecting the Seat track rail to the Crossbeam is considered, two different load paths exist. As shown in *Figure 14-53*, the total applied vertical load is shared between the Seat Track rail and the Shear clip according to their respective stiffness.

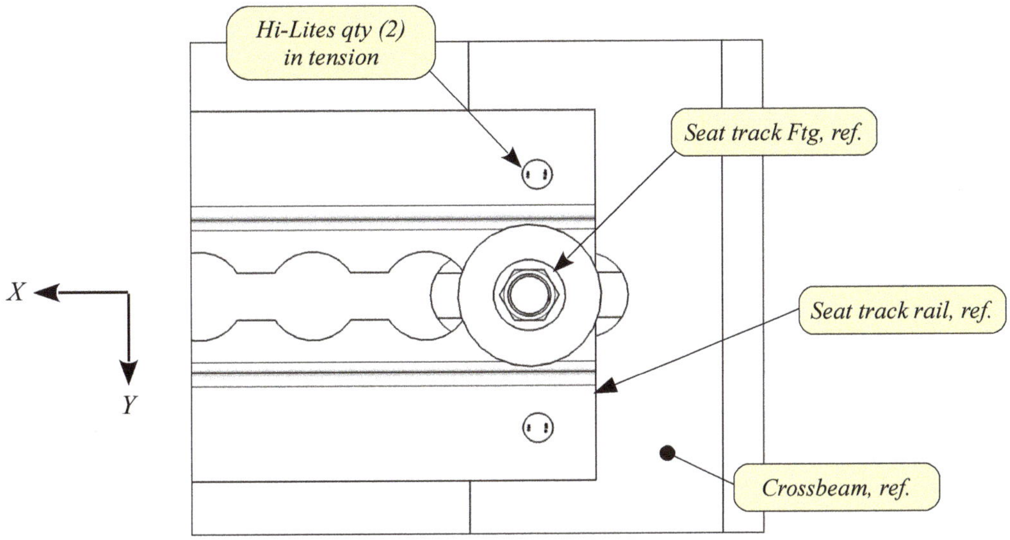

Figure 14-52 *– Seat track rail connection arrangement with qty(2) Hi-Lites and a Shear clip.*

The load path through the Seat track rail flanges can be seen in *Figure 14-49* and *Figure 14-50*, while the idealized load path through the Shear clip is shown in *Figure 14-54*. The stiffness of each of the two load paths is calculated in order to obtain the share of the vertical load that each load path carries.

1) Flange load path:

The flange load path is the shortest one but, since the flanges are only 0.070" thick, I expect it to be softer than the Shear clip load path. We'll see! Using the same assumptions as in *Figure 14-50* where the flange is considered as a *guided-fixed* beam, the deflection of the beam as:

$$\delta_1 = \frac{1/2\,W_1 L_1^3}{12\,EI_1}$$

Because there are two flanges, each flange will see half the load, so a ½ coefficient is added in front of W_1. Thus, the equation for the deflection of load path *1* is:

$$\delta_1 = \frac{1/2\,W_1(0.502)^3}{12\cdot 10\times10^6\cdot 2.8583\times10^{-5}} = 0.000018441\cdot W_1\,in.$$

For
$L_1 = 0.502\ in.$
$I_1 = 2.8583x\ 10^{-5}\ in^4$
$E = 10x10^6\ psi$

where, by considering an effective flange width of 1.0 *in.*, the moment of inertia is:

$$I_1 = \frac{bt^3}{12} = \frac{1.0(0.070)^3}{12} = 0.000028583\,in.^4$$

For
$b = 1.0\ in.$
$t = 0.070\ in.$

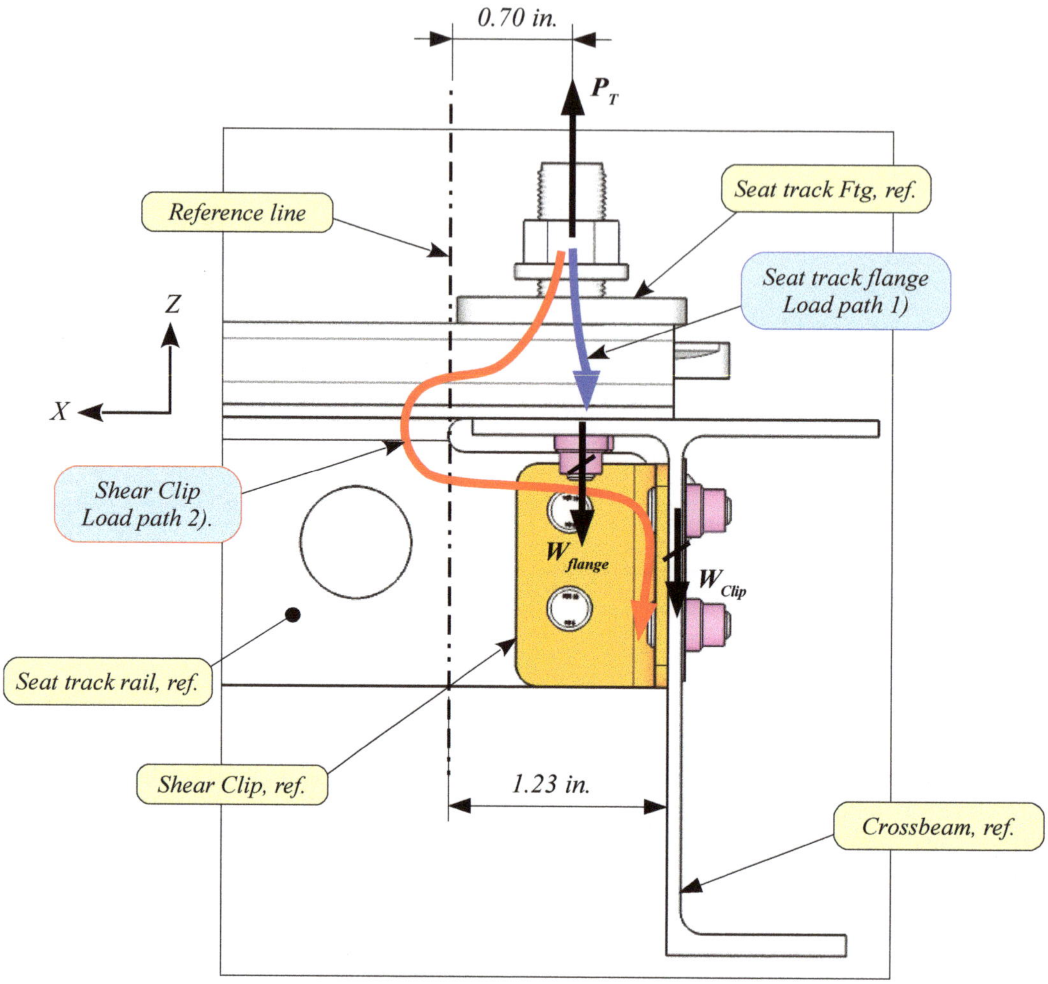

Figure 14-53 *– Two different load paths for the Vertical load.*

<u>2) Shear Clip load path:</u>

The Shear Clip load path is the longest one but the structure is sturdier hence I expect it to be stiffer than the Flange load path. Again, that's my guess at this time but the numbers will tell. As shown in *Figure 14-54*, the load path comprises of two *fixed-free* beams (cantilevered) of different dimensions and cross sections properties. The portion of the Seat track rail web being in tension (hatched rectangle in *Figure 14-54*) is neglected as its deflection would be very small compared to the beams in bending. The deflection equations of the two beams are calculated as follows.

For the top beam "*2a*", the Seat track rail section *B-B* from *Figure 14-20,* the following equation is used:

$$\delta_{2a} = \frac{W_2 L_{2a}^3}{3\,EI_{2a}}$$

hence

$$\delta_{2a} = \frac{W_2 \cdot 0.70^3}{3 \cdot 10 \times 10^6 \cdot 0.0132} = 8.66 \times 10^{-7} \cdot W_2 \; in.$$

For
L_{2a} = 0.70 in.
I_{2a} = 0.0132 in^4
$E = 10x10^6$ psi

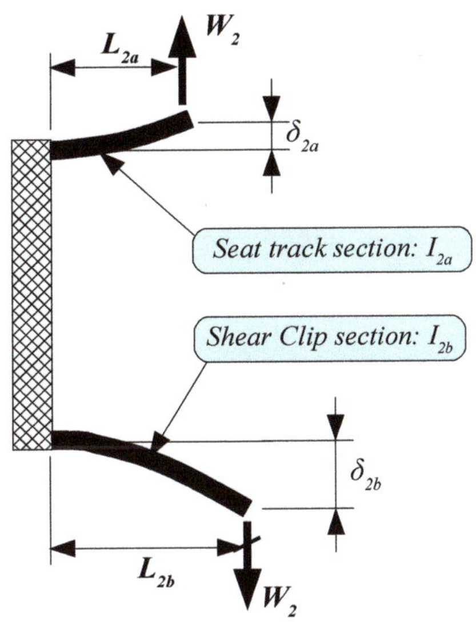

***Figure 14-54** – Idealized load path to the Shear clip.*

For the bottom beam *"2b"*, the section properties are shown in *Figure 14-55* and the deflection equation is:

$$\delta_{2b} = \frac{W_2 L_{2b}^3}{3\,EI_{2b}}$$

For
L_{2b} = 1.23 in.
I_{2b} = 0.032371 in^4
E = $10x10^6$ psi

hence

$$\delta_{2b} = \frac{W_2 \cdot 1.23^3}{3 \cdot 10 \times 10^6 \cdot 0.032371} = 1.916 \times 10^{-6} \cdot W_2 \; in.$$

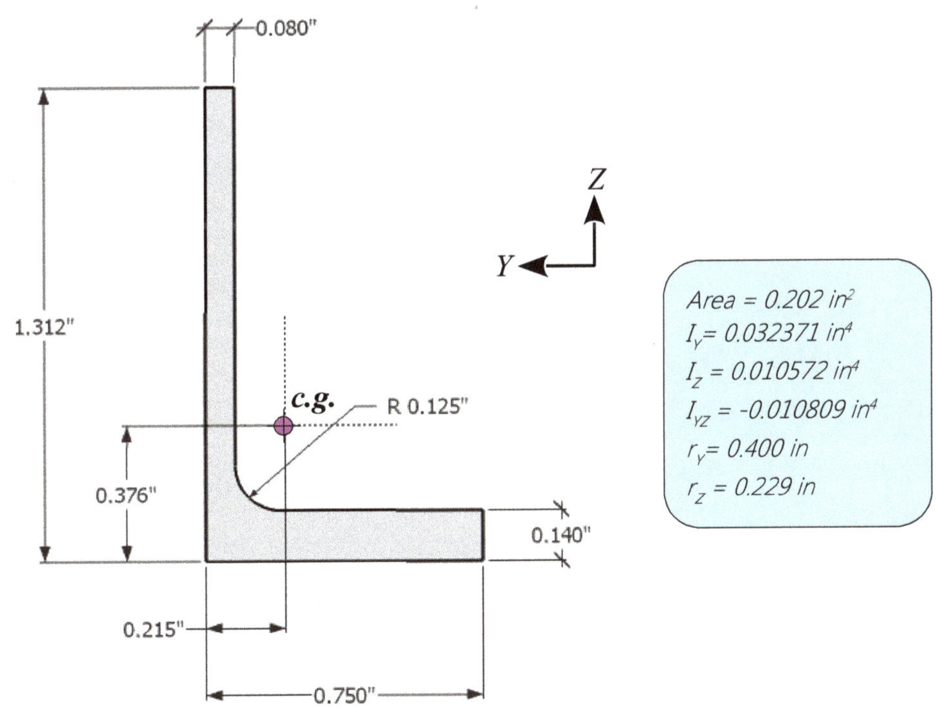

***Figure 14-55** – Seat track rail beam 2b section properties.*

Stiffness calculations

The stiffness "k" of a beam is the load "W" divided by the deflection "δ". Therefore, the stiffness of all the beams involved, using the deflections calculated previously, are:

For beam *1)*:

$$\delta_1 = 0.000018441 \cdot W_1 \ in.$$

hence

$$k_1 = \frac{W_1}{\delta_1} = \frac{W_1}{0.000018441 \cdot W_1} = 54{,}227 \ lb/in.$$

For beam *2a)*:

$$\delta_{2a} = 8.66 \times 10^{-7} \cdot W_2 \ in.$$

hence

$$k_{2a} = \frac{W_2}{\delta_{2a}} = \frac{W_2}{8.66 \times 10^{-7} \cdot W_2} = 1{,}154{,}734 \ lb/in.$$

For beam *2b)*:

$$\delta_{2b} = 1.916 \times 10^{-6} \cdot W_2 \ in.$$

hence

$$k_{2b} = \frac{W_2}{\delta_{2b}} = \frac{W_2}{1.916 \times 10^{-6} \cdot W_2} = 521{,}921 \ lb/in.$$

The system of stiffness can be idealized as the system of springs analogy as shown in *Figure 14-56*. As one can see, springs $\boldsymbol{k_{2a}}$ and $\boldsymbol{k_{2b}}$ are in series and those are in parallel with spring $\boldsymbol{k_1}$. At first, the equivalent spring stiffness for both springs in series is calculated as $\boldsymbol{k_2}$.

$$k_2 = \frac{1}{\dfrac{1}{k_{2a}} + \dfrac{1}{k_{2b}}} = \frac{1}{\dfrac{1}{1{,}154{,}734} + \dfrac{1}{521{,}921}} = 359{,}454 \ lb/in.$$

Since the relative displacement between the Seat track rail and the Crossbeam is the same for both $\boldsymbol{k_1}$ and $\boldsymbol{k_2}$ springs, then, so is their respective displacement. Therefore, the following relation is obtained:

$$\delta_1 = \delta_2 = \frac{W_1}{k_1} = \frac{W_2}{k_2}$$

$$W_1 = W_2 \cdot \frac{k_1}{k_2}$$

And since $W = W_1 + W_2$, thus:

$$W = W_2 \cdot \frac{k_1}{k_2} + W_2 = W_2 \left(\frac{k_1}{k_2} + 1 \right) = 852 \ lbs$$

solving for W_2:

$$W_2 = \frac{W}{\left(\dfrac{k_1}{k_2} + 1 \right)} = \frac{852}{\left(\dfrac{54{,}227}{359{,}454} + 1 \right)} = 740 \ lbs$$

and solving for W_1:

$$W_1 = W - W_2 = 852 - 740 = 112 \ lbs$$

That is of course 112 / 2 = 56 *lbs* of tension force per flush Hi-Lite fastener.

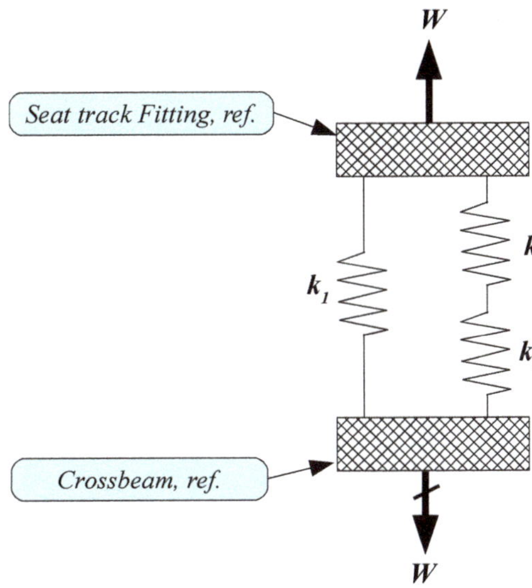

Figure 14-56 – *Load paths 1 & 2 spring system analogy.*

From the above analysis, as expected, the Shear Clip is a much better load path than the free flanges as it takes about 87% of the total applied vertical load. The bending stress analysis of the free flanges will then be re-done using the ***305 lbs*** of tension force from page *354*, at the location where no shear clip exists, instead of the 426 *lbs*!

In the previous analysis I tried to distribute the load in a more realistic way by taking into account the actual stiffness of the structure. This exercise demonstrates one thing. That is; to try to be closer and closer to the reality (truth), the analysis time required increases in an exponential way!

Using the same approach as shown in ***Figure 14-50*** and on page *353*, then the maximum bending moment for $a = 0$ (load applied at left end) is:

$$M_{Max} = \frac{Wl}{2}$$

For
W = 305 lbs.
l = 0.502 in.

then

$$M_{Max} = \frac{305 \times 0.502}{2} = 77\ in.-lb$$

Bending stress

For rectangular cross-sections, the bending stress equation is simplified as follows:

For
M = 77 in-lb
b = 1.0 in.
t = 0.070 in.

$$f_b = \frac{6M}{bt^2} = \frac{6 \times 77}{1.0 \times 0.070^2} = 94{,}286\ psi$$

The above calculated bending stress is still higher than the tensile strength of the material (F_{TU} = 80 *ksi*). Therefore, I will use the *plastic bending* approach to calculate the MS.

From page *353*, the plastic bending stress is: F_{pb} = 117,367 *psi*.

The bending stress ratio "R_b" is thus:

$$R_b = \frac{f_b}{F_{pb}} = \frac{94,286}{117,367} = 0.803$$

Shear stress

With the bending stress, a transverse shear exists that must be taken into account in the MS. With plastic bending, I consider the average shear stress through the beam section. For a cross-section area of: $A_S = b * t = (1.0)\ (0.070) = 0.070\ in^2$ and a vertical shear force of 305 *lbs*, the average shear stress "f_S" is:

$$f_S = \frac{V_Z}{A_S} = \frac{305}{0.07} = 4,357\ psi$$

For $F_{SU} = 40\ ksi$ (ref. *Table 14-15*), the shear stress ratio "R_S" is calculated as:

$$R_S = \frac{f_S}{F_{SU}} = \frac{4,357}{40,000} = 0.109$$

Margin of safety for the tension clip:

For bending stress combined with shear stress, the MS is calculated as follows:

$$MS = \frac{1}{\sqrt{R_b^2 + R_S^2}} - 1$$

hence

$$MS = \frac{1}{\sqrt{0.803^2 + 0.109^2}} - 1 = 0.23$$

However, the accepted general form of presenting the *Margins of Safety* is as follows:

Seat track rail Tension clip Ultimate	***0.23***
9g Fwd *Al Aly 7075-T6 Ext.*	*Plastic bending + Shear*

From the above analysis, it was found that the connection of the Seat track rail to the Crossbeam is more critical where there are *qty(4)* Hi-Lites and no shear clip. After a lap joint connection, Shear clips are always better at transferring shear loads while out-of-plane loads are always the worst and most flexible. Flexible joints lead to premature cracks and are therefore to be avoided as much as possible.

14.7 SEAT TRACK FITTINGS

The Seat track Fittings are often vendor parts and therefore, do not require any analysis other than making sure that the applied loads do not exceed the allowable providing they are applied as they should. In this example, the four *Seat track Fittings* used to attach the monument to the Seat track rails are the same Part Number and are made out of *CRES*. From *Figure 14-17*, they have an ultimate allowable load *"P_{V_all}"* of 6,000 *lbs.* vertically and *"P_{H_all}"* of 5,000 *lbs.* in any other horizontal direction. Hence:

$$P_{TU} = 6{,}000\ lbs$$
$$P_{SU} = 5{,}000\ lbs$$

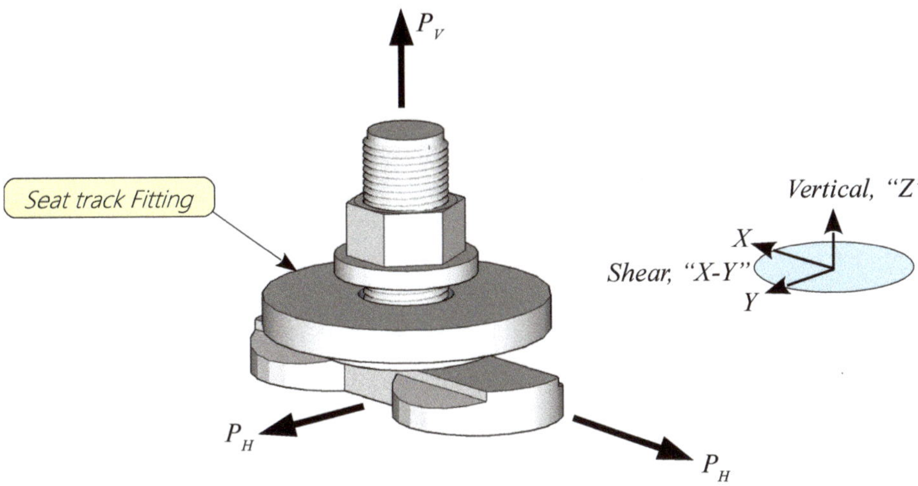

Figure 14-57 *– Seat track Fitting allowable loads.*

From *Table 14-14*, the maximum vertical reaction force "R_Z" (up or down) is 852 *lbs* at point ***1*** while the maximum shear load (in *X* or *Y* direction) is 1,367 *lbs* at point ***3***, both for the ***9g*** *Fwd LC*. Conservatively assuming that the maximum vertical and the maximum shear load occur both at the same point, for the same load case, the Margin of Safety is calculated by combining the load ratios as follows using a Fitting Factor of 1.15 as per (ref. 25. *FAR §25.625*).

$$MS = \frac{1}{FF\sqrt{R_T^2 + R_S^2}} - 1$$

Where R_T and R_S are respectively the tension and shear load ratios calculated as follows:

$$R_T = \frac{P_T}{P_{tu}} = \frac{|852|}{6{,}000} = 0.142$$
$$R_S = \frac{P_S}{P_{su}} = \frac{1{,}367}{5{,}000} = 0.273$$

Hence the *Seat track Ftg* MS is:

$$MS = \frac{1}{1.15\sqrt{0.142^2 + 0.273^2}} - 1 = 1.82$$

Seat Track Fitting Ultimate	***>1.00***
9g Fwd *CRES*	*Tension + Shear*

14.8 SHEAR CLIP

The Shear Clips are made out of *Al Aly 2024-T3 Clad* sheet 0.063 *in*. thick. They are located at *FS 498.5* and *FS 547.5* connecting the Seat track rails to the Crossbeams using *MS20470AD5* solid rivets. Shear Clips are added at those locations because the Seat track rails are discontinuous at those Crossbeams and the shear loads must be introduced adequately.

As shown in *Figure 14-58*, the Shear Clips are located on the positive *"Y"* side for ease of installation and because the location of the Seat track rail shear center *"SC"* is located on that same side, hence reducing the twisting effect.

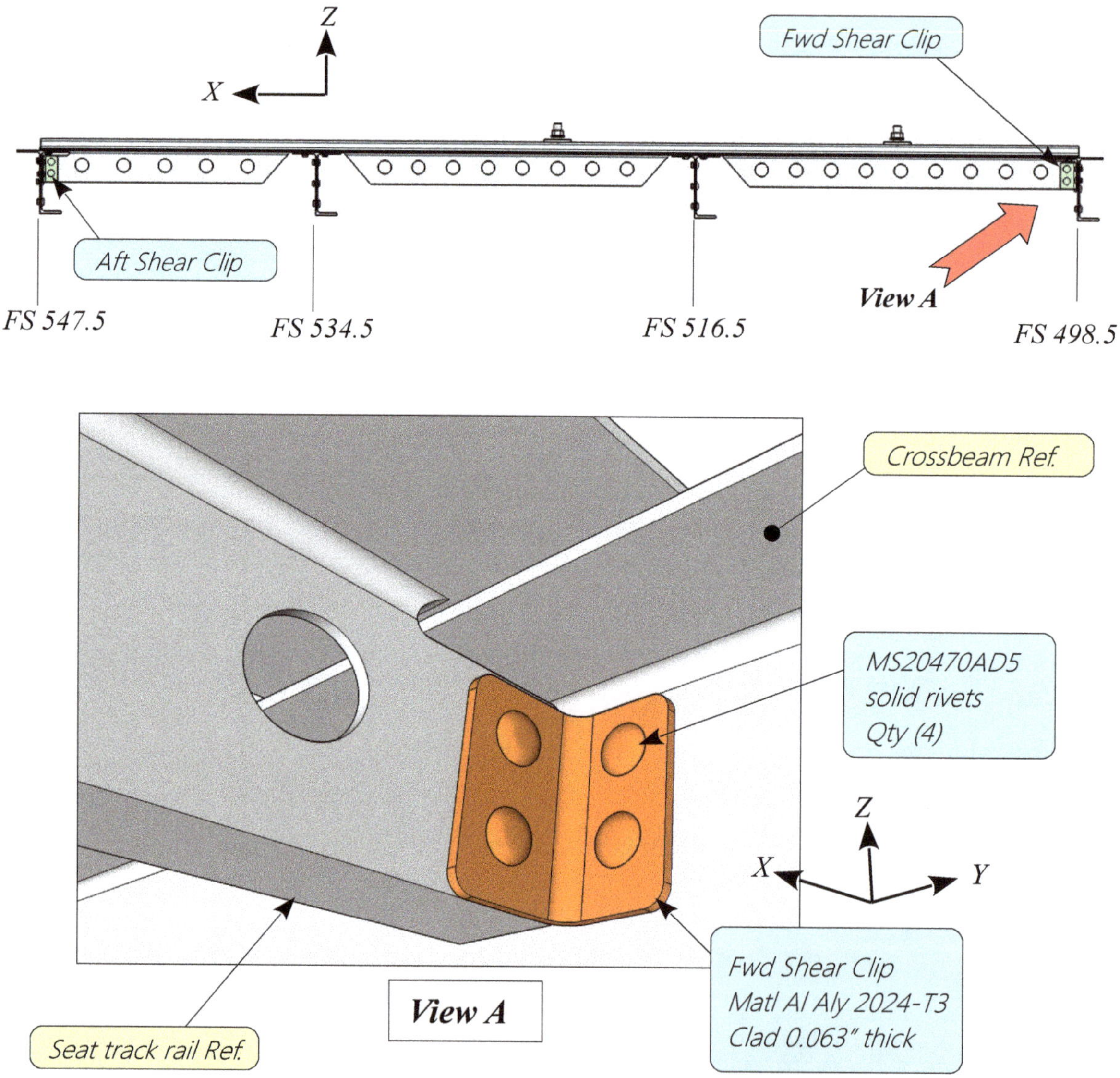

Figure 14-58 – *Shear Clip location and details.*

14.8.1 LOADS

The Free Body Diagram of the Shear Clip is shown in *Figure 14-59*. As in all shear clips, the shear load is considered applied at the faying surface of the flange and the shear clip is considered cantilevered off of the two (or more) rivets.

From *Table 14-14*, the maximum shear load "V_Z" is 852 *lbs.* at point *1* for the *9g Fwd LC*. It is conservatively assumed that the seat track Fitting is located right next to the Crossbeam to maximize the shear load reaction.

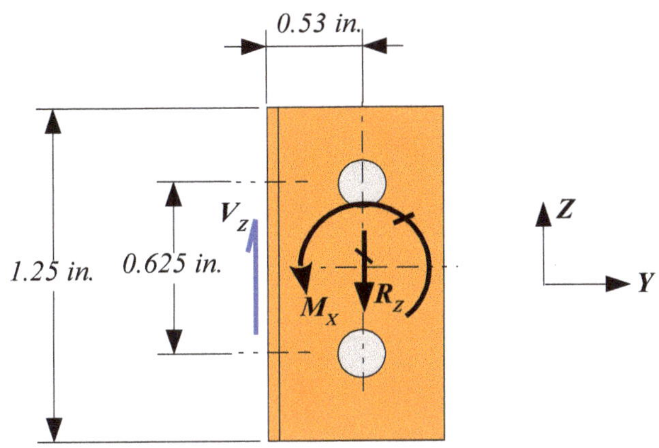

Figure 14-59 *– Shear Clip FBD.*

The shear clip is analyzed at the most critical location. That is through the fasteners where the moment "M_X" = 852 * 0.53 = 452 *in-lb*.

14.8.2 SECTION PROPERTIES

The cross section dimensions and properties across the fasteners are shown in *Figure 14-60*.

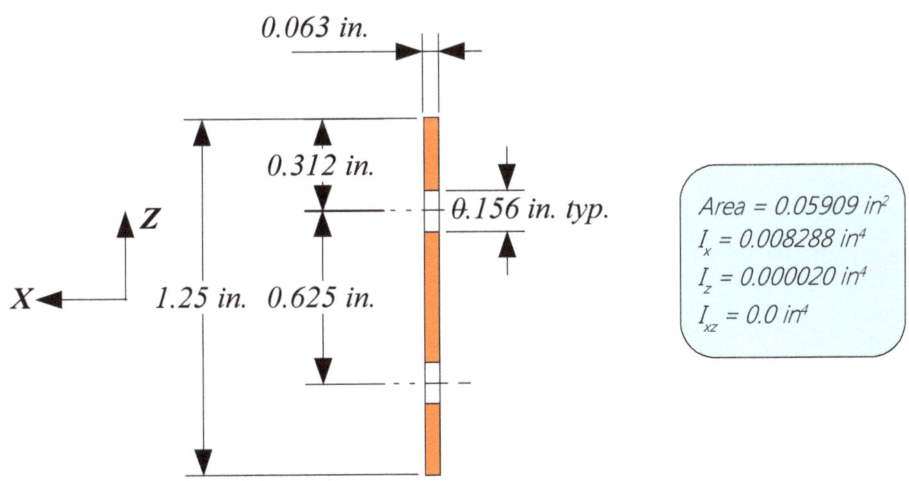

Figure 14-60 *– Shear Clip cross section.*

14.8.3 STRESS ANALYSIS

Bending stress

$$f_b = \frac{M_X\, c}{I_{XX}} = \frac{452\,(0.625)}{0.008288} = 34{,}086\ psi$$

$$f_S = \frac{V_Z}{A} = \frac{852}{0.05909} = 14{,}419\ psi$$

Shear stress

Combined stresses

The bending and shear stresses are combined according to the equation:

$$R_b^2 + R_S^2 = 1$$

Thus the MS is calculated as:

$$MS = \frac{1}{\sqrt{R_b^2 + R_S^2}} - 1$$

Where R_b and R_S are respectively the bending and shear stress ratios calculated as follows:

$$R_b = \frac{f_b}{F_{cy}} = \frac{|34{,}086|}{40{,}000} = 0.852$$

$$R_s = \frac{f_s}{F_{su}} = \frac{14{,}419}{35{,}000} = 0.412$$

F_{cy} is used in this case as the allowable since the maximum stress is in compression. It is a common practice to use F_{cy} whenever an ultimate stress in compression exists in relatively thin parts. Providing the offset between the applied load and the reaction plane would be large, a free flange buckling shall be done to get a more accurate compression allowable.

Hence the MS is:

$$MS = \frac{1}{\sqrt{0.852^2 + 0.412^2}} - 1 = 0.05$$

Shear Clip Ultimate	***0.05***
9g Fwd *Al Aly 2024-T3 Clad*	*Bending + shear*

14.8.4 CONNECTIONS

Shear Clip *MS20470AD5* solid rivets

Two rivets on either flanges of the Shear Clip are used. *Figure 14-61* shows the applied forces at the rivets from the *9g Fwd* LC.

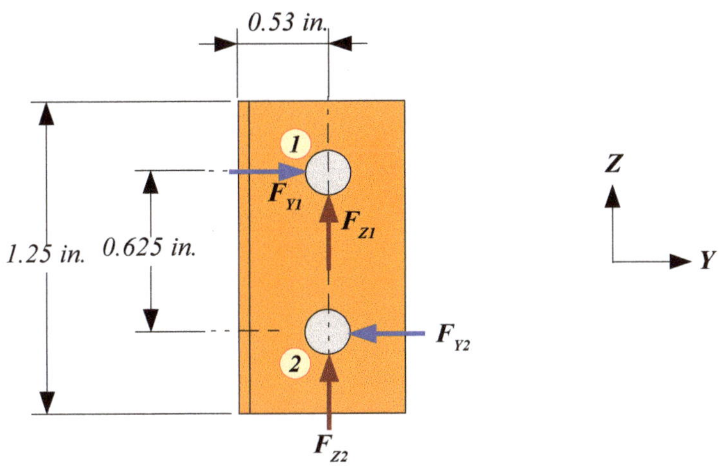

***Figure 14-61** – Applied forces on the Shear Clip rivets.*

Considering the maximum possible vertical shear load from *Table 14-14* of 852 *lbs*, and since $M_X = V_Z * e = 852 * 0.53 = 452$ *lbs*, the forces in the rivets are calculated as follows:

$$F_{Z1} = F_{Z2} = \frac{V_Z}{n} = \frac{852}{2} = 426\ lbs$$

$$F_{Y1} = -F_{Y2} = \frac{M_X}{pitch} = \frac{452}{0.625} = 723\ lbs$$

Combining the vectors, the resulting rivet load *"F"* is calculated as:

$$F = \sqrt{F_{Yi}^2 + F_{Zi}^2} = \sqrt{723^2 + 426^2} = 839\ lbs$$

Rivet shear:

As one can notice, the resulting shear load value *"F"* calculated above for one rivet greatly exceeds the shear allowable *"P_{SU}"* for an *MS20470AD5* of 596 *lbs* from *Table 14-57*. Therefore, another type of fastener shall be used. The rivets are changed for 5/32" Hi-Lites *HST10-5* having a shear allowable of 2,005 *lbs* as shown in *Table 14-26*.

A Fitting Factor *"FF"* of 1.15 (ref. 18 *FAR §25.625*) is used. No shear factor *"SF"* is required since the Crossbeam, the Seat track rail and the Shear clip are relatively thick. (ref. *Table 13-20*). The Margin of Safety for shear is thus calculated as follows:

$$MS = \frac{P_{su}}{FF \times F} - 1 = \frac{2{,}005}{1.15 \times 839} - 1 = 1.07$$

Bearing:

The bearing stress will be checked in all of the three parts involved in the joint. The bearing load allowable is calculated as follows:

$$P_{bru} = F_{bru} \times D \times t$$

Table 14-32 shows the resulting bearing *MS* for the final configuration and *Figure 14-62* shows a detailed view of the new configuration.

Shear clips should always be thinner than the parts they're connecting to so that they are a bit more flexible to avoid causing any undesirable residual stresses due to any slight misalignment at installation.

Table 14-32 *- Bearing stress MS table summary.*

	Material	*t*	*D*	*Fbru*	*Pbru*	*F*	*MS*
Part	*Al Aly*	*[in]*	*[in]*	*[ksi]*	*[lbs]*	*[lbs]*	*bearing*
Shear Clip	*2024-T3 Clad*	0.063	0.163	120	1232	839	0.47
Seat track	*7075-T6*	0.080	0.163	145	1891	839	1.25
Crossbeam	*7050-T7451*	0.080	0.163	105	1369	839	0.63

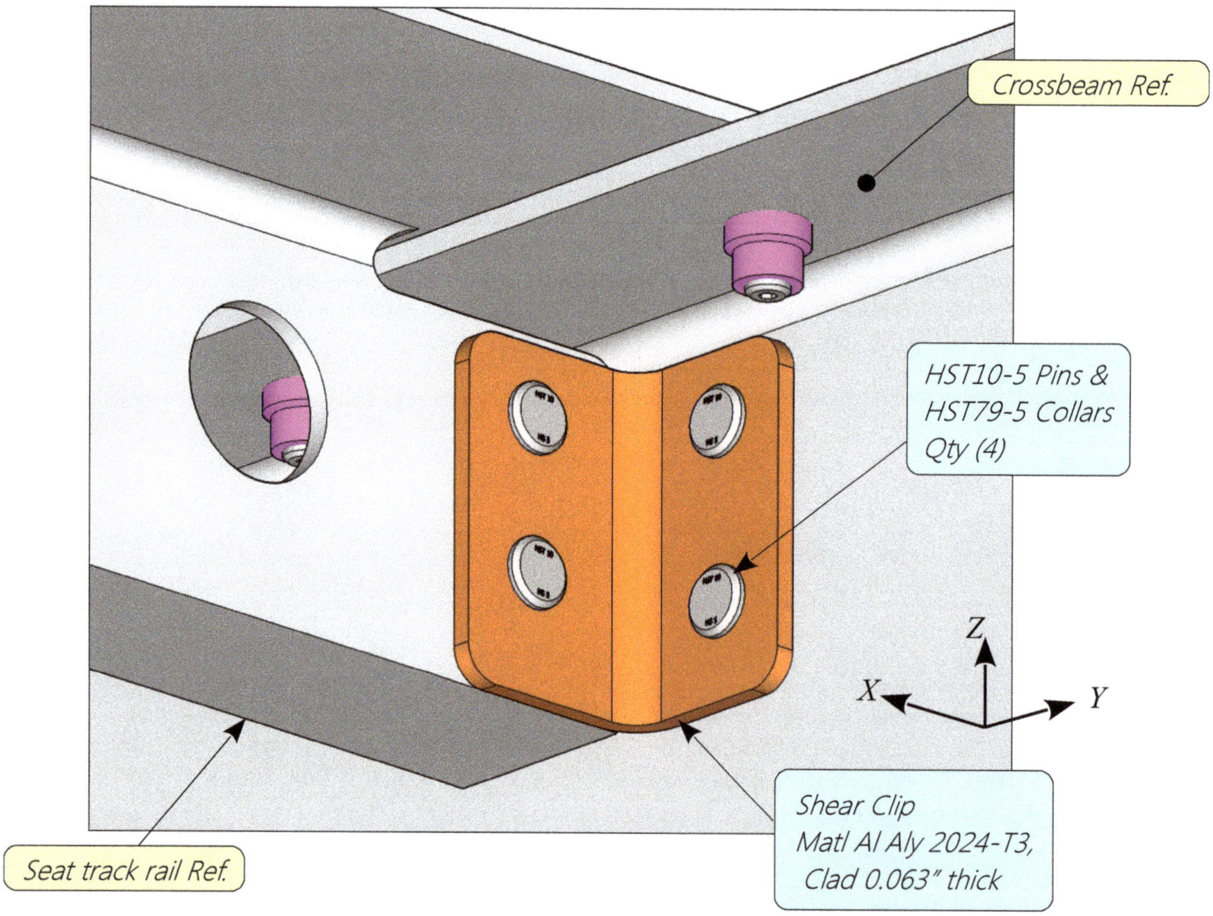

Figure 14-62 *– Detail view of the Shear clip and the new fasteners.*

A calculation example for the lowest bearing stress *MS* is done below:

The most critical part is the Shear Clip. The bearing ultimate allowable load "P_{bru}" in the Shear Clip made out of *Aluminum Alloy 2024-T3 Clad* 0.063 *in.* thick is calculated:

For: $F_{bru} = 120$ *ksi* (ref. *Table **14-19***), $e/D = 2.0$. And for $D = 0.163"$ and $t = 0.063"$ hence:

$$P_{bru} = 120{,}000 \times 0.163 \times 0.063 = 1{,}232 \; lbs$$

Margin of Safety for the Shear Clip connections

$$MS = \frac{P_{bru}}{F} - 1 = \frac{1{,}232}{839} - 1 = 0.47$$

Shear Clip *Ultimate*	***0.47***
9g Fwd *HST10-5 in Al Aly* *2024-T3 Clad*	*Bearing*

Since the *MS* in bearing is lower than the one in shear, the joint is thus bearing critical. For all materials involved, the ratio $F_{bru} \,/\, F_{bry}$ is smaller than 1.5. Therefore, no bearing check at limit (yield) is necessary.

"Don't worry, your ego WILL be scorched."

14.9 FLOOR PANELS

The Floor panels are honeycomb composite bonded panels and have two purposes. The first one is to distribute the floor pressure load to the Seat track rails and the Crossbeams. The second purpose is to transfer the in-plane shear loads between the Seat track rails and the Crossbeams. In that case. all of the longitudinal (fore-aft) loads from the monuments and seats must be routed (by shear) to the fuselage skin where those loads are reacted.

In our current cabinet installation, the Floor panels only react the shear load from the Seat track rails axial load principally for the *9g Fwd LC*.

For the in-plane shear load in a Floor panel, a shear buckling check and a connection check are going to be done. Our cabinet being located near the fuselage skin, chances are that the Floor panels have already some shear loads coming from other monuments or seats located inboard of that cabinet. In that case, the total shear load shall be considered in the analysis.

14.9.1 LOADS

Figure 14-63 shows a top view of the Floor panels involved in the cabinet installation. From *Table 14-14*, the longitudinal loads reacted at points *1 & 3* are: P_1 = 497 *lbs* and P_3 = 1,367 *lbs* respectively for the most critical loading condition (*9g Fwd LC*).

> Normally, the pressure distribution, determined to name a few, by the *6g Down Emergency landing LC*, the ultimate rapid decompression LC and, the ultimate vertical acceleration from flight loads, is plotted on a graph to be used for the analysis. The largest sized panel is checked for the maximum pressure thus passing the smaller panels by comparison. Providing all the panels have the same thickness and strength.
>
> As for the shear load, another graph is drawn that shows the maximum running load (shear flow), usually from the *9g Fwd LC*, that has to be distributed towards the fuselage skin.

For this example calculation, I will assume that the shear flow "q_i" coming from other items located inboard of the cabinet is 1,000 *lbs/ft* or 83 *lbs/in* also from the *9g Fwd LC*. Panel ***A*** is obviously the most critical panel as it is the largest in size and sees the highest load. The total shear flow load "q_A" going through that panel is:

$$q_A = q_i + \frac{P_1}{L} = 83 + \frac{497}{49} = 93\ lbs/in.$$

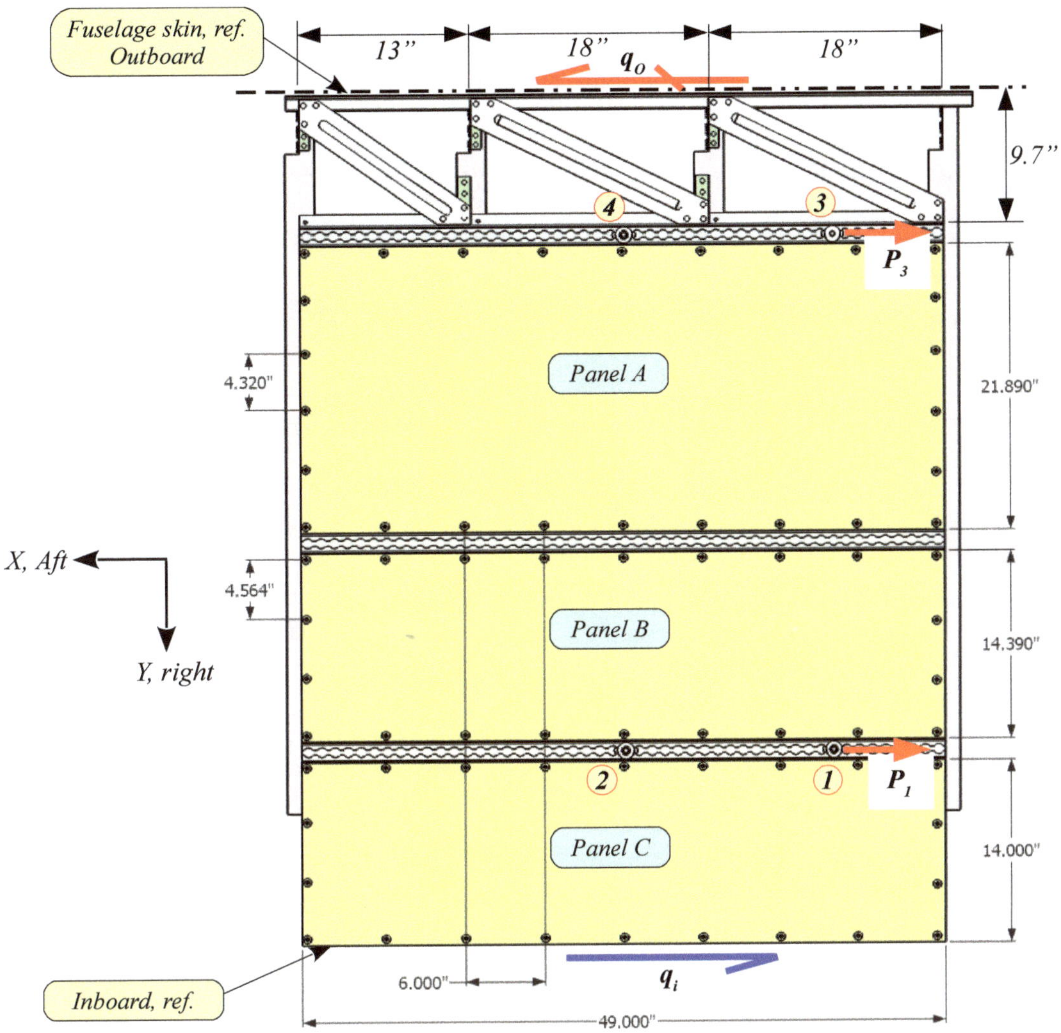

Figure 14-63 *– Top view of Floor panels and applied loads.*

14.9.2 PANEL CONSTRUCTION

The Floor panel is a sandwich bonded panel made out of GR/EP (Graphite-Epoxy) skins (facings) and a 0.50 *in.* thick of aramid fiber paper honeycomb core (similar to *Nomex®* from *Dupont*).

Both top and bottom facings are the same. In addition to the carbon fiber fabric plies, they both have an additional GL/EP (Glass-Epoxy) ply to protect the GR/EP from wear on the top and, to protect the contacting aluminum structure from corrosion on the bottom. Each face is composed of 1 ply of *0°* and 1 ply of *+45°* for a total thickness of 2 * 0.0077" = 0.0154 *in.* An adhesive ply of Epoxy is used to bond the faces to the core.

Each face has thus the following ply layout: [(0)/(45)]. The Glass/Epoxy plies having a lower Young's modulus are conservatively ignored in the analysis.

Whenever carbon (graphite) parts are in contact with metal, especially Aluminum and magnesium, it is very important to provide some sort of galvanic corrosion protection at the faying surface like glass fiber or some plastic/paint to prevent the metal from corroding.

Bonded panels are more complex to analyze than a simple aluminum sheet but we have all the tools necessary to substantiate them. In our case, the panel is loaded with pure shear but bonded panels can also be loaded in bending and transverse shear (due to a load normal to the surface) and, in edgewise compression. Of course, analysis tools made in a spreadsheet for example are quite handy and can save a lot of time while minimizing errors.

There are basically four checks to perform when analyzing bonded panels:

- General instability. That occurs when the whole panel buckles due to shear, compression or a combination of the two.
- Intracell buckling. That is when the faces (skins) buckle locally over a core cell length.
- Wrinkling. That is when the faces, considered as beams on an elastic foundation, buckle.
- Shear Crimping. This is a kind of general instability when the buckle wave length is small due to a soft core shear modulus.

All of the above checks have to be done while the lowest MS is used as the critical mode of failure.

14.9.3 GENERAL SHEAR BUCKLING

The largest panel "*A*" is assumed simply-supported on all sides and no longitudinal or lateral loads exist. The dimensions are a = 49 *in.* and b = 21.89 *in.* thus a/b = 49/21.89 = 2.24. The core ribbon direction is assumed along the long "a" direction as it is more conservative than in the short "b" direction.

The critical panel "*A*" is first analyzed by hand, and secondly, using *FEA* as a double check as will be demonstrated later. The equivalent quasi-isotropic material properties are taken from page *307*.

The general shear buckling allowable is calculated using (ref. 2.) as follows:

$$\frac{F_{SCR}}{\eta} = \frac{K_S}{S} \cdot \frac{U}{2t_F}$$

where

$$\eta = \frac{2E_T}{E+E_T} \qquad S = \frac{b^2 U}{\pi^2 D} \qquad F_i = \gamma_i F_{SCR}$$

$$\text{where } \gamma_i = \sqrt{3} \text{ for shear}$$

$$U = G'_C (c+2t_F) \qquad D = \frac{E t_F (c+t_F)^2}{2(1-\mu^2)}$$

therefore

$$U = 3{,}400\,(0.500+2 \cdot 0.0154) = 1{,}805 \; lb/in.$$

$$S = \frac{21.89^2 \cdot 1{,}805}{\pi^2 \cdot 14{,}898} = 5.9$$

$$D = \frac{6.67\times10^6 \cdot 0.0154\,(0.500+0.0154)^2}{2(1-0.29^2)} = 14{,}898 \; in.-lb$$

For
a = 49 in.
b = 21.89 in.
a/b = 2.24
E_{eff} = $6.67x10^6$ psi
μ = 0.29
t_F = 0.0154 in.
c = 0.500 in.
$G'_c = G_w$ = 3,400 psi

The shear buckling coefficient "*Ks*" for simply-supported flat honeycomb panels is obtained from *Figure B-5*. For a/b = 2.24 and S = 5.9, the value found for *Ks* is 3.5.

Therefore:

$$\frac{F_{SCR}}{\eta} = \frac{3.5}{5.9} \cdot \frac{1{,}805}{2 \cdot 0.0154} = 34{,}870 \ psi$$

$$\frac{F_i}{\eta} = \sqrt{3} \cdot 34{,}870 = 60{,}396 \ psi$$

Now to get the actual critical shear buckling stress, one has to determine the value for the plasticity coefficient "η". As there are generally no stress-strain curves for composite materials, hence the tangent Elastic modulus "E_T" cannot be determined. However, the critical shear buckling stress can be approximated by the following equations:

$$for: 0.0 \leqslant \frac{F_{CY}}{(F_i/\eta)} < 0.3 \rightarrow \frac{F_i}{F_{CY}} = 1$$

$$for: 0.3 \leqslant \frac{F_{CY}}{(F_i/\eta)} < 1.92 \rightarrow \frac{F_i}{F_{CY}} = -0.31 \frac{F_{CY}}{(F_i/\eta)} + 1.095$$

With a face compression strength of 70,000 *psi* for F_{CY}, the variable is calculated as:

$$\frac{F_{CY}}{F_i/\eta} = \frac{70{,}000}{60{,}396} = 1.2$$

Then the value for F_i/F_{CY} is found to be: 0.72. Therefore, $F_i = 0.72 * 70{,}000 = 50{,}400$ *psi*.

And finally, transforming the F_i stress back to shear, the critical shear buckling stress is:

$$F_{SCRi} = \frac{F_i}{\sqrt{3}} = \frac{50{,}400}{\sqrt{3}} = 29{,}098 \ psi$$

From the above result, the plasticity coefficient obtained is:

$$\eta = \frac{F_{SCRi}}{F_{SCR}} = \frac{29{,}098}{34{,}870} = 0.83$$

Shear stress

The shear flow "q_A" across the critical floor panel, calculated on page *368*, is 93 *lb/in*. Since that shear flow is taken by both the top and the bottom faces of the bonded panel, the shear stress "f_S" is calculated as follows:

$$f_S = \frac{q_A}{t_{total}} = \frac{93}{2 \cdot 0.0154} = 3{,}019 \ psi$$

Margin of Safety for the Floor panel in general shear buckling

$$MS = \frac{F_{SCRi}}{f_S} - 1 = \frac{29{,}098}{3{,}019} - 1 = 8.64$$

Floor panel A *Ultimate*	***> 1.0***
9g Fwd *GR/EP*	*Shear buckling*

This methodology is mainly used for bonded panels made out of CRES (Corrosion Resistant Steel) which shows, as most metals, some plasticity in the stress-strain curve.

Knowing that carbon fibers don't yield much before they break, I would expect that pretty much any stress value below the compression (or tension) strength would be in the elastic region and thus, a plasticity coefficient of $\eta = 1$ would be expected. Although any value lower than **one** would be conservative, I would still like to get a second "opinion" by using another approach.

For that reason, and because it's fun, I'll do a FEM of the panel and run it for buckling.

14.9.4 GENERAL BUCKLING USING FEA

Before I go ahead and start modeling large models, I sometimes like to perform a small test FEM in order to "de-bug" the model and to get a "feel" for the FEA results especially when the runs take a long time.

The software I am using performs buckling analyses only with "1D" elements like *line2* and "3D" elements like *hex8* or *tet10* for example. As a first trial, I decided to make a small test FEM, shown in *Figure 14-64,* by simply assuming that the two faces were joined together and no core existed. The small test FEM was a 6 *in.* x 6 *in.* x 0.0308 *in.* panel using *HEX8* elements as shown in *Figure 14-65.*

The panel is restrained vertically [3] with a sufficient quantity of grids around the periphery of the panel. Only one grid is retrained in [123] and another one in [23] thus blocking the panel from rigid body motion. A unit distributed load is applied on all four sides to create a unit shear load.

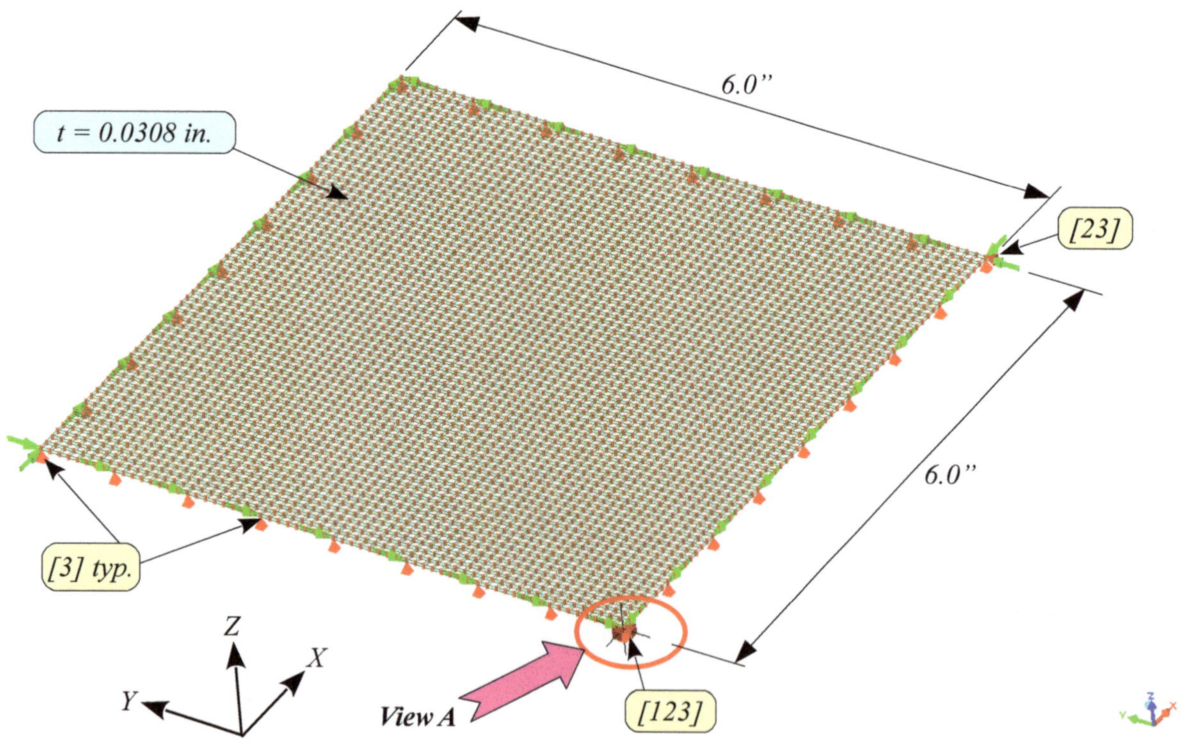

Figure 14-64 *– Small test FEM using Hex8 elements showing loads & B.C.*

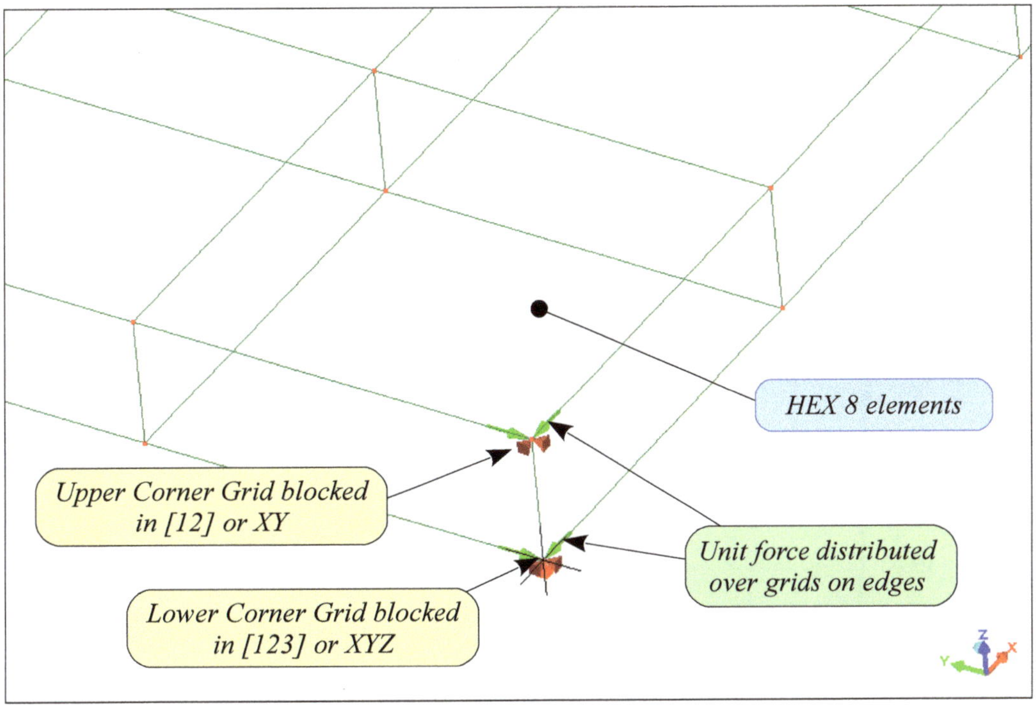

Figure 14-65 – *View A of Small test FEM using Hex8 elements showing loads & B.C.*

Estimating the shear buckling allowable

Before running the FEM, I calculated the critical shear buckling by hand. For a square panel ($a/b = 1$) simply supported on all edges, the buckling coefficient "k_S" obtained from *Figure A-2* is 9.7. Therefore, using *eq. (1)* from (ref. 20.), the hand calculated shear buckling allowable for this panel (without core) is calculated below using the material properties from page *307*.

$$F_{SCR} = \frac{\pi^2 \cdot k_S \cdot E_{eff}}{12(1-\mu^2)} \left(\frac{t}{b}\right)^2$$

$$F_{SCR} = \frac{\pi^2 \cdot 9.7 \cdot 6.67 \times 10^6}{12(1-0.29^2)} \left(\frac{0.0308}{6}\right)^2 = 1{,}531 \;\; psi$$

For
$a = 6$ in.
$b = 6$ in.
$a/b = 1$
$E_{eff} = 6.67x10^6$ psi
$\mu = 0.29$
$t = 0.0308$ in.
$k_s = 9.7$

The allowable shear flow "q_{CR}" is thus:

$$q_{CR} = F_{SCR} \times t = 1{,}531 \times 0.0308 = 47.1 \; lb/in.$$

and, the critical shear load "V_{CR}" is then:

$$V_{CR} = q_{CR} \times b = 47.1 \times 6.0 = 283 \; lbs$$

Therefore, using a unit shear load in the small *FEM*, I'm expecting to get a load factor "λ" of about 283.

14.9.4.1 FEA small test runs

Starting with 64 *Hex8* elements (8 x 8) and refining the *FEM* until the model had 16,384 elements (128 x 128), the results obtained for "λ" are shown in *Table 14-33*. To get a better idea of how bad the convergence was, a plot for those results can be seen in *Figure 14-66*.

Table 14-33 *- Load factor vs number of Hex8 elements.*

No. Elmts	λ
64	33688
256	8061
1024	2225
4096	801
16384	445

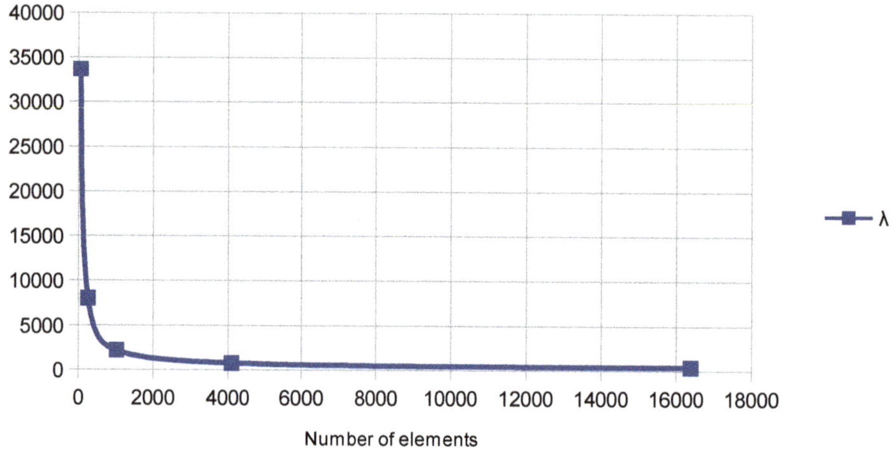

Figure 14-66 *– Load factor vs number of Hex8 elements.*

For the convergence test, I refined the model by splitting all the elements in two, ran the solver and recorded the load factor for each iteration. As I kept refining the model, I noticed that the results were always very different. So I plotted a graph (see ***Figure 14-66***) that shows the number of elements vs the load factor to see if this was going anywhere towards convergence. One can see that the curve is asymptotic and, before I could reach the final value, my computer ran out of memory...

As I couldn't refine the model any further due to an "out of memory" error message, I thought of changing the type of elements. The nomenclature of the *FEA* I used was mentioning that *Hex20* elements converged faster than *Hex8's*, so, I decided to give it a try.

So, I started over with relatively large elements but replaced the *Hex8*'s with *Hex20*'s. Again applying a unit shear load on all edges, the convergence occurred rapidly with 1,024 elements and a load factor of 275 was obtained. This result was pretty close to the hand calculations (283 *lbs*) within 3%. The results obtained for "λ" are shown in ***Table 14-34*** and the very rapid convergence can be see in the plot in ***Figure 14-67***.

Figure 14-69 shows the first mode for the shear buckling run of the small test panel.

Table 14-34 *- Load factor vs number of Hex20 elements.*

No. Elmts	λ
64	300
256	276
1024	275

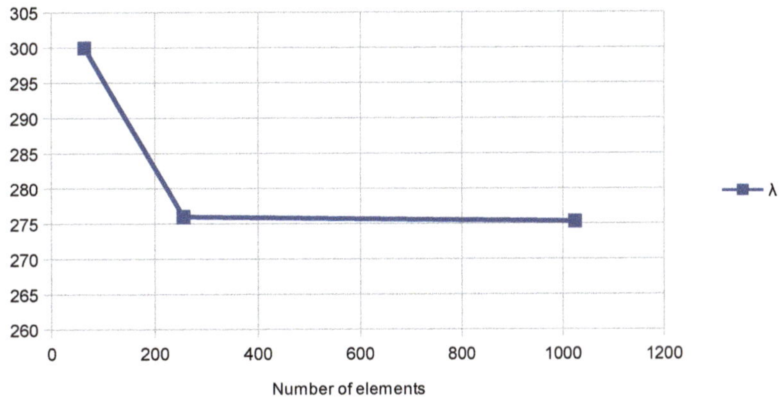

Figure 14-67 *– Load factor vs number of Hex20 elements.*

Therefore, *hex20*'s are definitely the type of elements to use for buckling using this *FEA* software. See *Figure 14-68* for a comparison between the *Hex8's* and *Hex 20's*. The *Hex8's* are created by extruding *Quad4's* and *Hex20's* are created by extruding *Quad8's*.

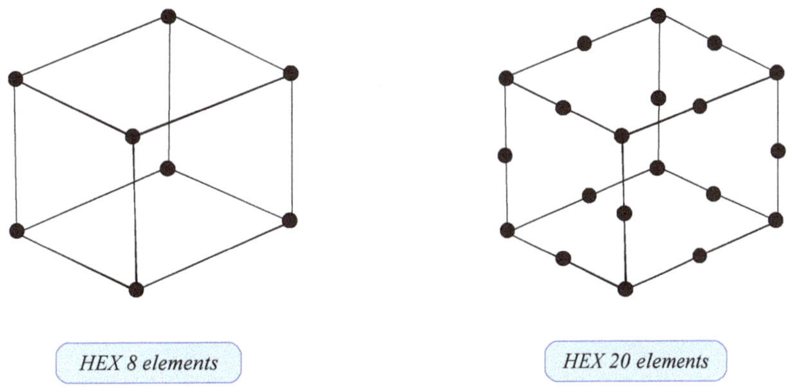

Figure 14-68 *– Hex8 vs Hex20 elements.*

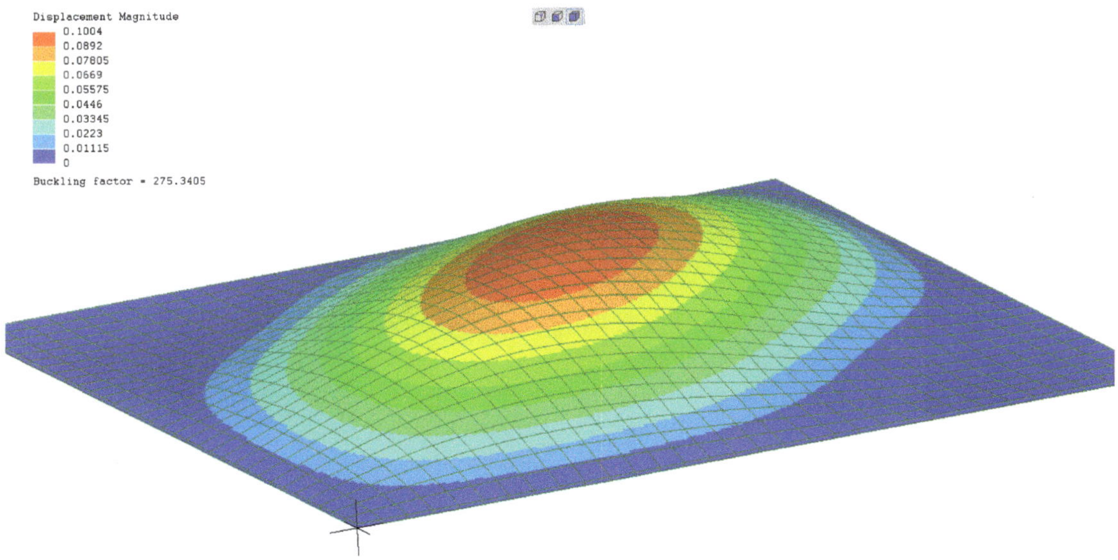

Figure 14-69 *– Small test panel FEM buckling first mode using Hex20 elements.*

14.9.4.2 FEA Floor panel runs

Now that I had determined that I needed to be using *Hex20* elements, I thus created a FEM for the actual Floor panel by meshing the two faces and the core using *hex20* elements. I then supported all the nodes on the lower edges in the "*Z*" direction, blocked one node at a corner in "*XYZ*" and another one at another corner in "*YZ*". I then applied the actual forces all along the edges in order to simulate the shear flow and ran the model recording the resulting load factors "λ" to test for the convergence. See *Figure 14-71* for the Floor panel *FEM*.

The material properties for the elements are shown in *Table 14-23* and *Table 14-24* by considering the faces as quasi-isotropic. For the core, I used the smallest shear modulus "G_W" that I converted into the Young's modulus as $E_{eff} = G_W(2(1+\mu)) = 3{,}400*(2(1+0.05)) = 6{,}800$ psi since I couldn't use orthotropic material properties for the buckling analysis using this software.

Starting with 96 elements, then 384, and finally 1,536 elements. From *Table 14-35* and *Figure 14-70* one can see that convergence was obtained quickly at 1,536 elements with a value for the load factor of $\lambda = 11.17$ compared to: $\lambda = MS + 1 = 8.64 + 1 = 9.64$ for the hand calculations.

This is a bit far from each other, so I decided to use another reference to get a third "opinion". I won't be showing the calcs here but, using (ref. 21.) for shear buckling of a sandwich panel with simply supported edges, I got a MS of 10.39 *($\lambda = 11.39$)*. This result is within 1% of the FEA!

Figure 14-72 shows the first mode of shear buckling for this sandwich floor panel.

Table 14-35 *- Load factor vs number of Hex20 elements for the Floor panel.*

No. Elmts	λ
96	12.59
384	11.25
1536	11.17

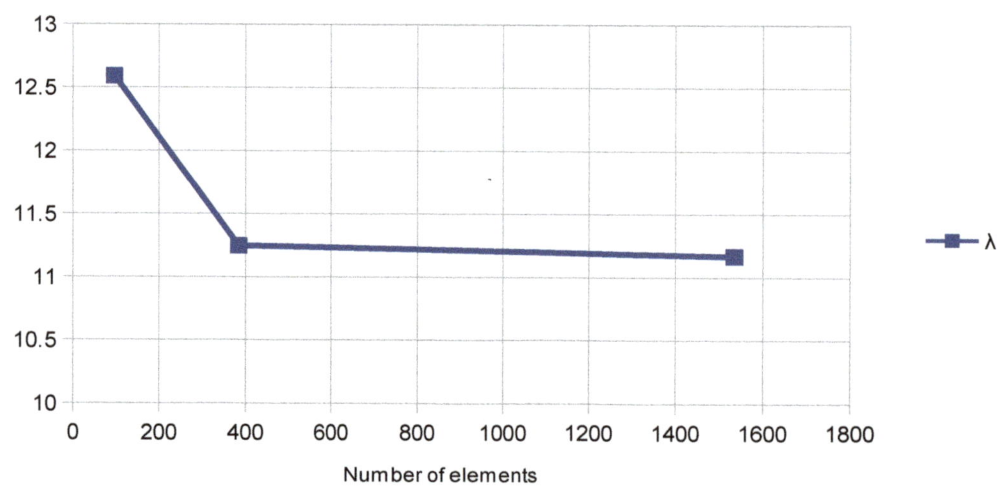

Figure 14-70 *– Load factor vs number of Hex20 elements for the Floor panel.*

The old hand calculation methodologies were developed along with some test articles. Therefore, they are empirical and pretty close to reality. When creating FEM's, we must always "tune" the model's boundary conditions, properties, etc. so that it matches the test article as close as possible. Tests are the actual reality that we're trying to simulate.

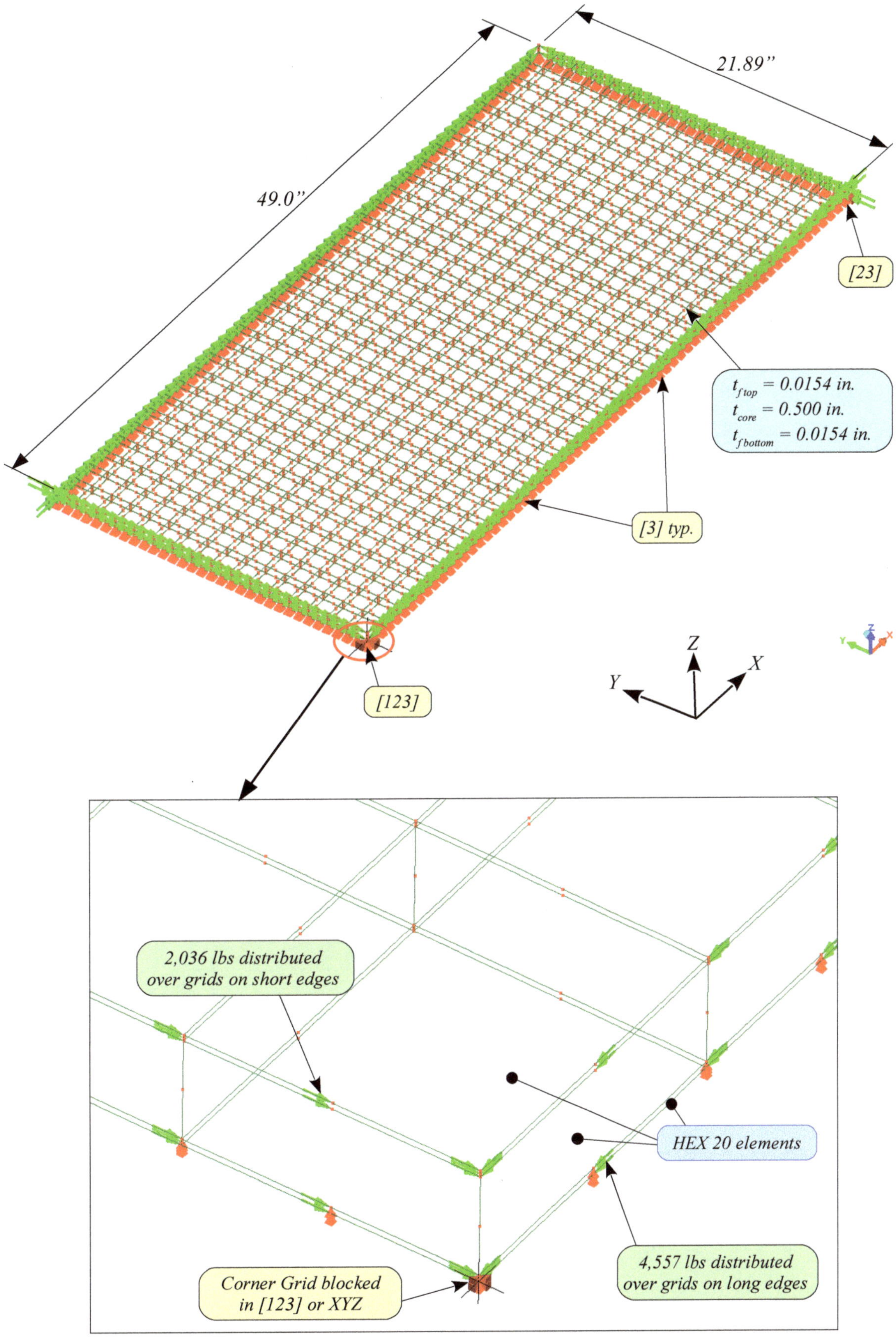

Figure 14-71 *– Floor panel FEM using Hex20 elements showing loads & B.C.*

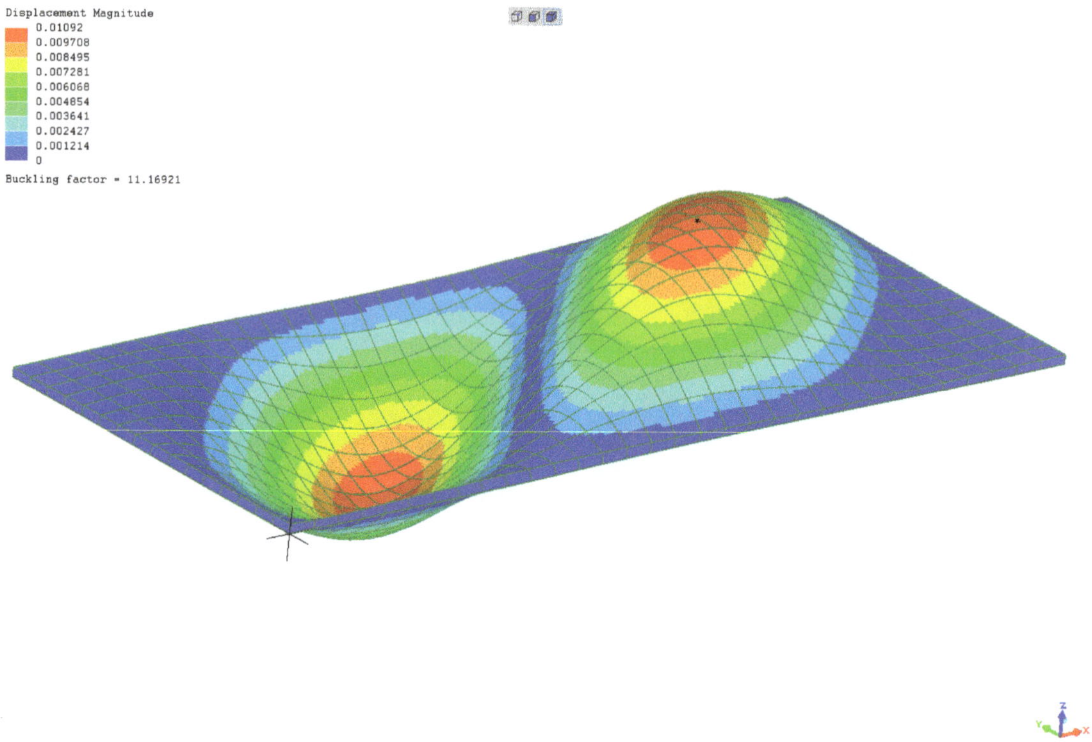

Figure 14-72 *– Floor panel FEM buckling first mode using Hex20 elements.*

From the results of the *FEA* and the ones from the two hand calculations from two different references, I got a very good confidence that the floor sandwich panel analyzed in shear buckling using those methodologies is structurally adequate. Only the more conservative analysis *(lowest MS)* was shown here. Therefore, the general instability MS shown on page *371* is considered good.

14.9.5 INTRACELL BUCKLING

This mode of failure of the faces is like a flat plate buckling with a span the length of a core cell "s". The intracell buckling due to shear is calculated below.

The intracell shear buckling allowable is calculated as follows:

$$\frac{F_{Si}}{\eta} = 0.6\,E_{eff}\left(\frac{t_F}{s}\right)^{3/2}$$

where

$$\eta = \frac{2\,E_T}{E+E_T} \qquad F_i = \gamma_i\,F_{Si} \quad \text{where } \gamma_i = \sqrt{3} \text{ for shear}$$

For
$E_{eff} = 6.67x10^6$ psi
$\mu = 0.29$
$t_F = 0.0154$ in.
$s = 3/16$ in.
$F_{CY} = 70{,}000$ psi

therefore

$$\frac{F_{Si}}{\eta} = 0.6 \cdot 6.67\times10^6\left(\frac{0.0154}{0.188}\right)^{3/2} = 93{,}840\ psi$$

$$\frac{F_i}{\eta} = \sqrt{3}\cdot 93{,}840 = 162{,}535\ psi$$

Now to get the actual critical shear buckling stress, one has to determine the value for the plasticity coefficient "η". As there are generally no stress-strain curves for composite materials, hence the tangent Elastic modulus "E_T" cannot be determined. However, the critical shear buckling stress can be approximated by the following equations:

$$for: 0.0 \leqslant \frac{F_{CY}}{(F_i/\eta)} < 0.3 \rightarrow \frac{F_i}{F_{CY}} = 1$$

$$for: 0.3 \leqslant \frac{F_{CY}}{(F_i/\eta)} < 1.92 \rightarrow \frac{F_i}{F_{CY}} = -0.31\frac{F_{CY}}{(F_i/\eta)} + 1.095$$

With a face compression strength of 70,000 *psi* for F_{CY}, the variable is calculated as:

$$\frac{F_{CY}}{F_i/\eta} = \frac{70{,}000}{162{,}535} = 0.4$$

Then the value for F_i/F_{CY} is found to be: 0.98. Therefore, $F_i = 0.98 * 70{,}000 = 68{,}600$ *psi*.

And finally, transforming the F_i stress back to shear, the critical shear buckling stress is:

$$F_{Sii} = \frac{F_i}{\sqrt{3}} = \frac{68{,}600}{\sqrt{3}} = 39{,}606 \ psi$$

From the above result, the plasticity coefficient obtained is:

$$\eta = \frac{F_{Sii}}{F_{Si}} = \frac{39{,}606}{93{,}840} = 0.42$$

Shear stress

The shear flow "q_A" across the critical floor panel, calculated on page *368*, is 93 *lb/in*. Since that shear flow is taken by both the top and the bottom faces of the bonded panel, the shear stress "f_S" is calculated as follows:

$$f_S = \frac{q_A}{t_{total}} = \frac{93}{2 \cdot 0.0154} = 3{,}019 \ psi$$

Margin of Safety for the Floor panel in intracell shear buckling

$$MS = \frac{F_{Sii}}{f_S} - 1 = \frac{39{,}606}{3{,}019} - 1 = 12.12$$

Floor panel A *Ultimate*	***> 1.0***
9g Fwd *GR/EP*	*Intracell shear buckling*

14.9.6 WRINKLING

Wrinkling is an instability problem similar to the case of a beam on an elastic foundation. The wrinkling due to in-plane shear is calculated below.

The wrinkling allowable is calculated as follows:

$$\frac{F_{SW}}{\eta_2} = 0.43\sqrt[3]{E_{eff} E'_C G'_C}$$

where

$$\eta_2 = \sqrt[3]{\frac{3E_T + E_S}{4E}}$$

$$F_i = \gamma_i F_{SW}$$
$$where\ \gamma_i = \sqrt{3}\ for\ shear$$

therefore

$$\frac{F_{SW}}{\eta_2} = 0.43\sqrt[3]{6.67\times10^6\cdot20{,}000\cdot3{,}400} = 33{,}040\ psi$$

$$\frac{F_i}{\eta_2} = \sqrt{3}\cdot33{,}040 = 57{,}226\ psi$$

For
$E_{eff} = 6.67x10^6\ psi$
$\mu = 0.29$
$t_C = 0.001\ in.$
$s = 3/16\ in.$
$F_{CY} = 70{,}000\ psi$
$E_C = 20{,}000\ psi$
$G'_C = G_W = 3{,}400\ psi$

To get the actual critical shear buckling stress, one has to determine the value for the plasticity coefficient "η". As there are generally no stress-strain curves for composite materials, hence the tangent Elastic modulus "E_T" cannot be determined. However, the critical shear buckling stress can be approximated by the following equations:

$$for: 0.0 \leqslant \frac{F_{CY}}{(F_{CW}/\eta_2)} < 0.67 \rightarrow \frac{F_{CW}}{F_{CY}} = 1$$

$$for: 0.67 \leqslant \frac{F_{CY}}{(F_{CW}/\eta_2)} < 3.15 \rightarrow \frac{F_{CW}}{F_{CY}} = -0.455\ \ln\frac{F_{CY}}{(F_{CW}/\eta_2)} + 0.83$$

where

$$\frac{F_{SW}}{\eta_2} = \frac{F_{CW}}{\eta_2}$$

$$\frac{F_{CY}}{F_i/\eta_2} = \frac{70{,}000}{57{,}226} = 1.2$$

Then the value for F_i/F_{CY} is found to be: 0.75. Therefore, $F_i = 0.75 * 70{,}000 = 52{,}293\ psi$.

And finally, transforming the F_i stress back to shear, the critical shear buckling stress is:

$$F_{SWi} = \frac{F_i}{\sqrt{3}} = \frac{52{,}293}{\sqrt{3}} = 30{,}191\ psi$$

From the above result, the plasticity coefficient obtained is:

$$\eta_2 = \frac{F_{SWi}}{F_{SW}} = \frac{30{,}191}{33{,}040} = 0.91$$

Shear stress

The shear flow "q_A" across the critical floor panel, calculated on page *368*, is 93 *lb/in.* Since that shear flow is taken by both the top and the bottom faces of the bonded panel, the shear stress "f_S" is calculated as follows:

$$f_S = \frac{q_A}{t_{total}} = \frac{93}{2\cdot0.0154} = 3{,}019\ psi$$

$$MS = \frac{F_{SWi}}{f_S} - 1 = \frac{30{,}191}{3{,}019} - 1 = 9.00$$

Margin of Safety for the Floor panel in wrinkling

14.9.7 SHEAR CRIMPING

Floor panel A *Ultimate*	***> 1.0***
9g Fwd *GR/EP*	*Wrinkling*

Shear crimping is a form of general instability problem where the wavelength of the buckles becomes small. The shear crimping due to in-plane shear is calculated below.

The shear crimping allowable is calculated as follows:

$$F_{CS\,max} = \frac{U}{2t_F} = \frac{G'_C\,(c+2t_F)}{2t_F}$$

therefore

$$F_{CS\,max} = \frac{3{,}400\,(0.500+2\cdot 0.0154)}{2\cdot 0.0154} = 58{,}595\ psi$$

For
$t_F = 0.0154$ in.
$c = 0.500$ in.
$G'_C = G_W = 3{,}400$ psi

Shear stress

The shear flow "q_A" across the critical floor panel, calculated on page *368*, is 93 lb/in. Since that shear flow is taken by both the top and the bottom faces of the bonded panel, the shear stress "f_S" is calculated as follows:

$$f_S = \frac{q_A}{t_{total}} = \frac{93}{2\cdot 0.0154} = 3{,}019\ psi$$

Margin of Safety for the Floor panel in shear crimping

$$MS = \frac{F_{CS\,max}}{f_S} - 1 = \frac{58{,}595}{3{,}019} - 1 = 18.41$$

Floor panel A *Ultimate*	***> 1.0***
9g Fwd *GR/EP*	*Shear crimping*

Since panels *B* and *C* are smaller and/or see lower loads, they are ***Passed by Comparison*** to panel *A*.

14.9.8 CONNECTIONS

The floor panels are attached to the airframe with screws. Nutplates (or more recently Clip-nuts) are usually fastened to the airframe structures and the screws are installed from the floor panel side through inserts. Those inserts are press-fit inserts and they are installed after the bonded panel has been manufactured. See *Figure 14-73* for a section view through the floor panel at the two-part insert.

Sometimes, when a high tension (or compression) out-of-plane load is applied to an insert in a bonded panel, higher density core or solid "hockey pucks" the same thickness as the core and made out of many GR/EP prepreg plies will be installed in the panel during its fabrication before the faces are bonded. This increases the transverse shear area of the core at the periphery of the "puck" to help transferring the load towards the edges of the panel.

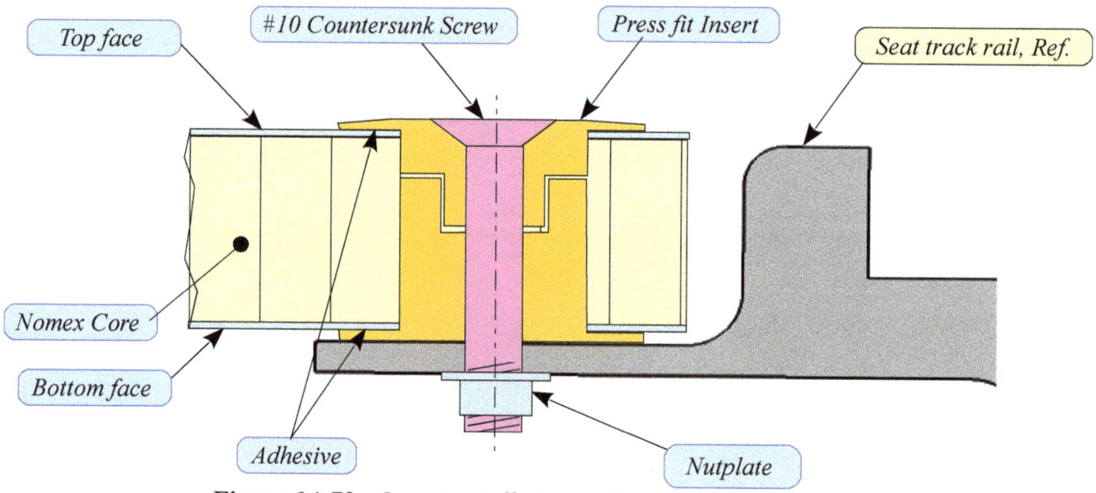

Figure 14-73 – *Insert installation and connection to structure.*

In our case, the floor panels only carry in-plane shear loads so no out-of-plane forces are applied to the inserts. Only the skins can carry in-plane loads while the core carries out-of-plane loads. Both faces are considered to react the in-plane shear load equally, especially since the critical loading condition is an emergency landing *(9g)*.

<u>Bearing</u>

The press-fit inserts are installed by first drilling a hole through the panel and then, by applying adhesive underneath the insert's flanges to be bonded to the faces of the panel and snap the two halves together from both sides of the panel. Although this adhesive would most likely transfer most (if not all) of the shear load between the insert and the panel faces, I will conservatively assume that no adhesive is used and that the shear is transferred by bearing between the shank of the insert and the edge of the holes in the panel's faces.

The capacity in bearing of GR/EP skins is usually based on tests. One can refer to the (ref. 3.) document for allowables on composite materials available on line. But, for our fictitious example, since the faces are considered quasi-isotropic, I will assume that the bearing strength "F_{bru}" is equal to the compression strength "F_{1C}" of the face material. So from *Table 14-23*, $F_{bru} = F_{1C} = 70$ *ksi*.

Since #10 countersunk screws (*AN509-3*) are used, they have a 3/16" diameter (0.190 *in.*). The insert diameter is 0.4375 *in.* So for face thicknesses of 0.0154 *in.* for both top and bottom faces, the total bearing load allowable "P_{br}" in the Floor panel is:

$$P_{br} = 2 \cdot F_{br} \cdot D_{insert} \cdot t = 2 \cdot 70{,}000 \cdot 0.4375 \cdot 0.0154 = 943 \ lbs$$

Note that the material allowables of composites are greatly affected by the environment in which the component is used. High temperature and high humidity are detrimental to composite parts in a significant way. That is why composite parts are normally designed using material allowables for ***hot-wet*** conditions. Sometimes, allowables from tests in those conditions are directly available and sometimes a knock down factor is used.

The same composite aircraft (especially if it's painted black!) parked in Florida won't have the same strength than parked in Alaska in winter. One could expect lower material properties at the former location. Therefore, it is important to design composite parts accordingly.

On the airframe side, the floor panels are fastened to the Crossbeams and the Seat track rails. Their respective bearing allowable are shown in *Table 14-36*.

***Table 14-36** - Airframe Bearing allowables.*

	Material	*t*	*D*	*Fbru*	*Pbru*
Part	*Al Aly*	*[in]*	*[in]*	*[ksi]*	*[lbs]*
Seat track	*7075-T6*	0.070	0.19	145	1929
Crossbeam	*7050-T7451*	0.090	0.19	105	1796

Fasteners

The screws are countersunk *AN509-3*'s having a shear strength "F_{SU}" of 75 *ksi*. Their ultimate single shear capacity "P_{SU}" in the shank is thus:

$$P_{SU} = F_{SU} \cdot \pi \frac{(D_{fast})^2}{4} = 75{,}000 \cdot \pi \frac{(0.19)^2}{4} = 2{,}125 \; lbs$$

Thus. the connection is thus critical in bearing in the Floor panel.

Margin of Safety for the Floor panel connections

The shear flow "q_A" across the critical floor panel (*A*) calculated on page *368*, is 93 *lb/in*. For a maximum fastener pitch "*s*" of 6.0 *inches*, as shown in *Figure 14-63*, the shear force in each fastener "*P*" is:

$$P = q_A \cdot s = 93 \cdot 6.0 = 558 \; lbs$$

No Fitting Factor is used in bearing.

$$MS = \frac{P_{br}}{P} - 1 = \frac{943}{558} - 1 = 0.69$$

Floor panel A *Ultimate*	***0.69***
9g Fwd *GR/EP*	*Bearing*

14.10 CRASH STRAPS

The Crash-straps (for lack of a better word) are the airframe components that ultimately transfer all of the forward load coming from the seat track rails to the fuselage skin. They work only in tension so stability is not a concern. A net section tension check and a connection check are the only analyses required to substantiate the Crash-straps. See *Figure 14-74* for their installation.

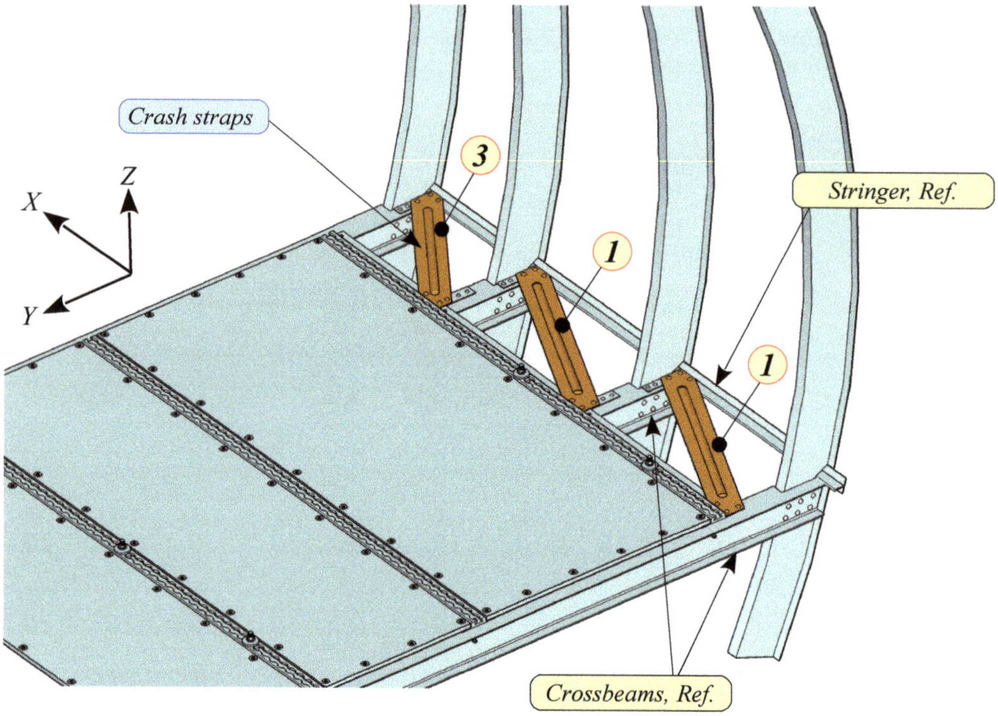

Figure 14-74 *– Crash straps installation.*

The Crash-straps are made out of *Al Aly 2024-T3 Clad* sheet 0.071 *in.* thick. They are connected to the Seat track rails at the inboard end and to the skin Stringer at the outboard end using *MS20470AD6* solid rivets. Crash-straps Items *(1)* have the same geometry while Item *(3)* is a little shorter and has a more obtuse angle.

> Item numbers or dash numbers are usually odd numbers on the left hand side and even on the right hand side when they are mirror parts. If the same part is installed on both sides then an odd number is used.

14.10.1 LOADS

The shear flow at the skin stringer is "q_o" as shown on *Figure 14-63*. From section *14.9.1*, the shear flow at the skin from the *9g Fwd* LC is calculated as follows:

$$q_O = q_i + \frac{P_1}{L} + \frac{P_3}{L} = 83 + \frac{497}{49} + \frac{1{,}367}{49} = 121\ lbs/in.$$

From *Figure 14-63*, there are two distinct frame pitches. For the two forward frame bays, the frame pitch "s_1" is 18 *in*. Hence, the shear load along the longitudinal axis "V_{X_1}" for those two bays is thus: $V_{X\_1} = qo * s_1 = 121 * 18 = 2{,}178$ *lbs* while, for the aft frame bay having a frame pitch "s_3" of 13 *in*., the shear load along the longitudinal axis "V_{X_3}" is thus: $V_{X\_3} = qo * s_3 = 121 * 13 = 1{,}573$ *lbs*.

The internal loads of the Crash-straps are shown in *Figure 14-75*. It is assumed that the connections at each end of the Crash-straps do not provide any restraint in in-plane *(X-Y)* rotation; thus no end moments only axial loads. This assumption makes sense since the Crash-straps are thin and their free flange would buckle on the compression side due to any bending.

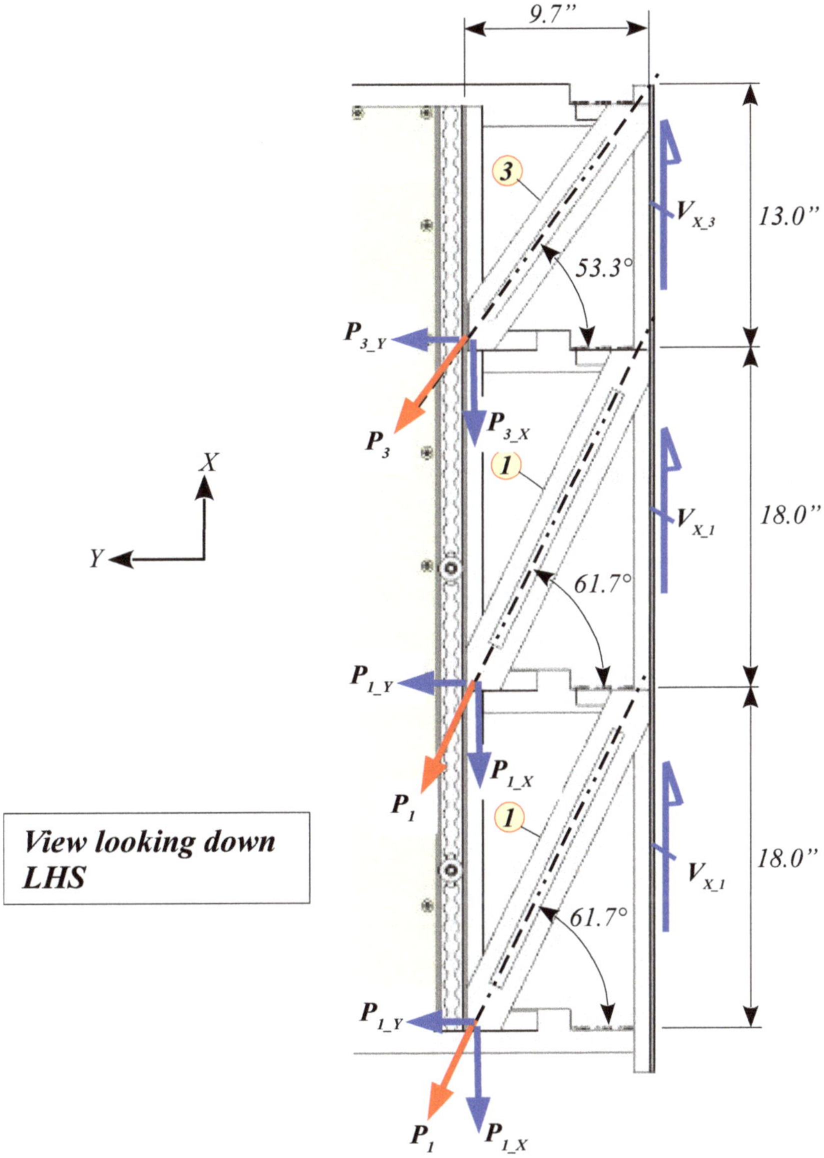

Figure 14-75 – *Crash-straps FBD.*

The internal loads for item *1* are calculated as follows:

$$P_{1X} = V_{X1} = 2{,}178 \ lbs.$$

$$P_1 = \frac{P_{1X}}{\sin\theta} = \frac{2{,}178}{\sin 61.7} = 2{,}474 \ lbs.$$

$$P_{1Y} = P_1 \cos\theta = 2{,}474 \cos 61.7 = 1{,}173 \ lbs.$$

The internal loads for item *3* are calculated as follows:

$$P_{3X} = V_{X3} = 1{,}573 \ lbs.$$

$$P_3 = \frac{P_{3X}}{\sin\theta} = \frac{1{,}573}{\sin 53.3} = 1{,}962 \ lbs.$$

$$P_{3Y} = P_3 \cos\theta = 1{,}962 \cos 53.3 = 1{,}173 \ lbs.$$

Item *(1)* has the highest load with an axial tension force of 2,474 *lbs*. Since the cross-section of both items *(1)* and *(3)* is the same, then only item *(1)* need to be analyzed passing item *(3)* by comparison. See *Figure 14-76* for the *FBD* of item (1) and the cross-sections locations.

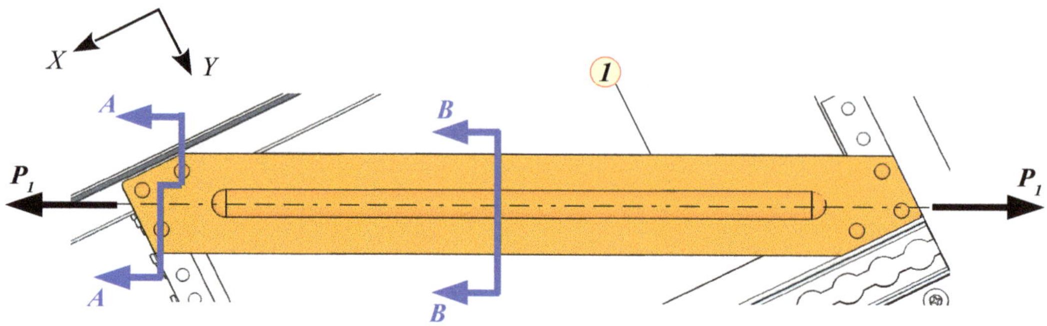

Figure 14-76 *– Internal axial tension load in Crash-strap item 1.*

14.10.2 SECTION PROPERTIES

The cross sections dimensions and properties for sections *A-A* and *B-B* of the Crash-straps are shown in *Figure 14-77*. No second moment of inertia is shown since they are not required as no bending moment exists in the parts.

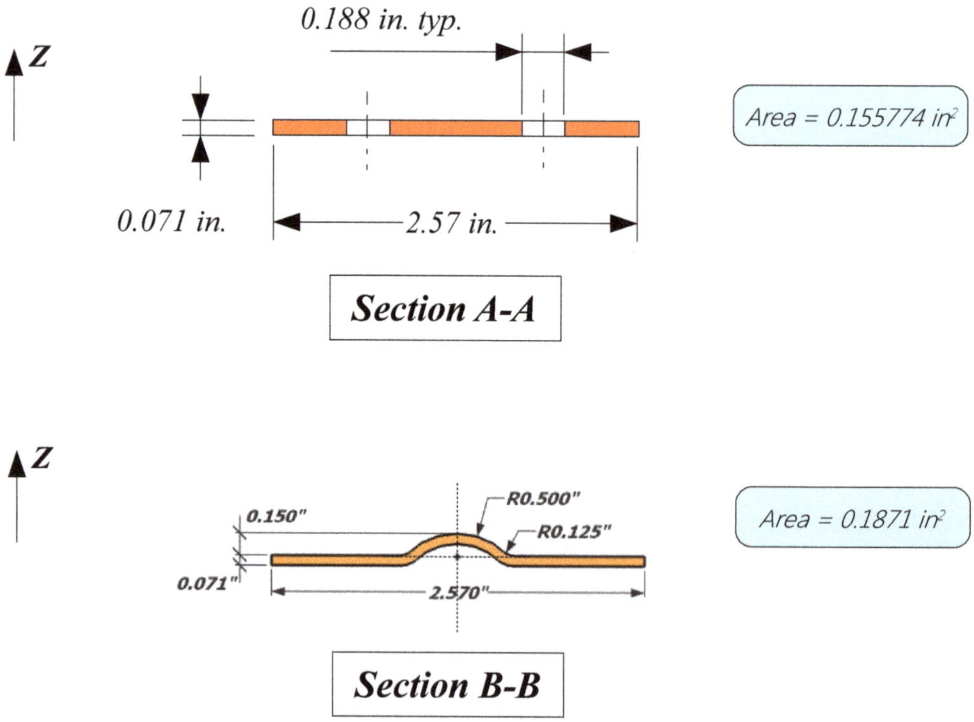

Figure 14-77 *– Crash-strap cross sections.*

Although the Crash-straps only have internal tension loads and thus are basically considered as cables, a bead was added in the design to give the part some out-of-plane stiffness. This would help prevent/reduce undesirable vibrations that could lead to premature cracks and/or cabin noise. Note that the bead is on the top side so that no water from possible condensation can be trapped.

14.10.3 STRESS ANALYSIS

Axial stress

Section *A-A* is the most critical location as its area is the smallest. Therefore, the net tension stress "f_a" through section *A-A* of item *(1)* is:

$$f_a = \frac{P}{A} = \frac{2{,}474}{0.155774} = 15{,}882 \ psi$$

Using the ultimate tensile allowable "F_{tu}" of 60 *ksi* from *Table 14-19*, the MS is calculated as:

$$MS = \frac{F_{tu}}{f_a} - 1 = \frac{60{,}000}{15{,}882} - 1 = 2.77$$

Crash-strap, Item 1 *Ultimate*	***> 1.0***
9g Fwd *Al Aly 2024-T3 Clad*	*Net tension*

14.10.4 CONNECTIONS

Crash strap *MS20470AD6* solid rivets

As shown in *Figure 14-78* and *Figure 14-79*, only two rivets from the installed three at each end provide a direct (stiff) load path. The total internal tension load of the Crash strap is assumed shared equally between the two rivets.

Note that, due to the high shear load at the rivets, as you'll see shortly, the type of fastener had to be changed and 5/32" *dia.* Hi-Lites are to be used. Therefore, the above analysis should be updated to account for the smaller holes. Or, a statement saying that the analysis is conservative because larger size holes were considered would probably do. Anyhow, the *MS* would still be >1.0.

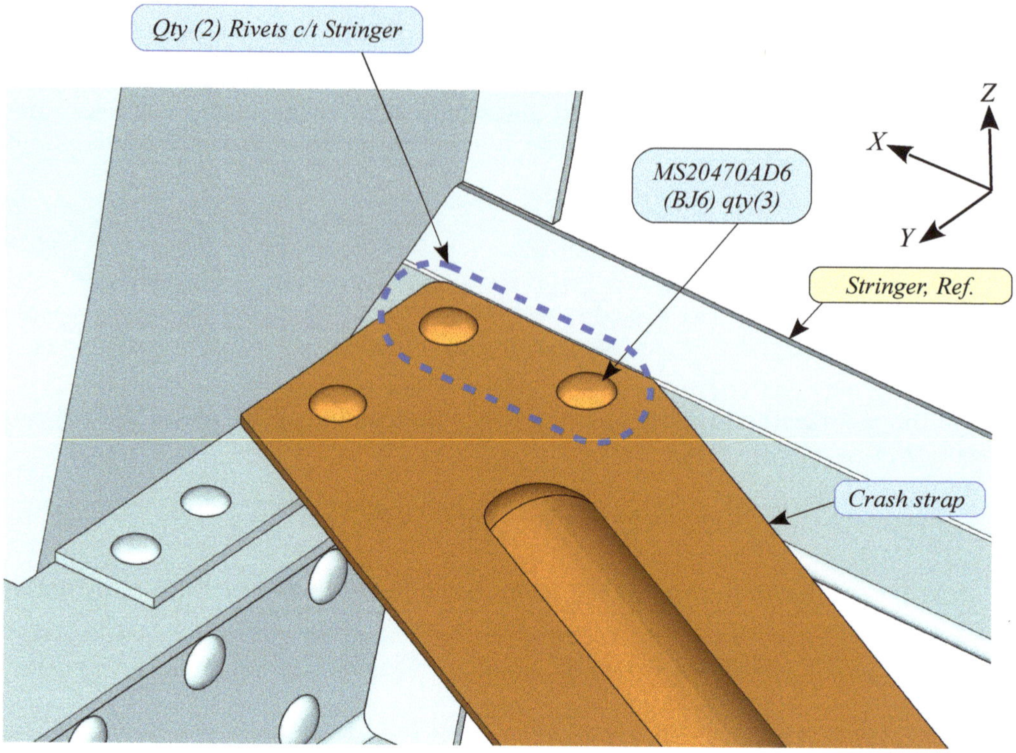

Figure 14-78 *– Crash strap outboard connection.*

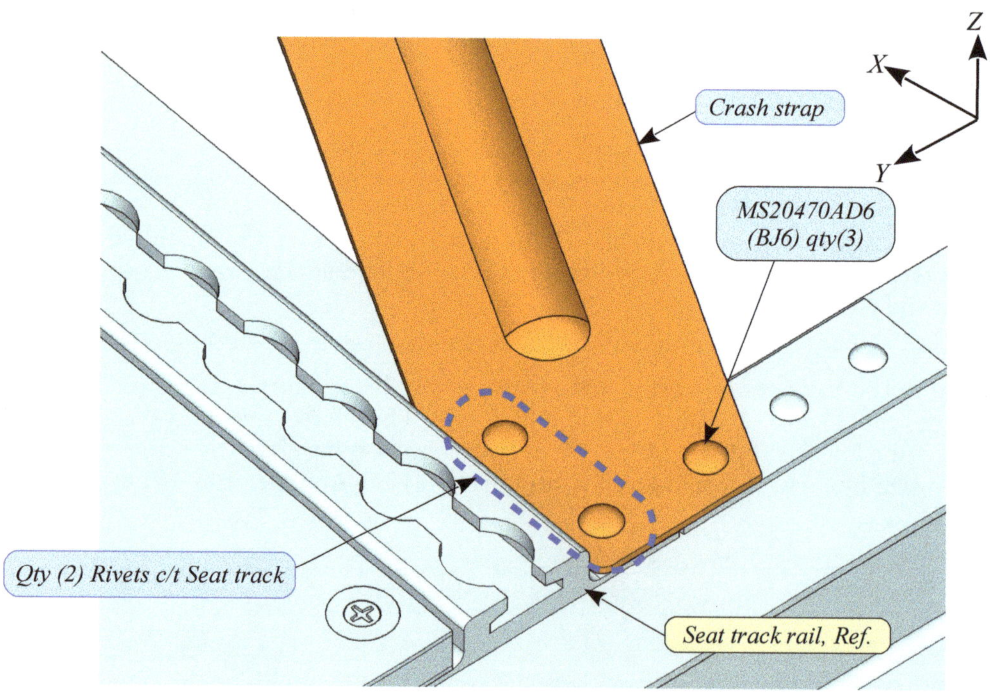

Figure 14-79 *– Crash strap inboard connection.*

Considering the maximum internal axial load between items *(1)* & *(3)* from page *385* of 2,474 *lbs*, the forces in each of the two rivets reacting that load are calculated as follows:

$$F = \frac{P_1}{n} = \frac{2{,}474}{2} = 1{,}237\ lbs$$

Rivet shear:

As one can notice, the resulting shear load value *"F"* calculated above for one rivet exceeds the shear allowable *"P_{SU}"* for an *MS20470AD6*. For a *BJ6* rivet having a single shear capacity of 862 lbs as shown in *Table 14-57*, it is clear that the strength of those solid rivets is insufficient. It is thus decided to change the fasteners from *MS20470AD6* to Hi-Lites *HST10-5* having a single shear strength of 2,005 *lbs* as shown in *Figure 14-25*.

Here, I prefer going down in diameter size (from -6 to -5) as it provides more flexibility in case of repairs. This is of course providing that the bearing capability of a smaller hole is still adequate. We'll see...

A Fitting Factor *"FF"* of 1.15 (ref. 25.. *FAR §25.625*) is used. No shear factor *"SF"* is required since the Crash-straps, the stringer and the Seat track rail are relatively thick. (ref. *Table 13-4*). The Margin of Safety for shear is thus calculated as follows:

$$MS = \frac{P_{su}}{FF \times F} - 1 = \frac{2{,}005}{1.15 \times 1{,}237} - 1 = 0.40$$

Bearing:

The bearing stress will be checked in all of the three parts involved in the joint. The bearing load allowable is calculated as follows:

$$P_{bru} = F_{bru} \cdot D \cdot t$$

Table 14-37 shows the resulting bearing MS and *Figure 14-80* shows a detailed view of the new configuration with the Hi-Lites fasteners.

***Table 14-37** - Bearing stress MS table summary.*

Part	Material Al Aly	t [in]	D [in]	Fbru [ksi]	Pbru [lbs]	F [lbs]	MS bearing
Crash-strap	2024-T3 Clad	0.071	0.163	120	1389	1237	0.12
Seat track	7075-T6	0.070	0.163	145	1654	1237	0.34
Stringer	7075-T6	0.080	0.163	145	1891	1237	0.53

Item *5* shown in *Figure 14-80* is a structural shim. Its purpose is to simply fill the gap between the Crossbeam and the Crash-strap. Since I'm assuming that all the axial load in the Crash-strap is transferred by only two fasteners, the third fastener that doesn't pick-up the Seat track rail sees no load. Hence, the shim and its connections are thus Passed by inspection.

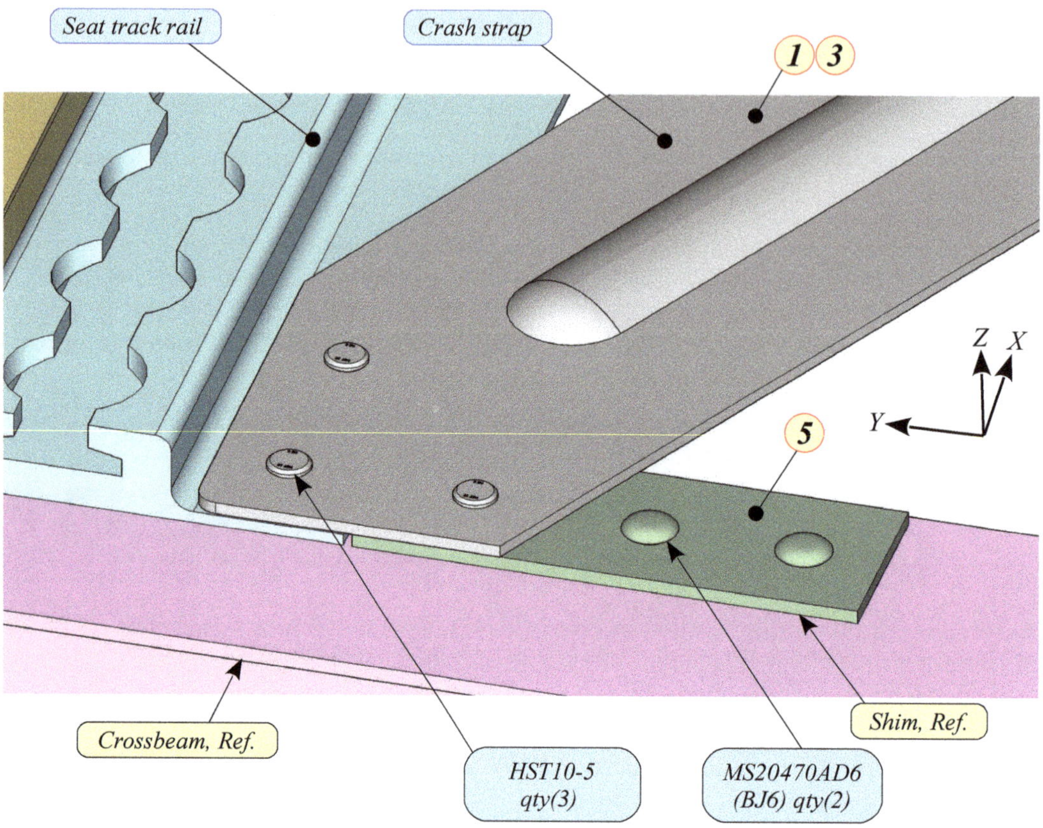

Figure 14-80 *– Detail view of the new Crash-strap connection configuration with the HST10-5's.*

A calculation example for the lowest bearing stress *MS* is done below:

The most critical part is the Crash-strap. The bearing ultimate allowable load "P_{bru}" in the Crash-strap made out of *Aluminum Alloy 2024-T3 Clad* 0.071 *in.* thick is calculated:

For: F_{bru} = 120 *ksi* (ref. *Table 14-19)*, $e/D = 2.0$. And for $D = 0.163$" and $t = 0.071$" hence:

$$P_{bru} = 120{,}000 \cdot 0.163 \cdot 0.071 = 1{,}389 \ lbs$$

No Fitting Factor is used in bearing.

Margin of Safety for the Crash-strap connections

$$MS = \frac{P_{bru}}{F} - 1 = \frac{1{,}389}{1{,}237} - 1 = 0.12$$

Crash-strap, Item 1 *Ultimate*	***0.12***
9g Fwd *HST10-5 in Al Aly* *2024-T3 Clad*	*Bearing*

Since the *MS* in bearing is lower than the one in shear, the joint is thus bearing critical. For all materials involved, the ratio $F_{bru} \, / \, F_{bry}$ is smaller than 1.5. Therefore, no bearing check at limit (yield) is necessary.

14.11 CROSSBEAMS

The Crossbeams were analyzed in section *13.1*. Hence, the analysis done here is to ensure that the Cabinet installation is not more critical than the other load cases used in that section. See *Figure 14-81* for a typical Crossbeam installation.

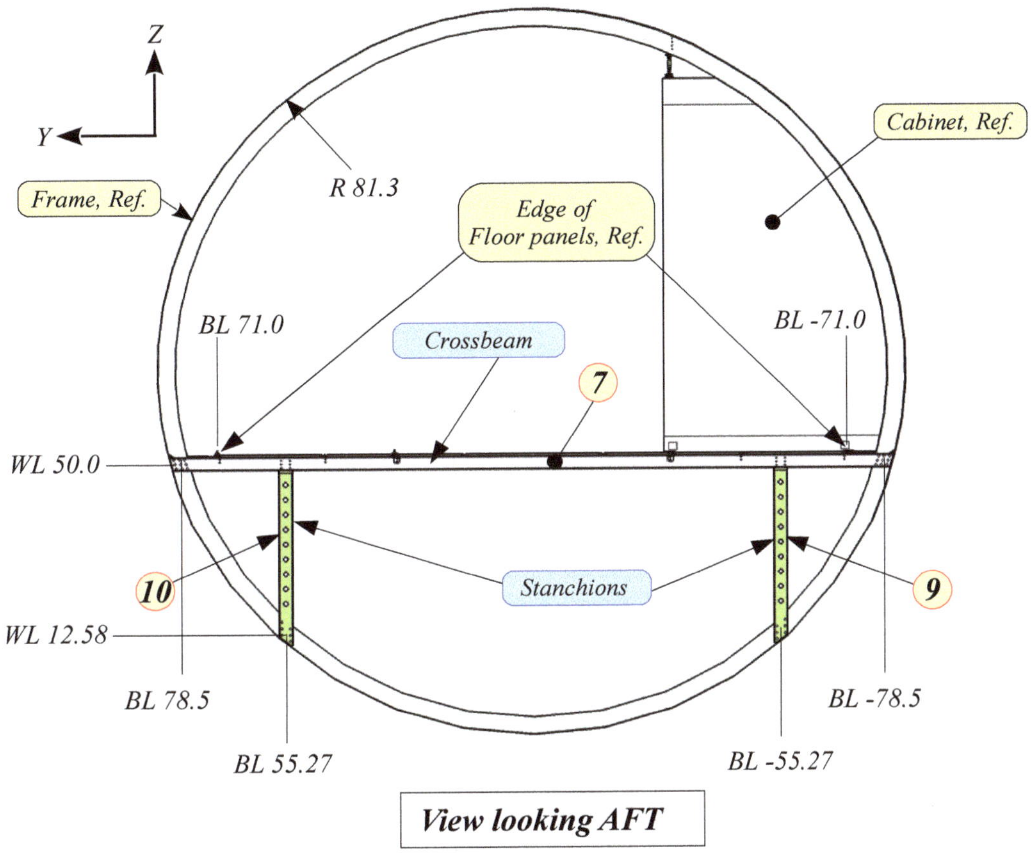

Figure 14-81 – *Cabinet and Crossbeam installation.*

14.11.1 LOADS

The analysis performed here is for the cabinet installation. The Crossbeams internal loads due to the Cabinet installation will simply be compared to the envelope load calculated in sub-section *13.1.2*. The internal loads due solely to the cabinet installation will be compared to that load envelope.

The same assumptions as in *Figure 14-4* are made. For instance, the Stanchions are assumed pinned at their Crossbeam attachment, ends are assumed simply supported and, the vertical reactions stiffness at the Frame and at the Stanchions are assumed the same.

By isolating the Crossbeam from the rest of the structure, the locations of the Cabinet points, its dimensions and *FBD* are shown in *Figure 14-82*. There is one Cabinet on each side.

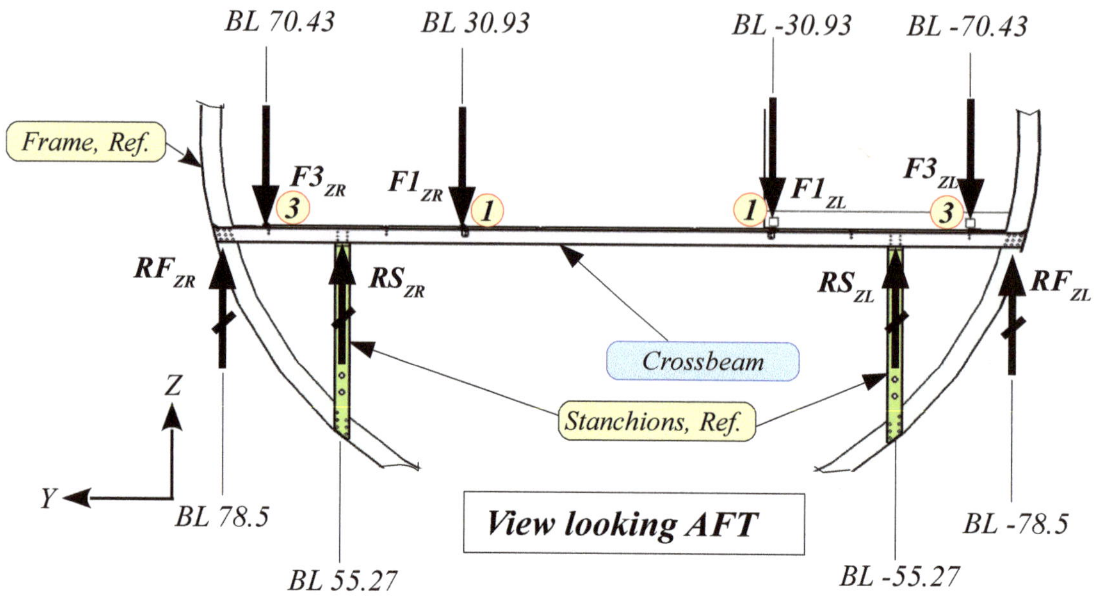

Figure 14-82 *– Crossbeam Dimensions and FBD.*

In order to determine which load case is the most critical, the shear and moment diagrams were calculated for each load case. Using the values for the Cabinet attachment points ***1*** *&* ***3*** from *Table 14-6*, *Table 14-9* and *Figure 14-12* for the *9g Fwd*, *3g Right* and *6g Down* load cases respectively, three sets of shear & moment diagrams are shown in *Figure 14-83* to *Figure 14-85*.

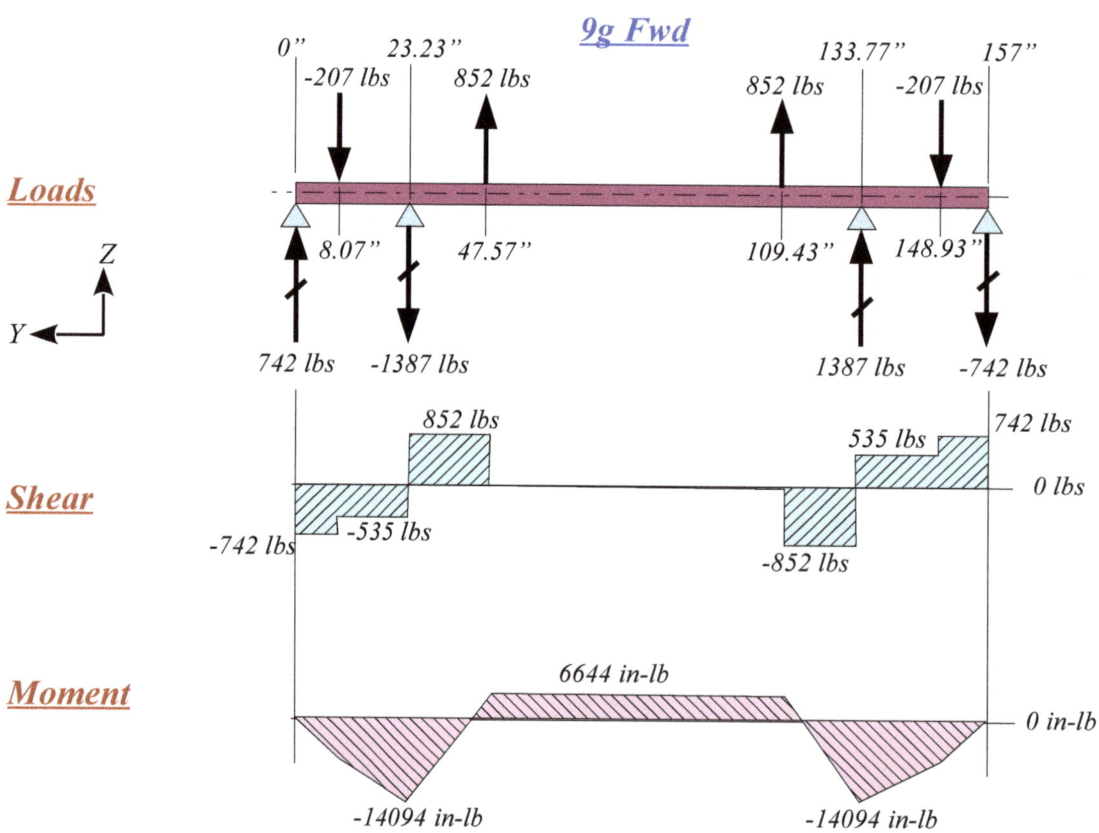

Figure 14-83 *– Crossbeam Shear & Moment diagram; 9g Fwd.*

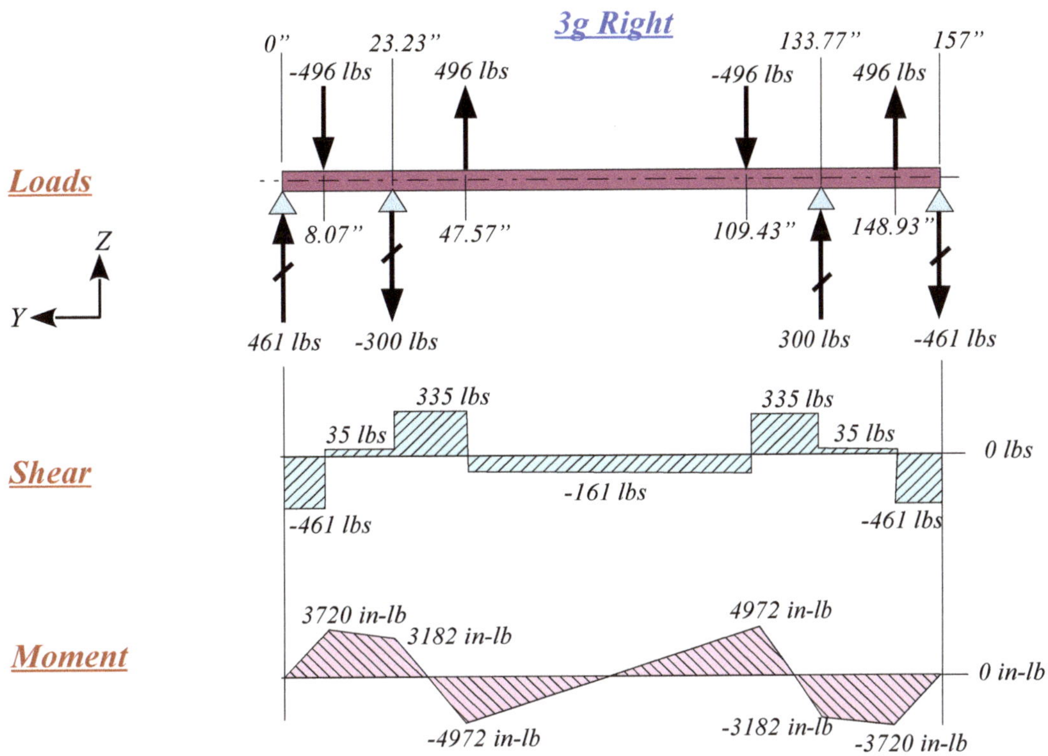

Figure 14-84 – *Crossbeam Shear & Moment diagram, 3g Right.*

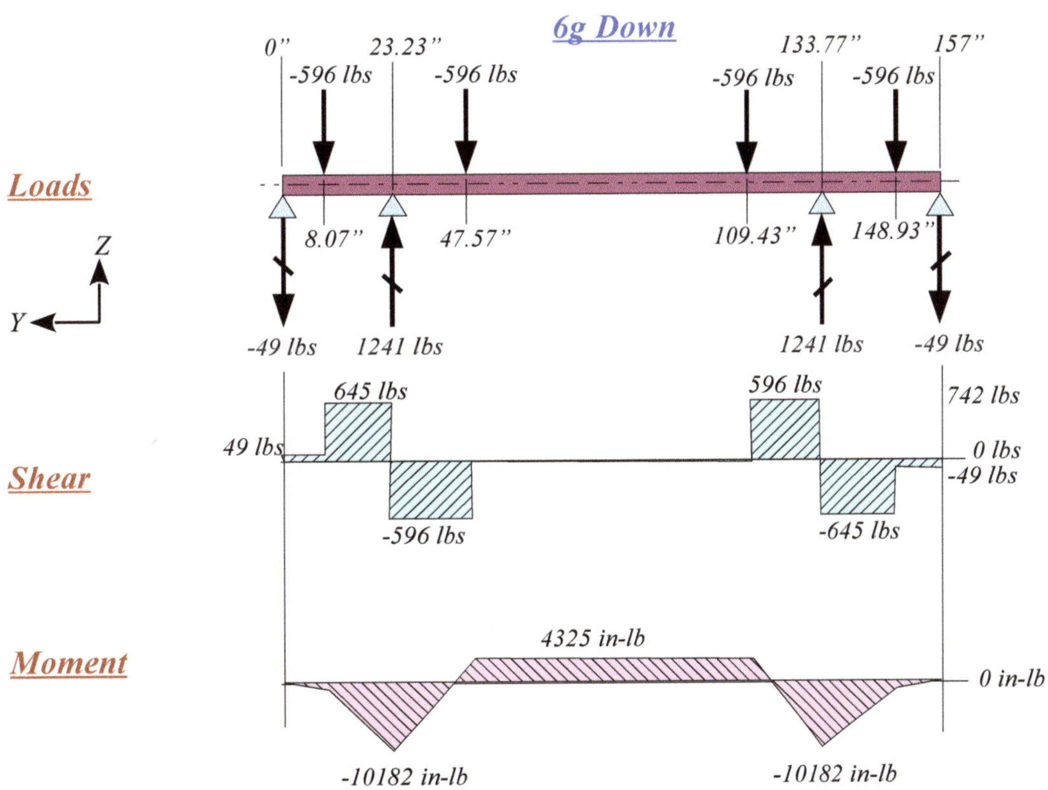

Figure 14-85 – *Crossbeam Shear & Moment diagram, 6g Down.*

One can notice that the critical condition is the *9g Fwd LC* with a maximum transverse shear of 852 *lbs* and a maximum bending moment of −14,094 *in-lb*. Comparing these results with the ones obtained in section *13.1.2* where the maximum shear and moment for the simply-supported condition of the *6g Down LC* (*Figure 14-4*) were 1,260 *lbs* and −20,515 *in-lb* respectively, the latter is found to be more critical.

There is no need to compare the conditions where the ends are fixed instead of simply-supported as they will still be higher for the floor loading case and we would still come up with the same conclusion. Omitting the shear & moment diagrams for the fixed ends was done to lighten the book, however, when writing an actual stress report, you may want to show all of the graphs to the satisfaction of your boss and to the certifying authority representative.

14.11.2 STRESS ANALYSIS

Since the internal loads used in section *13.1.2* are higher than the ones for the Cabinet installation, the condition from the Cabinet installation is thus **Passed by Comparison** to the floor loading under the *6g Down LC*. Hence, no further analysis is required for the Crossbeams.

Crossbeam Ultimate	***PBC***
6g Down *Al Aly 7050-T7451*	*Shear & Bending*

14.11.3 CONNECTIONS

<u>Crossbeams *MS20470AD6* solid rivets</u>

The Crossbeams are connected to the Frame with *qty(9) MS20470AD6* rivets and with *qty(6) MS20470AD6* also to the Stanchions. Since the end moments and the Stanchion loads are higher from the floor loading case analyzed in section *13.1.2*, no further analysis is required for the Cabinet installation.

14.12 LUG FITTINGS

The Lug fittings are used at each end of the struts located on top of the Cabinet to restrain it to the airframe at its upper end. They are machined out of *Al Aly* ***7050-T7451*** plates. There are four fittings from which the two attached to the Cabinet at points ***5 & 6*** are the same *"Fwd Ftgs"* while the two others, at points ***7 & 8,*** are different *"Aft inboard* and *Aft outboard Ftgs".* See *Figure* ***14-86*** for their locations.

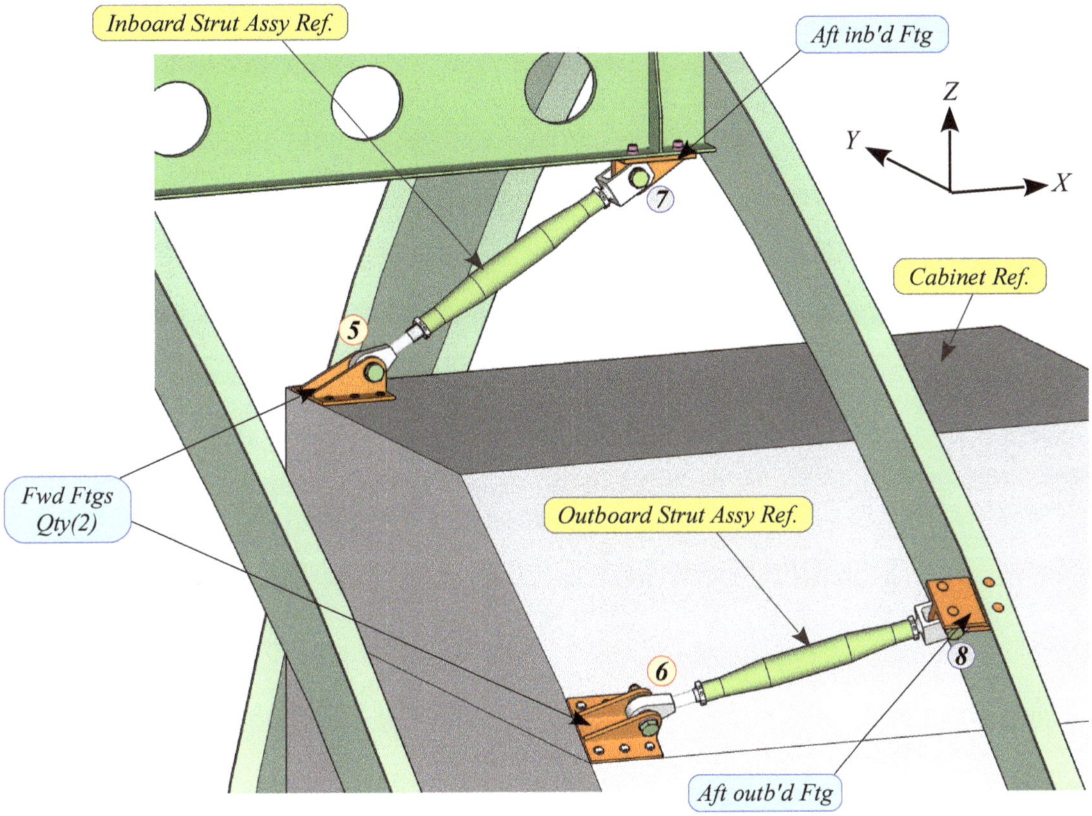

Figure 14-86 *– Lug Fittings locations.*

14.12.1 LOADS

For all Lug Fittings, the most critical load case is the ***9g*** *Fwd* emergency condition. Based on *Table* ***14-14***, the applied loads to the four Lug fittings are as shown in *Table* ***14-38***.

Table 14-38 *- Lug Fittings ultimate applied loads, 9g Fwd LC.*

Points	Name	P_{strut}	F_X	F_Y	F_Z	5% P_{strut}
		[lbs]	[lbs]	[lbs]	[lbs]	[lbs]
5	Fwd inb'd	784	668	0	411	39
6	Fwd outb'd	893	844	-178	234	45
7	Aft inb'd	784	-668	0	-411	39
8	Aft outb'd	893	-844	178	-234	45

From the above table, one can see that the outboard Lug fittings see more "out-of-plane" tension loads *"along Z"* while the inboard ones see more "in-plane" shear loads *"along X"*. So, every Lug fitting will be analyzed individually to ensure to capture everything.

Because of the lug geometry, the applied "P_X" and "$P_{Z'}$" loads to the fitting lugs must be transformed into an axial load "P_a" and a transverse load "P_t" as follows:

$$P_a = P_X \sin\theta + P_{Z'} \cos\theta$$
$$P_t = -P_X \cos\theta + P_{Z'} \sin\theta$$

Note that the axis "***Z'***" is the same as the "*Z*" axis for the Lug fittings at points *5 & 7* but is at an angle (rotated around the "*X*" axis by 37.2 *deg.*) for the Lug fittings at points *6 & 8*. See *Figure 14-11.*

The summary of all four Lug fittings transformed applied loads is shown in *Table 14-39.*

Table 14-39 *- Transformed applied loads along lug axes, 9g Fwd LC.*

Points	*Name*	P_X	$P_{Z'}$	P_a	P_t	Theta
		[lbs]	*[lbs]*	*[lbs]*	*[lbs]*	*[rad]*
5	*Fwd inb'd*	668	411	696	-361	31.0
6	*Fwd outb'd*	844	294	687	-572	31.0
7	*Aft inb'd*	668	411	678	-394	28.2
8	*Aft outb'd*	844	294	805	-389	45.0

A calculation example for the Fwd outb'd Lug fitting (*point 6*) is shown below.

$$P_a = 844\sin(31) + 294\cos(31) = 687\ lbs$$
$$P_t = -844\cos(31) + 294\sin(31) = -572\ lbs$$

Where

$$P_{Z'} = \sqrt{F_Z^2 + F_Y^2} = \sqrt{234^2 + (-178)^2} = 294\ lbs$$

And theta "θ" is the angle between the axial lug axis "P_a" and the "Z'" axis.

14.13 FORWARD LUG FITTINGS

The forward inboard and forward outboard Lug fittings are exactly the same part. Because of the different strut angles, the two fittings are loaded slightly differently.

The details of the Forward inboard (point *5*) and outboard (point *6*) Lug fittings are shown in *Figure 14-87* and *Figure 14-88.* As one can notice from the figure below, one hole is larger in a lug than the other. This is to allow for a sliding bushing in order to prevent "pinching" the two lugs when the strut bolt is torqued.

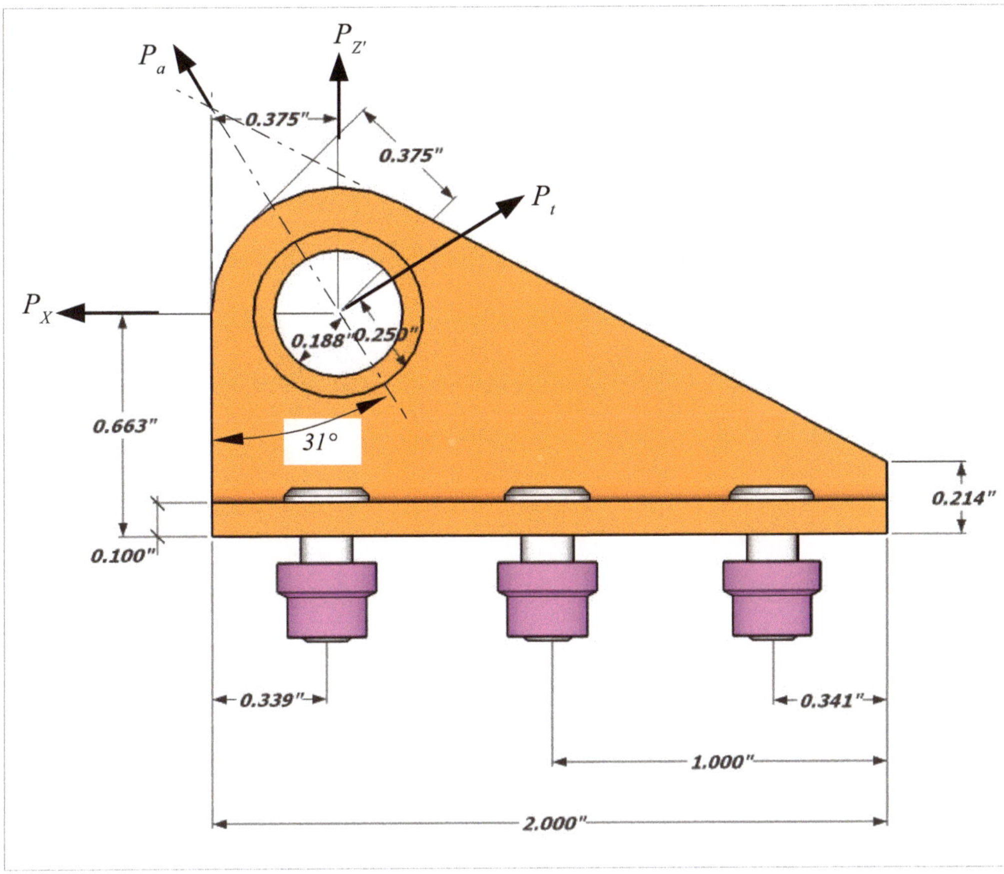

Figure 14-87 – Forward Lug Fitting dimensions – side view – points 5 & 6.

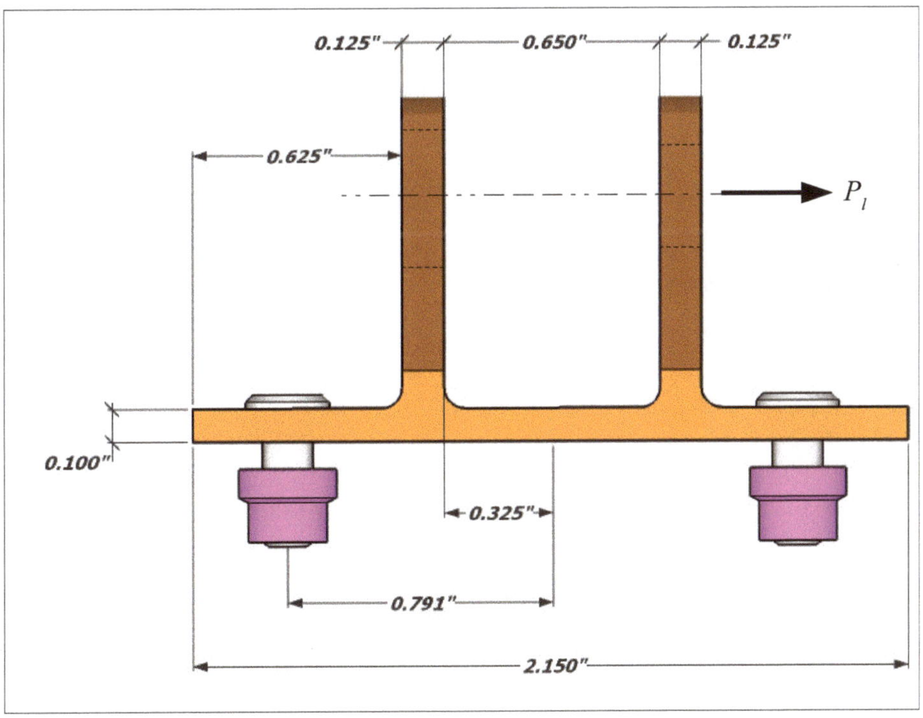

Figure 14-88 – Forward Lug Fitting dimensions – front view – points 5 & 6.

14.13.1 SECTION PROPERTIES

The critical cross-section are section *A-A* and section *B-B* as shown in *Figure 14-89*. At first, the exact position of section *B-B* is not known as I don't know yet where the most critical location is. It will be determined later as being through fastener *2* according to *Figure 14-94*.

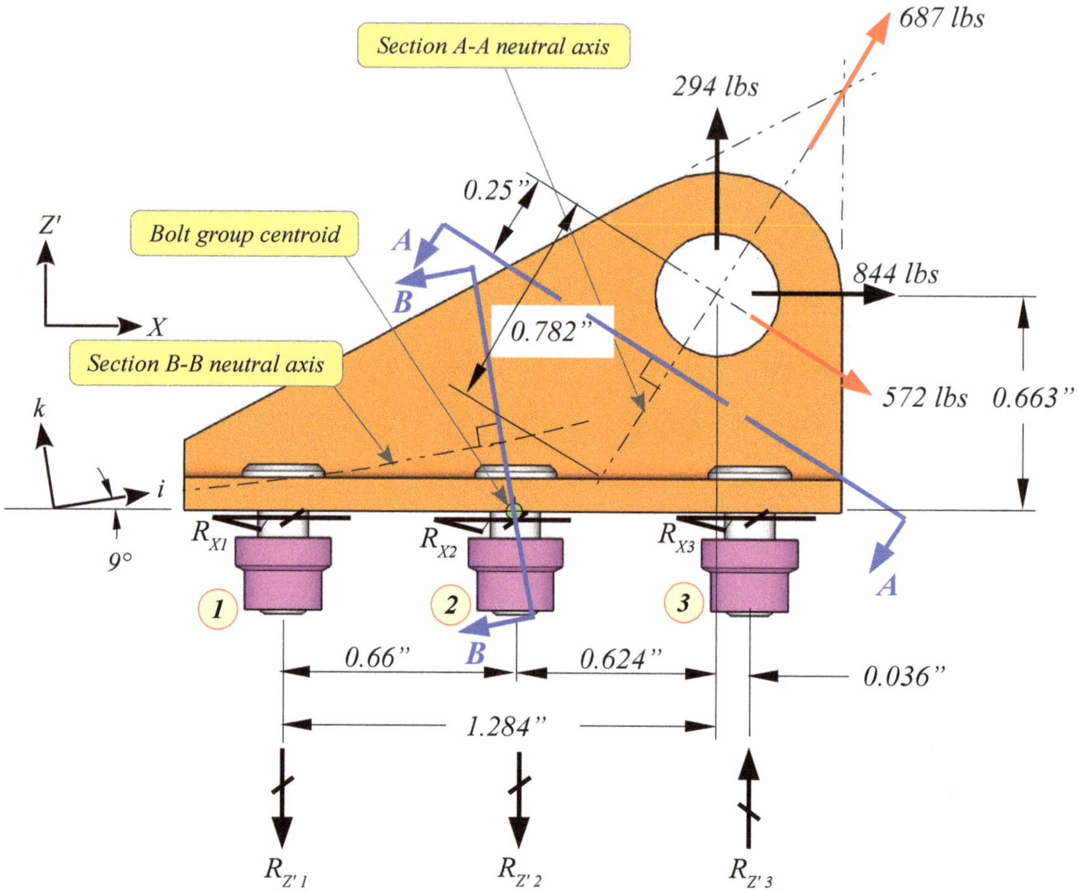

Figure 14-89 – *Fwd outboard Lug Fitting FBD – point 6.*

14.13.1.1 Through the Lugs (Section A-A)

Since we consider a side load applied to one of the two lugs and since a ***60/40*** load distribution is assumed, section *A-A* is for one lug only as shown in *Figure 14-90*.

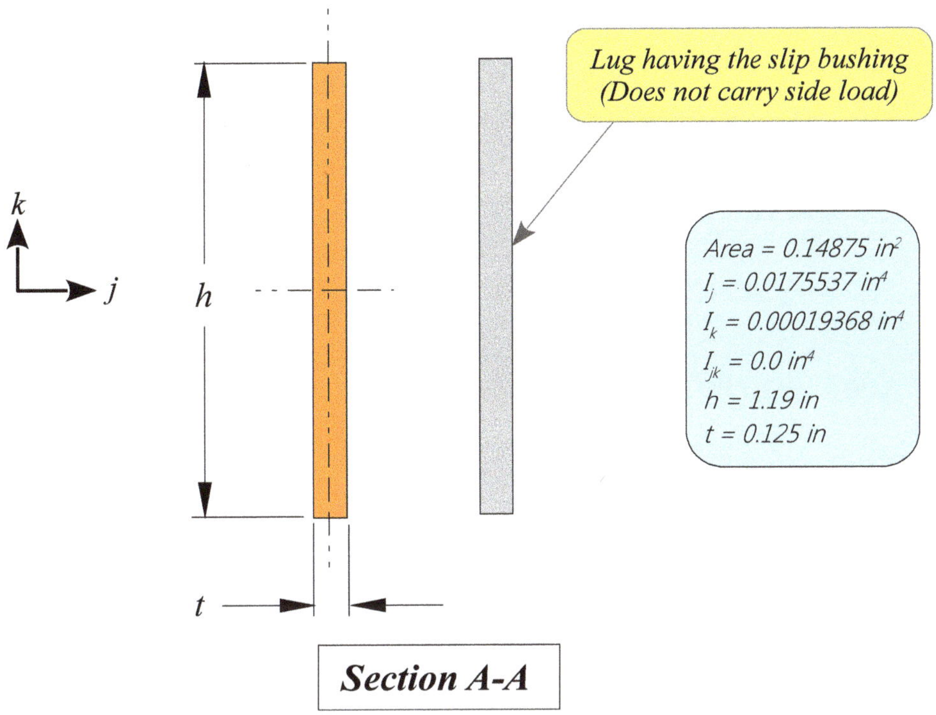

Figure 14-90 *– Lug fitting Section A-A.*

One would be tempted to consider the whole width of the integral stiffeners, having the same length as the fitting's base, as effective to react the side force. However, the portion of the stiffeners around fastener *1* being quite thin and far from the lug's elastic axis, this portion is expected to take a very small share of the load and is conservatively neglected.

Had the elastic axis been exactly at fastener 2 (as it should had been), then the whole width could have probably been used.

14.13.1.2 Through the base (Section B-B)

Many cross-sections across the Lug fitting base between fastener *1* and fastener *2* were made in order to determine the most critical location by trial and error. Since the section of the Lug fitting is not constant, the position of the sections' center of gravity varies along the *X* axis. By connecting the different center of gravity with a line, one obtain the "elastic axis" of the part. In our case, this elastic axis is slightly curved but, being almost straight, can be assumed a straight line having a *9 deg.* of inclination from the *X* axis. See *Figure 14-89.*

Now knowing the angle of the elastic axis, the cross-sections were redone but this time, perpendicular to that elastic axis. For conservatism, the thicknesses of the base and flanges were kept as their nominal thicknesses. Only the height of the integral stiffeners were affected by the inclination.

For the stress analysis, a new local coordinate system is created that has the "*i*" axis along the elastic axis and the "*k*" axis normal to it. The internal forces will be transformed accordingly of course.

The critical section *B-B* is shown in *Figure 14-91.*

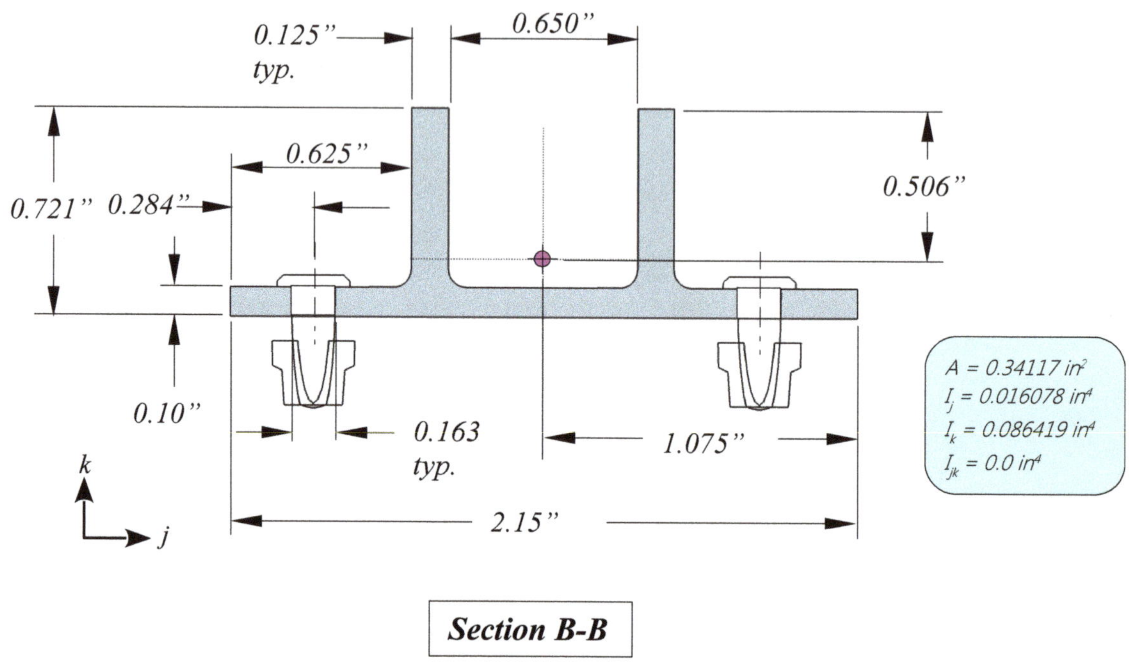

Figure 14-91 *– Lug fitting Section B-B.*

Although the section was cut at an angle of ***9*** *deg.*, the nominal base and integral stiffeners thicknesses are conservatively used instead of the projected thickness.

14.13.2 STRESS ANALYSIS

The following checks will be performed to analyze each Lug fitting:

- Lug analysis including out-of-plane Lug bending and hoop stress due to bushing interference fit.
- Flanged bushings interference fit
- Bending + tension + shear
- Tension Fitting
- Connections

The same part is used at point ***5*** and at point ***6*** but one has a slightly higher axial load while the other has a higher transverse load.

To be conservative (and faster), one could be tempted to use the maximum of both axial and transverse loads for the analysis thus substantiating both Forward inb'd and Forward outb'd parts at the same time. Although that would work for the lug analysis, it might not be conservative for the bending stress in the Lug fitting body as the axial and transverse forces could be acting against each other.

14.13.3 STRESSES AT SECTION A-A

Figure 14-89 shows the *FBD* of the *Forward outboard Lug fitting* for the *9g Fwd LC*. For conservatism, the load distribution between the two lugs is assumed as 60/40. Therefore, only one lug is analyzed with *60%* of the total applied load, plus, in combination, a side load corresponding to *5%* of the total applied load (strut load). The *FBD* of the Lug is shown in *Figure 14-92*.

The lug is considered as a cantilevered beam having the loads applied at the center of the hole and fixed at the base. For conservatism, section *A-A* is analyzed by using the bending moments "M_k" at the base. Following, is an example of the analysis through section *A-A* for the *Forward Outboard Lug fitting* under the *9g Fwd LC*.

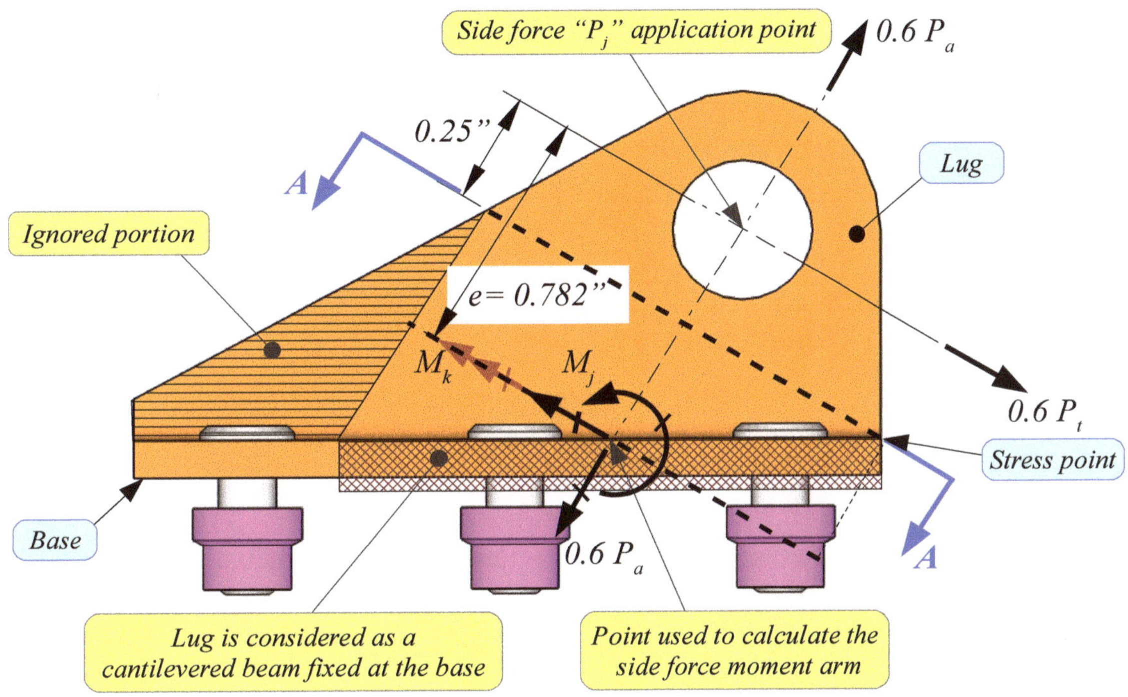

Figure 14-92 – *Lug FBD, Section A-A.*

Axial stress

The axial load going through the lug at section *A-A* is 60% of P_a. Hence, using *Figure 14-90* and *Table 14-39*, the axial stress f_a *is:*

$$f_a = \frac{0.60\,P_a}{A} = \frac{0.60\;687}{0.14875} = 2{,}770\ psi$$

Bending stress

Around the "*j*" axis:

Using *Figure 14-92*, the bending moment at the base "M_j" is calculated as:

$$M_j = 0.60\,P_t\cdot(0.782) = (0.60)572\cdot(0.782) = 268\ in.-lb$$

So, the section being a rectangle, the bending stress at the top and at the bottom is the same but with opposite signs.

$$f_{bj} = \frac{6\,M_j}{tb^2} = \frac{6\cdot 268}{0.125\cdot 1.19^2} = \pm 9{,}097\ psi$$

Around the "*k*" axis:

Again using *Figure 14-92*, the bending moment at the base "M_k" is calculated as:

$$M_k = P_j \cdot (0.782) = 45 \cdot (0.782) = 35 \ in.-lb$$

So, for a rectangle, the bending stress is:

$$f_{bk} = \frac{6M_k}{bt^2} = \frac{6 \cdot 35}{1.19 \cdot 0.125^2} = \pm 11{,}267 \ psi$$

Shear stress

The shear stress "f_S" is calculated through section *A-A* along both ***j*** and ***k*** directions due to P_j and P_t respectively.

Along the *j* axis:

$$f_{sj} = \frac{P_j}{A} = \frac{45}{0.14875} = 300 \ psi$$

Along the *k* axis:

$$f_{sk} = \frac{0.60 \, P_t}{A} = \frac{0.60 \cdot 572}{0.14875} = 2{,}307 \ psi$$

Adding vectorially the two shear stresses we get:

$$f_s = \sqrt{f_{sj}^2 + f_{sk}^2} = \sqrt{300^2 + 2{,}307^2} = 2{,}327 \ psi$$

Combined stresses

The axial, bending and shear stresses through section *A-A* are calculated at the "stress point"; shown in *Figure 14-92*. Those stresses are combined according to the interaction equation:

$$(R_a + R_b)^2 + R_S^2 = 1$$

In our case, the axial and the bending stresses are simply added up and a single axial stress ratio "R_a'" is calculated. Thus the MS is calculated as:

$$MS = \frac{1}{FF\sqrt{R_a'^2 + R_S^2}} - 1$$

Using the material allowables from *Table 14-16*, the stress ratios are calculated where R_a' and R_S are respectively the axial and shear stress ratios calculated as follows:

$$R_a' = \frac{f_a + f_{bj} + f_{bk}}{F_{tu}} = \frac{2{,}770 + 9{,}097 + 11{,}267}{75{,}000} = 0.308$$

$$R_s = \frac{f_s}{F_{su}} = \frac{2{,}327}{40{,}000} = 0.058$$

Hence, using a *FF* of 1.15, the MS at the stress point of section *A-A* is:

$$MS = \frac{1}{1.15\sqrt{0.308^2 + 0.058^2}} - 1 = 1.77$$

For additional conservatism, I could have used the Yield compression allowable "F_{CY}" instead of the tensile strength "F_{TU}" since the lug is relatively thin and could buckle under compression stresses. It would be interesting to see what the free flange buckling allowable due to bending would be. However, I suspect it to be close to F_{CY} which would still lead to an *MS* greater than 1.0.

The *Fwd inboard* (point *5*), *Fwd outboard* (point *6*) and the *Aft inboard* (point *7*) *Lug fittings* are very similar and are thus analyzed the same way. See *Figure 14-90* for the two *Fwd Ftgs* section properties and *Figure 14-105* for the *Aft inb'd Ftg* section properties. *Table 14-40* shows a summary of the properties, the stresses and the MS using a Fitting Factor *"FF"* of 1.15 for those three parts under the *9g Fwd LC*. The *Aft outboard Lug fitting* being different is analyzed separately in section *14.15.3*.

Table 14-40 *- Resulting stresses and MS through Section A-A Lug fittings at points 5, 6 & 7, 9g Fwd LC.*

Points	Name	e	b	h	c	A	I_j	I_k
		[in]	[in]	[in]	[in]	[in²]	[in⁴]	[in⁴]
5	Fwd inb'd	0.782	0.125	1.19	0	0.14875	0.017554	0.000194
6	Fwd outb'd	0.782	0.125	1.19	0	0.14875	0.017554	0.000194
7	Aft inb'd	0.548	0.25	1.352	0.58	0.193	0.047421	0.001005

Points	Name	%P	P_a	P_t	M_j	f_a	f_b	f_a+f_b	M_k	f_b	f_{Total}	R_A
		one lug	[lbs]	[lbs]	[in-lbs]	[psi]	[psi]	[psi]	[psi]	[psi]	[psi]	
5	Fwd inb'd	0.60	418	-217	169	2809	5740	8549	31	9892	18440	0.246
6	Fwd outb'd	0.60	412	-343	268	2770	9097	11867	35	11267	23134	0.308
7	Aft inb'd	1.00	678	-394	216	3512	3082	6594	21	2672	9266	0.124

Points	Name	P_a	P_t	f_{St}	f_{St}	f_S	R_S	MS
		[lbs]	[lbs]	[psi]	[psi]	[psi]		
5	Fwd inb'd	418	-217	1456	264	1479	0.037	2.50
6	Fwd outb'd	412	-343	2307	300	2327	0.058	1.77
7	Aft inb'd	678	-394	2044	203	2054	0.051	5.50

14.13.4 STRESSES AT SECTION B-B

FBD

For this *FBD* analysis, not knowing how stiff the cabinet is in the out-of-plane direction, the vertical reaction forces will all be assumed at the fasteners. There are *3* pairs of bolts having their centroid right at the center of fastener pair no. *2*. The reaction forces at the bolts are determined as follows. The sum of the moments at the bolt centroid is calculated (positive CW):

$$M_Y = 294\left(-0.624\right) + 844\left(0.663\right) = 376.1\ in.-lb$$

Since fastener *2* is right at the moment application point (bolts centroid), it can't contribute in reacting the moment; only fasteners *1 & 3* can react the moment as a couple. For a distance between them of 1.32 *in.*, the reaction forces due to the moment *"M_Y"* are:

$$R_{Z'1M} = -R_{Z'3M} = \frac{M_Y}{d} = \frac{376.1}{1.32} = 285\ lbs$$

Then, the sum of the forces at the bolt centroid is:

$$R_{Z'1V} + R_{Z'2V} + R_{Z'3V} = 294\ lbs$$

The total vertical force is shared evenly between all 3 pairs of fasteners. Therefore, the reaction forces due to the vertical force is:

$$R_{Z'1V} = R_{Z'2V} = R_{Z'3V} = \frac{294}{3} = 98\ lbs$$

The contribution of the moment and the vertical load is then combined to determine the resulting fastener loads.

$$R_{Z'1} = R_{Z'1M} + R_{Z'1V} = 285 + 98 = 383\ lbs$$
$$R_{Z'2} = R_{Z'2V} = 98\ lbs$$
$$R_{Z'3} = R_{Z'3M} + R_{Z'3V} = -285 + 98 = -187\ lbs$$

For the shear load, all 3 rows of fasteners are assumed to take an equal share of the total load. Therefore, the shear force in each fastener is:

$$R_{X1} = R_{X2} = R_{X3} = \frac{844}{3} = 281\ lbs$$

As we've seen previously, a new coordinate system "*C.S.*" is to be used to calculate the stress at the critical cross-section of the Lug fitting base. The internal loads are then transformed from the "*X-Z'* " *C.S.* to the "*i-k*" *C.S.* as follows:

$$P_i = P_X \cos(9) + P_{Z'} \sin(9)$$
$$P_k = -P_X \sin(9) + P_{Z'} \cos(9)$$

At the cross-section right at fastener *1*, the following internal loads exist: $F_X = R_{X1} = 281\ lbs$ and, $F_{Z'} = R_{Z'1} = 383\ lbs$. Thus transforming those forces in the "*i-k*" *C.S.* we get:

$$P_i = 281\cos(9) + 383\sin(9) = 337\ lbs$$
$$P_k = -281\sin(9) + 383\cos(9) = 334\ lbs$$

Due to the distance between the bolt shear plane and the section's center of gravity "*h*", a moment "M_{Y1}" exists and is calculated as follows. As shown in *Figure 14-93*, for $h = 0.095\ in.$ at the section through fastener *1*:

$$M_{Y1} = M_{j1} = P_X h = 281.3(0.095) = 26.7\ in.-lb$$

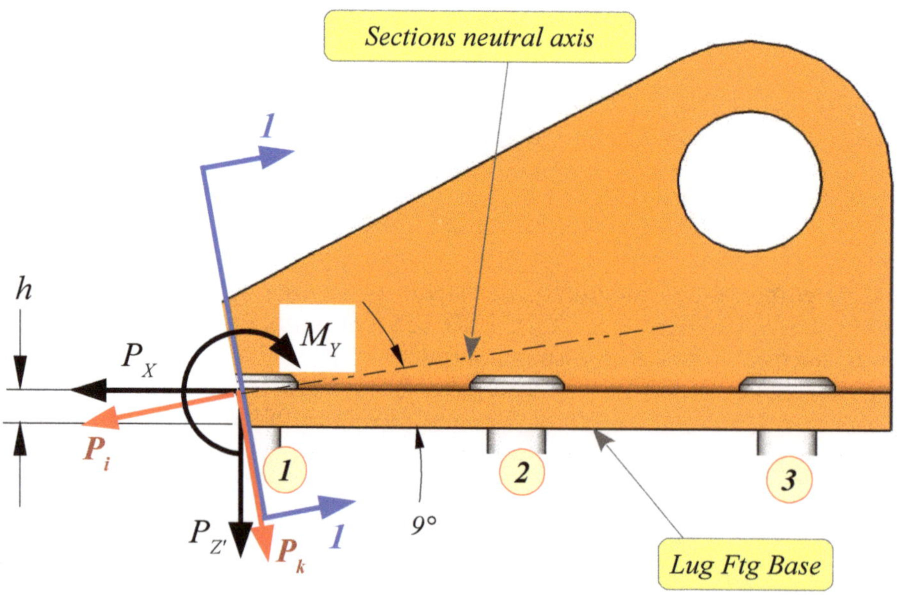

Figure 14-93 *– Internal loads at section through fastener 1.*

Shear and moment diagram for Section *B-B*

The shear and moment diagram of the Lug fitting is shown in *Figure 14-94* along the fitting's elastic axis ("*i*" axis). For conservatism, it is assumed that the attachment fasteners are pinned and therefore provide no restraining moment. In addition, the section modulus "$S = I/c$" is shown as well as the bending stress "$f_b = M/S$" in order to determine the most critical location. Only the portion between fastener *1* and just passed fasteners *2* was necessary to determine the most critical location, without taking the fasteners' holes into account, the critical location is right as fastener *2*.

Table 14-41 shows the internal loads, stresses and *MS* for the Forward inboard and outboard Lug fittings. The detailed calculations are shown further for the most critical one.

Table 14-41 *- Resulting stresses and MS through Section B-B Lug fittings at points 5 & 6, 9g Fwd LC.*

Points	Name	P_i	P_k	M_j	f_a	f_b	f_a+f_b	R_A
		[lbs]	[lbs]	[in-lbs]	[psi]	[psi]	[psi]	
5	Fwd inb'd	338	334	-197	990	6390	7380	0.098
6	Fwd outb'd	263	240	-134	772	4344	5116	0.068

Points	Name	P_i	P_k	f_S	R_S	MS
		[lbs]	[lbs]	[psi]		
5	Fwd inb'd	338	334	1854	0.046	6.99
6	Fwd outb'd	263	240	1331	0.033	10.46

Then the stresses at the most critical location through section *B-B* due to the *9g Fwd LC* are calculated below.

Axial stress

Using *Figure 14-94*, the axial load going through the part at section *B-B* is 338 *lbs*. The section properties of the section to be used is from *Figure 14-91* which takes into account the fastener holes. The axial stress "f_a" is thus:

$$f_a = \frac{P_i}{A} = \frac{338}{0.34117} = 991\ psi$$

Bending stress

Then again, from *Figure 14-94*, the maximum bending moment is −197 *in-lb* so the bending stress is calculated both at the top and at the base of the Fwd Lug Ftg.

$$f_{btop} = -\frac{M_j\, c_{top}}{I_j} = -\frac{-197\,(0.521)}{0.016078} = 6{,}390\ psi$$

$$f_{bbase} = -\frac{M_j\, c_{base}}{I_j} = -\frac{-197\,(-0.200)}{0.016078} = -2{,}451\ psi$$

Shear stress

The average shear stress "f_S" is calculated through section *B-B* but by removing the base flanges as those won't be carrying much vertical shear due to their vertical flexibility. Therefore, the remaining shear area is: $A_S = 2(0.721 * 0.125) = 0.18025\ in^2$. Hence, using the transverse shear load of 334 *lbs* just left of section *B-B*, we get:

$$f_S = \frac{V_k}{A_S} = \frac{334}{0.18025} = 1{,}853\ psi$$

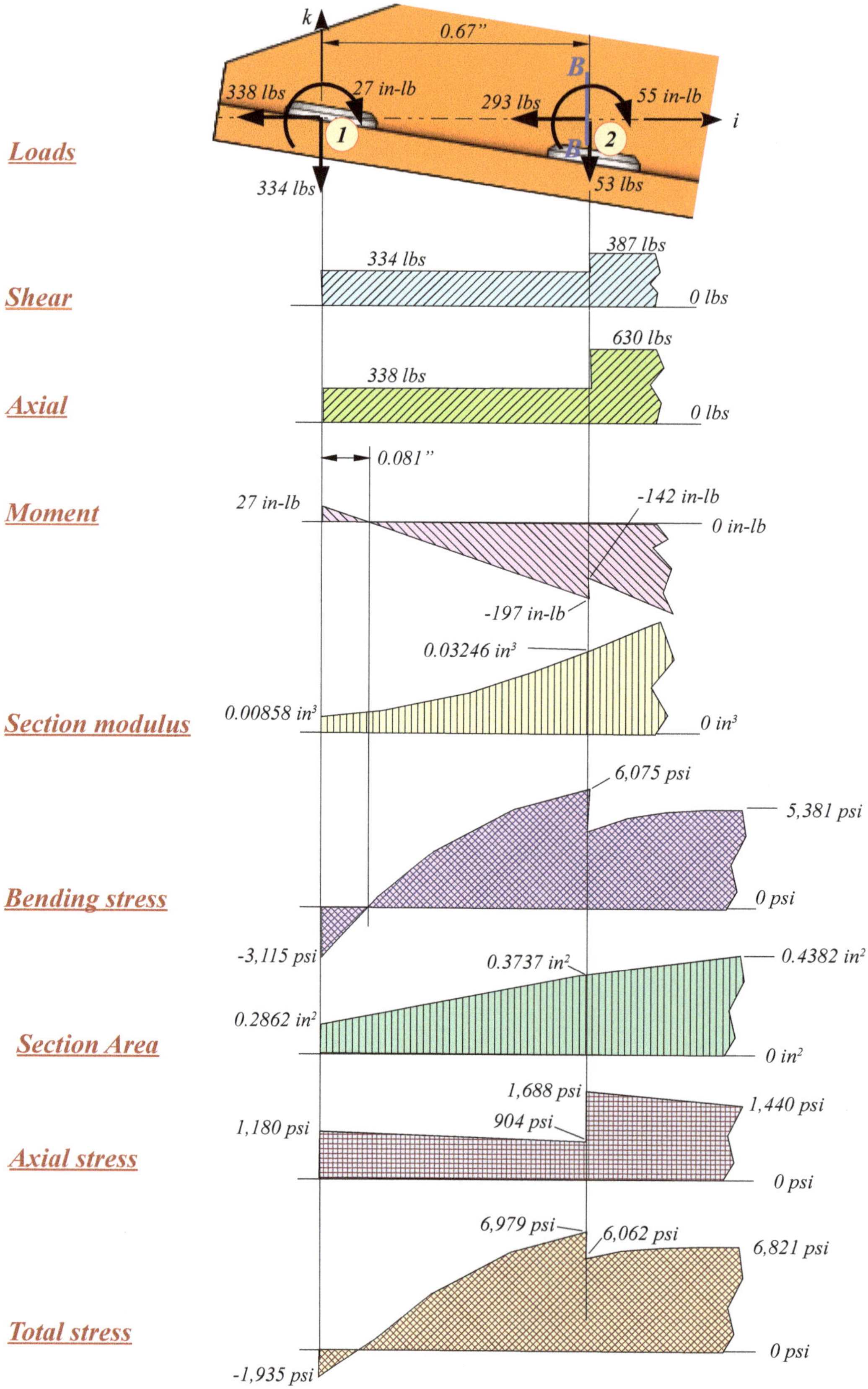

***Figure 14-94** – Shear & Moment and Section modulus & Stress graphs.*

Combined stresses

At the top

The axial, bending and shear stresses through section *B-B,* are combined according to the equation:

$$(R_a + R_b)^2 + R_S^2 = 1$$

In our case, because the stress is in tension and no plastic bending is used, the axial and the bending stresses are simply added up and a single axial stress ratio "R_a'" is calculated.

Thus the MS is calculated as:

$$MS = \frac{1}{FF\sqrt{R_a'^2 + R_S^2}} - 1$$

Using the material allowables from *Table 14-16*, the stress ratios are calculated where R_a' and R_S are respectively the axial and shear stress ratios calculated as follows:

$$R_a' = \frac{f_a + f_{btop}}{F_{tu}} = \frac{991+6{,}390}{75{,}000} = 0.098$$

$$R_s = \frac{f_s}{F_{su}} = \frac{1{,}853}{40{,}000} = 0.047$$

Hence, using a *FF* of 1.15, the MS at the top of section *B-B* is:

$$MS = \frac{1}{1.15\sqrt{0.098^2 + 0.047^2}} - 1 = 6.99$$

At the base

Using the same approach at the base but using the yield strength "F_{CY}" as the allowable from *Table 14-16,* the stress ratios are calculated as follows:

$$R_a' = \frac{f_a + f_{bbase}}{F_{CY}} = \frac{991 - 2{,}451}{60{,}000} = 0.024$$

$$R_s = \frac{f_s}{F_{su}} = \frac{1{,}853}{40{,}000} = 0.047$$

Hence, using a *FF* of 1.15, the MS at the base of section *B-B* is:

$$MS = \frac{1}{1.15\sqrt{0.024^2 + 0.047^2}} - 1 = 15.47$$

Margin of Safety of the Forward Lug Ftg

The above analysis showed that section *A-A* was more critical than at section *B-B*. The most critical MS for that fitting is then:

Fwd Lug Ftg *Section A-A, Ultimate*	***>1.0***
9g Fwd *Al Aly 7050-T7451*	*Tension + bending + shear*

14.13.5 LUG ANALYSIS

The most critical lug analysis is shown below as an example. The *Forward outboard Lug fitting* located at point ***6*** is the most critical one. Since there are bushings in the lugs, the hole diameters without the bushings have to be considered.

Before doing the longer lug analysis, a quick preliminary bearing check is done using ***Table 14-16*** and ***Figure 14-87*** and ***Figure 14-88***.

The bearing load allowable "P_{bru}" is calculated for the smaller hole as:

For:
$t = 0.125$ *in*
$e = 0.375$ *in*
$D = 0.375$ *in*
$e/D = 2$
$F_{bru} = 105$ *ksi*

$$P_{bru} = F_{bru} * D * t = 105{,}000 * 0.375 * 0.125 = 4{,}921\ lbs$$

Assuming a ***60/40*** applied load distribution between the two lugs, the maximum applied load to a lug is: 0.60 * 893 = 536 *lbs*. The bearing allowable is way higher than the ultimate applied load. Therefore the lug could be Passed by engineering judgment. But for academic purposes, a complete lug analysis is performed.

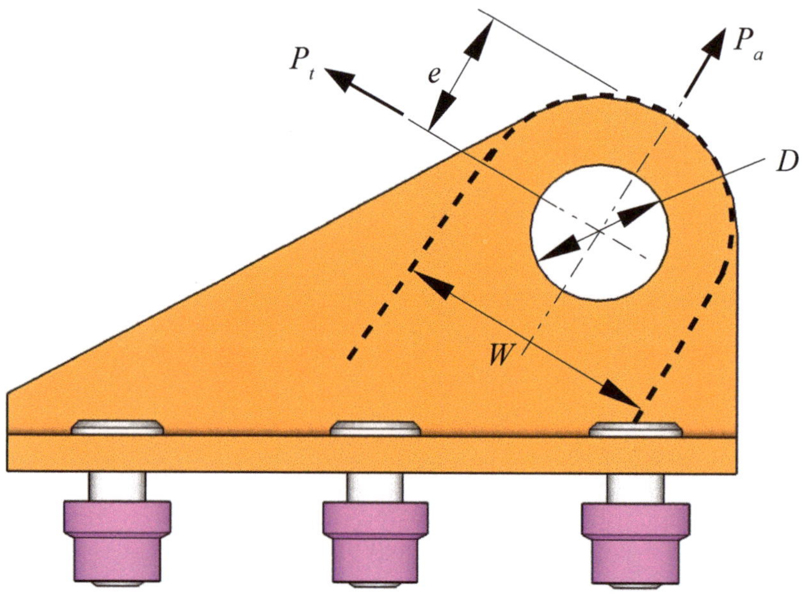

Figure 14-95 – *Forward Lug fitting idealized lug geometry.*

Lug axial net section strength

For conservatism and to simplify the analysis, the lug geometry is assumed as the dotted line shown in *Figure 14-95*. The largest hole "D_2" is used in the analysis.

> There might be requirements sometimes to perform the lug analysis by considering oversized holes. This allows for repairs without having to revisit the analysis.

The axial tension allowable "P_u" is calculated as:

$$P_u = K_t \, F_{tu} \, A_t$$

The tension efficiency factor "K_t" is obtained from *Figure B-6* using curve *2* For $W/D = 0.75 / 0.50 = 1.5$, we find a K_t of 0.86.

> For:
> $t = 0.125$ in
> $D_1 = 0.375$ in
> $D_2 = 0.50$ in
> $W = 0.75$ in

The net tension area "A_t" is calculated as:

$$A_t = (W - D)t$$

Thus $A_t = (0.75 - 0.50)\ 0.125 = 0.03125\ in^2$

For $F_{tu} = 75$ *ksi*, ref. *Table 14-16*. Therefore, the tension allowable is:

$$P_u = (0.86) * 75{,}000 * 0.03125 = 2{,}015\ lbs$$

Shear out bearing Strength

As for the axial net section check, to be consistent, the lug geometry is assumed as the dotted line shown on *Figure 14-95*.

Ultimate:

The shear out-bearing ultimate allowable "P_{bru}" is calculated as:

$$P_{bru} = K_{bru} \, F_{tu} \, A_{br}$$

The shear-bearing efficiency factor "K_{bru}" for axially loaded lugs is obtained from *Figure B-7*. For $D/t = 0.50 / 0.125 = 4.0$ and $e/D_2 = 0.375 / 0.50 = 0.75$, we find a K_{bru} of 0.50.

The bearing area "A_{br}" is calculated as: $A_{br} = D \ t$

Thus $A_{br} = 0.50 * 0.125 = 0.0625\ in^2$

For $F_{tu} = 75$ *ksi*, ref. *Table 14-16*. Therefore, the shear out-bearing allowable is:

$$P_{bru} = (0.50) * 75{,}000 * 0.0625 = 2{,}343\ lbs$$

Yield:

The shear bearing yield allowable "P_{bry}" is calculated as:

$$P_{bry} = K_{bry} \, F_{ty} \, A_{br}$$

The shear-bearing yield efficiency factor "K_{bry}" for axially loaded lugs is obtained from *Figure B-8*. For $t/D = 0.125 / 0.50 = 0.25$ and $e/D = 0.75$, we find a K_{bry} of 0.50.

For $A_{br} = 0.0625\ in^2$, $F_{ty} = 65\ ksi$, ref. *Table 14-16*.

Therefore, the shear bearing yield allowable is:

$$P_{bry} = (0.50) * 65{,}000 * 0.0625 = 2{,}031\ lbs$$

Since $P_{bru} / P_{bry} = 2{,}343 / 2{,}031 = 1.15$ and thus < 1.5, then no Yield *MS* needs to be calculated since it is less critical than the ultimate *MS*.

Lug Strength under transverse loading

Using the same lug geometry as the dotted line shown on *Figure 14-95* for consistency and conservatism.

Ultimate:

The transverse ultimate allowable "P_{tu}" is calculated as:

$$P_{tu} = K_{tu}\ F_{tu}\ A_{br}$$

The tension efficiency factor "K_{tu}" for transversely loaded lugs is obtained from *Figure B-9* using curve *3* for 7075-T6 plate > 0.5 *in*. For $A_{av}/A_{br} = 0.0207 / 0.0625 = 0.33$, we find a K_{tu} of 0.37.

The average area "A_{av}" is calculated using *Figure 14-96* as:

$$A_{av} = \frac{6}{\dfrac{3}{A_1} + \dfrac{1}{A_2} + \dfrac{1}{A_3} + \dfrac{1}{A_4}}$$

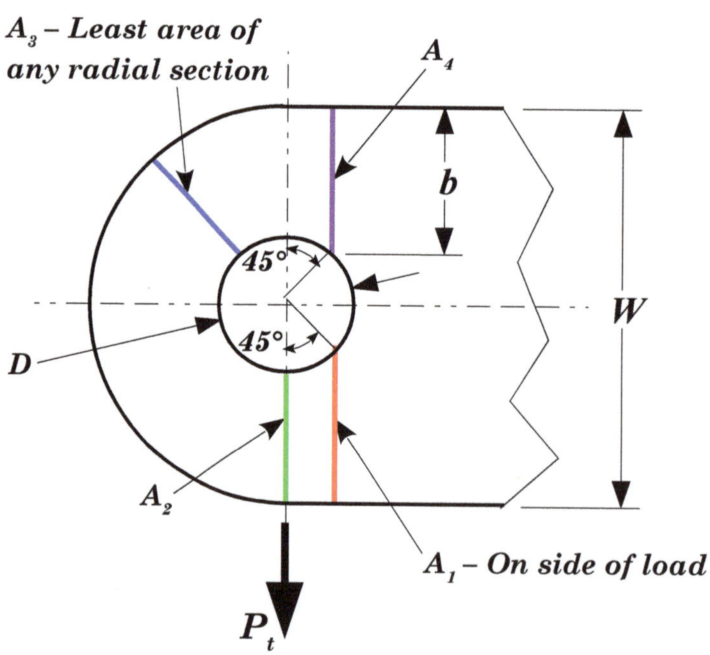

Figure 14-96 – *Lug section areas for transverse loading.*

Where, in our case;

$$A_2 = A_3 = \frac{(W-D)t}{2} = \frac{(0.75-0.50)(0.125)}{2} = 0.01562\ in.^2$$

$$A_1 = A_4 = \frac{(W-D\cos(45))t}{2} = \frac{(0.75-0.50\cos(45))(0.125)}{2} = 0.02478\ in.^2$$

Therefore;

$$A_{av} = \frac{6}{\frac{3}{0.02478}+\frac{1}{0.01562}+\frac{1}{0.01562}+\frac{1}{0.02478}} = 0.0207\ in.^2$$

For $A_{br} = 0.0625\ in^2$, $F_{tu} = 75\ ksi$ and $K_{tu} = 0.37$.

Therefore, the transverse ultimate allowable is:

$$P_{tu} = (0.37) * 75{,}000 * 0.0625 = 1{,}734\ lbs$$

Yield:

The transverse yield allowable "P_{ty}" is calculated as:

$$P_{ty} = K_{ty}\, F_{ty}\, A_{br}$$

The tension efficiency factor "K_{ty}" for transversely loaded lugs is obtained from *Figure B-9* using curve *1* for all Aluminum at yield. For A_{av}/A_{br} = 0.0207 / 0.0625 = 0.33, we find a K_{ty} of 0.41.

For $A_{br} = 0.0625\ in^2$, $F_{ty} = 65\ ksi$ and $K_{ty} = 0.41$.

Therefore, the transverse yield allowable is:

$$P_{ty} = (0.41) * 65{,}000 * 0.0625 = 1{,}665\ lbs$$

Since P_{tu}/P_{ty} = 1,734 / 1,665 = 1.04 and thus < 1.5, then no Yield MS would need to be calculated.

Margin of safety

Since we have a combination of axial and transverse loading (oblique loading), the following interaction equation is commonly used at yield and ultimate:

$$R_a^{1.6} + R_{tr}^{1.6} = 1$$

Thus the MS is calculated as:

$$MS = \frac{1}{FF\left(R_a^{1.6} + R_{tr}^{1.6}\right)^{0.625}} - 1$$

Where R_a and R_{tr} are respectively the axial and transverse load ratios calculated as follows. From *Table 14-39* for the *Fwd outb'd Lug fitting* (at point *6*), assuming a 60/40 load distribution between the two lugs, the axial load applied to one lug "P_a" is: 0.60 * 687 = 412 *lbs* and the transverse load applied to one lug "P_t" is: 0.60 * 572 = 343 *lbs* .

Ultimate:

$$R_a = \frac{P_a}{Min(P_u \ ; \ P_{bru})} = \frac{412}{Min(2{,}015 \ ; \ 2{,}343)} = 0.204$$

$$R_{tr} = \frac{P_t}{P_{tu}} = \frac{343}{1{,}734} = 0.198$$

Hence, considering the 1.15 Fitting Factor, the Ult. MS is:

$$MS = \frac{1}{1.15\left(0.204^{1.6} + 0.198^{1.6}\right)^{0.625}} - 1 = 1.80$$

As explained earlier, no Yield MS is necessary.

Table 14-42 shows the results of the lug analysis performed on all the Lug fittings.

***Table 14-42** - Resulting lug analysis stresses and MS for all Lug fittings, 9g Fwd LC.*

Points	Name	W	D	W/D	K_t	t	A_t	P_u	D/t	e	e/D	K_{bru}	A_{br}	P_{bru}
		[in]	[in]			[in]	[in²]	[lbs]		[in]			[in²]	[lbs]
5	Fwd inb'd	0.75	0.5	1.5	0.86	0.125	0.03125	2016	4	0.375	0.75	0.5	0.0625	2344
6	Fwd outb'd	0.75	0.5	1.5	0.86	0.125	0.03125	2016	4	0.375	0.75	0.5	0.0625	2344
7	Aft inb'd	1.034	0.656	1.58	0.85	0.25	0.0945	6024	2.6	0.517	0.79	0.58	0.164	7134
8	Aft outb'd	1.034	0.656	1.58	0.85	0.25	0.0945	6024	2.6	0.517	0.79	0.58	0.164	7134

Points	Name	A_{av}	A_1	A_2	A_3	A_4	A_{av}/A_{br}	K_{tu}	P_{tu}	P_a	R_a	$\lvert P_t \rvert$	R_t	MS
		[in²]	[in²]	[in²]	[in²]	[in²]			[lbs]	[lbs]		[lbs]		
5	Fwd inb'd	0.02073	0.024778	0.015625	0.015625	0.024778	0.33	0.37	1734	418	0.207	217	0.125	2.33
6	Fwd outb'd	0.02073	0.024778	0.015625	0.015625	0.024778	0.33	0.37	1734	412	0.204	343	0.198	**1.80**
7	Aft inb'd	0.060942	0.071267	0.04725	0.04725	0.071267	0.37	0.4	4920	678	0.113	394	0.080	4.80
8	Aft outb'd	0.060942	0.071267	0.04725	0.04725	0.071267	0.37	0.4	4920	805	0.134	389	0.079	4.20

Lug analysis is far from being an exact science and the methodology used here has to be used with caution. It is recommended to have a *MS* of at least 0.25 (maybe 0.15!) including any fitting factor or other factors. Also, try to avoid having the *(ST)* direction in plane with the lug; especially in the transverse direction. Always use a methodology that is approved by your boss and the *DER* and double check your results with a small *FEM* or some tests.

14.13.6 BUSHING INTERFERENCE FIT

Both *Fwd inboard* and *Fwd outboard Lug fittings* have the exact same bushings arrangement. As shown in *Figure 14-97*, one flanged bushing is installed with a slight interference in each of the fitting's lug, one being larger than the other to allow for a sliding bushing. This configuration prevents pinching of the two lugs when the bolt is torqued which would cause sustained tension (bending) stresses as the base of the lugs that could lead to premature cracking.

With this arrangement, only one lug "thrust lug" can react side loads (along the bolt's axis). This lug is sometimes designed thicker than the other one for that reason.

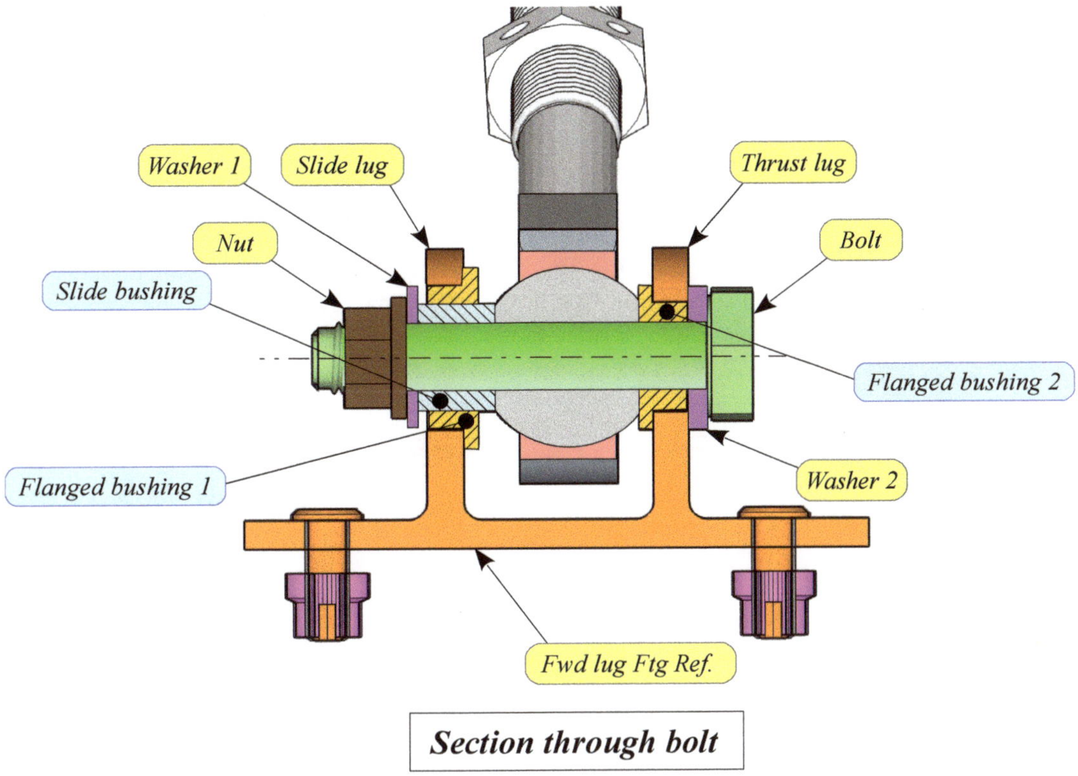

Figure 14-97 – *Fwd Lug Fittings bushings arrangement.*

The bushing interference analysis won't be done here as it is demonstrated in details in section *14.15.5*. Two important things in this design must be respected:

1. there must be a small clearance between the outside diameter of the Slide bushing and the inner diameter of the Flanged bushing to be able to slide.

2. the Slide bushing must be long enough so that, at the minimum tolerances of all parts involved (minimum material condition), the washer next to the nut "*washer1 as shown in the above figure*" must not touch the Slide lug.

14.13.7 CONNECTIONS

Based on the load distribution assumptions that were used for the *Fwd outb'd Lug Ftg.* in section *14.13.4*, *Figure 14-98* and *Figure 14-99* show the *FBD* in the *X-Z'* and *Y'-Z'* planes respectively with the reactions at the fasteners. *Figure 14-100* shows the *FBD* in an isometric view so that all the reaction forces can clearly be seen.

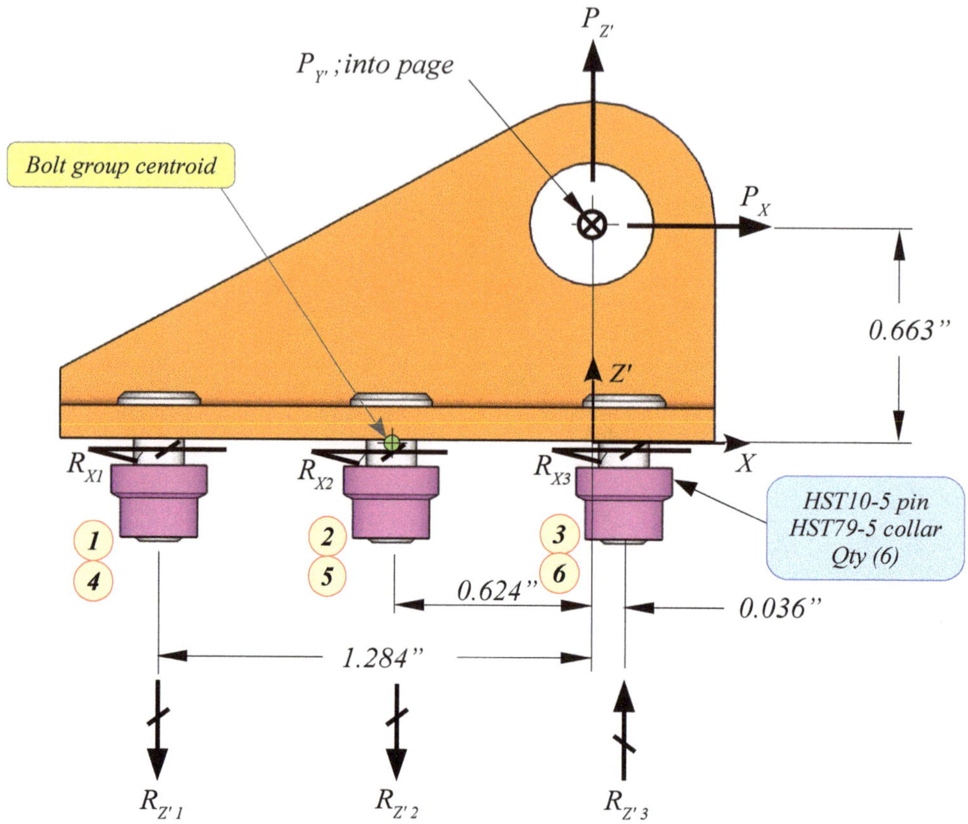

Figure 14-98 – *FBD in X-Z' plane showing the fasteners reactions.*

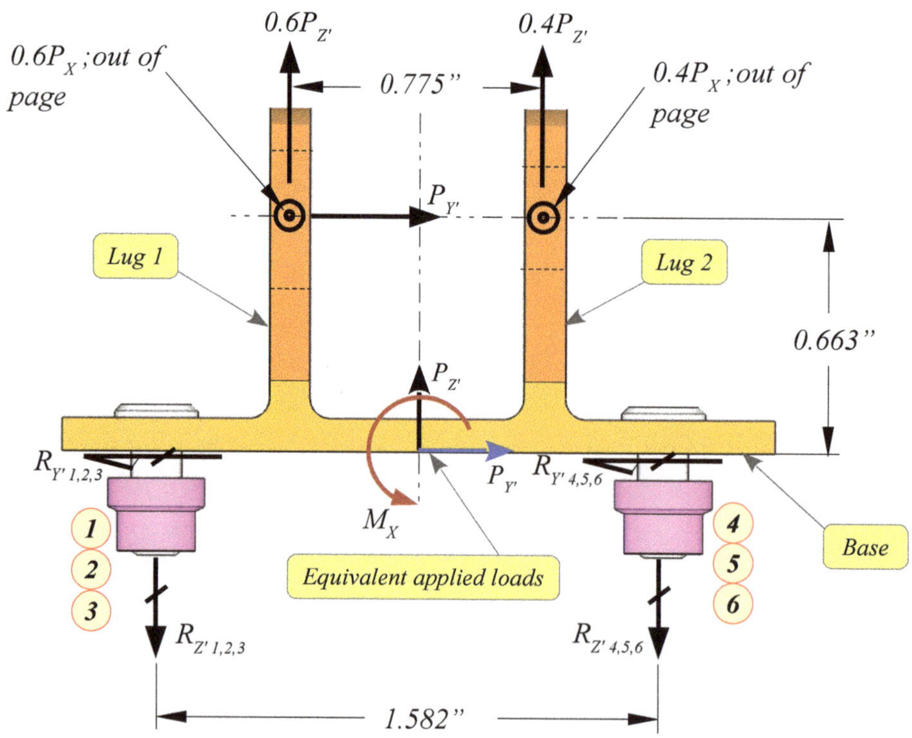

Figure 14-99 – *FBD in Y'-Z' plane showing the fasteners reactions.*

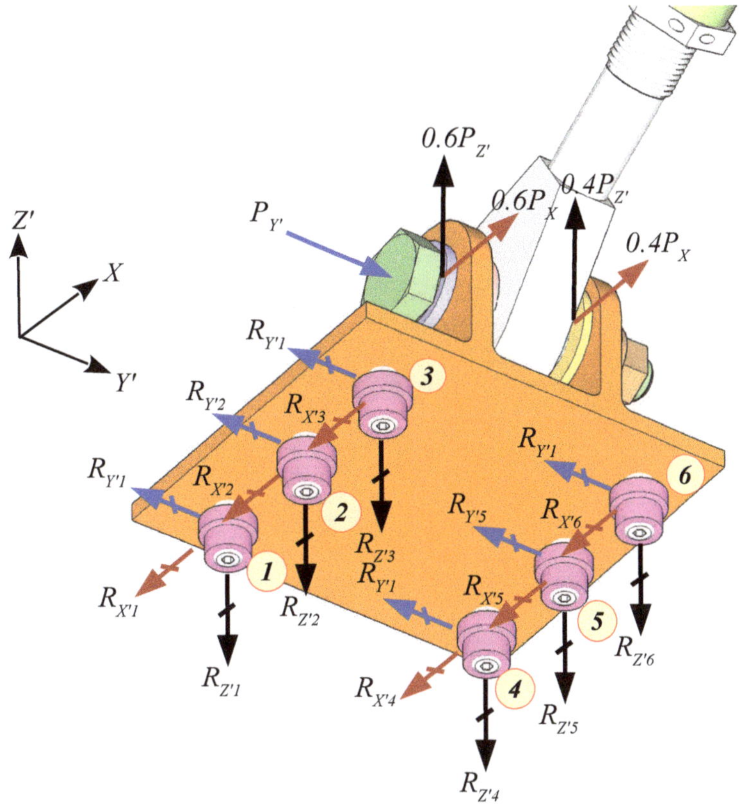

Figure 14-100 *– Fwd Lug Ftg Isometric view of the FBD with the fasteners reactions.*

<u>Fasteners forces</u>

The reactions forces at the fasteners had already been calculated in section ***14.13.4*** for the *Fwd outb'd lug ftg* for the ***9gFWD LC*** and are reproduced below in *Table 14-43*. However, the side load had been considered to calculate the bending stress at the root of the thrust lug but the reaction forces at the fasteners hadn't been done then as they weren't necessary. The effect of the side load and the 60/40 load distribution between the two lugs will now be super-imposed to the fasteners forces below.

Table 14-43 *- Fwd outb'd Lug Ftg. Fasteners summary forces from Strut load only.*

	R_X	$R_{Y'}$	$R_{Z'}$
fastener	*[lbs]*	*[lbs]*	*[lbs]*
Pair1	281	0	383
Pair2	281	0	98
Pair3	281	0	-187

Having 60% of the strut load applied to one lug and 40% to the other causes moments around the *"X"* and the *"Z' "* axes.

Because the Fwd Lug fitting is considered relatively soft in torsion, it is conservatively assumed that the moment "M_X" caused at the base of the fitting by the side load "P_Y" (applied at the center of the bearing) and the 60/40 lug distribution is reacted only by the two closest fasteners (*3 & 6*) as a couple.

On the other hand, the bolt pattern is pretty square and the fitting base is very rigid in the *X-Y'* plane. Therefore, the "$M_{Z'}$ " moment can be considered reacted by all of the six fasteners using a *2D bolt group* method as described in section ***10.6.2***.

For this *2D* analysis, I will consider all the loads that are applied at the center of the lug bearings to be applied at the base level in order to get rid of the "M_X" and "$M_{Y'}$" moments as they are taken care of in other parts of this chapter. Refer to *Figure 14-99* and *Figure 14-101* to see the *FBD* in the *Y'-Z'* and *X-Y'* planes respectively.

Since the fitting is not considered completely rigid, I couldn't simply use a *3D* bolt group tool. However, I like to verify quickly my hand results using that tool to see if they're in the "ball park". Doing a hand calculation to get the fasteners' reaction forces provides more control over the load paths.

A more accurate way to get the fastener reaction forces is to use *FEA*. However, as a warning, if you're tempted to do a detailed *FEM* to get the reaction forces at the fasteners, be sure to also model the back-up structure to where the analyzed part is connected as this will have a major impact on the load paths.

In general, when the structure is considered sturdy enough, we tend to assume the back-up structure to be infinitely rigid as it usually provides a quick and conservative way to get the fasteners loads. Just make sure that you describe in details in the report what are the assumed boundary conditions.

Shear loading

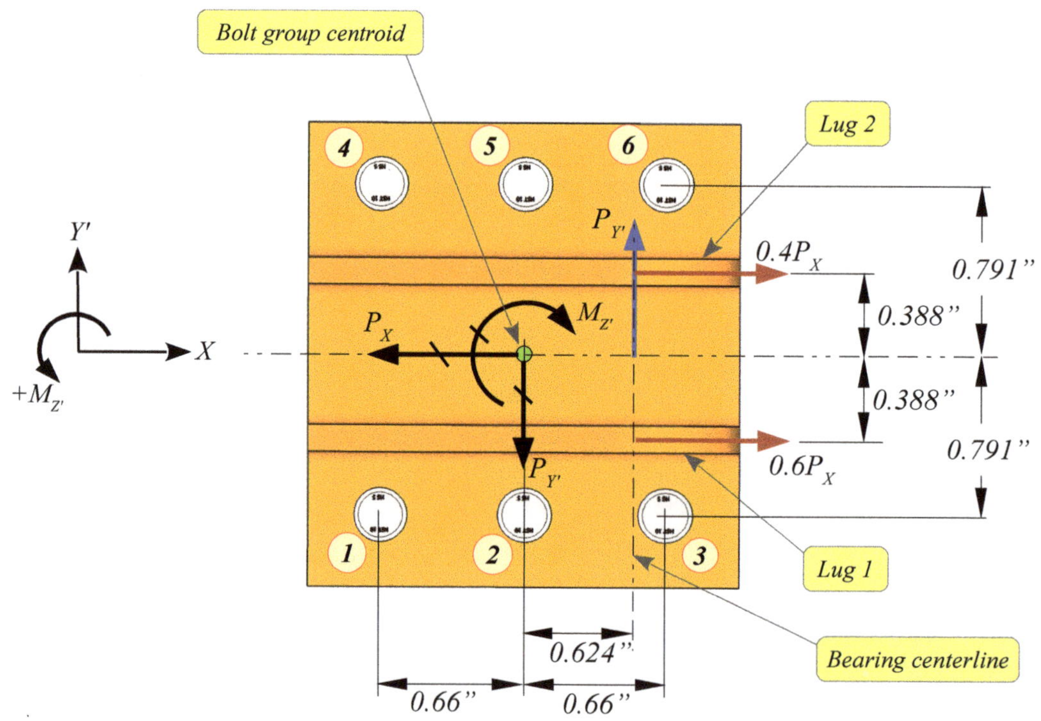

Figure 14-101 – *Fwd Lug Ftg FBD in X-Y' plane.*

Table 14-44 shows the position of the fasteners according to the local coordinate system, at the bolt group centroid, used in the *2D* bolt group tool as shown in *Figure 14-101*.

Table 14-44 *- Fwd Lug Ftg.; Positions of fasteners.*

fastener	x [in]	y' [in]	z' [in]
1	-0.66	-0.791	0
2	0	-0.791	0
3	0.66	-0.791	0
4	-0.66	0.791	0
5	0	0.791	0
6	0.66	0.791	0

As an example, the equivalent forces and moment applied at the bolt group cg for the *Fwd outb'd Ftg* (point *6*) under the *9g FWD LC* using *Table 14-37* are:

$$P_X = 0.6P_X + 0.4P_X = 844\,lbs$$

and from *Table 14-38*:

$$P_{Y'} = 45\,lbs$$

The moment around the Z' axis is thus:

$$\begin{aligned} M_{Z'} &= 0.6P_X(0.388) - 0.4P_X(0.388) + P_{Y'}(0.624) \\ &= 0.6(844)(0.388) - 0.4(844)(0.388) + 45(0.624) = 94\ in.-lb \end{aligned}$$

Using the strut loads from *Table 14-38* and *Table 14-39*, *Table 14-45* shows the loads in the *X-Y'* plane applied to each of the two lugs and their positions for the two *Fwd Lug ftgs* located at points *5* & *6*, for the *9g Fwd LC*. Note that *Lug 1* sees 60% of the total strut load and *Lug 2* sees 40%.

Table 14-45 *- Fwd inb'd (point 5) & outb'd (point 6) Lug Ftgs.; positions and applied loads.*

Point 5 Load	x [in]	y' [in]	z' [in]	P_X [lbs]	$P_{Y'}$ [lbs]
1	0	-0.388	0	401	39
2	0	0.388	0	267	0
			Σ	**668**	**39**

Point 6 Load	x [in]	y' [in]	z' [in]	P_X [lbs]	$P_{Y'}$ [lbs]
1	0	-0.388	0	506	45
2	0	0.388	0	338	0
			Σ	**844**	**45**

Table 14-46 shows the resulting applied fastener forces for the two *Fwd Lug Ftgs* (points *5* & *6*) from the *2D* bolt group analysis. The shear force in each fastener is calculated as follows:

$$F_S = \sqrt{(F_X^2 + F_{Y'}^2)}$$

Table 14-46 - *Fwd inb'd (point 5) & outb'd (point 6) Lug Ftgs.;resulting applied fasteners forces.*

Point 5	F_X	$F_{Y'}$	F_S
fastener	*[lbs]*	*[lbs]*	*[lbs]*
1	122	-3	122
2	122	7	123
3	122	16	123
4	100	-3	100
5	100	7	101
6	100	16	102
Σ	**668**	**40**	

Point 6	F_X	$F_{Y'}$	F_S
fastener	*[lbs]*	*[lbs]*	*[lbs]*
1	154	-4	154
2	154	8	154
3	154	19	155
4	127	-4	127
5	127	8	127
6	127	19	129
Σ	**844**	**46**	

Tension loading

From *Figure 14-99*, the M_X moment is calculated as:

$$M_X = -0.6P_Z(0.388) + 0.4P_Z(0.388) - P_{Y'}(0.663)$$
$$= -0.6(294)(0.388) + 0.4(294)(0.388) - 45(0.663) = -53\ in.-lb$$

Then, as assumed earlier, that moment is reacted by the fastener couple *3 & 6*. Thus, their axial load "$F_{Z'}$" is calculated as:

$$F_{Z'} = \frac{|M_X|}{1.582} = \frac{|-53|}{1.582} = 33\ lbs$$

Hence, since the resulting applied M_X is negative (clockwise in *Figure 14-99*), this adds 33 *lbs* of tension to fastener *3* and subtracts 33 *lbs* of tension to fastener *6*.

Resulting fasteners forces

Adding it all up from the above result, *Table 14-43* and *Table 14-46*, the resulting fastener forces in both inboard and outboard *Fwd Lug Ftgs* are summarized in *Table 14-47*.

Table 14-47 - *Fwd inb'd (point 5) & outb'd (point 6) Lug Ftgs.;resulting applied fasteners forces.*

Point 5	F_X	$F_{Y'}$	$F_{Z'}$	F_S	F_T
fastener	*[lbs]*	*[lbs]*	*[lbs]*	*[lbs]*	*[lbs]*
1	122	-3	139	122	139
2	122	7	69	123	69
3	122	16	35	123	35
4	100	-3	139	100	139
5	100	7	69	101	69
6	100	16	-39	102	0
Σ	**668**	**40**	**411**		

Point 6	F_X	$F_{Y'}$	$F_{Z'}$	F_S	F_T
fastener	*[lbs]*	*[lbs]*	*[lbs]*	*[lbs]*	*[lbs]*
1	154	-4	192	154	192
2	154	8	49	154	49
3	154	19	-60	155	0
4	127	-4	192	127	192
5	127	8	49	127	49
6	127	19	-127	129	0
Σ	**844**	**46**	**294**		

Then, the most critical fastener is no. ***1*** of the *Fwd outb'd Lug ftg* at point *6*.

Bearing

The bearing stress "f_{bru}" is checked in the *Fwd Lug Ftgs* and in the cabinet. The bearing load capability is calculated as:

$$P_{bru} = F_{bru} \cdot D \cdot t$$

The shear load used in the analysis is from *Table 14-47* fastener *3* for the *Fwd outb'd Lug Ftg* and is 155 *lbs*. The material used in the construction of the vendor cabinet may not be known but the allowable loads at the attachment points may be provided by the vendor's specifications. If that's the case, then no bearing check (or any other check) is necessary on the side of the vendor's cabinet. However, for the sake of the demonstration, let's assume that the cabinet is made out of Fiberglass-Epoxy (*GL/EP*) having an 0.063 *in.* thick embedded *Al Aly 6061-T6 aluminum alloy* doubler.

Because the *Al Aly* doubler is stronger than the *GL/EP*, most of the bearing load would be taken by the doubler. Therefore, the *GL/EP* is totally ignored as a load bearing material.

Their respective calculated bearing allowable are shown in *Table 14-48* based on allowables from *Table 14-16* and *Table 14-22*.

Table 14-48 *- Fwd outb'd Lug Ftg connections Bearing allowables and MS.*

Part	Material Al Aly	t [in]	D [in]	Fbru [ksi]	Pbru [lbs]	F [lbs]	MS bearing
Cabinet	6061-T6	0.063	0.163	85	873	155	4.63
Lug Ftg	7050-T7451	0.100	0.163	105	1712	155	10.04

The *MS* for the connections of the Forward inboard and outboard Lug fittings are covered by the Aft inboard Lug fitting *MS* shown on page *429*.

Looking back at this fictitious design, screws should have been used to connect the *Fwd Lug ftgs* to the monument instead of the Hi-Lites since the latter are considered permanent fasteners and that it would probably be interesting to be able to remove the cabinet from time to time.

14.13.8 TENSION CLIP

For the *9g Fwd LC*, since a substantial tension force of 192 *lbs* exists at fasteners *1 & 4* of the *Fwd outb'd Lug ftg*, this out-of -plane load will cause a bending stress at the root of the flanges. This arrangement can be considered similar to a tension clip. The flanges are conservatively idealized as cantilevered beams having one end fixed at the lug and one end free where the fastener is located. See *Figure 14-102* for details.

I could have idealized the flange as a fixed-guided beam as was done in section 14.6.12 but since I'm not sure of the stiffness of the back-up structure (the cabinet) I'd rather assume it soft such that the flange could pivot freely at the fastener.

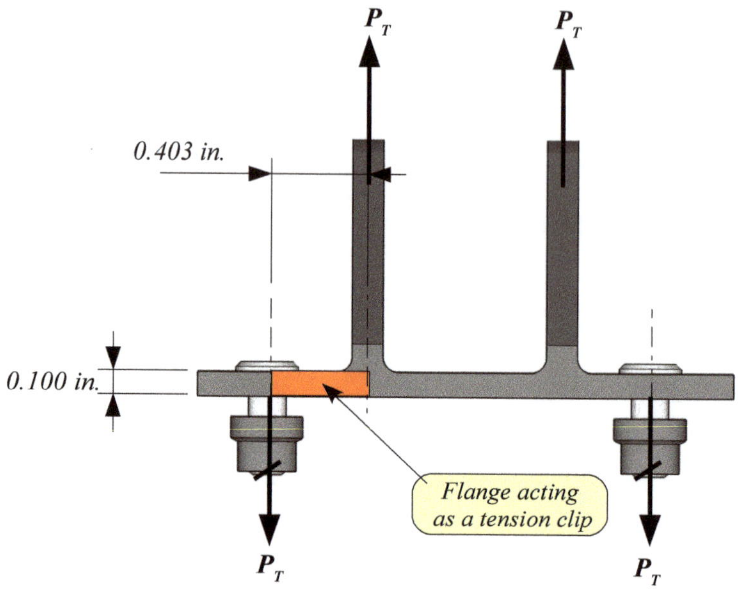

Figure 14-102 *– Tension load applied out of plane at the Fwd Lug Ftg – Cabinet connection.*

For the tension clip analysis, the Lug fitting flange is idealized as shown in ***Figure 14-103***. The flange is assumed fixed at the center of the lug and free at the center of the fastener. An effective flange width of 1.0 *inch* is considered.

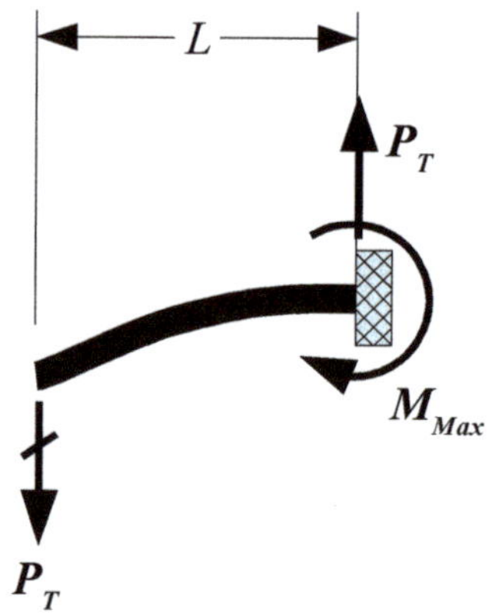

Figure 14-103 *– Idealized Lug fitting flange in out-of-plane bending.*

The maximum bending moment at the flange root is:

$$M_{Max} = P_T L$$

then

$$M_{Max} = 192 \cdot (0.403) = 78 \; in. - lb$$

For
P_T = 192 lbs.
L = 0.403 in.

Bending stress

For rectangular cross-sections, the bending stress equation is simplified as follows:

$$f_b = \frac{6M}{bt^2} = \frac{6 \times 78}{1.0 \times 0.100^2} = 46{,}800 \; psi$$

For
M = 78 in-lb
b = 1.0 in.
t = 0.100 in.

The bending stress is well below the tensile strength of the material of F_{TU} = 75 *ksi* (ref. *Table 14-16*). Providing that the bending stress would have been slightly higher than F_{TU}, I could have used a plastic bending allowable since we're dealing with ultimate loads; as demonstrated in section ***14.6.12***. However, in this example, I will conservatively be using F_{TU} as the allowable. Hence, the bending stress ratio "R_b" is thus:

$$R_b = \frac{f_b}{F_{TU}} = \frac{46{,}800}{75{,}000} = 0.624$$

Shear stress

With the bending stress, a transverse shear exists at the flange root that must be taken into account in the MS. The average shear stress through the beam section is used. For a cross-section area of: $A_S = b * t = (1.0)\ (0.100) = 0.100\ in^2$ and a vertical shear force of 192 *lbs*, the average shear stress "f_s" is:

$$f_s = \frac{V_Z}{A_S} = \frac{192}{0.100} = 1{,}920\ psi$$

For F_{SU} = 40 *ksi* (ref. *Table 14-16*), the shear stress ratio "R_S" is calculated as:

$$R_S = \frac{f_S}{F_{SU}} = \frac{1{,}920}{40{,}000} = 0.048$$

Margin of safety for the tension clip:

For bending stress combined with shear stress, using a *FF* of 1.15, the MS is calculated as follows:

$$MS = \frac{1}{FF\sqrt{R_b^2 + R_S^2}} - 1$$

hence

$$MS = \frac{1}{1.15\sqrt{0.624^2 + 0.048^2}} - 1 = 0.38$$

Fwd outb'd Lug Ftg Tension clip, Ultimate	***0.38***
9g Fwd *Al Aly 7050-T7451*	*Shear + bending*

The other Lug fittings have lower fastener tension loads and thicker flanges and thus are *Passed by Comparison* to the *Fwd outb'd Lug ftg.*

14.14 AFT INBOARD LUG FITTING

The Aft inboard Lug fitting is machined out of *Al Aly 7050-T7451*. *Figure 14-104* shows the details of the *Aft inboard Lug fitting* (point 7).

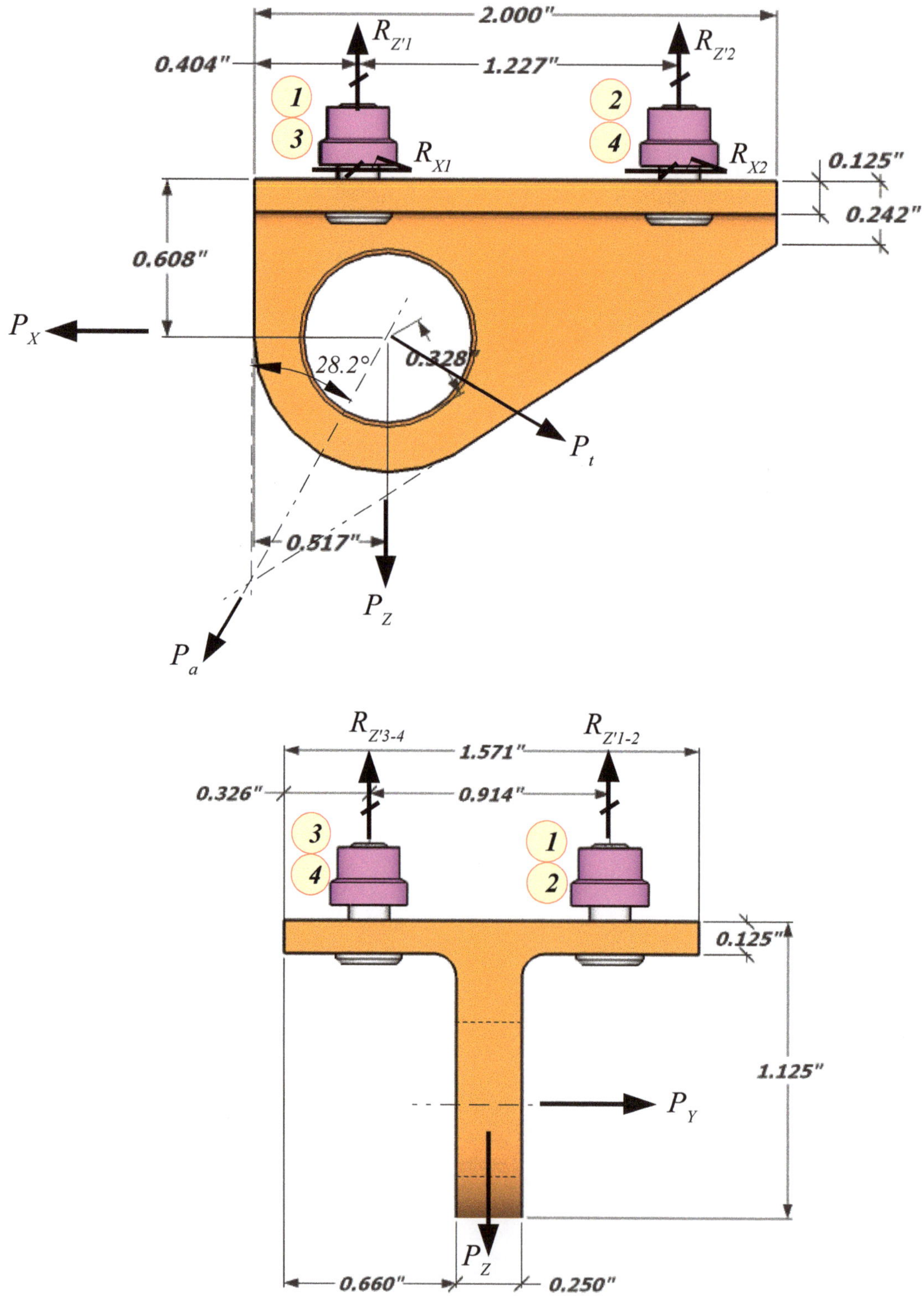

Figure 14-104 *– Aft inboard Lug Fitting dimensions – point 7.*

14.14.1 SECTION PROPERTIES

Figure 14-105 shows the cross-section properties of the Aft inboard Lug fitting at section *A-A* shown in *Figure 14-106*.

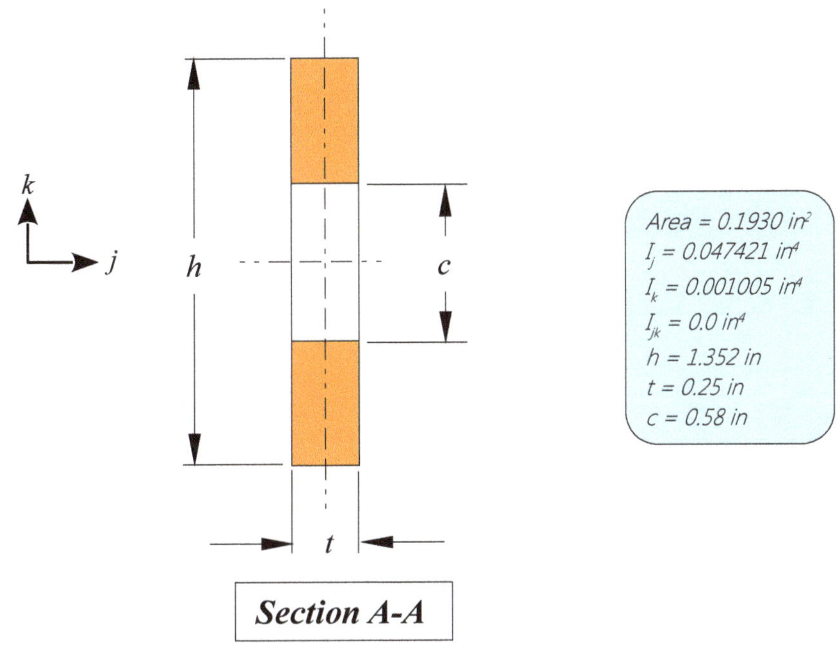

Figure 14-105 *– Aft inboard Lug Section A-A properties.*

14.14.2 STRESS ANALYSIS

The *Aft inb'd Lug Ftg* is analyzed in the same manner as for the *Fwd outb'd Lug Ftg* done in section *14.13*. See *Table 14-40* for the resulting stresses and MS.

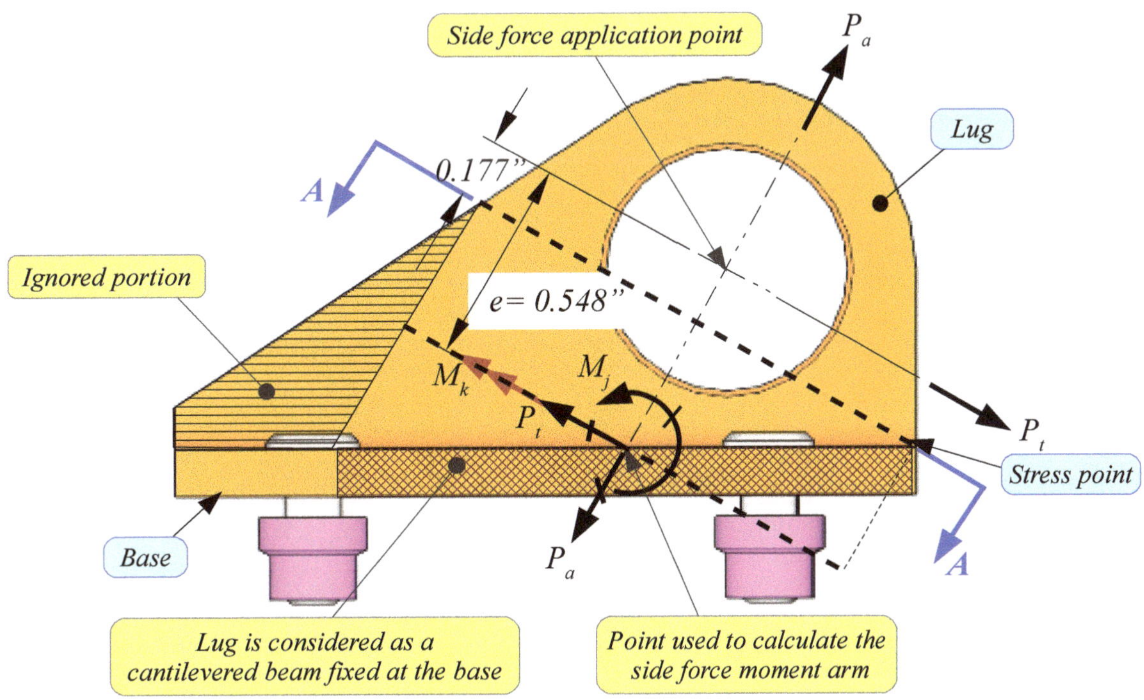

Figure 14-106 *– Aft inboard Lug FBD, Section A-A.*

14.14.3 LUG ANALYSIS

Figure 14-107 shows how the lug has been idealized for the *Aft inboard Lug fitting* . The fitting is analyzed in the same manner as for the *Fwd outb'd Lug Ftg* done in section *14.13.5*. See *Table 14-42* for the resulting stresses and MS.

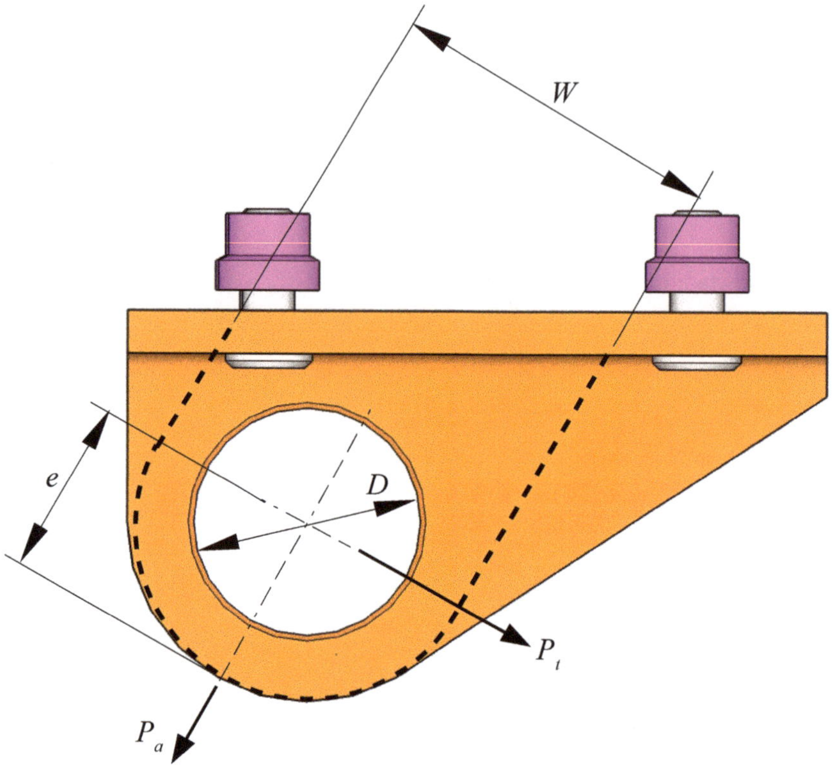

Figure 14-107 *– Aft inboard Lug Fitting lug idealization– point 7.*

14.14.4 CONNECTIONS

The *Aft inb'd Lug Ftg.* is connected (at point *7*) to the Intercostal that spans between two frames using *qty(4)* 5/32" Hi-Lites *Figure 14-108* shows the *FBD* in an isometric view so that all the reaction forces can clearly be seen. See *Figure 14-104* for the fastener's positions.

Fasteners forces

Along the *X* axis:

As assumed previously, the applied load P_X is evenly distributed among the four fasteners (*1, 2, 3 & 4*) common to the Intercostal. Using the P_X value of 668 *lbs* from *Table 14-39*, the fastener's reactions in the *X* direction are then:

$$R_{X1} = R_{X2} = R_{X3} = R_{X4} = \frac{P_X}{4} = \frac{668}{4} = 167\ lbs$$

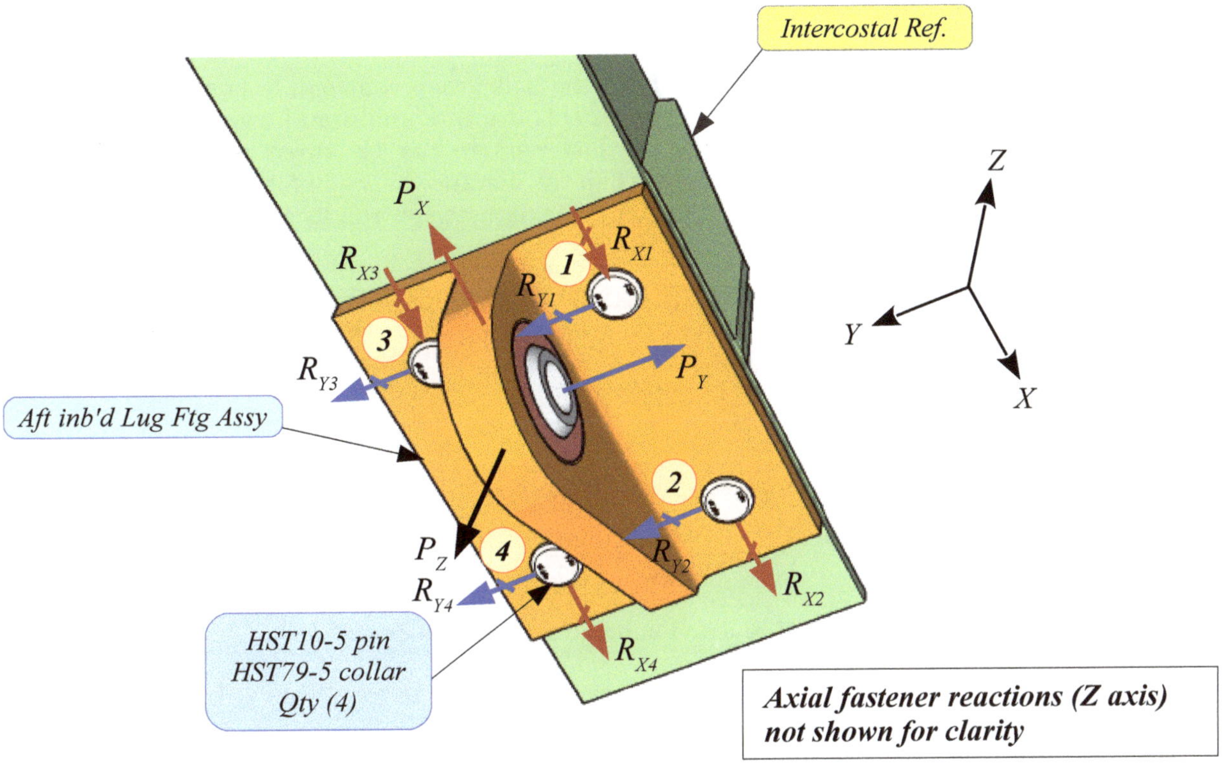

Figure 14-108 – *Aft inb'd Lug Ftg Isometric view of the FBD with the fasteners reactions.*

Along the *Y* axis:

Due to P_Y shear

Like for the load in *X*, the load along the *Y* axis is also reacted by the four fasteners (*1 to 4*). From *Figure 14-104*, the side load "P_Y" (from *Table 14-38*) is applied at a distance of $\boldsymbol{a}$ = 0.517 – 0.404 = 0.113 *in.* from the *1-3* fastener pair and, $\boldsymbol{b}$ = 1.227 – 0.113 = 1.114 *in.* from the *2-4* fastener pair.

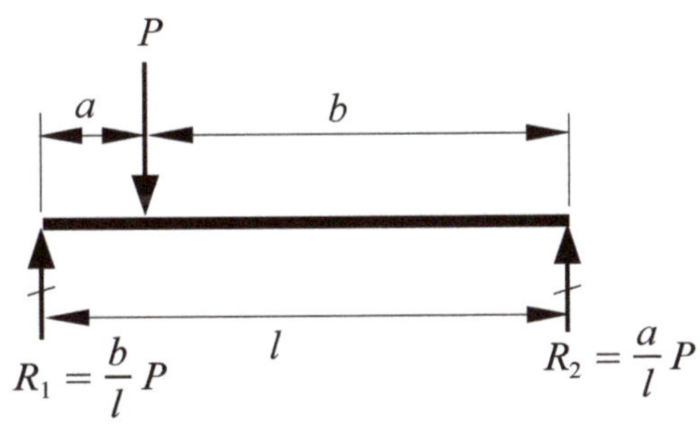

Figure 14-109 – *Shear load distribution.*

Using the quick shear distribution method, for a value l of 1.227 *in.*, as shown in *Figure 14-109*, the fastener's reactions in the "*Y*" direction are then:

$$R_{Y'1-3} = \frac{b}{l}P_Y = \frac{1.114}{1.227} \cdot 39 = 35\ lbs$$

$$R_{Y'2-4} = \frac{a}{l}P_Y = \frac{0.113}{1.227} \cdot 39 = 4\ lbs$$

Due to M_Z moment

The side load application point being away from the bolt group centroid, a small moment around the *Z* axis exists. Instead of doing a "complicated" *2D* bolt group analysis, I conservatively assumed that this moment is reacted as a couple by the fasteners in the "*Y*" direction since the distance between fasteners *1* and *2* (*1.227"*) is a bit larger than the distance between fasteners *1* and *3 (0.914")*. Based on *Figure 14-104*, the distance "*d*" along the *X* axis between the location where P_Y is applied and the bolt group c.g. is:

$$d = \frac{1.227}{2} + 0.404 - 0.517 = 0.500\ in.$$

Thus the moment "M_Z" to be reacted is:

$$M_Z = P_Y \cdot d = 39 \cdot 0.500 = 20\ in.-lb$$

The force couple "R_{YC}" reacted by the fasteners *1-3* pair and the *2-4* pair is:

$$R_{YC} = \frac{M_Z}{1.227} = \frac{20}{1.227} = 16\ lbs$$

This force is additive for fasteners *1 & 3* and subtractive for fasteners *2 & 4*.

Resulting forces in *Y*

Distributing the load evenly between the two fasteners in the same row, the resulting reaction forces in the four fasteners are:

$$R_{Y'1} = R_{Y'3} = \frac{35+16}{2} = 26\ lbs$$

$$R_{Y'2} = R_{Y'4} = \frac{4-16}{2} = -6\ lbs$$

Along the *Z* axis:

Due to P_X and P_Z

Using *Figure 14-104*, the reaction forces at the bolts are determined by calculating the sum of the forces and moments at fastener *2-4*.

The sum of the moments is calculated (*positive CCW*):

$$M_{Y2-4} = R_{Z1-3}(1.227) - P_Z(1.227+0.404-0.517) + P_X(0.608) = 0$$

For $P_X = 668\ lbs$ and $P_Z = 411\ lbs$, the force in the fasteners *1 & 3* is:

$$R_{Z1-3} = \frac{411(1.114) - 668(0.608)}{1.227} = 42\ lbs$$

Then, from the sum of the forces:

$$R_{Z2-4} = P_Z - R_{Z1-3} = 411 - 42 = 369\ lbs$$

Due to M_X

Since the fitting is short and stubby, I consider it to have a constant torsional stiffness and act as a torque beam such that the torque "M_X" caused at the fitting base by the side load "P_Y" is reacted by the *1-3* and *2-4* fasteners pairs. Using the exact same method as for the shear from *Figure 14-109*, the reacting moments are calculated as:

$$M_{X1-3} = \frac{b}{l} M_X = \frac{1.114}{1.227} \cdot 24 = 22\ in.-lb$$
$$M_{X2-4} = \frac{a}{l} M_X = \frac{0.113}{1.227} \cdot 24 = 2\ in.-lb$$

where

$$M_X = P_Y(0.608) = 39(0.608) = 24\ in.-lb$$

As expected, most of the torque is reacted by the *1-3* fastener pair. The couple forces caused in both fastener pairs are thus:

$$R_{ZC1-3} = \frac{M_{X1-3}}{0.914} = \frac{22}{0.914} = 24\ lbs$$
$$R_{ZC2-4} = \frac{M_{X2-4}}{0.914} = \frac{2}{0.914} = 2\ lbs$$

Resulting forces in Z

The contribution of all the forces is then combined to determine the resulting fastener loads. Therefore, the resulting axial forces in the fasteners are:

$$R_{Z1} = \frac{R_{Z1-3}}{2} + R_{ZC1-3} = \frac{42}{2} + 24 = 45\ lbs$$
$$R_{Z3} = \frac{R_{Z1-3}}{2} - R_{ZC1-3} = \frac{42}{2} - 24 = -3\ lbs$$
$$R_{Z2} = \frac{R_{Z2-4}}{2} + R_{ZC2-4} = \frac{369}{2} + 2 = 187\ lbs$$
$$R_{Z4} = \frac{R_{Z2-4}}{2} - R_{ZC2-4} = \frac{369}{2} - 2 = 183\ lbs$$

Table 14-49 shows the fasteners loads summary along with the shear force "F_S" and the tension force "F_T" in each of them.

$$F_S = \sqrt{F_X^2 + F_Y^2}$$

Table 14-49 *- Aft inb'd Lug Ftg. Fasteners summary forces.*

Point 7	F_X	F_Y	F_Z	F_S	F_Z
fastener	*[lbs]*	*[lbs]*	*[lbs]*	*[lbs]*	*[lbs]*
1	167	26	45	169	45
2	167	-6	187	167	187
3	167	26	-3	169	0
4	167	-6	183	167	183
Σ	**668**	**40**	**412**		

The most critical fastener is no. *2* .

Margin of safety for the Hi-Lites

Of all the Hi-Lites installed in the four Lug fittings, the most critical fastener is the no. **2** at *point 7* for the *Aft inb'd lug fitting* under the *9g Fwd LC* with an ultimate shear force of *167 lbs* and an ultimate tension force of *187 lbs*. The MS of the Pin and the Collar for the most critical fastener is calculated below substantiating all the Hi-Lites from all the Lug Ftgs.

Pin (HST10-5)

The interaction equation for bolts subjected to shear and tension loads is:

$$R_s^3 + R_t^2 = 1$$

See another example of that method regarding fasteners in (ref. 27.).

Where, using the fastener allowables from *Table 14-26*, the load ratios are:

$$R_s = \frac{P_s}{P_{su}} = \frac{167}{2{,}005} = 0.083 \qquad R_t = \frac{P_t}{P_{tu}} = \frac{187}{1{,}940} = 0.096$$

Then from the interaction equation above and using a Fitting Factor of 1.15, the Margin of Safety for the HST10-5 no. **2** is thus calculated as follows:

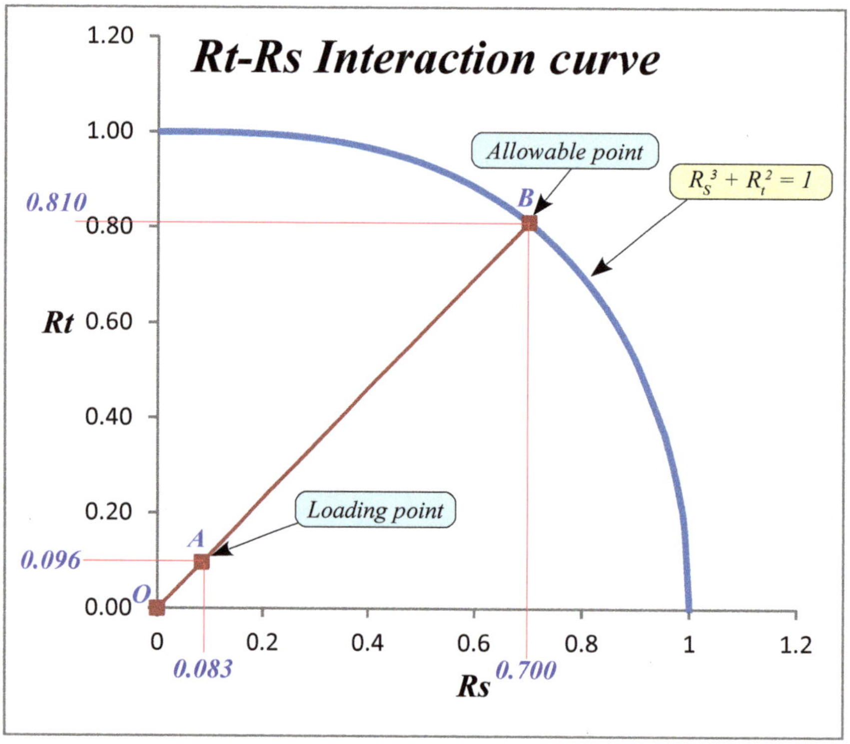

where

$$MS = \frac{\overline{OB}}{FF\,\overline{OA}} - 1 = \frac{R_{S\,All}}{FF\,R_S} - 1 = \frac{0.700}{1.15*0.083} - 1 = >1.0$$

or

$$MS = \frac{R_{t\,All}}{FF\,R_t} - 1 = \frac{0.810}{1.15*0.096} - 1 = >1.0$$

Collar (HST79-5)

The highest tension force occurs at fastener ***1*** of the *Fwd outb'd Lug ftg* with a value of 192 *lbs*. That tension force is compared to the tension allowable for the Collar from *Table 14-37* of 1,400 *lbs*.

The Margin of Safety for the *HST79-5 Collar* is thus calculated as follows using a Fitting Factor of 1.15:

$$MS = \frac{P_{tu}}{FF\,P_t} - 1 = \frac{1{,}400}{1.15 \times 192} - 1 = >1.0$$

Bearing

The bearing stress "f_{bru}" is checked in the *Aft inb'd Lug Ftg* as well as in the Intercostal. The bearing load capability is calculated as:

$$P_{bru} = F_{bru} \cdot D \cdot t$$

The shear load used in the analysis from fasteners *1 & 3* for the *Aft inb'd Lug Ftg* is 169 *lbs*. The calculated bearing allowables for both parts are shown in *Table 14-50*.

Table 14-50 *- Aft inb'd Lug Ftg to Intercostal connection Bearing allowables.*

Part	*Material Al Aly*	*t [in]*	*D [in]*	*Fbru [ksi]*	*Pbru [lbs]*	*F [lbs]*	*MS bearing*
Intercostal	*7050-T7451*	0.120	0.163	105	2054	169	11.15
Lug Ftg	*7050-T7451*	0.125	0.163	105	2139	169	11.66

Margin of Safety

The *MS* for the connections of the *Lug fittings* is:

Lug Ftgs Connections, Ultimate	***>1.00***
9g Fwd *HST10-5 in Al Aly 6061-T6*	*Shear and bearing*

Note that, since the *Aft inb'd Lug Ftg* is the most critical in the connections, this *MS* covers all the other Lug fittings' connections.

14.15 AFT OUTBOARD LUG FITTING

The *Aft outboard Lug fitting* (point *8*) is different from the other ones as it connects to two different parts; namely the skin and the Frame. *Figure 14-110* shows the details of the *Aft outboard Lug fitting*.

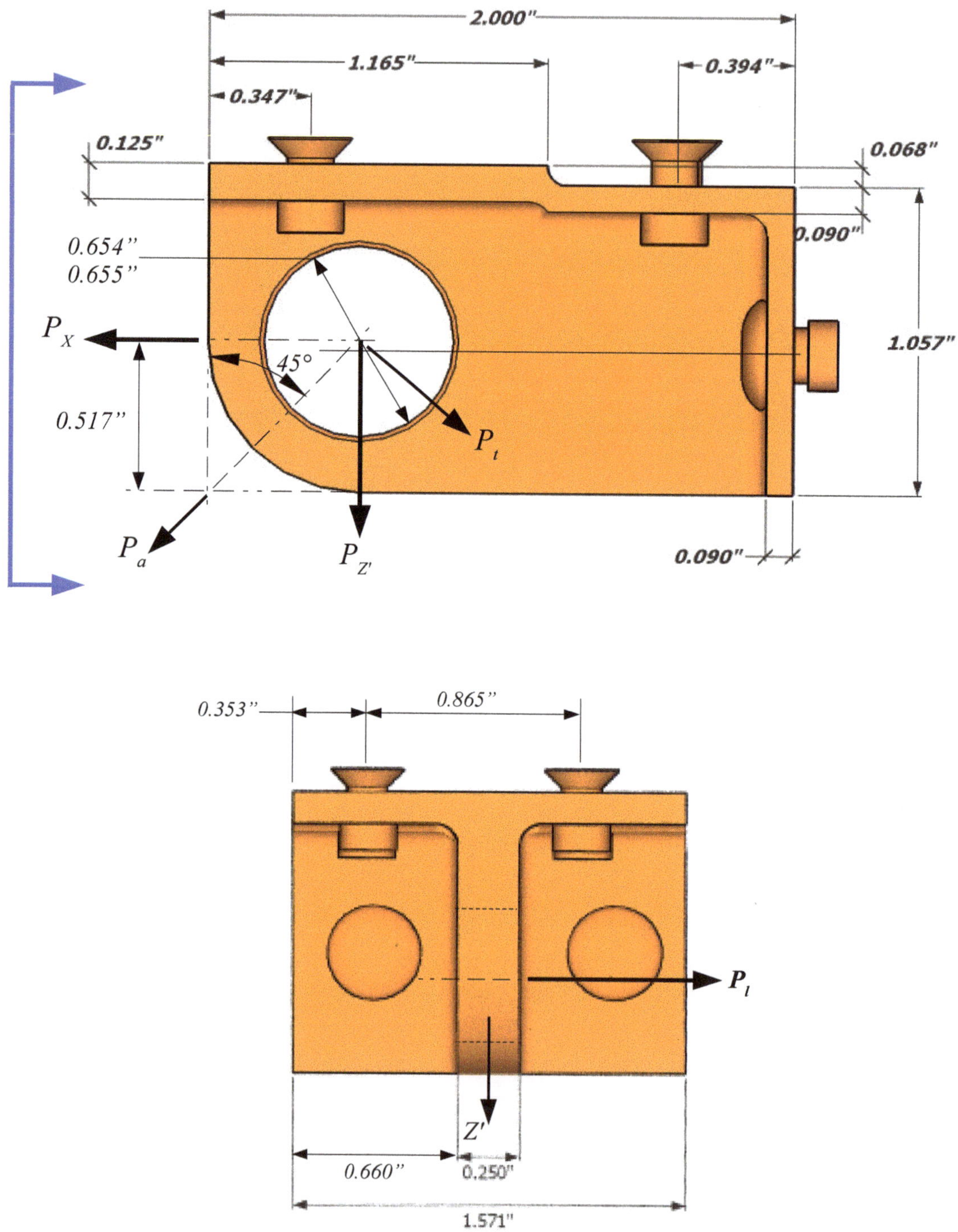

***Figure 14-110** – Aft outboard Lug Fitting dimensions – point 8.*

Figure 14-111 shows the *FBD* of the Lug fitting. As can be noticed, the line of force in the Strut doesn't exactly meet at the corner of the two shear planes where the load is reacted. Therefore, a small moment exists at that point that must be reacted by an out-of-plane force from one of the fasteners "$R_{Z'2}$".

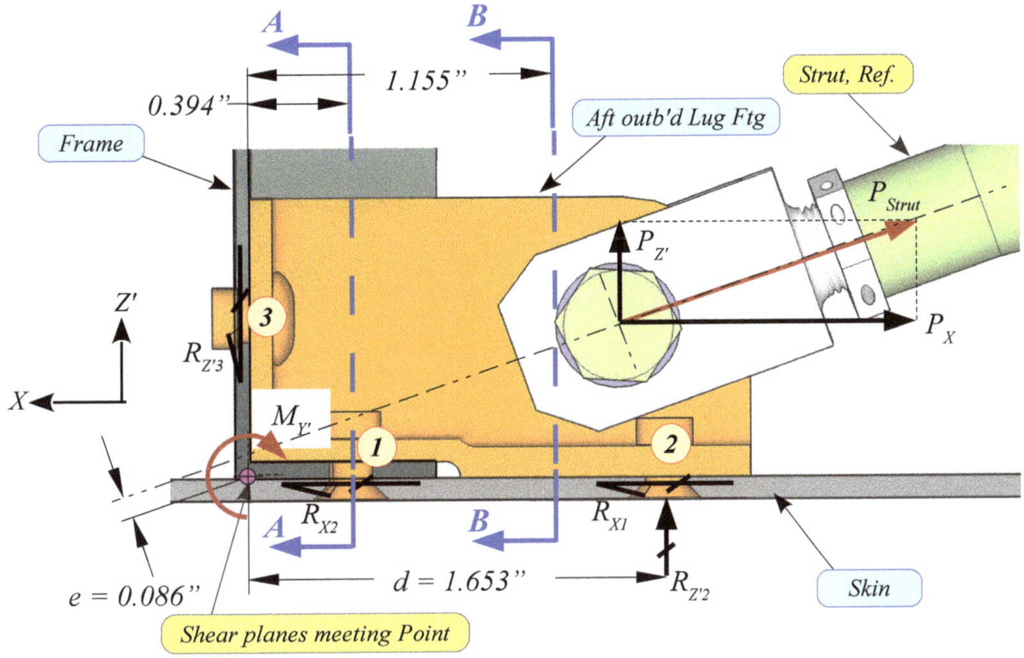

Figure 14-111 *– FBD Aft outboard Lug Ftg.*

If the designer had done his job right (or well guided by the stressman), the strut line of force would go exactly through the corner where the two shear planes meet and no moment "$M_{Y'}$" would exist.

The moment is assumed reacted by the out-of-plane force at fastener *2* since that fastener is the farthest from the shear planes meeting point. One could have used the very end of the Lug Ftg to react that moment, by contact as a compression force in this case, instead of fastener *2*. But, when the Strut load is reversed (due to another load case), a tension force instead of a compression would need to be reacted by fastener *2* as this can only be done by a fastener.

14.15.1 LOADS

From *Table 14-38*, having a Strut load of 893 *lbs* for the *9g FWD LC*, the sum of the moments "$M_{Y'}$" at the shear planes meeting point is calculated as:

$$M_{Y'} = P_{Strut} \cdot e = 893(0.086) = 77 \ in. - lb$$

Hence, since the moment at that point has to be 0, another counteracting moment has to exist. That moment is created as a couple between fastener *3* (in shear) and fastener *2* (in compression). The couple force "$R_{Z'2}$" is calculated as:

$$R_{Z'2} = \frac{M_{Y'}}{d} = \frac{77}{1.653} = 47 \ lbs$$

From the sum of the forces in Z', the vertical reaction force in fastener *3* is thus:

$$R_{Z'3} = P_{Z'} + R_{Z'2} = 294 + 47 = 341\ lbs$$

The shear reactions in fasteners *1 & 2* are assumed equally distributed. Hence:

$$R_{X1} = R_{X2} = \frac{P_X}{2} = \frac{844}{2} = 422\ lbs$$

> As always, we're trying to avoid to a minimum the out-of-plane loads in thin structures. To make the above *FBD*, since the Lug Fitting is attached to the thin skin and the thin Frame web, I always assume that the out-of-plane stiffness of those thin structures is very low, and therefore, is not an adequate load path. However, those thin members are very strong and stiff along their axes so the adequate load path is by shear from the fasteners.

14.15.2 SECTION PROPERTIES

A similar exercise as was previously done for the Forward outboard Lug Ftg was done to determine the most critical location. See the example shown in *Figure 14-94.* The critical location causing the highest stress in the lower "*TEE*" cap was found through fastener *1* (section *A-A*) while it was just at the edge of the bearing hole for the highest stress at the "top" (section *B-B*). See *Figure 14-111* for the location of the sections.

The internal forces at the centroid of section *A-A* (through fastener *1*) and section *B-B* are shown in *Figure 14-112* and *Figure 14-113* respectively.

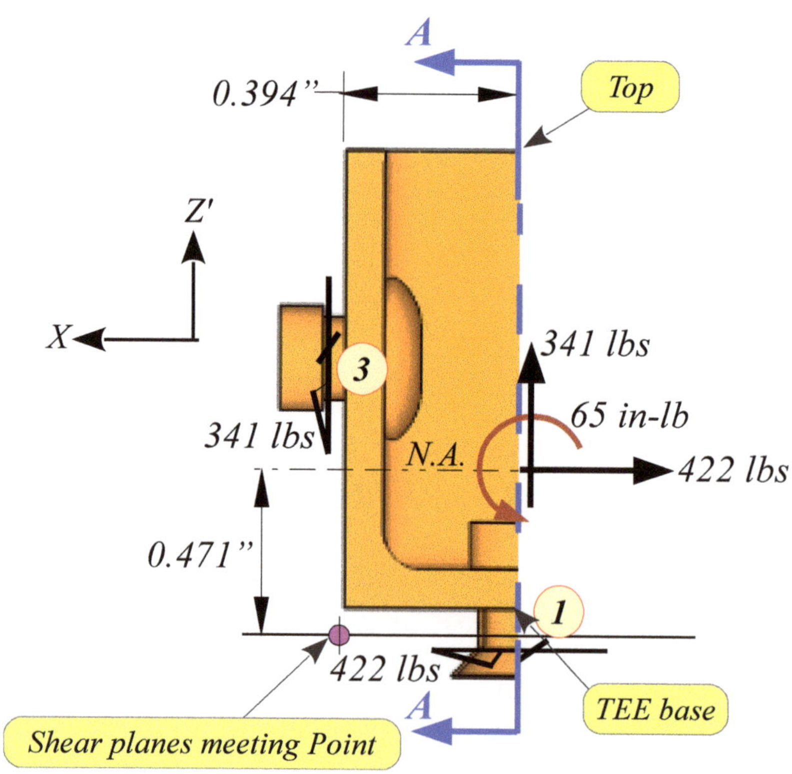

***Figure 14-112** – Internal forces, Section A-A.*

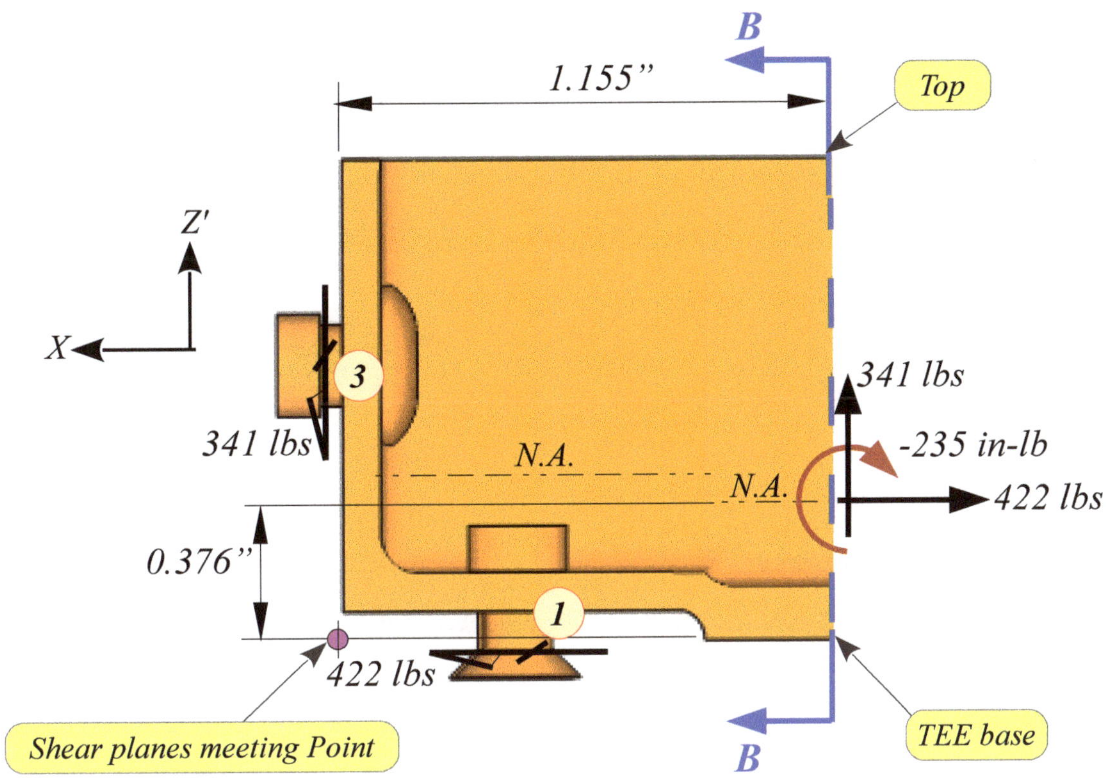

***Figure 14-113** – Internal forces, Section B-B.*

See *Figure 14-114 and Figure 14-115* for properties of section *A-A* and section *B-B* respectively.

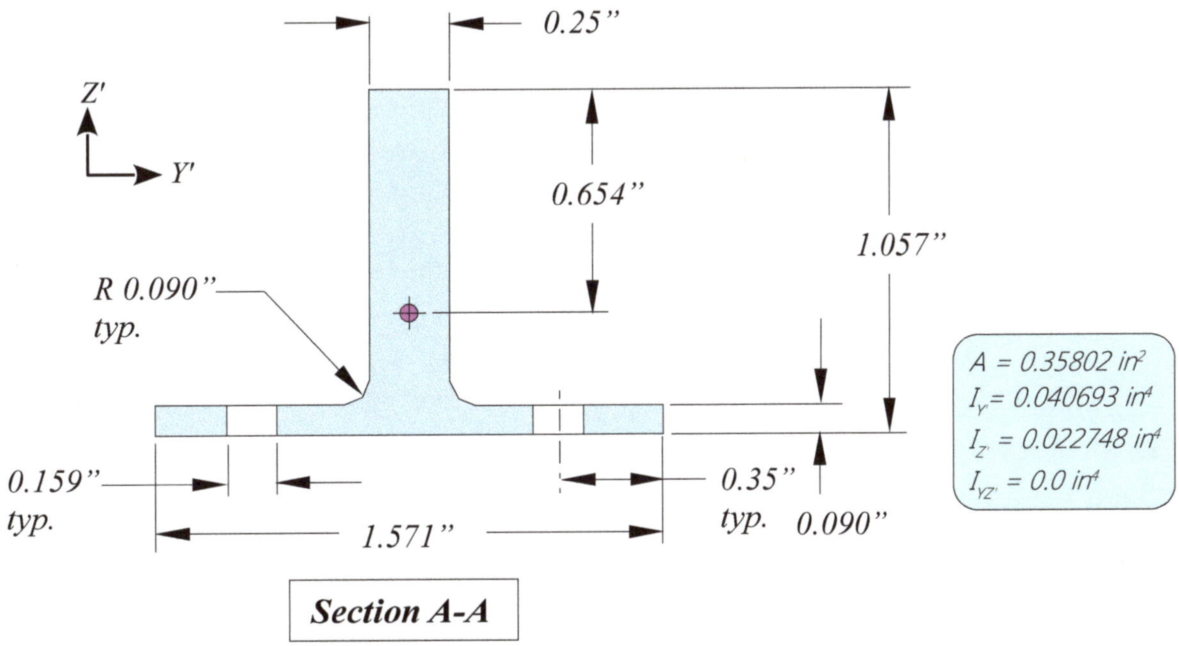

***Figure 14-114** – Aft outboard Lug Ftg., Section A-A.*

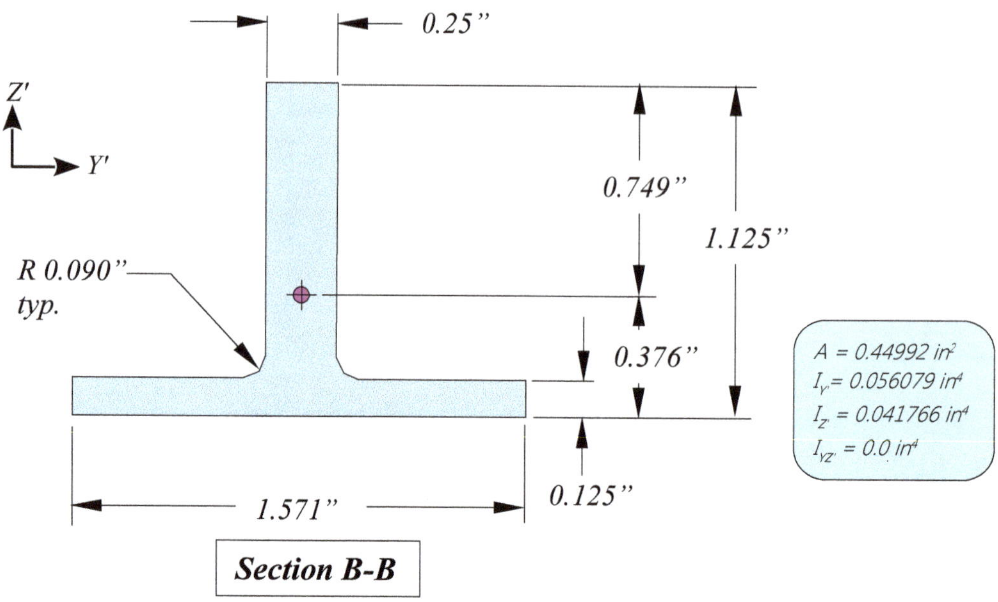

Figure 14-115 *– Aft outboard Lug Ftg., Section B-B.*

14.15.3 STRESS ANALYSIS

As the technique is the same for the stress calculations of both Sections, only the stresses at section *B-B* will be demonstrated.

Stresses due to Strut loads

Axial stress

The axial load going through the part at section *B-B* is half of P_X since fastener *2* took the other half. Hence, using *Table 14-39*, *Figure 14-113* and *Figure 14-114*, the axial load is $P_a = P_X / 2 = 844 / 2 = 422$ *lbs* and the axial stress f_a *is:*

$$f_a = \frac{P_X}{2A} = \frac{422}{0.44992} = 938 \ psi$$

Bending stress

Using *Figure 14-113*, the bending moment is −235 *in-lb* so the bending stress is calculated both at the top and at the base "*TEE cap*".

$$f_{btop} = -\frac{M_Y' \, c_{top}}{I_Y'} = -\frac{-235\,(0.749)}{0.056079} = 3{,}139 \ psi$$

$$f_{bbase} = \frac{M_Y' \, c_{base}}{I_Y} = \frac{-235\,(-0.376)}{0.056079} = -1{,}574 \ psi$$

Shear stress

The shear stress "$f_{XZ'}$" is calculated through section *B-B* but by removing the base flanges as those won't be carrying much vertical shear due to their vertical flexibility. Therefore, the remaining shear area is: $A_S = 1.125 * 0.25 = 0.28125 \ in^2$.

$$f_{XZ}' = \frac{P_Z' + R_{Z3}}{A_S} = \frac{341}{0.28125} = 1{,}212 \ psi$$

Stresses due to Side load at lug

The *Aft outb'd Lug Ftg* having a different geometry than the others is thus treated a bit differently. As can be seen in *Figure 14-116*, there are two possible load-paths to react the side load applied at the center of the bearing hole. The load-paths can be assumed as two cantilevered beams that are loaded at their free end. The most direct (stiffest) load-path "*1*" is towards the base common to the skin as its length is the shortest. Load-path "*2*" will be conservatively ignored in this analysis. Refer to the analysis on page *355* for an example on how to determine the load distribution between two load-paths.

To be thorough *"especially if a negative MS is obtained by considering only one load-path"*, one could evaluate the load distribution between the two load-paths. This is done by using the deflection equation of a cantilevered beam for each "beam" and equating their deflection.

In addition, one must take into consideration the stiffness of the back-up structure.

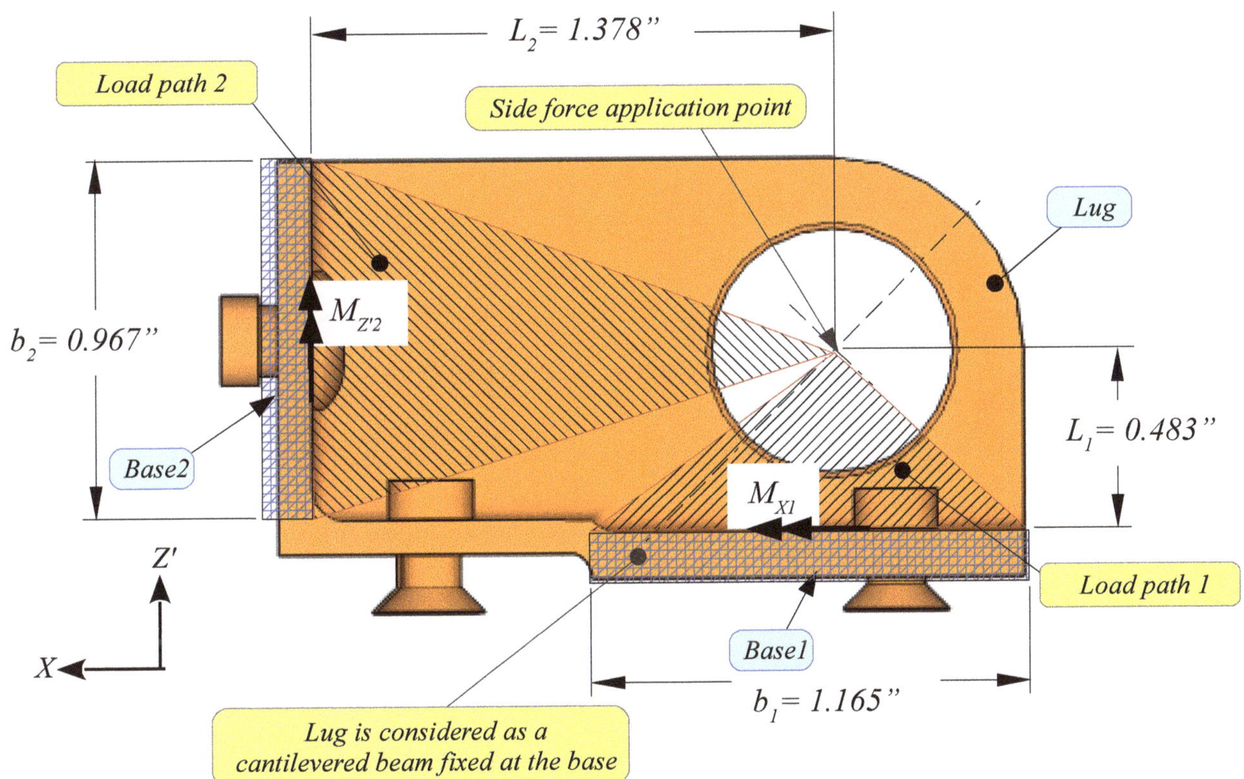

Figure 14-116 – *Aft outb'd Lug Ftg Load paths for the side load.*

Because the side load occurs at the same time as the Strut load, there are stress components along the "*X*" axis and along the "*Z'* " axis. The different stress components can be seen in *Figure 14-119* where the critical stress point location is shown as the *analysis point*.

Bending

The side load is reacted by the four fasteners attaching the Lug fitting to the skin. The *FBD* in the *X-Y'* plane due to the side load applied at the center of the bearing is shown in *Figure 14-117* along with the shear & moment diagram.

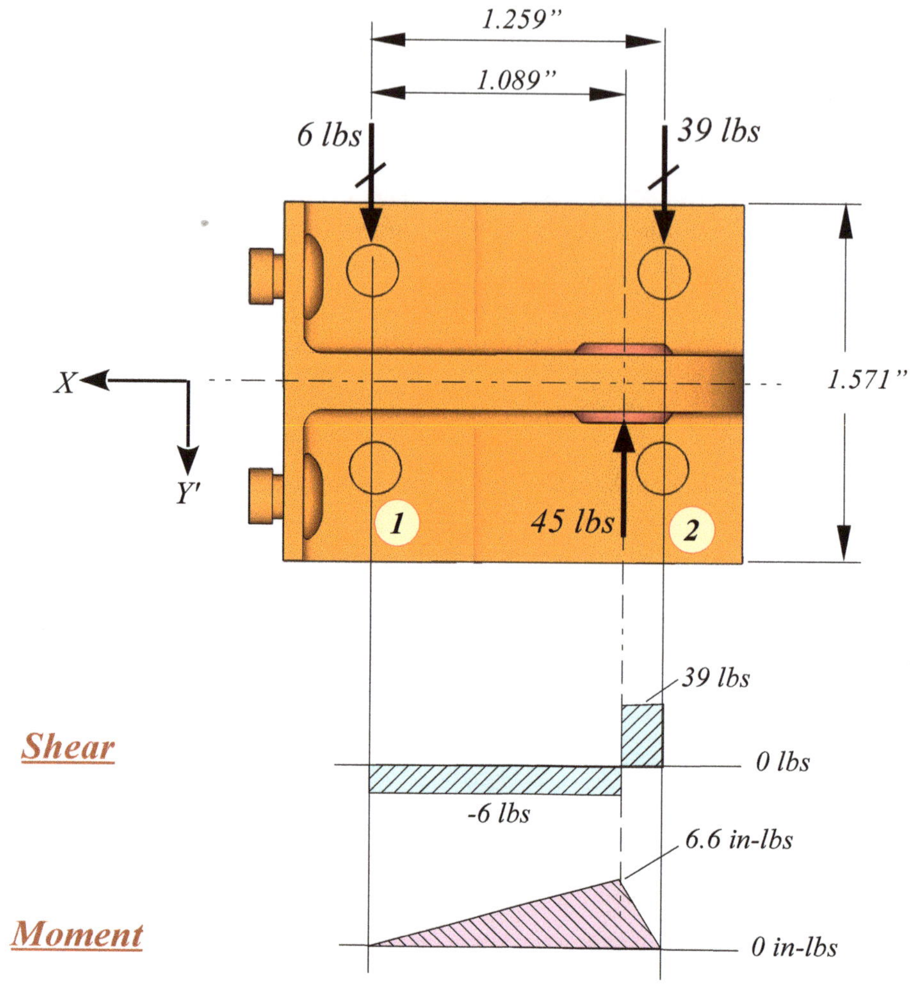

***Figure 14-117** – Aft outb'd Lug Ftg FBD in X-Y' plane due to side load.*

At the base of the lug

Using *Table 14-38* and *Figure 14-117*, the side load to apply at the center of the bearing hole is 45 *lbs*. For a length "L_1" of 0.483 *in.*, the bending moment "M_{X1}" is thus:

$$M_{X1} = P_{Side} \cdot L_1 = 45 \cdot 0.483 = 21.7\ in.-lb$$

Considering an effective base length "b_1" of 1.165 *in.* and the thickness of the lug of 0.25 *in.*, the bending stress "$f_{Z'}$" for this rectangular section at the "analysis point" is thus:

$$f_{Z'} = \frac{6M_{X1}}{b_1 t^2} = \frac{6 \times 21.7}{1.165 \cdot 0.25^2} = 1{,}788\ psi$$

At the edge of the base

The bending stress "f_X" is calculated at the edge of the base due to the side load. From *Figure 14-117*, the maximum bending moment "$M_{Z'}$" is 6.6 *in-lb*. The section properties are taken from *Figure 14-115*. Then:

$$f_X = \frac{M_Z' \cdot c}{I_Z'} = \frac{6.6 \cdot (0.786)}{0.41766} = 12\ psi \approx 0$$

For
$M_{Z'} = 6.6$ in-lb
$c = 0.786$ in.
$I_{Z'} = 0.041766$ in.

This stress component is so small that it is neglected.

Transverse shear

In the X-Y' plane, the average shear stress normal to the lug "$f_{XY'}$" is thus:

$$f_{XY}' = \frac{P_{Side}}{b_1 \cdot t} = \frac{45}{1.165 \cdot 0.25} = 155\ psi$$

In the Y'-Z' plane, the average shear stress "$f_{Y'Z'}$" by considering only the base is thus:

$$f_{Y'Z'} = \frac{V_Y'}{b \cdot t} = \frac{39}{1.571 \cdot 0.125} = 199\ psi$$

To be more accurate and more conservative, the shear stress "$f_{Y'Z'}$" is calculated using the shear stress equation:

$$f_s = \frac{VQ}{I \cdot t}$$

The first moment of Inertia "Q" is calculated below using *Figure 14-118.*

$$Q = A \cdot \bar{y} = \frac{0.66^2 \cdot 0.125}{2} = 0.02722\ in.^3$$

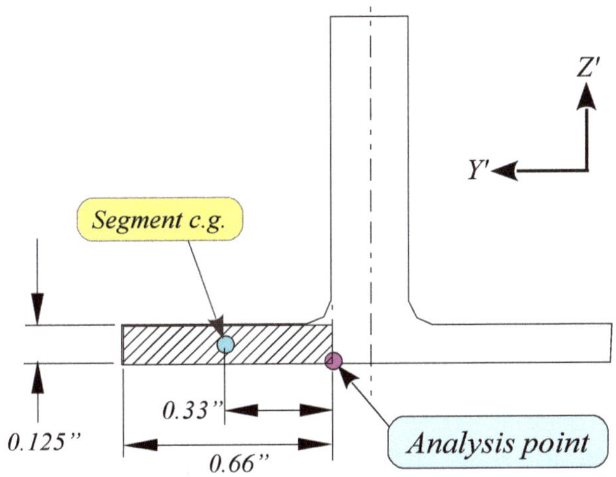

***Figure 14-118** – First moment of Inertia at analysis point.*

Then, the maximum shear stress "$f_{Y'Z'}$" at the analysis point is calculated as follows:

$$f_{Y'Z'} = \frac{V_Y'Q}{I_Z' \cdot t} = \frac{39(0.02722)}{(0.041766) \cdot (0.125)} = 203\ psi$$

For
$V_{Y'} = 39\ lbs$
$t = 0.125\ in.$
$I_{Z'} = 0.041766\ in.$
$Q = 0.02722\ in^3$

This is very close (in this case) to the previous faster approximation.

Combined stresses

At the top

At that location, the side load has no impact. Then, the axial, bending and shear stresses through section *B-B*, calculated previously from the Strut loads, are combined according to the equation:

$$(R_a + R_b)^2 + R_S^2 = 1$$

In our case, the axial and the bending stresses are simply added up and a single axial stress ratio "R_a'" is calculated. Thus the MS is calculated as:

$$MS = \frac{1}{\sqrt{R_a'^2 + R_S^2}} - 1$$

Using the material allowables from *Table 14-16*, the stress ratios are calculated where R_a' and R_S are respectively the axial and shear stress ratios calculated as follows:

$$R_a' = \frac{f_a + f_{b\,top}}{F_{tu}} = \frac{938 + 3{,}139}{75{,}000} = 0.055$$

$$R_s = \frac{f_s}{F_{su}} = \frac{1{,}212}{40{,}000} = 0.030$$

Hence, using a *FF* of 1.15, the MS at the top of section *B-B* is:

$$MS = \frac{1}{1.15\sqrt{0.055^2 + 0.030^2}} - 1 = 12.87$$

At the base:

Not only do we have to combine the axial, bending and shear stresses but we also have to include the bending and shear stresses due to the side load as it occurs at the same time. Although the bending stress due to the side load is at the top of the base, I will conservatively assume that they occur at the same point. Since the stresses "f_X" caused by the strut load and the one caused by the side-load "f_Z" are not in the same direction, they thus have to be combined vectorially. *Figure 14-119* shows an isometric view of both sets of stresses.

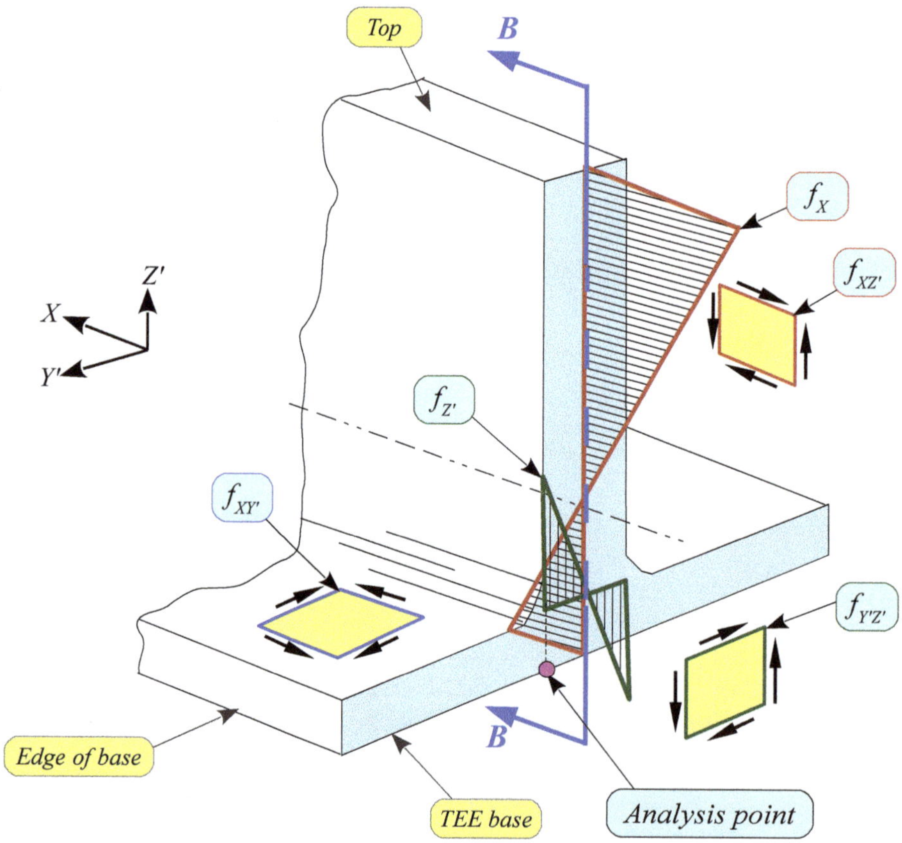

Figure 14-119 *– Stress combination at Section B-B.*

The *Von-Mises* effective stress "f_{eff}" for a tri-axial case is calculated as:

$$f_{eff} = \sqrt{\frac{1}{2}\left[(f_X - f_Y)^2 + (f_Y - f_Z)^2 + (f_Z - f_X)^2 + 6(f_{XY}^2 + f_{YZ}^2 + f_{XZ}^2)\right]}$$

The infinitesimal point of analysis is represented in *Figure 14-113*. Note that there is no stress "$f_{Y'}$" in the Y' direction in our case.

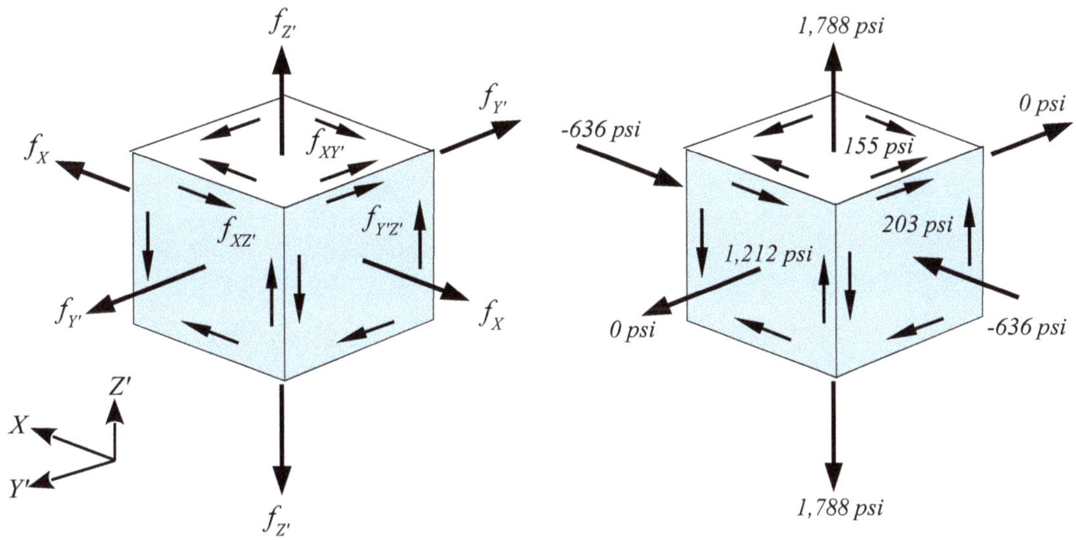

Figure 14-120 – *Infinitesimal element at analysis point.*

The *Von-Mises* effective stress calculated at the analysis point is thus:

$$f_X = f_a + f_{b\,base}$$

where,

$$f_X = 938 + (-1{,}574) = -636 \; psi$$

> For
> f_X = -636 psi
> $f_{Y'}$ = 0 psi
> $f_{Z'}$ = 1,788 psi
> $f_{XY'}$ = 155 psi
> $f_{Y'Z'}$ = 203 psi
> $f_{XZ'}$ = 1,212 psi

hence,

$$f_{eff} = \sqrt{\frac{1}{2}\left[(-636-0)^2 + (0-1{,}788)^2 + (1{,}788-(-636))^2 + 6(155^2+203^2+1{,}212^2)\right]} = 3{,}057 \; psi$$

Thus, using a *FF* of 1.15, the MS at the analysis point (base of section *B-B)* is calculated by conservatively using "F_{CY}" from *Table 14-16* as the allowable:

$$MS = \frac{F_{CY}}{FF\, f_{eff}} - 1 = \frac{60{,}000}{1.15 \cdot 3{,}057} - 1 = 16.07$$

The *Aft outb'd Lug Ftg* top of section *B-B* is therefore the most critical location. So the stress Margin of Safety is as follows:

Aft Outb'd Lug Ftg Section B-B, Ultimate	***>1.00***
9g Fwd *Al Aly 7050-T7451*	*Combined axial, bending & shear*

> Here again, F_{CY} was used as the allowable since there was compression stress at the base. No crippling allowable was calculated though since, as the ratio b/t of the base was low, I expected it to be equal to F_{CY} and also, since the internal axial stress of −636 *psi* was quite low, crippling was certainly not a concern.

14.15.4 LUG ANALYSIS

Figure 14-121 shows how the lug has been idealized for the *Aft outboard Lug fitting*. The fitting is analyzed in the same manner as for the *Fwd outb'd Lug Ftg* done in section *14.13.5*. See *Table 14-42* for the resulting stresses and MS.

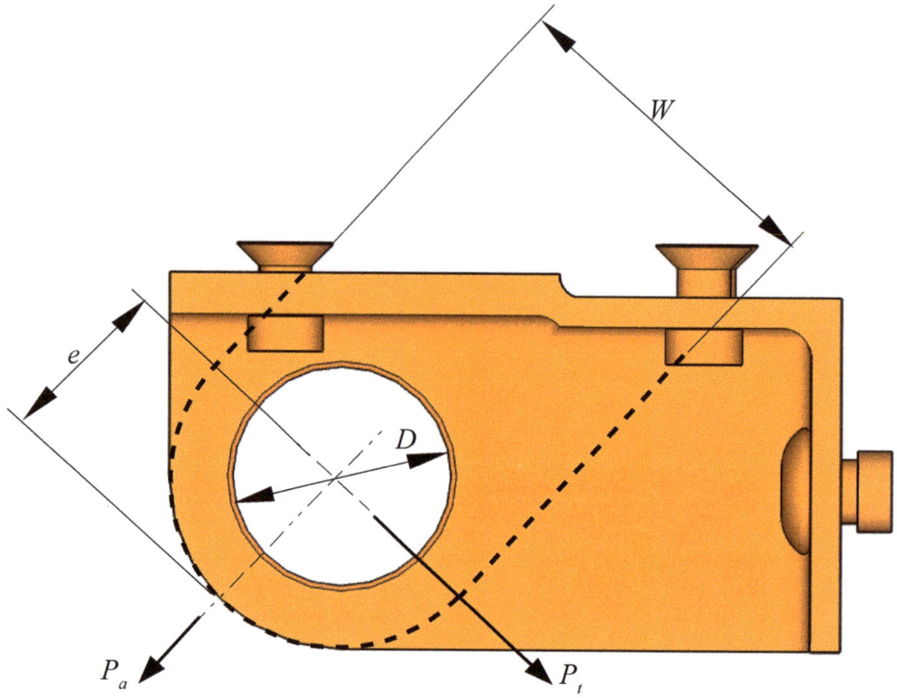

***Figure 14-121** – Aft outboard Lug Fitting lug idealization – point 8.*

14.15.5 BEARING INTERFERENCE FIT

Both *Aft inboard* and *Aft outboard Lug fittings* have the exact same spherical bearing installation. They also both have the same material and lug geometry. Therefore, this analysis covers both parts.

The spherical bearings are installed with a slight interference fit and, as a result, a tangential "hoop" tension stress will develop in the lug. Note that the spherical bearing manufacturer may specify a maximum (and minimum) interference fit range so that the bearing can function properly without being overly "squeezed".

In checking interference fit installations involving parts having different materials, I always like to do the analysis at the extreme temperatures to ensure that the stress levels are still adequate and that the bearing will be functioning as expected. This analysis is more a durability check than a static check as the thermal gradient will vary for each flight and cause cycling stresses.

I have seen instances where the temperature gradients were ignored which caused either premature durability issues (cracking) or knocking of parts due to loose fits. Remember that an aircraft can be parked on a tarmac at +50°C and could be flying at 40,000 *ft*. at −65°C with dissimilar materials having different thermal expansion coefficients! On top of that, those conditions could be occurring at every single flight (GAG cycle) which becomes a fatigue issue.

In our case, we have a Corrosion Resistant Steel "*CRES*" bearing outer race installed in an aluminum part. The aluminum having a larger thermal coefficient of expansion "α" than the *CRES*, will expand and shrink more than the bearing will for the same temperature gradient. So my concerns are: first, to make sure that the tension stress in the Ftg lug isn't too high at cold temperature and second, that the bearing doesn't become loose at elevated temperature.

The linear thermal coefficient of expansion for the materials involved are assumed as:

$$\alpha_{Alu} = 23.0 \times 10^{-6} in./in./C$$

$$\alpha_{CRES} = 11.0 \times 10^{-6} in./in./C$$

Since the Lug fittings are installed inside the airplane, they won't see as high a temperature gradient as if they were installed outside, like in the case of flight controls for example. I will thus do the analysis for the temperatures at −20°C and +40°C.

The following assumptions are made for this analysis:

- the temperature was +20°C when the parts were machined and installed.
- The lug is considered as a ring.
- The thermal coefficient of expansion is constant throughout the temperature range.
- Only the bearing outer race is considered having an inner radius equal to the average between the ball radius and the shoulder radius.
- The stresses caused by the emergency or flight load cases are not combined.
- The interference stress is considered limit.

Determination of the interference

At room temperature (20°C)

From the dimensions and taking into account their tolerances shown in *Figure 14-16* and *Figure 14-110*, the hole diameter range in the lug "D_{lug}" and the spherical bearing outer diameter range "D_{brg}" are:

$$D_{Lug} = 0.6540 : 0.6550 \; in.$$

$$D_{brg} = 0.6562 : 0.6557 \; in.$$

The minimum interference "e_{min}" on the radius is obtained from the largest hole with the smallest bearing, so:

$$e_{min} = \frac{0.6557 - 0.6550}{2} = 0.00035 \; in.$$

And, the maximum interference "e_{max}" on the radius is obtained from the smallest hole with the largest bearing, so:

$$e_{max} = \frac{0.6562 - 0.6540}{2} = 0.0011 \; in.$$

At low temperature (−20°C)

At that temperature, the temperature gradient "ΔT" is −40°C. The reduction in radius "ΔR" is thus:

$$\Delta R = \alpha(\Delta T)R$$

At the interface radius "R" of 0.328 *in.*, that value for each material is thus:

$$\Delta R_{Alu} = 23 \times 10^{-6}(-40)(0.328) = -0.000302 \; in.$$

$$\Delta R_{CRES} = 11 \times 10^{-6}(-40)(0.328) = -0.000144 \; in.$$

Assuming that we have the maximum interference as per drawing, then the final interference at −20°C "$e_{max\text{-}20}$" is thus:

$$e_{max-20} = e_{max} - \Delta R_{Alu} + \Delta R_{CRES}$$
$$e_{max-20} = 0.0011 - (-0.000302) + (-0.000144) = 0.001258 \; in.$$

That interference value is the one that will be used in the following calculations.

Interface pressure

The interface pressure "p" (radial stress) is calculated using the material properties from *Table 14-16* and *Table 14-18*. The pressure is calculated using the following equation:

$$p = \frac{e}{R\left[\frac{1}{E_o}\left(\frac{r_o^2+R^2}{r_o^2-R^2}+\mu_o\right)+\frac{1}{E_i}\left(\frac{R^2+r_i^2}{R^2-r_i^2}-\mu_i\right)\right]}$$

The dimensions of the idealized rings to use in the analysis are shown in *Figure 14-122*. The inner ring inner radius "r_i" is approximated as follows:

$$r_i = \frac{r_{ball}+r_{shoulder}}{2} = \frac{0.297+0.182}{2} = 0.240 \; in.$$

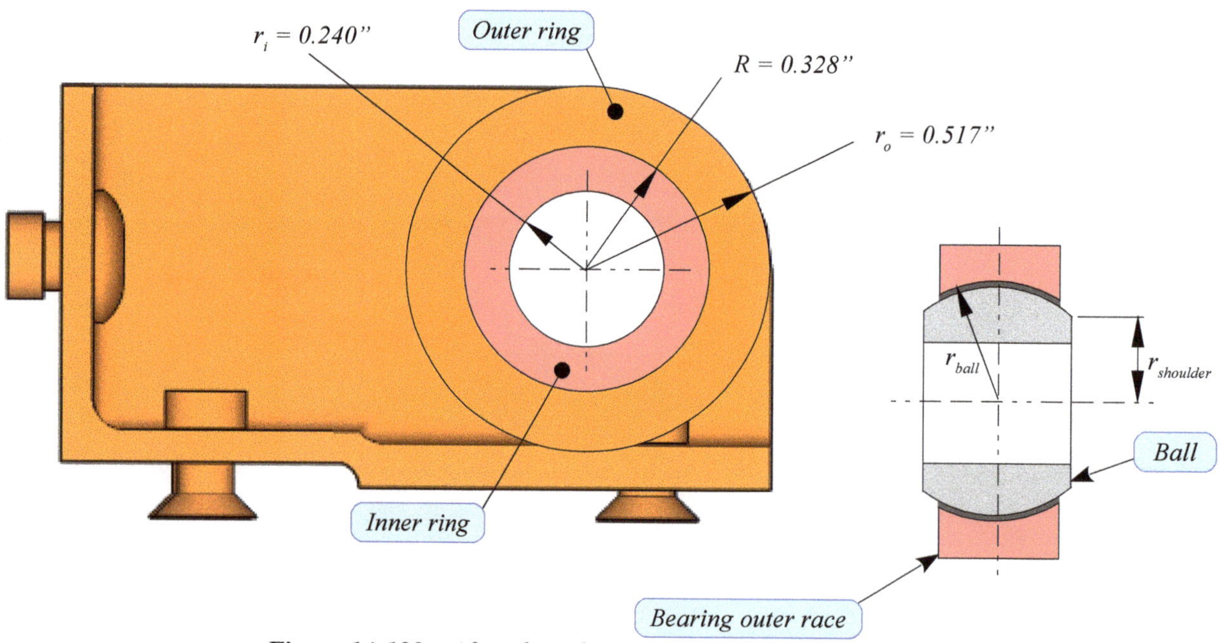

Figure 14-122 *– Aft outboard Lug Fitting lug rings idealization.*

Hence, the interface pressure is:

For			
$e = 0.001258$ in.	$r_i = 0.240$ in.	$E_{alu} = E_o = 10x10^6$ psi	$\mu_o = 0.33$
$R = 0.328$ in.	$r_o = 0.517$ in.	$E_{CRES} = E_i = 30x10^6$ psi	$\mu_i = 0.30$

$$p = \frac{0.001258}{0.328\left[\frac{1}{10\times10^6}\left(\frac{0.517^2+0.328^2}{0.517^2-0.328^2}+0.33\right)+\frac{1}{30\times10^6}\left(\frac{0.328^2+0.240^2}{0.328^2-0.240^2}-0.30\right)\right]} = 10{,}425 \; psi$$

As can be noticed above, the interface pressure is relatively high despite the fact that the interference is very small. It is interesting to note that the smaller the dimensions of the bearing and the lug, the higher the pressure for the same interference value.

The maximum limit tangential stress in the lug occurs at the interface and is calculated using the equation for cylinders having an internal pressure:

$$\sigma_{\theta\,max} = p\,\frac{r_o^2 + R^2}{r_o^2 - R^2}$$

hence:

$$\sigma_{\theta\,max} = 10{,}425\,\frac{0.517^2 + 0.328^2}{0.517^2 - 0.328^2} = 24{,}471\ psi$$

Using a *FF* of 1.15 and an ultimate load factor of 1.5, the MS in the *Lug fitting* is calculated below using the ultimate tensile strength from *Table 14-16.*

$$MS = \frac{F_{TU}}{1.5\,FF \cdot \sigma_{\theta\,max}} - 1 = \frac{75{,}000}{1.5 \cdot 1.15 \cdot 24{,}471} - 1 = 0.77$$

Aft Lug Ftg, Ultimate	***0.77***
Bearing Interference fit at -20°C Al Aly 7050-T7451	*Tangential stress*

Similarly, the maximum limit tangential stress in the bearing outer race occurs at the interface and is calculated using the equation for cylinders having an external pressure:

$$\sigma_{\theta\,max} = -p\,\frac{R^2 + r_i^2}{R^2 - r_i^2}$$

hence:

$$\sigma_{\theta\,max} = -10{,}425\,\frac{0.328^2 + 0.240^2}{0.328^2 - 0.240^2} = -34{,}453\ psi$$

Using a *FF* of 1.15 and an ultimate load factor of 1.5, the MS in the Lug fitting is calculated below by conservatively using the yield compression strength from *Table 14-18.*

$$MS = \frac{F_{CY}}{1.5\,FF \cdot \sigma_{\theta\,max}} - 1 = \frac{135{,}000}{1.5 \cdot 1.15 \cdot 34{,}453} - 1 = 1.27$$

Spherical bearing outer race, Ultimate	***>1.0***
Bearing Interference fit at -20°C CRES	*Tangential stress*

Temperature check

No fatigue check of the lug fitting was deemed necessary due to the variations in temperature since the Lug fittings are inside the plane and that the temperature inside is pretty much always constant at around +20°C. However, a static check is done at high temperature to ensure that a small interference still exists.

At elevated temperature (40°C)

At that temperature, the temperature gradient "ΔT" is 20°C. The increase in radius "ΔR" is thus:

$$\Delta R = \alpha(\Delta T)R$$

At the interface radius "R" of 0.328 *in.*, that value for each material is thus:

$$\Delta R_{Alu} = 23\times10^{-6}(20)(0.328) = 0.000151 \ in.$$

$$\Delta R_{CRES} = 11\times10^{-6}(20)(0.328) = 0.000072 \ in.$$

Assuming that we have the minimum interference as per drawing, then the final interference at 40°C "$e_{min\ 40}$" is thus:

$$e_{min\,40} = e_{min} - \Delta R_{Alu} + \Delta R_{CRES}$$
$$e_{min\,40} = 0.00035 - 0.000151 + 0.000072 = 0.000271 \ in.$$

Since the resulting interference is still positive, then the bearing doesn't come loose at 40°C. There is no need to evaluate the stress for the minimum interference since it is always lower than the stress at the highest interference. However, due to a side-load, we wouldn't want the bearing to slide out of the fitting. Therefore, one thing that must be verified is the axial force required to remove the spherical bearing to ensure that it is higher than any possible applied side force. The axial force "F_a" required is calculated with the equation:

$$F_a = \mu\, p_{min}\, \pi D t$$

Assuming a coefficient of friction "μ_f" between the aluminum and the CRES of 0.5 and calculating the minimum interface pressure "p_{min}" at elevated temperature for an interference "$e_{min\ 40}$" of 0.000271 *in.*, the axial force is calculated below.

Using the same equation for the interface pressure as done previously for the maximum interference at low temperature, an interface pressure of p_{min} = 2,246 *psi* is obtained. Thus the axial force is:

$$F_a = (0.5)2{,}246\,\pi(0.656)(0.25) = 578 \ lbs$$

For	
$e = 0.000271$ in.	$D = 0.656$ in.
$p_{min} = 2{,}246$ psi	$t = 0.25$ in.
$\mu = 0.50$	

This force being higher than the ultimate applied side load of 45 *lbs*, this installation is structurally acceptable.

14.15.6 CONNECTIONS

Based on the previous load distribution assumptions for the *Aft outb'd Lug Ftg.*, *Figure 14-123* shows the FBD in the Y'-Z' plane with the reactions at the fasteners. *Figure 14-124* shows the FBD in an isometric view so that all the reaction forces can clearly be seen.

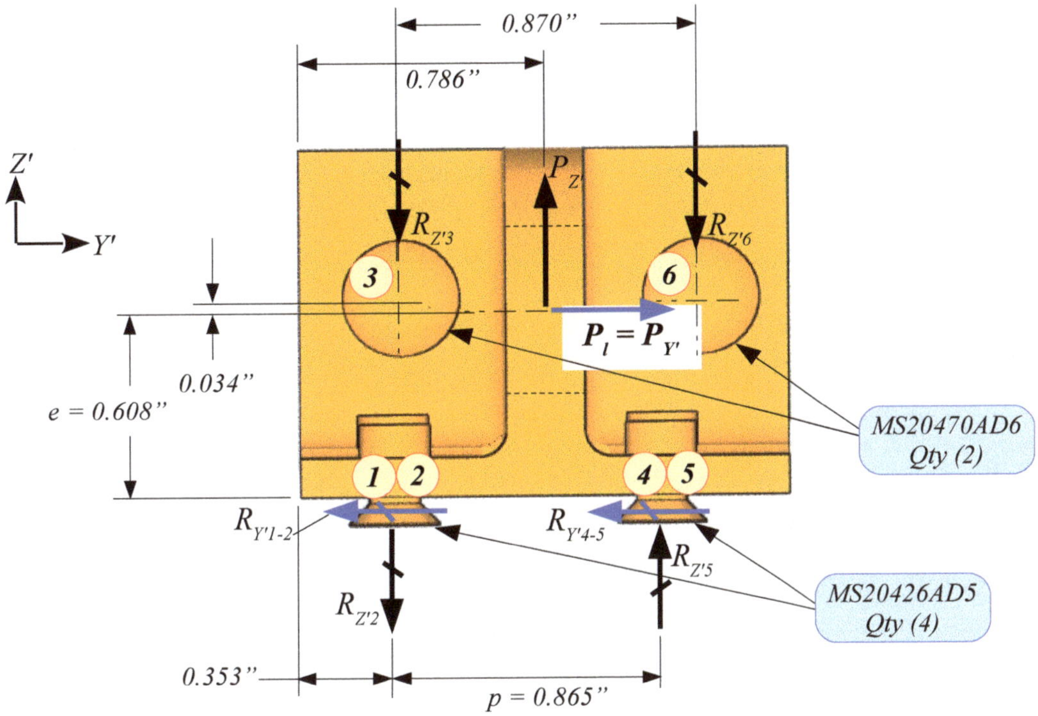

Figure 14-123 *– FBD showing the Aft outb'd Ftg fasteners reactions.*

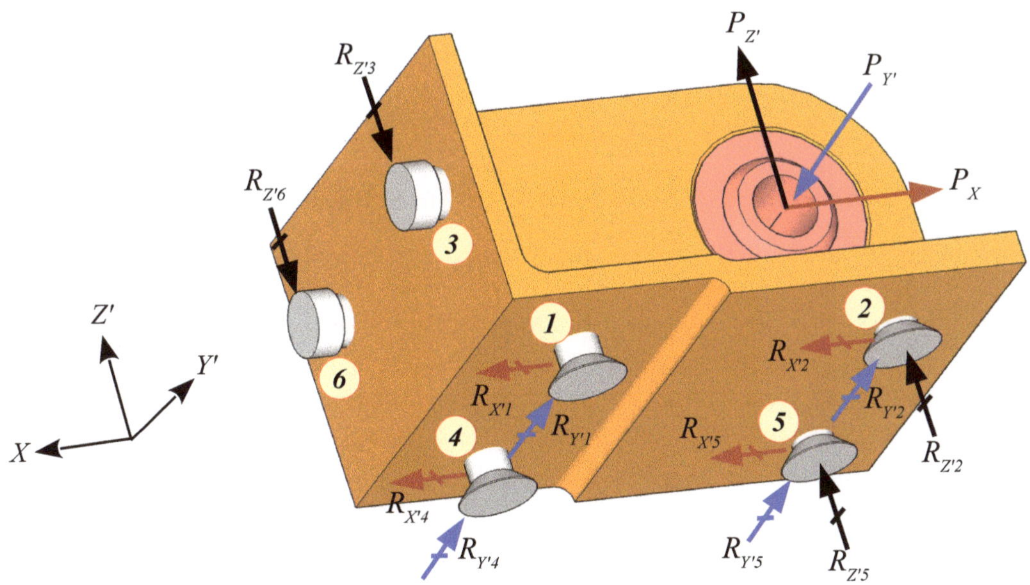

Figure 14-124 *– Isometric view of the Aft outb'd Ftg FBD with the fasteners reactions.*

The torque "M_X" around the X axis caused by the side load "$P_{Y'}$ "could have been assumed to be reacted in shear by the fasteners *3 & 6*. However, even though the fitting is short, it is still an open section "*TEE*" having a low stiffness in torsion. I then preferred assuming that moment to be reacted by the fasteners *2 & 5* since they are very close to the bearing where the side load is applied.

14.15.6.1 Fasteners forces

Along the *X* axis:

As assumed previously, the applied load P_X is evenly distributed among the four fasteners (*1, 2, 4 & 5*) common to the skin. Using the P_X value of 844 *lbs* from *Table 14-39*, the fastener's reactions in the *X* direction are then:

$$R_{X1} = R_{X2} = R_{X4} = R_{X5} = \frac{P_X}{4} = \frac{844}{4} = 211\ lbs$$

Along the *Y'* axis:

Like for the load in *X*, the load along the *Y'* axis is also reacted by the four fasteners (*1, 2, 4 & 5*). Using *Figure 14-117*, the reaction at the fastener rows are 6 *lbs* for the pair *1 & 4* and 39 *lbs* for the pair *2 & 5*. The fastener's reactions in the *Y* direction are then:

$$R_{Y'1} = R_{Y'4} = \frac{6}{2} = 3\ lbs$$

$$R_{Y'2} = R_{Y'5} = \frac{39}{2} = 20\ lbs$$

Along the *Z'* axis:

From section *14.15.1*, the reactions at the fastener pair *2 & 5* is $R_{Z2\text{-}5}$ = 47 *lbs* and it is $R_{Z3\text{-}6}$ = 341 *lbs* at the fastener *3 & 6* pair. Since it is assumed that the "torque" caused by the side load is only reacted by the fasteners *2 & 5* as a couple, the vertical load is considered evenly distributed between the *3 & 6* pair. The fastener's reactions in the *Z'* direction are then:

$$R_{Z'3} = R_{Z'6} = \frac{341}{2} = 171\ lbs$$

Using *Figure 14-123* and *Table 14-38*, the "torque" M_X" from the side load is calculated as:

$$M_X = P_{Y'} e = 45 \cdot 0.608 = 27.4\ in.-lb$$

Then, the couple force "R_{ZC}" in fasteners *2 & 5* to react that moment is:

$$R_{ZC} = \frac{M_X}{p} = \frac{27.4}{0.865} = 31.6\ lbs$$

Then, combining the forces in fasteners *2 & 5* we get:

$$R_{Z'2} = \frac{R_{Z2-5}}{2} - R_{ZC} = \frac{47}{2} - 31.6 = -8\ lbs$$

$$R_{Z'5} = \frac{R_{Z2-5}}{2} + R_{ZC} = \frac{47}{2} + 31.6 = 55\ lbs$$

Shear + Tension

Table 14-51 shows the fasteners loads summary along with the shear force "F_S" and the tension force "F_T" in each of them. The shear force in fasteners *3 & 6* is simply equal to the $R_{Z'}$ load while for fasteners *1, 2, 4 & 5,* it is calculated as follows:

$$F_S = \sqrt{R_X^2 + R_{Y'}^2}$$

Table 14-51 *- Aft outb'd Lug Ftg. Fasteners summary forces.*

	R_X	$R_{Y'}$	$R_{Z'}$	F_S	F_T
fastener	*[lbs]*	*[lbs]*	*[lbs]*	*[lbs]*	*[lbs]*
1	211	3		211	0
2	211	20	-8	212	8
3			171	171	0
4	211	3		211	0
5	211	20	55	212	-55
6			171	171	0

The forces in each fasteners could also have been calculated using a "3D bolt group" tool. This is the same approach as described in section *10.6.2* but in the 3 axes. When using such a tool, one has to use the proper stiffness at each fastener location. For our example above, fasteners *3 & 6* would have a stiffness of 0 in the *X* and *Y* axes but 1 along the *Z'* axis. But using a stiffness of 1 for all fasteners, one assumes that the part is infinitely rigid. This wouldn't be realistic for our Lug fitting and for a lot of other parts made with relatively thin flanges.

Bearing

The bearing stress "f_{bru}" is checked in the *Aft outb'd Lug Ftg* as well as in the skin and the Frame. The bearing load capability is calculated as:

$$P_{bru} = F_{bru} \cdot D \cdot t$$

The shear load used in the analysis is from fasteners *1 & 4* for the *Aft outb'd Lug Ftg* is 211 *lbs*, it is from fasteners *3 & 6* for the Frame web at 171 *lbs*, and is from fasteners *2 & 5* for the skin at 212 *lbs*.

Their respective calculated bearing allowable are shown in *Table 14-52*. Note that one should always be using the joint allowables from (ref. 8. chapter 8*)* or from the aircraft manufacturer whenever they are available, instead of calculating the shear and bearing allowables, as they are based on tests and are more accurate and usually more conservative. This is especially true for flush-head fasteners which have lower joint allowables.

Table 14-52 *- Aft outb'd Lug Ftg connections Bearing allowables.*

	Material	*t*	*D*	*Fbru*	*Pbru*	*F*	*MS*
Part	*Al Aly*	*[in]*	*[in]*	*[ksi]*	*[lbs]*	*[lbs]*	*bearing*
Frame	*7075-T7351*	0.070	0.191	135	1805	171	9.56
Lug Ftg	*7050-T7451*	0.090	0.159	105	1503	211	6.12
Skin	*2024-T3 Clad*	0.090	0.159	120	1717	212	7.10

The MS for the connections of the *Aft outboard Lug fitting* are covered by the *Aft inboard Lug fitting* MS shown on page *429*.

14.15.6.2 Non-structural shim

Fasteners *1 & 4* pick-up the Frame flange that is sandwiched between the *Aft Lug Ftg* and the skin. As the shear load in those two fasteners is really reacted by the skin and since there is only one row of rivets in the Frame flange, no shear is assumed reacted by the Frame flange in the longitudinal direction "*X*". The Frame could be considered to react the lateral load "*Y'* " but it would unnecessarily complicate things. Therefore, that flange is to be considered as a non-structural shim in both *X* and *Y'* directions.

The method that I normally use to substantiate rivets that are common to a non-structural shim is to simply consider it as a fixed-fixed beam. The effective length of the beam goes from the mid-thickness of the Aft Lug Ftg (in our case) to the mid-thickness of the skin. See *Figure 14-125* for a typical rivet joint idealization having a non-structural shim.

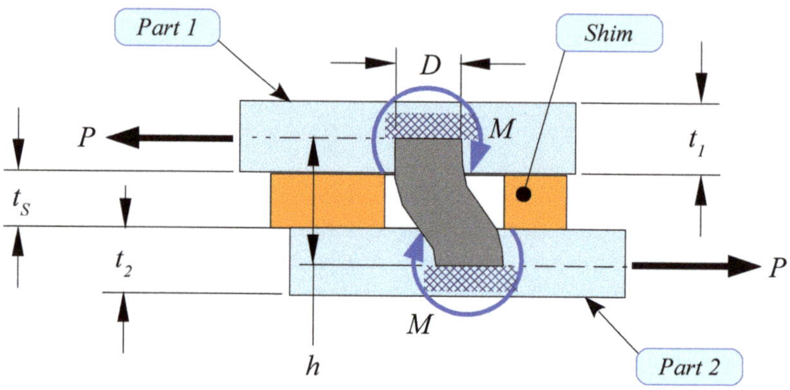

Figure 14-125 – *Idealization of a rivet joint with a non-structural shim.*

> In the case of bolts having a preload, I would consider the effective fastener length to be the sum of the thicknesses from all parts involved and, I would assume the end moments to be reacted as a couple between a tension load in the fastener shank and a contact force at the edge of the bolt's head/nut.

Using the fixed-fixed beam model, the end moment "*M*" is calculated as follows:

$$M = \frac{P \cdot h}{2}$$

In our example, the shear load "*P*" in fasteners *1 & 4* is 211 *lbs* from *Table 14-51*. The maximum end moment "*M*" is thus:

$$M = \frac{211 \cdot (0.16)}{2} = 16.9\, in. - lb$$

Where

$$h = \frac{t_1}{2} + t_S + \frac{t_2}{2} = \frac{0.090}{2} + 0.070 + \frac{0.090}{2} = 0.16\ in.$$

> *For:*
> $t_1 = t_{Ftg} = 0.090$ *in*
> $t_S = t_{frame} = 0.070$ *in*
> $t_2 = t_{skin} = 0.090$ *in*
> $D = 0.159$ *in*

The *MS20426AD5* rivets are made out of *Al Aly 2117-T3* having the following properties: F_{su} = 30 *ksi* while F_{tu} is evaluated as: F_{su} / 0.577 = 30 / 0.577 = 52 *ksi*.

Bending stress

For a rivet shank diameter of 0.159 *in.*, the maximum bending stress is calculated as:

$$f_b = \frac{M \cdot \frac{D}{2}}{\pi \frac{D^4}{64}} = \frac{32\text{M}}{\pi D^3} = \frac{32 \cdot 16.9}{\pi (0.159)^3} = 42{,}825 \ psi$$

and the bending stress ratio "R_b" is thus:

$$R_b = \frac{f_b}{F_{tu}} = \frac{42{,}825}{52{,}000} = 0.824$$

Shear stress

The average shear stress is calculated as:

$$f_s = \frac{P}{\pi \frac{D^2}{4}} = \frac{211 \cdot (4)}{\pi (0.159)^2} = 10{,}627 \ psi$$

and the shear stress ratio "R_S" is thus:

$$R_S = \frac{f_S}{F_{su}} = \frac{10{,}627}{30{,}000} = 0.354$$

Hence, for the *MS20426AD5* (*BB5* rivets *1 & 4*) using a *FF* of 1.15, the MS is calculated as:

$$MS = \frac{1}{FF\sqrt{R_b^2 + R_S^2}} - 1 \qquad MS = \frac{1}{1.15\sqrt{0.824^2 + 0.354^2}} - 1 = -0.03$$

Thus we find that the *BB5* rivets at those locations (*1 & 4*) just don't do the job. That confirms my suspicions that flush Hi-Lite fasteners should be used there. This fixed-fixed beam idealization of the rivet is most probably too conservative but I prefer doing that as I know that the same fasteners are also used in the joint between the Frame and the skin.

This latter joint analysis uses the difference in shear flow "Δq" in the adjacent skin panels obtained from the global FEM, see section *13.3* for an example. However, the critical load case for the *Lug Ftg* being the *9g Fwd LC*, that shear load is not going to be combined to any flight load case that would be loading those same fasteners.

Since fasteners 1 & 4 need to be changed from MS20426AD5 solid rivets to HST11-5 Hi-Lites, fasteners 2 & 5 will also be changed to the same fasteners in order to be uniform in the type of fasteners used.

Judging from the relatively small magnitude of the fasteners loads from *Table 14-51*, with their much higher shear and tension strengths, the Hi-Lites are thus Passed by engineering judgment.

If I were to write an actual stress report, I would most certainly show an MS for the fasteners. The method to use when there is a combination of shear and tension loads can be seen in section *14.6.11*. In our current example, I would be adding the tension and the bending stresses together to get the tension stress ratio. Then, I would proceed with the graph using the following interaction equation:

$$R_s^3 + R_t^2 = 1$$

14.15.7 SPHERICAL BEARING

A spherical bearing is installed press-fitted into the lug hole. They are usually rolled swaged to keep them into place. From *Figure 14-16*, the allowables in the radial and axial directions are 5,500 *lbs* and 400 *lbs* respectively. See *Figure 14-126* to see how the loads are applied to the bearing.

From *Table 14-38*, the Strut load "P_{strut}" at point *8* is 893 *lbs* and the side load "P_{side}" is 45 *lbs*. The MS is calculated using the interaction equation:

$$R_r^2 + R_a^2 = 1$$

The load ratios are calculated as follows:

$$R_r = \frac{P_{Strut}}{P_{radial}} = \frac{893}{5{,}500} = 0.162$$

$$R_a = \frac{P_{Side}}{P_{axial}} = \frac{45}{400} = 0.113$$

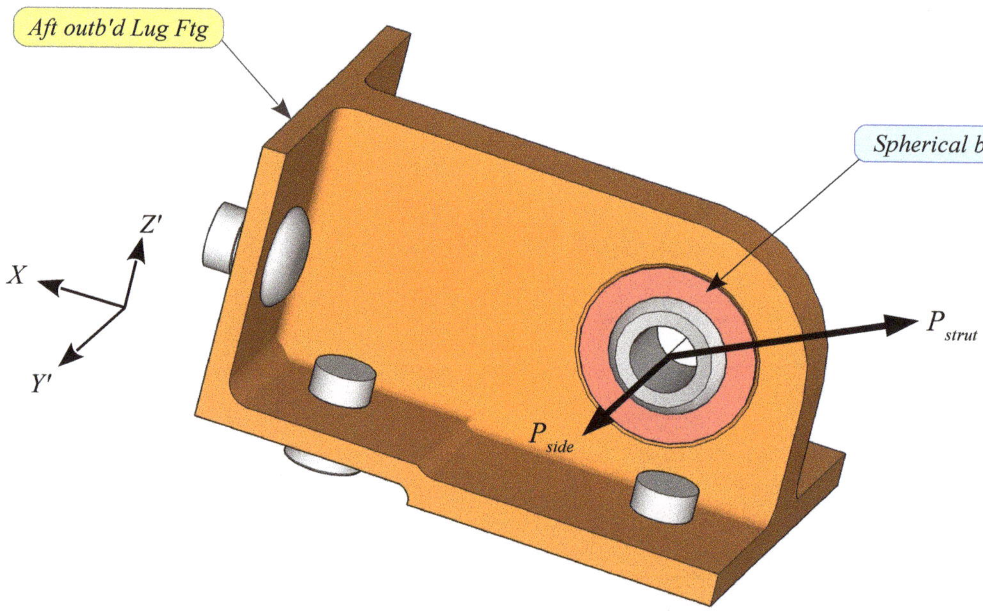

Figure 14-126 *– Aft outboard Lug Ftg assembly.*

Using a *FF* of 1.15, the MS for the spherical bearing is thus:

$$MS = \frac{1}{FF\sqrt{R_r^2 + R_a^2}} - 1$$

$$MS = \frac{1}{1.15\sqrt{0.162^2 + 0.113^2}} - 1 = 3.40$$

Aft Outb'd Lug Ftg Spherical bearing, Ult.	***>1.00***
9g Fwd *MS14101-4*	*Radial + axial*

14.15.8 COMMENTS

The *Aft outboard Lug fitting* is adequate for the emergency load cases by having *Hi-Lites* fasteners connecting to the skin. The *BJ6* fasteners common to the Frame web could be changed to *BJ5*'s thus allowing room for repairs.

14.16 STRUTS

To react the fore-aft loads, two Struts are used at the top of the cabinet. Both Struts are exactly the same even though their installed length is different by about a ¼ in. as their length is adjustable. Each Strut is made from an assembly of: ***1-*** a thin *Aluminum Alloy 7075-T62* tube, ***2-*** a rod-end made of *CRES* with a spherical bearing (ball joint) and, ***3-*** a rod-end also made of *CRES* called a clevis. The Struts are attached to the Lug Fittings with 0.25" dia. bolts. *Figure 14-2* shows the struts installation and *Figure 14-127* shows a Strut assembly.

For the *9g FWD LC,* a simple cable could have done the job but, it wouldn't have worked for the *1.5g Aft LC* which puts the Struts in compression.

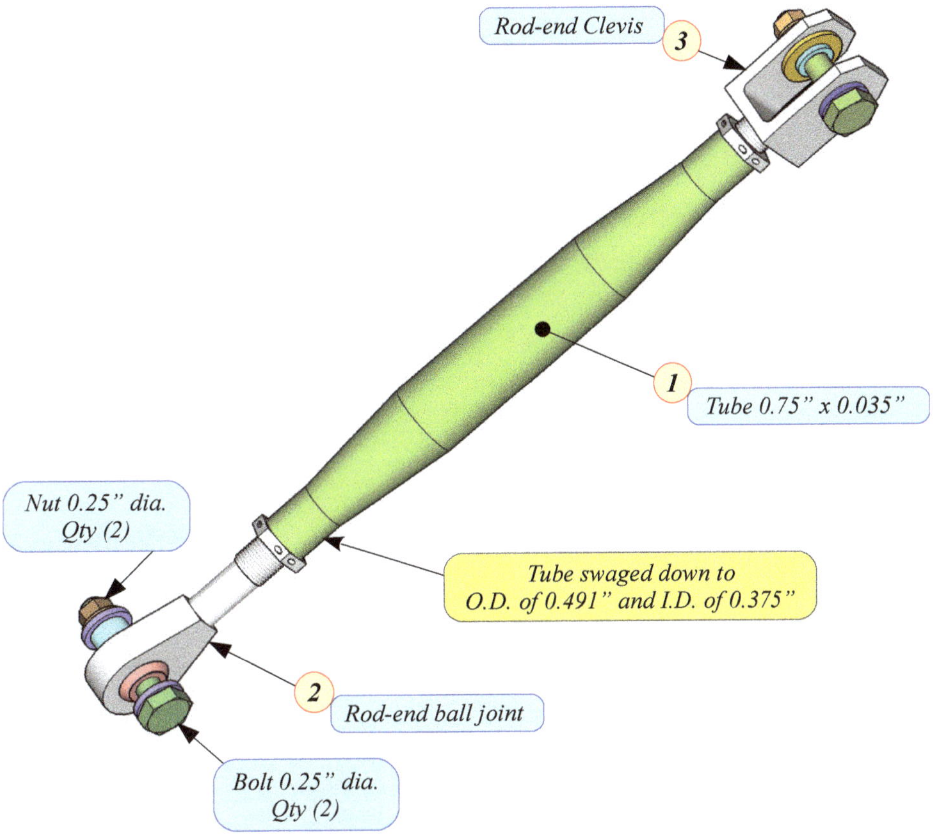

Figure 14-127 *– Strut assembly.*

14.16.1 LOADS

The struts are loaded in tension from the *9g Fwd LC* and in compression for the *1.5g Aft LC. Table 14-53* shows the ultimate loads for each strut. No other emergency load case loads-up the struts.

Table 14-53 *- Strut loads and lengths.*

		P 9g fwd	P 1.5g aft	Length
Strut	point	[lbs]	[lbs]	[in]
1	5	784	-131	8.69
2	6	893	-149	8.40

Struts are mainly sized according to the magnitude of the internal compression loads and their length as they are generally more critical due to the instability. For the purpose of the demonstration (although this is overly conservative), I will assume the loads to be fully reversible. In order to substantiate both struts, I will use the highest load and the largest length.

14.16.2 SECTION PROPERTIES

As can be seen in *Figure 14-128*, typical Struts are not simple columns as they have multiple cross-sections along their length. See *Figure 14-129* for the cross-sections dimensions and *Table 14-54* for the section properties of the three chosen cross-sections. Note that the outer diameter of the Rod-end "D_O" is based on the minor diameter of an external thread 0.375-24 UNJF according to the military standard *MIL-S-8879* (ref. 12.).

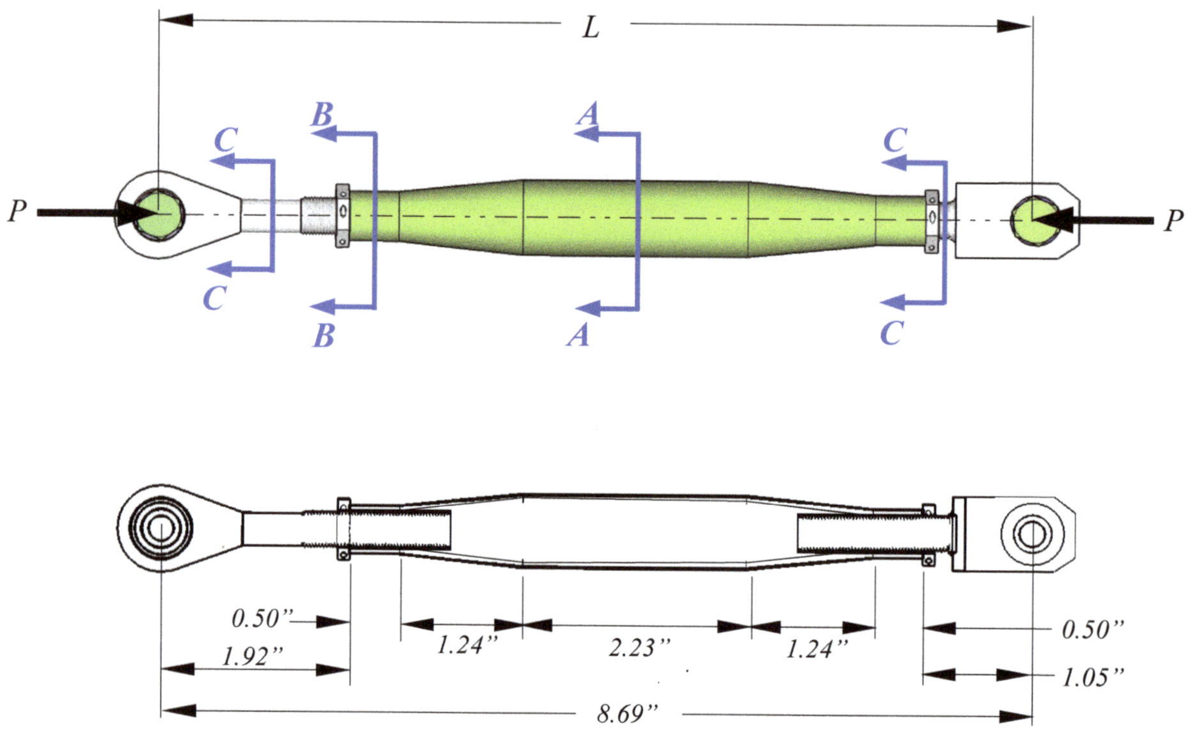

Figure 14-128 *– Strut assy dimensions and sections locations.*

Therefore, in order to analyze the struts in column buckling, one could use an *FEA* method or, in a quick , simple and conservative way, analyze it as a stepped column, as will be done in the next chapter.

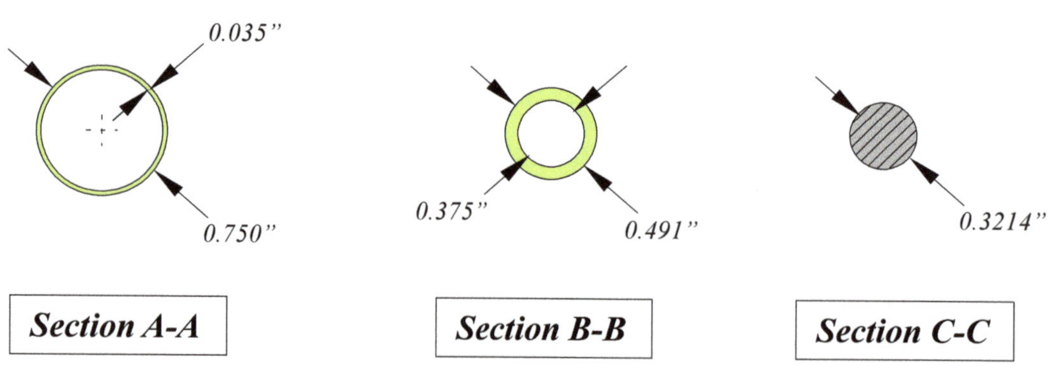

Figure 14-129 *– Strut assy sections properties.*

Table 14-54 *- Strut loads and lengths.*

Section	Do [in]	Di [in]	A [in²]	I [in⁴]	E [psi]	EI [lb-in²]
A-A	0.75	0.68	0.078618	0.005036	1E+007	50360
B-B	0.491	0.375	0.078898	0.001882	1E+007	18822
C-C	0.3214	0	0.08113	0.000524	3E+007	15714

14.16.3 COLUMN BUCKLING

The Strut is considered as a long stepped column pinned-pinned at the center of the bolts. I will conservatively consider the Rod-Ends (section *C-C*) to go all the way to the largest section of the tube (section *A-A*) thus idealizing the column as shown in *Figure 14-130*. Since the Rod-end and the tube are made from two different materials with different Young's modulus, *Figure B-4* is used for $EI_2 \neq EI_1$. As a result, section *B-B* won't be used in the analysis.

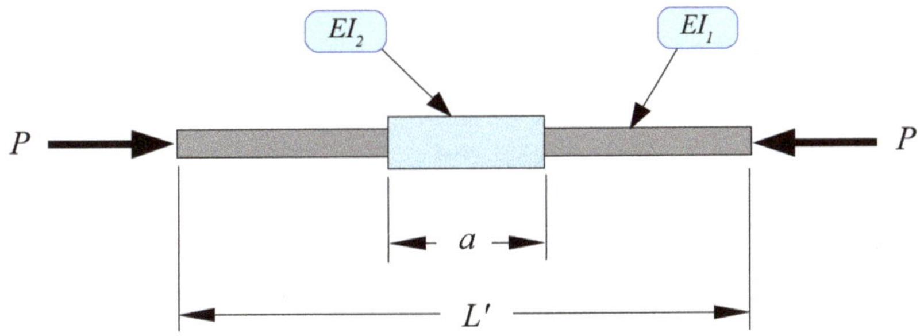

Figure 14-130 *– Strut idealization.*

The modified Euler compression allowable load is calculated as:

$$P_{CR} = \frac{B_2 (E_2 I_2)}{(L')^2}$$

where L' is the effective length calculated as: $L' = \frac{L}{\sqrt{c}}$

Where "*L*" is the column length and "*c*" is the End Fixity Coefficient. In the case shown in *Figure 14-130*, the End Fixity Coefficient "*c*" is obtained from *Figure B-2* and is equal to 1.0 for pinned-pinned columns. Hence, $L' = L = 8.69$ *in*.

Hence, using the values from *Table 14-54*, the ratio E_2I_2/E_1I_1 is calculated as:

$$\frac{E_2 I_2}{E_1 I_1} = \frac{50{,}360}{15{,}714} = 3.2$$

Then, for $a = L_2 = 2.23$ *in*. and $L' = 8.69$ *in*.:

$$\frac{a}{L'} = \frac{2.23}{8.69} = 0.25$$

From *Figure B-4* we find the value for B_2 of about 4.44.

Then the critical compression load for the strut is:

$$P_{CR} = \frac{4.44 \cdot 50{,}360}{(8.69)^2} = 2{,}961 \ lbs$$

The internal stresses across each of the three cross-sections due to P_{CR} are shown in *Table 14-55*.

Table 14-55 *- Internal compression stress at each section.*

Section	*fc* [psi]	*Fcy* [psi]
A-A	37673	65000
B-B	37540	65000
C-C	36507	135000

Since the stresses are well below the Yield compression stress allowable "F_{CY}", the critical load is within the elastic zone of the materials so no tangent modulus "E_T" calculation is necessary.

The critical column buckling compression load allowable *(2,961 lbs)* is found to be higher than the applied load (893 *lbs*) so no instability occurs. The column buckling MS is thus:

$$MS = \frac{P_{CR}}{P} - 1 = \frac{2{,}961}{893} - 1 = 2.32$$

I had assumed that the column was **long** right from the start. However, using (ref. 2. *Table C4.2*), the transitional L'/ρ value that differentiates between short and long columns would be calculated as 47.5 *in.* and since the current value is smaller with 34.3 *in*, the column would be considered short. As such, the critical compression stress would be calculated using (ref. 2. *eq. C4.6*) and a value of 53.6 *ksi* would be obtained. Multiplying this by the area of 0.786 in^2 we'd get a critical compression load of P_C = 4,216 *lbs*. This value being higher than the one obtained by assuming the column to be long, a higher *MS* would be obtained so, this is conservative.

14.16.4 STRESS ANALYSIS

Struts are pretty simple and yet, a fair amount of analyses are required to substantiate them. They are mainly critical in compression for stability but here's a list of checks that need to be done.

- Beam-column; including secondary bending
- Thread shear in tube (in the weakest material)
- Rod ends; bearing, lug, net section
- Bolts; bending + shear.

14.16.4.1 Beam column

We already checked it in column buckling but, even though struts normally only see axial loads (tension or compression), they also have some degree of bending in them due to manufacturing tolerances.

As shown in *Figure 14-131*, since no strut is perfect, it is common to consider a certain amount of offset between the line drawn between the center of the two bearings and, the centroid of the tube at mid-span. Depending on the company, the offset value could be the maximum between 0.050 *in.* and L/500.

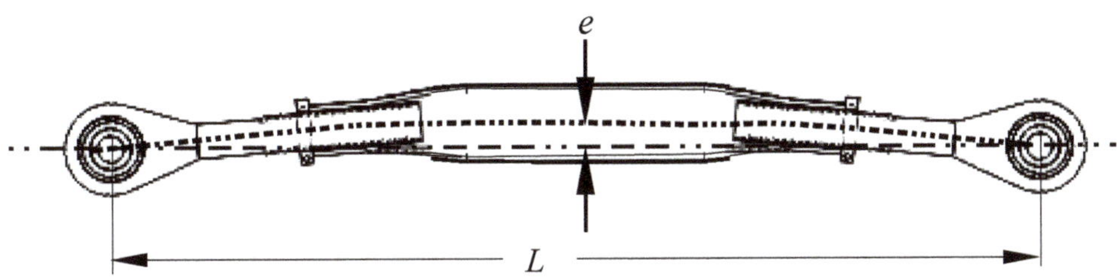

Figure 14-131 – *Strut offset.*

In our example, for $L = 8.69$ *in.*, the value for $L/500$ is 0.018 *in.* So a value of $e = 0.050$ *in.* will be used as the offset. Thus, the bending moment "M_O" caused by the offset is:

$$M_O = P \cdot e = 893(0.050) = 44.7 \; in.-lb$$

This moment exists whether the axial load is in tension or compression. However, when there is compression, the maximum bending moment that normally occurs at mid-span is amplified by the secondary bending. *Figure 14-132* shows the assumed idealization of the loaded beam-column.

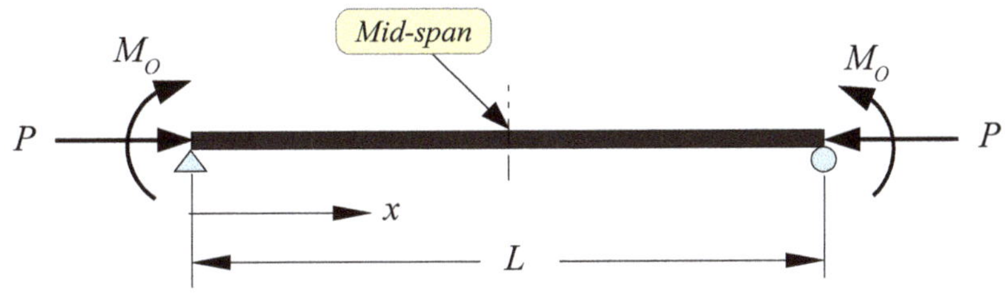

Figure 14-132 – *Strut Beam-column idealization.*

Of course, this is an idealization since, in reality, there is no end moments and the applied bending moment varies from 0 at the bearings to a maximum at mid-span. Some people will assume a single concentrated moment applied at mid-span. In both cases, an amplified moment is obtained which is used in the bending stress calculations.

The amplified bending moment can be quickly approximated with the equation:

$$M_{Max} = \frac{M_O}{1 - \frac{P}{P_{CR}}}$$

then

$$M_{Max} = \frac{44.7}{1 - \frac{893}{2{,}961}} = 64 \; in.-lb$$

For:
M_O = 44.7 in-lb
P = 893 lbs
P_{CR} = 2,961 lbs

or, it can be calculated more accurately using the beam column equations where:

$$M_{max} = \frac{M_o}{\cos\left(\frac{L}{2j}\right)} \qquad j = \sqrt{\frac{EI}{P}} \qquad j = \sqrt{\frac{10x10^6 \cdot 0.005036}{893}} = 7.51 \; in.$$

and

$$M_{max} = \frac{44.7}{\cos\left(\frac{8.69}{2 \cdot 7.51}\right)} = 53.3 \; in.-lb$$

For:
$E = 10 \times 10^6$ psi
$I = 0.005036$ in^4
P = 893 lbs
L = 8.69 in.

Since the applied loads aren't so high, I will conservatively use the result from the approximate method that provides the highest number such that $M_{max} = 64$ *in-lb*.

Here I could calculate the bending stress and simply add it to the axial stress and, compare that sum to a common allowable stress. However, since graphs for bending allowable for tubes exist and are based from tests data, I'd rather keep the axial and the bending stresses separate and calculate a stress ratio for each. The interaction equation to evaluate the MS is:

$$MS = \frac{1}{R_a + R_b} - 1$$

Since the bending moment is the greatest at mid-span, the axial stress "f_a" is calculated from section *A-A*. Thus,

$$f_a = \frac{P}{A} = \frac{893}{0.078618} = 11{,}359\ psi$$

The axial stress ratio "R_a" is calculated conservatively using F_{CY} from *Table 14-17*:

$$R_a = \frac{f_a}{F_{CY}} = \frac{11{,}359}{65{,}000} = 0.175$$

Then, using *Figure B-10* and the curve for 7075-T6 we find, for $D/t = 0.75/0.035 = 21$, a ratio for F_{bU}/F_{TU} of 1.17. Thus, the ultimate bending stress allowable: $F_{bU} = 1.17 * 75{,}000 = 87{,}000$ *psi*.

The bending stress at section *A-A* is:

$$f_b = \frac{M_{MAx}\ D}{2I}$$

hence

$$f_b = \frac{64\ 0.75}{2(0.005036)} = 4{,}760\ psi$$

For:
$D = 0.75$ in.
$I = 0.005036$ in^4
$M_{Max} = 64$ in-lb

The bending stress ratio is:

$$R_b = \frac{f_b}{F_{bU}} = \frac{4{,}760}{87{,}000} = 0.055$$

The MS in axial and bending in the *Strut Assy* is then:

$$MS = \frac{1}{0.175+0.055} - 1 = 3.34$$

Strut Assy, Ultimate	***>1.0***
9g Fwd *Al Aly 7075-T62*	*Axial + bending*

14.16.4.2 Shear in threads

The threads are machined (tapped) directly into the swaged ends of the Aluminum tube. The Rod-ends having the exact same thread are made out of *CRES* and thus the threads are critical on the aluminum side. The thread size is 0.375-24 UNJF-3A.

From the specification *MIL-S-8879*, the pitch diameter "D_P" for this thread is 0.3479 – 0.3516 *in.* Using the average of that (0.350 *in.*), the shear area "A_S" for a single thread is calculated as:

$$A_S = \frac{1}{2\cdot(threads\ per\ inch)}\cdot\pi\, D_P = \frac{1}{2\cdot 24}\cdot\pi\, 0.35 = 0.0229\ in.^2$$

For a shear strength "F_{SU}" of 45 *ksi* from *Table 14-17*, the shear force "P_{SI}" capacity of one thread is:

$$P_{SI} = A_s \cdot F_{SU} = 0.0229 \cdot 45{,}000 = 1{,}030 \; lbs/thread$$

Therefore, with an applied ultimate load of 893 *lbs* for the *9g Fwd LC*, only one thread would be sufficient (although not recommended). We typically want at least three full threads engaged.

This is fine for our design example as the Rod-ends are completely screwed into the swaged tube having a threaded length of 0.50 *in.*, thus 12 threads.

14.16.5 ROD ENDS

The Rod-ends are quite often vendor parts and therefore, are purchased off the shelf from a distributor. In that case, it is simply a matter of checking that the applied loads do not exceed the allowables provided by the vendor.

Ball joint Rod-end:

The Ball Rod-end is machined out of CRES with a spherical bearing swaged into place. The spherical bearing is the same as the one used in the *Aft Lug fittings* analyzed earlier. See section *14.15.7* for the substantiation.

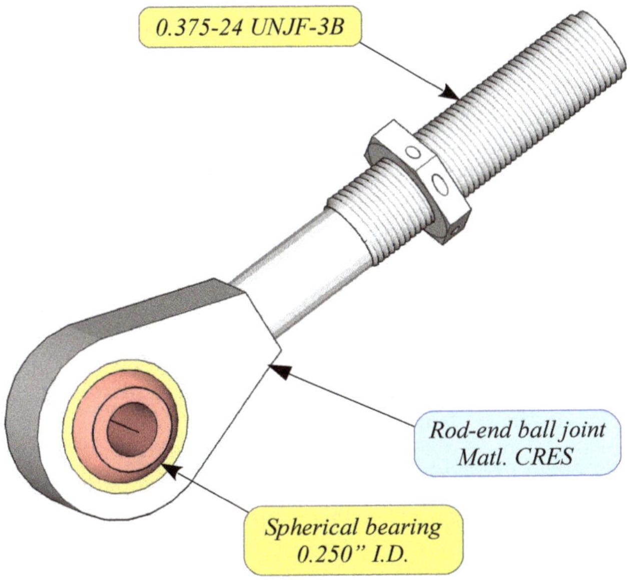

Figure 14-133 – *Ball joint Rod-end.*

Clevis Rod-end:

The *Clevis Rod-end* is machined out of CRES with two flanged bushings made out of Aluminum-Bronze press-fitted into the lugs. The bushings are somewhat sacrificial as they are softer than the CRES Clevis and the CRES bolt. They are meant to be replaced when they are worn out instead of having to replace the more expensive Clevis.

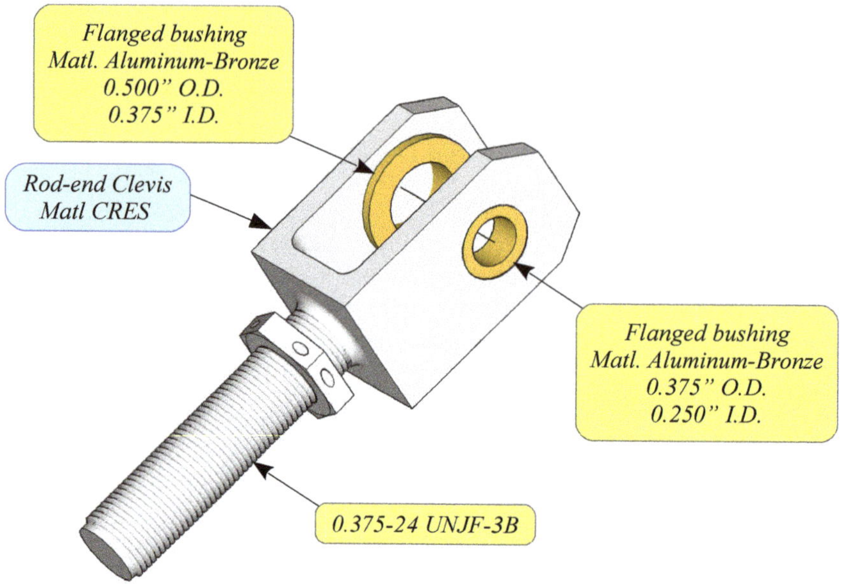

Figure 14-134 *– Clevis Rod-end.*

Note that for static analyses the bearings are always considered having no rotational resistance at all, that is no end moment. However, for fatigue analyses of flight control tubes for instance, one must take into account the small end moment caused by the bearing rotational friction as this could lead to fatigue cracks in the Rod-end's threaded section (or in the swaged aluminum tube).

14.16.6 BOLTS

The four *NAS6204* 0.250-28 hexagonal head bolts that connect the Strut assemblies to the Lug Fittings are all the same. They are made out of *CRES* having a tensile strength of 160 *ksi* and a shear strength of 95 *ksi*. They are checked in shear, tension and bending.

Figure 14-135 and *Figure 14-136* show the details for the bolts installations in the Fwd and Aft Lug Ftgs respectively. Since both configurations use the same bolt and same flanged bushings, the one having the largest span between the lugs, for the same applied load, is the most critical one. Hence, the configuration at the Forward end is thus the one to analyze passing the other one by comparison.

For the analysis, the bolt is idealized as shown in *Figure 14-137*. The following assumptions are made:

1. The bolt is considered as a simply supported beam,
2. The flanged bushings are ignored,
3. The spherical bearing does not contribute in helping the bolt in bending.

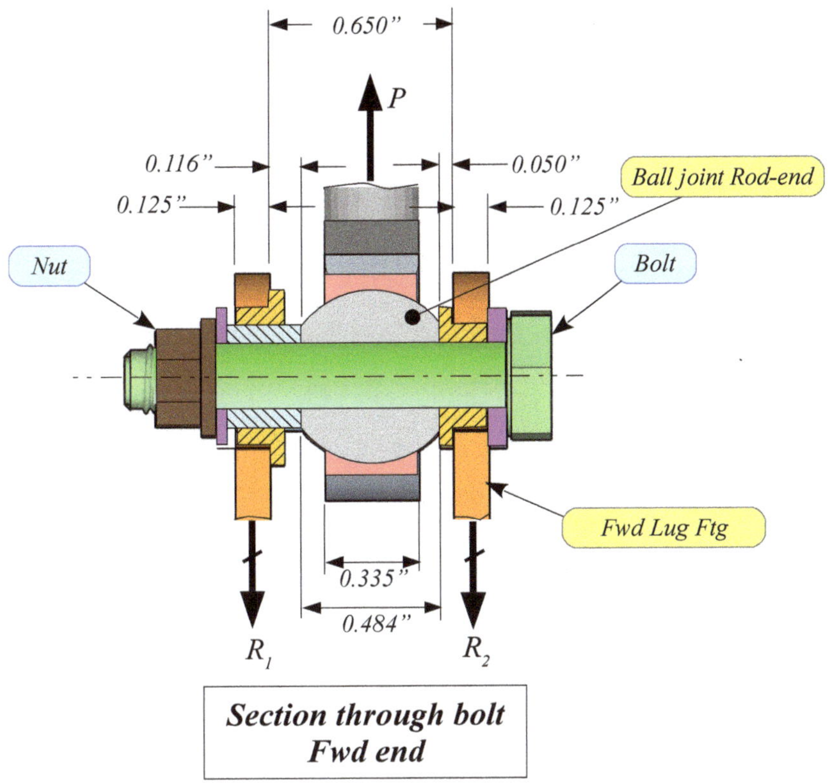

Figure 14-135 – *Bolt installation in Fwd Lug Ftgs.*

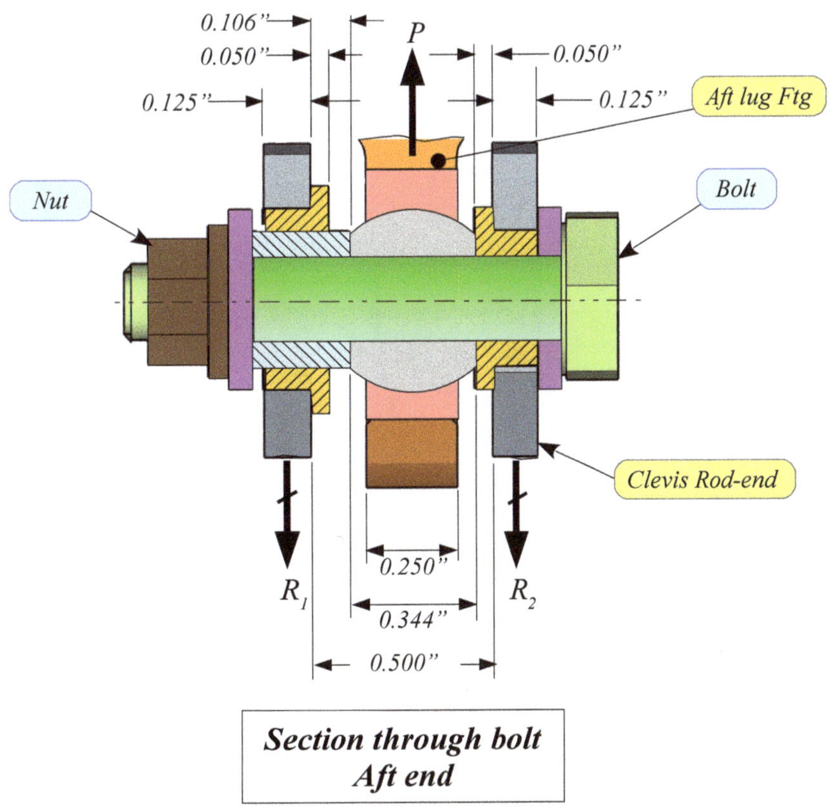

Figure 14-136 – *Bolt installation in Aft Lug Ftgs.*

In general, the following approximation is done in order to determine the shear and moment of the bolt.

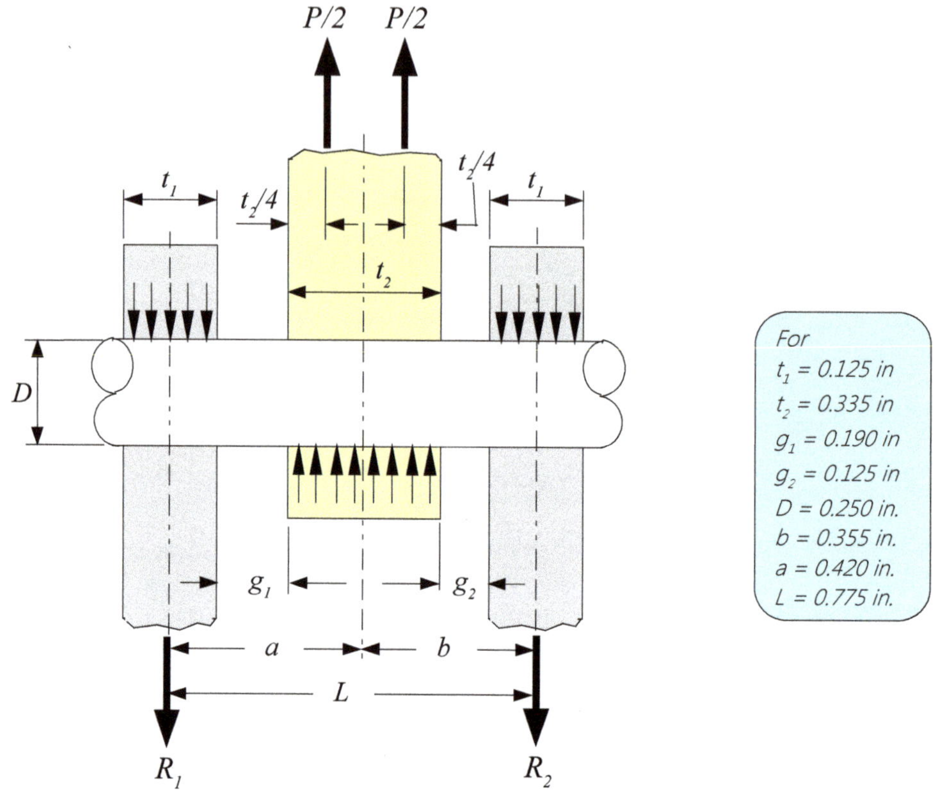

Figure 14-137 *– Bolt FBD idealization.*

The shear and moment diagrams for the bolt are shown in *Figure 14-138*.

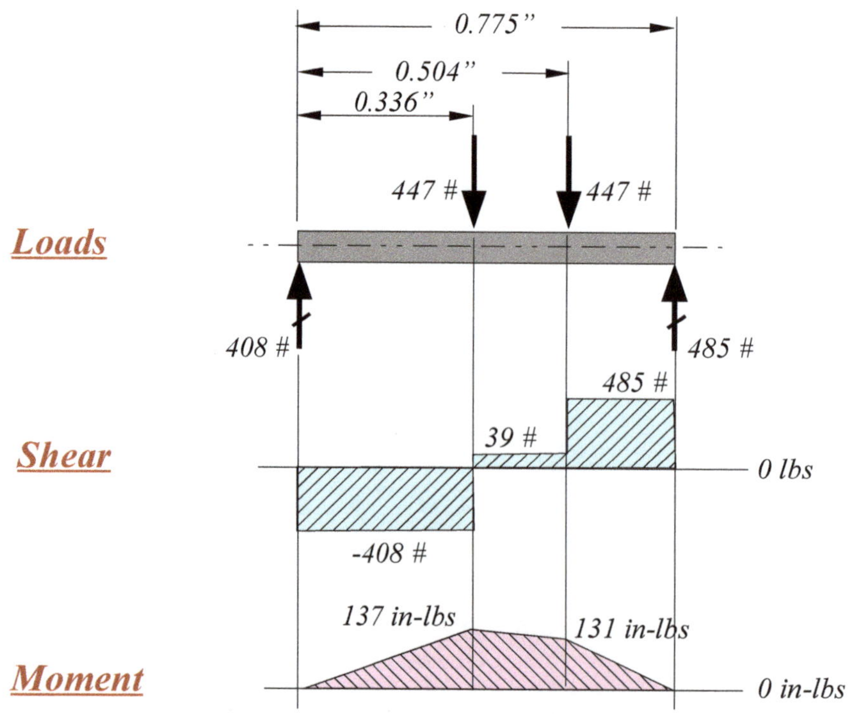

Figure 14-138 *– Bolt Shear and Moment diagrams.*

Shear:

Using *Figure 14-138*, because the bending moment at the location where the maximum shear load occurs is not much lower than the maximum bending moment, the maximum transverse shear force "*V*" of 485 *lbs.* is conservatively combined with the maximum bending moment as if they were at the same location. The average shear stress "f_S" is calculated below.

For
$$A = \frac{\pi D^2}{4} = \frac{\pi 0.250^2}{4} = 0.049087 \; in.^2$$
$$f_S = \frac{V}{A} = \frac{485}{0.049087} = 9{,}880 \; psi$$

With an ultimate shear strength of $F_{SU} = 95$ *ksi*, the shear load ratio "R_S" for the *9g Fwd LC* is thus:

$$R_S = \frac{f_S}{F_{SU}} = \frac{9{,}880}{95{,}000} = 0.104$$

Bending

The maximum bending moment is 137 *in-lb*. The bending stress is thus:

For
$$I = \frac{\pi D^4}{64} = \frac{\pi 0.250^4}{64} = 0.0001917 \; in.^4$$
$$f_b = \frac{M\,D}{2I} = \frac{137(0.25)}{2(0.0001917)} = 89{,}332 \; psi$$

With an ultimate tensile strength of F_{TU} = 160 *ksi*, one could simply be comparing the bending stress calculated above to F_{TU} and get a positive MS. This would be overly conservative as the bolt could take a lot more load by considering the plastic region. However, even if the current critical load case is an emergency landing, in general we do not want the bolt to deform so much from that ultimate load that we wouldn't be able to remove and replace it. Therefore, I like to calculate a **Yield** plastic bending allowable "M_{PY}" for bolts.

The plastic bending allowable "M_{PY}" is defined as: $M_{PY} = \sigma_Y Z$

Where, for solid circles, the plastic section modulus "*Z*" is:

$$Z = 1.333\,R^3 = 1.333(0.125)^3 = 0.0026035 \; in.^3$$

Hence, for $\sigma_Y = F_{TY} = 142$ *ksi*,

$$M_{PY} = 142{,}000 \cdot 0.0026035 = 369 \; in.-lb$$

The yield plastic bending moment "M_{PY}" is about 1.5 times greater than the maximum ultimate bending moment "M_{Ue}" that would have been obtained from the normal elastic equation:

$$M_{Ue} = F_{TU}\frac{I}{c} = 160{,}000 \cdot 0.0015336 = 245 \; in.-lb$$

Using the yield plastic bending allowable, the bending load ratio "R_B" for the *9g Fwd LC* is thus:

$$R_B = \frac{M_{Max}}{M_{PY}} = \frac{137}{369} = 0.371$$

In cases where the excessive deformation due to plastic bending is not a concern, then the ultimate bending stress allowable "F_{BU}" can be used. From (ref. 10. *Fig. 9.8.16*) for $D/t = 2$ and F_{TU} = 160 *ksi*, we find F_{BU} = 265 *ksi*. This leads to an ultimate plastic bending allowable "M_{PU}" of 406 *in-lb*. Thus, our approach is more conservative in this case.

Tension

There are discussions whether the tension preload in bolts should be considered or not when calculating the MS due to an ultimate static load condition (especially an emergency landing). Some say that, as soon as the bolt starts yielding under the ultimate applied load, the bolt will lose its preload and the tension will therefore disappear. While, some other people say that, as the bolt gets loaded and deforms in a semi-sine curve shape, an internal tension load would develop since the bolt is restrained by the clevis and what is in between.

As I'm not sure who's right, I'd rather be on the safe side and consider the preload. Since the bolt is in a shear application, the torque to apply to the bolt and nut should be relatively small. It is not uncommon to see a note on a drawing for this type of application next to a bolt that says: "*finger tight*". Meaning that the nut is turned until, using a torque wrench, the "*run-down torque*" starts increasing or, when any gap is just closed.

In our example, I will assume that a preload tension force "P_i" equivalent to about ***10%*** of the tensile capability of the bolt is used. From the ***MIL-S-8879*** military specification, for a thread of size 0.2500-28 UNJF-3A, the minor diameter "D_m" at the threads is 0.2041 in. Hence the cross section area "A_m" at that location is:

$$A_m = \frac{\pi\, D_m^2}{4} = \frac{\pi\, 0.2041^2}{4} = 0.032717\ in.^2$$

Thus, the ultimate tension bolt capability "P_{TU}" is:

$$P_{TU} = F_{TU} A_m = 160{,}000 \cdot 0.032717 = 5{,}235\ lbs$$

Therefore, 10% of P_{TU} would be 525 *lbs*, but I decided to round it up to 700 *lbs* conservatively. Using the well known general expression to calculate a bolt preload:

$$P_i = \frac{5\,T}{D}$$

For P_i = 700 *lbs* and D = 0.25 *in.* we obtain:

$$T = \frac{P_i\, D}{5} = \frac{700 \cdot 0.25}{5} = 35\ in. - lb$$

Note that the internal torsion stress initially caused in the bolt shank by the installation torque is considered to have disappeared and thus is not taken into account.

The tension stress "f_T" in the bolt shank is: $f_T = \frac{P_i}{A} = \frac{4\,P_i}{\pi\, D^2} = \frac{4 \cdot 700}{\pi\, 0.25^2} = 14{,}260\ psi$

In the technical report from *NASA* on aerospace threaded fasteners (ref. 28.) MS calculation with combined shear and tension loads, available for download, it was found that the tension stress ratio "R_T" calculated using the stress at the thread minor diameter better matched the tests results.

Therefore, recalculating the tension stress at the threads "$f_{T\,thread}$" we get:

$$f_{T\,thread} = \frac{P_i}{A_m} = \frac{700}{0.032717} = 21{,}396\ psi$$

Hence, with a tensile strength of F_{TU} = 160 *ksi*, the tension load ratio "R_T" for the ***9g*** *Fwd* LC is thus:

$$R_T = \frac{f_{T\,thread}}{F_{TU}} = \frac{21{,}396}{160{,}000} = 0.134$$

Margin of Safety

To my knowledge, there are no interaction equation for bolts that combine shear, tension and bending available in the public domain. However, since we're dealing with plastic bending where almost the whole bolt cross-section has reached the yield stress *(one half in tension and one half in compression)*, then it appears logical to simply add the tension and the bending stress ratios together as one tension stress ratio "$R_T{}'$". Therefore, the interaction equation for bolts subjected to shear, tension and bending loads becomes:

$$R_S^3 + R_T{}'^2 = 1 = R_S^3 + (R_T + R_B)^2$$

Where the load ratios are:

$$R_s = 0.104 \qquad R_T = 0.134 \qquad R_b = 0.371 \qquad R_{T'} = 0.134 + 0.371 = 0.505$$

Then from the interaction equation above and using a Fitting Factor of 1.15, the Margin of Safety due to the *9g Fwd* LC for the bolt is thus calculated as follows using *Figure 14-139.*

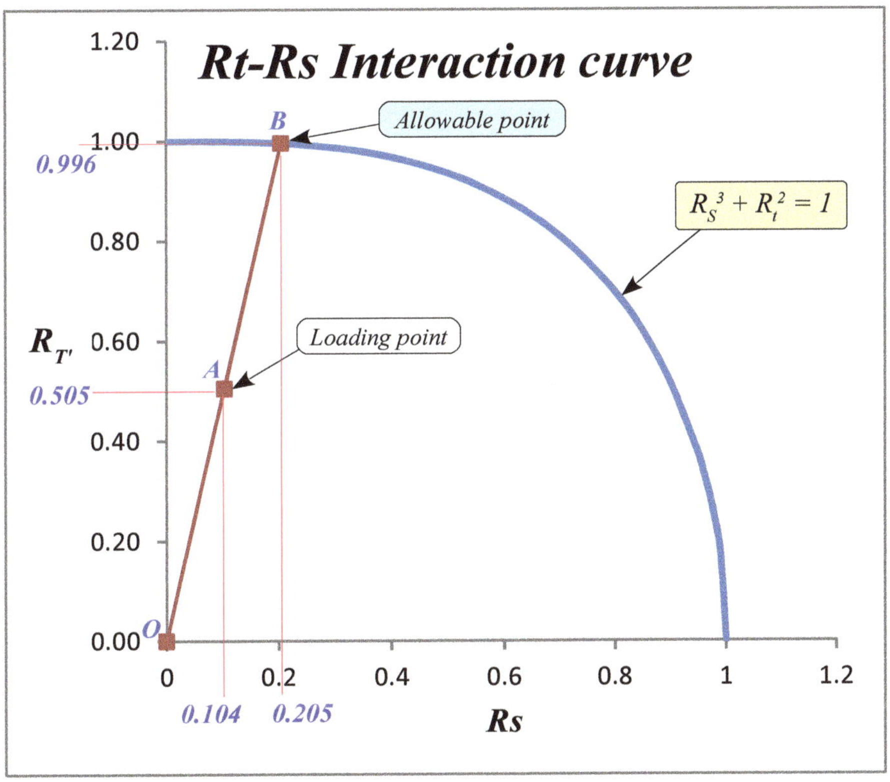

Figure 14-139 *– Bolt Shear and tension interaction curve.*

Where:

$$MS = \frac{\overline{OB}}{FF\,\overline{OA}} - 1 = \frac{R_{S\,All}}{FF\,R_S} - 1 = \frac{0.205}{1.15*0.104} - 1 = 0.71$$

$$MS = \frac{R_{t\,All}}{FF\,R_t{}'} - 1 = \frac{0.996}{1.15*0.505} - 1 = 0.71$$

Bolt, Ultimate	***0.71***
9g Fwd *Alloy Steel*	*Shear + tension + bending*

The above interaction equation is found to be more conservative (in this case anyway) compared to other more complicated interaction equations from some aircraft manufacturer companies.

14.17 INTERCOSTAL

The Intercostal is made out of *Al Aly* ***7050-T7451*** machined Plate 1.0 in. thick. *Figure* ***14-140*** shows the Intercostal installation. It is used to pick-up the loads from the *Aft inb'd Lug Ftg* and transfer them to the adjacent Frames and the skin.

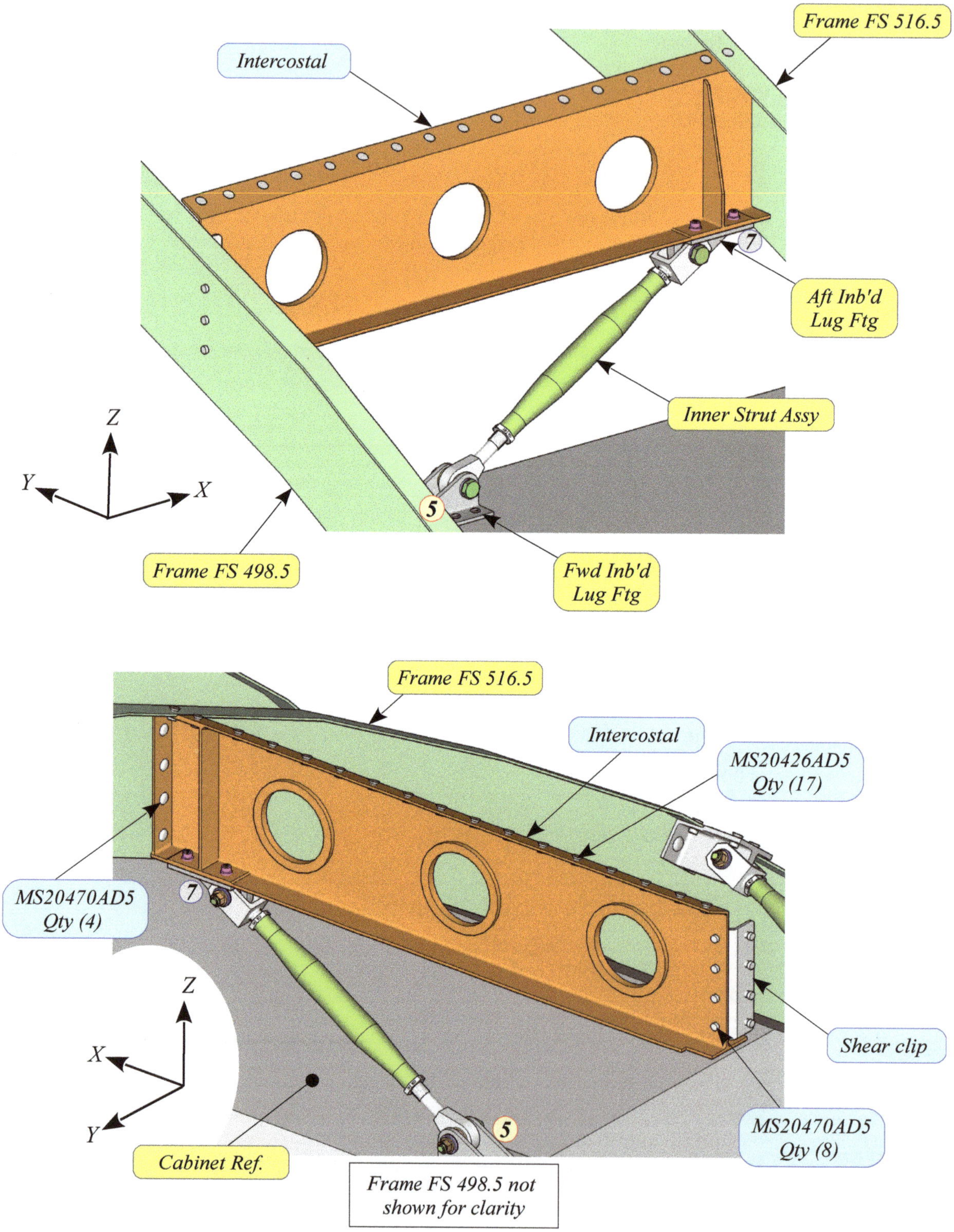

Figure 14-140 – *Intercostal installation details.*

Quite often, such Intercostal is not considered as part of the airframe primary structure. In which case it wouldn't be modeled in the global FEM so it would only be loaded by the emergency landing cases. In cases where the Intercostal is in the FEM, then all load cases (flight, ground, emergency landing) shall be considered for the analysis. *Figure 14-141* shows the dimensions of the Intercostal.

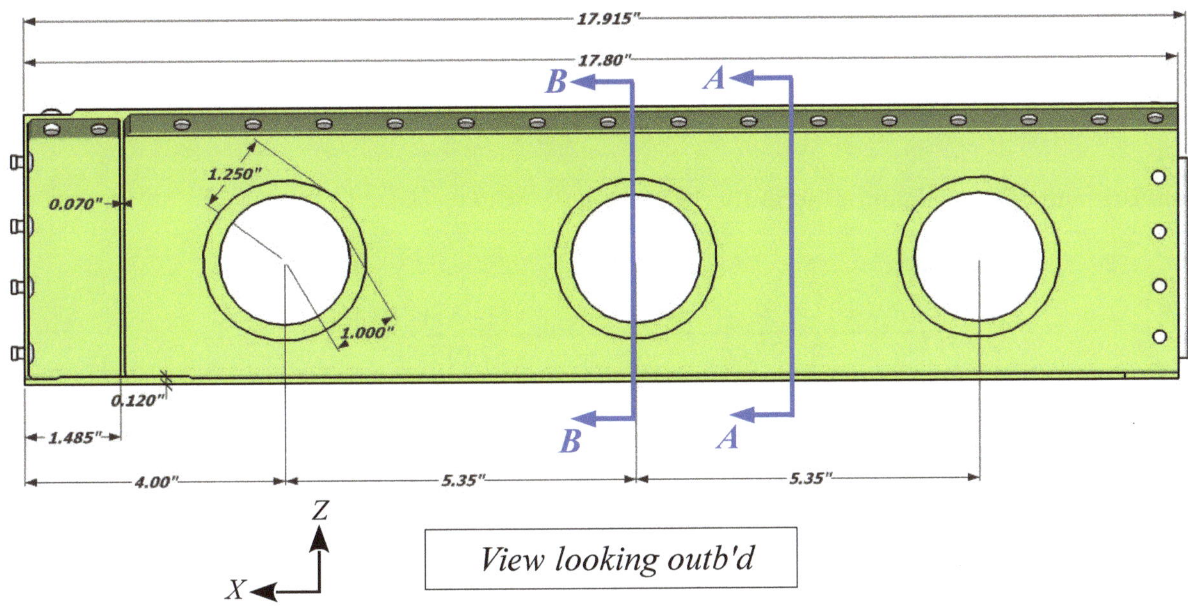

Figure 14-141 *– Intercostal dimensions and cross-sections locations.*

14.17.1 LOADS

14.17.1.1 X-Z plane

The Intercostal is loaded by the inboard Strut at point ***7***. *Figure 14-142* shows the *FBD* of the Intercostal in the *X-Z* plane when loaded by the Strut. The Intercostal is considered simply supported at the Frames while the *fore-aft* load in "*X*" is reacted in shear by the skin.

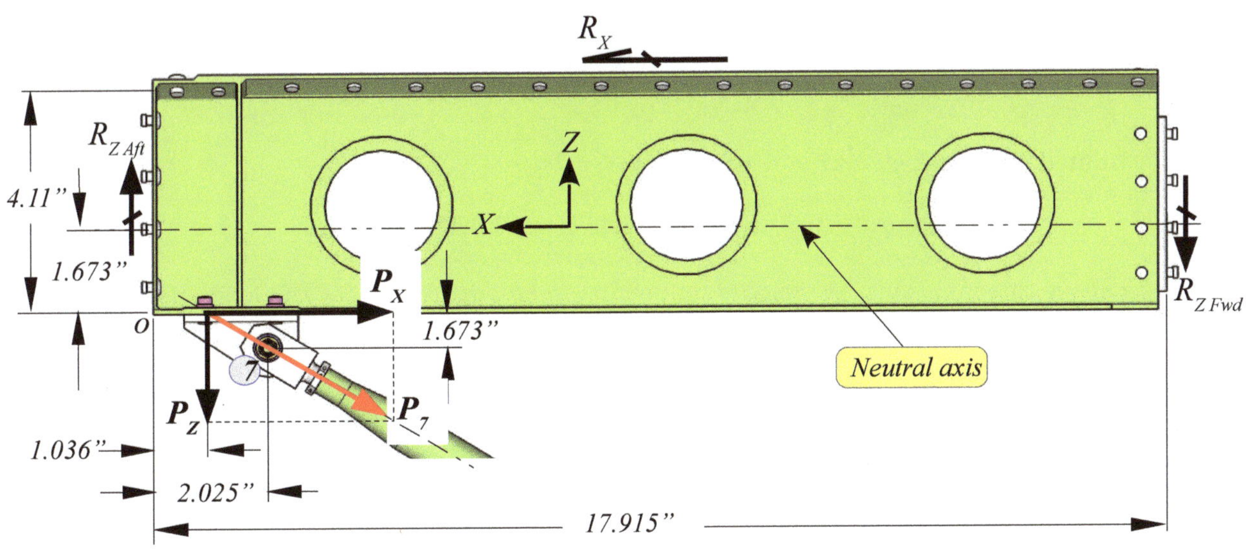

Figure 14-142 *– Intercostal assy F.B.D. In X-Z plane.*

To help in calculating the reaction forces, the strut load is applied at the point where the strut axis intersects the base of the Intercostal. The reaction forces are calculated below for the *9g FWD LC*.

From *Table 14-39*, the strut load at point *5* "P_5" is the same as "P_7" and is 784 *lbs*. The strut orthogonal components are:

$$P_X = 668\ lbs$$
$$P_Z = 411\ lbs$$

By the sum of the forces in "X":

$$R_X = P_X = 668\ lbs$$

From the sum of the moment around the "Y" axis at point *O (positive CCW)*:

$$\sum M_O = R_X(4.11) - P_Z(1.036) - R_{Z\,Fwd}(17.915) = 0$$

$$R_{Z\,Fwd} = \frac{R_X(4.11) - P_Z(1.036)}{17.915} = \frac{668(4.11) - 411(1.036)}{17.915} = 130\ lbs$$

And finally, from the sum of the forces in the "Z" axis:

$$\sum F_Z = R_{Z\,Aft} - P_Z - R_{Z\,Fwd} = 0$$

$$R_{Z\,Aft} = P_Z + R_{Z\,Fwd} = 411 + 130 = 541\ lbs$$

Shear and moment diagram

The shear and moment diagrams of the Intercostal in the **"X-Z"** plane can be seen in *Figure 14-143*. The applied forces are transformed onto the beam's neutral axis. Thus, the concentrated applied moment M_Y caused by the offset of the P_X force is calculated as:

$$M_Y = P_X(1.673) = 668(1.673) = 1{,}118\ in.-lb$$

Similarly, the shear flow at the skin has to also be transformed to the neutral axis as follows:

For a shear flow "q_X" at the skin of:

$$q_X = \frac{R_X}{17.915} = \frac{668}{17.915} = 37.287\ lbs/in.$$

The distributed moment "m_Y" at the neutral axis is:

$$m_Y = q_X(2.437) = 37.287\ (2.437) = 90.8736\ in.-lb/in.$$

With the above internal loads, the maximum bending moment "$M_{Y\,MAX}$" at the load application point is calculated from the forward end as:

$$M_{Y\,MAX} = q_X(17.915 - 1.036) - R_{Z\,fwd}(17.915 - 1.036)$$
$$= 90.87(16.879) - 130(16.879) = -652\ in.-lb$$

The maximum axial "F_X" (compression) internal load at the load application point is:

$$F_X = q_X(17.915 - 1.036) = -37.287\ (16.897) = -629\ lbs$$

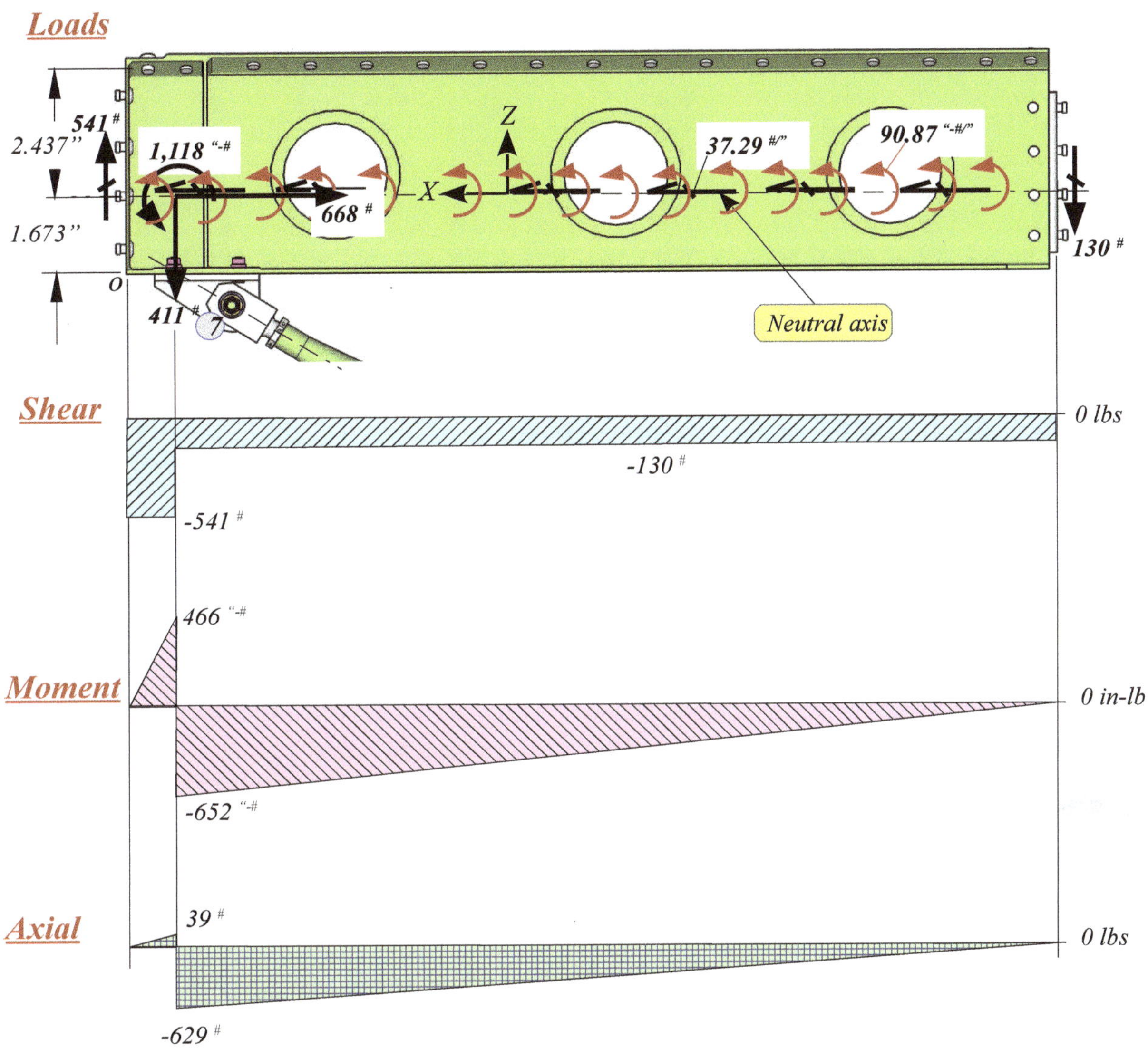

Figure 14-143 *– Intercostal Shear & Moment diagrams in X-Z plane for the 9g FWD LC.*

14.17.1.2 Y-Z plane

The Intercostal is loaded by the side load "P_Y" at point *7*. *Figure 14-144* shows the *FBD* of the Intercostal in the *Y-Z* plane. The Intercostal is considered simply supported at the Frames. The figure shows a section through point *7* at the *Aft inboard lug ftg* as if it were located at a lightening hole (section *B-B*). Of course this isn't the case and I'll explain why I do this.

The Intercostal being a thin opened section, it has a very weak torsional stiffness. Therefore, since the Side load would cause a torque through the Intercostal, I'm assuming that the Intercostal lower cap acts like a beam on its own; that is stabilized along the *Z* axis (vertically). The lower cap alone will react the side load and transfer the transverse shear to the adjacent Frames. Hence, using section *B-B* in the figure was simply to emphasize the fact that only the lower cap is reacting the side load.

But because there is an offset between the side load application point and the shear center of the lower cap (which is assumed at the same place as the centroid), a torque "*T*" still exists and must be reacted somehow.

The main purpose of the vertical integral ribs in the Intercostal located right at the load introduction point is to transfer the vertical shear efficiently into the web. In addition, the ribs being very stiff in the *Y-Z* plane (providing of course that they are non-buckling) can transfer efficiently any M_X moment (torque). This torque is then reacted as a couple of forces (in the Y axis) as R_T; one located at the lower cap c. of g. and the other at the rivets in the upper cap connecting to the skin. The total side load applied to the lower cap is thus the sum of R_Y (or P_Y) and R_T.

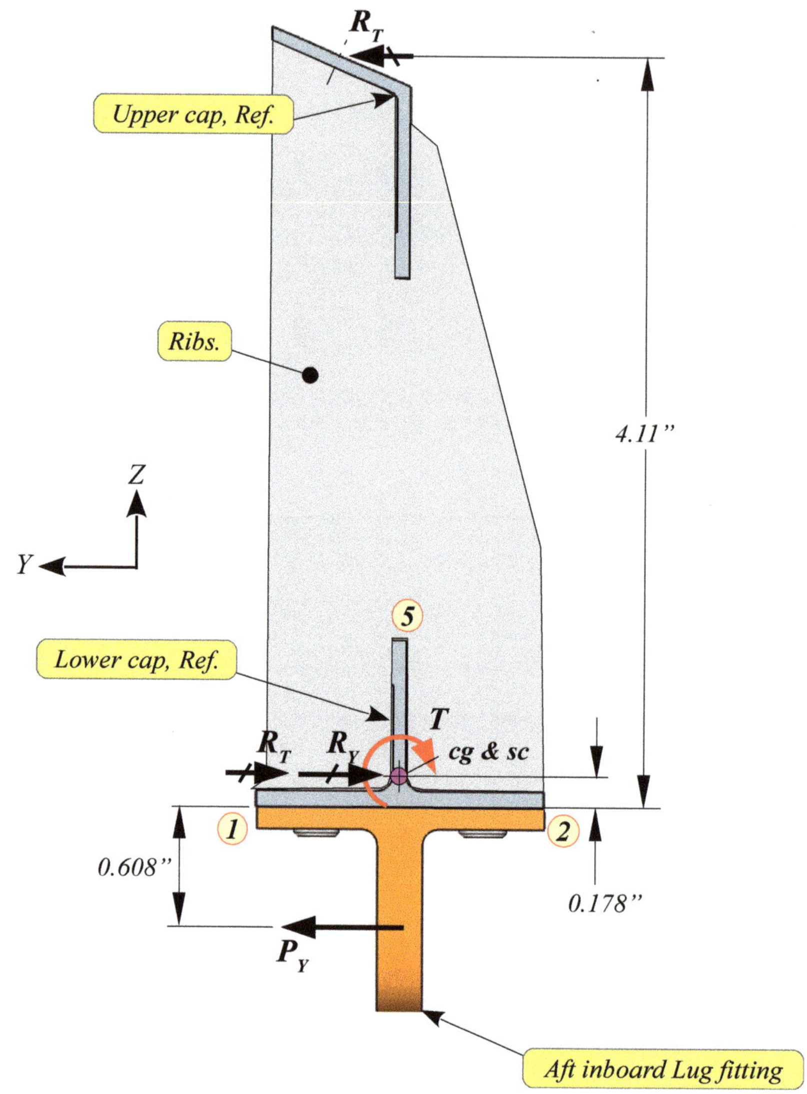

Figure 14-144 *– Intercostal assy F.B.D. in Y-Z plane.*

From *Table 14-38*, the side force P_Y at point *7* is 39 *lbs* for the *9g FWD LC*. The torque "*T*" applied at the centroid of the lower cap is thus:

$$T = P_Y \cdot e = 39(0.608 + 0.178) = 31 \; in. - lb$$

The couple force "R_T" is then:

$$R_T = \frac{T}{h} = \frac{31}{(4.11 - 0.178)} = 8 \; lbs$$

Therefore, the total applied load "$P_{Y\,tot}$" to the lower cap is:

$$P_{Y\,tot} = R_Y + R_T = 39 + 8 = 47 \; lbs$$

14.17.1.3 X-Y plane

An *FBD* of the Intercostal in the *X-Y* plane is shown in *Figure 14-145*. Applying the total side load "$P_{Y\,Total}$" at the same location in *X* as the center of the spherical bearing, the forward and aft reactions in the *Y* directions are calculated.

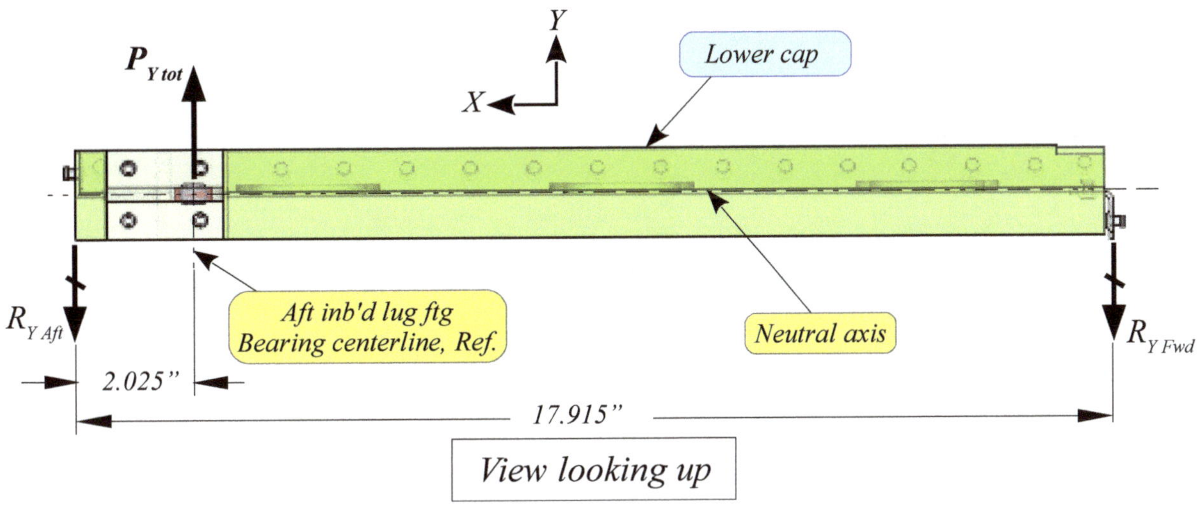

Figure 14-145 *– Intercostal Lower cap F.B.D. in X-Y plane.*

The forward "$R_{Y\,Fwd}$" and aft "$R_{Y\,Aft}$" shear reactions are calculated below as R_2 and R_1 respectively.

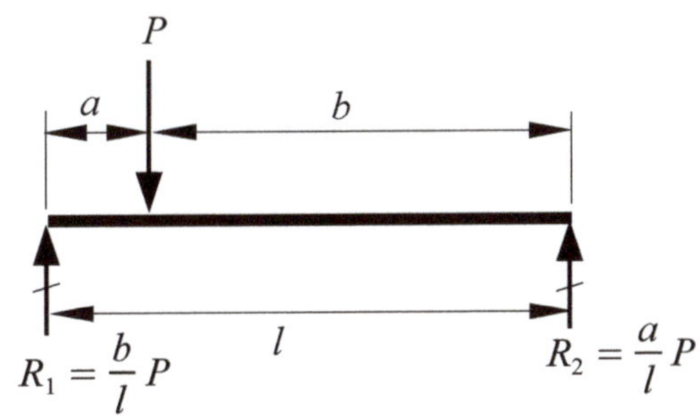

Figure 14-146 *– Shear load distribution for a simply supported beam.*

Using the quick shear distribution method, for l = 17.915", a = 2.025" and b = 15.89", as shown in *Figure 14-146*, the fastener's reactions in the "*Y*" direction are then:

$$R_{Y\,Aft} = \frac{b}{l} P_{Y\,tot} = \frac{15.89}{17.915} \cdot 47 = 42\ lbs$$

$$R_{Y\,Fwd} = \frac{a}{l} P_{Y\,tot} = \frac{2.025}{17.915} \cdot 47 = 5\ lbs$$

Shear and moment diagram

The shear and moment diagrams of the Intercostal in the "**X-Y**" plane can be seen in *Figure 14-147*. The maximum bending moment "$M_{Z\,max}$" at the load application point is:

$$M_{Z\,max} = R_{Y\,Aft} \cdot a = -42 \cdot 2.025 = -85\ in.-lb$$

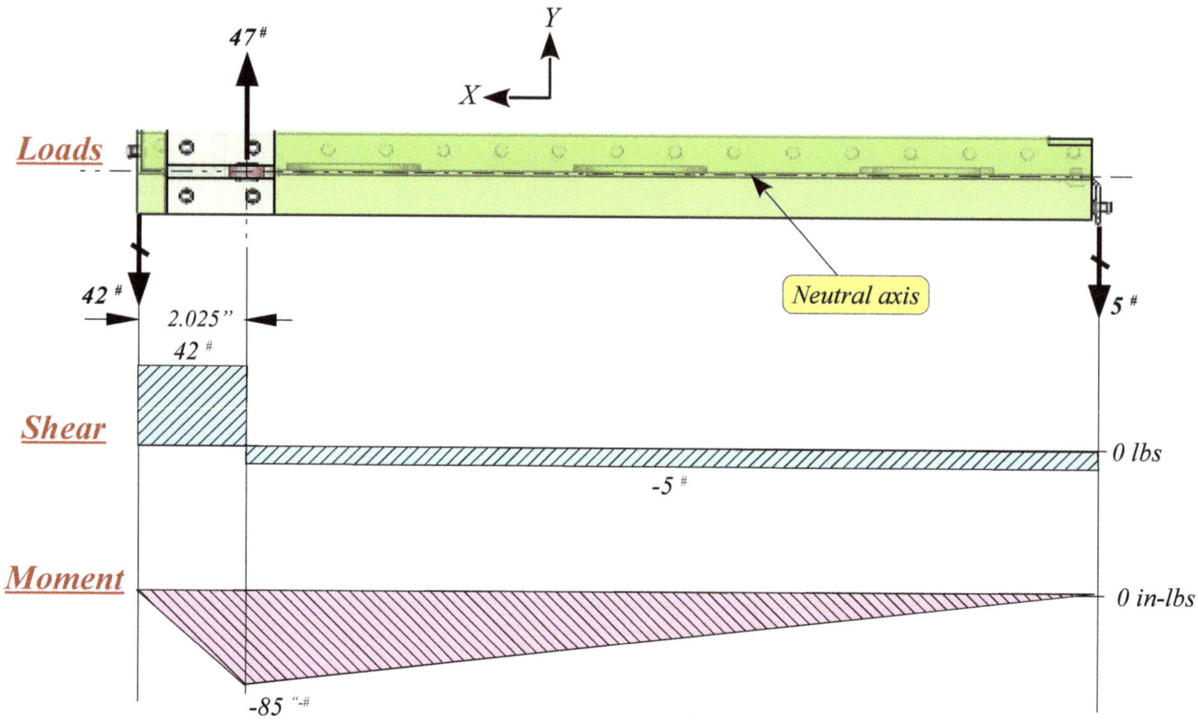

Figure 14-147 *– Intercostal Shear & Moment diagrams in the X-Y plane.*

14.17.2 SECTION PROPERTIES

Two critical cross-section locations were determined as shown in *Figure 14-141;* section *A-A* and section *B-B*. The section properties of both cross-sections are shown in *Figure 14-148* and *Figure 14-149*. The section properties are obtained from a *3D* drawing software. No effective skin was considered in the cross-sections for conservatism.

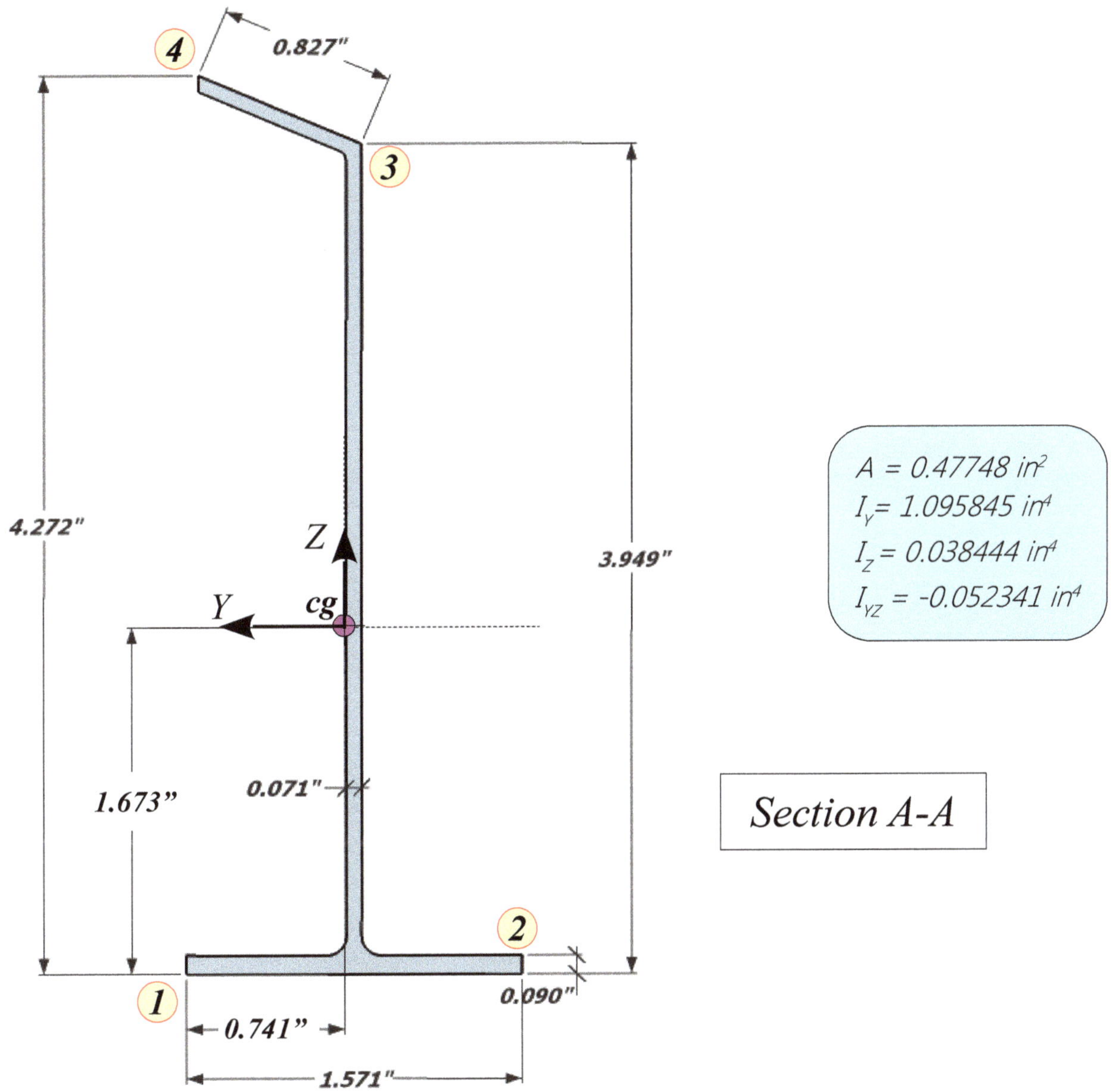

Figure 14-148 *– Intercostal section A-A.*

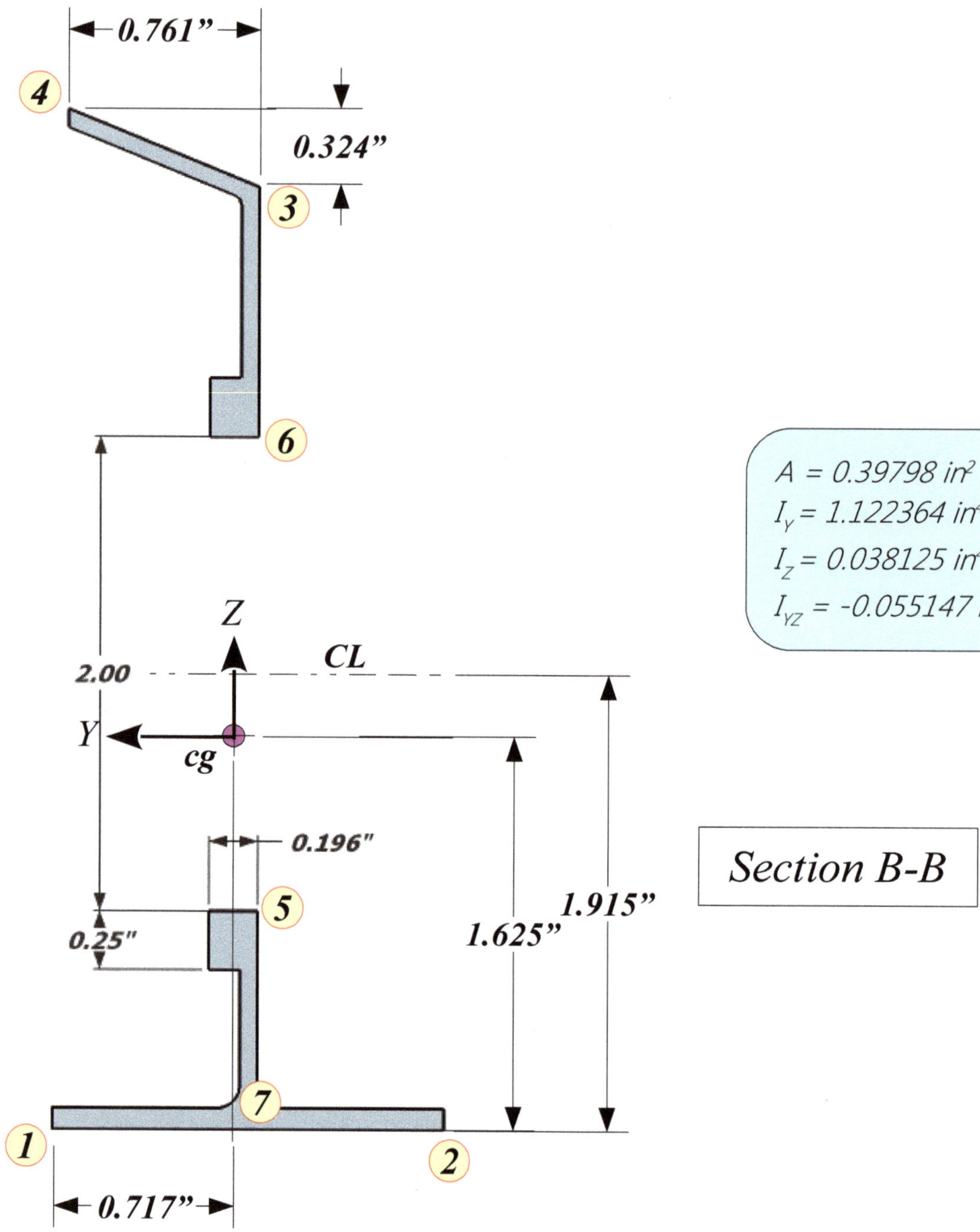

Figure 14-149 *– Intercostal section B-B.*

Lower Cap section properties

Since the upper cap of the Intercostal is fastened to the skin, it is inherently stable in compression. However, the lower cap not being stabilized laterally by anything between the two Frames, could be subjected to column instability. The lower cap dimensions are determined from section *B-B* at a lightening hole while the lip around the hole had been ignored. That section is assumed constant over the whole length of the Intercostal. The section properties of the lower cap are shown in *Figure 14-150*.

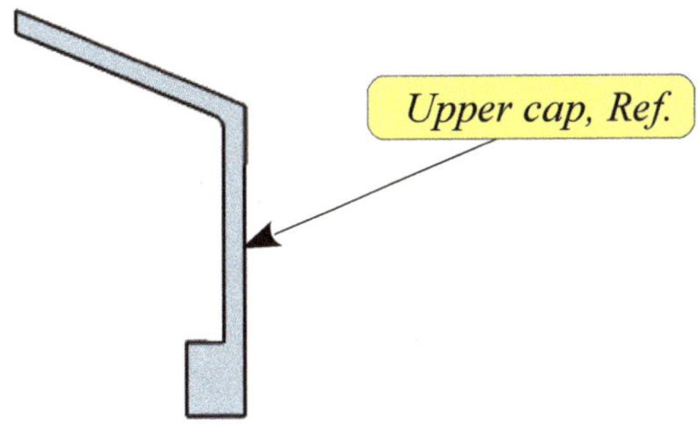

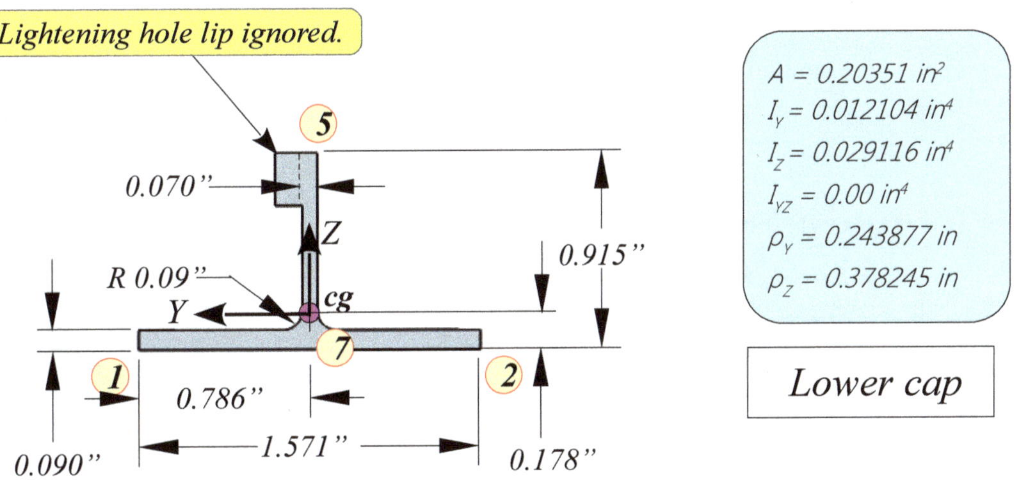

Figure 14-150 *– Intercostal lower cap section properties.*

14.17.3 STRESS ANALYSIS

In order to substantiate the Intercostal assembly, the following checks have to be performed:

- Crippling in lower cap.
- Column buckling
- Free-flange buckling
- Shear buckling of web
- Combined stresses
- Connections to skin and Frames
- Shear clip

Using *Figure 14-57* and *Figure 14-134*, the bending stress is calculated at the location where the bending moments are the largest which is at the load application point. Section *A-A* having a slightly lower second moment of inertia than section *B-B*, only the stresses at the points shown in *Figure 14-148* for section *A-A* will be calculated.

The stresses are calculated in two different planes:

1. Due to the moment M_Y and the axial load P_X (full machined Intercostal section *A-A*).
2. Due to the moment M_Z (Lower cap section only).

The resulting stresses obtained at points *1* through *5* are superimposed from the two calculations in order to get the total stresses. No complex bending is deemed necessary as the Intercostal is stabilized at the upper cap by the skin and is fairly short.

Due to M_Y and P_X:

The Intercostal full section *A-A* as shown in *Figure 14-148* is used. Section *A-A* is used conservatively at the load application point (where the moment is the highest) even if the lower cap flange at that location is slightly thicker. As explained earlier, for conservatism, no effective skin is considered.

The results of the axial and bending stresses calculated at the five stress points of section *A-A* are shown in *Table 14-56.*

Using the shear and moment diagram on *Figure 14-57*, the stresses are thus calculated as:

$$f_i = \frac{P_X}{A} + \frac{-M_Y(z_i)}{I_Y}$$

Following is a stress calculation example at point *2* of section *A-A*:

$$f_2 = f_a + f_{b2}$$
$$f_2 = -\frac{629}{0.47748} + \frac{-(-652)\cdot(-1.673)}{1.095845}$$
$$f_2 = -2{,}313\ psi$$

For:
P_X = -629 lbs I_Y = 1.095845 in^4
A = 0.47748 in^2 z_2 = -1.673 in
M_Y = -652 in-lb

The stresses at all stress points are:

Table 14-56 *- Axial and bending stresses at stress points of section A-A.*

Points	y [in]	z [in]	f_a [psi]	f_b [psi]	$f_a + f_{b\,A\text{-}A}$ [psi]
1	0.741	-1.673	-1,317	-995	-2,313
2	-0.830	-1.673	-1,317	-995	-2,313
3	-0.080	2.276	-1,317	1,354	37
4	0.682	2.600	-1,317	1,547	230
5	-0.045	0.915	-1,317	544	-773

Due to M_Z:

Now, the bending stresses is calculated due to the side load by considering the Lower cap only. The Intercostal lower cap section is shown in *Figure 14-150.*

Using the shear and moment diagram in *Figure 14-134*, the bending stress is thus calculated as:

$$f_{bi} = \frac{-M_Z(y_i)}{I_Z}$$

Following is a stress calculation example at point *2* of the lower cap by considering the moment amplification factor "*AF*" of 1.063 due to the secondary bending effect calculated in section ***14.17.6***:

$$f_{b2} = \frac{-(AF)M_Z(y_2)}{I_Z}$$

$$f_{b2} = \frac{(1.063)(-85)(-0.786)}{0.029116}$$

$$f_{b2} = -2{,}439\ psi$$

For:
M_Z = -85 in-lb I_Z = 0.029116 in⁴
AF = 1.063 y_2 = -0.786 in

The bending stresses at the stress points are shown in *Table 14-57*.

Table 14-57 *- Bending stresses at the stress points of the Lower cap due to M_Z.*

Points	y	z	$f_{b\ low\ cap}$
	[in]	[in]	[psi]
1	0.786	-0.178	2,438
2	-0.786	-0.178	-2,438
5	0.000	0.737	0

Total stresses:

The stress due to the axial load "P_X" and the bending moment "M_Y" applied to the full machined Intercostal section *A-A* is superimposed to the stress due to the bending moment M_Z applied to the Intercostal Lower cap section to get the total stress at each stress point. See *Table 14-58* for the results.

Table 14-58 *- Total Bending stresses at section A-A.*

Points	$f_a + f_{b\ A\text{-}A}$	$f_{b\ low\ cap}$	f_{total}
	[psi]	[psi]	[psi]
1	-2,313	2,438	125
2	-2,313	-2,438	-4,750
3	37	0	37
4	230	0	230
5	-773	0	-773

As seen above, the most critical location is at point *2* where the maximum compression stress exists. Since point *2* is in compression, a free flange buckling analysis will be performed.

Shear stress

Along the *Y* axis:

The maximum transverse shear stress "$f_{S\ max\ Y}$" in the "*Y*" direction for the Intercostal lower cap is calculated at the neutral axis (center), where the shear stress is maximum, using the following formula for a rectangular section:

$$f_{S\,max_Y} = \frac{3V_Y}{2A}$$

From *Figure 14-134*, the maximum shear load "V_Y" is 42 *lbs*. The flange area "*A*" of the lower cap flange hatched area from the *Y* axis shown on *Figure 14-151* is calculated as follows:

$$A = b \cdot t = 1.571 \cdot 0.090 * 0.372 = 0.14139\, in.^2$$

The web is neglected as it doesn't contribute much in taking shear in the *Y* direction.

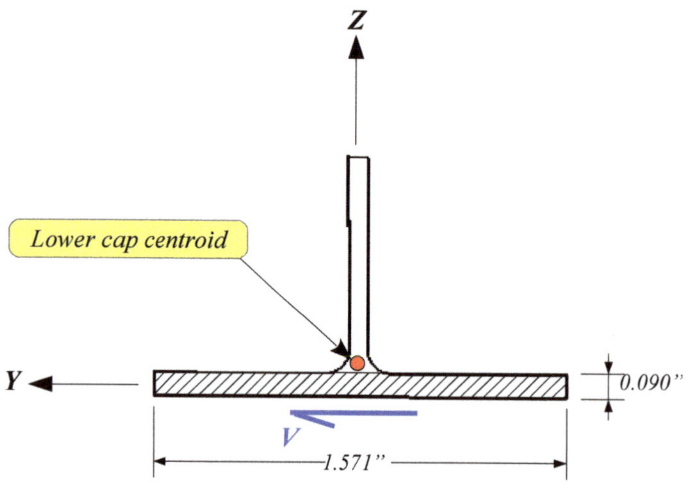

Figure 14-151 – *Lower cap flange dimensions.*

Then

$$f_{S\,max_Y} = \frac{3 \cdot 42}{2 \cdot 0.14139} = 446\ psi$$

Along the *Z* axis:

The average transverse shear stress "$f_{S\,Z}$" in the "***Z***" direction of the Intercostal is considered taken by the web of section *A-A* only.

$$f_{S_Z} = \frac{V_Z}{A_{web}}$$

From *Figure 14-57*, the average shear load "V_Z" is 541 *lbs*. Using *Figure 14-148*, the web area "A_{web}" of the Intercostal is calculated as:

$$A_{web} = h \cdot t_{web} = 3.949\,(0.070) = 0.27643\,in.^2$$

Then

$$f_{S_Z} = \frac{541}{0.27643} = 1{,}957\ psi$$

14.17.4 FREE FLANGE BUCKLING

Using the stresses from *Table 14-61*, a free-flange buckling analysis is performed since the Intercostal lower cap has the highest compression stress at the free edge (point ***2***). Therefore, considering the flange to start from the free edge (point ***2***) and stop at the web (half way between point ***1*** and point ***2,*** or point ***7***), as shown on *Figure 14-147*, the variation of the stress goes from a relatively high compression stress at the free edge to a smaller compression stress at the web.

The stress at the web (point *7*) is interpolated as follows:

$$f_{7web} = \frac{f_1 + f_2}{2} = \frac{(125) + (-4{,}750)}{2} = -2{,}313\ psi$$

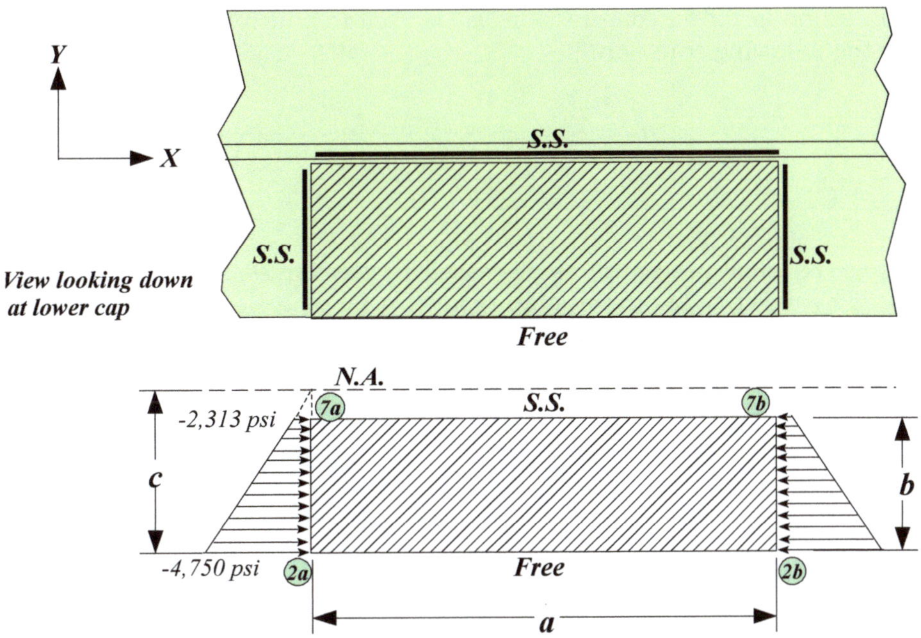

***Figure 14-152** – Free flange buckling in bending.*

For conservatism, the flange is considered simply supported at all sides except at the free edge. Using (ref. 20.) for flat plate buckling in bending, the bending stress allowable is calculated as:

$$F_{b\,cr} = \frac{k_b\,\eta_c\,\pi^2\,E}{12\left(1-\mu^2\right)}\left(\frac{t}{b}\right)^2$$

From *Figure 14-147,* the length "*a*" is arbitrarily shown since it is not required in this analysis. Since the stress varies along the Intercostal lower cap, the bending stress at point ***2a*** is different (higher) than the one at point ***2b***. The same for ***7a*** and ***7b***. The maximum (at points ***2a*** and ***7a***) are used for conservatism.

From *Table 14-61* we know that the total stress at point ***2*** is −4,750 *psi* and that it is −2,313 *psi* at point ***7*** located 0.785 *in*. away. For this analysis, we are interested to know the location where the stress is zero. That is *"c"* on *Figure 14-147*. By doing a linear extrapolation from points *2 & 7* we get $c = 1.530$ *in.*

In this case, $b = 0.785$ *in*. and $c = 1.53$ *in*. Thus $b/c = 0.785/\,1.53 = 0.51$. Hence from *Figure B-1*, the value for k_b read is: 0.46. The Young's modulus from *Table 14-16* is used.

From the above equation:

$$\frac{F_{b\,cr}}{\eta_c} = \frac{k_b\,\pi^2\,E}{12\left(1-\mu^2\right)}\left(\frac{t}{b}\right)^2$$

$$\frac{F_{b\,cr}}{\eta_c} = \frac{(0.46)\,\pi^2\,10*10^6}{12\left(1-0.33^2\right)}\left(\frac{0.090}{0.785}\right)^2 = 55{,}807\ psi$$

For
$k_b = 0.46$
$b = 0.785$ in
$t = 0.090$ in
$E = 10.0*10^6$ psi

This value is fairly close to F_{cy} *(60 ksi)*. Thus, I'll use the iteration process to determine the actual critical stress in free flange buckling by considering plasticity.

A plasticity correction factor for compression "η_c" is used to multiply the Young's modulus "E" and is calculated using the following equation:

$$\eta_C = \frac{E_t}{E} = \frac{1}{1+\left(\frac{0.002\ E\ n}{F_{cy}}\right)\left(\frac{F_{c\,cr'}}{F_{cy}}\right)^{n-1}}$$

Therefore, using the same procedure as described in section ***10.5.6***, the critical buckling bending stress is calculated by iteration by first assuming a value for $F_{c\,cr}'$.

A first value of 30,000 *psi* was assumed for $F_{c\,cr}'$ in the above equation, and from the iterative process, the value for "η_c"of 0.87727 was obtained. The final critical bending buckling stress allowable is thus;

For
$F_{cy} = 60{,}000\ psi$
$n = 20$
$E = 10.0*10^6\ psi$
$\mu = 0.33$
$t = 0.090\ in$
$b = 0.785\ in$
$k_b = 0.46$

$$F_{b\ cr} = \frac{k_b\ \eta_c\ \pi^2\ E}{12\left(1-\mu^2\right)}\left(\frac{t}{b}\right)^2 = \frac{0.46\,(0.87727)\ \pi^2\ 10*10^6}{12\left(1-0.33^2\right)}\left(\frac{0.090}{0.785}\right)^2 = 48{,}959\ psi$$

Free flange Margin of safety

Thus the MS in free flange buckling is calculated as:

$$MS = \frac{F_{b\,cr}}{f_2} - 1 = \frac{48{,}959}{|-4{,}606|} - 1 = >1.0$$

14.17.5 CRIPPLING

Since the lower cap of the Intercostal is in compression, a crippling check is performed. However, this check is somewhat redundant since that, as shown is section *14.17.4*, the free flange doesn't buckle. The flange buckling would be the first step to happen before the corner cripples therefore no crippling is expected.

But for the purpose of the demonstration, and since it is common to systematically perform a crippling check as soon as there is compression in a corner, the crippling stress allowable "F_{CC}" will be calculated below. This value is also required when the section analyzed is in compression and unsupported over a fairly long distance, as in the case of a column, where a *Johnson-Euler* analysis is necessary.

The crippling analysis is performed by considering the largest effective web width between *15t* (15*0.070 = 1.05 *in.*) from section *A-A* and, the upright flange length from the lower cap at section *B-B* (0.915 *in.*). The edge of all the flanges are considered free. See *Figure 14-153* for the effective segments used in the analysis.

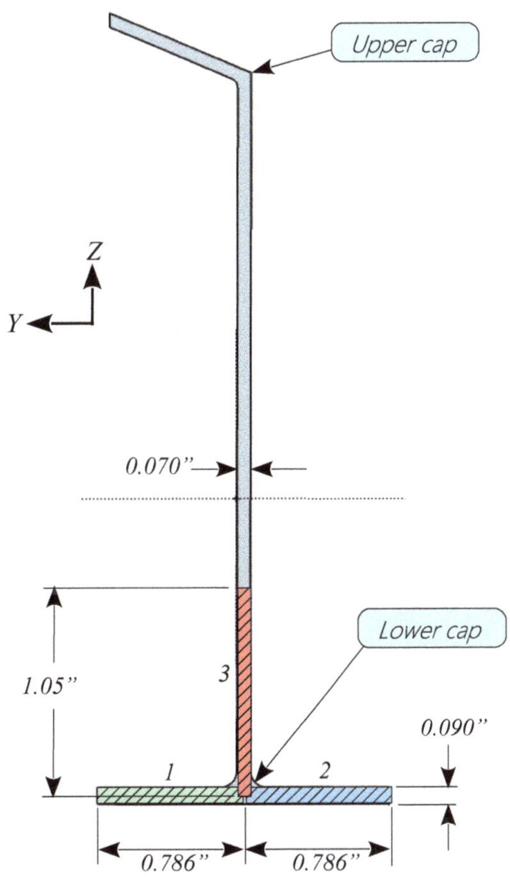

***Figure 14-153** – Effective segments for crippling at lower cap.*

The crippling allowable for each segment is calculated as shown below. Note that the crippling allowable "F_{CC}" of a segment can never be higher than F_{CY}. So the maximum value for F_{CCn} / F_{cy} is 1.0. Using *Figure 14-154*, the value at the abscissa is calculated as:

$$abscissa = \sqrt{F_{CY} / E}\,(b / t)$$

See *Figure 14-154* for typical crippling allowable curves.

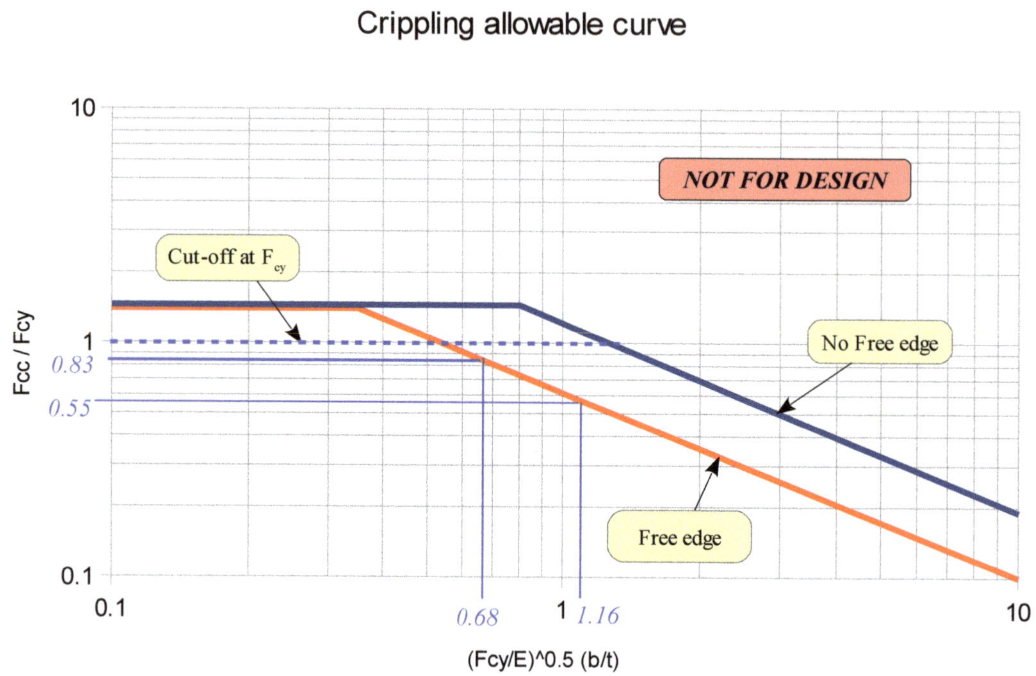

***Figure 14-154** – Typical Crippling curves.*

So for $E = 10.0*10^6$ *psi*, $F_{cy} = 60{,}000$ *psi*, the crippling allowable of each segment is calculated in *Table 14-59*.

***Table 14-59** - Crippling analysis at Intercostal lower cap.*

Segment	*Free edge*	b_n [in]	t_n [in]	$b_n t_n$ [in²]	b_n / t_n	$(F_{cy} / E)^{0.5}$	$(F_{cy} / E)^{0.5}(b_n/t_n)$	F_{CCn} / F_{cy}	F_{CCn} [ksi]	$t_n b_n F_{CCn}$ [lbs]
1	1	0.786	0.09	0.071	8.733	0.0775	0.68	0.83	49.8	3523
2	1	0.786	0.09	0.071	8.733	0.0775	0.68	0.83	49.8	3523
3	1	1.05	0.07	0.074	15.000	0.0775	1.16	0.55	32.7	2403
			$\sum b_n t_n =$	0.2150					$\sum b_n t_n F_{ccn} =$	9449

thus:

$$F_{CC} = \frac{\sum b_n t_n F_{CCn}}{\sum b_n t_n} = \frac{9{,}449}{0.215} = 43{,}954 \ psi$$

From *Table 14-28,* the stresses at points *1, 2* and *5* are −20, −4,606 and −773 *psi* respectively. An average stress could be calculated from those three points and compared to F_{CC} but, to be conservative, I will simply use the maximum compression stress at point *2*.

Margin of safety

Thus the crippling MS is calculated as:

$$MS = \frac{F_{CC}}{f_2} - 1 = \frac{43{,}954}{|-4{,}606|} - 1 = > 1.0$$

14.17.6 COLUMN BUCKLING

From *Figure 14-57* one can see that the Intercostal lower cap is subjected to a linearly varying compression load. Therefore, it is possible that the lower cap, considered as a column, could be unstable laterally, especially since there is a side load that contributes in causing a secondary bending moment. Since the load application point (at the *Aft inb'd Lug Ftg*) is very close to the aft end, I could simply ignore the small part aft of the loading point where the load is in tension and assume the whole lower cap length to be in compression.

The lower cap cross section properties as shown in *Figure 14-150* is used and considered constant throughout the length of the column. *Figure 14-155* shows how the lower cap (highlighted) is idealized as a column as shown in *Figure 14-156*. The column is assumed pinned-pinned.

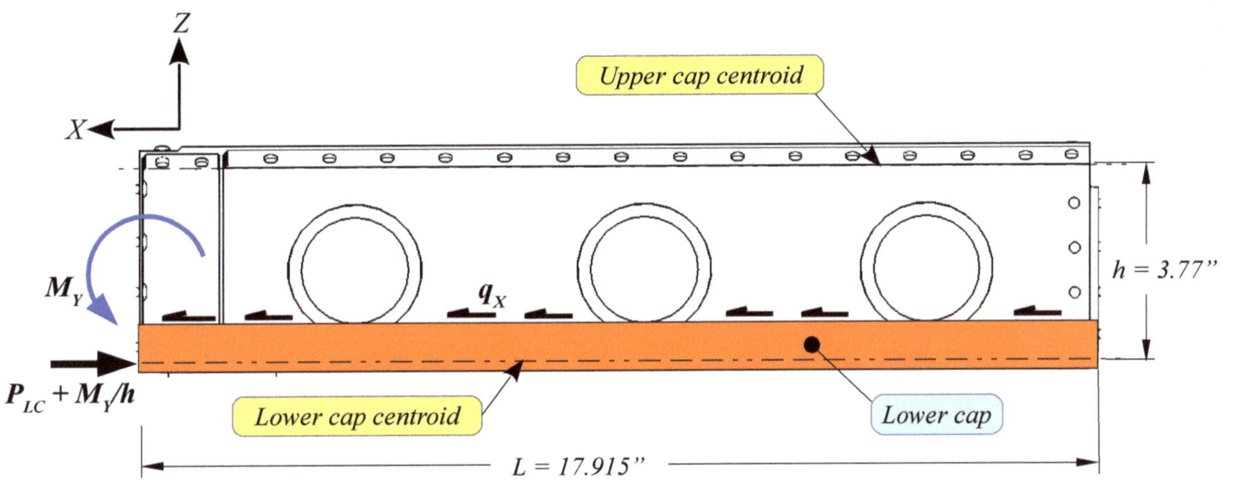

***Figure 14-155** – FBD on lower cap column.*

For

$$P_{LC} = \frac{P_X}{A} \cdot A_{LC} = -\frac{668}{0.47748} \cdot 0.20351 = -285 \ lbs$$

Using the values of $M_Y = -652$ *in-lb* (maximum bending moment), $L = 17.915$ *in.* (column length) and $h = 3.77$ *in.* (distance between the centroids of the upper and lower caps), then the total applied compression load is:

$$P_{total} = P_{LC} + \frac{M_Y}{h} = -285 + \frac{-652}{3.77} = -458 \ lbs$$

and the shear flow "q_X" is calculated as:

$$q_X = \frac{P_{total}}{L} = -\frac{458}{17.915} = 25.6 \ lb/in.$$

Hence, the column is idealized as shown below in *Figure 14-156*.

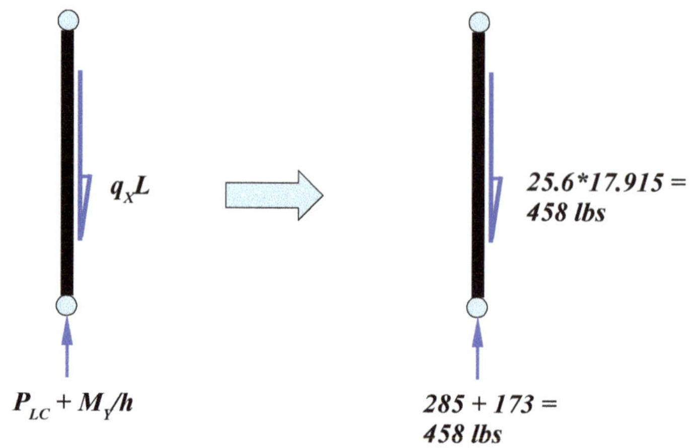

***Figure 14-156** – Idealized lower cap column.*

In order to determine the column buckling allowable, one needs to know the fixity coefficient *"C"* for the particular configuration. For a *pinned-pinned* uniform column having an applied end load and a shear flow, the ratio qL/P is evaluated as 25.6 (17.915)/ 458 = 1.0 and the *"C"* value is approximated as follows:

$$C = e^{(0.625\,qL/P)} = e^{(0.625*1.0)} = 1.87$$

Therefore, the effective column length "***L'***" is calculated as:

$$L' = \frac{L}{\sqrt{c}}$$

Hence, for L = 17.915 inches;

$$L' = \frac{17.915}{\sqrt{1.87}} = 13.1\ in.$$

Using the *Johnson-Euler* method, the critical column buckling stress is calculated as:

$$F_C = F_{CC} - \frac{F_{CC}^2}{4\pi^2 E_t}\left(\frac{L'}{\rho}\right)^2$$

then

$$F_C = 43{,}954 - \frac{43{,}954^2}{4\pi^2 10*10^6}\left(\frac{13.1}{0.378245}\right)^2 = 38{,}084\ psi$$

For
$E_t = E = 10.0*10^6$ psi
$\rho_z = 0.378245$ in
$L' = 13.1$ in.
$F_{CC} = 43{,}954$ psi

That value of F_C is sufficiently below F_{CY} that no plasticity should exist. However, just to make sure, I calculated the tangent modulus "E_t" using the *Ramberg-Osgood* relationship. That ratio is found by iteration using a *goal seek* function as follows:

$$\frac{E_t}{E} = \frac{1}{1 + \frac{3}{7} n \left(\frac{f}{F_{0.7}}\right)^{n-1}}$$

1. Knowing E and n, obtain the value for $F_{0.7}$.
2. Assume a value for f as equal to F_C calculated previously (38,084 *psi*).(in fact, any value will do).
3. Use the obtained value for E_t into the *Johnson-Euler* equation.
4. Calculate the difference between the initial value of f and the new value for F_C; it must be 0.

Determination of $F_{0.7}$

To get the $F_{0.7}$ value, a simple way to determine it is to use the stress-strain curve graphs approximated in *Figure 14-157* for the purpose of demonstration. Using the graph for the appropriate "n" value and the stress-strain curve in compression, I got the value at 70% of the primary E_C value.

For $E_C = 10.0 * 10^6\ psi$, then $E_{0.7} = 70\% * E_C = 7.0 * 10^6\ psi$. For a value of $n = 20$, I then found a value for $F_{0.7}$ of 56.0 *ksi*. See *Figure 14-157*.

From my iteration results, I obtained the following values:

$$\frac{E_t}{E} = \eta = 0.994$$

$$E_t = (0.994)10{,}000{,}000 = 9{,}944{,}801\ psi$$

$$\frac{f}{F_{0.7}} = 0.679$$

$$F_C = 38{,}051\ psi$$

This value, although slightly more conservative, is very close to the value of 38,084 *psi* by not considering plasticity. The lowest number of the two will be used into the critical compression load allowable P_{CR}.

$$P_{CR} = F_C \cdot A_{LC} = 38{,}051 \cdot 0.20351 = 7{,}743\ lbs$$

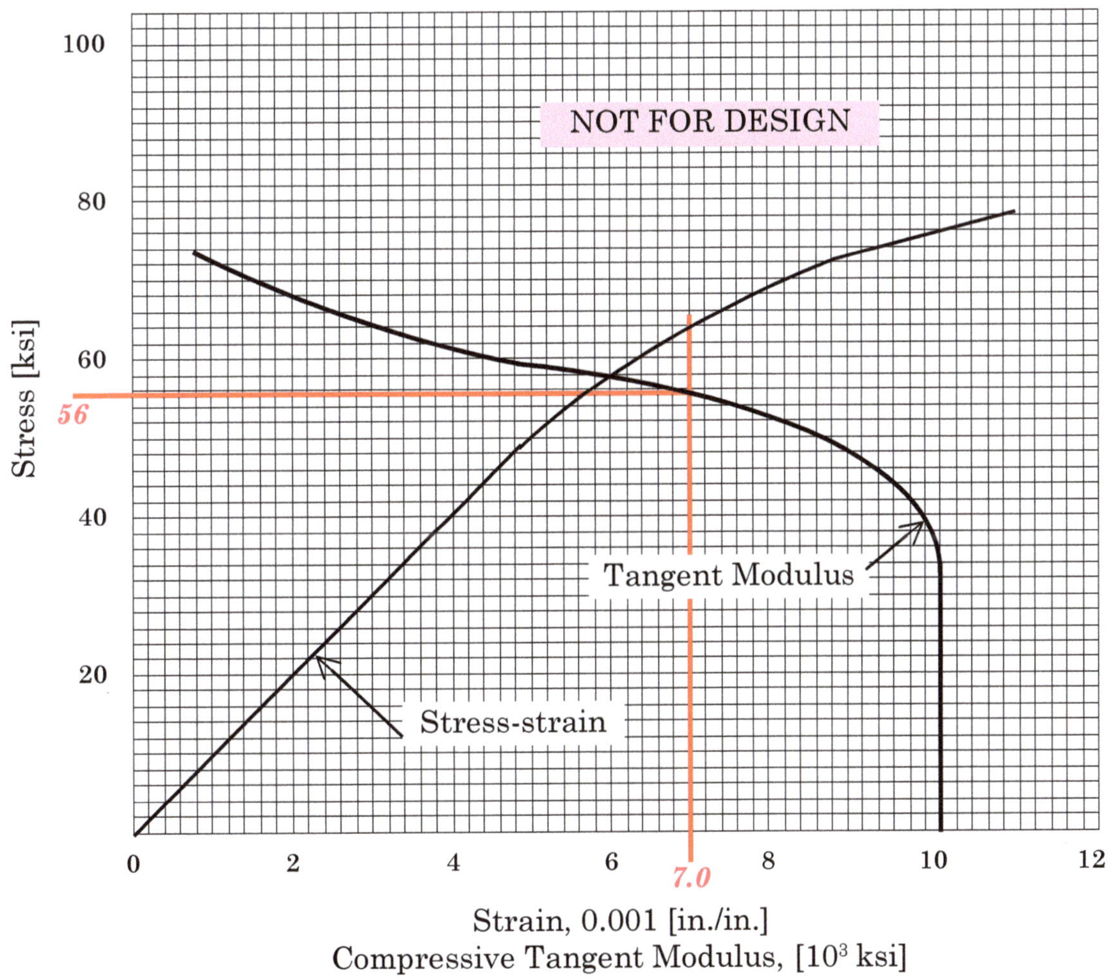

Figure 14-157 *– Approximation of Stress-strain and Tangent Modulus curves for Al Aly 7050-T7451.*

Compression load ratio

From the above analysis, the compression load allowable obtained was: $P_{CR} = 7{,}743\ lbs$.

Therefore, the Intercostal is not subject to general instability (column buckling) since the maximum applied compression load of 458 *lbs* is much lower than the allowable. The compression load ratio "R_C" is thus:

$$R_C = \frac{P_{total}}{P_{CR}} = \frac{458}{7{,}743} = 0.059$$

This value of R_C will be used later on along with the bending stress ratio "R_b" to calculate the MS.

To calculate the bending stress, because of the secondary bending effect, an "amplified" bending moment will be calculated using the approximate method as shown below. M_o is the primary bending moment while M is the "amplified" bending moment.

$$M = \frac{M_o}{1 - \dfrac{P}{P_{CR}}}$$

From the above equation, the "amplifying" factor "*AF*" is calculated as:

$$\frac{M}{M_o} = \frac{1}{1 - \dfrac{P}{P_{CR}}}$$

For $P = 458\ lbs$ and $P_{CR} = 7{,}743\ lbs$, hence:

$$\frac{M}{M_o} = \frac{1}{1 - \dfrac{458}{7{,}743}} = 1.063$$

Therefore, because the applied axial load is relatively low, the maximum bending moment is amplified only by about *6%*. This factor will be applied to the M_Z bending moment.

Margin of safety for column buckling

Thus the column buckling MS is calculated using the maximum compression stress at point *2*:

$$MS = \frac{P_{CR}}{P} - 1 = \frac{7{,}743}{458} - 1 = 15.90$$

Intercostal Lower cap Ultimate	***>1.00***
9g Fwd *Al Aly 7050-T7451.*	*Column buckling*

14.17.7 INITIAL BUCKLING OF WEB

The Intercostal web is checked for shear, compression and bending buckling. As shown in *Figure 14-152*, there are two panels in the web. Their boundaries are determined by the upper and lower flanges and by the vertical ribs and the lines of rivets. The panels are assumed simply supported on all sides. See *Table 14-60* for the results of the two shear panels using the iterative process due to plasticity.

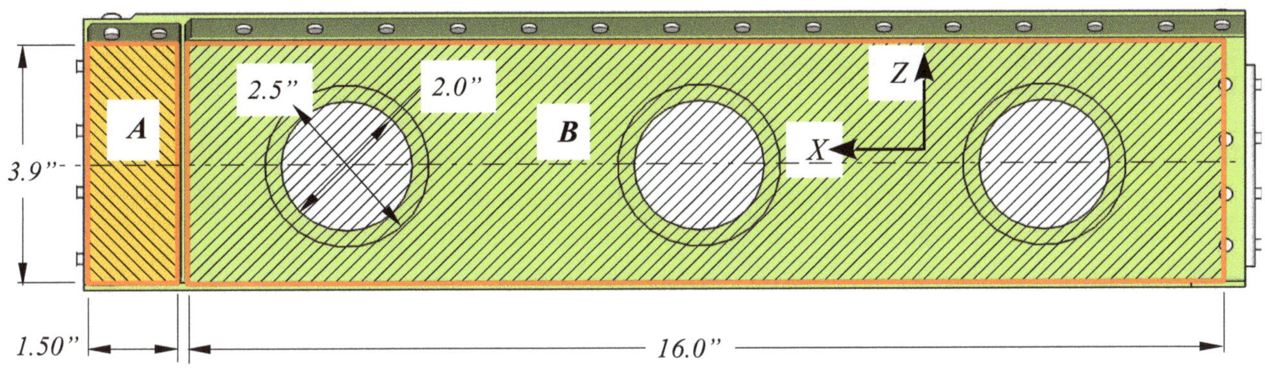

Figure 14-158 *– Intercostal web panels.*

14.17.7.1 Shear buckling

Although panel A is the most critical panel due to the higher transverse shear load, the calculations for panel B will be shown as an example since that panel has holes. From *Figure 14-143*, the transverse shear going through panel B of the Intercostal web is $V_Z = 130$ *lbs*. Since the web height is $h = 3.9$ *in*. and the thickness is $t = 0.070$ *in*, then the average shear stress is calculated as:

$f_s = V_Z/A_{web} = 130 / (3.9 * 0.070) = 476\ psi$. The shear flow is; $q = V_Z / h = 130 / 3.9 = 34\ lb/in$.

Table 14-60 *- Initial shear buckling analysis of Intercostal web.*

	a	b	a/b	k_S	t	η_s	F_{scr}	V_Z	f_s	R_S
Panel	*[in]*	*[in]*			*[in]*		*[psi]*	*[lbs]*	*[psi]*	
A	1.5	3.9	2.6	5.7	0.070	0.31511	**36103**	541	1982	0.055
B	16	3.9	4.1	5.8	0.070	1.00000	**17246**	130	476	0.028

Panel without a hole

Since there are three lightening holes in panel B, the shear buckling will be calculated in two steps. First, the shear buckling coefficient "k_s" is obtained, then, we get the knock down due to the holes. Using (ref. 20.) for flat plate buckling in shear, the shear stress allowable is calculated as:

$$F_{S\,cr} = \frac{k_S\,\eta_S\,\pi^2\,E}{12\left(1-\mu^2\right)}\left(\frac{t}{b}\right)^2$$

From *Figure 14-152*, the dimension "a" is 16.0 *in*. and the dimension "b" is 3.9 *in*. Note that for shear buckling, the value for "b" is always the smallest of the two dimensions. Thus $a/b = 16.0 / 3.9 = 4.1$. Hence from *Figure A-2*,, the value for k_s read is: 5.8. The primary Young's modulus in tension "E" is used from *Table 14-16*.

From the above equation:

$$\frac{F_{s\,cr}}{\eta_s} = \frac{k_s\,\pi^2\,E}{12\left(1-\mu^2\right)}\left(\frac{t}{b}\right)^2 = \frac{(5.8)\,\pi^2\,10*10^6}{12\left(1-0.33^2\right)}\left(\frac{0.070}{3.9}\right)^2 = 17{,}246\ psi$$

For
$k_s = 5.8$
$b = 3.9\ in$
$a/b = 4.1$
$t = 0.070\ in$
$E = 10.0*10^6\ psi$

Panel with holes

For this analysis, one non-reinforced hole is considered and it is assumed perfectly centered in a square panel. It think this is conservative since I'm not taking into account the lip around each hole. First the ratio D/b is calculated, that is 2.0 /3.9 = 0.51. Then the knock down ratio k/k_o can be approximated using the following equation.

$$k/k_o = 0.96 - \frac{D}{b} = 0.96 - 0.51 = 0.45$$

So using the shear buckling coefficient obtained previously (k_o = 5.8), the value for k having a hole is thus 0.45 * 5.8 = 2.61. Hence recalculating the shear buckling allowable we get:

$$\frac{F_{s\,cr}}{\eta_s} = \frac{(2.61)\,\pi^2\,10.0*10^6}{12\left(1-0.33^2\right)}\left(\frac{0.070}{3.9}\right)^2 = 7{,}761\ psi$$

As expected, this latter shear buckling allowable is much lower than the one calculated without the hole. Since this critical stress is relatively low compared to the yield shear stress allowable, the inelastic correction faction "η_s" is 1.0 in this case. The shear stress ratio for panel B with the holes is thus:

$$R_S = \frac{f_{S_z}}{F_{SCR}} = \frac{476}{7{,}761} = 0.061$$

14.17.7.2 Bending buckling

The stresses at the top *(3)* and bottom *(7)* of the web are obtained from *Table 14-56* and *Figure 14-149* respectively as; f_3 = 37 *psi* and f_7 = −2,313 *psi*. By extracting the pure bending and pure compression, we get the following stresses:

$$f_B = \frac{f_7 - f_3}{2} = \frac{-2{,}313 - 37}{2} = \pm 1{,}175\ psi$$

$$f_C = \frac{f_7 + f_3}{2} = \frac{-2{,}313 + 37}{2} = -1{,}138\ psi$$

Using *Figure B-1*, the buckling stress ratio in bending "R_B" is calculated for both simply supported panels. For pure bending, the ratio b/c is 2. See *Table 14-61* for the results.

Table 14-61 *- Initial bending buckling analysis of Intercostal web.*

Panel	a [in]	b [in]	b/c	k_B	t [in]	η_C	F_{Bcr} [psi]	f_B [psi]	R_B
A	1.5	3.9	2.0	23.2	0.070	0.74412	**51332**	1175	0.023
B	16	3.9	2.0	23.2	0.070	0.74412	**51332**	1175	0.023

Here, for bending and for compression cases, the plastic reduction factor η_C is calculated below.

$$\eta_C = \frac{1}{1 + \left(\frac{0.002\,E\,n}{F_{CY}}\right)\left(\frac{F_{CCR}}{F_{CY}}\right)^{n-1}}$$

For
$F_{cy} = 60{,}000\ psi$
$n = 20$
$E = 10*10^6\ psi$

By iteration, using the methodology described earlier in this book, the value of η_C of 0.74412 was obtained leading to the critical buckling stress in bending of 51,332 *psi*.

Using conservatively the same reduction coefficient due to the holes of 0.45 obtained previously in section *14.17.7.1* for the shear buckling analysis, the bending stress ratio for panel *B* with holes is thus:

$$R_B = \frac{k_o}{k}\frac{f_B}{F_{BCR}} = \frac{1{,}175}{(0.45)51{,}332} = 0.051$$

14.17.7.3 Compression buckling

Similarly, using *Figure A-1* curve *C,* for *simply-supported* edges, the buckling stress ratio in compression "R_C" is calculated for both panels. See *Table 14-62* for the results.

Table 14-62 *- Initial compression buckling analysis of Intercostal web.*

	a	b	a/b	k_C	t	η_C	F_{Ccr}	f_C	R_C
Panel	[in]	[in]			[in]		[psi]	[psi]	
A	1.5	3.9	0.4	5.9	0.070	1.00000	**17543**	1138	0.065
B	16	3.9	4.1	4	0.070	1.00000	**11894**	1138	0.096

Again, by iteration, the value of η_C of 1.0 was obtained for the panel *B* leading to the critical buckling stress in compression of 11,894 *psi*.

Using the same hole knock-down, the compression stress ratio for panel *B* with holes is thus:

$$R_C = \frac{k_o}{k}\frac{f_C}{F_{CCR}} = \frac{1{,}138}{(0.45)11{,}894} = 0.213$$

14.17.7.4 Margin of safety

As far as I know, there is no equation (in the public domain anyway) that I could use to directly determine the MS whenever, there is compression, bending and shear stress ratios involved at the same time. However, one solution would be to **not** segregate the bending and the compression and simply calculate a bending stress ratio "R_B" from that by using the proper b/c ratio. This value for R_B could be used along with R_S in the interaction equation shown below.

$$R_B^2 + R_S^2 = 1$$

I could redo the analysis using the above approach but, since the whole point of doing this analysis is to ensure that our initial assumption of considering the Intercostal web as non-buckling was correct, we don't care too much about the actual MS to the initial buckling as it is **not** a mode of failure. We simply want to know: "Does it buckle or not?"

For this, one could use the approximate following interaction equation:

$$R_B^2 + R_S^2 + R_C = 1$$

Therefore, in our case, for R_B = ***0.051***, R_S = ***0.061***, and R_C = ***0.213***, the interaction equation provides:

$$0.051^2 + 0.061^2 + 0.213 = 0.219$$

Hence, this value being much lower than 1.0, thus the panel won't buckle due to the *9g Fwd LC*.

Providing the panel would be found to buckle, then the section properties of the Intercostal would need to be recalculated by removing the web portion from sections *A-A* & *B-B* leaving only the upper and lower caps and their small effective web length of possibly *15t* .

Rib buckling check

As shown in *Figure 14-144*, the hypothesis is that the integral Intercostal rib is sufficiently stiff (doesn't buckle) in the *Y-Z* plane such that the torque can be reacted by a couple between the skin rivets and the lower cap. Having a moment "M_X" applied at its base and a simple shear reaction at the skin, the rib is idealized as a cantilevered beam where the highest bending stress is at the fixed end; that is at the lower cap flanges.

In this section, the integral rib free flange is checked in bending buckling. See *Figure 14-159* for the dimensions of the ribs' cross-section. The web is conservatively ignored from section *C-C* and thus, the section is simply a rectangle. For $T = M_X = 31$ *in-lb*, the bending stress "f_b" is thus:

$$f_b = \frac{6M}{b\,h^2} = \frac{6 \cdot 31}{0.070 \cdot 1.51^2} = \pm 1{,}166 \; psi$$

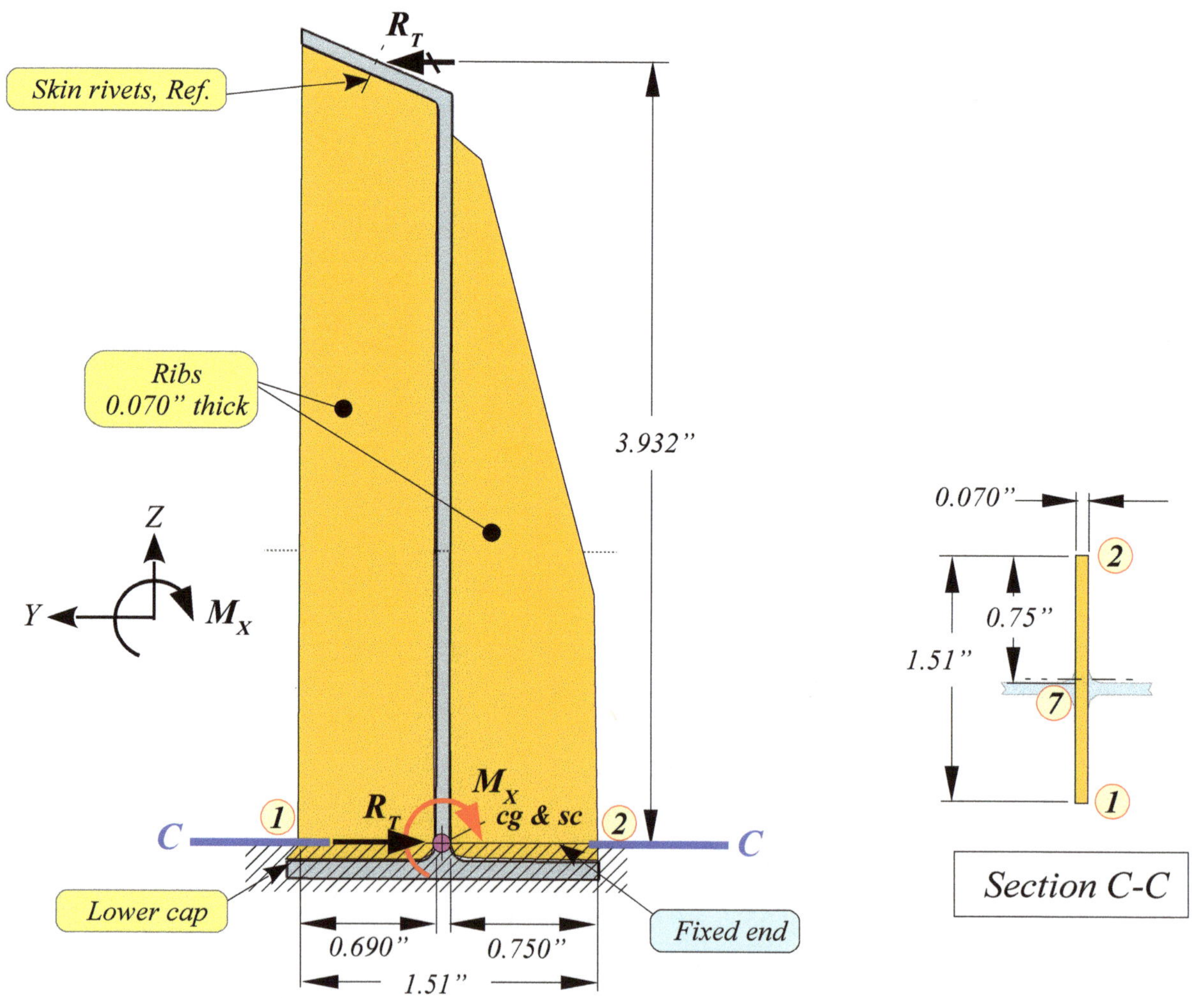

Figure 14-159 *– Intercostal integral ribs dimensions.*

Free flange buckling of rib

The stress at the longest flange base (point 7) being only 8 *psi* in tension, it is assumed to be 0 conservatively and, to simplify the calculations as $c = b = 0.75$ *in*. See *Figure 14-160*.

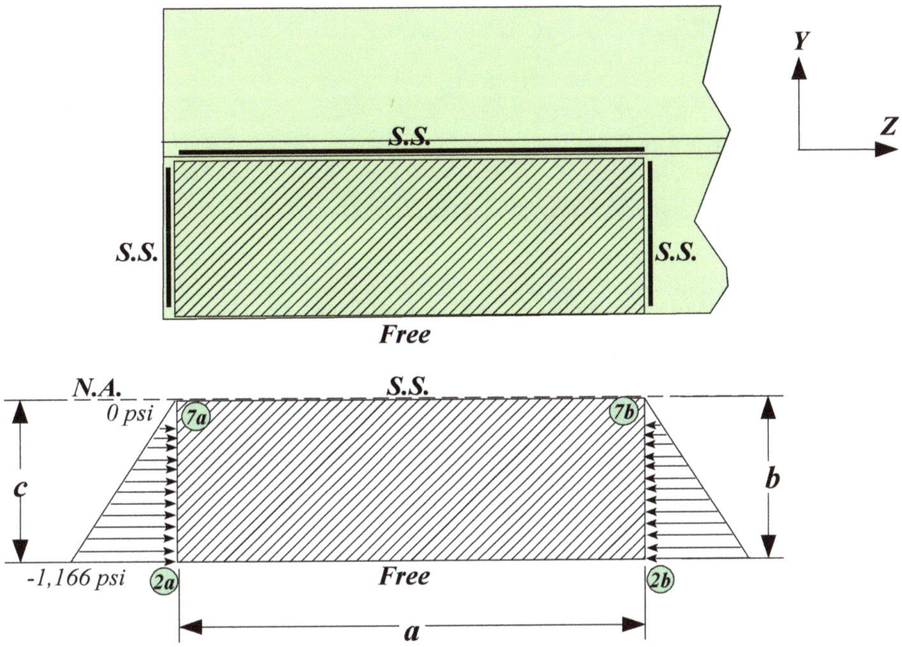

Figure 14-160 *– Free flange buckling in bending.*

For conservatism, the flange is considered simply supported at all sides except at the free edge. Using (ref. 20.) for flat plate buckling in bending, the bending stress allowable is calculated as:

$$F_{b\,cr} = \frac{k_b\,\eta_c\,\pi^2\,E}{12\left(1-\mu^2\right)}\left(\frac{t}{b}\right)^2$$

For b/c = 0.75/ 0.75 = 1.0, hence from *Figure B-1*, the value for $\boldsymbol{k_b}$ read is: 0.53. The Young's modulus from *Table 14-16* is used.

For
$k_b = 0.53$
$b = 0.750$ *in*
$t = 0.070$ *in*
$E = 10.0*10^6$ *psi*
$\mu = 0.33$

From the above equation:

$$\frac{F_{b\,cr}}{\eta_c} = \frac{k_b\,\pi^2\,E}{12\left(1-\mu^2\right)}\left(\frac{t}{b}\right)^2$$

$$\frac{F_{b\,cr}}{\eta_c} = \frac{(0.53)\,\pi^2\,10*10^6}{12\left(1-0.33^2\right)}\left(\frac{0.070}{0.75}\right)^2 = 42{,}613\ psi$$

As performed previously, the actual critical compression stress in bending is evaluated by iteration. We then find the following value:

$$F_{b\,cr} = \frac{k_b\,\eta_c\,\pi^2\,E}{12\left(1-\mu^2\right)}\left(\frac{t}{b}\right)^2 = \frac{0.53\,(0.99155)\,\pi^2\,10*10^6}{12\left(1-0.33^2\right)}\left(\frac{0.070}{0.75}\right)^2 = 42{,}253\ psi$$

The allowable is thus much higher than the bending stress in compression of 1,166 *psi*. Therefore, no buckling occurs in the ribs and our initial assumption is valid.

14.17.8 COMBINED STRESSES

The sum of the axial and bending stresses calculated at point *2* from *Table 14-58* is combined with the average shear stress along the *Y* axis calculated on page 476 in order to get the strength MS. The following interaction equation is normally accepted when combining an axial (or bending) stress with a shear stress at yield or ultimate:

$$R_a^2 + R_S^2 = 1$$

Thus the MS is calculated at point *2* as:

$$MS = \frac{1}{\sqrt{R_a^2 + R_S^2}} - 1$$

Where R_a and R_S are respectively the axial + bending and, shear stress ratios calculated as follows:

$$R_a = \frac{f_2}{F_{cy}} = \frac{|-4{,}750|}{60{,}000} = 0.080$$

$$R_s = \frac{f_{S_Y}}{F_{su}} = \frac{446}{40{,}000} = 0.012$$

Hence the MS is:

$$MS = \frac{1}{\sqrt{0.080^2 + 0.012^2}} - 1 => 1.0$$

14.17.9 CONNECTIONS

As shown in *Figure 14-161*, the forward and aft ends of the Intercostal are connected to the Frames at *FS* 498.5 via a Shear Clip and at *FS* 516.5 directly using *qty (4) MS20470AD5* solid rivets at each end. The connection at the forward end will be covered in the Shear clip section *14.18.4*. The Intercostal is also connected to the skin with *qty (17) MS20426AD5* solid flush rivets.

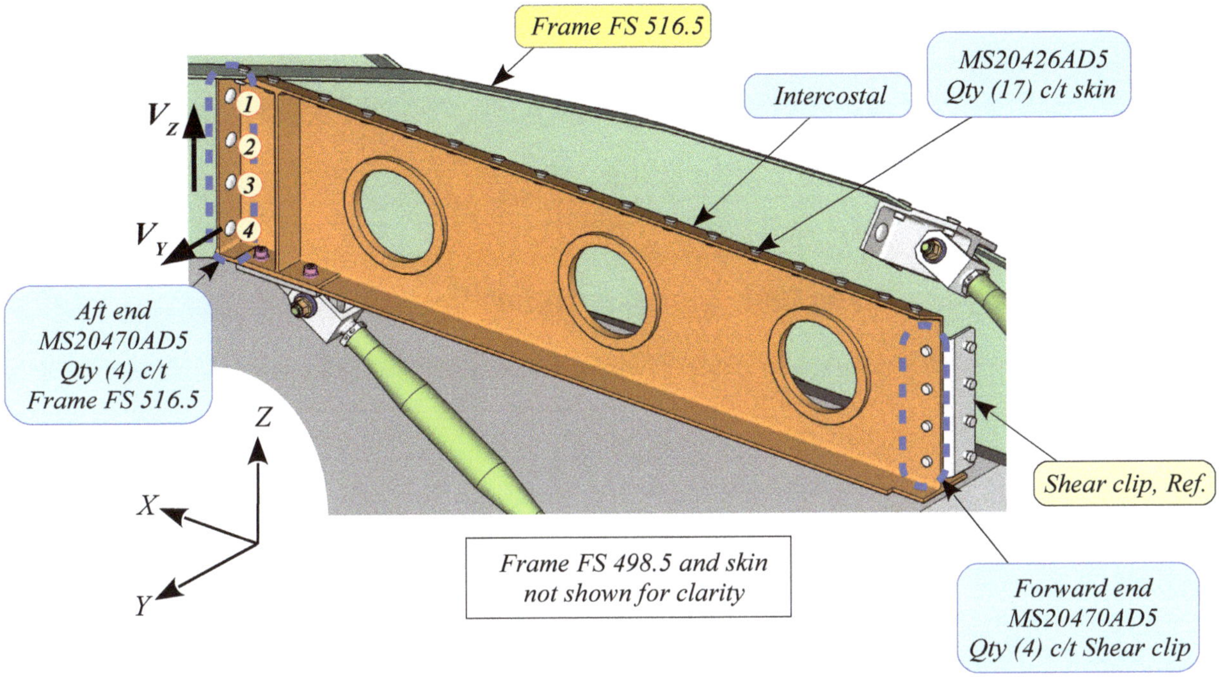

Figure 14-161 – *Intercostal connections.*

14.17.9.1 Aft end

The maximum shear load in the *Z* direction at the aft end obtained from *Figure 14-143* for the *9g Fwd LC* is 541 *lbs*. From *Figure 14-147*, the shear load at the aft end in the *Y* direction is 42 *lbs*.

The vertical load is thus assumed shared by all four rivets while, for conservatism, the side load is assumed taken by the bottom fastener only. Therefore, only the bottom rivet (*4*) is analyzed passing the other ones by comparison.

Along the *Z* axis:

$$R_{Z1} = R_{Z2} = R_{Z3} = R_{Z4} = \frac{P_Z}{4} = \frac{541}{4} = 136 \ lbs$$

Along the *Y* axis:

$$R_{Y4} = P_{Y\,total} = 42 \ lbs$$

The combined shear force "F_S" in fastener *(4)* is calculated as follows:

$$F_{S_4} = \sqrt{R_{Y_4}^2 + R_{Z_4}^2} = \sqrt{136^2 + 42^2} = 143 \ lbs$$

MS20470AD5 Rivet shear

A Fitting Factor "*FF*" of 1.15 (ref. *18. FAR §25.625*) is used. No shear factor "*SF*" is required since the Intercostal and the Frame web are relatively thick. (ref. *Table 13-4)*. The Margin of Safety for shear is thus calculated as follows using the allowables from *Table 14-25*:

$$MS = \frac{P_{su}}{FF \times F_{S_4}} - 1 = \frac{596}{1.15 \times 143} - 1 = 2.62$$

Bearing:

The bearing stress will be checked in both parts involved in the joint. The bearing load allowable is calculated as follows:

$$P_{bru} = F_{bru} \cdot D \cdot t$$

Table 14-63 shows the resulting bearing MS.

Table 14-63 *- Bearing stress MS table summary.*

Part	Material Al Aly	t [in]	D [in]	Fbru [ksi]	Pbru [lbs]	F [lbs]	MS bearing
Intercostal	7050-T7451	0.063	0.159	105	1052	143	6.36
Frame	7075-T7351	0.060	0.159	135	1288	143	8.01

No Fitting Factor is used in bearing.

The joint is thus shear critical. The Margin of Safety for the Intercostal aft end connection is:

$$MS = \frac{P_{bru}}{F} - 1 = \frac{1{,}052}{143} - 1 = 6.36$$

Intercostal Aft end Connection, Ultimate	***>1.00***
9g Fwd *MS20470AD5 in Al Aly 7050-T7451*	*Fastener shear*

14.17.9.2 Skin connection

The shear flow "q_X" in the X direction at the top of the Intercostal common to the skin is obtained from *Figure 14-142* for the *9g Fwd LC* and is 37.3 *lbs/in*. For a maximum fastener pitch "***s***" of 1.29 *in*., the maximum fastener shear force "F_X" is thus:

Along the *X* axis:

$$F_X = q_X \cdot s = 37.3 \times 1.29 = 48 \ lbs$$

Along the *Y* axis:

Similarly to the aft end connection, it is assumed that only one rivet at the Intercostal upper cap reacts the side load. From *Figure 14-144*, the lateral force R_T at the skin is 8 *lbs*. Note that the 23 *deg.* angle of the upper cap flange is neglected as the actual shear would be slightly smaller.

$$F_Y = R_T = 8 \ lbs$$

The combined shear force "F_S" is thus calculated as follows:

$$F_S = \sqrt{F_Y^2 + F_X^2} = \sqrt{8^2 + 48^2} = 49 \ lbs$$

MS20426AD5 Rivet shear

A Fitting Factor "*FF*" of 1.15 (ref. *18. FAR §25.625*) is used. For flush rivets, *Table 14-25* provides shear allowables for flush rivets according to the material thickness in which the countersink is performed. That would be the skin in our case. At that row of rivet, the skin is 0.090" thick so, from that table, the shear strength allowable is 575 *lbs*. The Margin of Safety for shear is calculated as follows:

$$MS = \frac{P_{su}}{FF \times F_S} - 1 = \frac{575}{1.15 \times 49} - 1 = 9.20$$

Bearing:

The bearing stress will be checked in both parts involved in the joint. The bearing load allowable is calculated as follows:

$$P_{bru} = F_{bru} \cdot D \cdot t$$

Table 14-64 shows the resulting bearing MS.

Table 14-64 *- Bearing stress MS table summary.*

Part	Material Al Aly	t [in]	D [in]	Fbru [ksi]	Pbru [lbs]	F [lbs]	MS bearing
Intercostal	7050-T7451	0.070	0.159	105	1169	49	22.85
Skin	2024-T3 Clad	0.090	0.159	120	1717	49	34.04

No Fitting Factor is used in bearing.

$$MS = \frac{P_{bru}}{F} - 1 = \frac{1{,}169}{49} - 1 = >1.0$$

The joint is thus shear critical.

"Professionalism = honesty."

14.18 SHEAR CLIP

The Shear Clip is made out of *Al Aly 2024-T3 Clad* sheet 0.063 *in.* thick. It is located at *FS 498.5* and connects the Intercostal to the Frame using *MS20470AD5* solid rivets as shown in *Figure 14-162*. The Shear clip has identical flanges, so only one flange needs to be analyzed by symmetry.

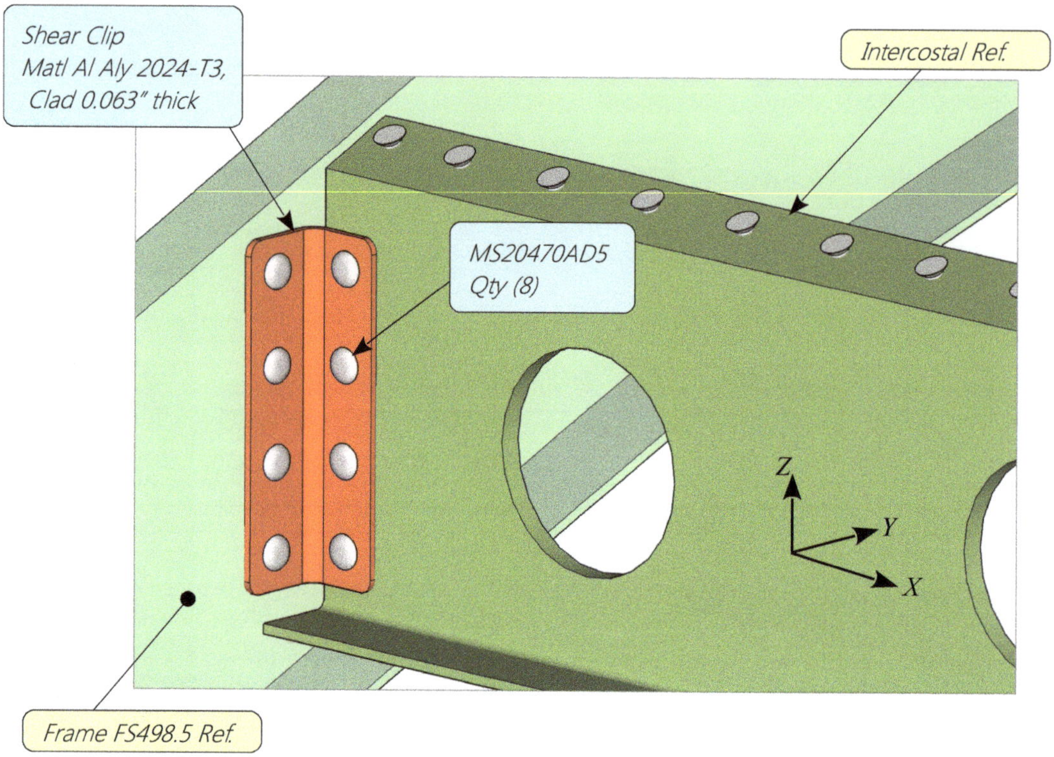

Figure 14-162 *– Shear Clip location.*

14.18.1 LOADS

The Free Body Diagram of the Shear Clip is shown in *Figure 14-163*. As in all shear clips, the shear load is considered applied at the faying surface of the flange and the shear clip is considered cantilevered off of the four rivets.

From *Figure 14-143*, the maximum shear load "V_Z" is 130 *lbs.* at the forward end for the *9g Fwd LC*.

With an offset "***e***" of 0.45" between the shear plane and the bolt group centroid position, the bending moment "M_Y" at the bolt group c. of g. is calculated as:

$$M_Y = V_Z \cdot e = 130 \cdot 0.45 = 59 \ in. - lb$$

and

$$R_Z = V_Z = 130 \ lb$$

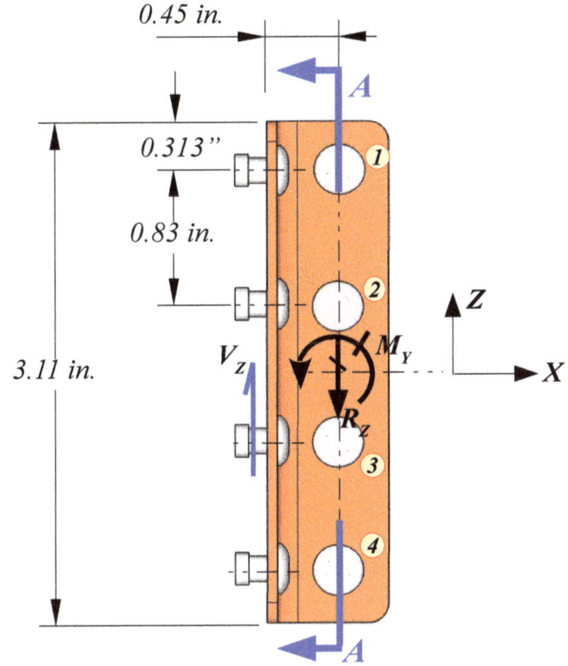

Figure 14-163 – Shear Clip FBD.

14.18.2 SECTION PROPERTIES

The cross section dimensions and properties across the fasteners are shown in *Figure 14-164*.

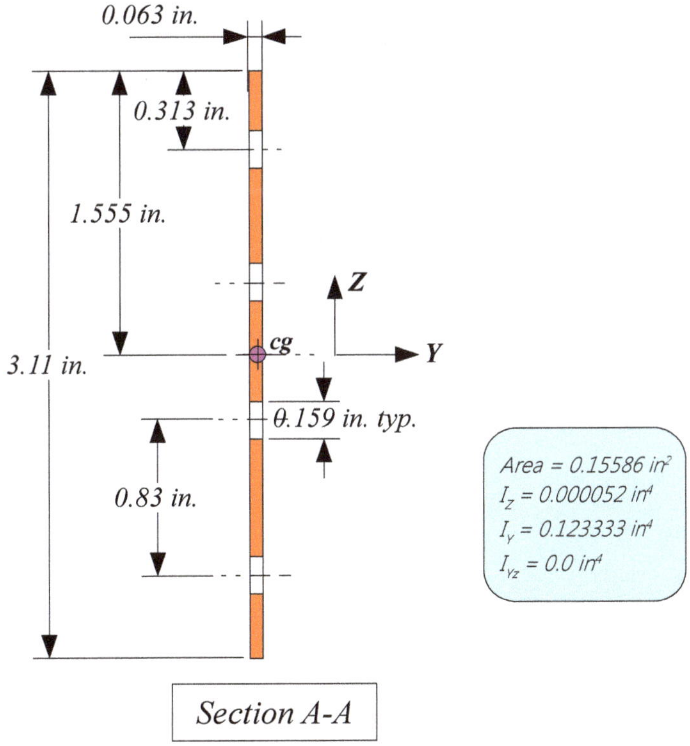

Figure 14-164 – Shear Clip cross section.

14.18.3 STRESS ANALYSIS

Bending stress

$$f_b = \frac{M_Y\, c}{I_Y} = \frac{59\,(1.555)}{0.123333} = 744\ psi$$

Shear stress

$$f_s = \frac{V_Z}{A} = \frac{130}{0.15586} = 834\ psi$$

Combined stresses

The bending and shear stresses are combined according to the equation:

$$R_b^2 + R_S^2 = 1$$

Thus the MS is calculated as:

$$MS = \frac{1}{\sqrt{R_b^2 + R_S^2}} - 1$$

Where R_b and R_S are respectively the bending and shear stress ratios calculated as follows using allowables from *Table 14-19* for *Al Aly 2024-T3 Clad* sheets:

$$R_b = \frac{f_b}{F_{cy}} = \frac{|744|}{40{,}000} = 0.019$$

$$R_s = \frac{f_s}{F_{su}} = \frac{834}{35{,}000} = 0.024$$

Hence the MS for the *Shear clip* is:

$$MS = \frac{1}{\sqrt{0.019^2 + 0.024^2}} - 1 = 31.66$$

Intercostal Shear Clip Ultimate	***>1.0***
9g Fwd *Al Aly 2024-T3 Clad*	*Bending + shear*

14.18.4 CONNECTIONS

Shear Clip *MS20470AD5* solid rivets

Four rivets on either flanges of the Shear Clip are used. *Figure 14-165* shows the applied forces at the rivets from the *9g Fwd* LC. Because the Shear clip is tall, it is conservatively assumed that only the top and bottom rivets react the moment. The vertical shear load is, of course, assumed distributed among all four rivets.

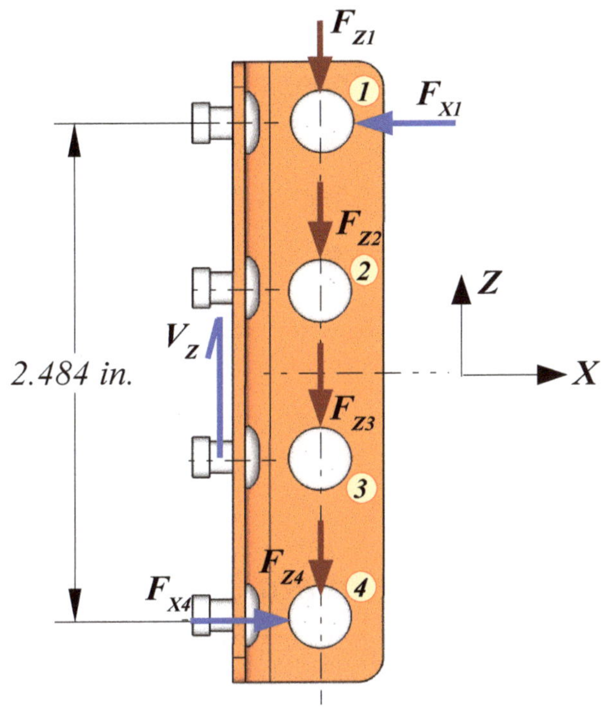

Figure 14-165 *– Applied forces on the Shear Clip rivets.*

For a shear load of V_Z = 130 *lbs* and a moment M_Y = 59 *in-lb* applied at the bolt group centroid, the forces in the rivets are:

$$F_{Z1} = F_{Z2} = F_{Z3} = F_{Z4} = \frac{V_Z}{n} = \frac{130}{4} = 33\ lbs$$

$$F_{X1} = -F_{X4} = \frac{M_Y}{pitch_{1-4}} = \frac{59}{2.484} = 24\ lbs$$

Combining the vectors, the resulting rivet force *"F"* in rivets *1 & 4* is calculated as:

$$F_1 = F_4 = \sqrt{F_{Xi}^2 + F_{Zi}^2} = \sqrt{24^2 + 33^2} = 41\ lbs$$

Margin of Safety for the Shear Clip connections

The force in the most critical rivets (*1 & 4*) is so low that the joint is *Passed by Engineering Judgment* in shear and in bearing.

14.19 CONCLUSION

As one could notice, the installation of the cabinet required a lot of analyses to substantiate the secondary structure as well as the Primary structure to which it is connected. This was a fairly good example that used pretty much all of the most common analysis techniques for hand calculations.

Normally, this analysis would be a *"first pass"* since, as you may have noticed, a lot of Margins of safety are high so there would be room for optimization by reducing the weight. For example, the Intercostal could be made thinner (out of sheet metal perhaps) and maybe taper with the smaller end at the forward end where the shear load is very small. A smaller and thinner Shear clip could also be used at that end along with *1/8"* rivets instead of *5/32"*.

This is why by doing the calculations in spreadsheets can speed-up the iterative optimization process immensely.

"An expert is someone who has made all the possible mistakes in a particular field and learned from them."

15. WING SPAR

For some reason, some people consider working on wings as more glamorous than any other part of the aircraft. Believe me, regarding stress analysis, there is nothing more important or more difficult to do on the wing. That fascination may be due to the fact that structural failures of a wing would be considered more catastrophic than a failure in the fuselage. Or maybe, that there is something mysterious or scientifically challenging regarding the advance design of a wing. I don't know. What I know however as a structural engineer is that a fuselage is a cylinder and that a wing is a rectangular box.

In this chapter, in order to show a complete example of a typical analysis, a portion of a wing *Rear Spar* (R/S) will be analyzed by showing all the necessary steps required to substantiate it, from: loads extraction, section properties, stresses and stability checks, and Margins of Safety.

But first, lets start by describing the structure of our wing example.

15.1 ANATOMY OF A WING

The wing box, as it is called, is a torsion cantilevered beam. It generally comprises of a Front and a Rear Spar (maybe more) and, top and bottom stiffened planks (skins). Ribs are spaced throughout the length of the wing. The leading and trailing edges are not considered as part of the wing box beam and thus are not part of the FEM. See *Figure 15-1* for a typical wing box.

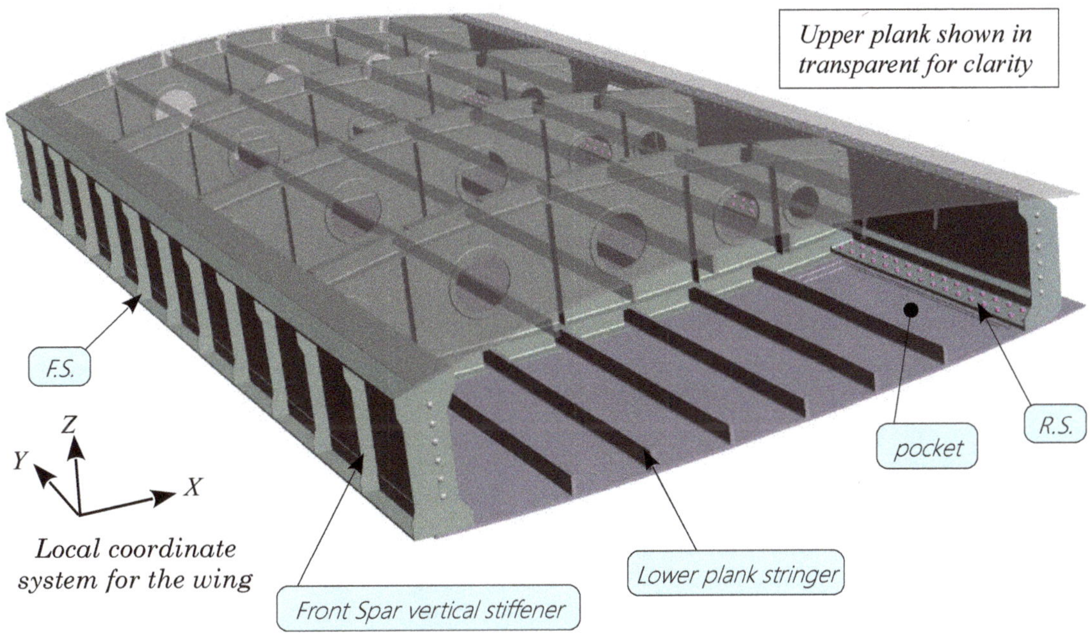

Figure 15-1 – *Typical wing box.*

The wing deflects the air downward hence, as a reaction, creates lift. It also supports the main landing gear and sometimes the engines. It also acts as reservoirs for the fuel. The portion of the wing containing the fuel is slightly pressurized due to the inertia of the fuel in dynamic loading.

For the example that follows, I created a simple cantilevered wing having a sweep angle of 30 *deg.* and, a wing profile ***747A315*** ***(15% thickness)*** from (ref. 1. *p.446*) was used with no taper (constant thickness and chord *"C"*). See *Figure 15-2* for a section of the wing structure showing its profile.

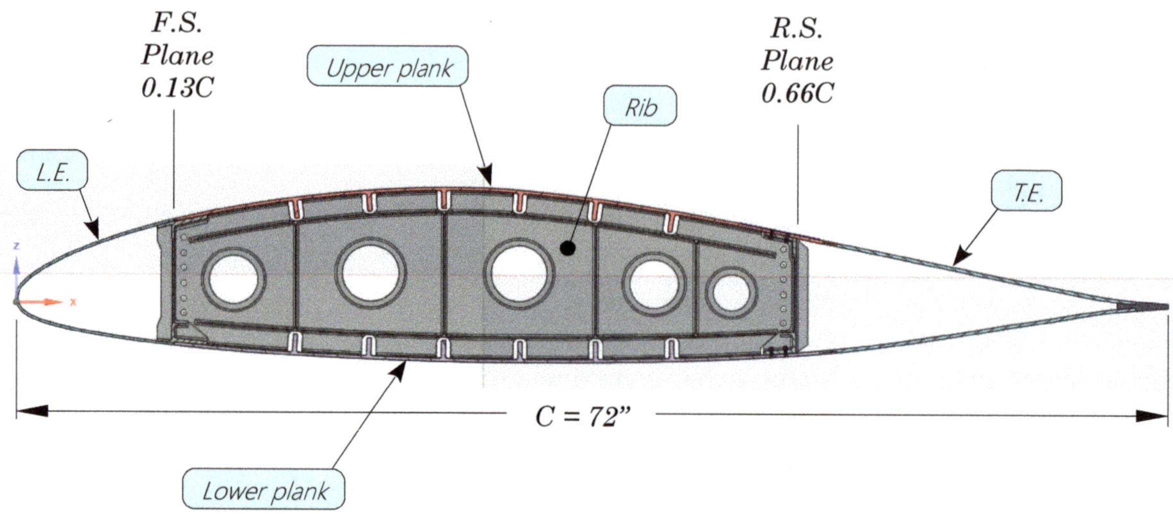

Figure 15-2 *– Wing section profile 747A315.*

15.1.1 WING PLANKS

In the past, the top and bottom of the wing box were made from skins and fastened stringers, just like for the fuselage. But in order to remove the maximum quantity of fastener holes, speed-up the manufacturing process and improve the strength and durability, the top and bottom planks are now machined out of Aluminum plates that are about 2.0 inches thick where fastened stringers are replaced by integral stiffeners.

Since the top plank is almost always in compression and the bottom plate is almost always in tension, each thus use different materials. The top plate usually uses *Aluminum **7000*** series alloy for its high compression strength, while the bottom plate uses *Aluminum **2000*** series alloy for its high durability (fatigue strength).

To fasten the Wing planks to the Spars and ribs, some manufacturers use conical shaped fasteners installed with interference fit by a robot (automatic riveting machine) which drills the holes and installs the fasteners in the same operation. This provides for very durable joints.

Helicopter main rotor blades, although they also provide lift, are loaded and constructed in a very different way. They are also analyzed in a very different manner as they are not part of the helicopter FEM.

This example is for a fictitious airplane wing modeled in a Finite Element loads Model. The upper plank is made out of *Al Aly **7075-T7351*** plate and the lower plank is made out of *Al Aly **2024-T351***. See *Figure 15-3* for the wing planks used.

In order to achieve a good compromise between the weight of the wing and its strength, airplane manufacturers will allow for a certain amount of buckling at **limit load**. In our example, we will allow buckling at *80%* of limit load except at cut-outs where no buckling is permitted at limit load.

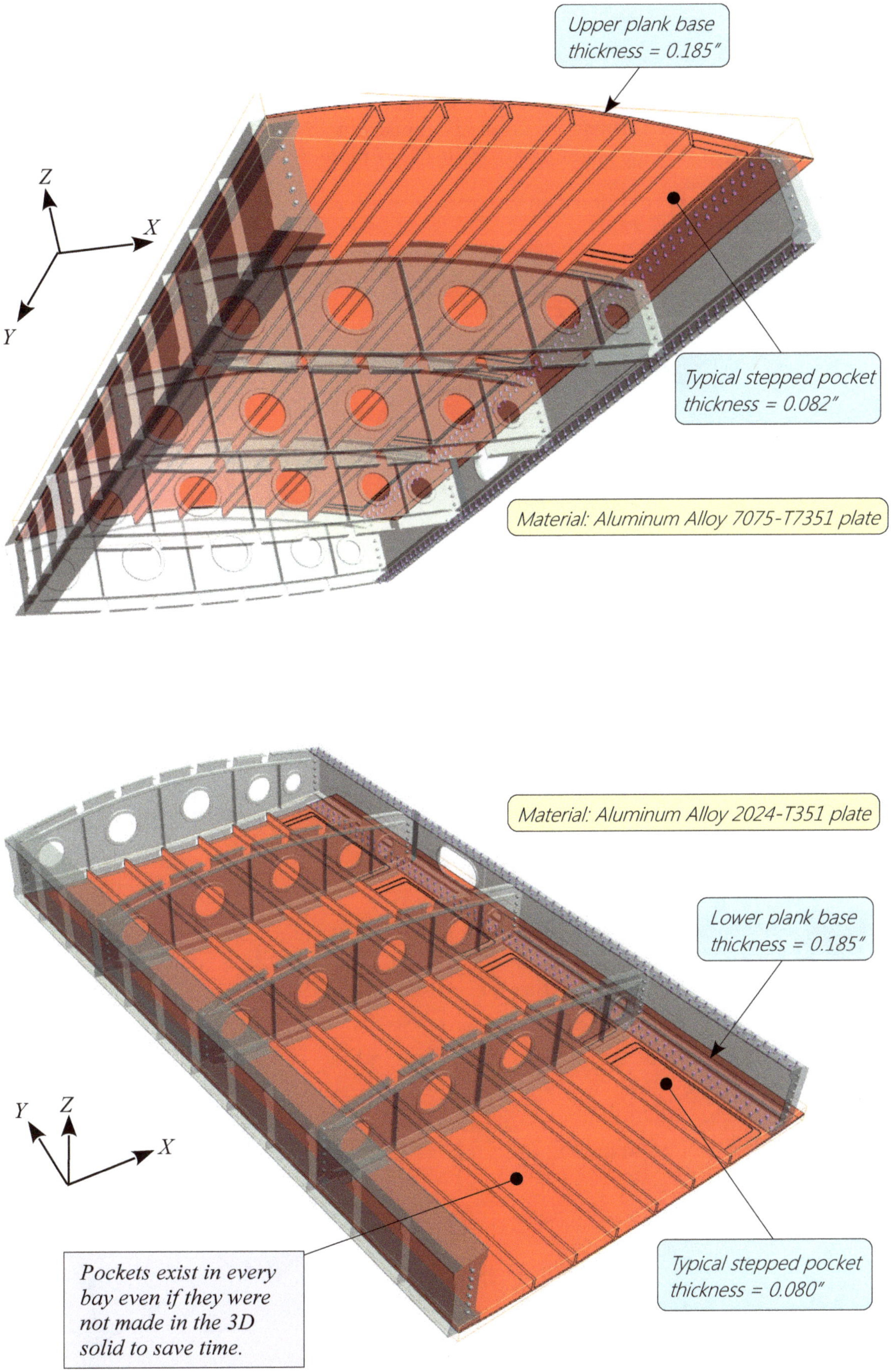

Figure 15-3 – *Example wing planks.*

15.1.2 SPARS

In addition to closing the front and back ends of the wing box, the spars also support in conjunction with the ribs; the control surfaces, the flaps, the slats, the de-icing system and the engine pylons. They are usually "***I***" or "***C***" beams sections machined out of the *Aluminum* ***7000*** series. In our example, the Rear Spar is made out of *Al Aly* ***7050-T7451*** plate.

Some effective "skin" is considered in the section properties. The Front and Rear Spars of our example are shown in *Figure 15-4.*

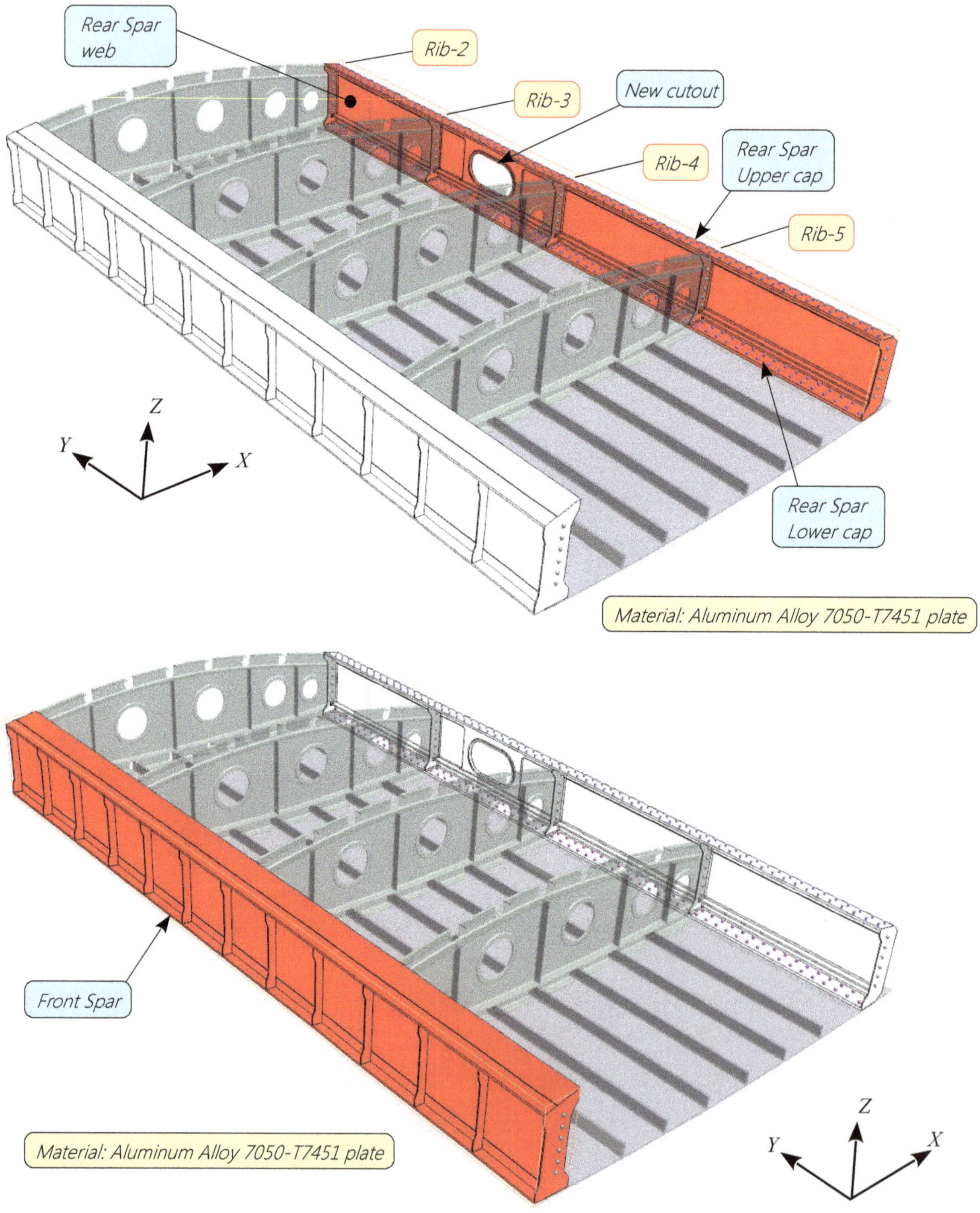

Figure 15-4 *– Example Rear Spar with cutout.*

15.1.3 RIBS

The wing planks are fastened onto the Ribs that determine the wing profile. The Ribs are then fastened to the Front and Rear Spars to which they transfer their vertical air loads and loads from any hard point loads like landing gear and engine nacelle drag struts. They act as panel breakers for the wing planks and baffles for the fuel reservoir.

In our example, the Ribs are machined out of *Al Aly* ***7050-T7451*** plates. Also, no fuel exists at the location of the analysis. See ***Figure 15-5*** for a picture of the ribs.

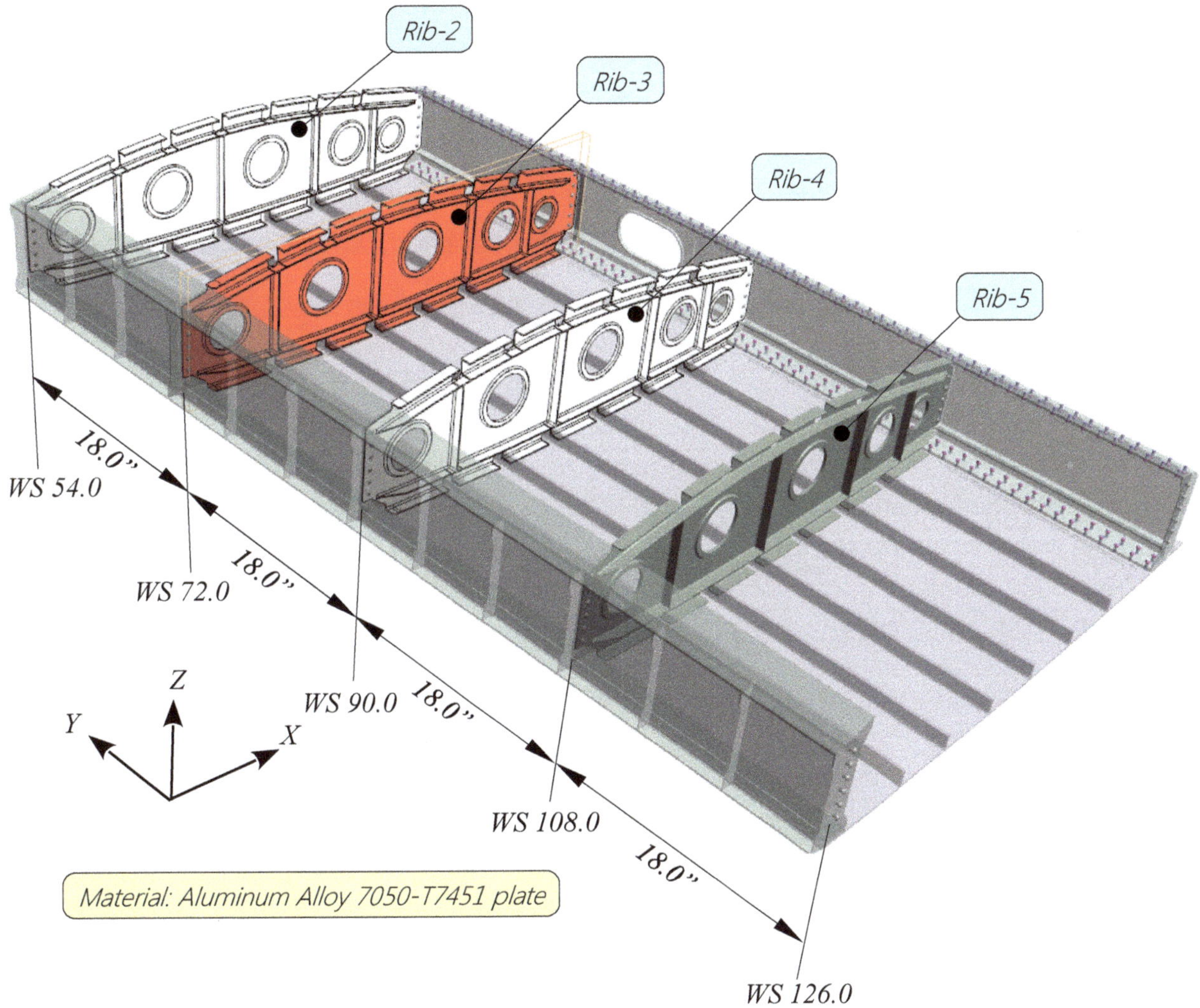

Figure 15-5 *– Example Ribs.*

15.2 DESCRIPTION

In this fictitious design example, I will assume that this analysis is required since the designer had added a cutout in the web, of the already analyzed Rear Spar, to allow clearance for some system. Therefore, only the portion of the Spar between the two ribs (*Rib-3 & Rib-4*) adjacent to, and, at the cutout, needs to be re-visited. See ***Figure 15-6*** for details of the cutout in the web of the Rear Spar and the locations of the most critical cross-sections.

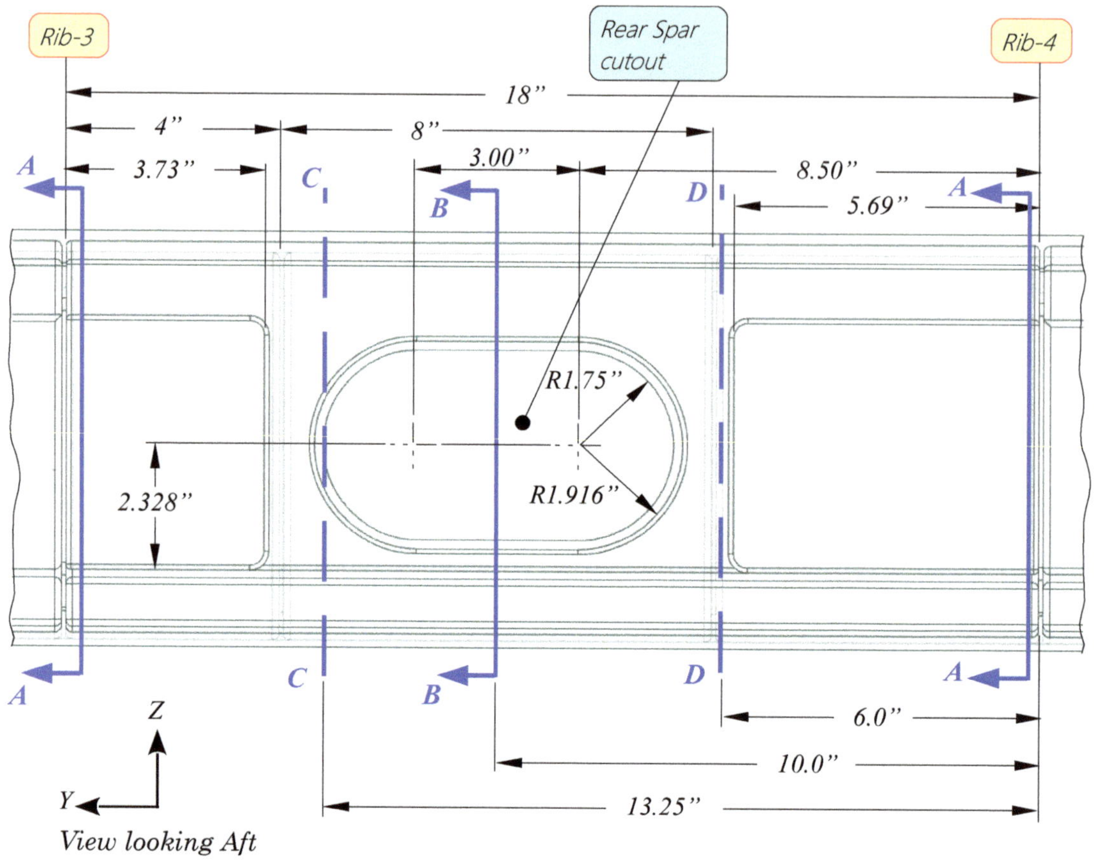

***Figure 15-6** – Rear Spar showing the new cutout and cross-sections.*

Here is a recommended list of the different steps that need to be done in order to substantiate the new *Rear Spar* with the new cutout.

1. Extract the internal loads at the junction of the Rear Spar and the Ribs *3 & 4*.
2. Perform the upper and lower wing plank buckling analysis.
3. Perform the Rear Spar web buckling analysis.
4. Evaluate the crippling allowables of the upper and lower Rear spar caps, through both sections *A-A* and *B-B*.
5. Evaluate the effective skin width.
6. Calculate the section properties near the Ribs (section *A-A*) and through the cutout (section *B-B*).
7. Perform preliminary stress calculations before secondary bending is considered.
8. Perform a slotted beam analysis at the cutout location (section *B-B & C-C*).
9. Perform a column buckling analysis of the *R/S* upper cap using the *Johnson-Euler* method.
10. Perform a beam-column analysis of the upper cap.
11. Calculate the final stresses and MS interaction.
12. Perform an Inter-rivet buckling analysis of the upper plank at the upper cap.
13. Perform a net tension analysis at the lower cap.
14. Perform a connections analysis for the joints between the Rear Spar and the upper and lower wing planks.

15.2.1 FEM

A *Global Finite Element Model* (GFEM) is generally done for all of the primary structure of an aircraft. This model is "rough" and it is considered a ***"Loads model"*** since it is used to extract loads (forces and moments) but NOT directly stresses.

GFEM*'s* have a relatively small number of elements and yet, they surprisingly match the actual tested structure very well – if well made that is. I believe it is more difficult to get a very good model made of fewer *1D* and *2D* elements than simply filling a *3D* solid geometry with millions of tinny *3D* elements. However, unless it changes soon (and I think it'll change eventually as the computing power of the machines increases and that excellent correlation is obtained with the real structure), all aerospace companies so far have used rough models to obtain the certification loads. See *Figure 15-7* for the *loads model* of our example.

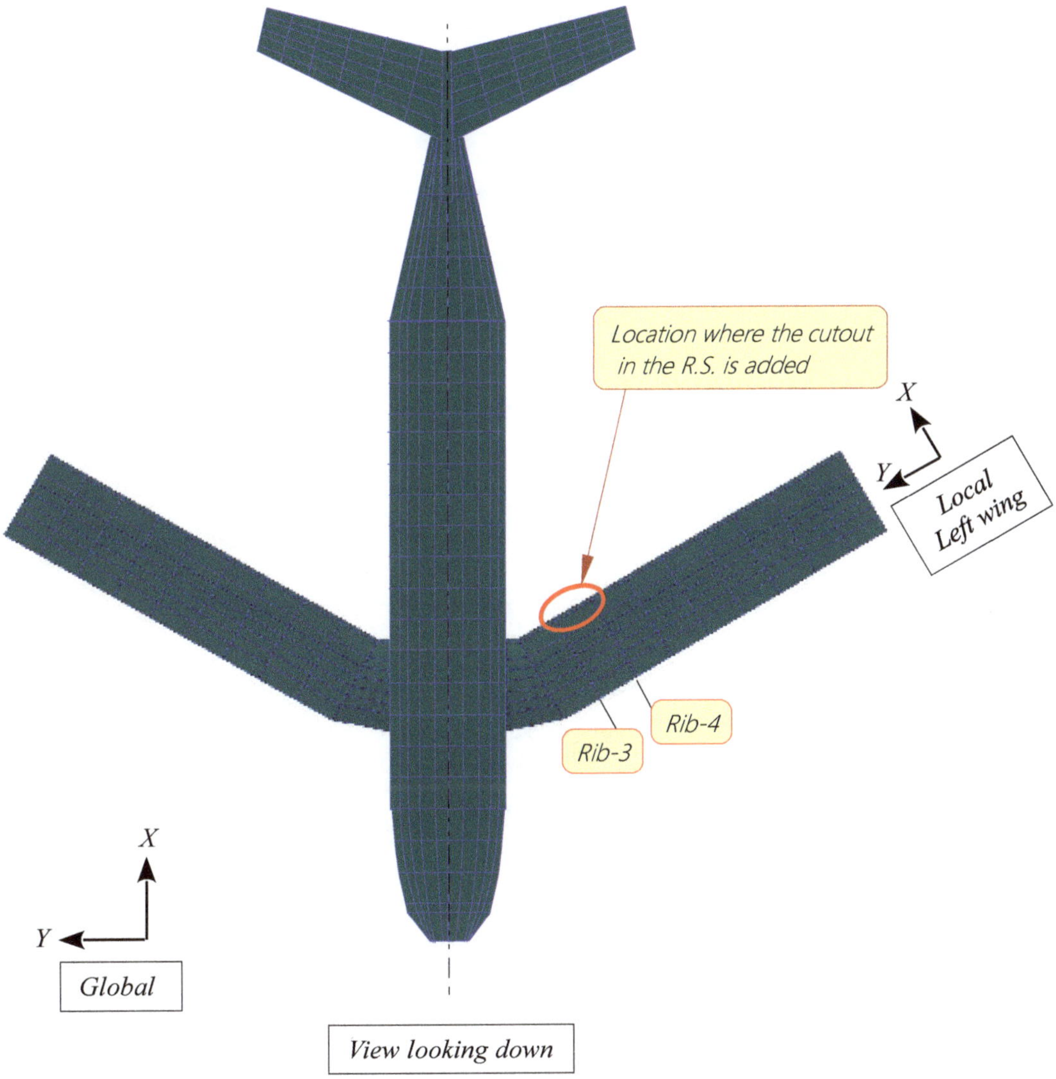

Figure 15-7 *– Loads FEM example.*

On the other hand, very detailed FEM*'s* (DFEM) using small *2D* and *3D* elements are created for critical areas, for instance, like around access panel or hard points, to better understand the load and stress distribution. But those DFEM's are generally not used (but can be) in the official stress report. To be used, they have to be validated against the actual test using strain gauges.

In this example of a FEM representing a fictitious *(and ugly)* commercial airplane, the *ROD–QUAD–ROD* approach is used. Every single bay, of the wing for instance, (framed by rib flanges and plank stringers) is modeled with one *QUAD* only.

To be conservative (and thorough), due to the large size of the cutout, normally I would have requested a change in the GFEM to better represent the new design by replacing the *CQUAD* element that models the Spar web where the large cutout is located and replace it with a *CSHEAR* element. I would also probably increase the upper and lower Spar caps section properties if necessary. The new FEM would then be re-run for the thousands of load cases and I would be extracting the new internal loads for this analysis.

However, it may be decided by the manager **not** to change the GFEM at this time for such a small change as it is quite time consuming. That's the assumption that I'll use in this example.

So I'll go ahead and use the existing internal loads and perform the analysis from that. The complete wing FEM is shown in *Figure 15-8* while a close-up of the FEM at the analysis location is shown in *Figure 15-9* that includes a superposition of the geometry.

For comparison purposes, I have done the said FEM change and re-ran. The results show changes in the internal loads of up to ±6% between having the web (where the cutout is) modeled with a *CQUAD* and one modeled with a *CSHEAR*. Therefore, it is good to keep that in mind when calculating the MS for the structure surrounding the cutout.

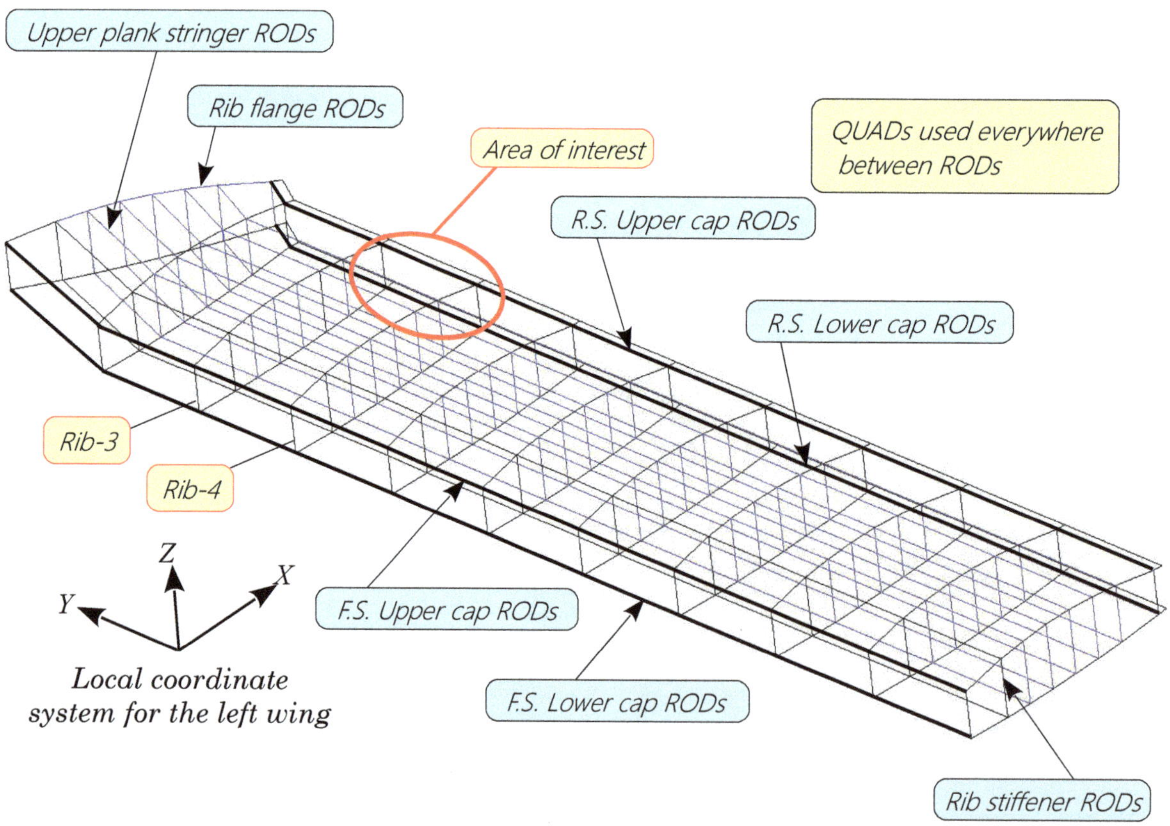

Figure 15-8 *– Complete Left wing global FEM and area of interest.*

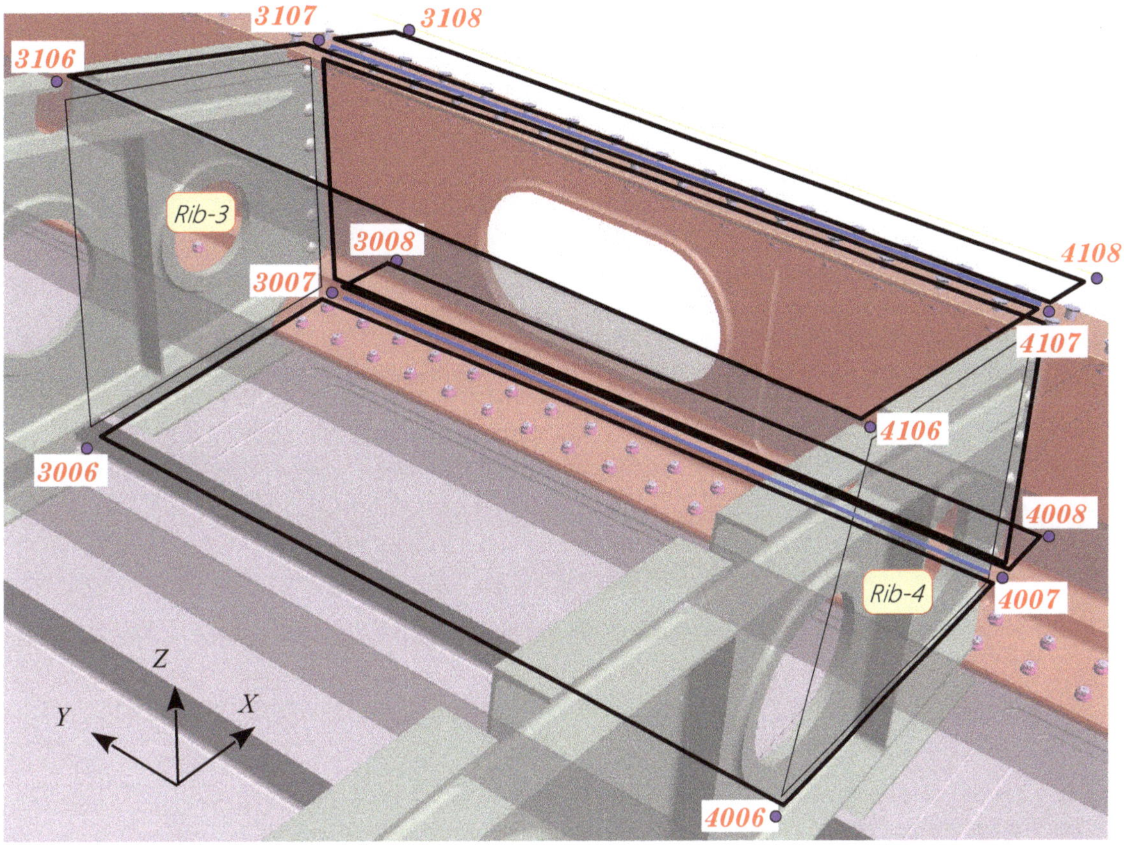

Figure 15-9 *– Superimposed geometry and FEM at R/S Cutout location.*

It is very important to superimpose the *3D* geometry onto the FEM to ensure that: ***1)*** the FEM matches the geometry in space and, ***2)*** the loads to be extracted from the *FEM* are at the correct location where the analysis is performed.

In my case, I drew the FEM by hand on top of the geometry since I couldn't perform the superposition with the software's I'm using. Try not to do like me and use the proper tools...

*Figure **15-10*** shows a close-up of the FEM at the area of interest along with grids and elements numbers. The *QUADS* are in black and the *RODS* are in blue. The green element numbers are for the *QUADS* representing the R/S web.

Note on the GFEM that I made:

The first iteration of the FEM that I did ran very well and the results seemed good. As I heard a couple of times: FEM always provides the correct answer, right?... Nothing could be further from the truth. Here's a good example.

As I started extracting loads for this example, I noticed something bizarre near the root of the wing that didn't make sense. There were very high loads in some RODS that represented the Upper flanges of some ribs.

I was puzzled since there were no redundant nodes, the loads were applied correctly, the boundary conditions were correct. After some time, I realized that the problem was simply inherent to the FEM construction. The elements making up the ribs were RODS for the flanges and vertical stiffeners and QUADS for the webs. Pretty standard you'll say but here's the thing. The problem was simply caused by the fact that I had put RODS only where actual vertical stiffeners existed and not at the other locations where the flange RODS had joints!

Since RODS (unlike BARS) have no end-moments, they act like linkages. Therefore, they need to be stabilized by other RODS (or BARS) in all planes. I then solved the problem by adding small RODS between each QUADS of the rib web.

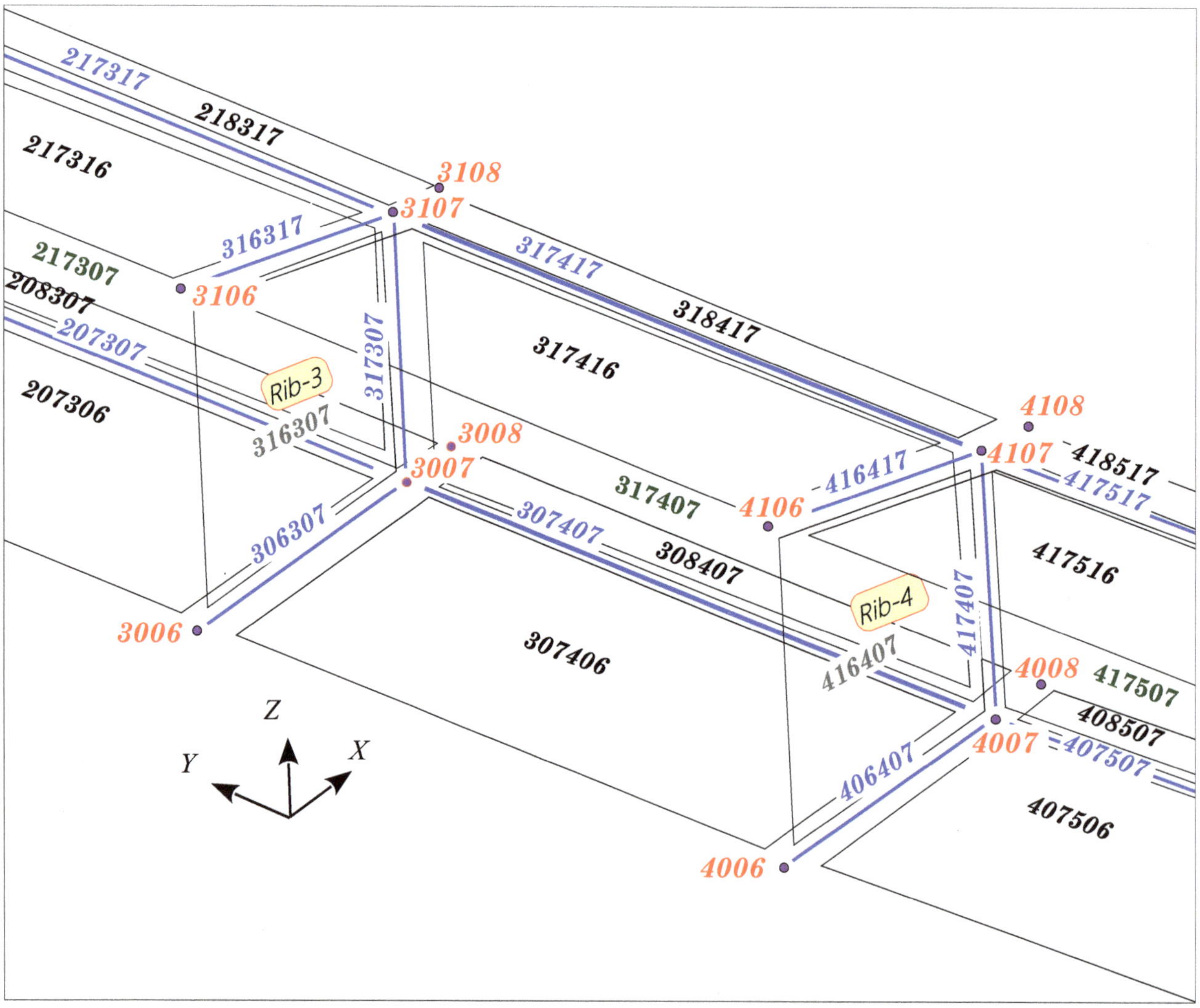

***Figure 15-10** – Close-up of the FEM with element numbers at the area of interest.*

15.3 SUMMARY OF MARGINS OF SAFETY

Table 15-1 shows the summary of minimum Margins of Safety for all parts analyzed in the Rear Spar modification.

Table 15-1 *- Rear Spar modification Summary of minimum Margins of Safety.*

Nomenclature	Description	Material	Mode of failure	Load case	MS	Page
Rear Spar	Upper cap	Al Aly 7050-T7451	Column buckling	2001	1.01	574
Rear Spar	Edge of cutout	Al Aly 7050-T7451	Axial + bending + shear	2001	0.01	588
Wing plank	Upper cap	Al Aly 7075-T7351	Inter-rivet buckling	2001	0.46	589
Rear Spar	Edge of cutout	Al Aly 7050-T7451	Net tension	2001	-0.03	593
Wing box Assembly	Connections at upper plank-R/S	HST11-6 Pin	Fastener Shear	2001	2.74	594

The margins of safety presented above should **NOT** be presented in a stress report since there is a negative MS and since the design hasn't been optimized.

This shows that the edge of the newly added cutout should be reinforced by increasing the web thickness and by increasing the width of the lip at its periphery in order to make a return flange.

"The computer says so, therefore it has to be the correct answer...."

15.4 LOADS

For this example, without getting into the details on how the external loads for this critical *Take-Off* load case (*LC 2001*) were applied to my fictitious wing design (as it is beyond the scope of this book), I'll simply say that I calculated some simple and rough linearly distributed load applied at the FEM *Grids* that connect the Front and Rear Spars to the upper plank. The idea was to get the FEM to provide some realistic internal loads to work with in our example. See ***Figure 15-11*** for the loaded FEM with the boundary conditions.

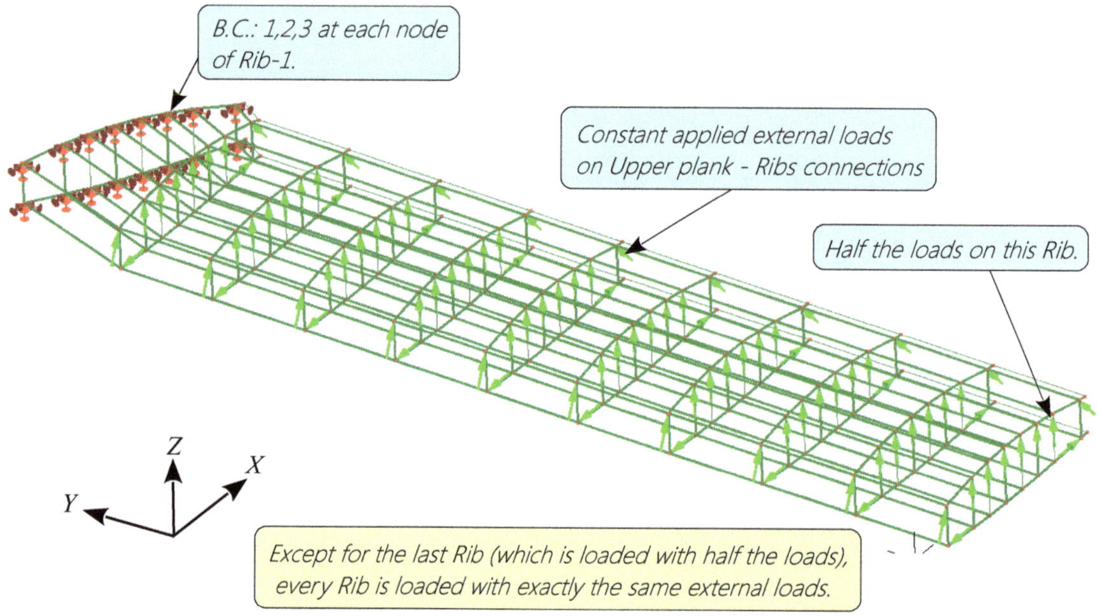

Figure 15-11 *– External loads and B.C. applied to the FEM, LC2001.*

To create the fictitious load case (*Take Off*; ***LC2001***) to load up the model, I simply considered a maximum lift coefficient for the wing profile of C_L = 1.5 and a speed at *Take-off* of 160 *knts*. The distributed load onto the wing was then multiplied by 1.5 to get the ultimate load condition. Note that this is just an example and that the most critical load case could be somewhere else in the aircraft flight envelope.

The actual applied external loads are not important here since that, as stress-engineers, we're mostly concerned about the extraction of the internal loads from the FEM results in order to perform our analysis. So, the applied loads are not shown.

What I am concerned about is that, due to the new large cutout, the internal loads that were flowing through the material that has been removed for the cutout will have to go somewhere else. That is why that, ideally, the element representing the Rear Spar web at the location of the cutout should be changed from a *QUAD* to a *SHEAR element* in the FEM as this would take care of the axial load re-distribution.

15.4.1 FEM VERIFICATION

But before starting extracting loads for the first time, I strongly suggest that you first verify the FEM to make sure it is correct! The creation of a FEM is beyond the goal of this book. However, unless you calculate your own loads from the emergency load cases and the like (*9g Fwd*, etc), the loads you'll be using will come from the Global FEM output. Therefore, it is essential to be able to understand the FEM output files in order to perform the analysis of any primary structure component.

Taking *MSC Nastran™* as an example for the solver — a very popular *FEA* package in aerospace — , generally, the output file extensions will be either an "***.op2***" or an "***.xdb***" file. Other formats are also available depending on the type of run performed. For the sizing phase, the FEM run will be a linear run static solution *SOL 101*. Those files are huge! Especially if a *GPFB* (Grid Point Force Balance) was requested. We usually can't import the whole results file data into a spreadsheet, we have to choose only the loads we need where the part to be analyzed is located.

Thus, for those few people that will be working directly on the global FEM, there are other books available (like ref. 17) and, also, the company for which you work for has most likely very detailed instructions for that.

Here's a non exhaustive list of things to check from the FEM for the other 95%+ of us stress engineers that are more FEM users:

1. Look at the deformed shape in the Post-processor to see if it makes sense for a typical load case like pressurization or take-off.
2. No gaps exist or coincident nodes. Note that coincident nodes may exist in some cases but generally, an equivalence should have been made.
3. Type of elements (*QUADS, RODS, BARS*, etc.) along with their respective material properties (material, thickness, section properties, orientation/normal, end fixity, etc).
4. The offset for the element's centroid is about right or as close as possible from what you will calculate. This is for *QUADS* and *BARS* as you can't offset *RODS*.

The information for the elements (material, section properties, etc) are normally all made available in a spreadsheet file by a stress analyst at the beginning of the program (*loop 0*) and given to the FEA group. That file is kept updated all along the duration of the program and provided for every run *loop* as the design evolves. For the certification, that information must match exactly the actual optimized structure geometry and this normally occurs for the final (certification) *loop 3* run. The stress reports will be based on those certification loads.

If you've been involved in this process from the beginning of the program then you know exactly how the FEM is. However, when you join the team somewhere in the middle of a program, you have no idea whether the FEM is correct or not. So get familiarized with it so you can trust the loads you'll be using.

There is always something to change whether it's the section properties of a *BAR* element or its offset, or the release of one end (that is for instance when a *BAR* is considered pinned instead of fixed) or, due to a negative MS that triggered the increase of a thickness for example. In which case, the FEA group/person needs to be notified through a "*FEM Change Request*" document.

This exercise is crucial as, not only the loads will be more accurate for your analysis but also —and mainly— will better represent the tested structure as far as strains and stiffness behavior. This will provide you, and the certification authority representative, a very good confidence in the model.

It is important to note that the section properties of the elements used in the FEM are fixed for all the thousands of runs (unless specific models are used for specific load cases). Therefore, if one analyzes a wing Spar, as shown in this example, one may want to provide the section properties for a member in compression for the upper cap and provide the section properties for a member in tension for the lower cap as it is the most common loading configuration and most likely the one which will be static tested at ultimate loads. Some aerospace companies may want to be more conservative (or for other reasons) by using the section properties for compression members for both upper and lower Spar caps.

One way to mitigate the fact that thin structures in compression have a smaller cross-section due to buckling than the ones in tension can be done by using *SHEAR* elements panels instead of *QUADS* for the skin and webs, therefore, assuming the panel to be buckled already from any compression loads.

Once you have a "warm and fuzzy feeling" (you're happy) with the FEM as the elements properties are good and the loads distribution makes sense, internal loads are then extracted for the thousands of load cases.

During my career, it appeared that the more powerful the computers became, the more load cases were run! Thus, it almost takes as much time as before to get the critical *MS*...

From my experience, analyses are performed in a spreadsheet using already-made templates with macros using popular software. Then, the macro is run for the thousands of load cases. Only the analysis for the most critical load case and having the lowest MS is detailed in the stress report.

In this fictitious example, I have asked the designer (me) to increase the thickness of the web around the cutout and add a lip, and also, move the two vertical small stiffeners to frame the cutout. This is a machined Spar so it is easy to add material; before manufacturing of the Spar has started that is!

I could have increased the Spar's upper and lower flanges as well but decided not to. Note that this design example is NOT optimized, otherwise, it would make sense to have thicker flanges at the cutout and thinner flanges everywhere else.

In the case that this kind of change would occur after a couple of Spars had been manufactured, those parts could possibly be retro-fitted by adding a Doubler for the web and Angle-doublers for the caps. It wouldn't be pretty but it would prevent scrapping the Spars.

15.4.2 INTERNAL LOADS EXTRACTION

There are three methods to extract internal loads from the results output file. Which method to use depends on the availability of certain results, how much time you have to do the job and, how accurate you want it. From the three, the preferred method being the first one and the least preferred is the last one. They are listed below:

1. *GPFB* (Grid Point Force Balance). This is the most accurate method and is obtained directly from the *FEA* run output file. It should always be used whenever possible. That results file is very large and requires a lot more time to run. Therefore, this option may not be available if time was a constraint for the *FEA* group.
2. *Element Forces*. This is a good method and may provide results that are pretty close to the *GPFB* in some cases. Since Element Forces are always available and require less time to run, the stress analyst may not have any other choice but to use this one.
3. *Axial load*. This method is the quickest and simplest as only the axial load "*P*" in the *ROD* or *BAR* is used. This method may provide a quick and rough avenue in order to quickly evaluate the axial stress "f_a", in a spar cap for example, by simply calculating $f_a = P/A$. This method is used sometimes to determine whether a cap is in tension or in compression. I don't recommend using that method to calculate margins of safety in a stress report.

For the analysis of the Rear Spar in this example, only the GPFB method will be shown.

The Grid Point Force Balance approach:

Figure 15-12 shows the FEM Grids and elements from where the *GPFB* are performed.

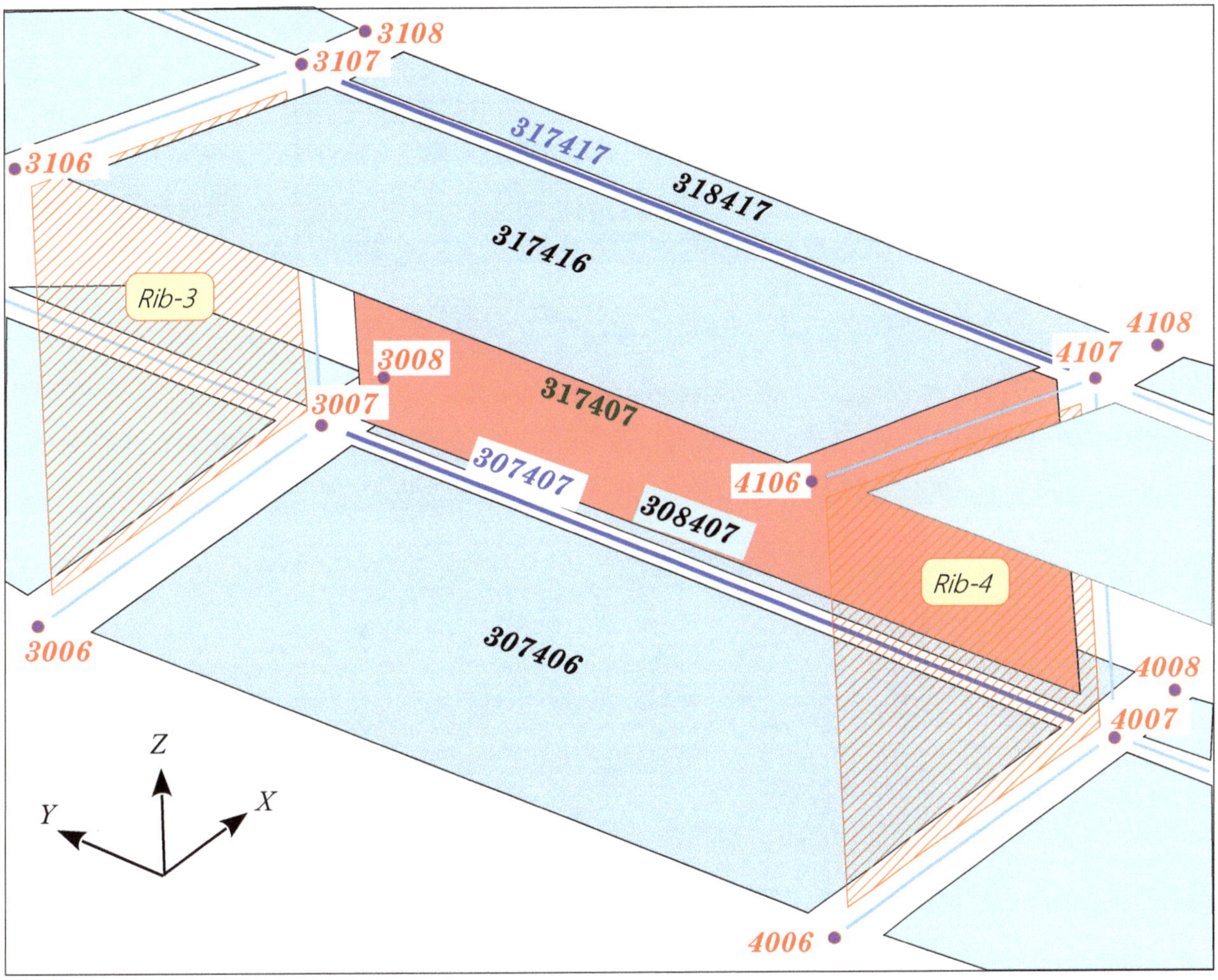

***Figure 15-12** – FEM Grids and elements used for the GPFB.*

What loads should I be extracting when looking at one particular Grid? In our example, since I'm concerned about the location of the R/S having the cutout, I will extract the loads "outboard" of Rib-3 (to the right on the figure) and "inboard" of Rib-4 (to the left on the figure). However, when analyzing the whole Spar, both sides "inboard" and "outboard" of any particular grid must be extracted and a MS calculated for both.

Figure 15-13, Figure 15-14 and *Figure 15-15* show the elements to use for extracting the internal loads along with the elements orientations and positive shear flow directions for the upper plank, Rear Spar, and lower plank respectively.

From those figures, one can see that I purposely showed the red section cuts slightly off from the grids. This is to emphasize the fact that I am extracting the loads from the elements just outside the grids and not from ALL the elements that connect to those grids. Hence, the elements that make-up the ribs for instance are not picked-up.

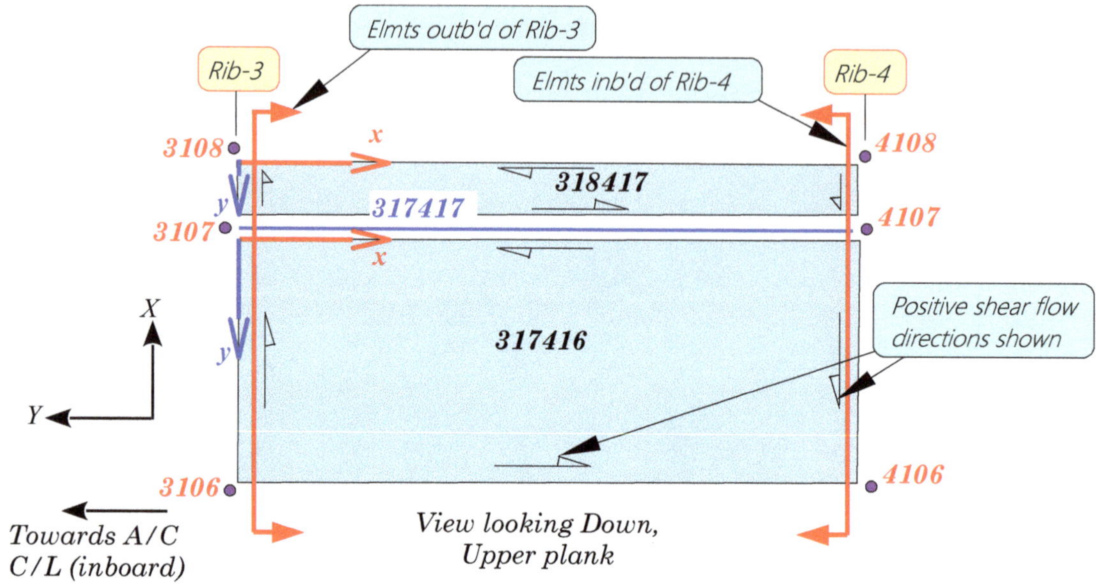

Figure 15-13 – Upper plank elements to be used with orientations.

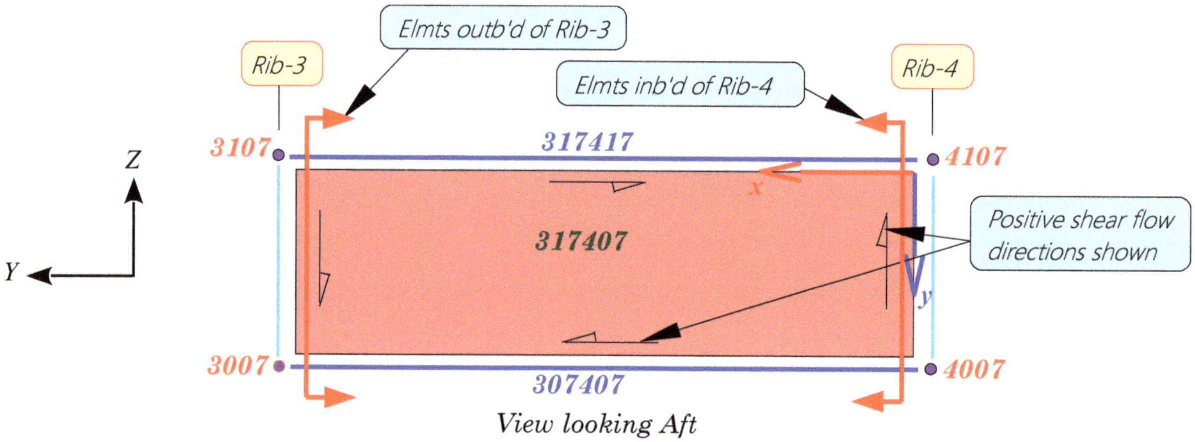

Figure 15-14 – Rear Spar elements to be used with orientations.

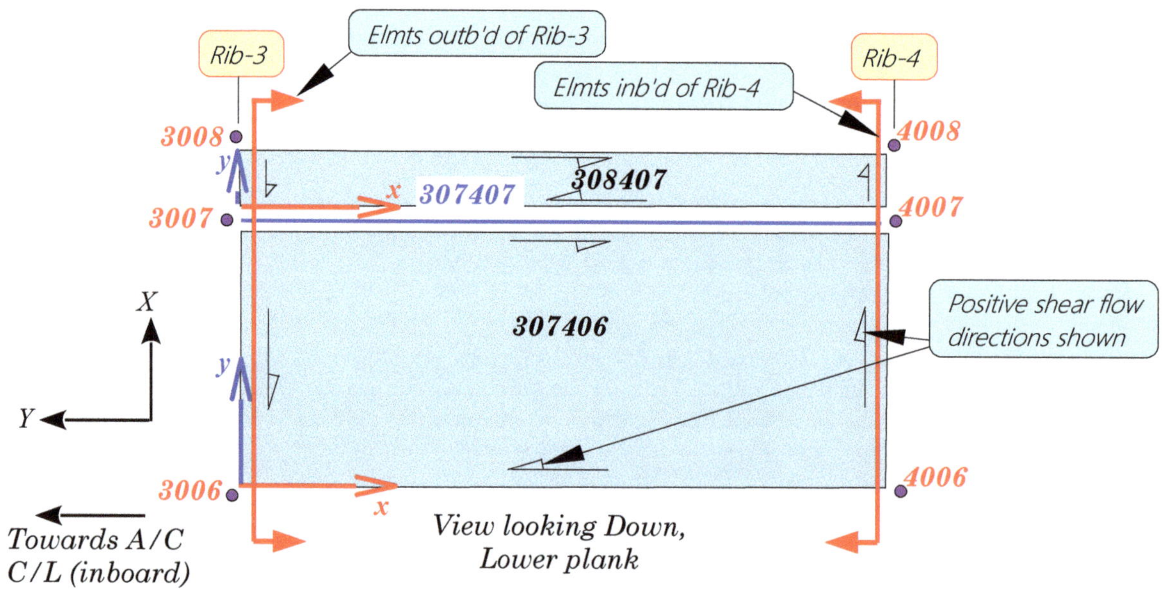

Figure 15-15 – Lower plank elements to be used with orientations.

From the FEM results file (*.f06* for Nastran), the GPFB loads for one load case from each element common to the analysis Grids are extracted as shown in *Table 15-2* to *Table 15-5*. Obviously, this load extraction would have to be done for the thousands of load cases. Like in every analysis, the *FBD* is essential. One must be able to determine the internal loads anywhere in a part by doing a *FBD*.

In the following tables, the rows with the pale background are the ones used to calculate the sum of the internal loads. The upper and lower caps are calculated separately. For the loads at Rib-3, the internal outboard loads at *GRIDS 3017 & 3018* are extracted for the upper cap as shown in *Table 15-2* and, at *GRIDS 3007 & 3008* for the lower cap as shown in *Table 15-3*. While at Rib-4, the internal inboard loads at *GRIDS 4017 & 4018* are extracted for the upper cap as shown in *Table 15-4* and, at *GRIDS 4007 & 4008* for the lower cap as shown in *Table 15-5*. Note that the loads are reactions loads.

Table 15-2 *- GPFB of R/S at Rib-3 Upper cap; LC2001.*

Rib 3	*Upper cap*	*GPFB*	*TX*	*TY*	*TZ*	*RX*	*RY*	*RZ*
Grid	*Element no.*	*Elmt type*	*[lbs]*	*[lbs]*	*[lbs]*	*[in-lbs]*	*[in-lbs]*	*[in-lbs]*
3107		APP-LOAD	-546	0	694	0	0	0
3106	317416	QUAD4	-1448	24213	409	698	34	-142
3107	217307	QUAD4	2	20112	-9831	265	6	-74
3107	217316	QUAD4	7005	-25713	-1509	-1379	-539	331
3107	218317	QUAD4	-3509	-7453	240	-357	331	32
3107	316307	QUAD4	2725	51	3783	175	-8	-15
3107	317407	QUAD4	-1	15797	5117	236	-33	-79
3107	317416	QUAD4	-7321	1380	1461	952	386	-183
3107	318417	QUAD4	2657	3067	-166	108	-226	-12
3107	217317	ROD	0	-15439	0	0	155	0
3107	316317	ROD	-1012	0	218	0	0	0
3107	317307	ROD	0	0	-7	0	0	0
3107	317417	ROD	0	8198	0	0	-72	0
3108	218317	QUAD4	2864	-675	-181	-165	-122	16
3108	318417	QUAD4	-2864	675	181	165	122	-16
		TOTALS	0	0	0	0	0	0
		Total Inboard	***8075***	***-29117***	***-7288***	***-1461***	***-176***	***291***
		Total Outboard	***-7529***	***29117***	***6594***	***1461***	***176***	***-291***

Table 15-3 *- GPFB of R/S at Rib-3 Lower cap; LC2001.*

Rib 3	*Lower cap*	*GPFB*	*TX*	*TY*	*TZ*	*RX*	*RY*	*RZ*
Grid	*Element no.*	*Elmt type*	*[lbs]*	*[lbs]*	*[lbs]*	*[in-lbs]*	*[in-lbs]*	*[in-lbs]*
3007		APP-LOAD	546	0	0	0	0	0
3006	307406	QUAD4	1124	-22837	108	902	41	40
3007	207306	QUAD4	-7691	23249	-352	-1015	-534	-91
3007	208307	QUAD4	4718	9175	300	-450	407	-51
3007	217307	QUAD4	-1	-18432	-9568	193	32	59
3007	307406	QUAD4	6581	-3444	254	794	396	5
3007	308407	QUAD4	-3548	-3954	-208	139	-282	0
3007	316307	QUAD4	-1589	-49	3247	174	3	5
3007	317407	QUAD4	0	-17358	6271	165	-23	73
3007	207307	ROD	0	24885	0	0	0	0
3007	306307	ROD	985	0	48	0	0	0
3007	307407	ROD	0	-14071	0	0	0	0
3007	317307	ROD	0	0	7	0	0	0
3008	208307	QUAD4	-3867	205	-230	-154	-179	-14
3008	308407	QUAD4	3867	-205	230	154	179	14
		TOTALS	0	0	0	0	0	0
		Total Inboard	***-7447***	***39032***	***-6548***	***-1252***	***-270***	***-92***
		Total Outboard	***6901***	***-39032***	***6548***	***1252***	***270***	***92***

Table 15-4 - *GPFB of R/S at Rib-4 Upper cap; LC2001.*

Rib 4	Upper cap	GPFB	TX	TY	TZ	RX	RY	RZ
Grid	Element no.	Elmt type	[lbs]	[lbs]	[lbs]	[in-lbs]	[in-lbs]	[in-lbs]
4107		APP-LOAD	-546	0	694	0	0	0
4107	317407	QUAD4	3	12274	-6132	143	19	-79
4106	317416	QUAD4	4766	-3360	-1114	-910	-265	211
4107	317416	QUAD4	4003	-22233	-756	-432	-219	130
4107	318417	QUAD4	909	-836	-64	-134	-74	11
4107	416407	QUAD4	166	-1	914	-12	-1	-6
4107	417507	QUAD4	-7	12563	4437	123	7	-36
4107	417516	QUAD4	-3551	50	744	168	208	-4
4107	418517	QUAD4	-477	387	26	145	9	-16
4107	317417	ROD	0	-8198	0	0	72	0
4107	416417	ROD	-500	0	108	0	0	0
4107	417407	ROD	0	0	30	0	0	0
4107	417517	ROD	0	5995	0	0	-21	0
4108	318417	QUAD4	-702	-2906	50	-65	16	6
4108	418517	QUAD4	702	2906	-50	65	-16	-6
		TOTALS	0	0	0	0	0	0
		Inboard w/o rib	***4212***	***-21900***	***-6902***	***-488***	***-187***	***68***
		Total Inboard	*3879*	*-21901*	*-5851*	*-501*	*-188*	*62*
		Total Outboard	*-3333*	*21901*	*5157*	*501*	*188*	*-62*

Table 15-5 - *GPFB of R/S at Rib-4 Lower cap; LC2001.*

Rib 4	Lower cap	GPFB	TX	TY	TZ	RX	RY	RZ
Grid	Element no.	Elmt type	[lbs]	[lbs]	[lbs]	[in-lbs]	[in-lbs]	[in-lbs]
4007		APP-LOAD	546	0	0	0	0	0
4006	307406	QUAD4	-4101	4625	-258	-732	-232	-43
4007	307406	QUAD4	-3603	21657	-104	-701	-175	-75
4007	308407	QUAD4	-1448	428	-100	-132	-109	-16
4007	317407	QUAD4	-2	-10712	-5257	163	27	72
4007	407506	QUAD4	3095	-1651	125	387	212	-20
4007	408507	QUAD4	820	-223	49	150	27	8
4007	416407	QUAD4	84	3	640	-11	5	0
4007	417507	QUAD4	6	-13508	4653	143	12	31
4007	307407	ROD	0	14071	0	0	0	0
4007	406407	ROD	502	0	25	0	0	0
4007	407507	ROD	0	-10063	0	0	0	0
4007	417407	ROD	0	0	-30	0	0	0
4008	308407	QUAD4	1128	3731	77	-98	39	-8
4008	408507	QUAD4	-1128	-3731	-77	98	-39	8
		TOTALS	0	0	0	0	0	0
		Inboard w/o rib	***-3924***	***29174***	***-5384***	***-767***	***-218***	***-28***
		Total Inboard	*-3338*	*29177*	*-4750*	*-778*	*-212*	*-28*
		Total Outboard	*2792*	*-29177*	*4750*	*778*	*212*	*28*

Table 15-6 shows a summary of the resulting GPFB of the Rear Spar at Rib-3 and Rib-4 upper and lower caps. The loads are reaction loads at the sections showed in *Figure 15-13*, *Figure 15-14* and *Figure 15-15* and are referenced to the wing coordinate system and their signs have nothing to do with whether they are tension or compression loads.

Note that the moments are simply the sum of the moments at the grids and not the total reacted moments for the whole section. Those will be calculated later.

Table 15-6 - Summary of the GPFB of R/S at Ribs 3&4 Upper & Lower caps.

	P_X [lbs]	P_Y [lbs]	P_Z [lbs]	M_X [in-lbs]	M_Y [in-lbs]	M_Z [in-lbs]
Rib-3 Upper cap	-7529	29117	6594	1461	176	-291
Rib-3 Lower cap	6901	-39032	6548	1252	270	92
Rib-4 Upper cap	4212	-21900	-6902	-488	-187	68
Rib-4 Lower cap	-3924	29174	-5384	-767	-218	-28

To perform the "beam" analysis of the R/S, we need the total applied loads at the section's c. of g. The reaction loads obtained above must then be summed-up and transformed accordingly. Since I don't know yet exactly where the centroid of the sections to be analyzed are located, I decided to do the sum of the forces and moments at a point located midway between the upper cap grid *(3107)* and the lower cap grid *(3007)* for Rib-3. The same is done at a point located midway between the upper cap grid *(4107)* and the lower cap grid *(4007)* for Rib-4.

The forces and moments, calculated at a summation point located half way between the upper cap and lower cap grids, could be used as a first analysis pass in order to size the Rear Spar. However, the actual location of the section's centroid has to be used to get the correct values for the internal moments in order to get the real stresses and optimize the design.

The sum of the forces and moments at that summation point is performed below for Rib-3 as an example while the same methodology is used for Rib-4. *Figure 15-16* shows the position of the summation point relative to the *GRIDS* from which the loads are extracted. Refer to *Table 15-2* and *Table 15-3* for the loads.

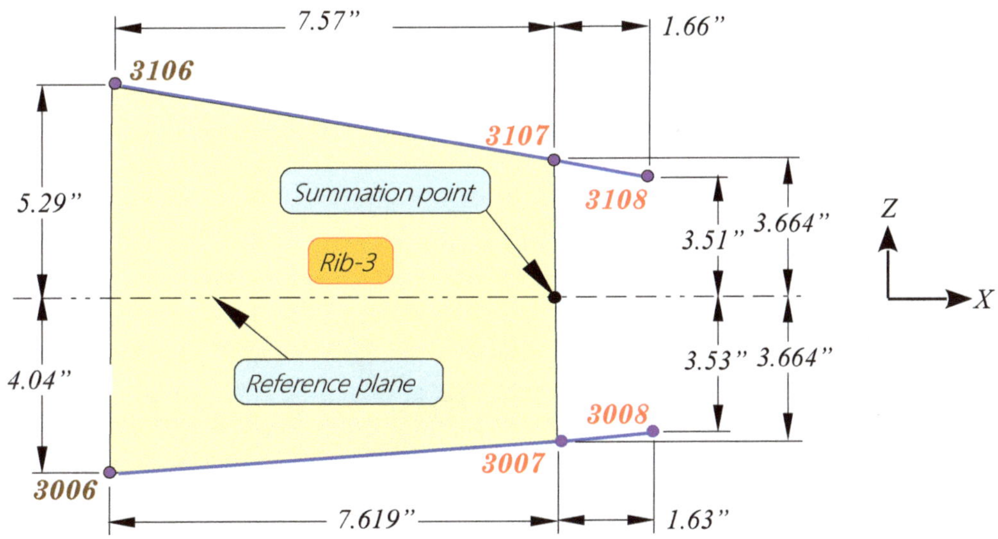

***Figure 15-16** – Position of the summation point relative to GRIDS at Rib-3.*

Forces in "X"

At GRID 3108; $F_X = -2{,}864\ lbs$

At GRID 3107; $F_X = -1 - 7{,}321 + 2{,}657 + 0 = -4{,}665\ lbs$

At GRID 3008; $F_X = 3{,}867\ lbs$

At GRID 3007; $F_X = 6{,}581 - 3{,}548 + 0 + 0 = 3{,}033\ lbs$

$$\sum F_X = -2{,}864 - 4{,}665 + 3{,}867 + 3{,}033 = -628 lbs.$$

Forces in "Y"

At GRID 3108; $F_Y = 675\ lbs$

At GRID 3107; $F_Y = 15{,}797 + 1{,}380 + 3{,}067 + 8{,}198 = 28{,}442\ lbs$

At GRID 3008; $F_Y = -205\ lbs$

At GRID 3007; $F_Y = -3{,}444 - 3{,}954 - 17{,}358 - 14{,}071 = -38{,}827\ lbs$

$$\sum F_Y = 675 + 28{,}442 - 205 - 38{,}827 = -9{,}915\ lbs.$$

Forces in "Z"

At GRID 3108; $F_Z = 181\ lbs$

At GRID 3107; $F_Z = 5{,}117 + 1{,}461 - 166 + 0 = 6{,}412\ lbs$

At GRID 3008; $F_Z = 230\ lbs$

At GRID 3007; $F_Z = 254 - 208 + 6{,}271 + 0 = 6{,}317\ lbs$

$$\sum F_Z = 181 + 6{,}412 + 230 + 6{,}317 = 13{,}141\ lbs.$$

Moments around "X"

At GRID 3108; $M_X = 165\ in\text{-}lb$

At GRID 3107; $M_X = 236 + 952 + 108 + 0 = 1{,}296\ in\text{-}lb$

At GRID 3008; $M_X = 154\ in\text{-}lb$

At GRID 3007; $M_X = 794 + 139 + 165 + 0 = 1{,}098\ in\text{-}lb$

Referring to *Figure 15-17*, the sum of the moments around the "X" axis is calculated as:

$$\sum M_X = 165 + 1{,}296 + 154 + 1{,}098 - 28{,}442(3.664) - 675(3.51) - 205(3.53) - 38{,}827(3.664)$$
$$= -246{,}854\ in.-lb.$$

By reversing the wing coordinate system, the reactions loads are now shown in *Figure 15-17* as applied loads at the section of the Rear Spar just outboard of Rib-3.

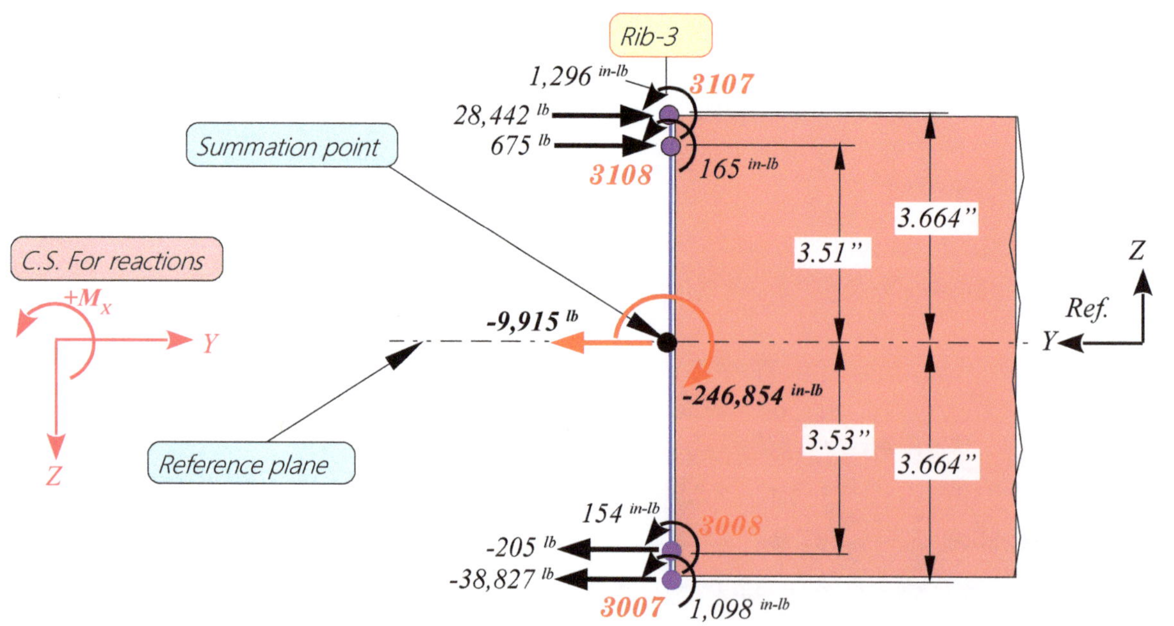

***Figure 15-17** – Sum of the forces and moments at the summation point; YZ plane.*

Moments around "Y"

At GRID 3108; $M_Y = 122$ *in-lb*

At GRID 3107; $M_Y = -33 + 386 - 226 - 72 = 55$ *in-lb*

At GRID 3008; $M_Y = 179$ *in-lb*

At GRID 3007; $M_Y = 396 - 282 - 23 + 0 = 91$ *in-lb*

Referring to *Figure 15-18*, the sum of the moments around the "Y" axis is calculated as:

$$\sum M_Y = 122+55+179+91-4{,}665(3.664)-2{,}864(3.51)-3{,}033(3.664)-3{,}867(3.53) \\ -181(1.66)-230(1.63) = -52{,}142 \ in.-lb.$$

Again, by reversing the wing coordinate system, the reactions loads are shown in *Figure 15-18* as applied loads at the section of the Rear Spar just outboard of Rib-3.

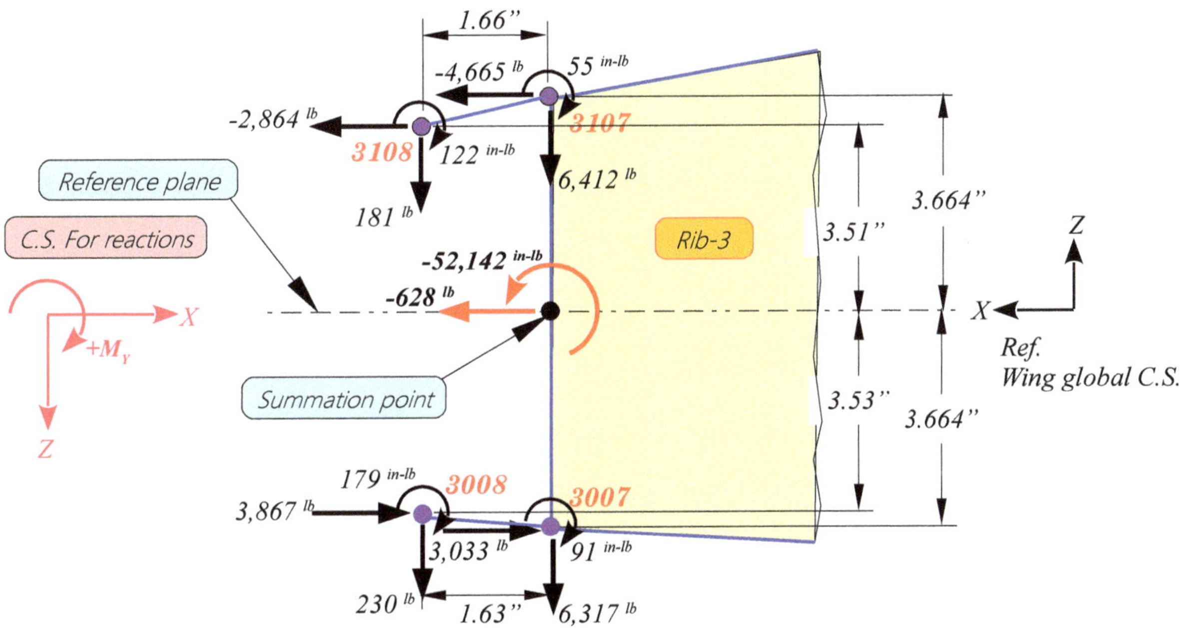

Figure 15-18 *– Sum of the forces and moments around Y at the summation point; XZ plane.*

Moments around "Z"

At GRID 3108; $M_Z = -16$ *in-lb*

At GRID 3107; $M_Z = -79 - 183 - 12 + 0 = -274$ *in-lb*

At GRID 3008; $M_Z = 14$ *in-lb*

At GRID 3007; $M_Z = 5 + 0 + 73 + 0 = 78$ *in-lb*

Referring to *Figure 15-19*, the sum of the moments around the "Z" axis is calculated as:

$$\sum M_Z = -16-274+14+78+675(1.66)-205(1.63) = 589 \ in.-lb.$$

Here also, by reversing the wing coordinate system, the reactions loads are shown in *Figure 15-19* as applied loads at the section of the Rear Spar just outboard of Rib-3.

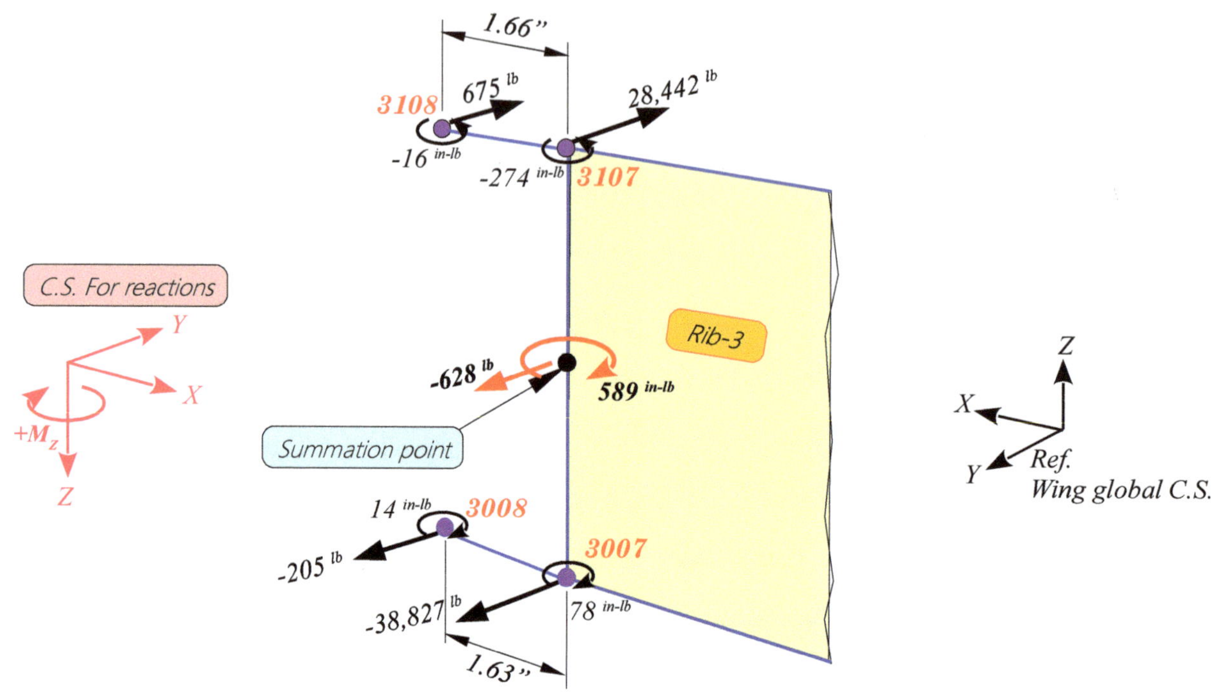

Figure 15-19 *– Sum of the forces and moments at the summation point; XY plane.*

Table 15-7 shows a summary of the sum of the forces and moments at the summation points at Ribs 3 & 4. The signs of the numbers are all changed to represent applied loads in the wing coordinate system. The *FBD* of the portion of the Rear Spar between Rib-3 and Rib-4 using the values from *Table 15-7* is shown in *Figure 15-21*.

Table 15-7 *- Applied forces & moments at the summation points of Ribs 3 & 4.*

	P_X	P_Y	P_Z	M_X	M_Y	M_Z
	[lbs]	[lbs]	[lbs]	[in-lbs]	[in-lbs]	[in-lbs]
Rib-3 @ Ref. Plane	628	9915	-13141	246854	52142	-589
Rib-4 @ Ref. Plane	288	7274	-12286	184932	29459	1299

The above loads correspond to the loads at the *Section A-A* locations, at the reference plane, as shown in *Figure 15-6* but we'll need to know what the loads will be at section *B-B* and section *C-C* for the analysis. A *FBD* for the portion of the R/S Between the Rib-3 and Rib-4 at the locations of interest is done using the sign conventions shown in *Figure 15-20* and the loads from *Table 15-7*. See *Figure 15-21* for a preliminary *FBD* of the *R/S* between Rib-3 and Rib-4.

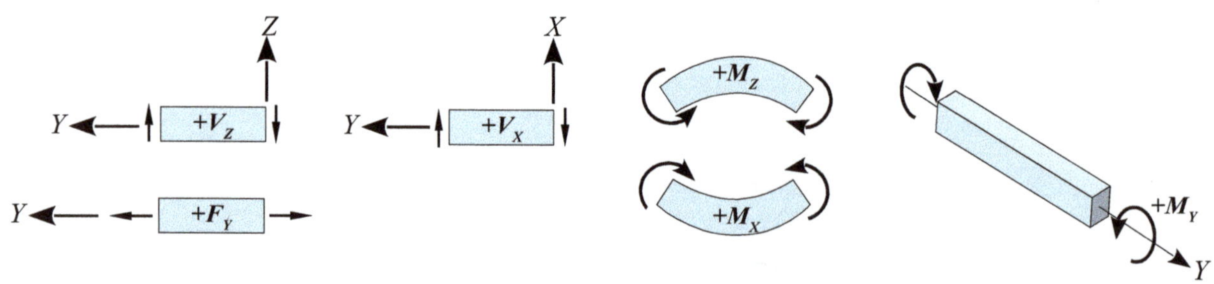

Figure 15-20 *– Sign convention for the applied loads.*

Since I'm analyzing the Rear Spar, I only want the loads at the Grids common to it. So, by extracting those loads, I'm taking into account, not only the loads from the elements making up the R/S but also part of the loads from the upper & lower planks. This is exactly what I want since I'm considering the wing planks to be buckled so that the loads that would normally (in the unbuckled state) got through the planks will have to go to the *R/S* upper and lower caps.

However, you may have noticed something wrong from *Figure 15-21*. **THE FREE BODY DIAGRAM DOES NOT BALANCE!** The reason being that, by not considering ALL the Grids (*Grids* ***3006, 3106, 4006 & 4106*** were excluded) from all the elements used for the loads extraction, some interaction from the *QUADS* representing the top and bottom planks (***QUADS 307406 & 317416***) was left out.

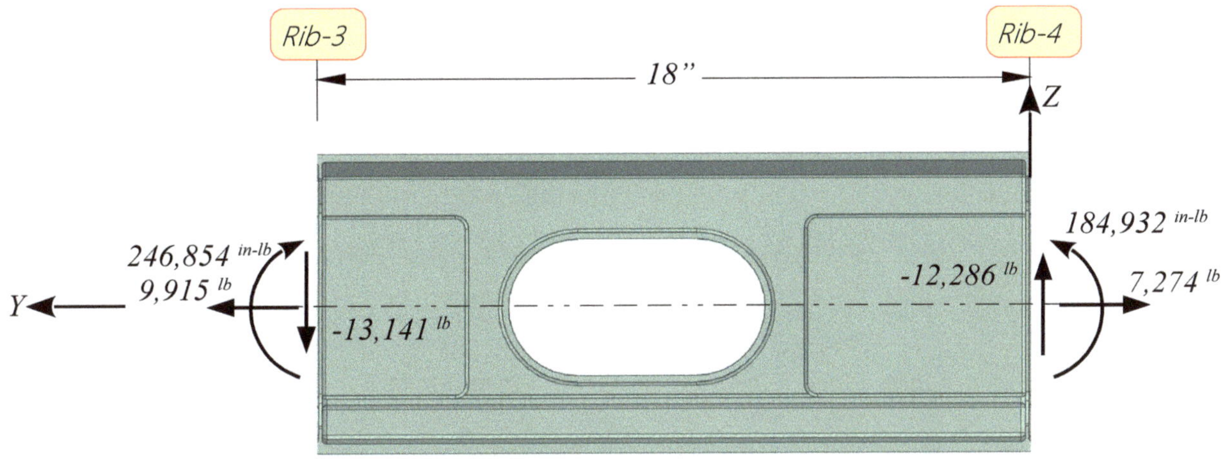

Figure 15-21 *– Preliminary FBD of the R/S Between Rib-3 & Rib-4.*

Although there is nothing wrong with the extracted loads themselves, there is some additional loading that must be taken into account from those two *QUADS*. The Rear Spar is NOT a simple beam! It is part of a box so it must not be considered by itself.

I could simply stop here and use the preliminary FBD (I call it preliminary because it is unbalanced) and I would get the right stresses. I could also interpolate the loads for the analysis of any intermediate sections but that wouldn't be the correct way of doing it. However, in my opinion (and in the opinion of most DERs), all FBD's MUST balance. This is the first thing that somebody verifying a stress report will do. If an FBD doesn't balance, then, chances are that the rest of the analysis is not even worth the cost of the paper and ink...

So let's see what needs to be done in order to get a balanced *FBD*.

To get a balanced FBD, as mentioned earlier, one must consider the contribution of the upper and lower plank *QUADS* ***(317416 & 307406)***. From *Table 15-2* to *Table 15-5*, the GPFB loads at *Grids* ***3006, 3106, 4006 & 4106*** are extracted and are used in the *FBD*. See *Figure 15-22* for a pictorial representation of the upper & lower planks reaction loads to be considered.

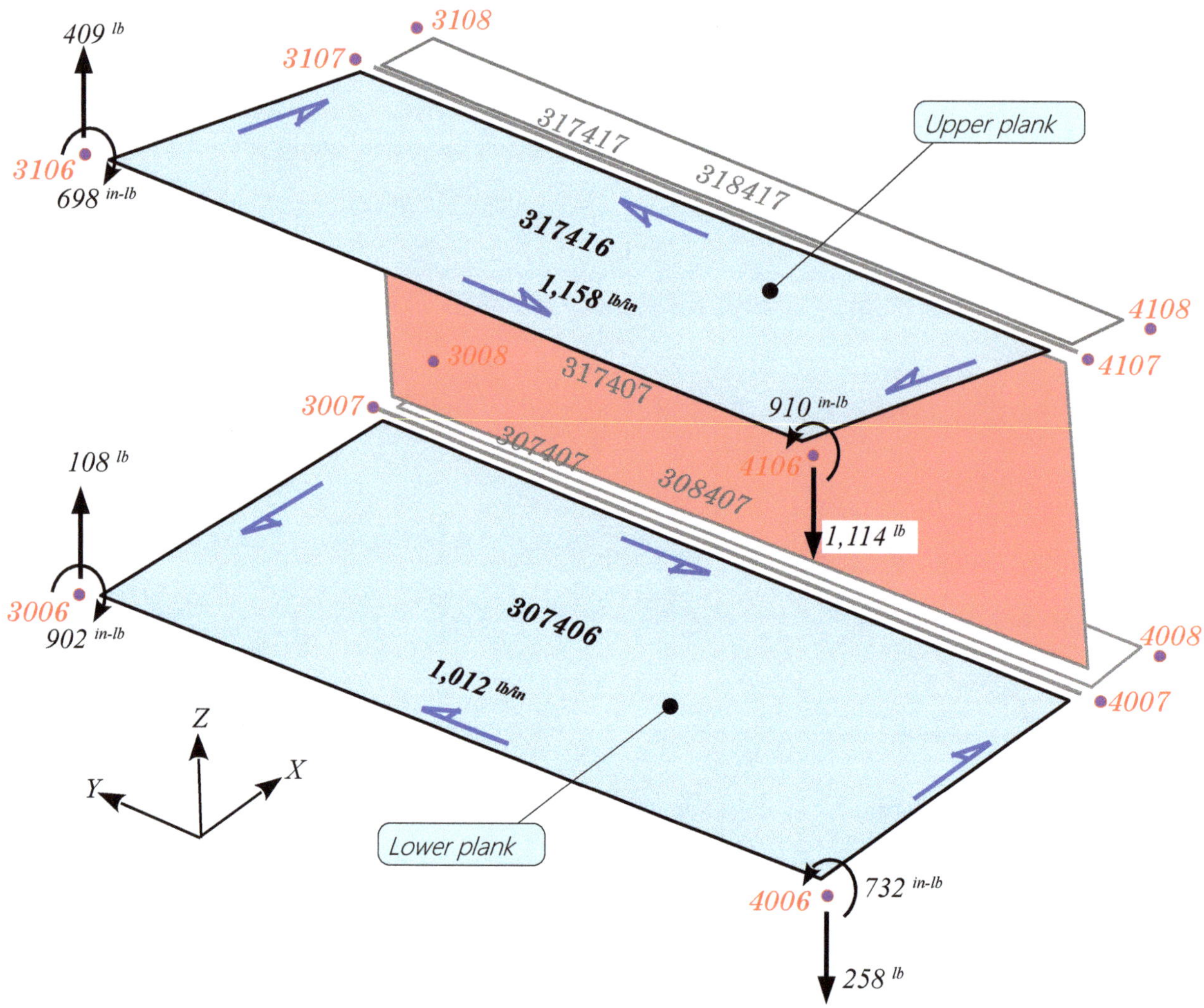

***Figure 15-22** – Additional contribution from the upper and lower plank for a balanced FBD.*

From the FEM output file, the shear flows "F_{XY}" in those two *QUADS* are:

- Element ***307406***; F_{XY} = 1,011.5 *lbs/in.*
- Element ***317416***; F_{XY} = 1,158.23 *lbs/in.*

As shown in ***Figure 15-16***, ***Grids 3106 & 4106*** are located 5.29 *in.* above the reference plane and ***Grids 3006 & 4006*** are located 4.04 *in.* below the reference plane. Therefore, the now balanced ***FBD*** of the R/S can be shown in ***Figure 15-23***. The loads are obtained from the ***GPFB Table 15-2*** to ***Table 15-5.***

Let's do the sum of the forces and moments using the wing coordinate system shown above in ***Figure 15-23*** to ensure the ***FBD*** is indeed balanced.

<u>Forces in "Y"</u>

$$\sum F_Y = 9{,}915 - 7{,}274 - 1{,}158.23(18) + 1{,}011.5(18) = 0 \; lbs.$$

<u>Forces in "Z"</u>

$$\sum F_Z = -13{,}141 - 409 - 108 + 12{,}286 + 1{,}114 + 258 = 0 \; lbs.$$

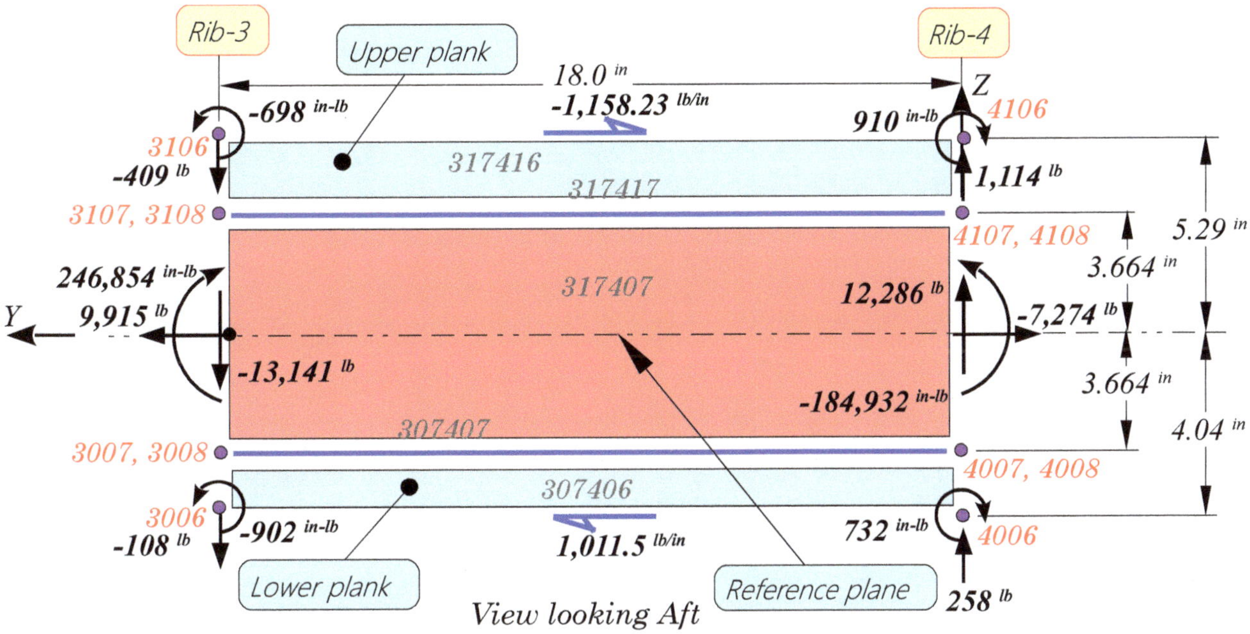

***Figure 15-23** – R/S Rib-3 – Rib-4 balanced FBD.*

<u>Moments around "X"</u>

$$\sum M_X = 246{,}854 - 184{,}932 - 698 - 902 + 910 + 732 + 1{,}158.23(18)(5.29) + 1{,}011.5(18)(4.04)$$
$$-(12{,}286 + 1{,}114 + 258)(18) = -37 \; in. - lb.$$

The results are very close to zero and, due to the round-off errors, are thus considered satisfactory.

Now that we have made the demonstration of a <u>balanced</u> FBD, we can obtain the internal loads at different locations along the elastic axis of the part, especially at the center of the cutout (section *B-B*) and at the inboard edge of the cutout (section *C-C*). Refer to *Figure 15-6* for the locations of the sections. Those FBD's are shown in *Figure 15-24* and *Figure 15-25* respectively. Since there are no external loads applied between Rib-3 and Rib-4, one could have simply obtained the internal loads at those intermediate sections by interpolation.

As explained earlier, those FBD's need the contribution of the upper and lower planks to be balanced. The detailed analysis are not shown for those two sections but it is the same approach as shown above but with lengths of 10 *in.* and 13.25 *in.* instead of 18 *in.* Note that, since there are no grids at the sections, the forces and the moments at the edge of the upper and lower planks have been interpolated.

The internal loads from the upper and lower planks are shown in blue.

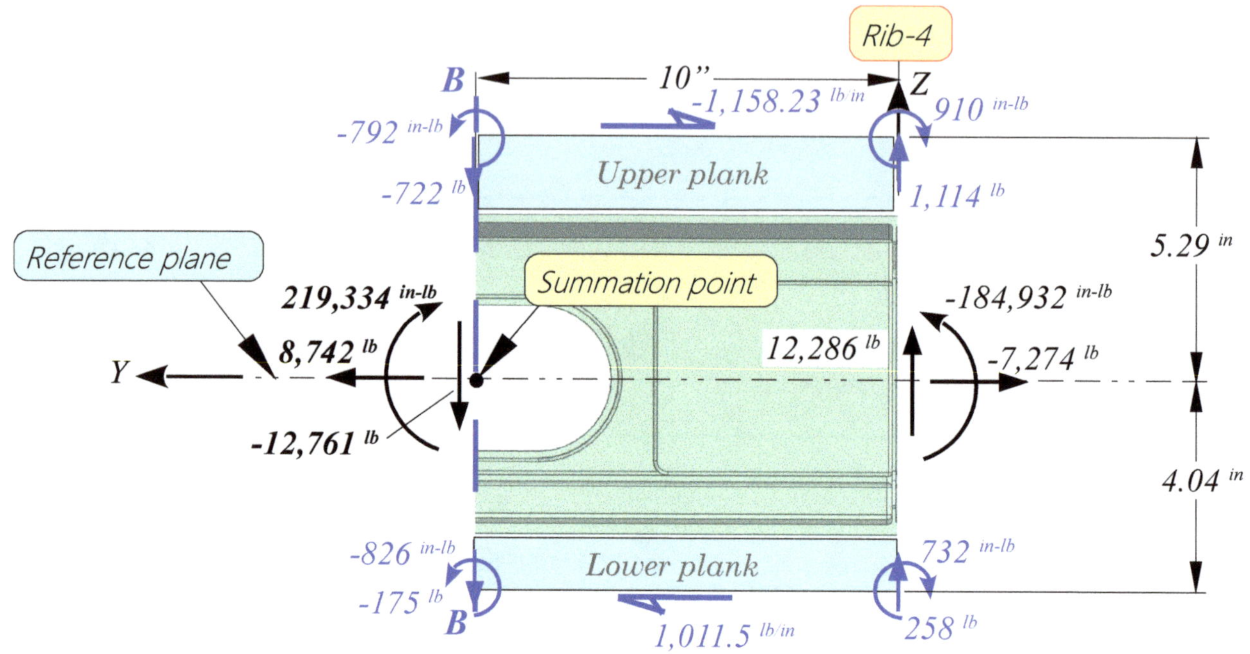

Figure 15-24 – R/S Section B-B - Rib-4 FBD.

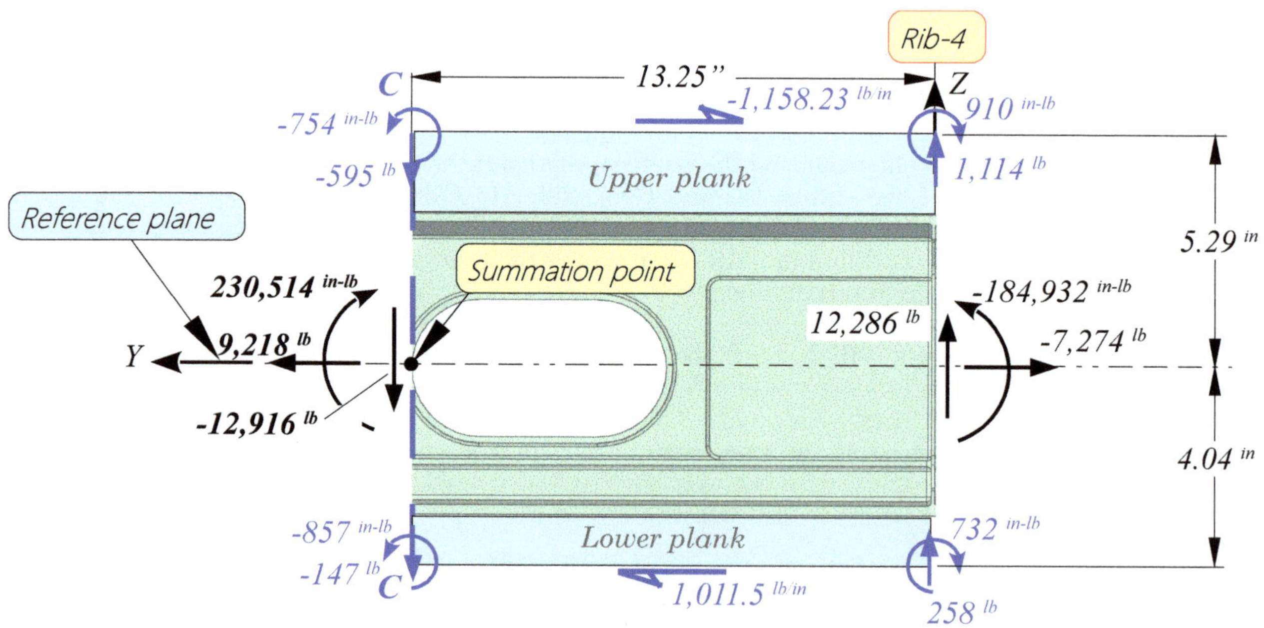

Figure 15-25 – R/S Section C-C - Rib-4 FBD.

Table 15-8 shows a summary of the internal at the different critical sections located at a distance "*y*" from Rib 4 along the reference axis.

Table 15-8 *- Summary of internal loads at the critical sections.*

At section:	*A-A (Rib3)*	*C-C*	*B-B*	*A-A (Rib4)*
y [in] =	*18*	*13.25*	*10*	*0*
Vx [lb] =	628	538	477	288
Py [lb] =	9915	9218	8742	7274
Vz [lb] =	-13141	-12916	-12761	-12286
Mx [in-lb] =	246855	230514	219334	184932
My [in-lb] =	52142	46156	42061	29459
Mz [in-lb] =	-589	-91	250	1299

Now that we have the internal loads at all locations of interest, we can thus start the analysis by first verifying our assumption regarding the buckled panels of the wing planks adjacent to the R/S See section *15.7* for the analysis.

The applied loads at the reference axis will be transformed later to the actual neutral axis of the sections for the stress analysis done in section *15.14*.

I consider the loads extraction to be the most important/critical step in the whole analysis. This is where I've seen most of the mistakes being done. Mistakes of signs, of element numbers and of elements orientation to list a few. To my opinion, it is worth spending the time making sure everything is correct for one particular load case by doing the process by hand before creating a macro that will do it for the thousands of load cases.

15.5 MATERIAL ALLOWABLES

The actual and latest material allowables for all metallic materials can be obtained from the latest edition of (ref. 17). The **Rear Spar**, is machined out of *Al Aly 7050-T7451* plate between 1.00 *in.* and 2.50 *in.* thick. *Table 15-9* shows fictitious values for that material.

Table 15-9 *- Al Aly 7050-T7451 Plate 2.50 in. fictitious material allowables.*

Aluminum Alloy 7050-T7451 Plate
Fictitious Values, NOT FOR DESIGN

	2.50” [ksi]
F_{tu}	75
F_{ty}	65
F_{cy}	60
F_{su}	40
F_{bru} $(e/D = 2.0)$	105
F_{bry} $(e/D = 2.0)$	85
$E, 10^3$	10
$G, 10^3$	3.8
μ	0.33
n	20

The **wing Upper plank**, is machined out of *Al Aly 7075-T7351* plate between 1.00 *in.* and 2.00 *in.* thick. *Table 15-10* shows fictitious values for that material.

Table 15-10 *- Al Aly 7075-T7351 Plate 2.00 in. fictitious material allowables.*

Aluminum Alloy 7075-T7351 Plate
Fictitious Values, NOT FOR DESIGN

	2.00” [ksi]
F_{tu}	70
F_{ty}	60
F_{cy}	55
F_{su}	40
F_{bru} $(e/D = 2.0)$	135
F_{bry} $(e/D = 2.0)$	85
$E, 10^3$	10
$G, 10^3$	3.8
μ	0.33
n	20

The **wing Lower plank**, is machined out of *Al Aly 2024-T351* plate between 1.00 *in.* and 2.00 *in.* thick. *Table 15-11* shows fictitious values for that material.

Table 15-11 *- Al Aly 2024-T351 Plate 2.00 in. fictitious material allowables.*

Aluminum Alloy 2024-T351 Plate
Fictitious Values, NOT FOR DESIGN

	2.00" [ksi]
F_{tu}	61
F_{ty}	41
F_{cy}	37
F_{su}	36
F_{bru} $(e/D = 2.0)$	111
F_{bry} $(e/D = 2.0)$	85
$E, 10^3$	10
$G, 10^3$	3.8
μ	0.33
n	20

15.6 FASTENERS ALLOWABLES

Flush Hi-Lites 3/16" are the only type of fasteners installed in this example. *Table 15-12* shows the ultimate single shear and tension allowables for 6/32" flush head *Hi-Lites HST11-6* Titanium Pins and *HST79-6* Aluminum Collars from (ref. 24).

Table 15-12 *- HST11 & HST79 Hi-Lites ultimate shear and tension allowables.*

	P_{tu} [lbs]	P_{su} [lbs]
Pin HST11-6	2,000	2,690
Collar HST79-6	1,600	N/A

15.7 WING PLANKS BUCKLING

As mentioned earlier, the upper wing plank is expected to buckle under the ultimate compression loads. It is in fact designed to buckle at a much lower load in order to save weight. For our example, the upper wing plank is designed to buckle at *80%* of limit load. This corresponds to about half the ultimate loads *(53%)*.

In this example, I'm only considering one fictitious load case which puts the upper plank into compression. Therefore, I've tailored the thickness of the upper plank pockets to buckle at the desired load level but the lower plank wasn't since that, in this load case, the lower plank only sees tension. Note that the buckling analysis is performed by considering the combined effect of the compression and the shear loads.

This analysis would normally appear in another stress report chapter that covers the wing planks and not in the Rear Spar analysis. However, for our purpose, I'm showing it here.

Upper plank

The upper wing plank is machined out of Aluminum Alloy ***7075-T7351*** plate 2.0 inches thick. The choice of material here is simply based on the fact that it is better to use the ***7000*** Aluminum series material due to its higher compression capability but other materials could do. ***Figure 15-26*** and ***Figure 15-27*** show the dimensions of the upper plank pocket adjacent to the Rear Spar upper cap.

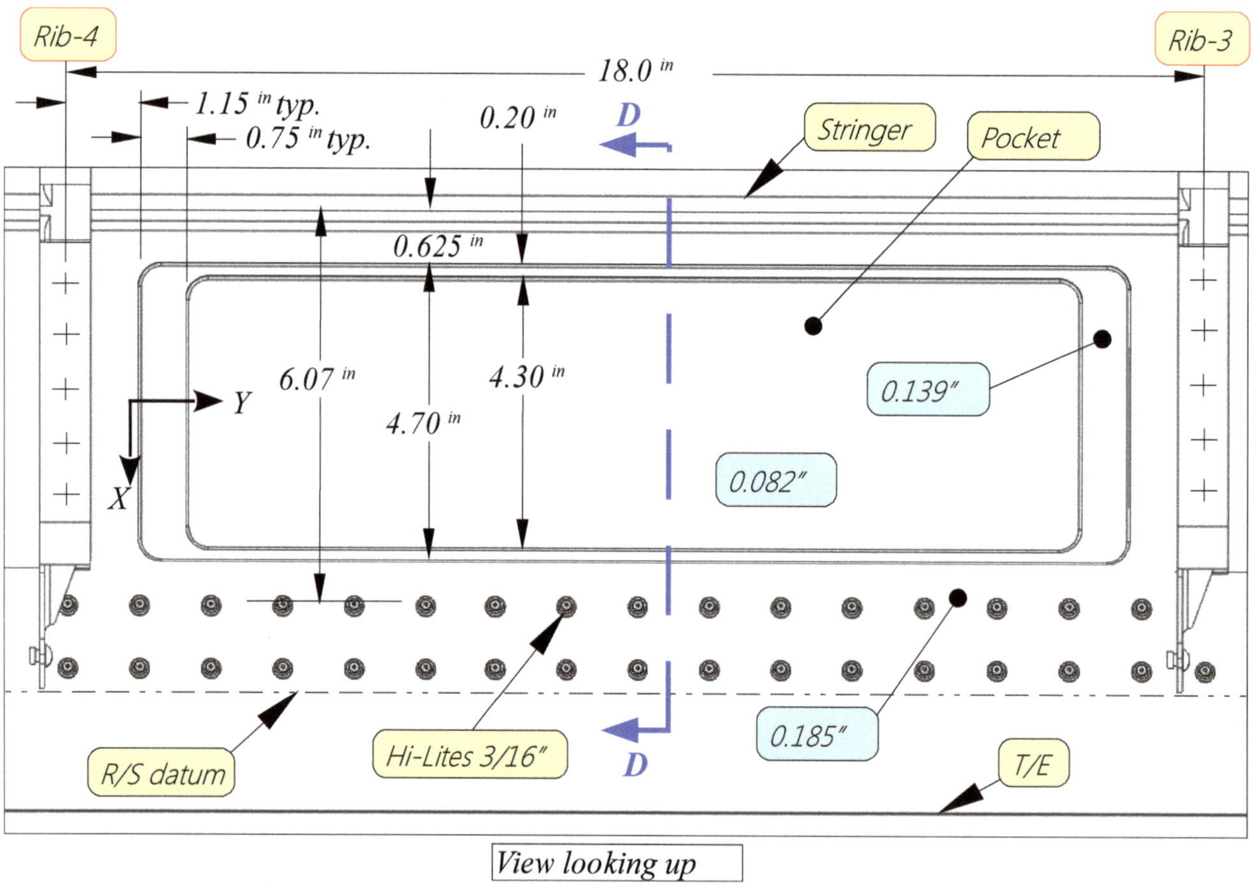

Figure 15-26 *– Upper wing plank pocket dimensions.*

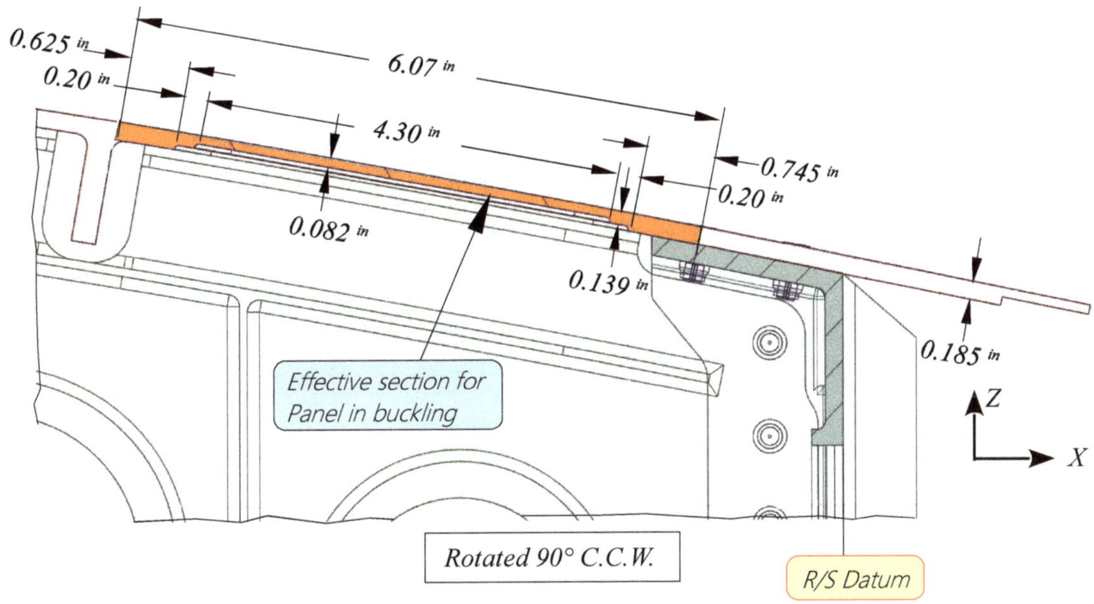

Figure 15-27 *– Section D-D Upper wing plank pocket dimensions.*

For the buckling analysis, since the thickness of the plank pocket isn't constant, one approach is to calculate an equivalent thickness "t_{eq}" that will be assumed uniform through-out the panel. For that, I calculate the total volume of material making the panel and divide it by its surface.

The volume; *Vol* is:

$$Vol = 18\times6.07(0.185) - 15.7\times4.70(0.185-0.139) - 14.2\times4.30(0.139-0.082) = 13.33834 \ in.^3$$

The equivalent thickness to use for the buckling analysis; t_{eq} is thus:

$$t_{eq} = \frac{Vol}{A} = \frac{13.33834}{18\times6.07} = 0.122 in.$$

This equivalent thickness can be different than the equivalent thickness used in the *FEM* for the *QUAD* representing this panel depending on the wing plank portion included in the *RODS* as part of the Rear Spar cap and stringer. See calculations further down.

The iterative procedure is to determine the minimum pocket thickness (0.082 *in.* in our case) by iteration in order to get a MS of 0 in combined compression and shear buckling for a load at 80% of limit load.

The panel is conservatively assumed simply-supported on all sides. As shown in *Figure 15-28*, the edges of the panel are considered at the center of the first row of fasteners and at the edge of the stringer.

<u>**Note:**</u> The study done in section ***15.9*** using a small FEM has demonstrated that it was probably better to consider the smallest pocket thickness as constant through out the panel for the buckling analysis than calculating an equivalent thickness based on the panel volume. It would be interesting here to do the same to see what equivalent thickness should be used.

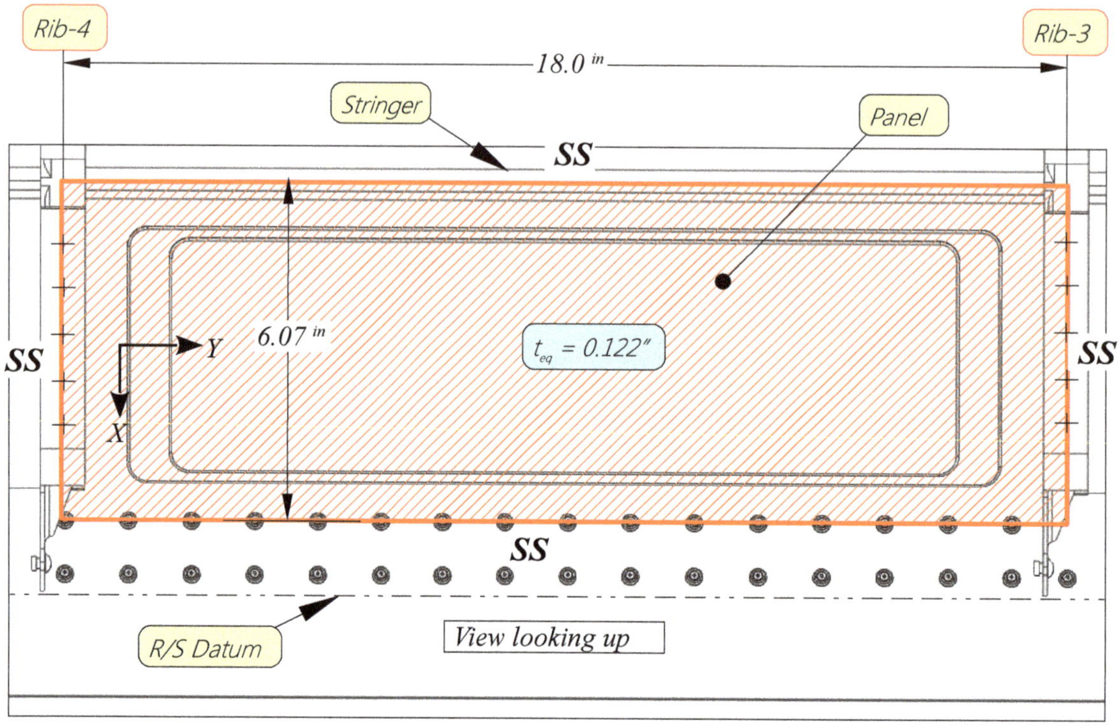

Figure 15-28 *– Upper wing plank panel (A) geometry for buckling analysis.*

From the FEM results file, the internal loads in the two *QUADS* making up the upper *(A)* and lower *(B)* panels are extracted and showed in *Table 15-13*. To evaluate the stress going through the panel, I'm using the element forces and dividing them by the cross section area based on the equivalent thickness that I used for those *QUADS*.

The following calculation is how I determined the equivalent thickness "t_{QUAD}" to assign to the *QUADS* in the FEM representing the upper plank. Refer to *Figure 15-29* for the dimensions of the upper plank. The cross-section is assumed constant between Rib-3 & Rib-4.

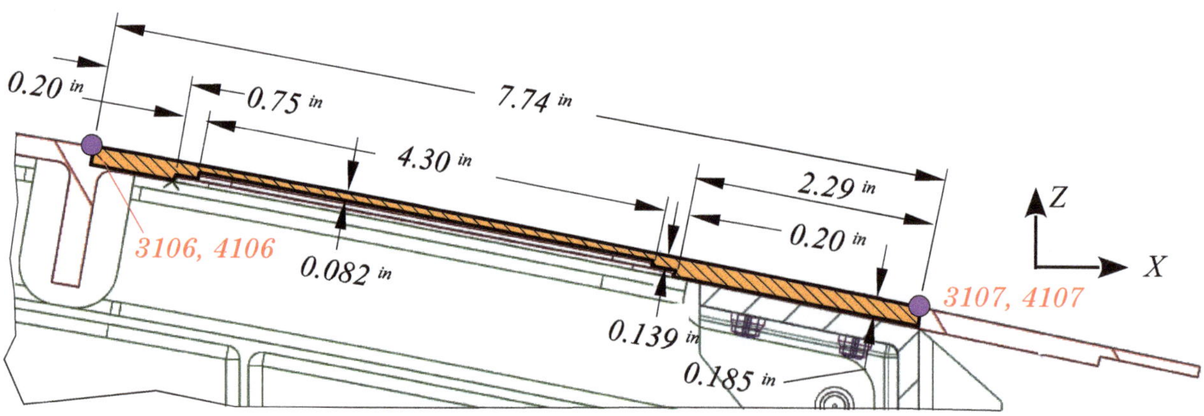

Figure 15-29 *– Section D-D Upper wing plank pocket dimensions used in QUAD 317416.*

$$t_{QUAD} = \frac{7.74\times(0.185) - 4.70(0.185-0.139) - 4.30(0.139-0.082)}{7.74} = 0.125\ in.$$

$$A_{QUAD} = t_{QUAD}\times b = (0.125)\times 7.74 = 0.9706\ in.^2$$

Note that the *QUAD* representing the upper plank was offset in the FEM by $t_{QUAD}/2 = 0.062\ in.$ inward of the wing box (toward the lower plank).

Lower plank

The lower wing plank is machined out of Aluminum Alloy ***2024-T351*** plate 2.0 inches thick. The choice of material here is simply based on the fact that it is better to use the ***2000*** Aluminum series material due to its higher durability in fatigue.

Here, since the load case used introduced tension into the lower plank, the buckling of the lower plank is not critical in compression as only shear can cause some buckling. The minimum pocket thickness was set to 0.080 *in.* and is kept at that value. Since the geometry is pretty much the same except for the width of the panel and its thickness, there is no need to repeat the images and the procedure.

Buckling check of upper and lower planks

See *Table 15-13* for a summary of the FEM ultimate loads and stresses for the upper and the lower panels *QUADS 317416 & 307406* respectively. The acceptable initial buckling loads are the ultimate loads divided by 1.5 then multiplied by 0.80.

Table 15-13 - *FEM ultimate and 80% limit loads and stresses.*

Ult load factor: 1.5

Element	F_X	F_Y	F_{XY}	d_Y	d_X	P_X	P_Y	A_{QUAD}	t_{QUAD}	f_X	f_{XY}	f_{A_pocket}	f_{S_pocket}
	[lb/in]	[lb/in]	[lb/in]	[in]	[in]	[lbs]	[lbs]	[in²]	[in]	[psi]	[psi]	[psi]	[psi]
317416	-3305	188	1158	7.74	18	-25581	3384	0.9706	0.125	-26356	9234	-34521	9486
307406	3445	-166	1012	7.63	18	26285	-2988	0.93605	0.123	28081	8249	37397	8573

Limit load factor: 0.80

Element	F_X	F_Y	F_{XY}	d_Y	d_X	P_X	P_Y	A_{QUAD}	t_{QUAD}	f_X	f_{XY}	f_{A_pocket}	f_{S_pocket}
	[lb/in]	[lb/in]	[lb/in]	[in]	[in]	[lbs]	[lbs]	[in²]	[in]	[psi]	[psi]	[psi]	[psi]
317416	-1763	100	618	7.74	18	-13643	1805	0.9706	0.125	-14056	4925	-18411	5059
307406	1837	-89	540	7.63	18	14019	-1594	0.93605	0.123	14977	4400	19945	4572

Table 15-14 shows the shear buckling MS summary for the wing box upper and lower panels adjacent to the R/S. The MS obtained is 0.01 (close enough to zero) for initial buckling of the upper panel under the 80% Limit load. This is not an actual MS but it is simply a target to size the panel pocket thickness.

Table 15-14 - *Wing upper and lower panels compression + shear buckling MS summary.*

Shear *Ref. Bruhn C5.11* *All sides SS*

		a	b	a/b	k_S	t_{eq}	η_s	F_{scr}	f_s	Rs	F_{scr}'	$F_{scr}-F_{scr}'$
	Panel	[in]	[in]			[in]		[psi]	[psi]		[psi]	[psi]
Upper	A	18	6.07	2.97	5.8	0.122	0.99601	**21567**	4925	**0.23**	21567	0.000
Lower	B	18	5.95	3.02	5.8	0.118	0.99760	**20994**	4400	**0.21**	20994	0.000

Compression *Ref. Bruhn C5.2* *All sides SS; case C*

		a	b	a/b	k_C	t_{eq}	η_C	F_{Ccr}	f_C	R_L	F_{Ccr}'	$F_{ccr}-F_{Ccr}'$
	Panel	[in]	[in]			[in]		[psi]	[psi]		[psi]	[psi]
Upper	A	18	6.07	2.97	4	0.122	1.00000	**14933**	14056	**0.94**	14933	0.000
Lower	B	18	5.95	3.02	4	0.118	1.00000	**14513**	-14977	**-1.03**	14513	0.000

	Panel	R_S	R_L	MS
Upper	A	0.23	0.94	**0.01**
Lower	B	0.21	-1.03	23.43

An example of calculation is shown below for the upper *QUAD "Panel A"*. The lateral running load "F_Y" is neglected as it is very small. The goal is to determine the minimum pocket thickness by iteration in order to get a MS of 0 in combined compression and shear buckling.

<u>Analysis example for the most critical panel "**A**" for element *317416*</u>

Stress calculations @ 80% limit:

$$P_X = F_X \times d_Y = -1{,}763 \times 7.74 = -13{,}643\ lbs$$

$$f_X = f_C = \frac{P_X}{A_{QUAD}} = \frac{-13{,}643}{0.9706} = -14{,}056\ psi$$

$$f_{XY} = f_S = \frac{F_{XY}}{t_{QUAD}} = \frac{618}{0.125} = 4{,}925\ psi$$

<u>Shear buckling</u>

Using (ref. 17) and *Figure A-2* with the *Hinged edges curve* for flat plate buckling in shear having simply supported edges, the shear stress allowable is calculated as:

$$F_{s\,cr} = \frac{k_s\, \eta_s \pi^2 E}{12\left(1-\nu^2\right)}\left(\frac{t}{b}\right)^2$$

From *Figure 15-28,* the dimension *"a"* is 18 *in.* and the dimension "*b*" is 6.07 *in.* Thus a/b = 18 / 6.07 = 2.97. Hence from *Figure A-2*, the value for k_S is: 5.8. The primary Young's modulus in tension "*E*" is used from *Table 15-10.* So from the above equation:

$$\frac{F_{s\,cr}}{\eta_s} = \frac{k_s\, \pi^2 E}{12\left(1-\mu^2\right)}\left(\frac{t_{eq}}{b}\right)^2$$

$$\frac{F_{s\,cr}}{\eta_s} = \frac{5.8\ \pi^2 10*10^6}{12\left(1-0.33^2\right)}\left(\frac{0.122}{6.07}\right)^2 = 21{,}653\ psi$$

For
k_s = 5.8
b = 6.07 in
t_{eq} = 0.122 in
μ = 0.33
*E = 10.0*10⁶ psi*

Although the above value is a bit far from the shear strength of the material "F_{su}", the plastic reduction factor in shear "η_S" will be evaluated.

Therefore, using the same procedure as described in section *11.2.7*, the critical shear buckling stress is calculated by iteration by first assuming a value for $F_{S\,cr}'$. A plasticity correction factor for shear "η_S" is used to multiply the Young's modulus "***E***" and is calculated below.

$$\eta_S = \frac{1}{1+\left(\frac{0.00346\ G}{F_{sy}}\right)\left(\frac{F_{scr}}{F_{sy}}\right)^{n-1}}$$

For
F_{sy} = 30,250 psi
n = 20
*G = 3.76*10⁶ psi*

Where

$$F_{sy} = 0.55\,F_{cy} = (0.55)\ 55{,}000 = 30{,}250\ psi$$

This equation is solved by iteration until $F_{s\,cr}'$ is equal to the value obtained from the equation:

$$F_{s\,cr} = \frac{k_s\, \eta_s \pi^2 E}{12\left(1-\mu^2\right)}\left(\frac{t}{b}\right)^2$$

From the iterative process, the value for "η_s" of 0.99926 was obtained. The final initial shear buckling allowable is thus;

$$F_{s\,cr} = \frac{5.8\ \pi^2 (0.99926)\,10*10^6}{12\left(1-0.33^2\right)}\left(\frac{0.122}{6.07}\right)^2 = 21{,}637\ psi$$

For
k_s = 5.8
*E = 10.0*10⁶ psi*
t = 0.122 in
b = 6.07 in.
μ = 0.33

Compression buckling

Using *Figure A-1* curve *C* for flat plate buckling in compression with all edges simply supported, the compression stress allowable is calculated as:

$$F_{C\,cr} = \frac{\pi^2 k_C\, E}{12(1-\nu^2)}\left(\frac{t_{eq}}{b}\right)^2$$

For
$k_C = 4.0$
$b = 6.07\ in$
$a/b = 2.97$
$t_{eq} = 0.122\ in$
$E = 10.0*10^6\ psi$
$\mu = 0.33$

hence

$$F_{C\,cr} = \frac{\pi^2(4.0)\times 10{,}000{,}000}{12(1-0.33^2)}\left(\frac{0.122}{6.07}\right)^2 = 14{,}933\ psi$$

This value is low enough that no plasticity occurs (η = ***1.0***).

Margin of safety

In the case of a panel having a longitudinal and a shear load, the equation for the MS equation is obtained from the interaction equation in (ref. 17) as follows for: $R_L + R_S^2 = 1$

$$MS = \frac{2}{R_L + \sqrt{R_L^2 + 4\ R_S^2}} - 1$$

where the stress ratios are calculated as:

$$R_L = R_C = \frac{f_C}{F_{CCR}} = \frac{14{,}056}{14{,}933} = 0.940$$

$$R_S = \frac{f_S}{F_{SCR}} = \frac{4{,}925}{21{,}637} = 0.228$$

hence

$$MS = \frac{2}{0.940 + \sqrt{0.940^2 + 4(0.228)^2}} - 1 = 0.007$$

With this MS, an inner pocket thickness of t_{pocket} = 0.082 *in.* for the upper panel *"A"* was obtained and applied in the design of the upper wing plank. The pocket thickness is used in order to get the equivalent panel thickness "t_{eq}" of 0.122 *in.*

Now, using the ultimate loads, a post-buckling analysis will be performed in the next sub-chapter.

One thing to be aware of is that, in order to simplify the equation, some references (like ref. 10. for instance) use graphs for a buckling coefficient variable *"K"* that is derived from a Poisson coefficient of 0.3. However, note that the original source for the buckling equation is (ref. 20.) for the buckling analysis of flat plates and the correct Poisson value should be used.

Just know that *"K"* is equal to: $\boldsymbol{k}\ \pi^2\ /\ 12(1\text{-}\mu^2)$ and is using a value of 0.3 for the *Poisson* coefficient "μ". Since we're using Aluminum Alloys having a "μ" of 0.33 most of the time, then the ***K*** value obtained from the figures, (ref.10.) for instance, is slightly under-evaluated and is thus slightly more conservative if not corrected for.

15.8 DIAGONAL TENSION FIELD IN WING PLANK

Since the upper plank panel buckles at 80% of limit loads, it will be far into post buckling at ultimate loads. When a panel gets into post buckling, the buckled panel can be imagined as a bunch of cables, acting at an angle from the longitudinal direction, pulling onto the stringers and causing additional compression loads to be added into the components framing that panel.

This is why one must be doing this analysis before analyzing the Rear Spar caps as we want to take into account all the loading components. Those loads comprise the load into the cap "P_P", obtained from the FEM *GPFB,* and the load added by the diagonal tension effect of the panel "P_{DT}". Hence , the total stringer (*R/S upper cap*) load "P_{STR}" is thus:

$$P_{STR} = P_P + P_{DT}$$

Many analysis methodologies exist but, for this example, I am using the approach described in (ref. 19.) for diagonal tension fields in curved web systems since it takes into account shear AND compression loads. It is thus more versatile since it can be used for the fuselage and any other curved panels as well. Although the wing box planks are slightly curved, the panel we're interested in is pretty flat. The curvature radius is thus considered infinite in this case.

In order to have a better comprehension of the diagonal tension analysis methodology, I strongly recommend the reader to read the *NACA paper TN-2661 (Part I)* (ref. 19), as well as *NACA paper TN-2662 (Part 2)* for actual test results on structures in diagonal tension. Note that *Bruhn* (ref. 2) is based on those papers.

The diagonal tension analysis example details that follow are only for the upper panel *(A)* but the results of both upper *(A)* and lower *(B)* panels are shown in *Table 15-15*. Refer to *Figure 15-28* for the dimensions of panel *(A)*.

Table 15-15 *- Wing upper and lower panels diagonal-tension field parameters.*

Panel	f_S [psi]	$F_{S\,CR}$ [psi]	f_C [psi]	$F_{C\,CR}$ [psi]	A	B	R_C	$f_{S\,CR\_C}$ [psi]	R_t	$f_{S\,CR\_t}$ [psi]
A	9234	21637	-26356	-14933	-0.690	-2.854	0.229	4958	–	–
B	8249	21035	28081	-14513	-0.690	3.404	–	–	1.967	41385

Panel	R [in]	d [in]	h [in]	d/h	300td/Rh	$f_S/f_{S\,CR}$	k
A	999999	18	6.070	2.000	0	1.86	0.13
B	999999	18	5.954	2.000	0	0.20	0.0

Upper panel *(A)* diagonal tension analysis

1) Stresses and allowables

To start, we have to calculate the compression and shear stresses going through the panel and the critical compression and shear initial buckling stress allowables.

From *Table 15-13,* the compression "f_C" and shear "f_S" ultimate stresses in *element 317416* are:

$$f_C = f_X = -26{,}356\,psi$$

$$f_S = f_{XY} = 9{,}234\,psi$$

The compression over shear stress ratio "*B*" is calculated as:

$$B = \frac{f_C}{f_S} = -\frac{26{,}356}{9{,}234} = -2.854$$

From *Table 15-14*, the compression "F_{C_CR}" and shear "F_{S_CR}" buckling stress allowables in the upper panel are:

$$F_{C_{CR}} = -14{,}933\ psi$$

$$F_{S_{CR}} = 21{,}637\ psi$$

The compression over shear allowables ratio "*A*" is calculated as:

$$A = \frac{F_{C_{CR}}}{F_{S_{CR}}} = -\frac{14{,}933}{21{,}637} = -0.690$$

Then, the actual shear stress "$f_{S\ CR}$" at which the panel buckles due to the presence of the compression stress is:

$$f_{S_{CR}} = F_{S_{CR}}\left[\frac{-\frac{B}{A} + \sqrt{\left(\frac{B}{A}\right)^2 + 4}}{2}\right]$$

hence, considering

$$f_{S_{CR}} = F_{S_{CR}}\ R_C$$

then

$$R_C = \left[\frac{-\frac{-2.854}{-0.690} + \sqrt{\left(\frac{-2.854}{-0.690}\right)^2 + 4}}{2}\right] = 0.229$$

and thus

$$f_{S_{CR}} = 21{,}637\ (0.229) = 4{,}955\ psi$$

2) Diagonal tension factor; ***k***

The factor "***k***" varies from 0 (when there is no buckling) to 1 (when the panel is fully in diagonal tension). Therefore, it is an index on the severity of the buckling. It is calculated as follows:

$$k = \tanh\left[\left(0.5 + 300\frac{td}{Rh}\right)\log_{10}\frac{f_S}{f_{S_{CR}}}\right]$$

For $R = \infty$, then

$$k = \tanh\left[(0.5 + 0)\log_{10}\frac{9{,}234}{4{,}955}\right] = 0.13$$

From this number, one can notice that "*k*" is closer to 0 than 1. Therefore, the post buckling severity is low.

As mentioned earlier, the stringers (the R/S upper cap) will see additional compression loads from the effect of the diagonal tension. This additional load simply depends on the severity of the buckling, thus, on the "*k*" factor and the angle "*α*" of the web waves. The angle value is unknown and is found by iteration.

As a first approximation, I can assume an angle *"α"* of 45 *deg.* *(π/4 rad.)* to calculate the additional load in the stringer due to the diagonal tension effect *"$P_{D.T.}$"* using the following equation.

$$P_{DT} = -k\ q\ h \cot\alpha$$

For
$k = 0.13$
$q = 1158\ lb/in$
$h = 6.07\ in$
$\alpha = 0.7854\ rad$

In this case, the additional load would be:

$$P_{DT} = -(0.13)(1{,}158)(6.07)\cot\pi/4 = -944\ lbs$$

Otherwise, the angle *"α"* is calculated from the equation:

$$\tan^2\alpha = \frac{\varepsilon - \varepsilon_{ST}}{\varepsilon - \varepsilon_{RG} + \dfrac{1}{24}\left(\dfrac{h}{R}\right)^2}$$

Where *"ε"* is the strain in the skin/web panel, *"ε_{ST}"* is the strain in the stringer/Spar cap, and *"ε_{RG}"* is the strain in the ring/rib flange. Each of them is calculated by first assuming a value for the angle and then converging to a final value by iteration until the angle obtained from the above equation is equal to the assumed value.

When running a macro that performs this kind of calculation for thousands of elements and thousands of load cases, sometimes the stress group lead will simply assume an angle of 45 *deg.* conservatively for all panels instead of having to perform the more accurate but complex and time consuming iterative process.

For our example, the resulting angle obtained was: ***0.9975 rad. (57.2 deg.)*** and thus:

$$P_{DT} = -(0.13)(1{,}158)(6.07)\cot(0.9975) = -610\ lbs$$

This is only 335 *lbs* lower than the value calculated with an angle of 45 *deg.* but still about 50% less. The higher angle obtained is mainly due to the fact that the panel is flat *(R = ∞)* and that the contribution of the compression stress is higher than from the shear stress. For conservatism, the *45 deg.* angle will be kept. The calculations to get the actual angle of *57.2 deg.* is shown below as a reference.

3) Stringer (R/S) loads, stresses and strains

For
$P_p = -30{,}431\ lbs$
$A_{STR} = 0.58755\ in^2$
$A_{e\ SKIN} = 0.72705\ in^2$
$k = 0.13$
$f_s = 9{,}234\ psi$
$h = 6.07\ in$
$t = 0.125\ in$
$\alpha = 0.9975\ rad$
$R_C = 0.229$
$E_{STR} = 10x10^6\ psi$

The strain in the stringer *"ε_{ST}"* is calculated as follows:

$$\varepsilon_{ST} = \frac{f_{STR}}{E_{STR}}$$

where the stress in the stringer is

$$f_{STR} = \frac{P_P}{A_{STR} + A_{e\ SKIN}} + \frac{-k\ f_S \cot\alpha}{\dfrac{A_{STR}}{ht} + 0.5(1-k)R_C}$$

then

$$f_{STR} = \frac{-30{,}431}{0.58755+0.72705} + \frac{-(0.13)\ 9{,}234\ \cot(0.9975)}{\dfrac{0.58755}{6.07\times 0.125} + 0.5(1-0.13)(0.229)} = -24{,}068\ psi$$

and

$$\varepsilon_{ST} = \frac{-24{,}068}{10\times 10^6} = -2{,}407\times 10^{-6}\ in./in.$$

4) Ring (Ribs) loads, stresses and strains

The strain in the ring "ε_{RG}" that is attached to the skin is calculated as follows:

$$\varepsilon_{RG} = \frac{f_{RG}}{E_{RG}}$$

where the stress in the ring is:

$$f_{RG} = \frac{k\, f_S \tan\alpha}{\dfrac{A_{RG}}{d\,t} + 0.5(1-k)}$$

For
$A_{RG} = 0.5735\ in^2$
$k = 0.13$
$f_s = 9{,}234\ psi$
$d = 18\ in$
$t = 0.125\ in$
$\alpha = 0.9975\ rad$
$E_{RG} = 10x10^6\ psi$

then

$$f_{RG} = \frac{(0.13)\ 9{,}234\ \tan(0.9975)}{\dfrac{0.5735}{18(0.125)} + 0.5(1-0.13)} = -2{,}797\ psi$$

and

$$\varepsilon_{RG} = \frac{-2{,}797}{10\times10^6} = -280\times10^{-6}\ in./in.$$

5) Skin (plank) strains

For
$k = 0.13$
$f_s = 9{,}234\ psi$
$\mu = 0.33$
$\alpha = 0.9975\ rad$
$E_{skin} = 10x10^6\ psi$

The strain in the skin "ε" is calculated as follows:

$$\varepsilon = \frac{f_S}{E_{Skin}}\left[\frac{2k}{\sin 2\alpha} + \sin 2\alpha(1-k)(1+\mu)\right]$$

then

$$\varepsilon = \frac{9{,}234}{10\times10^6}\left[2\frac{(0.13)}{\sin(2\times0.9975)} + \sin(2\times0.9975)(1-0.13)(1+0.33)\right] = 1{,}241\times10^{-6}\ in./in.$$

6) Determination of the buckling angle "α"

Using the following equation mentioned earlier, the buckling is found by iteration. The result is shown below:

$$\tan^2\alpha = \frac{1{,}241 - (-2{,}406)}{1{,}241 - (-280) + \dfrac{1}{24}\left(\dfrac{6.07}{\infty}\right)^2} = 2.39863$$

and thus

$$\alpha = atan\sqrt{2.39863} = 0.9975\ rad. = 57.15\ deg.$$

As explained earlier, the angle of ***45 deg.*** is used conservatively and the –944 *lbs* load will be added to the existing compression loads in the R/S upper cap. But then, we have to see if the web of the R/S also gets into post-buckling as it will too add a compression load into the R/S upper cap, as demonstrated in section ***15.10***.

7) Allowable shear stress in skin

In order to conclude with the wing upper plank, a check for the allowable shear stress in the skin is done. From (ref. 19. eq (53)), the following equation is used:

$$F_{S\,ALL} = F'_{S\,ALL}(0.65 + \Delta)$$

where

$$\Delta = 0.3\ \tanh\frac{A_{RG}}{dt} + 0.1\ \tanh\frac{A_{STR}}{ht}$$

so

$$\Delta = 0.3\ \tanh\frac{0.5735}{18(0.125)} + 0.1\ \tanh\frac{0.58755}{6.07(0.125)} = 0.0925$$

For
$A_{RG} = 0.5735\ in^2$
$A_{STR} = 0.58755\ in^2$
$d = 18\ in$
$h = 6.07\ in$
$t = 0.125\ in$

Then, before we can obtain a value for "$F'_{S\ ALL}$", we must first evaluate the maximum allowable shear stress "F_S" using the equation:

$$F_S = 0.344\ F_{TU} \cdot k^{(-0.08)} \leqslant MIN(0.45\ F_{TU}\ ;\ F_{SU})$$

Thus, for $k = 0.13$ and $F_{TU} = 70\ ksi$, a value of 28,280 *psi* is obtained for the allowable gross shear stress "F_S". This value is smaller than (0.45)*70,000 = 31,500 *psi* and, $F_{SU} =$ 40,000 *psi* so it's fine.

Then, using ***Figure A-6*** with the radius $R = \infty$ and the "R_R" and "R_S" parameters, a value of ***40 deg.*** is obtained for the angle in the case of Pure Diagonal Tension "a_{PDT}" The parameters are calculated as follows:

$$R_R = \frac{dt}{A_{RG}} = \frac{18(0.125)}{0.5735} = 3.94 \qquad R_S = \frac{ht}{A_{STR}} = \frac{6.07(0.125)}{0.58755} = 1.30$$

thus, the curve for the following parameter is chosen. The curve ***0.5*** is used in our case.

$$\frac{1 + R_S}{1 + R_R} = \frac{1 + 1.30}{1 + 3.94} = 0.47$$

For
$E = 10X10^6\ psi$
$F_S = 28{,}280\ psi$
$R_R = 3.94$
$h = 6.07\ in$
$R = \infty$

For the ordinate of the graph, the following parameter is calculated:

$$\frac{\frac{h}{R}\sqrt{\frac{E}{F_S}}}{\sqrt{1+R_R}} = 0$$

Instead of using the figures, the Pure Diagonal Tension angle "a_{PDT}" could have also been obtained by iteration by using the (ref. 19. eq. (44a)):

$$(1 + R_R)\tan^4\alpha + A\tan^3\alpha = 1 + R_S$$

where

$$A = \frac{1}{24}\left(\frac{h}{R}\right)^2 \frac{E}{f_S} = 0$$

For
$E = 10X10^6\ psi$
$f_S = 9{,}234\ psi$
$R_R = 3.94$
$R_S = 1.30$
$h = 6.07\ in$
$R = \infty$

then, by iteration, we find $a_{PDT} = 39.6$ deg. This is very close to 40 *deg.* and, using *Figure A-7* for *Al Aly 7075-T6* with a F_{TU} of 72 *ksi*, with $k = 0.13$ and "a_{PDT}" = 40 *deg.*, we find a value for $F'_{S\ ALL} = 31{,}500\ psi$. Since our fictitious material has a F_{TU} of 70,000 *ksi*, the value for $F'_{S\ ALL}$ is corrected as:

$$F'_{S\ ALL} = \frac{70{,}000}{72{,}000} \times 31{,}500 = 30{,}625\ psi$$

finally, the allowable shear stress in the skin is:

$$F_{S\ ALL} = 30{,}625(0.65 + 0.0925) = 22{,}740\ psi$$

The MS is thus:

$$MS = \frac{F'_{S\ ALL}}{f_S} - 1 = \frac{22{,}740}{9{,}234} - 1 = 1.46$$

Hence, the wing upper plank is adequate for the critical ultimate load case analyzed.

15.9 REAR SPAR WEB BUCKLING

The Rear Spar web is also checked for shear buckling. Although shear buckling is not a mode of failure, we need to verify if the web is non-buckling. As shown on *Figure 15-30*, there are two *(2)* web panels. The panel boundaries are determined by the upper cap flange and the lower stringer and, the vertical integral ribs that act as panel breakers (dividers). The panels are assumed simply supported on all sides.

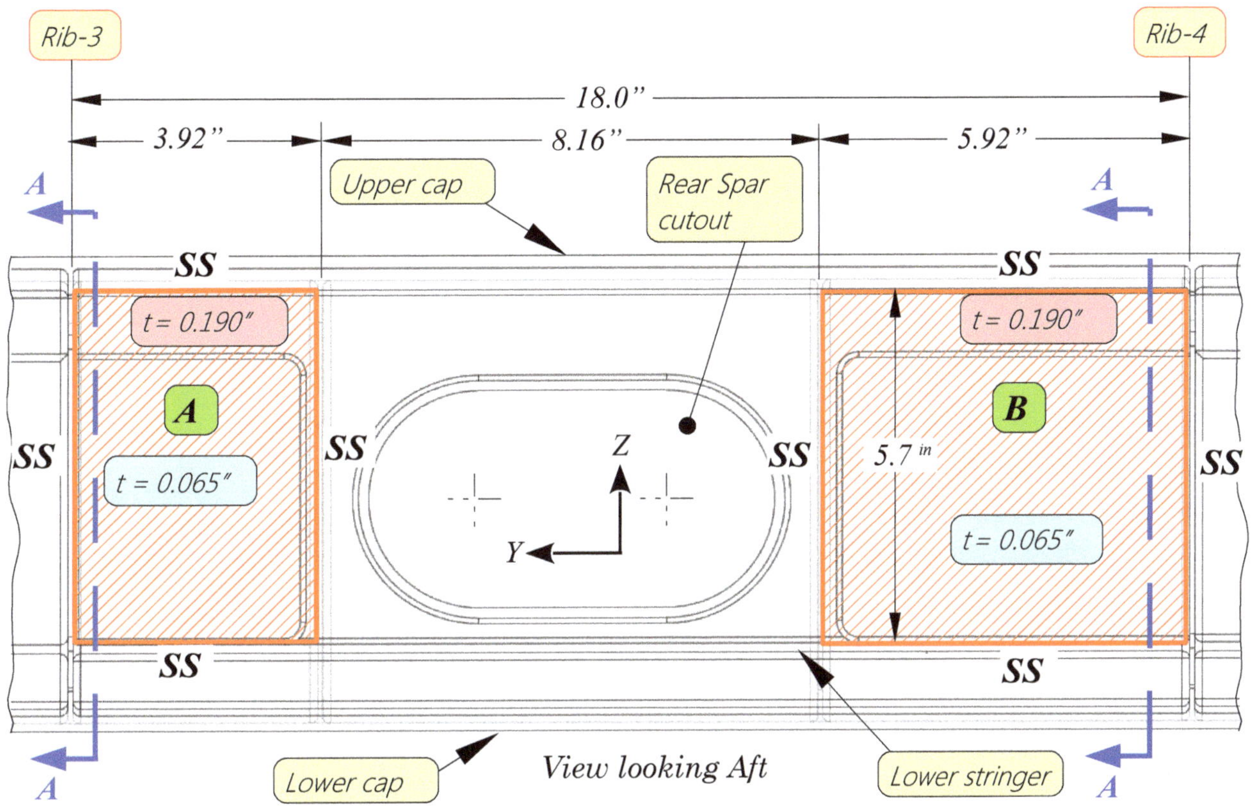

Figure 15-30 *– Rear Spar web panels.*

Each panel is checked using the most conservative transverse shear load within that panel. So from *Figure 15-23* the maximum shear force *"S"* through panel *"A"* is 13,141 *lbs* while, by doing a linear interpolation, the maximum shear force *"S"* through panel *"B"* is 12,567 *lbs*.

Although the whole shear force is assumed going through the Rear Spar vertical web effective area – as shown in *Figure 15-31* as section *A-A* – only the pocket needs to be checked for buckling. The average shear stress is used for each panel.

<u>Effective web shear area</u>

The Rear Spar flanges and the skins are considered to be inefficient to react vertical shear loads. Hence, referring to *Figure 15-31*, the effective shear area; "A_S" is:

$$A_S = 1.164(0.190) + 4.72(0.065) + 1.277(0.200) = 0.78336 \; in.^2$$

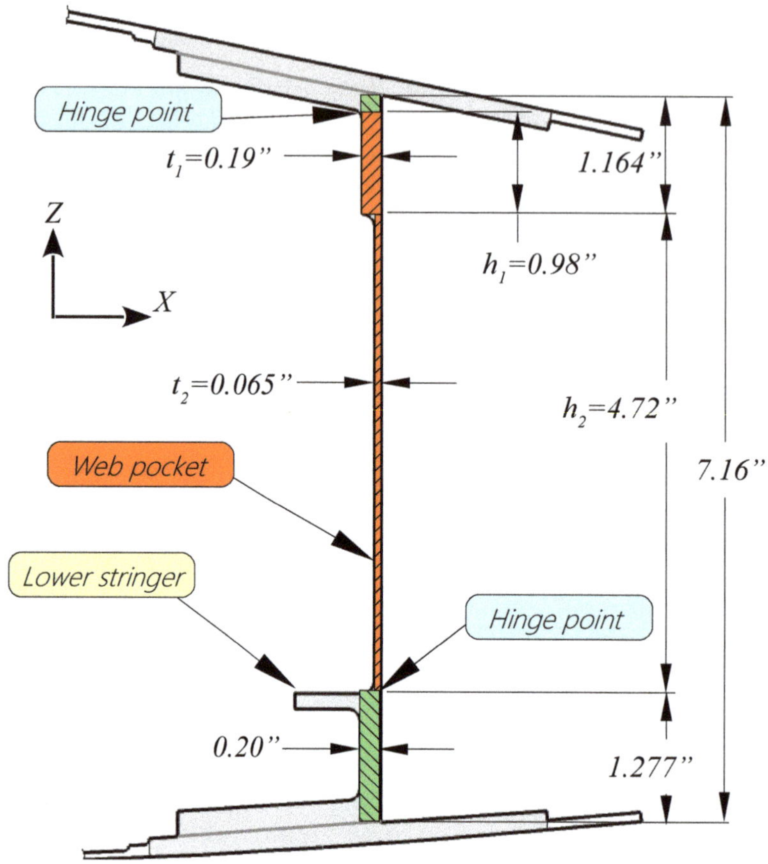

Figure 15-31 *– Section A-A - Rear Spar web effective vertical shear area.*

Table 15-16 shows the shear buckling MS summary for the Rear Spar web panels. From the results, panel ***A*** is shear buckling resistant for the ultimate load while panel ***B*** is not.

Table 15-16 *- Rear Spar panels ultimate shear buckling MS summary.*

	a	b	a/b	k_S	t	η_s	F_{scr}	V_Z	f_s	MS
Panel	[in]	[in]			[in]		[psi]	[lbs]	[psi]	
A	3.92	5.7	1.45	7.1	0.065	1.00000	**18018**	13141	16775	0.07
B	5.92	5.7	1.04	9.5	0.065	1.00000	**11402**	12567	16043	-0.29

Panel ***B*** was allowed to buckle in this example to demonstrate the methodology. However, it is not advisable to have a buckling panel right next to a cutout. Panel ***B*** should have its thickness increased to prevent buckling at least at limit loads.

Analysis example for the most critical panel "***B***"

The MS from *Table 15-16* were all calculated following the same procedure. Panel ***B*** is the most critical panel with an MS of –0.29. This MS does not mean that the panel fails but simply that it buckles under the ultimate loads and that a post buckling analysis is required. Using (ref. 20.) and *Figure A-2* for flat plate buckling in shear having simply supported edges, the shear stress allowable is calculated as:

$$F_{s\,cr} = \frac{k_s\,\eta_s\,\pi^2 E}{12\left(1-\nu^2\right)}\left(\frac{t}{b}\right)^2$$

From *Figure 15-30* for panel *B*, the dimension *"a"* is 5.92 *in.* and the dimension *"b"* is 5.7 *in.* Note that for shear buckling, the value for *"b"* is always the smallest of the two dimensions. Thus a/b = 5.92 / 5.7 = 1.04. Hence from *Figure A-2*, the value for k_S is: 9.5. The primary Young's modulus in tension *"E"* is used from *Table 15-9*.

Because the web pocket thickness is not constant, see *Figure 15-31*, I decided to use the smallest thickness as the effective panel thickness to be used in the buckling analysis. Not only this is a conservative assumption but, as you'll see, this is a pretty accurate one as well.

To verify this, I made a small FEM representing panel ***B*** using ***HEX20*** solid elements and ran a buckling solution by applying a 1 *lb* transverse shear force along the inboard and outboard edges of the panel. A shear force of 1.0386 *lb* is applied to the top and bottom edges for balance. See *Figure 15-32* for the FEM details and *Figure 15-33* for the results.

By using a unit load, the buckling factor "λ" obtained will correspond to the critical shear force "P_{S_CR}". In this case, I obtained 5,546 *lbs* for the first mode. By dividing that load by the pocket area "A_{SP}", we obtain the critical shear stress; "$F_{SCR\ FEM}$".

For

$$A_{SP} = t_1 h_1 + t_2 h_2 = (0.19)0.98 + (0.065)4.72 = 0.493\, in.^2$$

then

$$F_{SCR\,FEM} = \frac{P_{SCR}}{A_{SP}} = \frac{5{,}546}{0.493} = 11{,}249\ psi$$

Now going back to the theoretical approach, the initial shear buckling allowable is calculated as follows. There is no plasticity as the allowable stress is well below the yield shear stress.

$$F_{s\,cr} = \frac{k_s\, \eta_s\, \pi^2\, E}{12(1-\nu^2)} \left(\frac{t}{b}\right)^2$$

$$F_{s\,cr} = \frac{9.5\ 1.0\ \pi^2\ 10*10^6}{12(1-0.33^2)} \left(\frac{0.065}{5.7}\right)^2 = 11{,}402\ psi$$

For
k_s = 9.5
η_s = 1.0
b = 5.7 *in*
t = 0.065 *in*
E = 10.0*10^6 *psi*

This value is within 1% from the one obtained from the small *FEM*. Therefore, using the minimum thickness for the whole panel is a very good assumption in this case.

For comparison purposes, I calculated the critical shear stress by only considering the thin portion of the web with b = 4.72" and t = 0.065" and obtained F_{SCR} = 14,003 psi. Then, I did it by using b = 5.7" but calculating an equivalent thickness based on the volume of the panel where t_{eq} = 0.086" and obtained 19,960 *psi*.

This study clearly demonstrates that the smallest thickness should be used. However, I do recommend the analysts to do their own study for the geometry being analyzed.

All edge nodes are blocked in *X* while one node at the upper-inner corner is also blocked in *Y* and, one node at the lower-inner corner is blocked in all three directions *(XYZ)*.

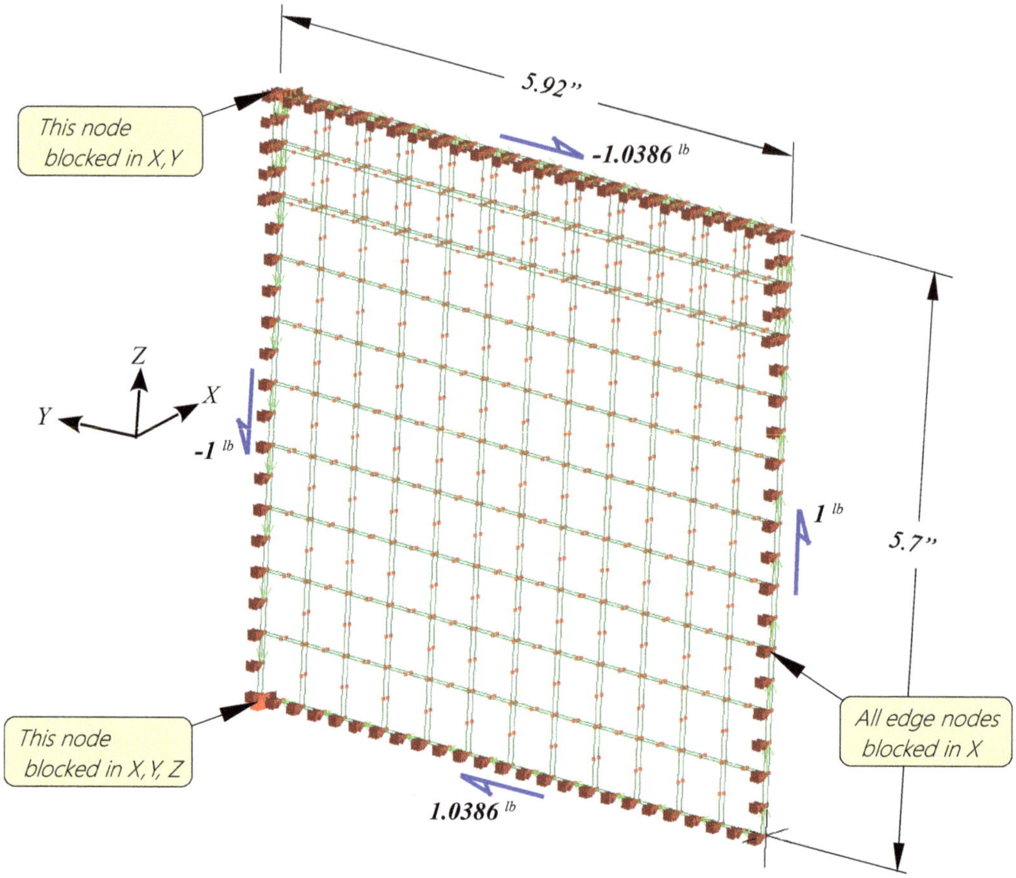

Figure 15-32 – *Panel B pocket FEM details.*

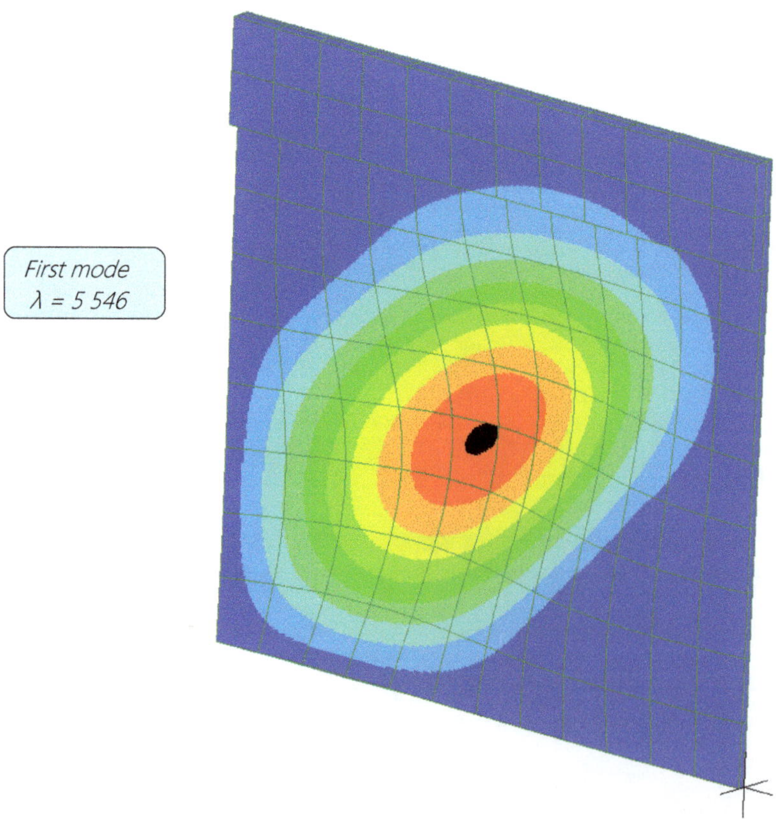

Figure 15-33 – *Panel B pocket FEM results*

Margin of safety for initial shear buckling

The maximum transverse shear load at the location of panel "*B*" is; V_Z = 12,567 *lbs*. The shear stress is thus; $f_S = V_Z / A_S$ = 12,567 / (0.78336) = 16,043 *psi*. Thus the MS in initial shear buckling of the Rear Spar web panel ***B*** due to the ultimate load is calculated as:

$$MS = \frac{F_{s\,cr}}{f_S} - 1 = \frac{11{,}402}{16{,}043} - 1 = -0.29$$

15.10 DIAGONAL TENSION FIELD IN REAR SPAR PANEL B

Just like for the upper wing plank panel analyzed earlier, the buckling of panel *B* in the Rear Spar web will introduce an additional compression load into the Rear Spar upper cap; "P_{DT}". For this analysis, since no compression buckling exists, the methodology based on (ref. 19.) for diagonal semi-tension field for flat webs will be used. See ***Figure 15-34*** for the geometry of panel *B* .

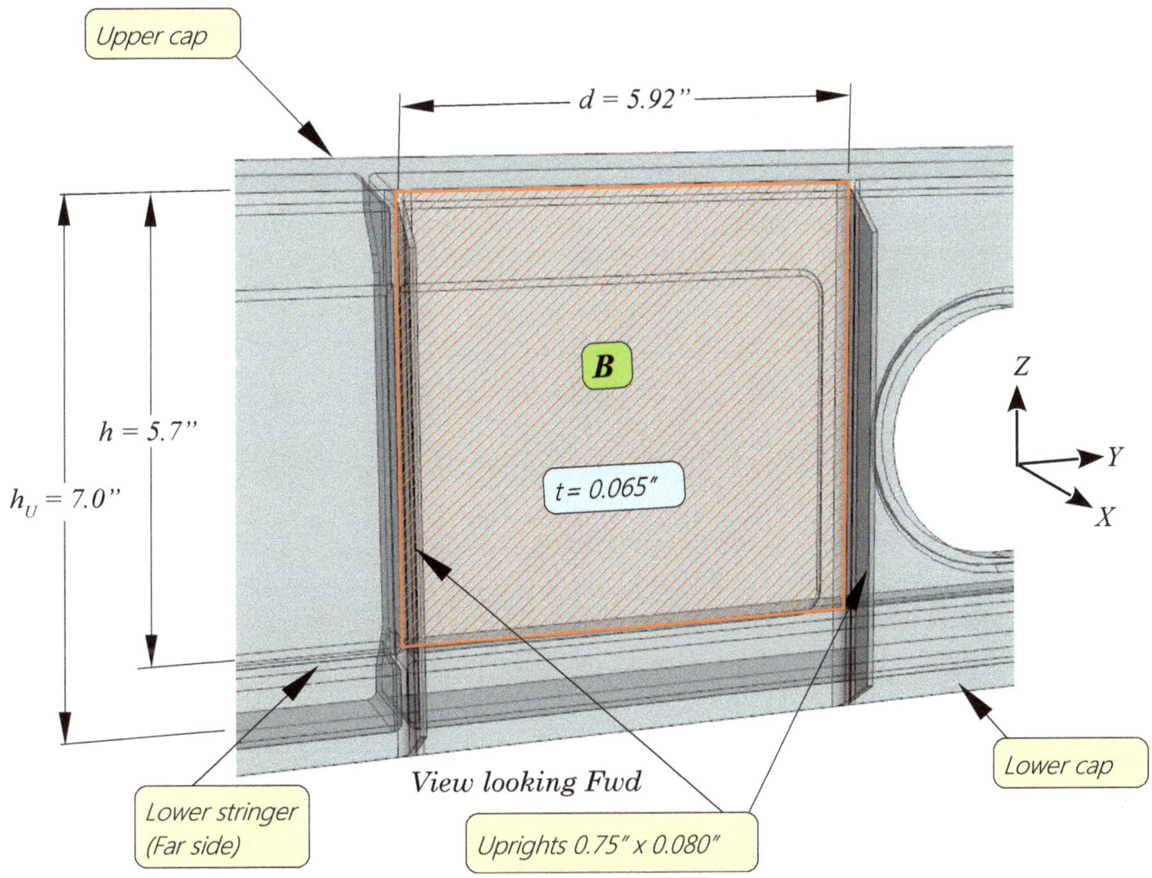

Figure 15-34 – *Panel B geometry for the diagonal tension analysis.*

The following parameters are calculated to compare with the methodology limitations:

> For
> t_U = 0.080 in
> $d = d_C$ = 5.92 in
> $h = h_C$ = 5.7 in
> t = 0.065 in

t_U/t = 0.080 / 0.065 = 1.23 > 0.6 *ok*

d/h = 5.92 / 5.7 = 1.04 ≈ 1.0 *ok*

h/t = 5.7 / 0.065 = 88 << 200, hence, this could lead to un-conservative predictions as this is a relatively thick beam.

1) Loading ratio

The loading ration is the ratio of the applied average shear stress over the critical shear stress. For $F_{S\,CR}$ = 11,402 *psi*, calculated in *section 15.9*, and f_S = 16,043 *psi*, the load ratio is:

$$\frac{f_s}{F_{s\,cr}} = \frac{16{,}043}{11{,}402} = 1.41$$

2) Diagonal tension factor; ***k***

The diagonal tension factor "*k*" is obtained from *Figure A-8* using the "*300td/Rh = 0*" curve for flat panels where $R = \infty$. Hence, with the loading ratio of 1.41, we find ***k = 0.09***.

3) Single upright

Since the upright is integral to the Rear Spar, its cross-section is considered as shown in *Figure 15-35* along with its section properties.

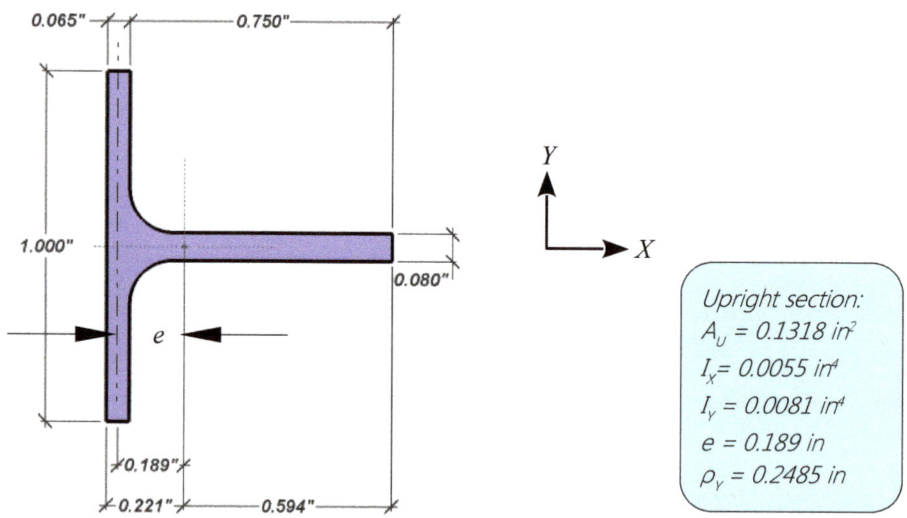

Figure 15-35 *– Upright section dimensions and properties.*

Average stress in upright

Due to the lack of symmetry from the web plane, the effective cross-section area of the upright "A_{Ue}" is calculated from (ref. 19. eq. (22)) as follows:

$$A_{U\,e} = \frac{A_U}{1 + \left(\frac{e}{\rho}\right)^2} = \frac{0.1318}{1 + \left(\frac{0.189}{0.2485}\right)^2} = 0.08350\,in.^2$$

Using *Figure A-9*, with the ordinate A_{ue} / *dt* = 0.0835 / (5.92 * 0.065) = 0.22, and *k* = 0.09 as the abscissa, we find f_U/f_S = 0.14. Thus:

$$f_U = 16{,}043(0.14) = -2{,}246\,psi$$

Because there are no upright on each side of the web, the upright is thus loaded in compression eccentrically causing the stress in the upright to vary from the web to the free edge. The maximum stress in the upright is calculated with the equation:

$$f_{U\,max} = f_U\left(\frac{f_{U\,max}}{f_U}\right)$$

Using *Figure A-10,* with d/h_U = 5.92 / 7.0 = 0.85 as the ordinate and k = 0.09 as the abscissa, we find $f_{U\,max}/f_U$ = 1.21. Thus:

$$f_{U\,max} = -2{,}246(1.21) = -2{,}718\,psi$$

Buckling angle; α

The buckling angle is then obtained from *Figure A-3*, with the abscissa f_u/f_S= 2,246 / 16,043 = 0.14, and k = 0.09 for the curve, we find ***tan α*** = 0.96. So the buckling angle is thus:

$$\alpha = atan(0.96) = 43.83\,deg = 0.765\,rad$$

4) Allowable stress in upright

Column failure

The uprights "vertical stiffeners" are checked for column instability due to the effect of the diagonal tension in the rear spar web. The effective column length is calculated using the following equation.

$$L_e = \frac{h_U}{\sqrt{1 + k^2\left(3 - 2\dfrac{d}{h}\right)}}$$

then

$$L_e = \frac{7.0}{\sqrt{1 + 0.09^2\left(3 - 2\,\dfrac{5.92}{5.7}\right)}} = 6.974\ in.$$

For
h_U = 7.0 in
d = 5.92 in
h = 5.7 in
k = 0.09

The Johnson-Euler method will be used to calculate the allowable column stress "F_C" using the equation:

$$F_C = F_{CS} - \frac{F_{CS}^2}{4\,\pi^2\,E}\left(\frac{L_e}{\rho}\right)^2$$

where the crippling stress allowable "F_{CS}" is obtained for *TEES* using the equation:

$$\frac{F_{CS}}{F_{CY}} = (0.67)\left[\sqrt{\left(\frac{g\,t^2}{A}\right)\left(\frac{E}{F_{CY}}\right)}\right]^{0.40}$$

hence

$$\frac{F_{CS}}{F_{CY}} = (0.67)\left[\sqrt{\left(\frac{3(0.065)^2}{0.1318}\right)\left(\frac{10\times10^6}{60{,}000}\right)}\right]^{0.40} = 0.731$$

For
F_{CY} = 60 ksi
g = 3 (for TEES)
t = 0.065 in
A = 0.1318 in²
E = 10,000 ksi

This value is within the limitation of 0.8 for *TEES*. Therefore, the crippling stress allowable is: F_{CS} = (0.731) 60,000 = 43,834 *psi*.

Now the column stress allowable is thus:

$$F_C = 43{,}834 - \frac{43{,}834^2}{4\,\pi^2\,10\times10^6}\left(\frac{6.974}{0.2485}\right)^2 = 40{,}001\ psi$$

For
F_{CS} = 43,834 psi
L_e = 6.974 in
ρ = 0.2485 in
E = 10,000 ksi

The column allowable stress is then compared to the average compression stress in the upright "f_U" to calculate the MS.

$$MS_{Column} = \frac{F_C}{f_U} - 1 = \frac{40{,}001}{2{,}246} - 1 \gg 1.0$$

Forced crippling

The upright being only on one side of the web sees some bending stress in addition to the axial compression stress. This caused a higher stress "$f_{U\,max}$" of −2,718 *psi* calculated earlier. This stress is compared to the allowable stress obtained from (ref. 19. eq. (37b)) for single uprights made out of ***Al Aly 7075-T6*** with $F_{CY} = 68$ *ksi* as follows:

$$F_{U\,7075} = 32{,}500\,k^{2/3}\left(\frac{t_U}{t}\right)^{1/3}$$

For
$t_U = 0.080$ in
$t = 0.065$ in
$k = 0.09$

hence

$$F_{U\,7075} = 32{,}500(0.09)^{2/3}\left(\frac{0.080}{0.065}\right)^{1/3} = 6{,}995\ psi$$

Since we're using a different material ***(7050-T7451 with F_{CY} = 60 ksi)***, then the actual forced crippling allowable is factored as follows:

$$F_{U\,7050} = F_{U\,7075}\left(\frac{\sqrt{F_{CY\,7050}}}{\sqrt{F_{CY\,7075}}}\right) = 6{,}995\left(\frac{\sqrt{60}}{\sqrt{68}}\right) = 6{,}570\ psi$$

The MS in forced crippling is thus:

$$MS_{F.C.} = \frac{F_{U\,7050}}{f_{U\,max}} - 1 = \frac{6{,}570}{2{,}718} - 1 = 1.41$$

From the above results, the uprights are adequate since they are sufficiently strong and stable to withstand the diagonal tension effect. Therefore, the uprights in the rear spar web can adequately support the *R/S* upper cap in the vertical direction subjected to column buckling.

5) Web max shear

The web is checked for the maximum *(peak)* shear stress "$f_{S\,max}$" in the panel subjected to diagonal semi-tension field. The methodology is from (ref. 19. §3.7) using the equation *(33a)*:

$$f_{S\,max} = f_S\,(1 + kC_1)(1 + kC_2)$$

Where C_1 is obtained from *Figure A-4* using *tan α* = 0.96, we then get C_1 = 0.002. We then obtain C_2 from *Figure A-5* using ***wd*** = 2.01 as the abscissa, then we get C_2 = 0.08. The parameter "*wd*" is calculated as follows:

$$wd = 0.7\,d\,\sqrt[4]{\frac{t}{(I_C + I_T)h_e}}$$

For
$d = 5.92$ in
$h_e = 5.7$ in
$t = 0.065$ in
$I_C = 0.0641$ in^4
$I_T = 0.1431$ in^4

so

$$wd = 0.7(5.92)\sqrt[4]{\frac{0.065}{(0.0641 + 0.1431)5.7}} = 2.01$$

where I_C and I_T are the moment of inertia of the *R/S* upper and lower caps respectively (Compression & Tension) without considering the planks nor the web. Therefore, the maximum shear stress is:

$$f_{S\,max} = 16{,}043\ (1 + (0.09)(0.002))(1 + (0.09)(0.08)) = 16{,}161\ psi$$

For
$f_S = 16{,}043$ psi
$k = 0.09$
$C_1 = 0.002$
$C_2 = 0.08$

Since the buckling coefficient is very small, the maximum shear stress is not much higher than the average shear stress f_S of 16,043 *psi*.

Margin of safety

Since there are no rivets connecting the web to the upper and lower caps —as the rear spar is a one piece machined part— no fasteners connecting the web to the stringers need be checked. Thus, the gross shear allowable "F_S" is obtained from the following equation for $k = 0.09$ and $F_{TU} = 75$ *ksi*.

$$F_S = 0.344\ F_{TU} \cdot k^{(-0.08)} = (0.344) \cdot 75{,}000 \cdot (0.09)^{(-0.08)} = 31{,}280\ psi$$

Therefore, the MS is calculated as follows:

$$MS_{Shear.} = \frac{F_S}{f_{S\,max}} - 1 = \frac{31{,}280}{16{,}161} - 1 = 0.93$$

6) Web permanent buckling

As a last check, we need to verify that the panel doesn't get into permanent buckling at limit load. The limit shear stress "$f_{S\,lim}$" is thus:

$$f_{Slim} = \frac{f_S}{Ult\ L.F.} = \frac{16{,}043}{1.5} = 10{,}695\ psi$$

The ratio of the permanent buckling stress over the critical shear stress "$F_{S\,P.B.}/F_{SCR}$" is obtained from the following equation:

$$\frac{F_{S\,P.B.}}{F_{SCR}} = 10^{(-0.7861\ \log x\ +\ 0.66437)} \quad for\ x < 7$$

$$= 1 \quad for\ x \geq 7$$

$$where \quad x = \frac{4\,F_{SCR} \times 10^8}{E_C\,F_{CY}} = \frac{4 \cdot 11{,}402 \times 10^8}{10 \times 10^6 \cdot 60{,}000} = 7.6$$

For
$F_{S\,CR} = 11{,}402\ psi$
$F_{CY} = 60{,}000\ psi$
$E_C = 10{,}000\ ksi$

therefore, since x > 7:

$$\frac{F_{S\,P.B.}}{F_{SCR}} = 1.0$$

thus

$$F_{S\,P.B.} = (1.0)\,11{,}402 = 11{,}402\ psi$$

Finally, the MS is calculated as:

$$MS_{P.B.} = \frac{F_{S\,P.B.}}{f_{Slim}} - 1 = \frac{11{,}402}{10{,}695} - 1 = 0.06$$

Since the MS is positive, no permanent buckling will occur at limit load. However, as mentioned earlier, because there is a cutout in the adjacent panel, it would be preferable to increase the thickness of the web (panel *B*) to make it shear resistant for ultimate loads.

Additional load in upper cap due to D.T.

The additional load in the stringer (upper cap) due to the diagonal tension effect is calculated using the following equation:

$$P_{DT} = -\frac{k\ S}{2} \cot\alpha$$

In this case, the additional load "P_{DT}" would be:

$$P_{DT} = -\frac{(0.09)\ 12{,}567}{2} \cot(0.765) = -589\ lbs$$

For
$k = 0.09$
$S = 12{,}567\ lbs$
$\alpha = 0.765\ rad$

15.11 CRIPPLING

Before we can compute the section properties of the rear spar cross sections, we need to know the effective skin/plank width. The first step in calculating that is to calculate the crippling stress allowable of the Rear Spar upper and lower caps (without the effective skin) for the two critical sections.

15.11.1 SECTION A-A

The crippling stress allowable *"F_{CC}"* will be calculated below for the section *A-A*. The section is divided into segments as shown in ***Figure 15-36***. The web is not taken into account as it is considered buckled. All segments have one edge free except segment *4*. A crippling stress allowable for each upper cap and lower cap are calculated as they are separate.

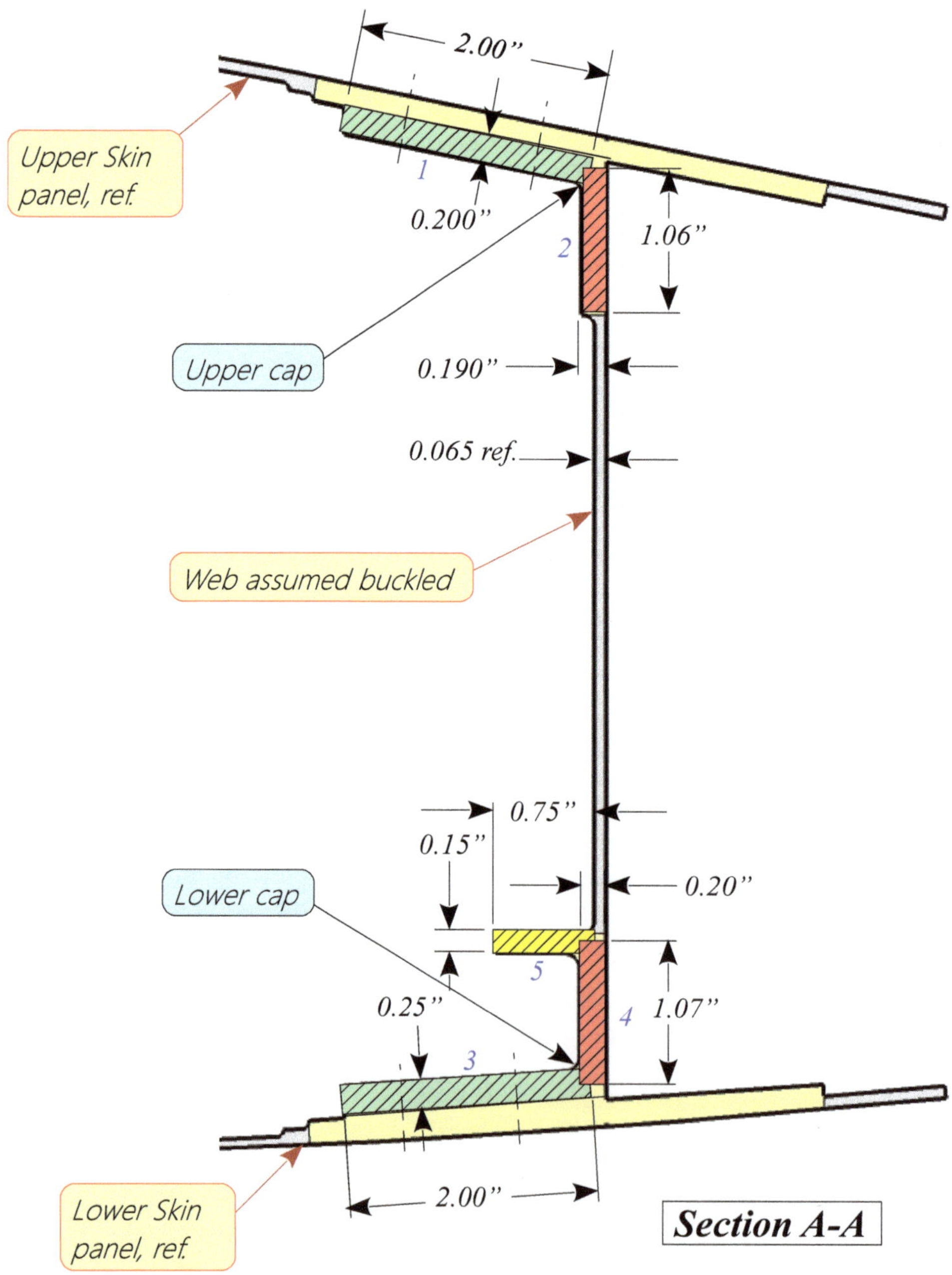

Figure 15-36 *– Effective segments for crippling at section A-A.*

The crippling allowable for each segment is calculated as shown below. Note that the crippling allowable "F_{CC}" of a segment can never be higher than F_{CY}. So the maximum value for F_{CCn} / F_{cy} is 1.0. Using *Figure 15-37* for section *A-A*, the value at the abscissa is calculated as:

$$abscissa = \sqrt{F_{CY}/E}(b/t)$$

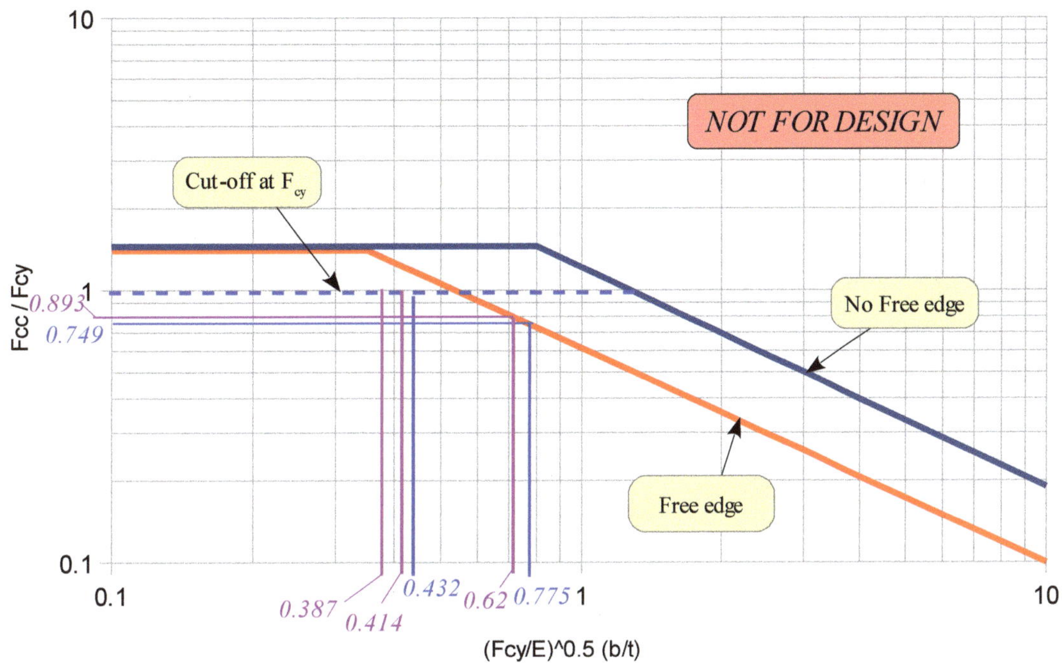

Figure 15-37 *– Typical Crippling curves section A-A.*

Note that, although the graph for extruded (machined) sections shows a cutoff at about *1.4* for the ratio F_{CC} / F_{CY}, we typically put the limit at F_{CY} anyway (ratio of 1.0). That higher than 1 cutoff ratio comes from F_{TU} / F_{CY} which in our case would be: *75 / 60 = 1.25*.

So for $E = 10.0*10^6$ *psi*, F_{cy} = 60,000 *psi* and from the dimensions shown in *Figure 15-36*, the crippling allowable of each segment is summarized in *Table 15-17* for the upper cap and in *Table 15-18* for the lower cap.

Table 15-17 *- Crippling analysis for the Rear Spar Section A-A upper cap.*

Upper cap RS section A-A

Segment	Free edge	b_n	t_n	$b_n t_n$	b_n / t_n	$(F_{cy} / E)^{0.5}$	$(F_{cy} / E)^{0.5}(b_n/t_n)$	F_{CCn} / F_{cy}	F_{CCn}	$t_n b_n F_{CCn}$
		[in]	[in]	[in²]					[ksi]	[lbs]
1	1	2.00	0.200	0.400	10.000	0.0775	0.775	0.749	44.9	17976
2	1	1.06	0.19	0.201	5.579	0.0775	0.432	1.000	60	12084
			$\sum b_n t_n =$	0.6014				$\sum b_n t_n F_{ccn} =$		30060

Thus the crippling allowable stress for section *A-A* upper cap is:

$$F_{CC} = \frac{\sum b_n t_n F_{CCn}}{\sum b_n t_n} = \frac{30{,}060}{0.6014} = 49{,}983 \ psi$$

***Table 15-18** - Crippling analysis for the Rear Spar Section A-A lower cap.*

Lower cap RS section A-A

Segment	Free edge	b_n	t_n	$b_n t_n$	b_n / t_n	$(F_{cy} / E)^{0.5}$	$(F_{cy} / E)^{0.5}(b_n/t_n)$	F_{CCn} / F_{cy}	F_{CCn}	$t_n b_n F_{CCn}$
		[in]	[in]	[in²]					[ksi]	[lbs]
3	1	2	0.250	0.500	8.000	0.0775	0.620	0.893	53.6	26790
4	0	1.07	0.2	0.214	5.350	0.0775	0.414	1.00	60	12840
5	1	0.75	0.15	0.113	5.000	0.0775	0.387	1.00	60	6750
			$\Sigma b_n t_n =$	0.8265					$\Sigma b_n t_n F_{ccn} =$	46380

Thus the crippling allowable stress for *Section A-A* lower cap is:

$$F_{CC} = \frac{\sum b_n t_n F_{CCn}}{\sum b_n t_n} = \frac{46{,}380}{0.8265} = 56{,}116 \ psi$$

15.11.2 SECTION B-B

The crippling stress allowable *"F_{CC}"* will be calculated below for the section *B-B*. The section is divided into segments as shown in *Figure 15-38*. All segments have one edge free except segment *4*. The small flange (lip) around the cutout is too small to be considered as a support for their adjacent segments. A crippling stress allowable for each upper cap and lower cap is calculated as they work separately.

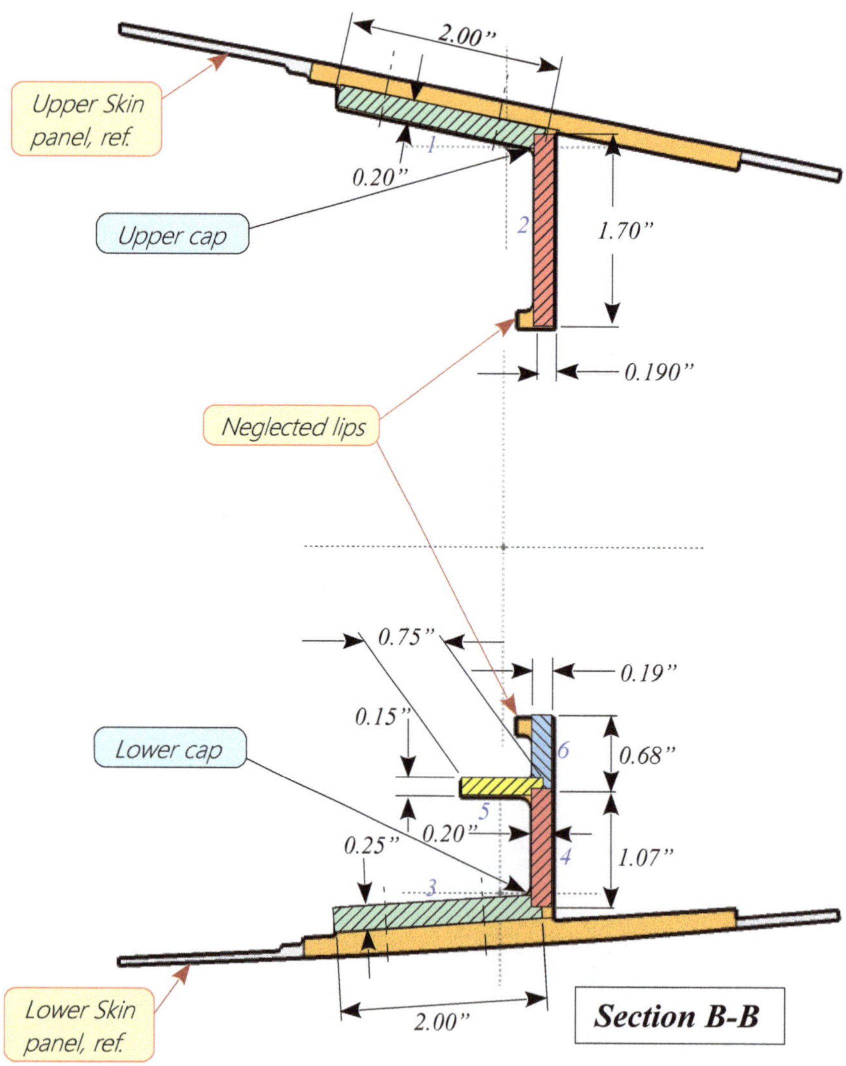

***Figure 15-38** – Effective segments for crippling at section B-B.*

The crippling allowable for each segment is calculated as shown below. Note that the crippling allowable "F_{CC}" of a segment can never be higher than F_{CY}. So the maximum value for F_{CCn} / F_{cy} is 1.0. Using *Figure 15-39* for section *B-B*, the value at the abscissa is calculated as:

$$abscissa = \sqrt{F_{CY}/E}(b/t)$$

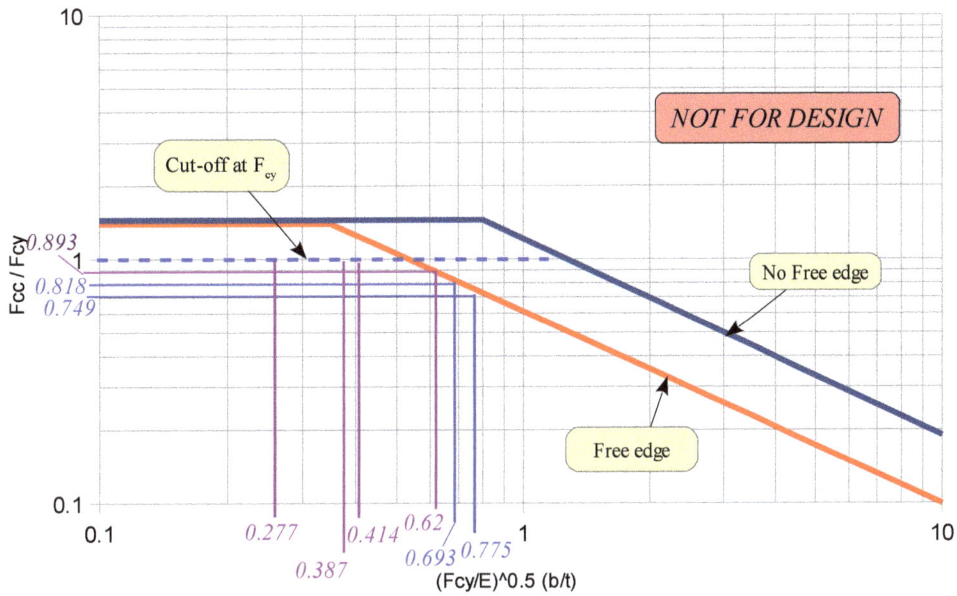

***Figure 15-39** – Typical Crippling curves section B-B.*

So for $E = 10.0*10^6$ *psi*, $F_{cy} = 60{,}000$ *psi* and from the dimensions shown in *Figure 15-38*, the crippling allowable of each segment is summarized in *Table 15-19* for the upper cap and in *Table 15-20* for the lower cap.

***Table 15-19** - Crippling analysis for the Rear Spar Section B-B upper cap.*

Rear Spar upper cap section B-B

Segment	*Free edge*	b_n	t_n	$b_n t_n$	b_n / t_n	$(F_{cy} / E)^{0.5}$	$(F_{cy} / E)^{0.5}(b_n/t_n)$	F_{CCn} / F_{cy}	F_{CCn}	$t_n b_n F_{CCn}$
		[in]	*[in]*	*[in²]*					*[ksi]*	*[lbs]*
1	1	2.00	0.200	0.400	10.000	0.0775	0.775	0.749	44.9	17976
2	1	1.7	0.19	0.323	8.947	0.0775	0.693	0.818	49.1	15853
			$\sum b_n t_n =$	0.72300					$\sum b_n t_n F_{ccn} =$	33829

Thus the crippling allowable stress for section *B-B* upper cap is:

$$F_{CC} = \frac{\sum b_n t_n F_{CCn}}{\sum b_n t_n} = \frac{33{,}829}{0.723} = 46{,}790 \ psi$$

***Table 15-20** - Crippling analysis for the Rear Spar Section B-B lower cap.*

Rear Spar Lower cap section B-B.

Segment	*Free edge*	b_n	t_n	$b_n t_n$	b_n / t_n	$(F_{cy} / E)^{0.5}$	$(F_{cy} / E)^{0.5}(b_n/t_n)$	F_{CCn} / F_{cy}	F_{CCn}	$t_n b_n F_{CCn}$
		[in]	*[in]*	*[in²]*					*[ksi]*	*[lbs]*
3	1	2	0.250	0.500	8.000	0.0775	0.620	0.89	53.6	26790
4	0	1.07	0.2	0.214	5.350	0.0775	0.414	1.00	60	12840
5	1	0.75	0.15	0.113	5.000	0.0775	0.387	1.00	60	6750
6	1	0.68	0.19	0.129	3.579	0.0775	0.277	1.00	60	7752
			$\sum b_n t_n =$	0.95570					$\sum b_n t_n F_{ccn} =$	54132

Thus the crippling allowable stress for section *B-B* lower cap is:

$$F_{CC} = \frac{\sum b_n t_n F_{CCn}}{\sum b_n t_n} = \frac{54{,}132}{0.9557} = 56{,}641 \ psi$$

15.12 EFFECTIVE SKIN WIDTH

Because the upper and lower wing planks are fastened to the Rear Spar upper and lower caps, a portion of those planks will effectively work with the Spar in taking a share of the axial and bending loads. In order to assume that there is indeed an effective portion of the plank that works with the Spar, there shall be NO inter-rivet buckling. Since we don't know yet what the stresses are in the plank, we can only assume that no inter-rivet buckling occurs for now. It will be verified in section *15.19*.

Figure 15-40 shows the dimensions of the upper plank at the Rear Spar connection that are used in the effective skin width calculations.

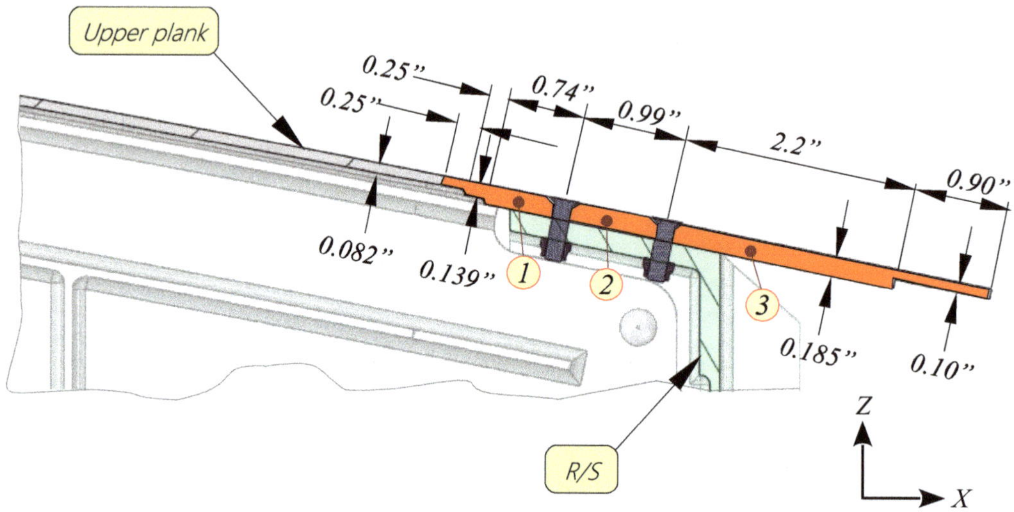

***Figure 15-40** – Upper plank effective skin width dimensions.*

Both plank and Rear spar are considered to have the same material properties in the elastic range. The skin/plank material properties are used.

There are three segments to consider in our example; the forward portion, the portion between the two fastener rows and, the trailing edge portion. The effective width of each portion is calculated and will be summed-up to get the total effective skin width.

1) For the portion forward of the first fastener row, the effective width "w" is:

$$w = 1.9t\sqrt{(E/F_{CY})}$$

Since there are three different skin thicknesses along that portion, as shown in *Figure 15-40*, an equivalent thickness "t_{eq}" is calculated to be used in the equation above. Since we know that the upper plank pocket buckles (see section *15.7*), an effective pocket length of only 0.25 *in.* is chosen.

$$t_{eq} = \frac{0.74(0.185) + 0.25(0.139) + 0.25(0.082)}{0.74 + 0.25 + 0.25} = 0.155\ in.$$

Note that some references use the crippling stress allowable "F_{CS}" instead of the compression yield stress "F_{CY}". This is indeed more conservative if you allow the crippling stress to be higher than the yield compression stress. However, since our crippling stress is limited to "F_{CY}", I prefer using "F_{CY}" in the equation instead of "F_{CS}".

As the effective width "*w*" is for a total width for a single row of fasteners and since we are only interested in one side (forward portion only) of that fastener row, the calculated width is divided by 2.

$$w_1 = w/2 = \frac{1.9}{2}(0.155)\sqrt{(10{,}000/55)} = 1.988\ in.$$

For
$t_{eq} = 0.155\ in$
$F_{CY} = 55{,}000\ psi$
$E = 10{,}000\ ksi$

2) The effective width of the portion between the two fastener rows is simply the pitch; $w_2 = 0.99\ in.$

3) The portion aft of the second fastener row representing the trailing edge is calculated using the equation for skins having one edge free.

$$w_3 = 0.62\,t\sqrt{(E/F_{CY})}$$

The equivalent thickness of portion (3) is calculated as:

$$t_{eq} = \frac{2.2(0.185) + 0.90(0.10)}{2.2 + 0.90} = 0.160\ in.$$

For
$t_{eq} = 0.160\ in$
$F_{CY} = 55{,}000\ psi$
$E = 10{,}000\ ksi$

hence

$$w_3 = 0.62(0.160)\sqrt{(10{,}000/55)} = 1.34\ in.$$

The total effective skin width is thus:

$$w_{Tot} = w_1 + w_2 + w_3 = 1.988 + 0.99 + 1.34 = 4.318\,in.$$

The corresponding area of the effective skin "A_{eff}" is:

$$\begin{aligned} A_{eff} &= w_1\,t_{eq1} + w_2\,t_{eq2} + w_3\,t_{eq3} \\ &= 1.988\,(0.155) + 0.99\,(0.185) + 1.34\,(0.160) = 0.706428\,in.^2 \end{aligned}$$

Now, because I expect the thinner segment of the pocket and also the thinner trailing edge portion to buckle under the compression stress, I decided to simply consider the effective skin width as the length of the plate base thickness (*0.185"*) between the machined steps. See ***Figure 15-41***. The effective width and area are then:

$$w_{eff} = (0.74 + 0.99 + 2.2) = 3.93\,in.$$

$$A_{eff} = w_{eff}(t_{base}) = 3.93(0.185) = 0.72705\,in.^2$$

Whether I use the lengthy approach or the assumed width having the constant base thickness, both effective areas are very close (within 3%) which is reassuring.

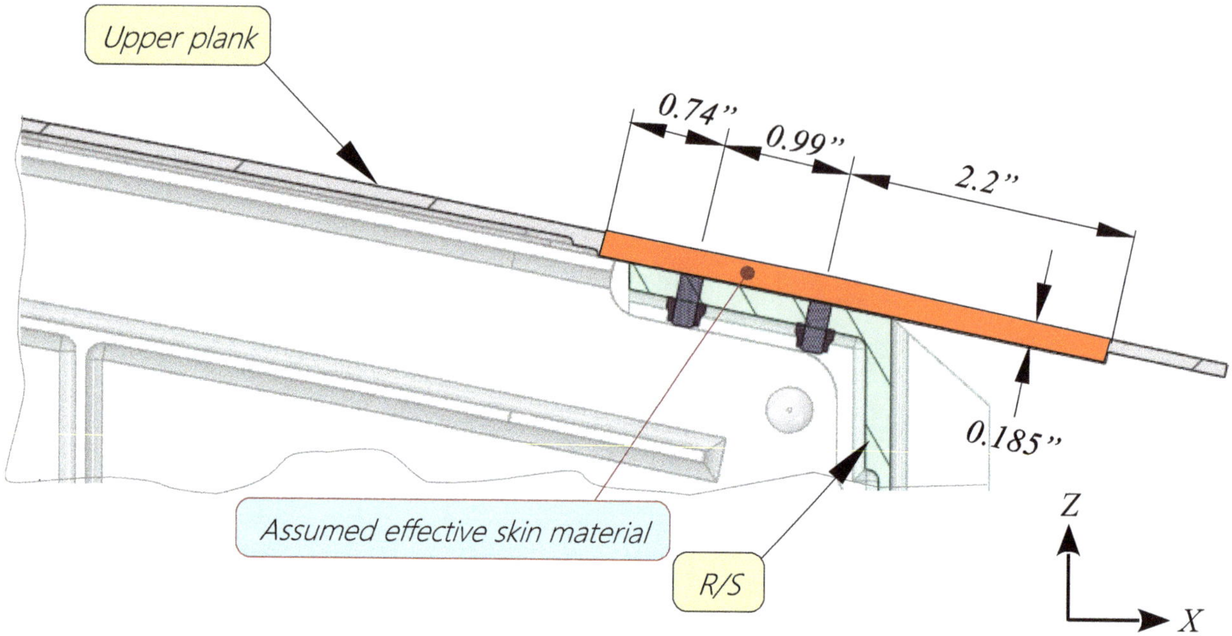

***Figure 15-41** – Assumed Upper plank effective skin width.*

The same assumption was made for the lower plank effective skin material but isn't demonstrated here as it would be redundant. Those effective skin material will be considered in the calculation of the section properties.

15.13 SECTION PROPERTIES

Referring to *Figure 15-6*, the section properties at sections *A-A* and *B-B* are calculated by including the effective skins. It was decided, for conservatism and simplicity, to consider the effective skin in compression for both upper and lower caps. This allows for a program to perform the stress analysis of all the thousands of load cases when hundreds of them may be putting the lower cap into compression.

Figure 15-42 and *Figure 15-43* show the dimensions and section properties for sections *A-A* & *B-B* respectively. The effective material is shown in orange while the considered buckled segments are in pale gray.

In both figures, the centroids of the whole sections are shown as well as the centroids of the upper and lower caps by themselves.

Even if the upper plank, the lower plank and the Rear Spar use different Aluminum alloys, their modulus of elasticity "*E*" is considered the same for the whole section. Otherwise, equivalent areas for the effective skin for instance would have to be calculated.

However, their respective stress allowables will be used according to the locations of the stress points. For instance, points *1, 2 & 3* of section *A-A* are located in the upper plank made out of *Al Aly 7075-T7351* having an F_{CY} of 55 *ksi* while points *6, 7 & 8* are located in the lower plank made out of *Al Aly 2024-T351* having an F_{CY} of 37 *ksi*. Points *4 & 5* are located in the Rear Spar made out of *Al Aly 7050-T7451* having an F_{CY} of 60 *ksi*.

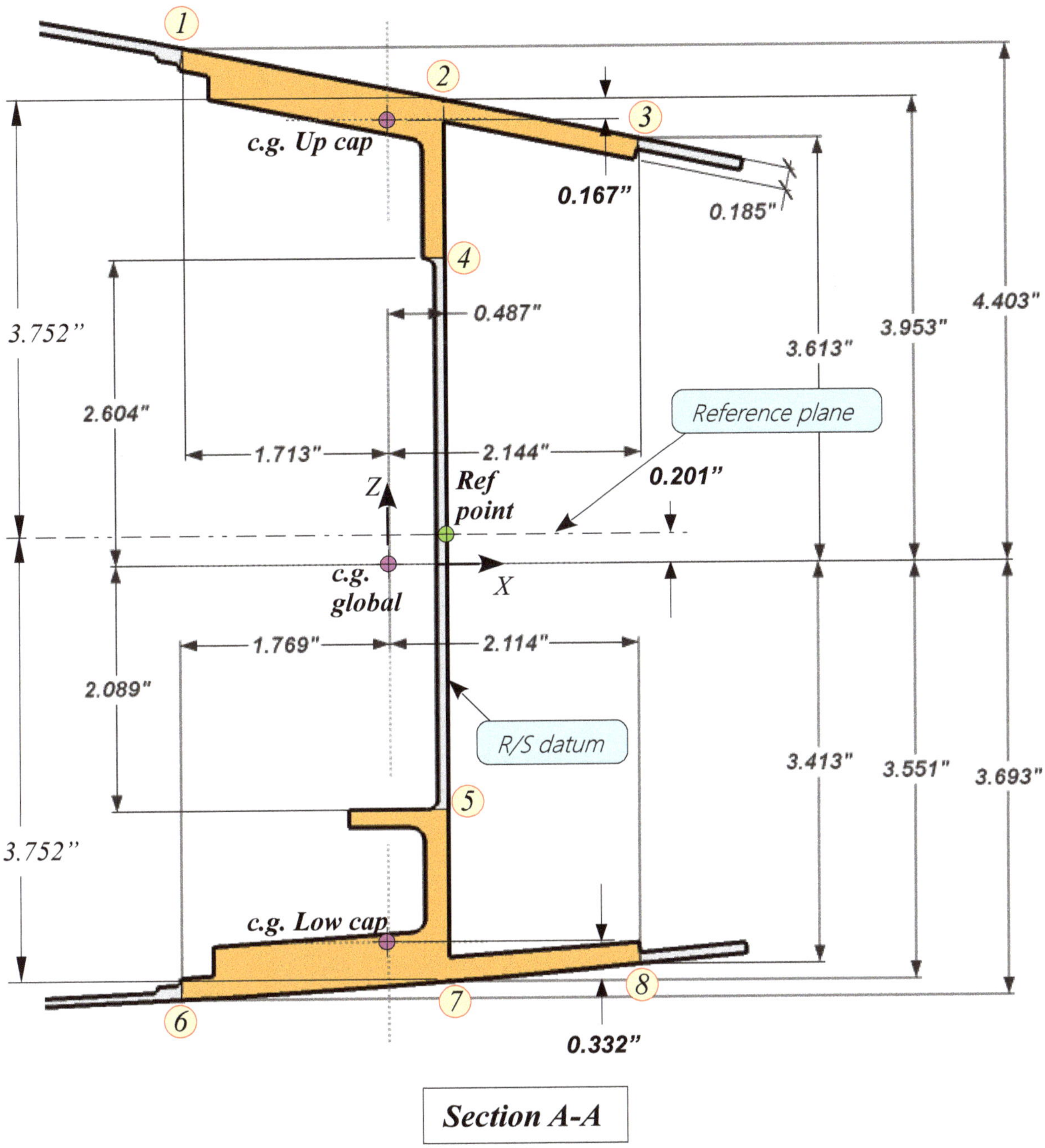

Global section:
Area = 2.8648 in^2
I_x = 35.4309 in^4
I_z = 2.4820 in^4
I_{xz} = 0.0380 in^4

Upper cap section:
Area = 1.3146 in^2
I_x = 0.1660 in^4
I_z = 1.1946 in^4
I_{xz} = 0.2593 in^4

Lower cap section:
Area = 1.5502 in^2
I_x = 0.2676 in^4
I_z = 1.2742 in^4
I_{xz} = 0.1129 in^4

***Figure 15-42** – Section A-A dimensions and section properties.*

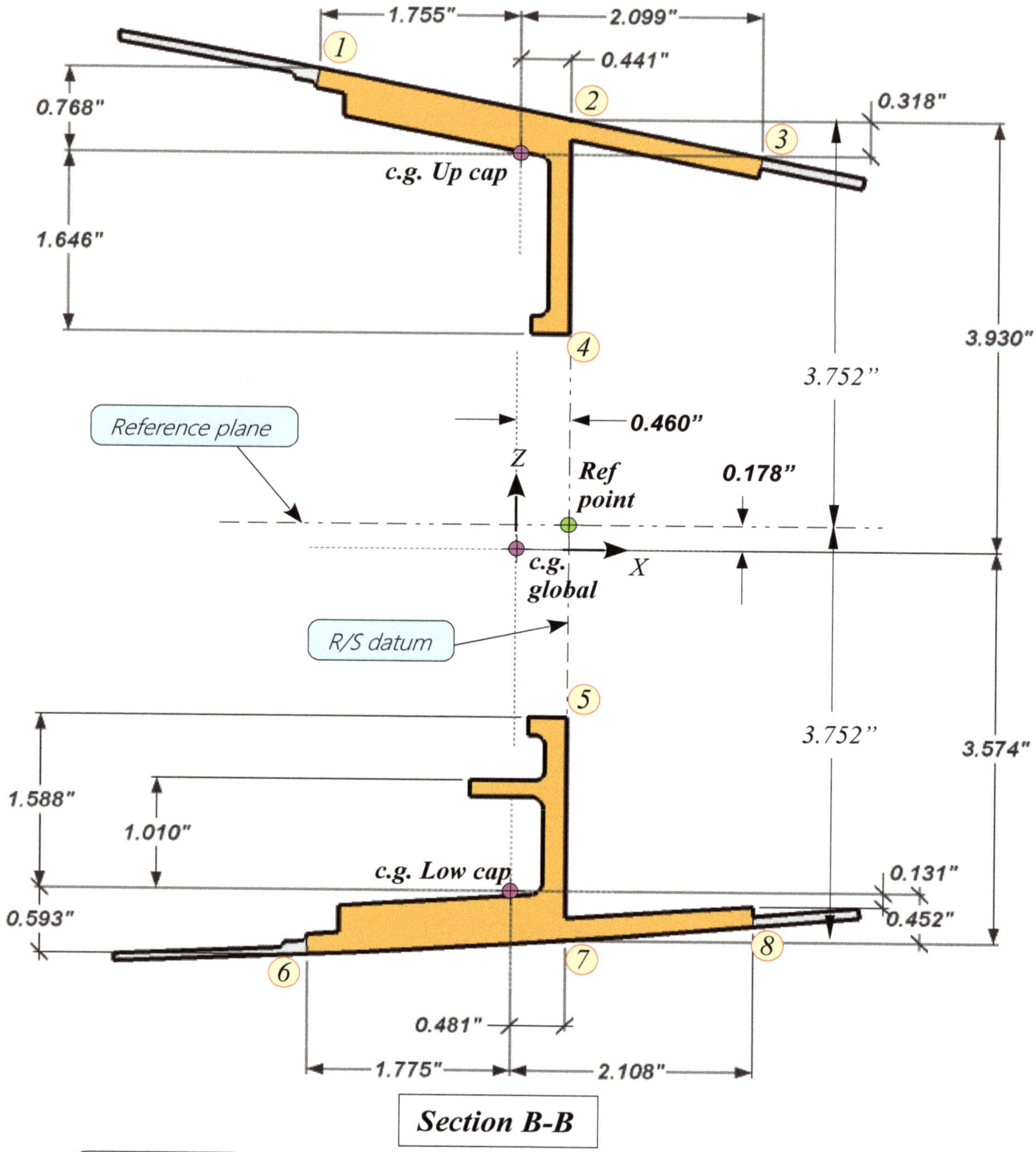

Global section:
Area = 3.1461 in^2
I_x = 36.5096 in^4
I_z = 2.5093 in^4
I_{xz} = 0.0630 in^4

Upper cap section:
Area = 1.4580 in^2
I_x = 0.4738 in^4
I_z = 1.2107 in^4
I_{xz} = 0.3275 in^4

Lower cap section:
Area = 1.6881 in^2
I_x = 0.5397 in^4
I_z = 1.2937 in^4
I_{xz} = 0.1828 in^4

Figure 15-43 – *Section B-B dimensions and section properties.*

15.14 PRELIMINARY STRESS ANALYSIS

The preliminary stresses are calculated for sections *A-A* & *B-B* from the hypothetical critical load case *LC2001*. The internal loads that were extracted at the reference plane, as shown in the FBD *Figure 15-21*, need to be transformed to each section's centroid.

The stresses calculated in this section are preliminary since they will be changed due to the column buckling effect and the slotted beam effect at the cutout. See section *15.18* for the actual stresses and MS.

15.14.1 SECTION A-A

In order to do the stress analysis of the R/S at section *A-A*, the sum of the forces and moments have to be transformed to the centroid of the section, as shown in *Figure 15-42*, by −0.487 *in.* in *X* and by −0.201 *in.* in *Z*. The new sum of the forces and moments are calculated below at the section near Rib-3 and the transformed FBD for section *A-A* is shown in *Figure 15-44*.

Of course, the forces remain unchanged from *Table 15-8* but the moments are re-calculated as follows:

$$M_X = M_{X\,ref} + P_Y(d_Z)$$
$$= 246{,}855 + 9{,}915(-0.201) = 244{,}862\ in.-lb$$

$$M_Y = M_{Y\,ref} - V_X(d_Z) + V_Z(d_X)$$
$$= 52{,}142 - 628(-0.201) + (-13{,}141)(-0.487)$$
$$= 58{,}668\ in.-lb$$

$$M_Z = M_{Z\,ref} - P_Y(d_X)$$
$$= -589 - 9{,}915(-0.487) = 4{,}240\ in.-lb$$

For
$V_{X\,ref} = 628\ lbs$
$P_{Y\,ref} = 9{,}915\ lbs$
$V_{Z\,ref} = -13{,}141\ lbs$
$M_{x\,ref} = 246{,}855\ in\text{-}lb$
$M_{Y\,ref} = 52{,}142\ in\text{-}lb$
$M_{Z\,ref} = -589\ in\text{-}lb$
$d_X = -0.487\ in$
$d_Z = -0.201\ in$

The same thing was done for section *A-A* near Rib-4 and for section *D-D* but are not demonstrated. Note that the section properties at section *D-D* are the same as section *A-A*.

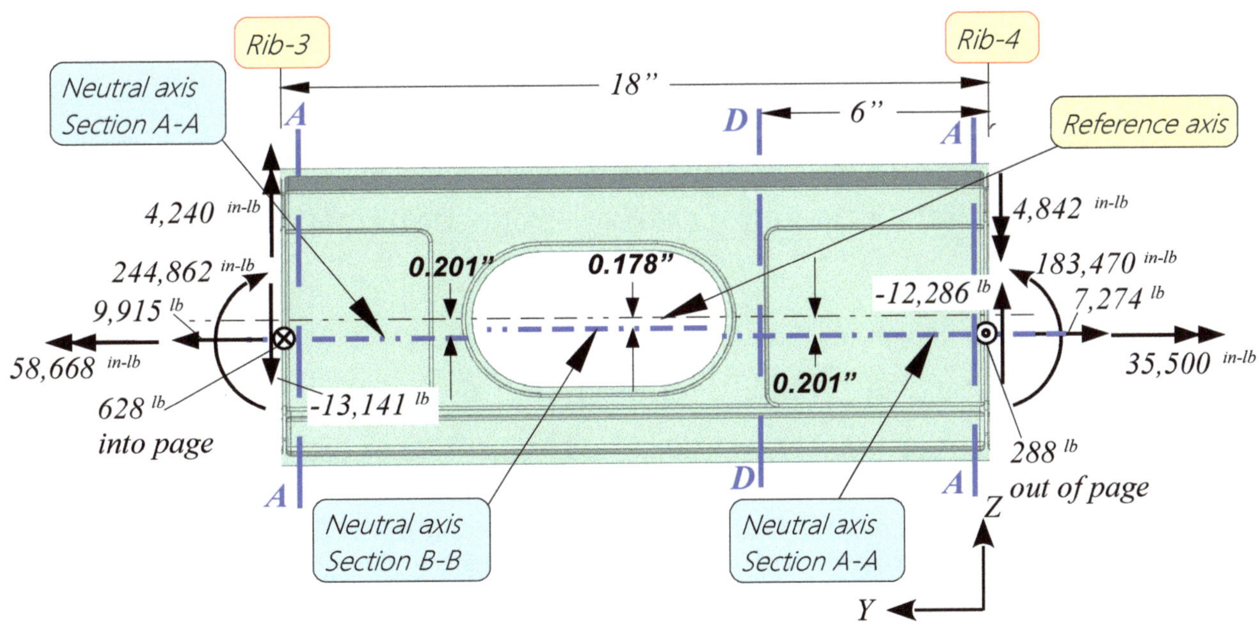

Figure 15-44 – *Transformed FBD at section A-A c of g at Ribs-3 & 4 and section D-D.*

The resulting transformed applied loads at Rib-3, section *D-D*, and Rib-4 for section *A-A* are summarized in *Table 15-21* for *LC 2001*.

Table 15-21 *- Summarized applied loads LC2001 at neutral axis of section A-A at Ribs-3 & 4 and section D-D.*

At section:	A-A (Rib3)	D-D	A-A (Rib4)
y [in] =	18	6	0
Vx [lb] =	628	401	288
Py [lb] =	9915	8155	7274
Vz [lb] =	-13141	-12571	-12286
Mx [in-lb] =	244862	203948	183470
My [in-lb] =	58668	43223	35500
Mz [in-lb] =	4240	4641	4842

The stresses at the qty*(8)* stress points, as shown in *Figure 15-42*, are calculated by conservatively using the complex bending equation even though the Rear Spar is fastened to the upper and lower wing planks and is somewhat stabilized.

<u>Axial</u>

The axial stress trough the section *A-A* is calculated as:

$$f_{ay} = \frac{P_Y}{A} = \frac{9{,}915}{2.8648} = 3{,}461 \ psi$$

For
$P_Y = 9{,}915$ lbs
$A = 2.8648$ in

<u>Bending</u>

The complex bending stress is thus calculated as:

$$f_{by} = -\frac{(M_X I_Z + M_Z I_{XZ})z + (M_Z I_X + M_X I_{XZ})x}{I_X I_Z - I_{XZ}^2}$$

Calculation example at point ***1*** of section *A-A* due to the ***LC 2001*** at Rib-3:

For
$M_X = 244{,}862$ in-lb $\quad I_Z = 2.482$ in^4 $\quad I_X = 35.4309$ in^4 $\quad x = -1.713$ in
$M_Z = 4{,}240$ in-lb $\quad I_{XZ} = 0.038$ in^4 $\quad z = 4.403$ in

$$f_{by1} = -\frac{(244{,}862(2.482) + 4{,}240(0.038))(4.403) + (4{,}240(35.4309) + 244{,}862(0.038))(-1.713)}{35.4309 \times 2.482 - 0.038^2}$$
$$= -33{,}545 \ psi$$

The total stress "$f_{total\,1}$" at point ***1***, without considering the shear yet, is:

$$f_{total1} = f_a + f_b = 3{,}461 + (-33{,}545) = -30{,}084 \ psi$$

Table 15-22 shows the axial and bending stresses through section *A-A* of the Rear Spar at Rib-3 for all stress points along with the position of each stress point relative to the c.g.

Table 15-22 *- Bending stresses at all stress points of section A-A near rib-3.*

Stress points	*x* [in]	*z* [in]	*fa* [psi]	*fb* [psi]	*fa + fb* [psi]
1	-1.713	4.403	3,461	-33,545	**-30,084**
2	0.466	3.953	3,461	-26,481	-23,020
3	2.144	3.613	3,461	-21,087	-17,626
4	0.466	2.604	3,461	-17,156	-13,695
5	0.466	-2.089	3,461	15,286	18,747
6	-1.769	-3.693	3,461	22,320	25,781
7	0.466	-3.551	3,461	25,393	28,854
8	2.114	-3.413	3,461	27,429	30,890

Table 15-23 shows the axial and bending stresses through section *D-D* of the Rear Spar for all stress points of section *A-A* along with the position of each stress point relative to the c.g.

Table 15-23 *- Bending stresses at all stress points of section A-A, through section D-D.*

Stress points	*x* [in]	*z* [in]	*fa* [psi]	*fb* [psi]	*fa + fb* [psi]
1	-1.713	4.403	2,847	-28,708	**-25,862**
2	0.466	3.953	2,847	-21,850	-19,004
3	2.144	3.613	2,847	-16,607	-13,760
4	0.466	2.604	2,847	-14,082	-11,236
5	0.466	-2.089	2,847	12,942	15,788
6	-1.769	-3.693	2,847	17,802	20,648
7	0.466	-3.551	2,847	21,360	24,207
8	2.114	-3.413	2,847	23,793	26,639

Table 15-24 shows the axial and bending stresses through section *A-A* of the Rear Spar at Rib-4 for all stress points along with the position of each stress point relative to the c.g.

Table 15-24 *- Bending stresses at all stress points of section A-A near rib-4.*

Stress points	*x* [in]	*z* [in]	*fa* [psi]	*fb* [psi]	*fa + fb* [psi]
1	-1.713	4.403	2,539	-26,287	**-23,748**
2	0.466	3.953	2,539	-19,532	-16,993
3	2.144	3.613	2,539	-14,364	-11,825
4	0.466	2.604	2,539	-12,544	-10,005
5	0.466	-2.089	2,539	11,768	14,307
6	-1.769	-3.693	2,539	15,540	18,079
7	0.466	-3.551	2,539	19,342	21,881
8	2.114	-3.413	2,539	21,973	24,512

Vertical Shear; $f_{s\,z}$

The vertical shear "$f_{s\,z}$" is assumed to be transferred only through the web. The flanges (and the wing planks), being almost horizontal, are assumed inefficient in the vertical direction. See *Figure 15-31* and *Table 15-16* for the average vertical shear stress through section *A-A*.

The maximum stress is obtained near rib-3 (in panel *A* of the R/S) and is; $f_{s\,z}$ = 16,775 *psi*. Referring to *Figure 15-42*, this shear stress exists at stress points *2, 4, 5 & 7*.

Fore-aft Shear: $f_{S\,X}$

The shear stress "$f_{S\,X}$" in the *fore-aft* direction is considered transferred by the upper and lower effective skin and the R/S flanges. Not only this stress is caused by the transverse shear in the X direction "V_X" but also by the torsion moment "M_Y" acting upon the R/S as a couple between the centroids of the upper and the lower flanges. See *Figure 15-45* for the FBD of the fore-aft shear loads.

> The torsion moment "M_Y" is considered reacted as a force couple applied at the upper and lower caps centroid for simplicity. However, the couple should really be applied at the shear center instead of the centroid. Since the shear center is harder to calculate and since it would be very close to the centroid (in this case), the centroid will do.

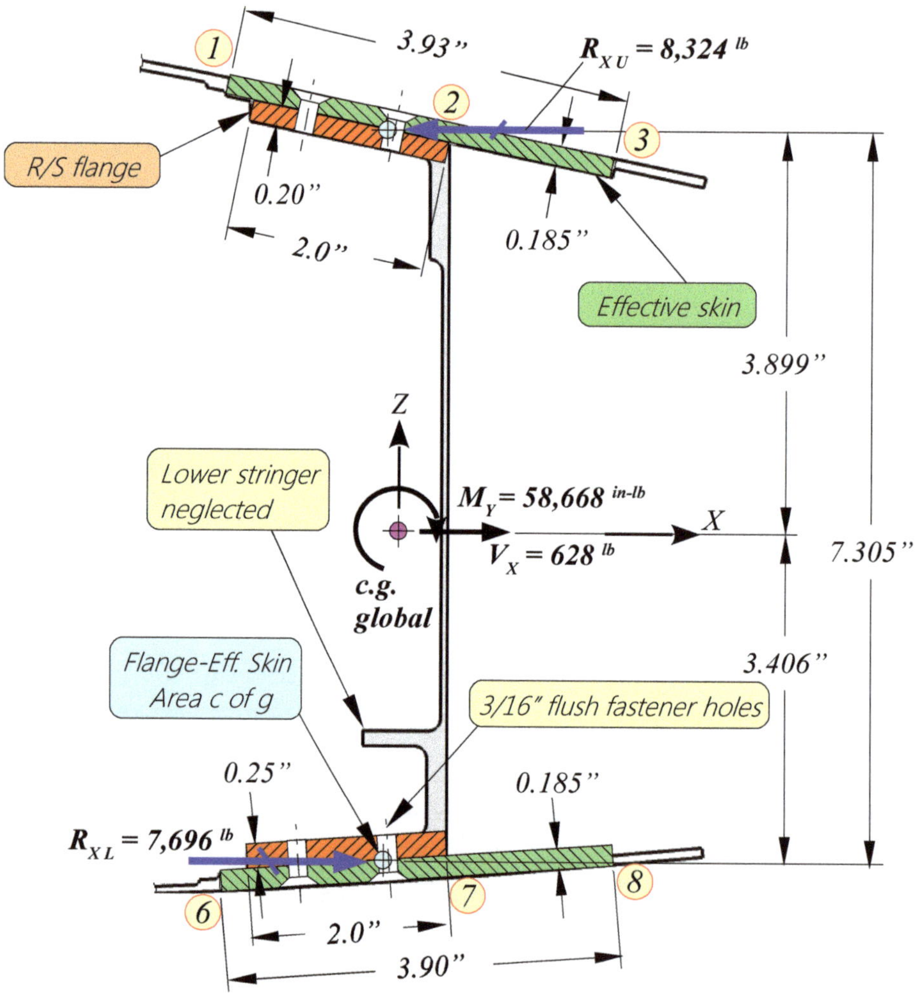

Figure 15-45 *– FBD of Section A-A due to fore-aft "X" shear loads.*

Using the loads applied at the centroid of section *A-A* from *Table 15-21*, the reactions forces at the centroid, in the directions shown, of the upper and lower fore-aft shear areas are calculated as follows:

$$R_{X\,U} = \frac{M_Y}{h} + V_X\,\frac{h_l}{h} = \frac{58{,}668}{7.305} + 628\,\frac{3.406}{7.305} = 8{,}324\ lbs$$

$$R_{X\,L} = \frac{M_Y}{h} - V_X\,\frac{h_u}{h} = \frac{58{,}668}{7.305} - 628\,\frac{3.899}{7.305} = 7{,}696\ lbs$$

Note here that I decided to conservatively neglect the 1,400 *lbs* reaction load applied to the upper cap in the *fore-aft* direction from the diagonal tension effect (See *Figure 15-58*). The reason being that this load is in the opposite direction and therefore would be a relief to the force calculated above. Providing the lower plank would get into diagonal tension, then the load from the *D.T.* would have to be added to get the total shear load in the lower cap.

The shear areas, by subtracting the flush fastener holes, are calculated as follows:

$$A_{XU} = 3.93(0.185) + 2.0(0.20) - 2\left[0.189(0.185+0.2) + \frac{(0.30-0.189)}{2}(0.047)\right] = 0.97626\ in.^2$$

$$A_{XL} = 3.90(0.185) + 2.0(0.25) - 2\left[0.189(0.185+0.25) + \frac{(0.30-0.189)}{2}(0.047)\right] = 1.05186\ in.^2$$

Then, the average shear stresses in the "X" direction, applicable to points *1, 2 & 3*, are calculated in the upper cap as follows:

$$f_{SXU} = \frac{R_{XU}}{A_{XU}} = \frac{8{,}324}{0.97626} = 8{,}526\ psi$$

Similarly, the shear stresses in the "X" direction, applicable to points *6, 7 & 8*, are calculated in the lower cap as follows:

$$f_{SXL} = \frac{R_{XL}}{A_{XL}} = \frac{7{,}696}{1.05186} = 7{,}317\ psi$$

15.14.2 SECTION B-B

Like for section *A-A*, in order to do the stress analysis of the R/S at section *B-B*, the sum of the forces and moments have to be transformed to the centroid of the section, as shown in *Figure 15-42*, by −0.460 *in.* in *X* and by −0.178 *in.* in *Z*. The new sum of the forces and moments is calculated below at the inboard edge of the cutout (shown as section *C-C* in *Figure 15-6)* and the transformed FBD for section *B-B* is shown in *Figure 15-46*. The cross-section of section *C-C* is assumed exactly the same as section *B-B* for conservatism.

Of course, the forces remain unchanged from *Table 15-8* but the moments are re-calculated as follows:

$$\begin{aligned} M_X &= M_{X\,ref} + P_Y(d_Z) \\ &= 230{,}514 + 9{,}218(-0.178) = 228{,}873\ in.-lb \end{aligned}$$

$$\begin{aligned} M_Y &= M_{Y\,ref} - V_X(d_Z) + V_Z(d_X) \\ &= 46{,}156 - 538(-0.178) + (-12{,}916)(-0.460) \\ &= 52{,}193\ in.-lb \end{aligned}$$

$$\begin{aligned} M_Z &= M_{Z\,ref} - P_Y(d_X) \\ &= -91 - 9{,}218(-0.460) = 4{,}150\ in.-lb \end{aligned}$$

For
$V_{X\,ref} = 538\ lbs$
$P_{Y\,ref} = 9{,}218\ lbs$
$V_{Z\,ref} = -12{,}916\ lbs$
$M_{x\,ref} = 230{,}514\ in\text{-}lb$
$M_{Y\,ref} = 46{,}156\ in\text{-}lb$
$M_{Z\,ref} = -91\ in\text{-}lb$
$d_X = -0.460\ in$
$d_Z = -0.178\ in$

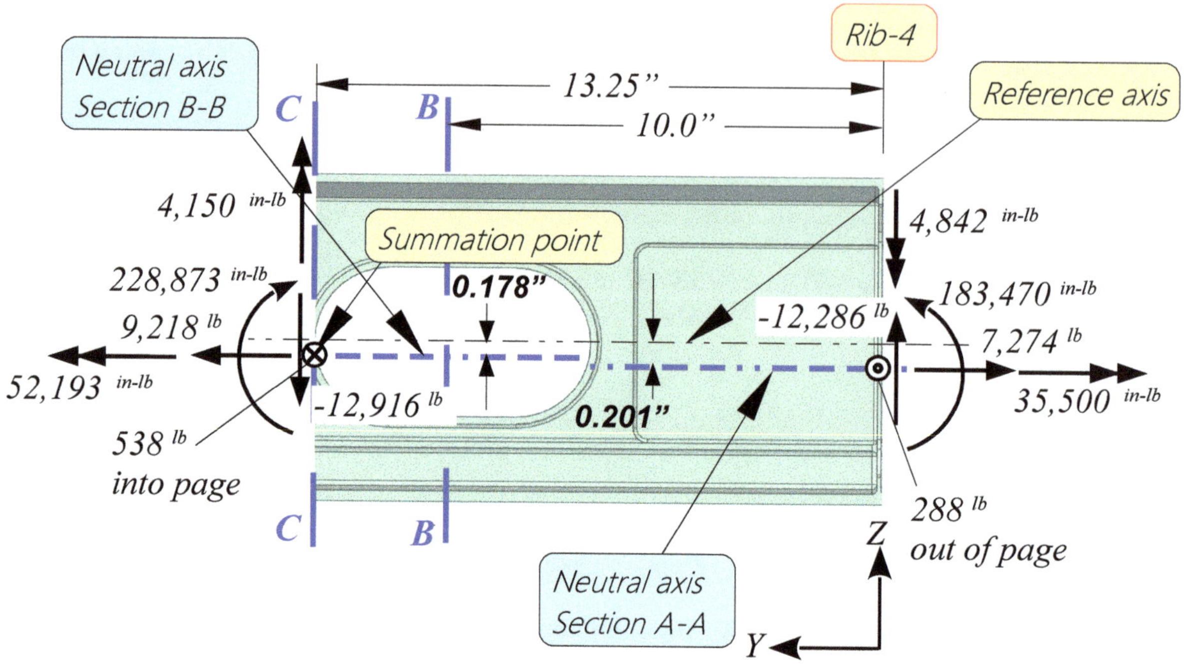

***Figure 15-46** – Transformed FBD at section B-B c.of g located at section C-C.*

The resulting transformed applied loads at section *C-C* (*13.25 inches* from Rib-4), and at the center of the cutout (section *B-B)* are summarized in *Table 15-25* for *LC 2001*.

***Table 15-25** - Summarized applied loads LC2001 at neutral axis of section B-B at two locations.*

At section:	C-C	B-B
y [in] =	13.25	10
Vx [lb] =	538	477
Py [lb] =	9218	8742
Vz [lb] =	-12916	-12761
Mx [in-lb] =	228873	217778
My [in-lb] =	52193	48016
Mz [in-lb] =	4150	4271

The stresses at the qty*(8)* stress points, as shown in *Figure 15-43*, are calculated by conservatively using the complex bending equation even though the Rear Spar is fastened to the upper and lower wing planks.

Axial

The axial stress trough the section *B-B* is calculated as:

$$f_{ay} = \frac{P_Y}{A} = \frac{9{,}218}{3.1461} = 2{,}930 \ psi$$

For
$P_Y = 9{,}218$ lbs
$A = 3.1461$ in

Bending

The complex bending stress is thus calculated as:

$$f_{by} = -\frac{(M_X I_Z + M_Z I_{XZ})z + (M_Z I_X + M_X I_{XZ})x}{I_X I_Z - I_{XZ}^2}$$

Below is shown a calculation example at point ***1*** of section *B-B* due to the ***LC 2001*** at section *C-C*:

For
$M_X = 228{,}873$ in-lb $\quad I_Z = 2.5093$ in⁴ $\quad I_X = 36.5096$ in⁴ $\quad x = -1.755$ in
$M_Z = 4{,}150$ in-lb $\quad I_{XZ} = 0.0630$ in⁴ $\quad z = 4.380$ in

$$f_{by1} = -\frac{(228{,}873(2.5093) + 4{,}150(0.063))(4.38) + (4{,}150(36.5096) + 228{,}873(0.063))(-1.755)}{36.5096 \times 2.5093 - 0.0630^2}$$

$$= -30{,}650\ psi$$

The total stress "$f_{total\ 1}$" at point *1* (without considering the shear yet) is:

$$f_{total1} = f_a + f_b = 2{,}930 + (-30{,}650) = -27{,}720\ psi$$

*Table **15-26*** shows the axial and bending stresses through section *B-B* of the Rear Spar at the section *C-C* location for all stress points along with the position of each stress point relative to the c.g.

Table 15-26 *- Bending stresses at all stress points of section B-B at inb'd edge of cutout.*

Stress points	x [in]	z [in]	fa [psi]	fb [psi]	fa + fb [psi]
1	-1.755	4.38	2,930	-30,650	**-27,720**
2	0.46	3.93	2,930	-23,816	-20,886
3	2.099	3.612	2,930	-18,853	-15,923
4	0.46	1.966	2,930	-11,498	-8,567
5	0.46	-1.534	2,930	10,454	13,384
6	-1.775	-3.715	2,930	20,085	23,016
7	0.46	-3.574	2,930	23,249	26,179
8	2.108	-3.438	2,930	25,381	28,311

*Table **15-27*** shows the axial and bending stresses through section *B-B* of the Rear Spar at the center of the cutout for all stress points along with the position of each stress point relative to the c.g.

Table 15-27 *- Bending stresses at all stress points of section B-B at center of cutout.*

Stress points	x [in]	z [in]	fa [psi]	fb [psi]	fa + fb [psi]
1	-1.755	4.38	2,779	-29,391	**-26,612**
2	0.46	3.93	2,779	-22,603	-19,824
3	2.099	3.612	2,779	-17,669	-14,891
4	0.46	1.966	2,779	-10,881	-8,103
5	0.46	-1.534	2,779	10,007	12,786
6	-1.775	-3.715	2,779	18,884	21,663
7	0.46	-3.574	2,779	22,182	24,961
8	2.108	-3.438	2,779	24,423	27,201

Vertical Shear: $f_{s\,z}$

The vertical shear "$f_{s\,z}$" is assumed to be transferred only through the vertical components of the section. The flanges (and the wing planks), being almost horizontal, are assumed inefficient in the vertical direction.

Effective web shear area

The Rear Spar flanges and the skins as well as the lower stringer are considered to be inefficient to react vertical shear loads. Hence, referring to *Figure 15-47,* the effective shear area; "$A_{S\,Z}$" is:

$$A_{SZ} = h_1 t_1 + h_2 t_2 + h_3 t_3$$

$$A_{SZ} = 1.164(0.190) + 0.578(0.190) + 1.277(0.200) = 0.58638\ in.^2$$

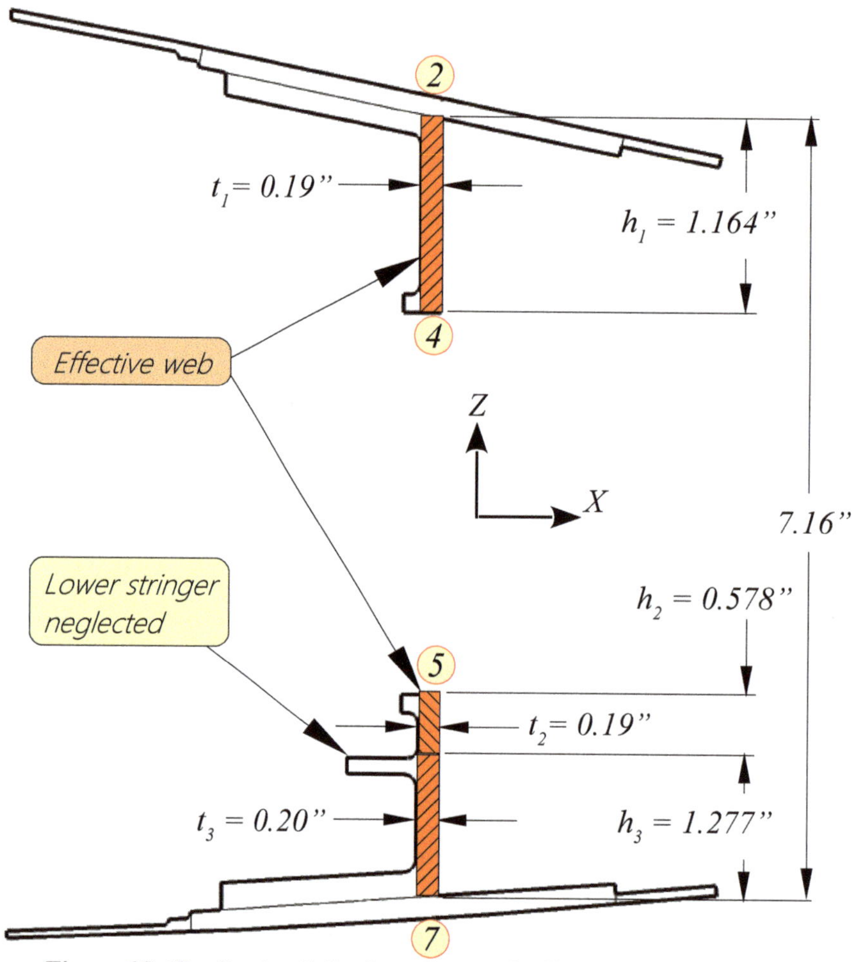

Figure 15-47 *– Section B-B - Rear Spar web effective vertical shear area.*

From *Table 15-25*, the vertical shear load "V_Z" at section *C-C* is: 12,916 *lbs*.

The average shear stress through the effective web area is thus:

$$f_{SZ} = \frac{V_Z}{A_{SZ}} = \frac{12{,}916}{0.58638} = 22{,}027\ psi$$

This shear stress exists at stress points *2, 4, 5 & 7*.

Fore-aft Shear: $f_{S\,X}$

In the same way as in section *A-A*, the shear stress "$f_{S\,X}$" in the *fore-aft* direction at section *C-C* is considered transferred by the upper and lower effective skin and the R/S flanges. Not only this stress is caused by the transverse shear in the *X* direction "V_X" but also by the torsion moment "M_Y" acting upon the R/S as a couple between the centroids of the upper and the lower flanges. See *Figure 15-48* for the *FBD* of the fore-aft shear loads.

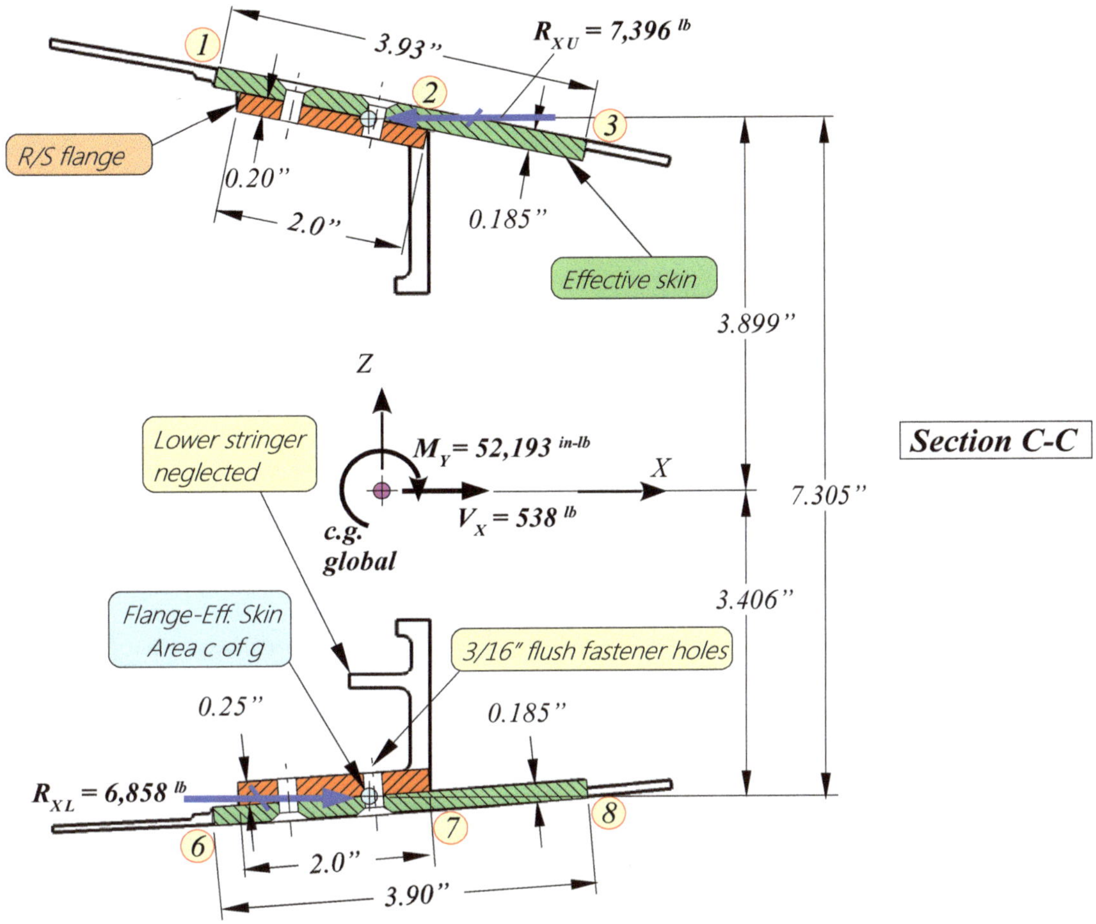

Figure 15-48 – *FBD of Section C-C due to fore-aft "X" shear loads.*

Using the loads applied at the centroid of section *C-C* from *Table 15-25,* the reactions forces at the centroid, in the directions shown, of the upper and lower fore-aft shear areas are calculated as follows:

$$R_{XU} = \frac{M_Y}{h} + V_X \frac{h_l}{h} = \frac{52{,}193}{7.305} + 538\,\frac{3.406}{7.305} = 7{,}396\ lbs$$

$$R_{XL} = \frac{M_Y}{h} - V_X \frac{h_u}{h} = \frac{52{,}193}{7.305} - 538\,\frac{3.899}{7.305} = 6{,}858\ lbs$$

The shear areas are exactly the same as for section *A-A*. Hence:

$$A_{XU} = 0.97626\ in.^2$$

$$A_{XL} = 1.05186\ in.^2$$

Then, the average shear stresses in the "X" direction, applicable to points *1, 2 & 3*, are calculated in the upper cap as follows:

$$f_{SXU} = \frac{R_{XU}}{A_{XU}} = \frac{7{,}396}{0.97626} = 7{,}576\ psi$$

Then, the shear stresses in the "X" direction, applicable to points *6, 7 & 8*, are calculated in the lower cap as follows:

$$f_{SXL} = \frac{R_{XL}}{A_{XL}} = \frac{6{,}858}{1.05186} = 6{,}520\ psi$$

Lateral Shear: $f_{S\,Y}$

Upper plank lateral shear.

Here we are interested in evaluating the shear stress, in the plank and in the R/S flange, going through the row of fasteners. That load comes from the wing upper and lower planks and is assumed transferred to the Rear Spar by the first row of fasteners. Since the thickness of the planks are constant between rib-3 & 4, the lateral shear stress is also constant. Since the top plank gets into post buckling, the increased shear flow "q_{DT}" due to the diagonal semi-tension field is used.

From page *534*, the shear stress in the upper plank is: $f_S = 9{,}234\ psi$.

From (ref. 20.), the equation to calculate the increased shear flow is:

$$q_{DT} = f_S\, t \left[1 + k\left(\frac{1}{\cos\alpha} - 1 \right) \right]$$

hence, using the standard buckling angle of 45 degrees:

$$q_{DT} = 9{,}234\ (0.125) \left[1 + 0.13\left(\frac{1}{\cos 45} - 1 \right) \right] = 1{,}222\ lb/in.$$

For
$f_S = 9{,}234\ psi$
$t = 0.125\ in$
$k = 0.13$
$\alpha = 45\ deg$

Figure 15-49 shows the shear flows in the lateral direction between the wing planks and the Rear Spar.

Referring to *Figure 15-13*, for the upper plank, one can see that the shear flow from element no. *318417* should be taken into account since the wing plank is continuous in the aft direction over the two rows of fasteners. From the FEM output file, we find a shear flow of $F_{XY} = 124\ lb/in$ in that element. Since their positive shear flow directions are opposite, we would be subtracting one from another to get the resulting shear flow taken by the row of fasteners.

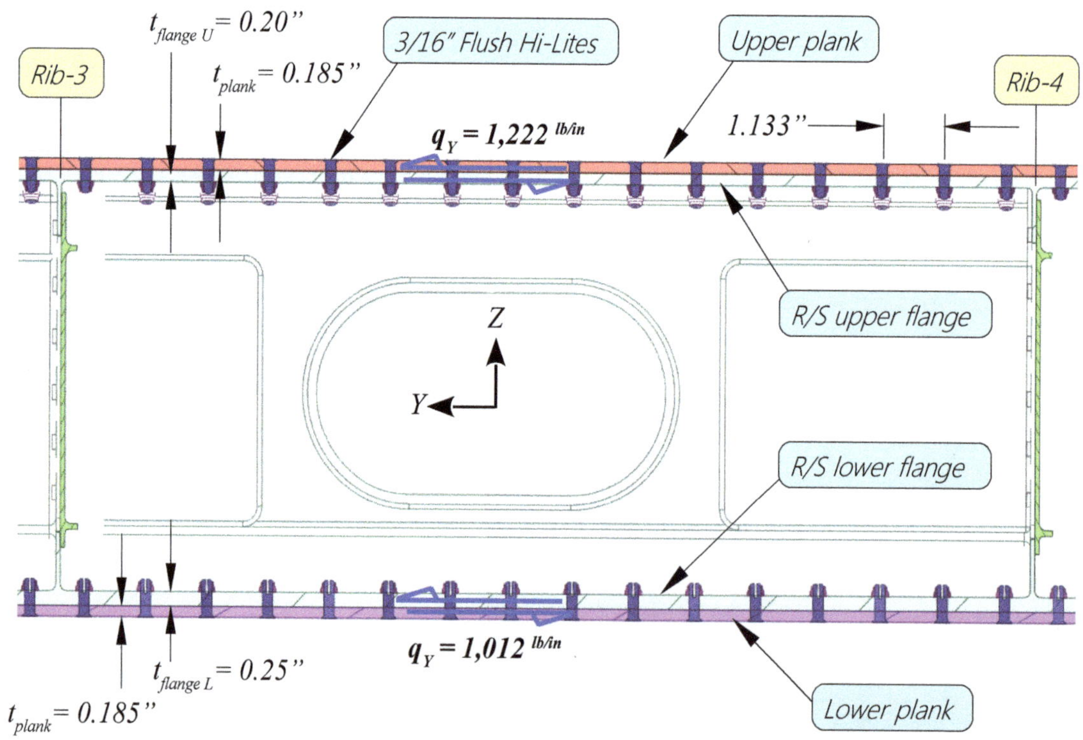

Figure 15-49 *– Shear flow along Y between wing planks and R/S.*

However, since there are two rows of fasteners, I will assume that the forward row only reacts the wing plank shear flow and the aft row reacts the trailing edge shear flow. See *Figure 15-50* for the shear flow distribution between the two rows of fasteners.

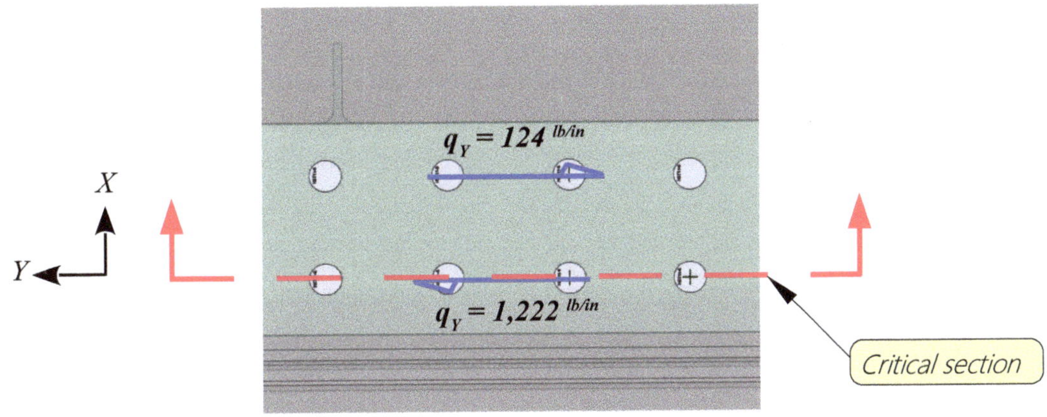

***Figure 15-50** – Shear flow distribution between the two rows of fasteners.*

The shear force for one fastener pitch "s" is; $V_Y = q\ s = 1{,}222\ (1.133) = 1{,}385\ lbs.$

The shear area in the wing plank "$A_{S\ plank}$" is calculated for one fastener pitch:

$$A_{S\ plank} = 1.133(0.185) - \left[0.189\ (0.185) + \frac{(0.30-0.189)}{2}(0.047)\right] = 0.172031\ in.^2$$

The shear stress "$f_{S\ Y plank}$" is thus:

$$f_{SY\ plank} = \frac{V_Y}{A_{S\ plank}} = \frac{1{,}385}{0.172031} = 8{,}051\ psi$$

The shear area in the Rear Spar upper flange "$A_{S\ flange}$" is calculated for one fastener pitch:

$$A_{S\ flange} = 1.133(0.20) - 0.189\ (0.20) = 0.1888\ in.^2$$

The shear stress "$f_{S\ Y flange}$" is thus:

$$f_{SY\ flange} = \frac{V_Y}{A_{S\ flange}} = \frac{1{,}385}{0.1888} = 7{,}336\ psi$$

This shear stress is considered to exist at stress point *1*.

Lower plank lateral shear

Since the lower plank doesn't get into diagonal tension *(k = 0)*, then the shear flow of the *QUAD* element *no. 307406* remains the same with $F_{XY} = q = 1{,}012\ lb/in$. The shear flow being smaller and the R/S flange of the lower cap being thicker, there is no need to demonstrate the shear stress calculations for the lower plank.

15.15 SLOTTED BEAM

Due to the large elongated cutout that was added (or cut out) in the Rear Spar web, the vertical shear *"V_Z"* will cause some additional bending stresses, at section *C-C*, at the upper and lower caps. See *Figure 15-51* for the dimensions and applied loads at the center of the slot (section *B-B*). This approach considers an inflection point *(M = 0)* at the center of the slot.

Naming the upper cap as member *"U"* and the lower cap as member *"L"*, the bending stresses are calculated using the following equations:

$$f_{iU} = -\frac{M_B}{I/c_i} - \frac{V_B\, x\, I_U/(I_U + I_L)}{(I_U/z_i)} \quad compression\,@\,point\,(i)\,upper\,cap$$

$$f_{iL} = \frac{M_B}{I/c_i} + \frac{V_B\, x\, I_L/(I_U + I_L)}{(I_L/z_i)} \quad tension\,@\,point\,(i)\,lower\,cap$$

The left hand part of the equation is for the stress *"$f_{bending\ global}$"* in the global cross section and was already calculated in *Table 15-26* for section *B-B* and *Table 15-27* for section *C-C*. The right hand part of the equation is for the additional stresses caused by the slot. Note that only the additional loads in the *YZ* plane are considered in this slot analysis as the loads in the *XY* plane are not affected by the slot.

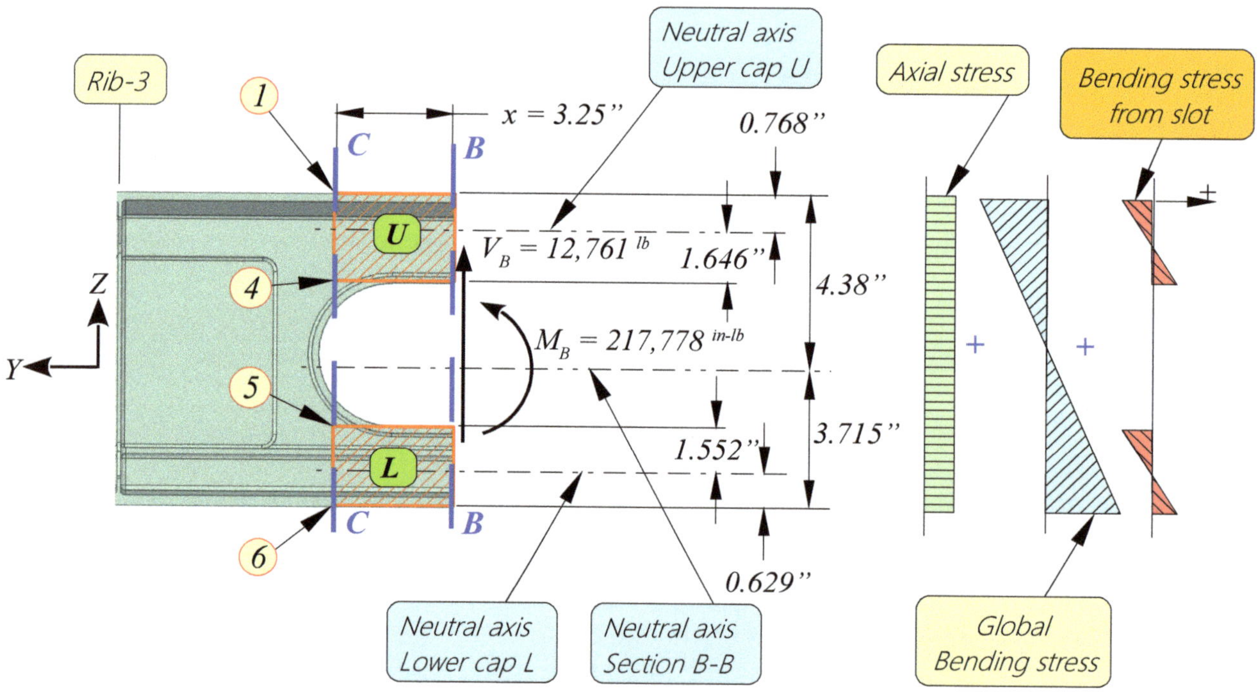

Figure 15-51 *– Rear Spar slot dimensions and applied loads.*

However, from the above picture, one cannot directly use the transverse shear "V_B" as is in the equations. As explained previously and shown in *Figure 15-23*, the *FBD* is quite complex since the contribution from the wing upper and lower planks must be taken into account. Therefore, in order to get the bending moments at section *C-C* in the upper and lower caps, an equivalent vertical shear force "V_{EQ}" is calculated from the already known *"M_X"* moments at sections *B-B "M_B"* & *C-C "M_C"* from *Table 15-25*.

Note that the distance *"x"* is local to the slot and is from its center.

Since the bending moment varies linearly, the equivalent shear force is therefore:

$$V_{EQ} = \frac{M_C - M_B}{x}$$

For
$M_B = 217{,}778$ in-lb
$M_C = 228{,}873$ in-lb
$x = 3.25$ in

therefore

$$V_{EQ} = \frac{228{,}873 - 217{,}778}{3.25} = 3{,}414\ lb$$

Note that this equivalent shear is NOT to be used to calculate the actual shear stress through the section but to simply evaluate the bending moments at the caps.

Thus, the right side of the equation "$f_{bending\ slot}$" is calculated for all qty(8) stress points of section *B-B* (see *Figure 15-43*) as shown in the summary *Table 15-28* and described in details for point *1* in the upper cap as follows:

$$f_{1\ slot} = -\frac{3{,}414\,(3.25)\,(0.4738)\,/\,(0.4738 + 0.5397)}{0.4738\ /\ 0.768} = -8{,}408\ psi$$

For
$V_{EQ} = 3{,}414$ lbs
$x = 3.25$ in
$I_U = 0.4738$ in^4
$I_L = 0.5397$ in^4
$z_1 = 0.768$ in

From *Table 15-26,* for $f_{1\ bending\ global} = -30{,}650\ psi$ and $f_{1\ axial\ global} = 2{,}930\ psi$, then:

$$f_1 = f_{1\ bending\ global} + f_{1\ axial\ global} + f_{1\ slot}$$
$$= -30{,}650 + 2{,}930 + (-8{,}408) = -36{,}128\ psi$$

Table 15-28 *- Stress summary at section C-C in upper & lower caps due to cutout.*

Upper cap Points	z_i [in]	$f_{bending\_global}$ [psi]	V_{EQ} [lbs]	$V_{EQ} * x$ [in-lbs]	$I_U/(I_U+I_L)$	I_U/z_i [in^3]	$f_{bending\_slot}$ [psi]	f_{axial} [psi]	f_{total} [psi]
1	0.768	-30650	3414	11096	0.467	0.617	-8408	2930	**-36128**
2	0.318	-23816	3414	11096	0.467	1.490	-3481	2930	-24367
3	-0.021	-18853	3414	11096	0.467	-22.562	230	2930	-15693
4	-1.646	-11498	3414	11096	0.467	-0.288	18020	2930	9453

Lower cap Points	z_i [in]	$f_{bending\_global}$ [psi]	V_{EQ} [lbs]	$V_{EQ} * x$ [in-lbs]	$I_L/(I_U+I_L)$	I_L/z_i [in^3]	$f_{bending\_slot}$ [psi]	f_{axial} [psi]	f_{total} [psi]
5	1.588	10454	3414	11096	0.533	0.340	-17385	2930	-4001
6	-0.593	20085	3414	11096	0.533	-0.910	6492	2930	29508
7	-0.452	23249	3414	11096	0.533	-1.194	4948	2930	31128
8	-0.313	25381	3414	11096	0.533	-1.724	3427	2930	**31738**

A good thing to notice here is that, at the edge of the cutout (points *4 & 5*), there is either tension or a small compression stress. Therefore, no buckling of the free edge will occur. Although, as we'll see later, those points are the most critical ones.

See section *15.18* for the MS calculations.

15.16 COLUMN BUCKLING

Above everything else, one must make sure that no general instability of the members in compression (like the Rear Spar upper cap) occurs. Any other MS are meaningless if the structure collapses due to general instability! Therefore, the Rear Spar upper cap must be checked in column buckling by taking into account the internal compression load from the FEM plus the additional compression loads from the diagonal tension effect of the upper plank and the Rear Spar web panel *"B"* calculated earlier.

But first, we must understand the possible behavior of that column due to an axial compression force from its geometry and its restraints in both *YZ* and *XY* planes. ***Figure 15-52*** shows the diagonal semi-tension field effect on the R/S upper cap in the *XY* plane, while ***Figure 15-53*** shows diagonal semi-tension field effect on the R/S upper cap in the *YZ* plane.

Since there is no way for the R/S upper cap to move outboard of the wing box (in the positive *X* direction) or upward (in the positive *Z* direction), then, the R/S upper cap can only be assumed fixed at its supports. It is thus considered fixed at every rib in the *XY* plane and fixed at the uprights in the *YZ* plane.

The uprights are used as supports because, as determined in section ***15.10***, they are non buckling.

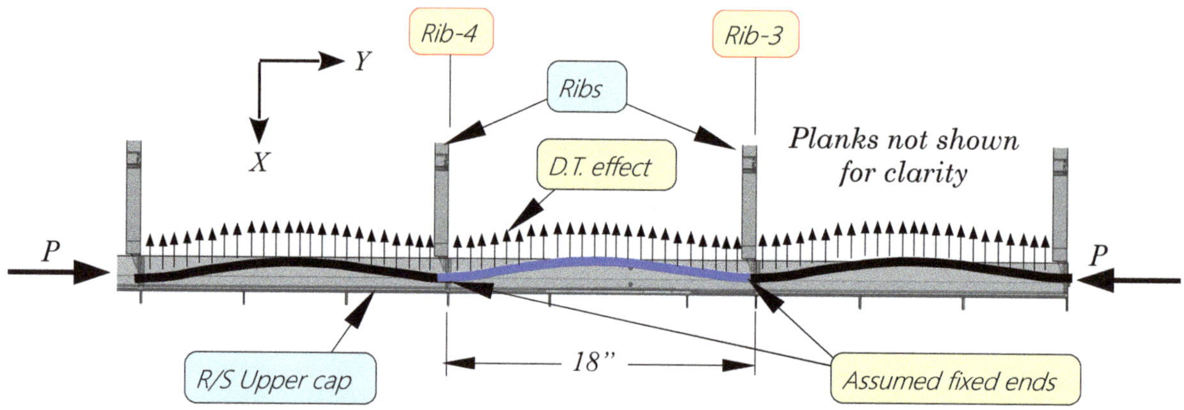

Figure 15-52 *– Diagonal tension effect on R/S upper cap; XY plane.*

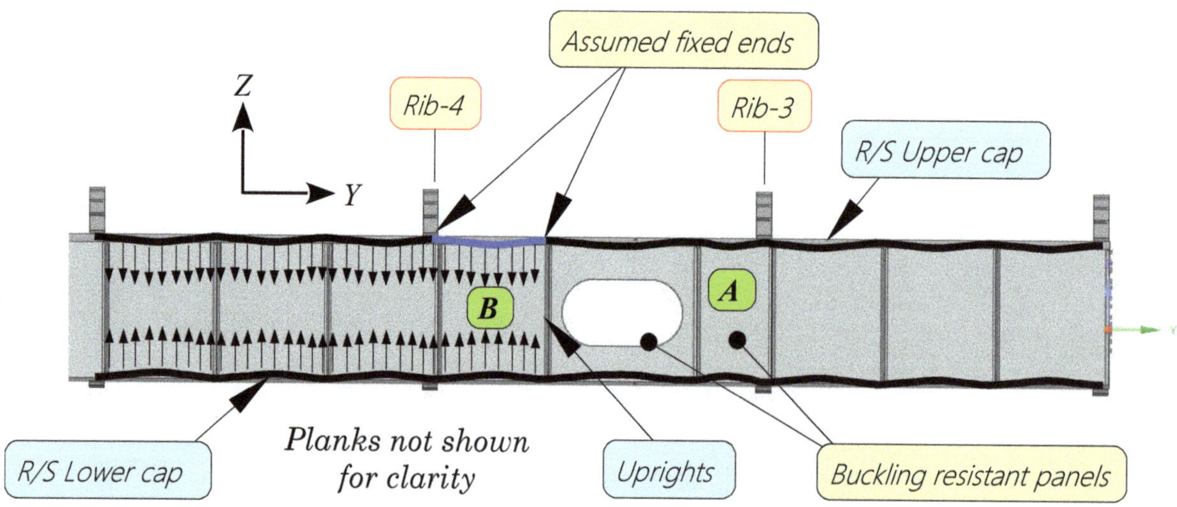

Figure 15-53 *– Diagonal tension effect on R/S upper cap; YZ plane.*

Upper cap loads

Using the resulting stresses from *Table 15-22* at stress points *1, 3 & 4* and the positions of the upper and lower caps centroids from *Figure 15-55,* the equivalent axial force and moments going through the upper cap only are calculated. *Figure 15-54* shows the R/S upper cap between rib-3 and rib-4 with its equivalent applied loads at both ends applied along its neutral axis.

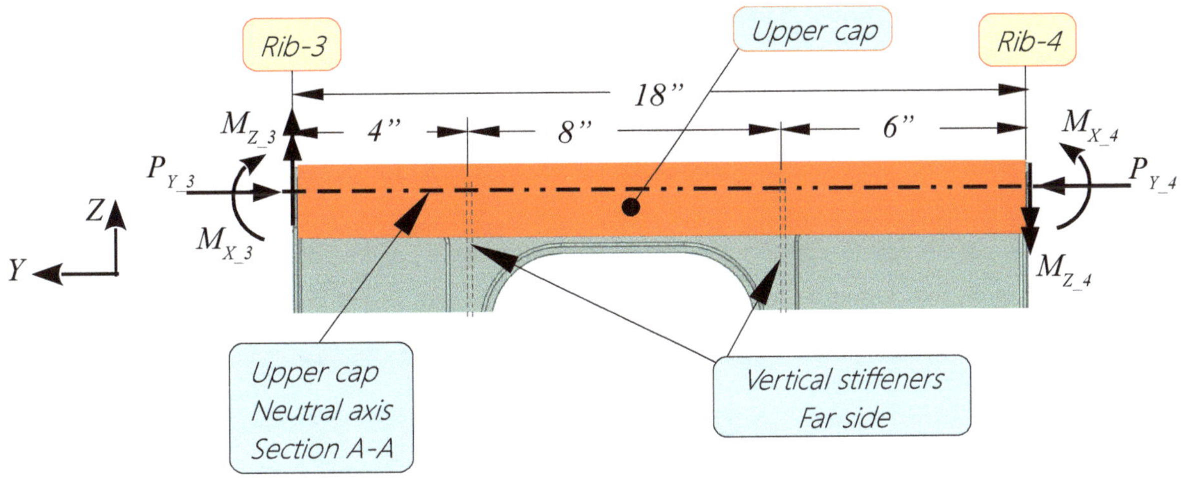

Figure 15-54 *– Upper cap equivalent applied loads.*

There are three unknowns that can be derived from the stresses at the three critical points *1, 3 & 4,* namely: the axial load "P_Y", the moment "M_X" and, the moment "M_Z". Therefore, we need three equations to solve the problem and they are the stress equations applicable to each stress point based on the following equation:

$$f_i = \frac{P_Y}{A} - \frac{M_X z_i}{I_X} + \frac{M_Z x_i}{I_Z}$$

For point 1:
$A = 1.3146\ in^2$ $x_1 = -1.723\ in$
$I_X = 0.1660\ in^4$ $z_1 = 0.618\ in$
$I_Z = 1.1946\ in^4$

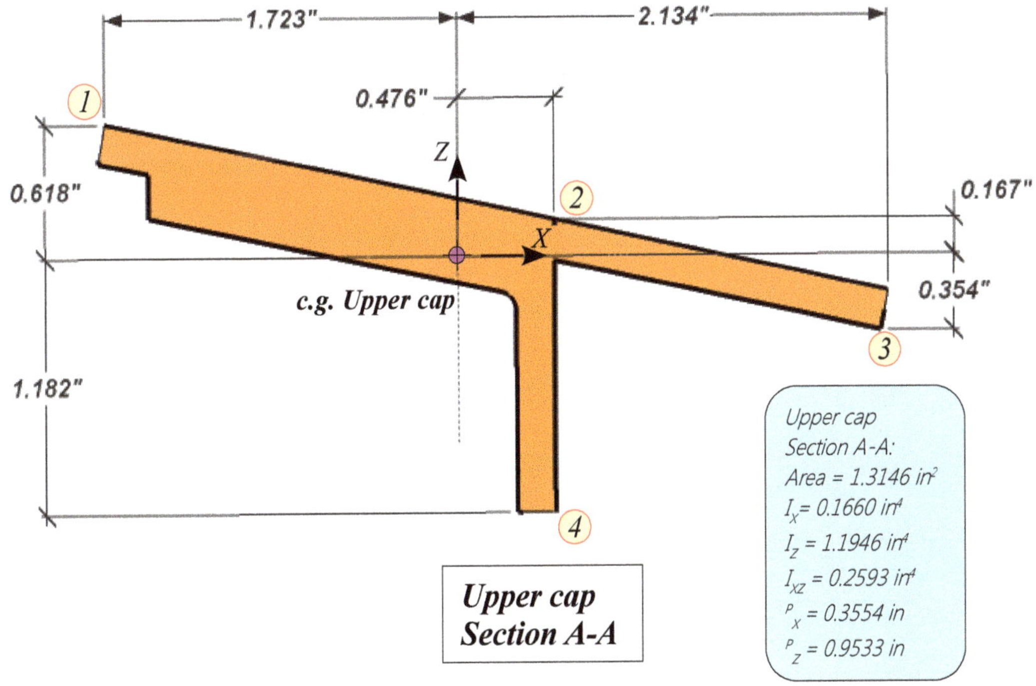

Figure 15-55 *– Section A-A Upper cap dimensions and section properties.*

The complex bending equation is not used here (we're assuming $I_{XZ} = 0$) not only because the upper cap is part of the Rear Spar and is somewhat stabilized and refrained from rotation but the idea here is to simply find equivalent internal loads that can reproduce the stresses at the critical locations. Therefore, the stress equation for point *1* becomes:

$$f_i = \frac{P_Y}{1.3146} - \frac{M_X(0.618)}{0.166} + \frac{M_Z(-1.723)}{1.1946}$$

doing it using the stresses at Rib-3 from *Table 15-22*, this reduces to:

$$f_1 = (0.7607)P_Y - 3.7229\,M_X - 1.4423\,M_Z = -30{,}084\ psi$$

then, similarly for points *3 & 4* we get:

$$f_3 = (0.7607)P_Y + 2.1325\,M_X + 1.7864\,M_Z = -17{,}626\ psi$$
$$f_4 = (0.7607)P_Y + 7.1205\,M_X + 0.3985\,M_Z = -13{,}695\ psi$$

And using matrices, the unknown loads at the upper cap centroid of section *A-A* Rib-3 are solved as shown in *Table 15-29*. The same thing is done at section *C-C* and at section *D-D*, as shown in *Table 15-30* and *Table 15-31* respectively.

Table 15-29 *- Equivalent section A-A upper cap loads at rib-3.*

Rib 3
Sect A-A

[A]	P	M_X	M_Z	[B] f
	[lb]	[in-lb]	[in-lb]	[psi]
Point 1	0.7606876616	-3.722891566	-1.44232379	-30084
Point 3	0.7606876616	2.1325301205	1.7863720074	-17626
Point 4	0.7606876616	7.1204819277	0.3984597355	-13695

	[F]
P [lb] =	-30431
M_X [in-lb] =	1237
M_Z [in-lb] =	1614

$[A]^{-1}$			
	0.6439760963	0.4766910951	0.1939328086
	-0.05727745	-0.075966896	0.1332443459
	-0.2058467	0.4474928265	-0.241646126

$[F] = [A]^{-1} \times [B]$

The stresses at section *C-C* are taken from the slotted beam analysis *Table 15-28.*

Table 15-30 *- Equivalent section B-B upper cap loads at section C-C.*

At cutout
Sect C-C

[A]	P	M_X	M_Z	[B] f
	[lb]	[in-lb]	[in-lb]	[psi]
Point 1	0.6858710562	-1.620937104	-1.449574626	-36128
Point 3	0.6858710562	0.0443224989	1.7337077724	-15693
Point 4	0.6858710562	3.4740396792	0.3642520856	9453

	[F]
P [lb] =	-28812
M_X [in-lb] =	8185
M_Z [in-lb] =	2138

$[A]^{-1}$			
	0.6635687212	0.491084989	0.3033462899
	-0.10376034	-0.137429256	0.2411895951
	-0.259861361	0.3860340472	-0.126172686

$[F] = [A]^{-1} \times [B]$

The stresses at section *D-D* are taken from *Table 15-23.*

Table 15-31 *- Equivalent section A-A upper cap loads at section D-D.*

Sect D-D	[A]	P	M_X	M_Z	[B] f
		[lb]	[in-lb]	[in-lb]	[psi]
	Point 1	0.7606876616	-3.722891566	-1.44232379	-25862
	Point 3	0.7606876616	2.1325301205	1.7863720074	-13760
	Point 4	0.7606876616	7.1204819277	0.3984597355	-11236

$[A]^{-1}$			
	0.6439760963	0.4766910951	0.1939328086
	-0.05727745	-0.075966896	0.1332443459
	-0.2058467	0.4474928265	-0.241646126

	[F]
P [lb] =	-25393
M_X [in-lb] =	1030
M_Z [in-lb] =	1881

$[F] = [A]^{-1} \times [B]$

Then, with those loads at each end of the upper cap (column), the column analysis can be performed. The maximum load between both ends is conservatively used and assumed constant along the span. The column allowables will be calculated for both *XY* and *YZ* planes.

15.16.1 XY PLANE

The R/S upper cap is idealized as shown in *Figure 15-56* with both ends fixed at Ribs 3 & 4.

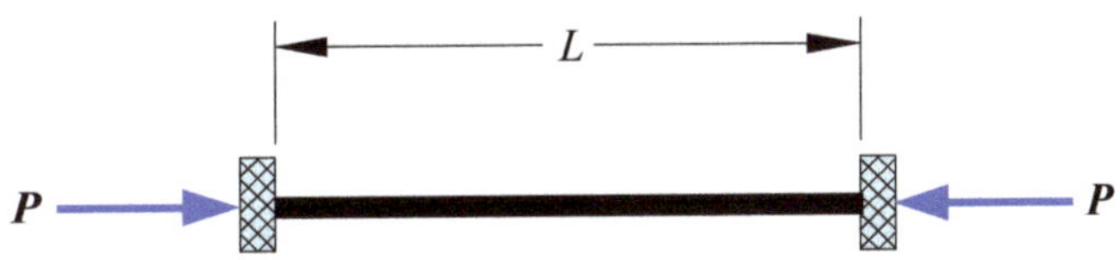

Figure 15-56 *– Idealized upper cap column.*

For conservatism and simplicity, the applied load *"P"* is considered the same at both ends using the highest compression force at Rib-3, and, the section properties of the upper cap of section *A-A* is assumed constant along the span. This is reasonable since the moment of inertia around the *"Z"* axis (which concerns us here) for the upper caps of sections *A-A & B-B* are almost identical.

For fixed-fixed column, the buckling coefficient *"c"* from *Figure B-2* is: *4.* Therefore, the effective column length "***L'***" is calculated as:

$$L' = \frac{L}{\sqrt{c}}$$

Hence, for $L = 18$ inches;

$$L' = \frac{18}{\sqrt{4}} = 9.0\,in.$$

Using the *Johnson-Euler* method, the critical column buckling stress is calculated as:

$$F_C = F_{CC} - \frac{F_{CC}^{\;2}}{4\pi^2 E_t}\left(\frac{L'}{\rho}\right)^2$$

then

$$F_C = 49{,}983 - \frac{49{,}983^2}{4\pi^2 5.929 * 10^6}\left(\frac{9.0}{0.9533}\right)^2 = 49{,}032\ psi$$

> *For*
> $E_t = 5.929*10^6$ *psi*
> $\rho_Z = 0.9533$ *in*
> $L' = 9.0$ *in.*
> $F_{CC} = 49{,}983$ *psi*

Where the radius of gyration "ρ_Z" is obtained from *Figure 15-55* and the allowable crippling stress from section *15.11*. The tangent modulus "E_t" is calculated using the *Ramberg-Osgood* relationship.

That ratio is found by iteration using a *goal seek* function as follows:

$$\frac{E_t}{E} = \frac{1}{1+\frac{3}{7}n\left(\frac{f}{F_{0.7}}\right)^{n-1}}$$

The $F_{0.7}$ value was previously calculated for the monument Intercostal made out of the same material from *Figure 14-157* and is 56 *ksi*. By iteration, the following values were obtained:

$$\frac{E_t}{E} = \eta = 0.593$$

$$E_t = (0.593)10{,}000{,}000 = 5{,}929{,}639\ psi$$

$$\frac{f}{F_{0.7}} = \frac{49{,}032}{56{,}000} = 0.876$$

$$F_C = 49{,}032\ psi$$

Compression load ratio; R_C

From the above analysis, the compression stress allowable obtained was: F_{CR} = 49,032 *psi*, and therefore, the critical load is; $P_{CR} = F_{CR} * A$ = 49,032 * 1.3146 = 64,458 *lbs*.

From *Table 15-29,* the maximum upper cap load is −30,431 *lbs* at rib-3 and is conservatively assumed constant between rib-3 & rib-4. However, because the upper wing plank panel adjacent to the R/S upper cap gets into diagonal semi-tension field, as well as the R/S web panel ***B***, (for almost half of the length of the span) the additional compression loads from those two diagonal tension effects must be taken into account.

From the wing upper plank: $P_{D.T.\_plank}$ = −944 *lbs*, ref. page *536*.

From the R/S panel *B:* $P_{DT\ R/S}$ = −589 *lbs,* ref. page *547*.

The total compression load in the upper cap "P_{Total}" is:

$$P_{Total} = P_P + P_{DT\ plank} + P_{DT\ R/S}$$
$$= -30{,}431 - 944 - 589 = -31{,}965\ lbs$$

Therefore, the R/S upper cap is not subject to general instability (column buckling) in the *XY* plane since the maximum applied compression load of 31,965 *lbs* is lower than the allowable. The compression load ratio "R_C" is thus:

$$R_C = \frac{P_{Total}}{P_{CR}} = \frac{31{,}965}{64{,}458} = 0.496$$

Margin of safety for column buckling

Thus the column buckling MS is calculated as:

$$MS = \frac{P_{CR}}{P_{Total}} - 1 = \frac{64{,}458}{31{,}965} - 1 = 1.01$$

Rear Spar upper cap Ultimate	***> 1.0***
LC2001 *Al Aly 7050-T7451*	*Column buckling*

15.16.2 YZ PLANE

As shown in *Figure 15-30*, there are three spans to look at. Panel *A* having the same dimensions as panel *B* but with a shorter span, is passed by comparison to the latter. Therefore, only the span over panel *B* and over the cutout will be looked at.

Here, since the vertical stiffeners are non-buckling, they thus act as efficient supports to the upper cap. Again, the R/S upper cap is idealized as shown in *Figure 15-56* with both ends fixed.

Panel B span

Using the section *A-A* upper cap properties, the span of that column is 6 inches. For fixed-fixed column, the buckling coefficient *"c"* from *Figure B-2* is: 4. Therefore, the effective column length ***"L'"*** is calculated as:

$$L' = \frac{L}{\sqrt{c}}$$

Hence, for $L = 6$ inches;

$$L' = \frac{6}{\sqrt{4}} = 3.0\ in.$$

Using the *Johnson-Euler* method, the critical column buckling stress is calculated as:

$$F_C = F_{CC} - \frac{F_{CC}^2}{4\pi^2 E_t}\left(\frac{L'}{\rho}\right)^2$$

then

$$F_C = 49{,}983 - \frac{49{,}983^2}{4\pi^2 5.77*10^6}\left(\frac{3.0}{0.3554}\right)^2 = 49{,}202\ psi$$

For
$E_t = 5.770*10^6\ psi$
$\rho_X = 0.3554 in$
$L' = 3.0\ in.$
$F_{CC} = 49{,}983\ psi$

Where the radius of gyration *"ρ_X"* is obtained from *Figure 15-55* and the allowable crippling stress from section *15.11*. The tangent modulus "E_t" is calculated using the *Ramberg-Osgood* relationship.

That ratio is found by iteration using a *goal seek* function as follows:

$$\frac{E_t}{E} = \frac{1}{1 + \frac{3}{7}n\left(\frac{f}{F_{0.7}}\right)^{n-1}}$$

The $F_{0.7}$ value is the same as before with 56 *ksi*. By iteration, the following values were obtained:

$$\frac{E_t}{E} = \eta = 0.577$$
$$E_t = (0.577)10{,}000{,}000 = 5{,}770{,}336\ psi$$
$$\frac{f}{F_{0.7}} = \frac{49{,}202}{56{,}000} = 0.879$$
$$F_C = 49{,}202\ psi$$

<u>Compression load ratio: R_C</u>

From the above analysis, the compression stress allowable obtained was: F_{CR} = 49,202 *psi*, and therefore, the critical load is; $P_{CR} = F_{CR} * A =$ 49,202 * 1.3146 = 64,681 *lbs*.

From *Table 15-31*, the maximum upper cap load at section *D-D* is −25,393 *lbs* and is assumed constant along the span. However, because the upper wing plank panel adjacent to the R/S upper cap gets into diagonal semi-tension field, as well as the R/S web panel ***B***, the additional compression loads from those two diagonal tension effects must be taken into account.

From the wing upper plank: $P_{D.T.\_plank} = -944\ lbs$, ref. page *536*.

From the R/S panel *B*: $P_{DT\ R/S} = -589\ lbs$, ref. page *547*.

The total compression load in the upper cap "P_{Total}" is:

$$\begin{aligned} P_{Total} &= P_P + P_{DT\ plank} + P_{DT\ R/S} \\ &= -25{,}393 - 944 - 589 = -26{,}926\ lbs \end{aligned}$$

Therefore, the R/S upper cap is not subject to general instability (column buckling) in the *YZ* plane since the maximum applied compression load of 26,926 *lbs* is lower than the allowable. The compression load ratio "R_C" is thus:

$$R_C = \frac{P_{Total}}{P_{CR}} = \frac{26{,}926}{64{,}681} = 0.416$$

Cutout span

Using the section *B-B* upper cap properties, the span of that column is 8 inches. For fixed-fixed column, the buckling coefficient "*c*" to use is: *4*. Therefore, the effective column length "***L'***", for $L = 8$ inches, is calculated as:

$$L' = \frac{L}{\sqrt{c}} = \frac{8}{\sqrt{4}} = 4.0\ in.$$

Using the *Johnson-Euler* method, the critical column buckling stress is calculated as:

$$F_C = F_{CC} - \frac{F_{CC}^2}{4\pi^2 E_t}\left(\frac{L'}{\rho}\right)^2$$

then

$$F_C = 46{,}790 - \frac{46{,}790^2}{4\pi^2 8.0287 * 10^6}\left(\frac{4.0}{0.5701}\right)^2 = 46{,}449\ psi$$

For
$E_t = 8.0287*10^6\ psi$
$\rho_X = 0.5701\ in$
$L' = 4.0\ in.$
$F_{CC} = 46{,}790\ psi$

Where the radius of gyration "ρ_X" is obtained from *Figure 15-57* and the allowable crippling stress from section *15.11*. The tangent modulus "E_t" is calculated using the *Ramberg-Osgood* relationship. That ratio is found by iteration using a *goal seek* function as follows:

$$\frac{E_t}{E} = \frac{1}{1 + \frac{3}{7} n \left(\frac{f}{F_{0.7}}\right)^{n-1}}$$

The $F_{0.7}$ value is the same as before with 56 *ksi*. By iteration, the following values were obtained:

$$\frac{E_t}{E} = \eta = 0.80$$

$$E_t = (0.80)10{,}000{,}000 = 8{,}028{,}674\ psi$$

$$\frac{f}{F_{0.7}} = \frac{46{,}449}{56{,}000} = 0.829$$

$$F_C = 46{,}449\ psi$$

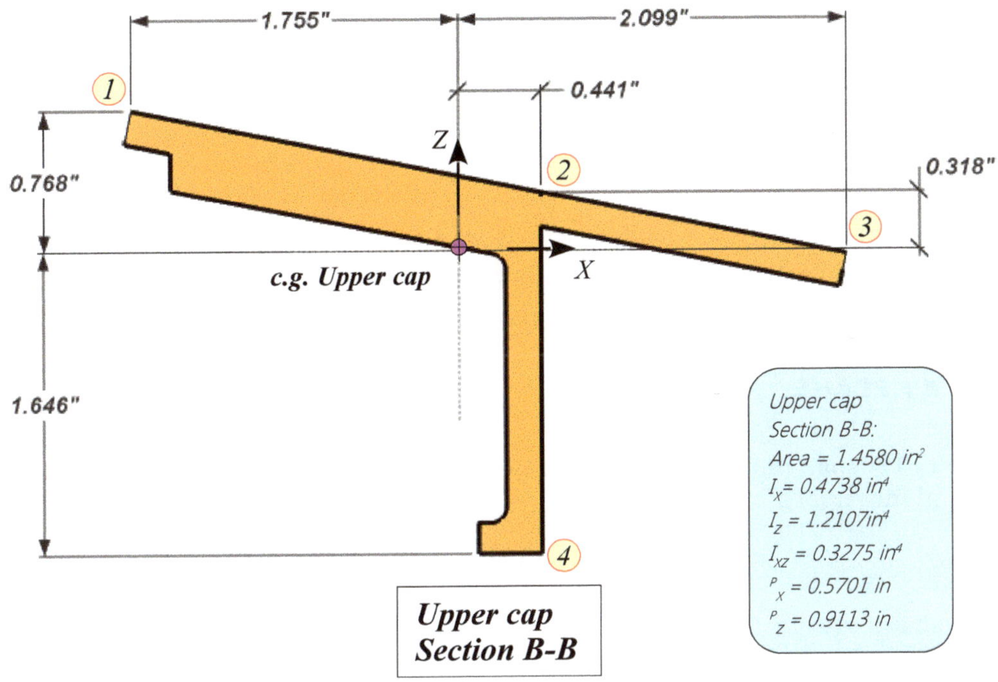

Figure 15-57 *– Upper cap dimensions and properties at section B-B.*

Compression load ratio: R_C

From the above analysis, the compression stress allowable obtained was: F_{CR} = 46,449 *psi*, and therefore, the critical load is; $P_{CR} = F_{CR} * A =$ 46,449 * 1.458 = 67,723 *lbs*.

From *Table 15-30*, the maximum upper cap load section *C-C* is −28,812 *lbs* and is assumed constant along the span. However, because the upper wing plank panel adjacent to the R/S upper cap gets into diagonal semi-tension field, the additional compression load from the diagonal tension effect must be taken into account.

From the wing upper plank: $P_{D.T.\_plank} = -944$ *lbs*, ref. page *536*.

The total compression load in the upper cap "P_{Total}" is:

$$P_{Total} = P_P + P_{DT\,plank} + P_{DT\,R/S}$$
$$= -28{,}812 - 944 = -29{,}756\ lbs$$

Therefore, the R/S upper cap is not subject to general instability (column buckling) in the *YZ* plane since the maximum applied compression load of 29,756 *lbs* is lower than the allowable. The compression load ratio "R_C" is thus:

$$R_C = \frac{P_{Total}}{P_{CR}} = \frac{29{,}756}{67{,}723} = 0.439$$

Margin of safety for column buckling

Since this compression stress ratios are smaller in the *YZ* plane than in the *XY* plane, then the MS of the longer span column calculated previously is the most critical one and drives the others.

15.17 BEAM-COLUMN

The Diagonal Semi-Tension field effect, in the wing upper plank and in panel *B* of the Rear Spar, pulls onto the Rear Spar upper cap causing a bending moment in it. This bending moment then gets amplified by the large compression load in the upper cap. This amplification (called secondary bending moment) is calculated later on.

First, the primary bending moment must be calculated. The Rear Spar upper cap, is considered as a beam fixed at both ends having a uniformly distributed load *"w"*. This distributed load is calculated for the "end-bay" effect in diagonal tension as follows.

$$w = k\, f_S t \tan\alpha$$

15.17.1 XY PLANE

So, for the long 18" span in the *XY* plane, the distributed load *"w_X"* due to the diagonal tension in the wing upper plank from section *15.8* is:

$$w_X = -(0.13)\,9{,}234\,(0.125)\tan 45 = -155.5\ lb/in.$$

For
$k = 0.13$
$f_s = 9{,}234$ psi
$t = 0.125$ in.
$\alpha = 45$ deg.

Figure 15-58 shows the idealized *FBD* and the shear & moment diagrams in the *XY* plane of the R/S upper cap.

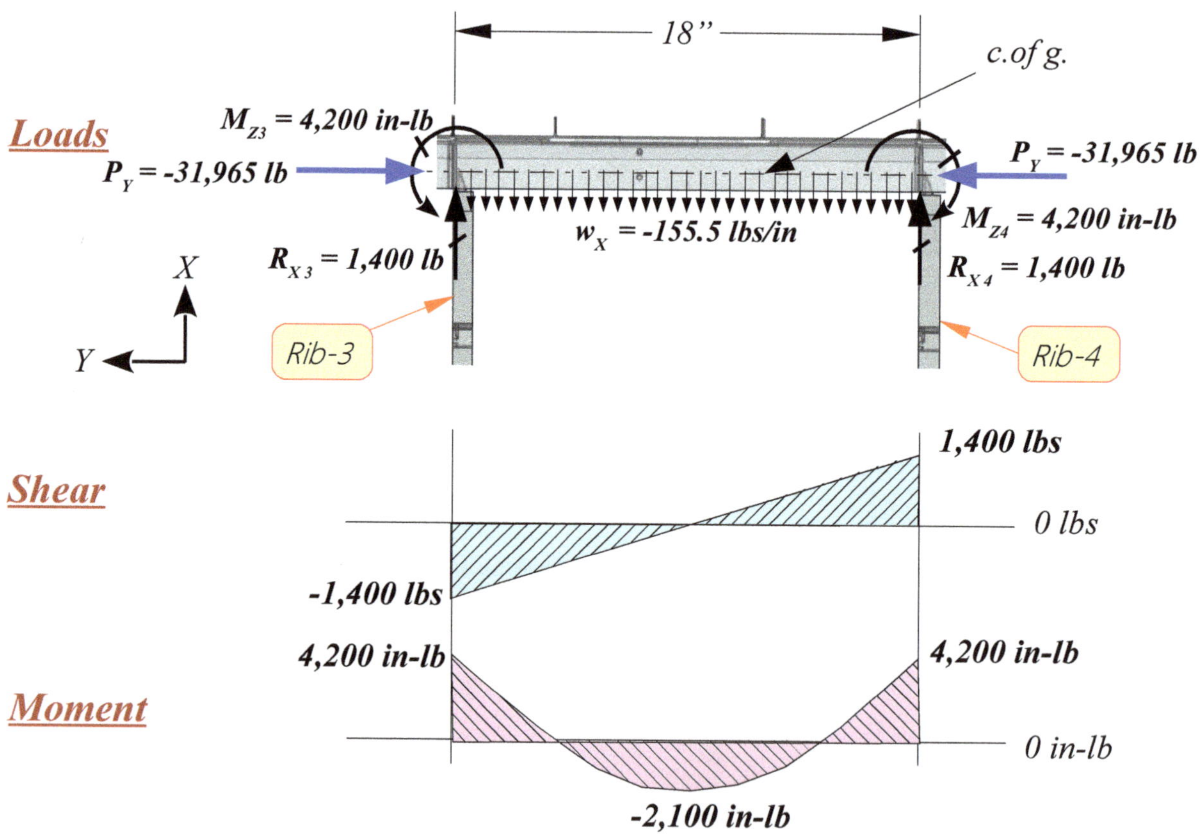

***Figure 15-58** – Idealized FBD and Shear & moments diagrams of the upper cap beam, XY plane.*

The shear reactions at the ribs and the maximum bending moment at the center are calculated for a constant distributed load of $w_X = -155.5\ lb/in$.

where

$$R_{X3} = R_{X4} = \frac{w_X\, l}{2} = \frac{-155.5 \times 18}{2} = 1{,}400\ lbs$$

$$M_{Z3} = M_{Z4} = -\frac{w_X\, l^2}{12} = -\frac{-155.5 \times 18^2}{12} = 4{,}200\ in.-lb$$

$$M_{Z\,CTR} = \frac{w_X\, l^2}{24} = \frac{-155.5 \times 18^2}{24} = 2{,}100\ in.-lb$$

Then, the secondary bending moments are calculated for a fixed-fixed beam having an evenly distributed load and an axial compression force. Referring to *Figure 15-59*, the increased bending moments at the ends and at mid-span are calculated as follows:

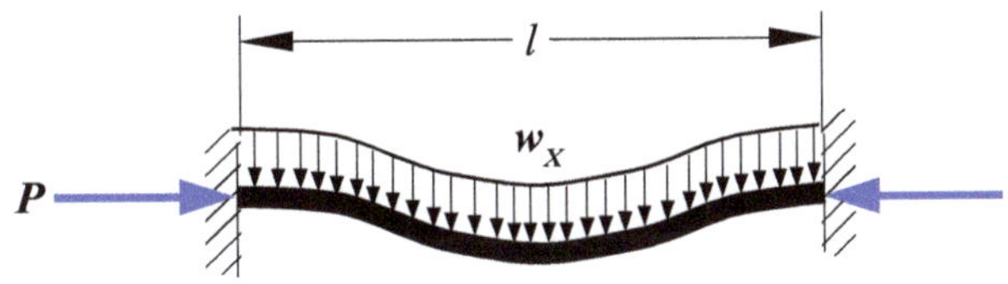

Figure 15-59 *– Beam-column with left and right ends fixed.*

Note that the approximate method used to calculate the approximate "amplified" bending moment generally gives good results when beams are simply supported but it is not applicable when the beams are fixed at both ends. The reason being that this equation assumes a parabolic deflection curve; which is not the case for fixed-fixed beams.

$$M = \frac{M_o}{1 - \dfrac{P}{P_{CR}}}$$

For conservatism and simplicity, the maximum axial compression force in the R/S upper cap between the two ends is considered constant along the beam with $P = 31{,}376$ *lbs*. The critical compression load being; $P_{CR} = 64{,}458$ *lbs*. Hence, for example, from the above equation, the "amplifying" factor "*AF*" in the *XY* plane would be:

$$\frac{M}{M_o} = \frac{1}{1 - \dfrac{31{,}376}{64{,}458}} = 1.95$$

This would mean that the end moments as well as the mid-span moment would be almost twice as much as the *FBD* moments. I don't recommend using this equation in this case as it is way too conservative.

The secondary bending moments equations for a beam-column having both ends fixed are:

$$M_A = M_B = -\frac{w_X}{k^2}\left[1 - \frac{kl/2}{\tan(kl/2)}\right]$$

$$M_{CTR} = \frac{w_X}{k^2}\left[\frac{kl/2}{\sin(kl/2)} - 1\right] \quad at\ y = \frac{l}{2}$$

where

$$k = \sqrt{\frac{P}{EI}} = \sqrt{\frac{31{,}376}{10\times10^6\,(1.1946)}} = 0.05 \quad 1/in.$$

therefore

$$M_A = M_B = -\frac{(-155.5)}{(0.05)^2}\left[1 - \frac{(0.05)18/2}{\tan((0.05)18/2)}\right] = 4{,}259 \; in.-lb$$

$$M_{CTR} = -\frac{155.5}{(0.05)^2}\left[\frac{(0.05)\,18/2}{\sin((0.05)\,18/2)} - 1\right] = -2{,}153\, in.-lb \quad at\ y = \frac{l}{2}$$

For
a = 0 in
w_x = -155.5 lb/in
l = 18 in.
P = 31,376 lbs
E = 10x106 psi
I_z = 1.1946 in^4

The above results show that the secondary bending moment effect on this fixed-fixed beam is quite small with an amplification factor of somewhere between 1% and 3%.

15.17.2 YZ PLANE

So, for the short 6" span in the *YZ* plane, the distributed load *"wz"* due to the diagonal tension in panel *B* of the Rear Spar web from section *15.10* is:

$$w_Z = -(0.09)\,16{,}043\,(0.065)\tan 43.8 = -90.1 \; lb/in.$$

For
k = 0.09
f_s = 16,043 psi
t = 0.065 in.
α = 43.8 deg.

Figure 15-60 shows the idealized *FBD* and the shear & moment diagrams in the *YZ* plane of the R/S upper cap.

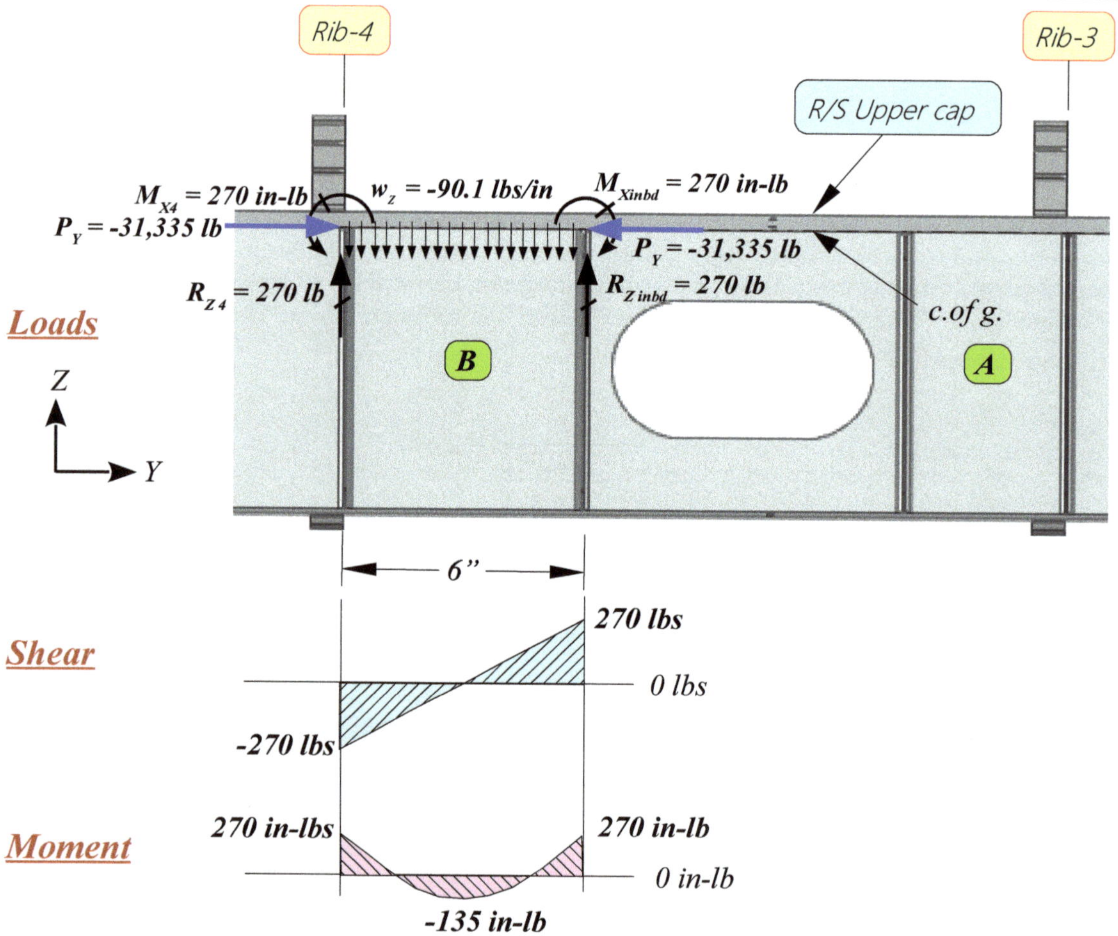

***Figure 15-60** – Idealized FBD and Shear & moments diagrams of the upper cap beam, YZ plane.*

The shear reactions at the ribs and the maximum bending moment at the center are calculated for a constant distributed load of $w_Z = -90.1\ lb/in.$.

where

$$R_{Z\,inbd} = R_{Z4} = \frac{w_Z\, l}{2} = \frac{-90.1\times 6}{2} = 270\ lbs$$

$$M_{X\,inbd} = M_{X4} = -\frac{w_Z\, l^2}{12} = -\frac{-90.1\times 6^2}{12} = 270\ in.-lb$$

$$M_{X\,CTR} = \frac{w_Z\, l^2}{24} = \frac{-90.1\times 6^2}{24} = 135\ in.-lb$$

Then, using the same methodology to determine the secondary bending moments as in the *XY* plane, the bending moments equations are:

$$M_A = M_B = -\frac{w_Z}{k^2}\left[1 - \frac{kl/2}{\tan(kl/2)}\right]$$

$$M_{CTR} = \frac{w_Z}{k^2}\left[\frac{kl/2}{\sin(kl/2)} - 1\right]\quad at\ y = \frac{l}{2}$$

where

$$k = \sqrt{\frac{P}{EI}} = \sqrt{\frac{31{,}335}{10\times 10^6\,(0.166)}} = 0.14\quad 1/in.$$

For
$a = 0\ in$
$w_Z = -90.1\ lb/in$
$l = 6\ in.$
$P = 31{,}335\ lbs$
$E = 10x106\ psi$
$I_x = 0.166\ in^4$

therefore

$$M_A = M_B = -\frac{(-90.1)}{(0.14)^2}\left[1 - \frac{(0.14)6/2}{\tan((0.14)6/2)}\right] = 273\ in.-lb$$

$$M_{CTR} = -\frac{90.1}{(0.14)^2}\left[\frac{(0.14)\,6/2}{\sin((0.14)\,6/2)} - 1\right] = -138\ in.-lb\quad at\ y = \frac{l}{2}$$

Here again, the above results show that the secondary bending moment effect on this fixed-fixed beam is quite small with an amplification factor of somewhere between 1% and 3%.

> *"Unlike doctors, engineers can't bury their mistakes"*
>
> *(Jean-Guy Sicotte M.D., P.E.)*

15.18 STRESS ANALYSIS AND MS INTERACTION

The stresses that were calculated in section *15.14* were preliminary because they didn't take into account the secondary bending moment (although small) from the diagonal tension effect and, the effect of the cutout as a slotted beam.

The summary tables for the stresses and the MS in the upper and lower caps are shown for section *A-A* at Rib-3, section *C-C* (on the inboard edge of the cutout) and, at section *A-A* at Rib-4. See *Table 15-32, Table 15-33* and *Table 15-34* respectively. As can be seen, the most critical location for *LC2001* is at stress point *(4)* of section *C-C*. An example of the detailed analysis for the most critical location is demonstrated below.

15.18.1 INTERACTION EQUATIONS

For the analysis of the upper cap, three MS interaction equations are used for comparison while only one is used for the lower cap as it is in tension.

The first equation "MS_1" is combining: bending "R_b", flexural shear "R_S", and axial compression "R_a". This equation is very common and found to be conservative. It is used in both compression and tension.

$$MS_1 = \frac{1}{\sqrt{(R_a + R_b)^2 + R_S^2}} - 1$$

The second equation "MS_2" is for round tube columns with combined; bending "R_b", torsional shear "R_{ST}", and axial compression "R_C". This equation is to be used mainly for control tubes, struts or the like. I am using it here simply for comparison purposes. Although it provides lower MS in our case, I find it inappropriate for the R/S upper cap so its MS is ignored.

$$MS_2 = \frac{1}{R_C + \sqrt{R_b^2 + R_{ST}^2}} - 1$$

The third equation "MS_3" is for the analysis of Stringers due to Diagonal Tension. Although this equation uses the axial and bending stresses, it ignores the flexural shear. As this equation was developed empirically, I shall use it to check the R/S upper cap since diagonal tension occurs in the wing upper plank.

$$MS_3 = \frac{1}{\frac{f_a + f_b}{F_{CC}} + \frac{f_{ST_{MAX}}}{F_{ST}}} - 1$$

***Table 15-32** - Stress summary at section A-A @ Rib3, Upper and Lower caps.*

Upper cap

Section A-A Stress points	x [in]	z [in]	f_a [psi]	f_{b_1} [psi]	f_{b_2nd} [psi]	f_b [psi]	f_a+f_b [psi]	$R_a=R_C$	R_B	R_1	R_2	MS_3
1	-1.723	0.618	-24,315	-6,935	-6,145	-13,080	**-37,395**	0.496	0.238	0.748	0.138	0.13
2	0.476	0.167	-24,315	-602	1,698	1,096	-23,219	0.496	0.016	0.465	0.138	0.66
3	2.134	-0.354	-24,315	5,523	7,611	13,133	-11,182	0.496	0.188	0.224	0.138	1.77
4	0.476	-1.182	-24,315	9,454	1,698	11,152	-13,164	0.496	0.149	0.263	0.138	1.49

Section A-A Stress points	f_{S_X} [psi]	f_{S_Y} [psi]	f_{S_Z} [psi]	f_{S_XY} [psi]	f_{S_XZ} [psi]	F_{SU} [psi]	R_S	MS_1	MS_2
1_{plank}	8,527	8,048	0	11,725	8,527	40,000	0.293	0.27	0.14
1_{flange}	8,527	7,333	0	11,246	8,527	40,000	0.281	0.27	0.16
2_{plank}	8,527	0	0	8,527	8,527	40,000	0.213	0.80	0.41
2_{flange}	8,527	0	16,775	8,527	18,818	40,000	0.470	0.44	0.03
3	8,527	0	0	8,527	8,527	40,000	0.213	0.40	0.28
4	0	0	16,775	0	16,775	40,000	0.419	0.30	0.06

$$MS_1 = \frac{1}{\sqrt{(R_a + R_b)^2 + R_S^2}} - 1$$

$$MS_2 = \frac{1}{R_C + \sqrt{R_b^2 + R_{ST}^2}} - 1$$

Lower cap

Section A-A Stress points	x [in]	z [in]	f_a [psi]	f_{b_1} [psi]	f_{b_2nd} [psi]	f_b [psi]	f_a+f_b [psi]	R_a	R_B
5	0.512	1.129	25,657	-6,910	0	-6,910	18,747	0.342	0.126
6	-1.744	-0.474	25,657	124	0	124	25,781	0.367	0.002
7	0.512	-0.332	25,657	3,235	0	3,235	28,893	0.367	0.046
8	2.139	-0.194	25,657	5,232	0	5,232	30,890	0.367	0.075

$$MS_3 = \frac{1}{\dfrac{f_b + f_p}{F_{CC}} + \dfrac{f_{ST\,Max}}{F_{ST}}} - 1 = \frac{1}{R_1 + R_2} - 1$$

Section A-A Stress points	f_{S_X} [psi]	f_{S_Y} [psi]	f_{S_Z} [psi]	f_{S_XY} [psi]	f_{S_XZ} [psi]	F_{SU} [psi]	R_S	MS
5	0	0	16,775	0	16,775	40,000	0.419	0.59
6_{plank}	7,317	6,665	0	9,897	7,317	40,000	0.247	1.25
6_{flange}	7,317	4,858	0	8,783	7,317	40,000	0.220	1.33
7_{plank}	7,317	0	0	7,317	7,317	40,000	0.183	1.21
7_{flange}	7,317	0	16,775	7,317	18,302	40,000	0.458	0.62
8	7,317	0	0	7,317	7,317	40,000	0.183	1.09

$$MS = \frac{1}{\sqrt{(R_a + R_b)^2 + R_S^2}} - 1$$

Table 15-33 - *Stress summary at section C-C, Upper and Lower caps.*

Upper cap

Section C-C Stress points	x [in]	z [in]	f_a [psi]	f_{b_1} [psi]	f_{b_2nd} [psi]	f_b [psi]	f_a+f_b [psi]	$R_a=R_C$	R_B	R_1	R_2	MS_3
1	-1.755	0.768	-20,453	-16,366	3,122	-13,244	**-33,698**	0.440	0.241	0.720	0.138	*0.17*
2	0.441	0.318	-20,453	-4,715	-784	-5,499	-25,953	0.440	0.100	0.555	0.138	*0.44*
3	2.099	-0.021	-20,453	4,069	-3,734	335	-20,118	0.440	0.005	0.430	0.138	*0.76*
4	0.441	-1.646	-20,453	29,214	-784	28,430	7,976	0.440	0.379	0.106	0.138	*3.10*

Section C-C Stress points	f_{S_X} [psi]	f_{S_Y} [psi]	f_{S_Z} [psi]	f_{S_XY} [psi]	f_{S_XZ} [psi]	F_{SU} [psi]	R_S	MS_1	MS_2
1_{plank}	7,576	8,048	0	11,053	7,576	40,000	0.276	*0.36*	*0.24*
1_{flange}	7,576	7,333	0	10,544	7,576	40,000	0.264	*0.37*	*0.25*
2_{plank}	7,576	0	0	7,576	7,576	40,000	0.189	*0.75*	*0.53*
2_{flange}	7,576	0	22,026	7,576	23,292	40,000	0.582	*0.26*	*-0.03*
3	7,576	0	0	7,576	7,576	40,000	0.189	*1.07*	*0.59*
4	0	0	22,026	0	22,026	40,000	0.551	***0.01***	*-0.10*

$$MS_1 = \frac{1}{\sqrt{(R_a + R_b)^2 + R_S^2}} - 1$$

$$MS_2 = \frac{1}{R_C + \sqrt{R_b^2 + R_{ST}^2}} - 1$$

Lower cap

Section C-C Stress points	x [in]	z [in]	f_a [psi]	f_{b_1} [psi]	f_{b_2nd} [psi]	f_b [psi]	f_a+f_b [psi]	R_a	R_B
5	0.481	1.588	22,509	-26,509	0	-26,509	-4,001	0.300	0.442
6	-1.775	-0.593	22,509	6,999	0	6,999	29,508	0.322	0.100
7	0.481	-0.452	22,509	8,669	0	8,669	31,177	0.322	0.124
8	2.108	-0.313	22,509	9,229	0	9,229	31,738	0.322	0.132

$$MS_3 = \frac{1}{\frac{f_b + f_p}{F_{CC}} + \frac{f_{ST\,Max}}{F_{ST}}} - 1 = \frac{1}{R_1 + R_2} - 1$$

Section C-C Stress points	f_{S_X} [psi]	f_{S_Y} [psi]	f_{S_Z} [psi]	f_{S_XY} [psi]	f_{S_XZ} [psi]	F_{SU} [psi]	R_S	MS
5	0	0	22,026	0	22,026	40,000	0.551	*0.08*
6_{plank}	6,519	6,665	0	9,323	6,519	40,000	0.233	*1.08*
6_{flange}	6,519	4,858	0	8,131	6,519	40,000	0.203	*1.14*
7_{plank}	6,519	0	0	6,519	6,519	40,000	0.163	*1.11*
7_{flange}	6,519	0	22,026	6,519	22,970	40,000	0.574	*0.38*
8	6,519	0	0	6,519	6,519	40,000	0.163	*1.08*

$$MS = \frac{1}{\sqrt{(R_a + R_b)^2 + R_S^2}} - 1$$

***Table 15-34** - Stress summary at section A-A @ Rib4, Upper and Lower caps.*

Upper cap

Section A-A / Stress points	x [in]	z [in]	f_a [psi]	f_{b_1} [psi]	f_{b_2nd} [psi]	f_b [psi]	f_a+f_b [psi]	$R_a=R_C$	R_B	R_1	R_2	MS_3
1	-1.723	0.618	-18,085	-6,351	-7,160	-13,511	**-31,595**	0.368	0.246	0.632	0.138	*0.30*
2	0.476	0.167	-18,085	-128	1,423	1,295	-16,790	0.368	0.018	0.336	0.138	*1.11*
3	2.134	-0.354	-18,085	5,572	8,192	13,764	-4,320	0.368	0.184	0.086	0.138	*3.46*
4	0.476	-1.182	-18,085	7,392	3,639	11,032	-7,053	0.368	0.147	0.141	0.138	*2.59*

Section A-A / Stress points	f_{S_X} [psi]	f_{S_Y} [psi]	f_{S_Z} [psi]	f_{S_XY} [psi]	f_{S_XZ} [psi]	F_{SU} [psi]	R_S	MS_1	MS_2
1_{plank}	5,115	8,048	0	9,536	5,115	40,000	0.238	*0.52*	*0.41*
1_{flange}	5,115	7,333	0	8,941	5,115	40,000	0.224	*0.53*	*0.43*
2_{plank}	5,115	0	0	5,115	5,115	40,000	0.128	*1.46*	*1.01*
2_{flange}	5,115	0	15,684	5,115	16,497	40,000	0.412	*0.77*	*0.28*
3	5,115	0	0	5,115	5,115	40,000	0.128	*0.77*	*0.69*
4	0	0	15,684	0	15,684	40,000	0.392	*0.55*	*0.27*

$$MS_1 = \frac{1}{\sqrt{(R_a + R_b)^2 + R_S^2}} - 1$$

$$MS_2 = \frac{1}{R_C + \sqrt{R_b^2 + R_{ST}^2}} - 1$$

Lower cap

Section A-A / Stress points	x [in]	z [in]	f_a [psi]	f_{b_1} [psi]	f_{b_2nd} [psi]	f_b [psi]	f_a+f_b [psi]	R_a	R_B
5	0.512	1.129	19,153	-4,845	0	-4,845	14,307	0.255	0.088
6	-1.744	-0.474	19,153	-1,074	0	-1,074	18,079	0.274	0.018
7	0.512	-0.332	19,153	2,772	0	2,772	21,924	0.274	0.040
8	2.139	-0.194	19,153	5,359	0	5,359	24,512	0.274	0.077

$$MS_3 = \frac{1}{\frac{f_b + f_p}{F_{CC}} + \frac{f_{ST\,Max}}{F_{ST}}} - 1 = \frac{1}{R_1 + R_2} - 1$$

Section A-A / Stress points	f_{S_X} [psi]	f_{S_Y} [psi]	f_{S_Z} [psi]	f_{S_XY} [psi]	f_{S_XZ} [psi]	F_{SU} [psi]	R_S	MS
5	0	0	15,684	0	15,684	40,000	0.392	*0.92*
6_{plank}	4,474	6,665	0	8,027	4,474	40,000	0.201	*1.83*
6_{flange}	4,474	4,858	0	6,605	4,474	40,000	0.165	*1.98*
7_{plank}	4,474	0	0	4,474	4,474	40,000	0.112	*2.01*
7_{flange}	4,474	0	15,684	4,474	16,309	40,000	0.408	*0.94*
8	4,474	0	0	4,474	4,474	40,000	0.112	*1.72*

$$MS = \frac{1}{\sqrt{(R_a + R_b)^2 + R_S^2}} - 1$$

15.18.2 CRITICAL STRESS ANALYSIS EXAMPLE.

Referring to *Figure 15-6* and *Figure 15-57,* the stress analysis at point *(4)* of section *C-C* is demonstrated in detail as it is the most critical location. Note that section *C-C* is the same as section *B-B* but located at the inboard edge of the cutout, the stress point *(4)* is located right at the edge of the cutout and is considered as part of the upper cap.

As shown in *Figure 15-51,* the Rear Spar was considered as two distinct beams; the upper and the lower caps. The upper cap is the one we're mainly interested in as it is in compression and is therefore more critical in static analysis.

As shown in page *572*, an equivalent axial force *"P"* and bending moments *"M_X & M_Z"* applied to the upper cap are derived from the stresses obtained in the slotted beam analysis section *15.15* where both moments puts point *(1)* into compression; the equivalent loads are:

$P = -28{,}812\ lbs$

$M_X = 8{,}185\ in\text{-}lb$

$M_Z = 2{,}138\ in\text{-}lb$

Then, because the upper wing plank panel adjacent to the R/S upper cap gets into diagonal semi-tension field, the additional compression loads from the diagonal tension effect is taken into account. The additional compression force from the diagonal tension field is constant within that bay regardless of the location of the section cut.

From the wing upper plank: $P_{D.T.\_plank} = -944\ lbs$, ref. page *536,*

Then the total axial load "P_{Total}" that includes the contribution of the diagonal tension to section *B-B* is:

$$P_{Total} = P + P_{DT\ plank}$$
$$= -28{,}812 - 944 = -29{,}756\ lbs$$

Note that the R/S web that slightly gets into diagonal tension only in the bay aft of the cutout (panel *B*) is neglected. See section *15.10* for the diagonal tension analysis of the web. Therefore the additional compression load caused by it is not considered in the bay of the cutout (section *B-B*).

Axial stress

$$f_a = \frac{P_{Total}}{A} = \frac{-29{,}756}{1.458} = -20{,}409\ psi$$

Because the axial stress is in compression, the axial stress ratio *"R_a"* is obtained by comparing the above stress with the column buckling allowable stress *"F_C"* for the upper cap span over the cutout from page *576*. Hence:

$$R_a = \frac{|f_a|}{F_C} = \frac{|-20{,}409|}{46{,}449} = 0.439$$

Bending stress

The equivalent bending moments shown above are the primary moments and will cause the primary bending stress *"f_{b1}"* while the bending moment from the beam-column analysis is the secondary bending moment and will cause the secondary bending stress *"f_{b2}"*. The secondary moment at the center of the span is used since that is where the cutout is located.

From the primary equivalent moments due to the slotted-beam: $M_X = 8{,}185$ *in-lb*, $M_Z = 2{,}138$ *in-lb*, ref. *Table 15-30.*

From the secondary beam-column at center: $M_{Z\,CTR} = -2{,}153$ *in-lb*, ref. page ***580.***

Then the primary and secondary bending stresses at point *(4)* are calculated below:

for the primary bending:

$$f_{b1-4} = \frac{-M_X z_4}{I_X} + \frac{M_Z x_4}{I_Z}$$

For
$M_X = 8{,}185$ in-lb
$M_Z = 2{,}138$ in-lb
$z_4 = -1.646$ in
$x_4 = 0.441$ in
$I_X = 0.4738\ in^4$
$I_Z = 1.2107\ in^4$

then

$$f_{b1-4} = -\frac{8{,}185(-1.646)}{0.4738} + \frac{2{,}138(0.441)}{1.2107} = 29{,}214\ psi$$

for the secondary bending:

$$f_{b2-4} = \frac{M_{Z\,CTR} x_4}{I_Z}$$

For
$M_{Z\,CTR} = -2{,}153$ in-lb
$x_4 = 0.441$ in
$I_Z = 1.2107\ in^4$

then

$$f_{b2-4} = -2{,}153\frac{(0.441)}{1.2107} = -784\ psi$$

the total bending stress at point *(4)* is thus:

$$f_{b\ Total} = f_{b1-4} + f_{b2-4} = 29{,}214 + (-784) = 28{,}430\ psi$$

Because the total bending stress is in tension, the bending stress ratio "R_B" is obtained by comparing the above stress with the tensile strength "F_{TU}" of the R/S *(7050).*

I used F_{TU} in the bending stress ratio calculation for conservatism since I consider this structure as thin. In some cases, where the flanges are relatively thick —and thus considered stable— the plastic bending allowable F_{BU} could be used instead. Note that the compression stress capacity of a free flange is usually the limiting factor.

hence

$$R_B = \frac{f_B}{F_{TU}} = \frac{28{,}430}{75{,}000} = 0.379$$

Shear stress

The average shear stress through the section is calculated instead of using the actual shear stress equation $\tau = VQ/It$. The reasons for that are, that not only this is simpler but one have to remember that the Rear Spar is part of the wing box and that we're only looking at a small part of the actual complete wing box section. Therefore, one can confidently assume that the shear stress is constant over small lengths since that the shear loads we obtain from the global *FEM*, at the elements we're interested in, has already been distributed correctly according to the above equation.

For all stress points, the shear stresses are calculated along the three axes (or in the three planes if you prefer). In all cases, it is assumed that the shear can only be transferred in the "strong" direction of the part; those are in-plane loads. It is considered that the shear loads that act normal to a flange or a web cannot effectively be reacted since those relatively thin flange or web are too flexible in the out-of-plane "normal" direction.

At stress point *(4)*, since this point is in the Rear Spar vertical web, only the vertical shear "in Z" is considered. Referring to ***Figure 15-47***, the vertical shear stress "f_{SZ}" is 22,027 *psi*.

> I chose to call the shear stress here as "f_{SZ}" where S is for Shear and Z is for the vertical load direction. However, most textbooks would call it "f_{YZ}" since shear always acts within a plane. Whatever nomenclature you choose, try to be consistent through out your analysis and, as far as I'm concerned, as long as it is clear and that the nomenclature makes sense, it's all that matters.
>
> Providing that a shear stress from another direction would co-exist at the same stress point (as it is the case in some other locations), both shear stresses would have to be vectorially added. Note that, for thin structures, there shouldn't be more than two shear stresses acting at the same point, otherwise, I suggest you simplify your assumptions accordingly.

Since point *(4)* is located in the Rear Spar, the shear stress ratio "R_S" is obtained by comparing the above stress with the shear strength "F_{SU}" of the R/S *(7050)*.

hence

$$R_S = \frac{f_{SZ}}{F_{SU}} = \frac{22{,}027}{40{,}000} = 0.551$$

Margin of safety in axial + bending + shear

The lowest MS is obtained at stress point *(4)* located at the upper edge of the cutout through section *C-C* for the fictitious ***LC2001*** load case. As explained earlier, I am using the following interaction equation to calculate the MS in the Rear Spar.

$$MS_1 = \frac{1}{\sqrt{(R_a + R_b)^2 + R_S^2}} - 1$$

thus

$$MS_1 = \frac{1}{\sqrt{(0.439 + 0.379)^2 + 0.551^2}} - 1 = 0.01$$

For
$R_a = 0.439$
$R_B = 0.379$
$R_S = 0.551$

Rear Spar upper cap *Section C-C point (4)* *Ultimate*	***0.01***
LC2001 *Al Aly 7050-T7451*	*Axial + bending + shear*

15.19 INTER-RIVET BUCKLING

The wing planks are fastened to the Rear Spar flanges using flush 3/16" *Hi-Lites* fasteners. When I drew section **A-A** and section **B-B,** I assumed that an effective skin *(plank)* was considered working with the Rear Spar upper cap as shown in *Figure 15-42* and *Figure 15-43.*

However, in order to validate this assumption, it is necessary to verify that no buckling in the plank indeed occurs in between the fasteners, otherwise no effective skin could be considered in the cross sections. As shown in *Figure 15-50*, the pitch "***p***" of the fasteners is 1.133" corresponding to ***6D*** (or maximum). Based on the Euler equation, the inter-rivet buckling allowable is calculated as:

$$F_{ir}=\frac{\pi^2 E_t}{\left(\frac{p}{\sqrt{C}}\frac{\sqrt{12}}{t}\right)^2}$$

The maximum compression stress at the forward edge (stress point ***(1)***) is conservatively assumed constant across the effective skin.

The end fixity coefficient "***C***" for flush type fasteners is ***1.5*** and the tangent elastic modulus "E_t" is found by iteration using a *Goal Seek* tool. A plasticity correction factor for compression "η_c" is used to multiply the Young's modulus "***E***" and is calculated as follows. No correction for Clad material is used since the plank is machined.

$$\eta_C=\frac{E_t}{E}=\frac{1}{1+\left(\frac{0.002\ E\ n}{F_{cy}}\right)\left(\frac{F_{ccr'}}{F_{cy}}\right)^{n-1}}$$

Therefore, using the same procedure as described in section ***11.2.7***, the critical inter-rivet buckling stress is calculated by iteration by first assuming a value for F_{ccr}' (F_{ir}' in this case).

The plastic reduction factor "η_C" obtained from the iteration is 0.16411 and thus, the tangent modulus is: $E_T = 0.16411 * 10\times10^6 = 1.64\times10^6$ *psi*.

hence

$$F_{ir}=\frac{\pi^2\ 1.64\times10^6}{\left(\frac{1.133}{\sqrt{1.5}}\frac{\sqrt{12}}{(0.185)}\right)^2}=53{,}979\ psi$$

For
F_{cy} = 55,000 psi
n = 20
*E = 10.0*10^6 psi*
p = 1.133in
t = 0.185 in
*E_t = 1.64*10^6 psi*

Margin of safety in inter-rivet buckling

With the maximum compression stress at point ***1*** of section *A-A* near Rib-3 of −36,945 *psi* from *Table 15-32,* the MS in inter-rivet buckling is calculated as:

$$MS=\frac{F_{ir}}{|f_{c1}|}-1=\frac{53{,}979}{36{,}945}-1=0.46$$

Hence, no inter-rivet buckling occurs confirming that the assumption to consider an effective plank width is valid.

15.20 NET TENSION

At the beginning of the analysis, it was decided to define the cross-sections of the Rear Spar by considering the effective skin widths as if the upper AND the lower caps were in compression. This was a conservative approach and would also cover load cases that would put the lower cap in compression.

Therefore, for this net tension analysis, instead of re-calculating new cross-sections by considering the full effective skin in tension of the lower cap and adding the fastener holes, I'll simply add the fastener holes in the lower cap that is the most critical in tension and assume that the centroid location of the whole section is unchanged.

The rational behind this assumption is based on the fact that the fastener holes *(1D)* exist every *(6D)* of distance. Therefore, no holes exist over 83% of the length. So it is closer stiffness-wise to assume that there aren't any holes on 100% of the length instead of supposing that the two rows of holes are considered as two continuous slots over the whole length of the beam.

This thus considers that the internal loads in the lower cap are exactly the same even with the fastener holes but will be transformed to the new centroid location of the lower cap having the holes.

From the previous stress analyses shown in *Table 15-33*, the highest tension stress occurs at section *C-C* due to the slotted beam effect and is 31,738 *psi* at point *(8)*. Therefore, this is where the two fastener holes will be added to get the highest tension stress possible. The dimensions and section properties of the lower cap in tension of section *B-B* is shown in *Figure 15-61*.

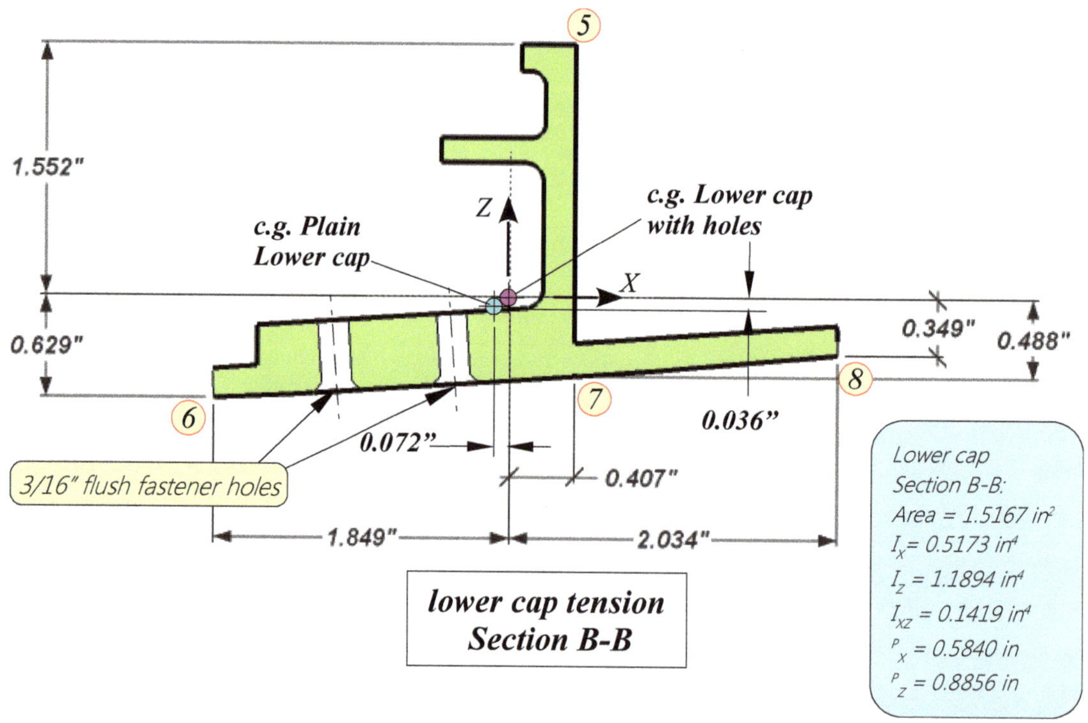

***Figure 15-61** – Lower cap dimensions and properties of section B-B in tension.*

Comparing the tension section shown above and the compression section from *Figure 15-43*, one can notice that, with the addition of the holes, not only the area and moments of inertia are slightly reduced but that, the *c.g.* moved up by 0.036" and aft by 0.072". Therefore, the internal loads in the lower cap at section *C-C*, derived in the same way as for the upper caps in section *15.16*, are shown in *Table 15-35* and are transformed as follows:

Table 15-35 *- Equivalent section B-B lower cap loads at section C-C.*

Sect C-C — ***At cutout***

[A]	P [lb]	M_X [in-lb]	M_Z [in-lb]	[B] f [psi]
Point 5	0.5923819679	-2.942375394	0.3718018088	-4001
Point 6	0.5923819679	1.0987585696	-1.372033702	29508
Point 8	0.5923819679	0.5799518251	1.629434954	31738

dz = 0.036 *in*

dx = 0.074 *in*

	[F]	*Transformed loads*	
P [lb] =	37997	P [lb] =	37997
M_X [in-lb] =	9307	M_X' [in-lb] =	10674
M_Z [in-lb] =	2352	M_Z' [in-lb] =	-460

$[A]^{-1}$			
	0.388925586	0.7534722821	0.5457021318
	-0.267400391	0.1120423476	0.1553580431
	-0.046220415	-0.313803602	0.3600240164

$[F] = [A]^{-1} \times [B]$

The axial load remains unchanged of course at P = 37,997 *lbs* but the new bending moments to apply at the *cg* of the lower cap section having holes are calculated as:

$$M_X' = M_X + P\,dz = 9{,}307 + 37{,}997\,(0.036) = 10{,}674\ in.-lb$$

$$M_Z' = M_Z - P\,dx = 2{,}352 - 37{,}997\,(0.074) = -460\ in.-lb$$

Then, the stresses at points *(5)* to *(8)* are calculated in the same way as in section *15.18* but using the perforated lower cap section *B-B*. The results at those stress points are shown in *Table 15-36*.

Table 15-36 *- Stress summary at section C-C, tension Lower cap.*

Lower cap in net tension

Section C-C / *Stress points*	x [in]	z [in]	f_a [psi]	f_{b_1} [psi]	f_{b_2nd} [psi]	f_b [psi]	f_a+f_b [psi]	R_a	R_B
5	0.407	1.552	25,052	-32,183	0	-32,183	-7,131	0.334	0.536
6	-1.849	-0.629	25,052	13,695	0	13,695	38,747	0.358	0.196
7	0.407	-0.488	25,052	9,912	0	9,912	34,965	0.358	0.142
8	2.034	-0.349	25,052	6,415	0	6,415	31,467	0.358	0.092

Section C-C / *Stress points*	$f_{S\,X}$ [psi]	$f_{S\,Y}$ [psi]	$f_{S\,Z}$ [psi]	$f_{S\,XY}$ [psi]	$f_{S\,XZ}$ [psi]	F_{SU} [psi]	R_S	MS
5	0	0	22,026	0	22,026	40,000	0.551	*-0.03*
$\mathbf{6}_{plank}$	6,519	6,665	0	9,323	6,519	40,000	0.233	*0.67*
$\mathbf{6}_{flange}$	6,519	4,858	0	8,131	6,519	40,000	0.203	*0.70*
$\mathbf{7}_{plank}$	6,519	0	0	6,519	6,519	40,000	0.163	*0.90*
$\mathbf{7}_{flange}$	6,519	0	22,026	6,519	22,970	40,000	0.574	*0.31*
8	6,519	0	0	6,519	6,519	40,000	0.163	*1.09*

Example of stress analysis at stress point *(5)* in the tension lower cap at Section *C-C*.

The most critical location is at the stress point *(5)* located right at the edge of the cutout and is considered as part of the lower cap.

The loads applied at the cg for the lower cap with holes, as shown in *Figure 15-61*, are:

$P = 37{,}997\ lbs$

$M_X' = 10{,}674\ in\text{-}lb$

$M_Z' = -460\ in\text{-}lb$

Axial stress

$$f_a = \frac{P}{A} = \frac{37{,}997}{1.5167} = 25{,}052\ psi$$

For
$A = 1.5167\ in^2$

The axial stress ratio "R_a" is obtained by comparing the above stress with the tensile strength "F_{TU}" for the Rear Spar *(Al Aly 7050)* from *Table 15-9.*

hence

$$R_a = \frac{f_a}{F_{TU}} = \frac{25{,}052}{75{,}000} = 0.334$$

Bending stress

Of course, no diagonal tension effect occurs here since the plank is in tension and its buckling coefficient "*k*" is 0.

Then the bending stress at point *(5)* is calculated below:

$$f_{b5} = -\frac{M_X' z_5}{I_X} + \frac{M_Z' x_5}{I_Z}$$

then

$$f_{b5} = -\frac{10{,}674(1.552)}{0.5173} + \frac{-460(0.407)}{1.1894} = -32{,}183\ psi$$

For
$M_X' = 10{,}674$ in-lb
$M_Z' = -460$ in-lb
$z_5 = 1.552$ in
$x_5 = 0.407$ in
$I_X = 0.5173\ in^4$
$I_Z = 1.1894\ in^4$

Because the total bending stress is in compression, the bending stress ratio "R_B" is obtained by conservatively comparing the above stress with the compression yield stress allowable "F_{CY}" of the R/S *(7050)*. Hence:

$$R_B = \frac{|f_{b5}|}{F_{CY}} = \frac{|-32{,}183|}{60{,}000} = 0.536$$

Note that, for more conservatism, the Free Flange buckling allowable could be calculated and used as the compression allowable. I wouldn't be comfortable in using the plastic bending allowable for this section, unless maybe that the lip around the cutout would be made larger thus making the web at point *(5)* more stable and more capable in compression due to bending.

Shear stress

The average shear stress had already been calculated using the net section between the fastener holes. Therefore, the same shear values for the lower cap from *Table 15-33* are used.

At stress point *(5)*, since this point is in the Rear Spar vertical web, only the vertical shear "*in Z*" is considered. Referring to *Figure 15-47*, the vertical shear stress "f_{SZ}" is 22,027 *psi.*

Since point *(5)* is located in the Rear Spar, the shear stress ratio "R_S" is obtained by comparing the above stress with the shear strength "F_{SU}" of the R/S *(7050)*. Hence:

$$R_S = \frac{f_{SZ}}{F_{SU}} = \frac{22{,}027}{40{,}000} = 0.551$$

Margin of safety in net tension

The lowest MS is obtained at stress point *(5)* located at the lower edge of the cutout through section *C-C* for the fictitious *LC2001*. I am using the following interaction equation to calculate the MS in the Rear Spar.

$$MS_1 = \frac{1}{\sqrt{(R_a + R_b)^2 + R_S^2}} - 1$$

For
$R_a = 0.334$
$R_B = 0.536$
$R_S = 0.551$

thus

$$MS_1 = \frac{1}{\sqrt{(0.334 + 0.536)^2 + 0.551^2}} - 1 = -0.03$$

Recommendation: Increase the web thickness of section *B-B* from 0.19" and 0.20" to 0.213" throughout.

I would also increase the width of the lip around the cutout to make it a decent return flange, somewhere between 0.5 and 0.75 in. total width.

15.21 CONNECTIONS

The upper and lower wing planks are fastened to the Rear Spar upper and lower flanges with two rows of 3/16" flush *Hi-Lites*. The fasteners are spaced (pitched) at 6D (*s = 1.133"*). The pins used are *HST11-6* along with shear application collars *HST79-6*. See *Table 15-12* for the fasteners allowables. See *Figure 15-50* for the details of the joints.

The loads are taken from the FEM output file for the *QUADS* that model the upper and lower plank panels adjacent to the R/S. The loads that are to be transferred from the wing planks to the Rear Spar are the running (distributed) shear flow along the global wing *Y* axis *"Fxy"* and, the load normal to the joint along the global *X* axis which corresponds to *"Fy"* from the local element orientation.

In addition to *"Fy"*, there is another distributed load that we have to consider caused by the diagonal tension effect on the upper plank; that is "w_{DT}". This running load was previously calculated on page *578* as being: $w_{DT} = w_X = 155.5$ *lb/in*. See *Table 15-37* for the applied loads to the upper and lower joints.

***Table 15-37** - Applied loads to the wing planks-R/S joints.*

	Element	*(In global X direction)* *Fy* [lb/in]	w_{DT} [lb/in]	q_{DT} [lb/in]	*Fres* [lb/in]	*s* [in]	*Ps/row* [lbs]
Upper plank	317416	188	155.5	1222	1270	1.133	1439
Lower plank	307406	-166	0	1012	1026	1.133	1162

The diagonal tension effect also adds loads to the shear flow *"Fxy"* obtained from the FEM. As calculated previously on page *566*, the resulting shear flow caused by a buckling angle of 45 *deg.* is: q_{DT} = 1,222 *lb/in* for the upper joint. Since the lower plank doesn't get into post buckling, the shear flow and the normal distributed running loads are directly taken form the FEM output file.

The most critical fastener load calculation is done here for a pair of fasteners (row) as an example. The normal running load and shear flow are vectorially combined as follows to get the resulting distributed load "F_{RES}" applied to the two rows as it is assumed that both rows share the ultimate loads equally. See *Figure 15-62* for the total applied loads to the two fastener rows of the upper plank joint to the Rear Spar.

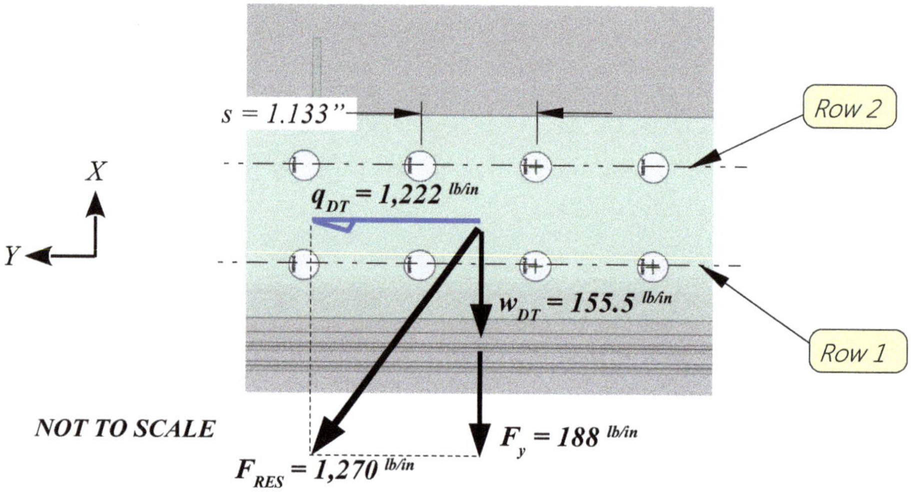

***Figure 15-62** – Total applied loads to the upper plank two rows of fasteners due to LC2001.*

$$F_{RES} = \sqrt{(F_Y + w_{DT})^2 + q_{DT}^2}$$
$$= \sqrt{(188 + 155.5)^2 + 1{,}222^2} = 1{,}270\ lb/in.$$

And the load per row is thus:

$$P_{S/row} = F_{RES} \cdot s = 1{,}270 \cdot 1.133 = 1{,}439\ lbs/row$$

The Margins of Safety in shear for the 3/16" *Hi-Lites* fasteners and in bearing for the upper and the lower wing planks as well as for the R/S flanges are calculated. They are summarized in *Table 15-38* and *Table 15-39.*

***Table 15-38** - MS calculation in shear for the upper and lower planks.*

Part	Material Al Aly	t [in]	Psu [lbs]	rows	Psu_tot [lbs]	P_S [lbs]	MS shear
Upper Plank	7075-T7351	0.185	2690	2	5380	1439	2.74
Lower Plank	2024-T351	0.185	2690	2	5380	1162	3.63

A MS calculation example is demonstrated below for the most critical joint in shear.

$$MS_{Shear} = \frac{P_{SU} \cdot rows}{P_S} - 1 = \frac{2{,}690 \cdot 2}{1{,}439} - 1 = 2.74$$

As on page 569, I could have considered only one row of fasteners as effective. However, it is common to use all fasteners when ultimate loads are applied.

Table 15-39 - *MS calculation in bearing for the upper and lower planks and the R/S.*

Rear Spar-Upper plank

Part	Material Al Aly	t [in]	D [in]	Fbru [ksi]	Pbru [lbs]	rows	P_S [lbs]	MS bearing
Upper Plank	7075-T7351	0.185	0.191	135	4770	2	1439	5.63
Rear Spar	7050-T7451	0.200	0.191	105	4011	2	1439	4.58

Rear Spar-Lower plank

Part	Material Al Aly	t [in]	D [in]	Fbru [ksi]	Pbru [lbs]	rows	P_S [lbs]	MS bearing
Lower Plank	2024-T351	0.185	0.191	111	3922	2	1162	5.75
Rear Spar	7050-T7451	0.250	0.191	105	5014	2	1162	7.63

A calculation example is demonstrated below for the most critical joint in bearing. It occurs in the Rear Spar upper flange.

$$MS_{BRG} = \frac{F_{BRU} \cdot D \cdot t \cdot rows}{P_S} - 1 = \frac{105{,}000\,(0.191)(0.200)\cdot 2}{1{,}439} - 1 = 4.58$$

As can be noticed, the joint is shear critical and only one row of fasteners would be sufficient to carry the ultimate applied loads. The joints are thus satisfactory.

As always, it is preferable to use joints allowables instead of using the fastener shear allowable and calculating the bearing capability. This is especially true when flush fasteners are used. The joint allowables are obtained from tests that involve the right fastener along with the right sheet material and thickness. Those can be obtained from (ref. 8) for many fastener arrangements.

15.22 COMMENTS

Since the MS is **negative** in net tension, it cannot be the final MS that you'll put in the stress report. It is simply an indication that the design (or the analysis approach) needs to be revised. Hence, that drawing cannot be signed yet!

Looking at the stress ratios from the analyses of the sections through the cutout, I find that the shear stress ratio "R_S" is relatively high. This is to be expected in areas where cutouts exist since there isn't much material left to carry that transverse shear load. To solve this, the web thickness around the cutout (section *B-B*) should be increased and the whole analysis redone accordingly.

Since the MS is also very low at point *(4)* for section *A-A* with 0.01, the increased web thickness will improve that MS also.

The reader most realize that this fictitious design was my first draft and that at least one other iteration of the analysis shall be performed in order to optimize it, or at least get a positive MS.

In order to speedup the iteration process, I do not recommend you get your section properties from a *CAD* software (like I did!) but to idealize the sections as made out of many rectangles, or other simple shapes, and use parameters: (t_1, h_1, t_2, etc). Then, providing you use a spreadsheet, all the parameters can be linked for all analyses so that you'll know right away the impact on all the MS.

16. CONCLUSION

The stress report is the product that aircraft stress engineers deliver. This is usually the program end game for the design/stress group. At the time you're writing the stress report, designers will already have been assigned to other projects or, let go! Then, it'll be your turn once you hand out the stress report(s) and it gets approved. At that time, I always got mixed feelings. I was happy and proud of what the group had achieved by working together on this great new aircraft for the past *3-4* years (or more) but, at the same time, this is a time for change and uncertainty.

Some would jump straight into this brand new hot program, some would be laid-off and, some would retire or change career or, go raise families. For example, a stress group could go from 40 analysts at the peak of a program down to only 3 at the end to support manufacturing. Just like the GAG "Ground-Air-Ground" cycle of an airplane, the job cycle of the stress engineer has a similar "Start-Cruise-End" cycle lasting from a few months to many years.

Writing reports can be considered boring for some people as they prefer the sizing phase and the interaction with other people. The report writing phase is a lonely job. There is not much interaction with other people except your boss and the DER. Personally, I always enjoyed that phase. Not only because I love writing but also because I know that the work I do will remain for others to see and be referred to for the generations to come.

I remember, during one particular assignment, looking at some stress reports on the *747,* that had been hand written (what else) when I was 5 years old, and be very impressed on the quality; even without the help of computers. As some would say: *"it's not the instrument that makes the musician".* Computers, *FEA* and software are simply tools for the stress engineer. The most important thing is and always will be your knowledge and judgment. The capacity to be able to analyze situations and find solutions is essential and irreplaceable by machines.

> *"The instrument doesn't make the musician"*

It is quite possible that this book might, in a near future, be "read" by *A.I.'s* (Artificial Intelligence Entities) to learn from it and eventually replace the structural analysts (as well as many other professionals). The hand calculations are, to my humble opinion, towards the end of their glory days. Quite often, stress analysts nowadays spend more time doing coding or manipulating columns and rows of numbers in a spreadsheet. They are more like accountants working on a balance sheet.

As the computers get more and more powerful and the number of degrees of freedom in a Finite Element Model gets larger and larger, stress analysis will eventually be reduced to extracting the maximum stress at the most critical location (the red spot on the pretty picture) of a very fine meshed model and simply comparing this value to the allowable to get the Margin of Safety.

Twenty years ago, I thought that this was going to happen any time but, here we are, almost a generation later and companies still use a crude loads FEM along with spreadsheet tools to process the always increasing number of load cases. In 2018, the FEMs look pretty much the same as they were back in 1987 and we're still using the same FEA products. The only difference now is that, instead of having books and binders filling the shelves at our desks we have a computer with two screens with all of the references we need in electronic format. This is beginning to look more and more like the paperless consoles of a spaceship.

Every year I tell myself: *That's it!* Some start-up company is going to develop this super computer program/code that will be replacing me. But this day hasn't arrived yet. Probably because aircraft manufacturing companies and the FAA are not ready for such a change. For all I know, maybe a machine could replace us for the static, fatigue & DTA analyses and the flight tests but I doubt that it will ever replace humans for our inventiveness, creativity and judgment, and certainly not for passion. But hey, I could be wrong.

Over my generation long career, very little things changed from the structure of an airplane (Actually, not much has changed on the structure side since *WW2*). We're still using the same aluminum alloys, the same corrosion resistant steels. Of course composites have become more common but aluminum is still the most popular material in commercial aircraft. The only changes maybe are that machined parts have replaced extrusions, forgings and castings. Titanium fasteners have replaced mostly steel and CRES fasteners and, Aluminum-Lithium alloy is gaining in popularity especially for fuselage skins. However, even if the basic configuration of an airplane hasn't changed for many generations, somehow, for some reason, we seem to be re-inventing the wheel on every new program.

The main reason for that is the lack of knowledge continuity. Regardless on the size or the reputation of a company, it is always the people working inside that building that have the expertise and do the job. If and when those people leave, they bring their expertise with them. Knowledge is a very volatile resource. Without its employees, any company is an empty shell. Whether or not they meet the *ISO* standards and have all the instructions to start over in the case that everybody was to be replaced by people unfamiliar to the field, it could never be the same. It is like replacing the chefs and all of their assistants by a bunch of new cooks unfamiliar with that restaurant's specialties. The books for all the recipes could still be there with the same pots and stoves, but I can assure you that it'll never taste the same! Unless, of course, in the case that the apprentice is mentored on a one to one basis by the master – as it should be!

Of course every aircraft company, like restaurants, have their own specialties, their own way of doing things, their own philosophy. There are also unwritten rules. However, as a good number of designers and analysts go from one company to another to surf the work wave, (like wolves following the migrating elks) those people tend to "standardize" how things are done. Therefore when people ask me: What is the best airplane? Brand *"A"* or brand *"B"*? I tell them they're all very good and that it is quite possible that, in some cases, the very same people worked on all of them and chances are, you probably couldn't tell the difference simply by looking at the airframe of each one.

Almost four years ago, when I started writing down ideas about a book that would help bridge the gap between what was taught in school and what is actually required in the industry of an accomplished aircraft stress engineer, I was far from imagining what was to hit us in our 2020 world. This book turns out to be more necessary than what I could have ever expected.

As I'm writing these lines, more and more engineers are working from their home office and, although working from home has definite advantages, it is lacking the close proximity of experienced people. Although this book doesn't replace a good mentor, it certainly comes in handy when the unsupervised stress analysts can't just turn around and ask their neighbors on how they would analyze a particular aircraft structure.

With this book, my intention is to try to keep that know-how alive by sharing it with the engineers that, unlike me, haven't worked in all of those companies out there and haven't had a mentor to help them develop and strengthen their skills. I sure hope that I was able to give you, the reader, some confidence in your art enough to be autonomous and being able to analyze any type of aircraft structure.

Flying machines are awesome and so are the aerospace engineers creating them!

"We can always make money later
but we can't make time later."

(Stef)

APPENDIX A

The following figures are used in this book and were extracted from the NACA Technical Notes (ref. 19., 20.) that are in the public domain. Those are available at the following web site:

https://ntrs.nasa.gov/search

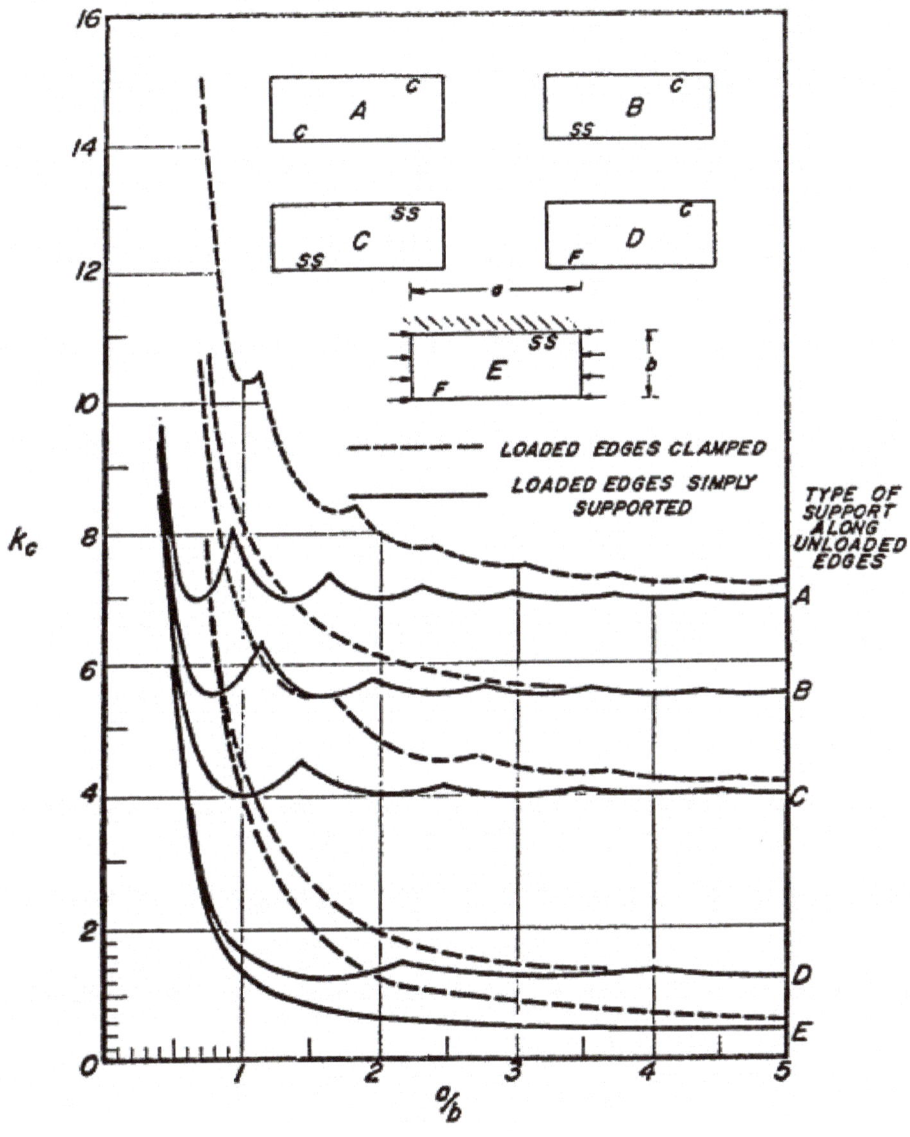

Figure A-1 *– Compressive-buckling coefficient for flat rectangular plates. Ref. 20. fig.14.*

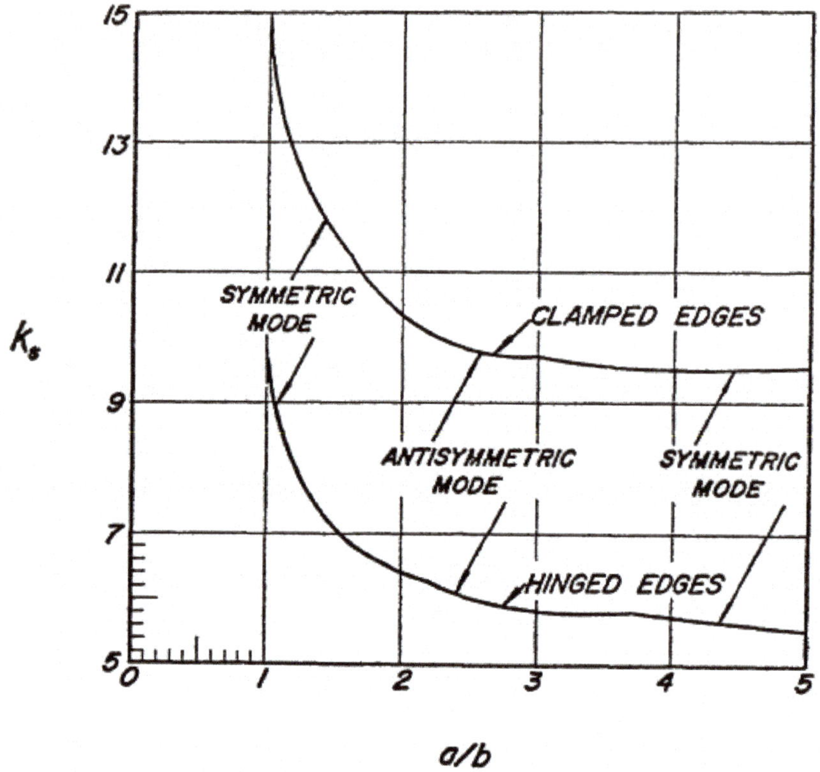

Figure A-2 – *Shear-buckling coefficient for flat rectangular plates. Ref. 20. fig.22.*

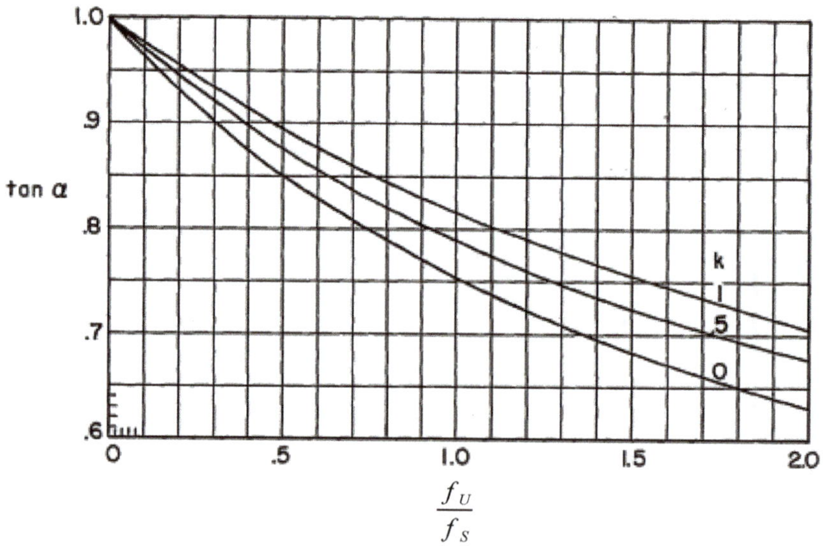

Figure A-3 – *Incomplete diagonal tension, Ref. 19. fig.16a.*

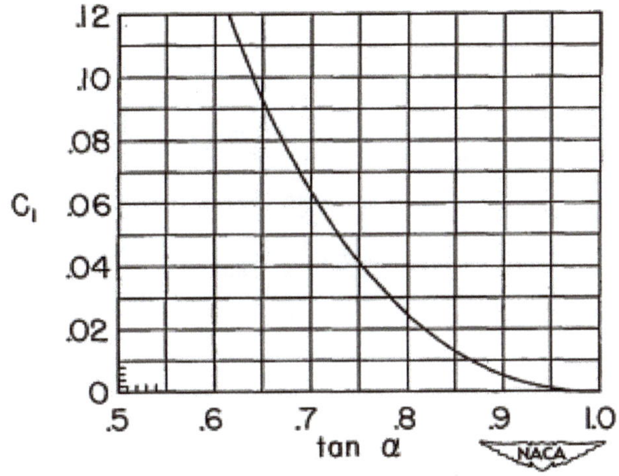

Figure A-4 *– Angle factor C1, Ref. 19. fig.17.*

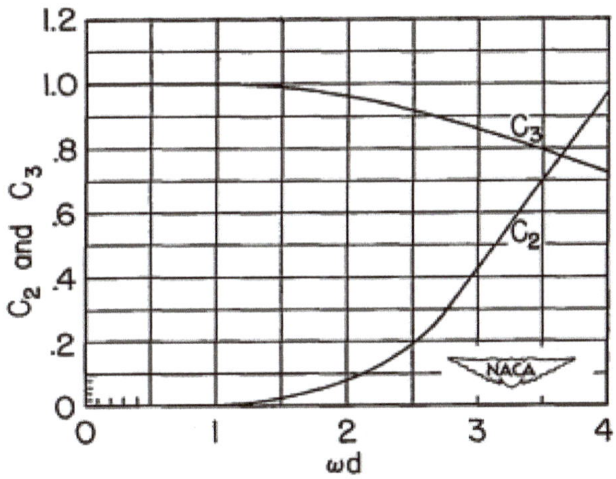

$$wd = 0.7\text{d}\sqrt[4]{\frac{t}{\left(I_C + I_T\right)h_e}}$$

Figure A-5 *– Stress concentration factors C2 and C3, Ref. 19. fig.18.*

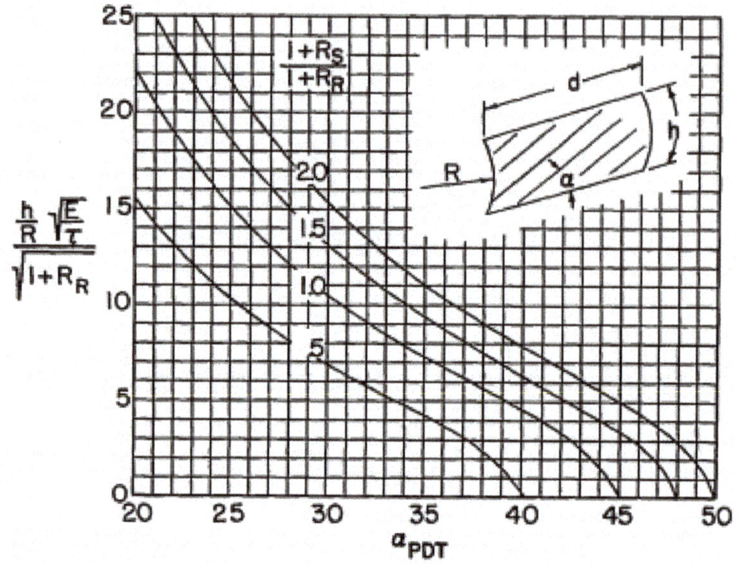

$$R_R = \frac{d\,t}{A_{RG}}\,;\, R_S = \frac{h\,t}{A_{ST}}$$

Figure A-6 *– Buckling Angle at Post Diagonal Tension, Ref. 19. fig.28a.*

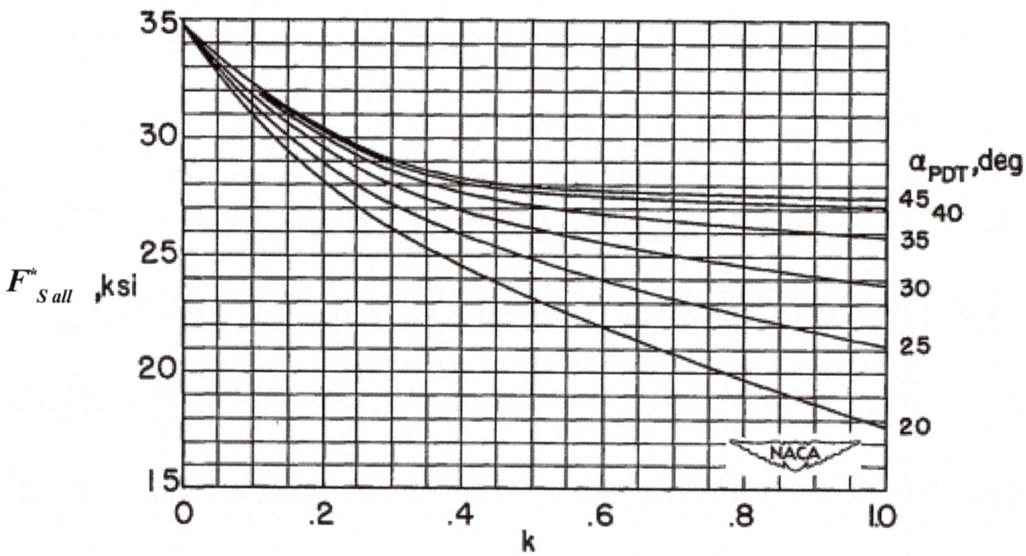

Figure A-7 *– Allowable shear stress for 7075-T6, Ftu = 72 ksi, Ref. 19. fig.19b.*

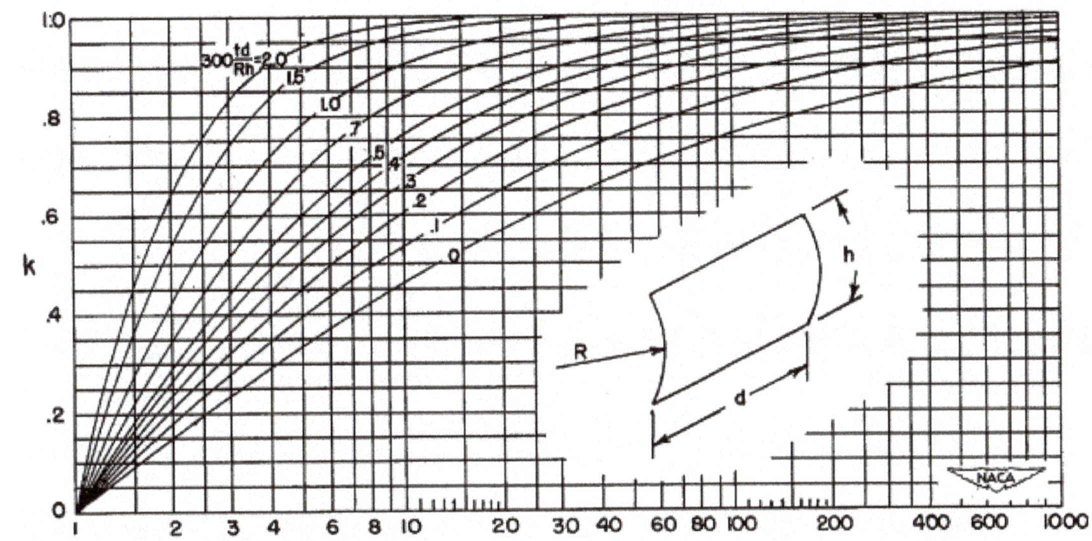

Use 0 curve for flat sheets f_S/F_{S_CR}

$$\left(\text{If } h>d\text{, replace} \frac{td}{Rh} \text{ by } \frac{th}{Rd}\text{; if } \frac{d}{h}\left(\text{or. } \frac{h}{d}\right) > 2\text{, use } 2\right)$$

Figure A-8 *– Diagonal-tension factor k according to the panel radius, Ref. 19. fig.13.*

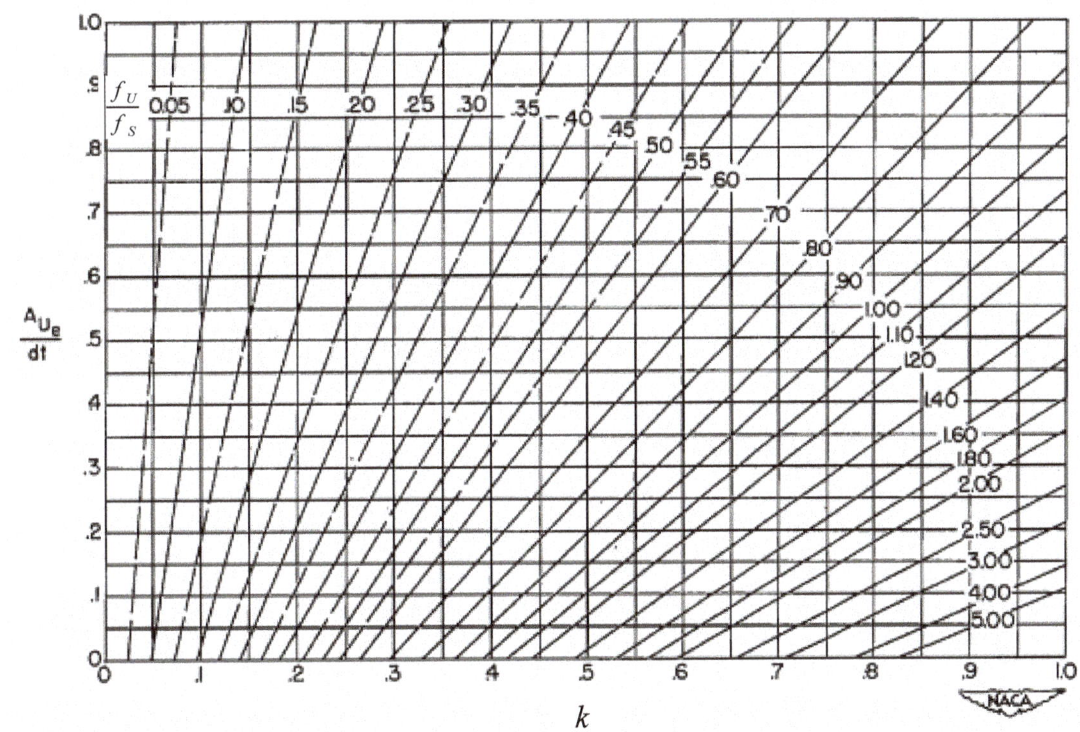

Figure A-9 *– Diagonal-tension analysis chart, Ref. 19. fig.14.*

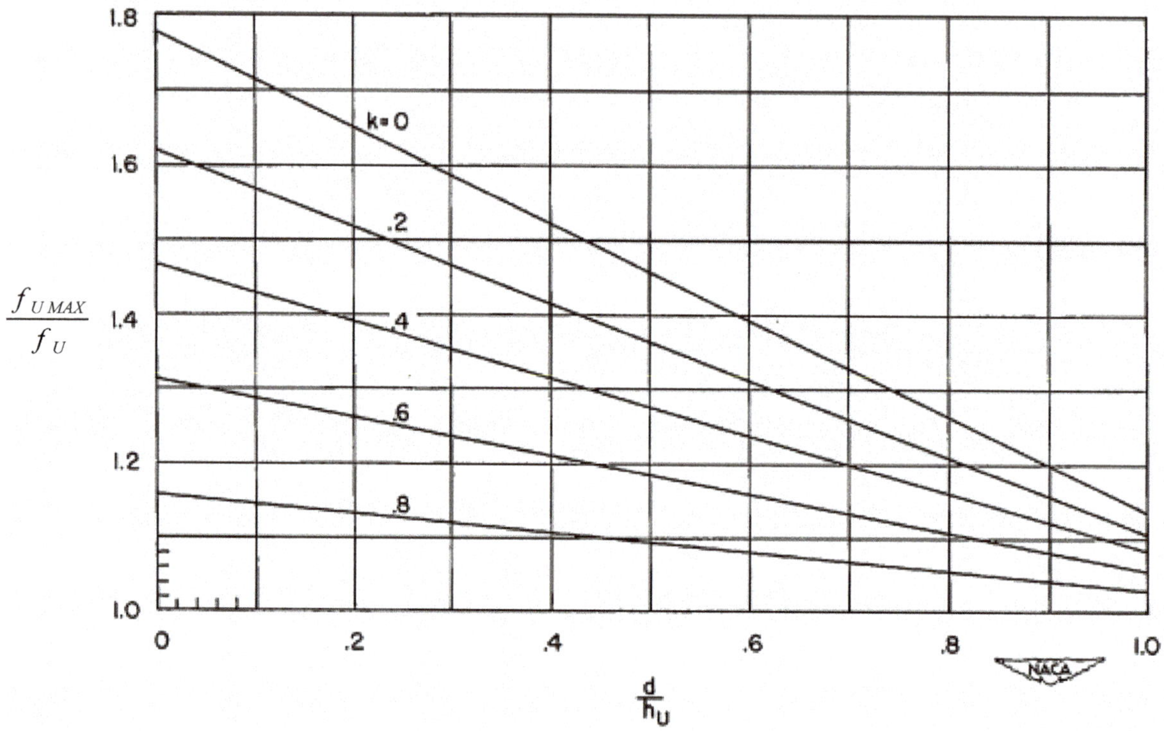

NOTE: When using on curved panels:

For Rings: Read abscissa as $\frac{d}{h}$*; for Stringers: Read abscissa as* $\frac{h}{d}$.

Figure A-10 *– Ratio of maximum stress over average stress in Web stiffener, Ref. 19. fig.15.*

APPENDIX B

The following figures are fictitious and not to be used for design. They are based on graphs used in the industry and in some popular references and are simply shown here to demonstrate the analysis steps.

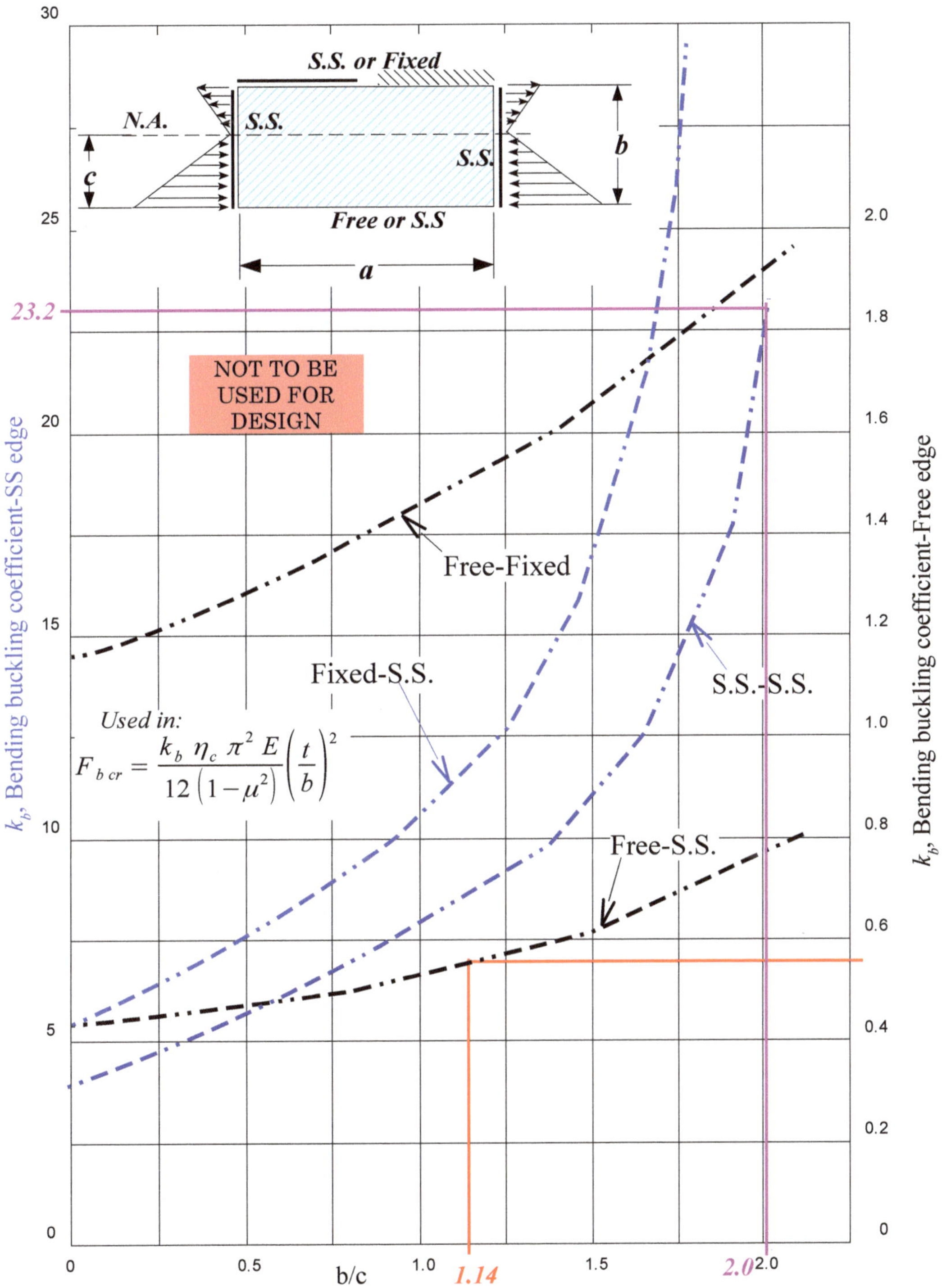

Figure B-1 *– Fictitious general bending buckling coefficient k_b for a/b>1, one edge free.*

Loading and end fixity		End fixity coefficient *C*
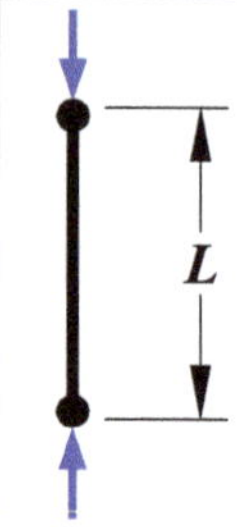	-Uniform column -axial load -pinned ends	***1***
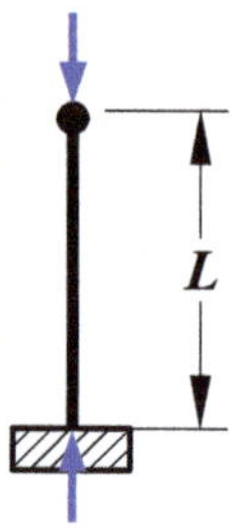	-Uniform column -axial load -pinned-fixed ends	***2.05***
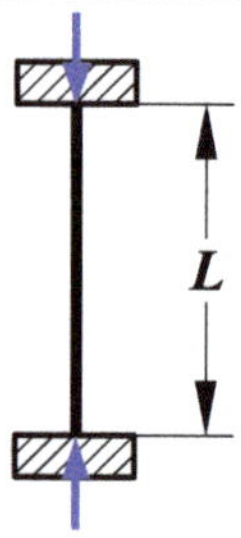	-Uniform column -axial load -fixed ends	***4***
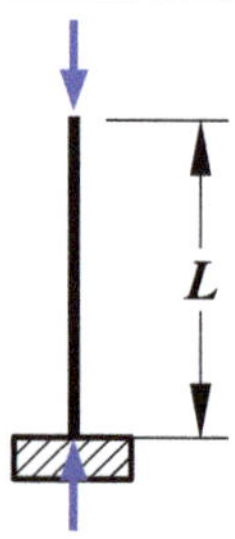	-Uniform column -axial load -fixed-free ends	***0.25***

Loading and end fixity		End fixity coefficient *C*
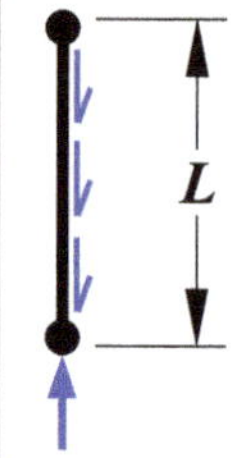	-Uniform column -distributed load -pinned ends	***1.87***
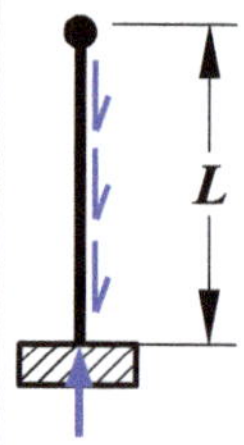	-Uniform column -distributed load -pinned-fixed ends	***3.55***
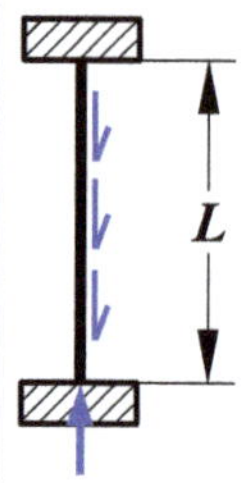	-Uniform column -distributed load -fixed ends	***7.5***
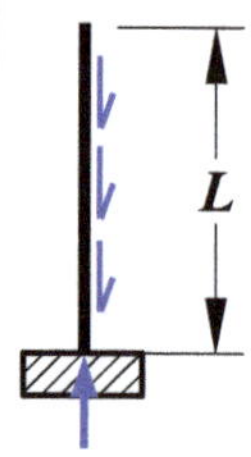	-Uniform column -distributed load -fixed-free ends	***0.794***

Figure B-2 *– Column end-fixity coefficient "C".*

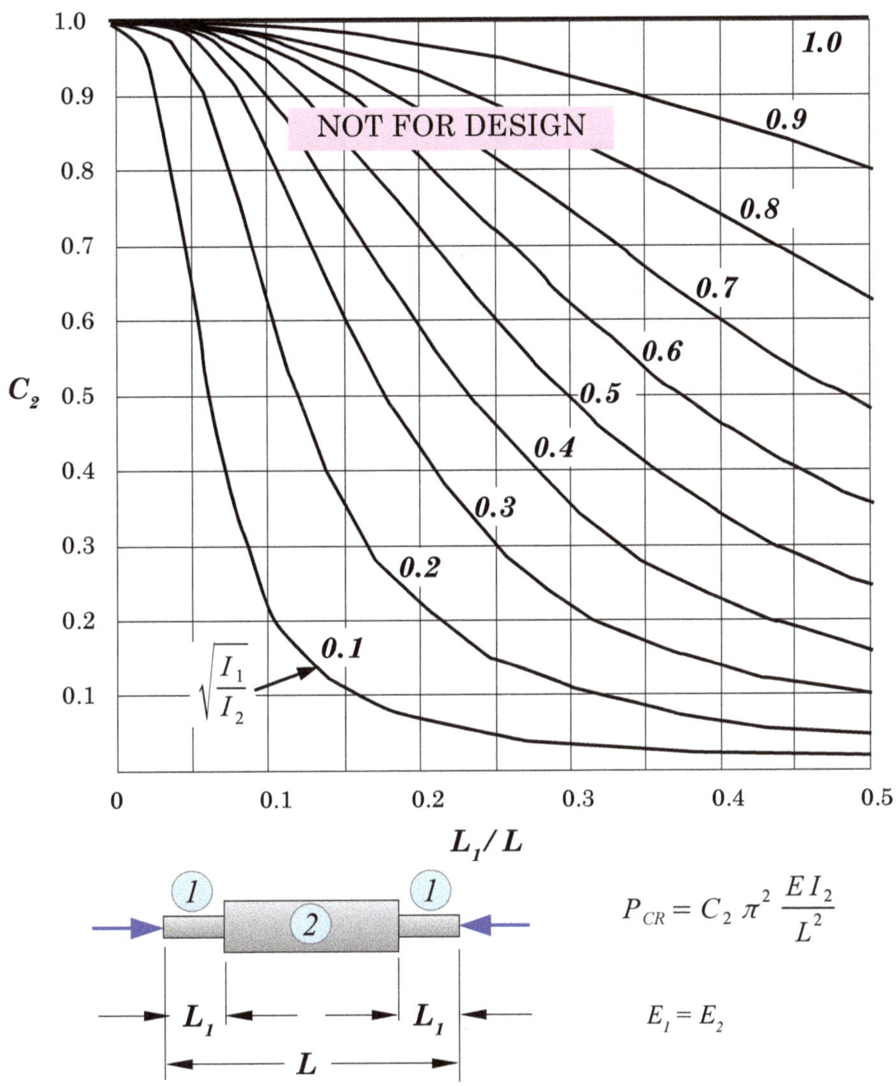

***Figure B-3** – Symmetrical stepped column "C_2" coefficient.*

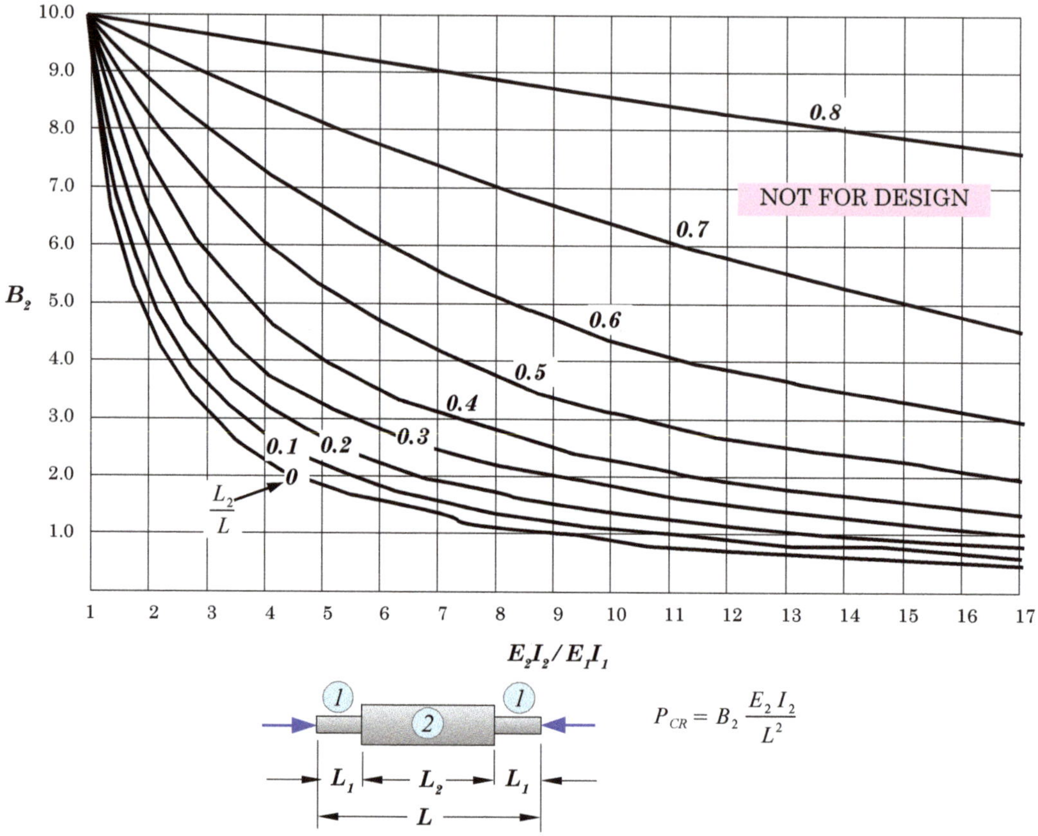

Figure B-4 – *Symmetrical stepped column "B_2" coefficient for $E_1 \neq E_2$.*

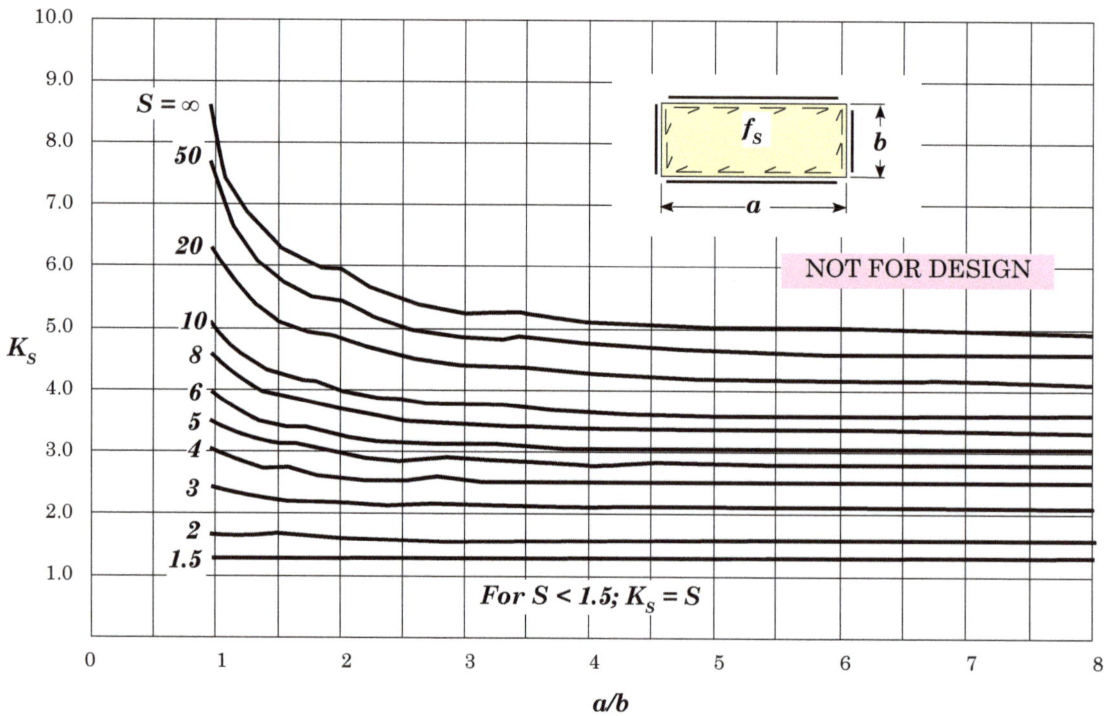

Figure B-5 – *Shear buckling coefficient for simply supported honeycomb bonded panels.*

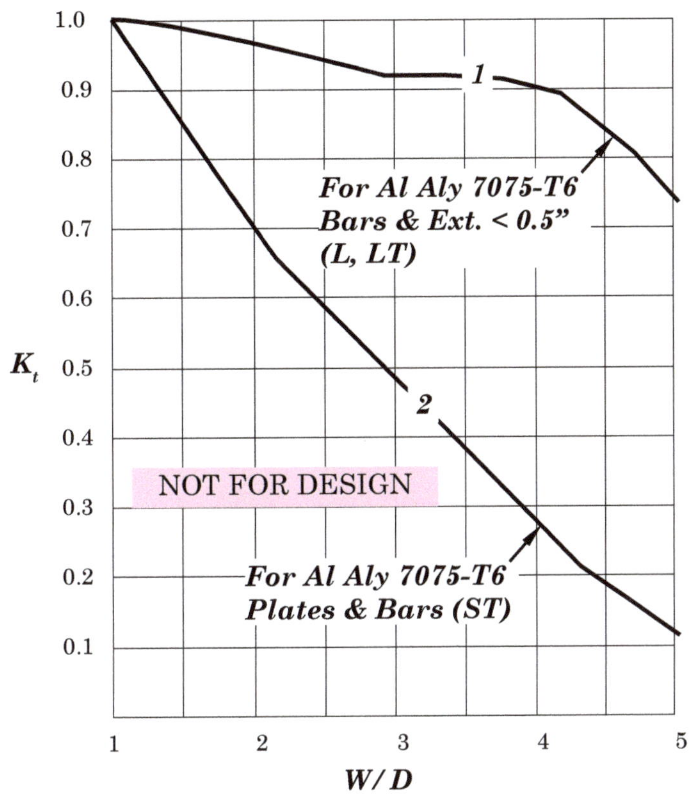

Figure B-6 *– Lug tension efficiency factor "K_t" for axially loaded lugs.*

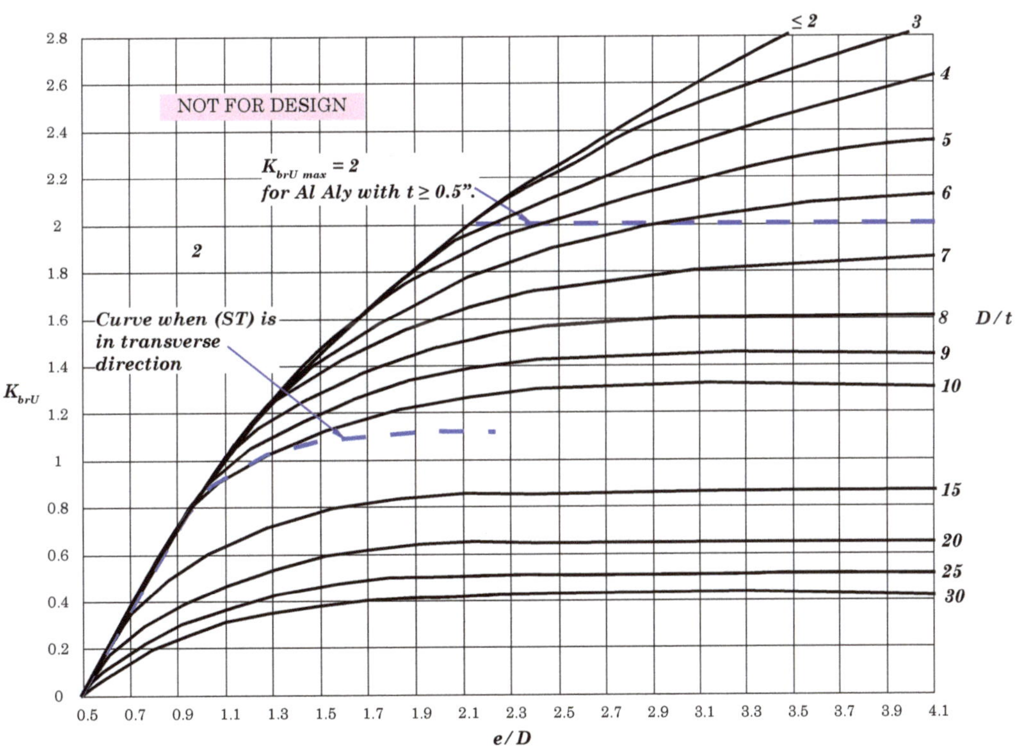

Figure B-7 *– Lug ultimate bearing coefficient "K_{brU}" for axially loaded lugs.*

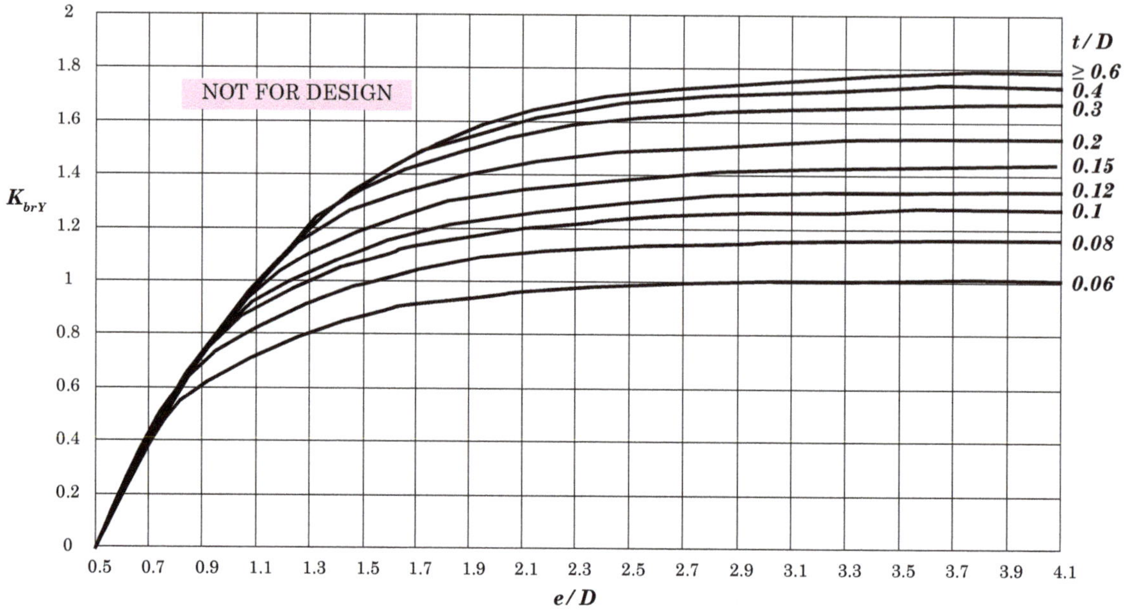

Figure B-8 *– Lug Yield bearing coefficient "K_{brY}" for axially loaded lugs.*

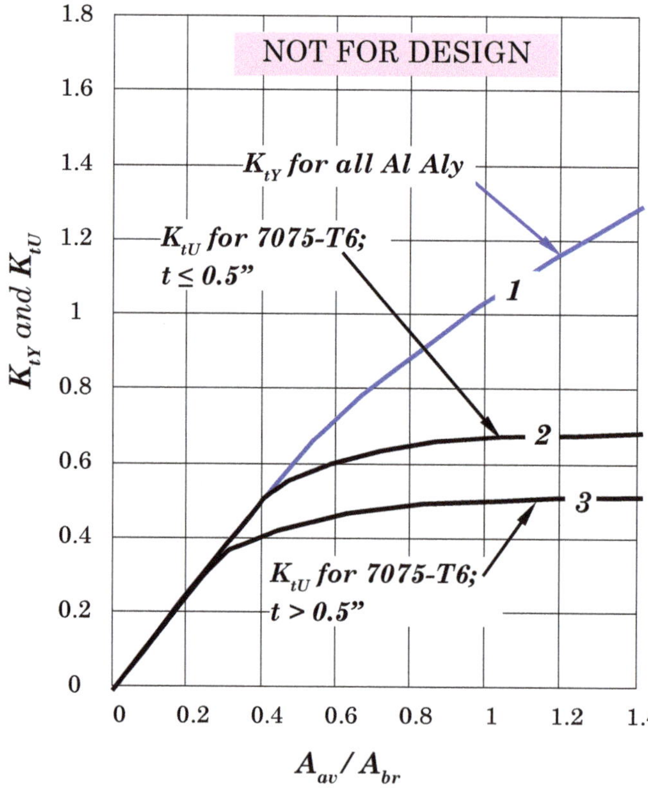

Figure B-9 *– Lug tension coefficients "K_{tY}" & "K_{tU}" for transversely loaded lugs.*

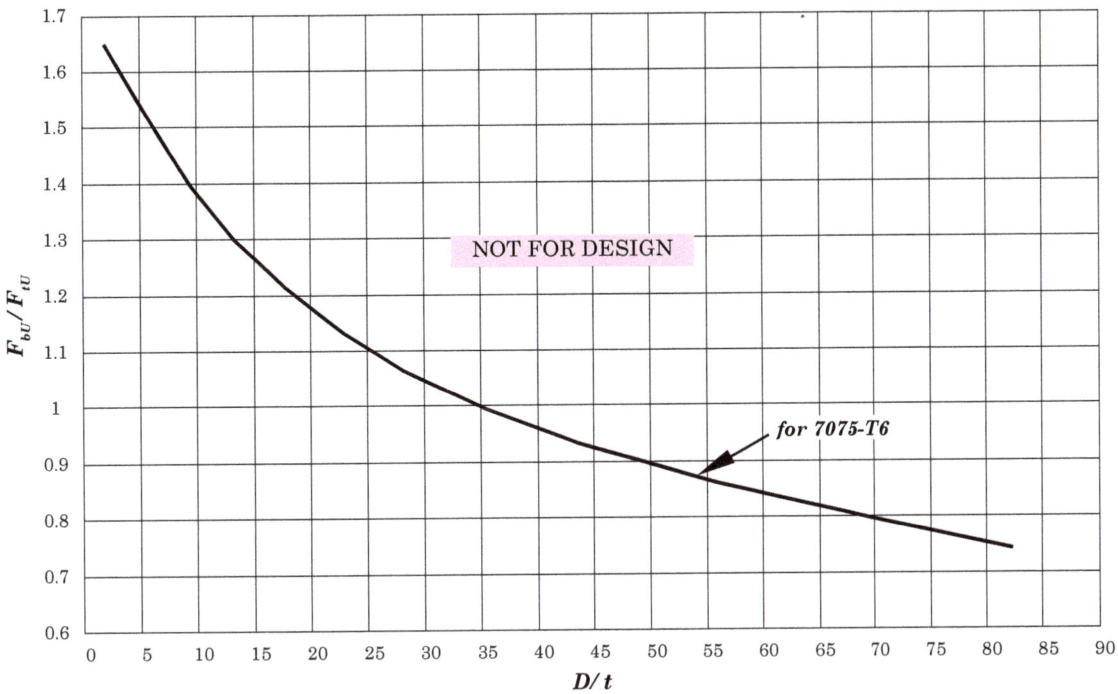

Figure B-10 *– Bending modulus "F_{bU}/F_{tU}" for Al Aly 7075-T6 tubes.*

INDEX

F

G

H

I

J

L

M

www.ingramcontent.com/pod-product-compliance
Lightning Source LLC
LaVergne TN
LVHW070829220926
848486LV00029B/228